Lecture Notes in Mechanical Engineering

Lecture Notes in Mechanical Engineering (LNME) publishes the latest developments in Mechanical Engineering—quickly, informally and with high quality. Original research or contributions reported in proceedings and post-proceedings represents the core of LNME. Volumes published in LNME embrace all aspects, subfields and new challenges of mechanical engineering.

To submit a proposal or request further information, please contact the Springer Editor of your location:

Europe, USA, Africa: Leontina Di Cecco at Leontina.dicecco@springer.com
China: Ella Zhang at ella.zhang@cn.springernature.com
India, Rest of Asia, Australia, New Zealand: Swati Meherishi at swati.meherishi@springer.com

Topics in the series include:

- Engineering Design
- Machinery and Machine Elements
- Mechanical Structures and Stress Analysis
- Automotive Engineering
- Engine Technology
- Aerospace Technology and Astronautics
- Nanotechnology and Microengineering
- Control, Robotics, Mechatronics
- MEMS
- Theoretical and Applied Mechanics
- Dynamical Systems, Control
- Fluid Mechanics
- Engineering Thermodynamics, Heat and Mass Transfer
- Manufacturing Engineering and Smart Manufacturing
- Precision Engineering, Instrumentation, Measurement
- Materials Engineering
- Tribology and Surface Technology.

Indexed by SCOPUS, EI Compendex, and INSPEC.

All books published in the series are evaluated by Web of Science for the Conference Proceedings Citation Index (CPCI).

To submit a proposal for a monograph, please check our Springer Tracts in Mechanical Engineering at https://link.springer.com/bookseries/11693.

Arun Kumar R. · G. Rajesh · Jaywant H. Arakeri ·
Hardik Kothadia
Editors

Proceedings of Fluid Mechanics and Fluid Power (FMFP) 2023, Vol. 6

Fluid Mechanics

Springer

Editors
Arun Kumar R.
Department of Mechanical Engineering
Indian Institute of Technology Jodhpur
Jodhpur, Rajasthan, India

G. Rajesh
Department of Aerospace Engineering
Indian Institute of Technology Madras
Chennai, India

Jaywant H. Arakeri
Jawaharlal Nehru Centre for Advanced Scientific Research
Bengaluru, Karnataka, India

Hardik Kothadia
Department of Mechanical Engineering
Indian Institute of Technology Jodhpur
Jodhpur, Rajasthan, India

ISSN 2195-4356 ISSN 2195-4364 (electronic)
Lecture Notes in Mechanical Engineering
ISBN 978-981-96-9599-7 ISBN 978-981-96-9600-0 (eBook)
https://doi.org/10.1007/978-981-96-9600-0

This Springer imprint is published by the registered company Springer Nature Singapore Pte Ltd.
The registered company address is: 152 Beach Road, #21-01/04 Gateway East, Singapore 189721, Singapore

Contents

About the Editors

Dr. Arun Kumar R. is working as an associate professor in the Department of Mechanical Engineering, Indian Institute of Technology, Jodhpur. He joined IIT Jodhpur in 2019 and prior to that, he worked as Inspire Faculty at IISc Bangalore—Aerospace department, from 2018 to 2019. Dr. Arun completed his Ph.D. from IIT Madras in 2018 from the Department of Aerospace Engineering. Dr. Arun works in the fields of high-speed aerodynamics, shock waves, and fluidic actuators using experimental and computational techniques. He has published more than 30 articles in reputed international journals including the Journal of Fluid Mechanics and Physics of Fluids and National and International Conference proceedings. He is a life member of the Society for Shockwave Research, India.

Prof. G. Rajesh is working as a professor in the Department of Aerospace Engineering, Indian Institute of Technology, Madras. Before joining IIT Madras in 2013, he worked in various positions at Keimyung University, South Korea, Indian Institute of Space Science and Technology, Indian Space Research Organization (ISRO) and Government Engineering Colleges in Kerala, India. He works in computational and experimental shock wave dynamics, interior, intermediate and exterior ballistics, blast attenuation, and armour development. He has published more than 80 articles in reputed international journals including Journal of Fluid Mechanics and Physics of Fluids and National and International Conference proceedings. He is a life member of International Shockwave Institute, International Ballistic Society and a life member and governing council member of the Society for Shockwave Research, India.

Prof. Jaywant H. Arakeri was in the Faculty of the Mechanical Engineering Department, the Centre for Product Design and Manufacture at the Indian Institute of Science, Bengaluru, till July 2022. He is currently an honorary professor at JNCaSR, Bengaluru, and visiting professor at IIT, Jodhpur. All his education has been in aeronautical engineering, B.Tech. (IIT, Madras), M.E. (IISc), and Ph.D. (Caltech). His research is primarily focused on the fundamental understanding of various phenomena in fluid mechanics. Some of the questions being addressed in his lab relate to the role of turbulence in condensation and droplet growth in clouds; flows around flexible surfaces like fish tails and heart valves; instability of unsteady flows, including those with curvature, like those found in arteries; heat and moisture loss from soils and leaves; thermal management in polyhouses and aeroponic chambers. He is the president of the National Society for Fluid Mechanics and Fluid Power.

Dr. Hardik Kothadia is currently working as Associate Professor in the Department of Mechanical Engineering, IIT Jodhpur. He received his Ph.D. from IIT Bombay in the area of multiphase flows and heat transfer. He is recipient of the National Award for Best M.Tech. Thesis from Indian Society for Technical Education. He received excellence in Thesis Work Award for Outstanding Research Contributions in Ph.D. work from IIT Bombay. He received teaching excellence award from IIT Jodhpur. His research area includes compact heat exchanger, direct steam generator, thermal energy storage, desalination, and multiphase flows analysis. He has published more than 40 articles in reputed international journals and national and international conference proceedings. He is a life member of the National Society for Fluid Mechanics and Fluid Power and Indian Society for Heat and Mass Transfer.

Nonlinear Numerical Study of Annular Fin of Different Profiles

Harishchandra Thakur

Abstract The current article performs heat transfer analysis on the annular fin of different profiles. This nonlinear analysis considers temperature-dependent thermal conductivity, variable heat transfer coefficient, and surface radiation. The meshless local Petrov–Galerkin (MLPG) method, regarded as a truly meshfree method is applied to create discrete equations. The function approximation is done using moving least square (MLS) scheme and essential boundary condition is imposed with the penalty method. The nonlinearity of the problem is handled with a simple iterative method. The performance of different profiles based on their respective volume in nonlinear cases is evaluated and compared.

Keywords Radial fin · MLPG method · Nonlinear analysis · Meshfree method

1 Introduction

Applying fin increases surface area to enhance heat transfer from the hot base surface by convection or by convection-radiation. Annular/radial fins are applied on heat exchangers, engines, compressors, electrical devices and control systems. But, application of fin invites addition of extra material into the thermal system. It poses a new kind of problem to achieve maximum heat transfer enhancement using the least amount of material. This helps to produce thermal system cost-effective, compact and lighter. This has attracted attention of researchers to try different profiles of annular fin. The difference in temperature between the fin surface and the surrounding fluid is maximum near the base of the fin. It decreases towards the tip. Hence, a fin profile which becomes thinner towards the tip is preferred. This improves the heat transfer rate per unit volume of the fin. Another aspect which must be taken into account is the dependency of thermal properties i.e. thermal conductivity and

H. Thakur (✉)
Department of Mechanical Engineering, School of Engineering, Gautam Buddha University, Greater Noida, India
e-mail: harish@gbu.ac.in

A. Kumar R. et al. (eds.), *Proceedings of Fluid Mechanics and Fluid Power (FMFP) 2023, Vol. 6*, Lecture Notes in Mechanical Engineering,
https://doi.org/10.1007/978-981-96-9600-0_1

heat transfer coefficient on temperature and thermal radiation particularly when fin operates under a large temperature difference. This introduces nonlinearity of different order. Finding exact solution of resulting differential equation is often not possible. Hence, this nonlinear heat transfer problem of annular fin is solved using semi-analytical methods—Adomian Decomposition Method, Differential Transform Method, Homotopy Perturbation Method, Homotopy Analysis Method, Lie Symmetry Method and Variational Iteration Method are few among them. Numerical methods such as Finite Difference Method, Finite Volume Method, Finite Element Method and Meshfree Methods such as Element Free Galerkin Method and Meshless Local Petrov Galerkin (MLPG) Method are also used to solve nonlinear fin problem.

2 Literature Review and Objective

As discussed, fins that have thinner profile towards the tip are preferred due to higher value of heat transfer per unit volume; therefore, fins with trapezoidal, triangular, hyperbolic, and parabolic profiles are often chosen for thermal analysis in the literature. Besides these step-fin and dimensionally optimized rectangular profile annular fin are also found good in number. Mokheimer [1] has solved annular fin of rectangular, hyperbolic, triangular and parabolic profiles using finite difference method. Thermal efficiency of the fin is obtained using variable heat transfer coefficient for a wide range of radius ration and thermo-geometric parameter. Arauzo et al. [2] have used power series method to address hyperbolic profile annular fin and obtained temperature distribution and heat transfer rate. Campo and Cui [3] have achieved exact solution of temperature variation in a hyperbolic annular fin using the mean value of theorem for integration. Kundu [4] has studied rectangular step-fin under dehumidification. An iterative approach is used to calculate the tip temperature. Kang and Look [5] have used a new analytical approach to obtain optimized set of geometric parameters of a trapezoidal profile annular fin. Khani and Aziz [6] have selected annular fin of rectangular and convex parabolic profiles mounted on a rotating shaft and presented an analysis using the Homotopy Method (HAM). Darvishi et al. [7] have studied porous radial fin of rectangular profile using the Homotopy Analysis Method. The variation of temperature and rate of heat transfer with different dependent parameters are studied in detail. Ndlovu and Moitsheki [8] have performed analysis on radial fins of rectangular and hyperbolic profiles using the Variational Iteration Method and obtained temperature distribution and heat transfer under different varying parameters. Celentano and Campo [9] have solved annular fin of quarter circle profile. A two-dimensional finite element analysis is performed to obtain temperature distribution and heat transfer. Kundu and Yook [10] have tried to derive a common mathematical expression which will be applicable for porous longitudinal fins, spines and radial fins together. They included all sorts of nonlinearities and obtained optimized performance of the fin for a given volume of the material.

This article investigates annular fin with rectangular, trapezoidal, triangular, hyperbolic and half-elliptical profile. The material of the fin is Aluminium. The thermal conductivity and heat transfer coefficient of the fin material are assumed depending on temperature. The thermal radiation is also included in the analysis. With the help of the MLPG method and a simple iterative technique, performance of annular fin with different profiles based on their volume used are presented. Forthcoming sections discuss about the numerical formulations of the problem followed by a detail discussion on results obtained.

3 The MLPG Method and the Problem Formulation

The MLPG method was proposed by Atluri and Zhu [11, 12]. The Meshfree methods have emerged in late nineties as a group of new numerical methods which were aimed to reduce the need of meshing of the computational domain. Few among them have completely removed the requirement of mesh at any stage of computation. The MLPG method is one among them. It performs interpolation based on nodal information and performs integration in a local domain. Any node based method can be used for function interpolation, the current analysis uses moving least square (MLS) scheme. The MLS approximants are obtained using three components: a weight function related to each node, a basis function and a set of coefficients which are not constant. Integration is performed in a conveniently chosen local domain using Gauss–Legendre quadrature rule. This method has a lot of flexibility in application. This method is successfully extended to solve fin problems [13–16] where stability and convergence of the method are also tested and found excellent.

The current analysis considers heat transfer in steady-state condition in an annular fin which has a temperature-dependent thermal conductivity $k(T)$. Different profiles of the radial fin are considered which are shown by a generalized representation in Fig. 1. The fin profile is confined between inner radius r_i and outer radius r_o. A differential element of radial thickness dr at radius r is selected for to apply the energy balance equation as highlighted in Fig. 1. The thickness $\delta(r)$ is the thickness parallel to the base surface of the fin at radius r. The fin has cross-sectional area $A(r) = 2\pi r\ \delta(r)$ and perimeter $P(r) = 2\pi r$. The fin loses heat by convection and radiation together to its surrounding maintained at temperature T_a. The heat transfer coefficient of the governing differential equation is a function of temperature.

The temperature of sink is assumed to be the same for convection and radiation to keep formulation simple.

Energy balance is applied on the differential element to derive governing differential equation. The heat conducted into the element will be equal to the combined heat transfer going out of the element due to conduction, convection and radiation in steady state condition. The final form of the governing equation domain Ω can be expressed as

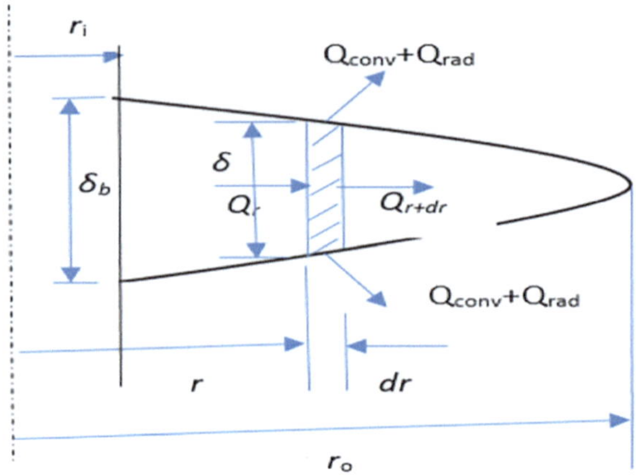

Fig. 1 Profile in general of annular fin and energy interactions on element

$$\frac{d}{dr}\left(A(r)k(T)\frac{dT}{dr}\right) - 2h_c(T)P(r)(T - T_a) - 2\varepsilon P(r)\sigma(T^4 - T_a^4) = 0 \quad (1)$$

If δ_b is the thickness at base and δ_t at tip, the thickness of different profiles at radius r is given by

Profile	Expression for Thickness, δ(r)
Rectangular	$\delta_t = \delta_b$
Triangular	$\delta(r) = \delta_b - \left(\frac{\delta_b}{r_o - r_i}\right)(r - r_i)$
Trapezoidal	$\delta(r) = \delta_b - \left(\frac{\delta_b - \delta_t}{r_o - r_i}\right)(r - r_i)$
Hyperbolic	$\delta(r) = \delta_b\left(\frac{r_b}{r}\right)$
Half - elliptical	$\delta(r) = 2 \times \frac{\delta_b}{(r_o - r_i)} \times \sqrt{(r_o - r_i)^2 - (r - r_i)^2}$

(2)

The Eq. (1) can also be written as

$$\frac{d}{dr}\left(A(r)k(T)\frac{dT}{dx}\right) - h(T)P(r)(T - T_a) = 0 \quad (3)$$

where $h(T)$ represents heat transfer coefficient and includes the effects of convection and radiation both.

Boundary Conditions: The cylindrical surface on which the fin is mounted is maintained at temperature T_b. There is no heat transfer from the tip of the fin i.e. it represents an adiabatic tip condition of the fin. These boundary conditions can be written as

$$T = T_b \quad \text{at} \quad r = r_i \quad \text{and} \quad \frac{dT}{dr} = 0 \quad \text{at} \quad r = r_o \quad (4)$$

The current paper uses moving least square scheme for interpolation of function. It has a drawback of not possessing Kronecker delta function property which makes imposition of essential boundary condition difficult. Additional method is used for this. The current article uses the penalty method for this purpose.

Equation (3) in terms of weighted residual form and after the inclusion of the penalty parameter α can be expressed as

$$\int_{\Omega_Q} \left\{ \frac{d}{dr}\left(A(r)k(T)\frac{dT}{dr} \right) - 2h(T)P(r)(T - T_a) \right\} v \, d\Omega - [\alpha(T - T_b)v]_{r=r_b} = 0 \tag{5}$$

where Ω_Q is the local domain chosen to perform the integration and v is the test function. The first term of the above equation is reduced to weak form by performing integration by parts which is desirable to apply the MLPG method. Equation (5) can now be written as

$$A(r)vk(T)\frac{dT}{dr}\mathbf{n}\Big|_{\partial\Omega_Q} - \int_{\Omega_Q} \left[A(r)k(T)\frac{dT}{dr}\frac{dv}{dr} + 2h(T)P(r)Tv - 2h(T)P(x)T_a v \right] d\Omega - \alpha(T - T_b)v]_{r=r_b} = 0 \tag{6}$$

where $\partial\Omega_Q$ is the boundary of the local domain Ω_Q and $\mathbf{n}$ is the unit normal to the boundary. Unlike Galerkin formulation, Petrov–Galerkin chooses trial and test function from different spaces and the current research picks MLS weight function as the test function, v. It attains a zero value at the boundary of local domain. Hence the first term of the above equation yields non-zero value only when the local domain intersects the global domain. Finally, The MLS trial function can be written as:

$$T(\mathbf{x}) = \sum_{i=1}^{n_s} \Phi_i T_i = \mathbf{\Phi T} \tag{7}$$

where n_s is the number of nodes in the support domain those will participate in formulation of trial function at arbitrary point $\mathbf{x}$ in the computational domain. The weak form of governing Eq. (6) is integrated over the local domain of each node of the domain after substituting Eq. (7) into it. A set of non-linear equations in following form is achieved.

$$\mathbf{K(T)T} = \mathbf{F(T)} \tag{8}$$

The matrix elements of Eq. (8) can be elaborated as:

$$K_{ij} = \int_{\Omega_Q} \begin{pmatrix} A(r)\frac{\mathrm{d}v}{\mathrm{d}r}k(T)\frac{\mathrm{d}\Phi_j}{\mathrm{d}r} + v_i 2h(T)P(r)\Phi_j \end{pmatrix} \mathrm{d}\Omega \\ -\left[A(r)v_i\frac{\mathrm{d}\Phi_j}{\mathrm{d}r}n_x\right]_\Gamma + \left[\alpha v_i \Phi_j\right]_{r=r_i} \tag{9}$$

where Γ is the global boundary.

$$F_i = \int_{\Omega_Q} (2h(T)P(r)T_a)v_i \mathrm{d}\Omega + [\alpha v_i T_b]_{r=r_i} \tag{10}$$

The annular fin chosen in the current problem has a temperature dependent thermal conductivity, $k(T)$. The heat transfer coefficient is also taken a function of temperature. This makes set of equations represented by Eq. (8) nonlinear. This nonlinearity of the problem is handled using an iterative method. This iterative method can be explained as follow:

$$\mathbf{K}(\mathbf{T}_p^{n+1})\mathbf{T}_{p+1}^{n+1} = \mathbf{F}(\mathbf{T}_p^{n+1}) \tag{11}$$

where $\mathbf{T}_0^{n+1} = \mathbf{T}^n$. p is the iteration counter. In the successive stages of iteration results start converging and termination of the loop is exercised when user defined condition of accuracy $\left(\left\|\mathbf{T}_{p+1}^{n+1} - \mathbf{T}_p^{n+1}\right\| < \in\right)$ is achieved. Here $\in$ is the set convergence criteria.

Calculation of fin efficiency is done to understand performance of the fin under different conditions. Fin efficiency is defined as,

$$\eta_{\text{fin}} = Q_{\text{fin}}/Q_{\text{max,fin}} \tag{12}$$

$$Q_{fin} = \int_{r_i}^{r_o} h(T)(T - T_a)P(r)\mathrm{d}r \tag{13}$$

and

$$Q_{\text{max,fin}} = \int_{r_i}^{r_0} h(T)(T_b - T_a)P(r)\mathrm{d}r = h_a(T_b - T_a)\int_{r_i}^{r_o} P(r)\mathrm{d}r \tag{14}$$

4 Results and Discussion

The MLPG method is applied to solve nonlinear annular fin problem. In this section results are presented and discussed in sequence: validation of result at first and then effects of different parameters on the variation of temperature and performance of the fin. A geometrically optimized fin of rectangular profile under convection is selected as reference problem [17] for validation. Geometric parameters of the fin and temperature conditions related to the current computational analysis are: Sink temperature = 35 °C, Fin base temperature = 135 °C, fin thickness at base = 0.00329 m, height of the fin over the base = 0.06 m and radius of the cylindrical base = 0.06 m. An annular fin with rectangular profile assuming thermal conductivity and heat transfer coefficient constant ($k = 228$ W/m °C, $h = 150$ W/m^2 °C) is chosen to solve initially. The exact solution of the problem is available and are plotted with the MLPG result shown in Fig. 2. An excellent agreement is observed.

Keeping the base thickness constant, the problem is solved using MLPG method selecting different profiles one by one. The ratio of the thickness at the tip to its base in case of the trapezoidal profile fin is taken 0.5. Thermal conductivity and heat transfer coefficient are assumed temperature-dependent and radiation effect is included for nonlinear analysis. For trial function the MLS scheme is used with linear base. All the line local domains are covered with ten number of gauss points placed appropriately to perform integration numerically. Related codes are written and executed on Dev C++ platform.

In the next analysis, keeping all parameters same as of the previous one, the thermal conductivity of the fin material is assumed varying as $k(T) = k_0(1 + \beta T)$, $k_0 = 228$ W/m °C and $\beta = -0.001$ i.e. for the case when it decreases with increasing temperature. The variation of temperature in different profiles are shown in Fig. 3. Temperature in all profiles decreases monotonically towards the tip of the fin. Among all the profiles the rectangular profile fin observes highest temperature at a given location followed by half-elliptic, trapezoidal, hyperbolic and triangular in decreasing order.

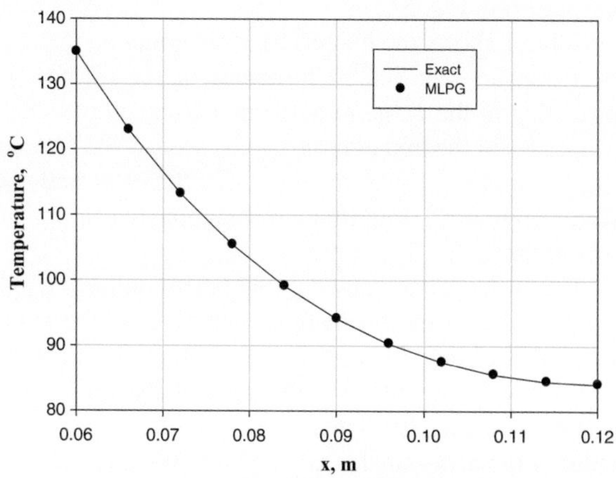

Fig. 2 Validation of result

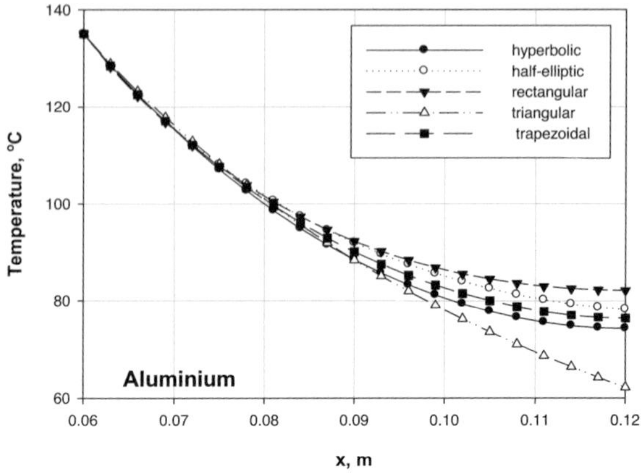

Fig. 3 Temperature Variation in radial fin of different profiles when thermal conductivity varies with temperature

Figure 4 shows temperature variation in the all selected profile radial fins under the condition of temperature-dependent heat transfer coefficient. A power law is used for variation of heat transfer coefficient given by $h(T) = h_0\left[\frac{T-T_a}{T_b-T_a}\right]^{nc}$, $h_0 = 150$ W/m^2 °C and n_c = ¼ and 2 which appropriately represents different flow condition and surrounding conditions. Values of n_c relates different physical conditions such as convection with constant heat transfer coefficient ($n_c = 0$), natural convection under laminar flow ($n_c = 1/4$), natural convection under turbulent flow ($n_c = 1/3$), laminar film boiling or condensation ($n_c = -$ 1/4), are to be mentioned a few. Only two values are selected here in the current analysis. A higher positive value of n_c represents weaker strength of convection and this allows fin to be maintained higher average temperature. A comparison of tip temperature of the fin in two parts of Fig. 4 is in the same line of explanation.

Finally, the effect of radiation is included in the analysis with two different values of surface emissivity. The higher is the value of emissivity, the higher will be the heat transfer rate due to radiation. The variation of temperature in different profiles are shown in Fig. 5.

Table 1 shows the fin performance under all the above conditions. Fin efficiency, heat transfer rate and performance of the fin based on the volume adjusted are presented in the Table to perform a comparative study. Performance with constant properties in the first column can be seen as the reference. For this condition performance of the profiles in terms of efficiency and heat transfer rate are in order from maximum to minimum: rectangular, half-elliptical, trapezoidal, hyperbolic and triangular.

This order is maintained for all other nonlinear conditions as well. As the thermal conductivity of the fin material decreases in the case of variable thermal conductivity, the average fin surface temperature also decreases, resulting in a decrease in heat transfer rate and fin efficiency of the given profile. With a variable convection exponent n_c, its higher value decreases the rate of convection heat transfer and reduces the heat transfer rate and fin efficiency. The effect of radiation is observed

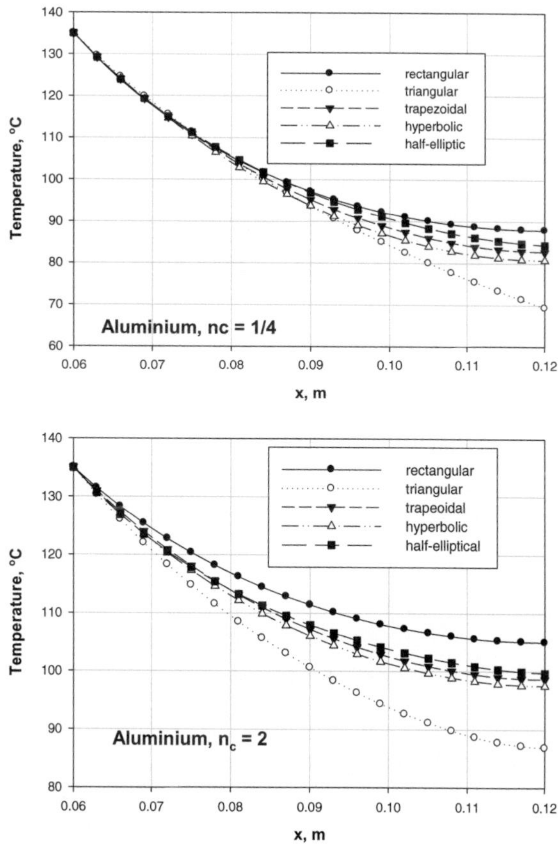

Fig. 4 Temperature variation in radial fin of different profiles when heat transfer coefficient varies with temperature

to be minimal due to lower temperature range; however, it is still significant to be included in the analysis. The last part of the Table 1 shows performance of different profile annular fin under the adjusted volume. The rate of heat transfer per unit volume from an annular fin of a rectangular profile with constant properties is chosen as the reference and that of other profiles under different operating conditions is calculated as its multiplier. It may be a wiser way to select a fin profile to make it economical and lighter. It is observed that a triangular profile outperforms all others in this category. Fins with trapezoidal and half-elliptical profiles perform nearly same and performance of hyperbolic profile fin lies above them. Even for a given profile, this ratio has a significant variation under different operating conditions. Hence, it is very appropriate to perform nonlinear analysis to improve the accuracy of the prediction. So based on the application, the appropriate profile can be selected. But, the selection of fin is judiciously done keeping in mind the cost of material, the ease of manufacturing, and most importantly, the requirements of the situation. In a situation when it is required to release heat at the maximum rate, the rectangular profile is the obvious choice.

Fig. 5 Temperature variation in radial fin of different profiles when radiation is included

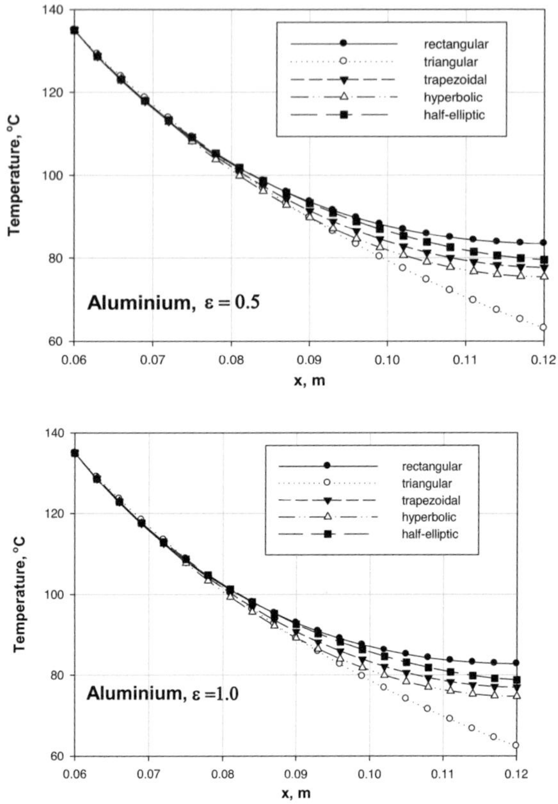

5 Conclusions

Current article has performed numerical analysis on annular fin for nonlinear condition using a truly meshfree method. The MLS scheme, the penalty method and an iterative approach are used for function approximation, essential boundary condition imposition and to handle nonlinearities arising due to temperature dependent material properties and other conditions, respectively. Numerical experiments are carried out on annular fin with five different profiles with a view to understand their performances in the convective and convective-radiative environments. Based on the requirement of higher rate of heat transfer, rectangular profile proved to be most suitable. Other profiles considered in this article are useful to reduce the overall weight and cost thermal system. Performance based on their volume gives a reasonable picture of comparison among all the profiles. Analysis in the nonlinear condition includes performance in actual working conditions. This can improve the accuracy of thermal system.

Table 1 Performance of different profile annular fin under different conditions

Profile	Constant properties	Variable k $\beta = -0.001$	Variable h $n_c = 1/4$	Variable h $n_c = 2$	With radiation $\varepsilon = 0.5$	With radiation $\varepsilon = 1.0$
Efficiency of the fin						
Rectangular	0.62	0.60	0.58	0.48	0.61	0.60
Triangular	0.54	0.53	0.51	0.33	0.54	0.53
Trapezoidal	0.59	0.57	0.55	0.42	0.58	0.57
Hyperbolic	0.57	0.56	0.54	0.40	0.56	0.55
Half-elliptical	0.60	0.59	0.57	0.43	0.59	0.59
Heat Transfer from the fin, W						
Rectangular	626.90	608.67	589.35	491.71	639.14	651.12
Triangular	553.94	537.36	520.90	333.55	564.26	574.39
Trapezoidal	598.06	580.25	561.81	424.39	609.41	620.52
Hyperbolic	582.55	564.97	547.01	409.58	593.42	604.07
Half-elliptical	614.71	596.72	577.81	433.65	626.61	638.25
Heat Transfer per unit volume relative to that of rectangular profile with constant properties						
Rectangular	1.00	0.97	0.94	0.78	1.02	1.04
Triangular	1.99	1.93	1.87	1.06	2.03	2.06
Trapezoidal	1.32	1.28	1.24	0.94	1.35	1.37
Hyperbolic	1.39	1.35	1.31	0.98	1.42	1.45
Half-elliptical	1.32	1.28	1.24	0.93	1.35	1.37

Nomenclature

$A(r)$ Fin cross-sectional area at radius r (m^2)
h Heat transfer coefficient (W/m^2 °C)
k Thermal conductivity (W/m °C)
m Thermo-geometric parameter ($/m^2$)
n_x Outward unit normal (–)
n_c Convection parameter (–)
P Perimeter of the fin (m)
Q_{fin} Rate of heat transfer from fin (W)
r Arbitrary radius (m)
T_a Temperature of sink (°C)
T_b Temperature of the fin base (°C)
v Test function (–)
α Penalty parameter (–)
β Thermal conductivity coefficient (/°C)
Γ Boundary of global domain (–)
δ_r Thickness of the fin at radius r (m)

δ_b Thickness of the fin base (m)
η_f Fin efficiency (–)
Φ Shape function (–)
Ω Global computational domain (–)
Ω_Q Local domain of integration (–)

References

1. Mokheimer EM (2002) Performance of annular fins with different profiles subject to variable heat transfer coefficient. Int J Heat Mass Transf 45(17):3631–3642
2. Arauzo I, Campo A, Cortés C (2005) Quick estimate of the heat transfer characteristics of annular fins of hyperbolic profile with the power series method. Appl Therm Eng 25(4):623–634
3. Campo A, Cui J (2008) Temperature/heat analysis of annular fins of hyperbolic profile relying on the simple theory for straight fins of uniform profile. J Heat Transfer 130(5):054501
4. Kundu B (2009) Analysis of thermal performance and optimization of concentric circular fins under dehumidifying conditions. Int J Heat Mass Transf 52(11–12):2646–2659
5. Kang HS, Look DC Jr (2009) Optimization of a trapezoidal profile annular fin. Heat Transf Eng 30(5):359–367
6. Khani F, Aziz A (2010) Thermal analysis of a longitudinal trapezoidal fin with temperature-dependent thermal conductivity and heat transfer coefficient. Commun Nonlinear Sci Numer Simul 15(3):590–601
7. Darvishi MT, Khani F, Aziz A (2016) Numerical investigation for a hyperbolic annular fin with temperature dependent thermal conductivity. J Propul Power 5(1):55–62
8. Ndlovu PL, Moitsheki RJ (2019) Analysis of temperature distribution in radial moving fins with temperature dependent thermal conductivity and heat transfer coefficient. Int J Thermal Sci 145:106015
9. Celentano D, Campo A (2021) Numerical heat transfer in annular fins of curved profile formed with the intersection of two equal circles. Int J Numer Meth Heat Fluid Flow 31(11):3343–3355
10. Kundu B, Yook SJ (2021) An accurate approach for thermal analysis of porous longitudinal, spine and radial fins with all nonlinearity effects–analytical and unified assessment. Comput Appl Math 402:126124
11. Atluri SN, Zhu T (1998) A new meshless local Petrov-Galerkin (MLPG) approach in computational mechanics. Comput Mech 22:117–127
12. Atluri SN, Zhu T (1998) A new meshless local Petrov-Galerkin (MLPG) approach to nonlinear problems in computer modeling and simulation. CMES Comput Modeling Eng Sci 3:187–196
13. Vyas P, Thakur H, Darji V (2019) Nonlinear analysis of convective-radiative longitudinal fin of various profiles. Int J Numer Meth Heat Fluid Flow 30(6):3065–3082
14. Safarpoor M, Shirzadi A (2019) A localized RBF-MLPG method for numerical study of heat and mass transfer equations in elliptic fins. Eng Anal Boundary Elem 98:35–45
15. Garg R, Thakur H, Tripathi B (2020) Numerical simulation of two-dimensional fins using the meshless local Petrov–Galerkin method. Eng Comput 37(8):2913–2938
16. Thakur H (2021) Nonlinear heat transfer analysis of spines using MLPG method. Eng Anal Boundary Elem 131:15–26
17. Kraus D, Aziz A, Wetlty J (2001) Extended surfaces heat transfer. Wiley, NY

Experimental and Numerical Investigation of Heat Transfer and Fluid Flow Through a Rectangular Duct with Pulsating Flow

Mohammad Kamran, Amir Yousf Sofi, and Adnan Qayoum

Abstract The present study investigates the heat transfer and fluid flow characteristics numerically as well as experimentally in a rectangular duct with pulsating velocity at the inlet. The numerical analysis is carried out using Finite Element Method (FEM) using k-ω turbulence model. Air (Pr = 0.71) is used as working fluid and is assumed to be incompressible and inviscid. The experimental setup features a test section that is 130 cm long with an aspect ratio of 1.2. During the experiments, the Reynolds number of the working fluid is adjusted to span from 12,800 to 35,000. Additionally, the pulsation frequency of the fluid is varied, ranging from 0 to 60.5 Hz. The experimental and simulation results depict enhancement of heat transfer. A maximum heat transfer augmentation of 19.5% is observed at a Reynolds Number of 29,000 and at a frequency of 15.5 Hz. The local Nusselt Number downstream of the duct first increases and then decreases at the end of the heated plate.

Keywords Heat transfer · Fluid flow · Pulsating velocity · Finite Element Method (FEM) · Pulsation frequency

1 Introduction

Heat transfer augmentation in rectangular ducts is of paramount importance in the field of heat and fluid flow analysis [1]. Rectangular ducts are commonly employed in various heat transfer systems, such as cooling channels in combustion chambers, compact heat exchangers, nuclear reactors and cooling networks for gas turbines [2–6]. In thermal science and fluid dynamics, forced turbulent heat convection in a rectangular duct is one of the basic issues. Many scholars began working as early as the first decades of the previous century, studying thermo-hydrodynamics of single-phase pulsing flow in channels or pipes ranging from conventional to mini/micro

M. Kamran · A. Y. Sofi · A. Qayoum (✉)
Department of Mechanical Engineering, National Institute of Technology Srinagar, Hazratbal, Srinagar, India
e-mail: adnan@nitsri.ac.in

A. Kumar R. et al. (eds.), *Proceedings of Fluid Mechanics and Fluid Power (FMFP) 2023, Vol. 6*, Lecture Notes in Mechanical Engineering,
https://doi.org/10.1007/978-981-96-9600-0_2

level. They did this experimentally, analytically and computationally. Pulsating fluid flow and heat transfer are common in various industrial applications, including pulse combustors used in both civilian and military contexts, and reciprocating engines. The periodic variations in fluid flow lead to unique flow and heat transfer characteristics. To investigate the impact of pulsation, Velazquez et al.'s study [7] used a pulsing flow to improve laminar heat transmission ahead of a step that turns backward. They showed an enhancement of 55% in the Nusselt Number. The forced pulsating convective flow in a parallel-plate channel with two porous connected strip heaters at the bottom wall was explored numerically by Huang et al. [8]. They discovered that a technique for augmenting heat transfer and cooling electronic equipment involves combining flow pulsation with a limited-sized fibre permeable heat sink. The effect of pulsation on heat transfer in a heat exchanger for parallel and counter water flows was investigated by Zohir et al. [9]. They found heat transfer enhancement of 20% for parallel flow and 90% for counter flow. The pulsing laminar flow of an incompressible, viscous liquid in a rectangular conduit was examined by Yakhot et al. [10]. A pulsating pressure disparity is applied to produce motion. They demonstrated that the induced velocity exhibits a phase lag relative to the applied pressure oscillations, ranging from no lag at very slow oscillations to a 90-degree lag at high oscillation frequencies. In an expanded rectangular channel, Yoshikawa et al. [11] computed coupled flow and heat transmission numerically in three dimensions. Studies have demonstrated that when the Reynolds number exceeds a threshold point, the symmetric channel flow becomes uneven and deflects to a single side of the wall. They also discovered that the channel flow is greatly influenced and heat transmission on the wall is increased by the inlet flow, which has a pulsating variation in the Strouhal number, which varies between 0.05 and 0.1. Mulcahey et al. [12] investigated ordered 2-D pulsation flow over a homogeneous linear design of heated square tubes at 0–80 Hz pulsation frequencies for 50–200 Reynolds numbers. They demonstrated how heat transmission in tube arrays may be improved by flow pulsation and how this improvement rises with pulsation period and average inflow Reynolds Number. Using both experimental and numerical methods, the main goal of this work is to investigate the effects of Reynolds number and pulsation frequency on heat transfer in a rectangular duct. The objective is to investigate how varying pulsation frequency and Reynolds number can enhance heat transfer, with the ultimate aim of utilizing pulsating flow for this purpose.

2 Experimental Setup

The experimental test section is 330 cm long with an aspect ratio of 1:2 and is shown in Fig. 1. The heated section is 100 cm long. Temperature measurements are obtained by 10 K-type thermocouples installed in the test section. The Reynolds number of the working fluid is varied from 1.28×10^4 to 3.5×10^4 by means of a suction motor and pulsation frequency is varied from 0 to 60.5 Hz. The heated section consists of a copper electric heater plate supplying a uniform heat flux. A 10 W 190 mm diameter

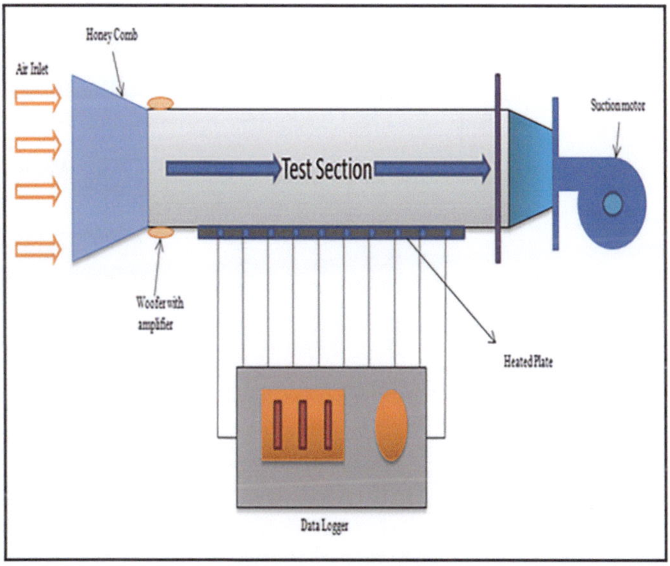

Fig. 1 Experimental setup

acoustic woofer with an impedance of 8 Ω and an amplifier is used to generate and control the pulsating flow.

3 Simulation Methodology

The simulation domain consists of a single-pass rectangular duct 100 cm in length with a height of 10 cm and is shown in Fig. 2. The numerical analysis is carried out using Finite Element Method (FEM) for an externally heated duct with a pulsating inflow along with k-ω turbulence model. Air (Pr $=$ 0.71) is used as working fluid and is assumed to be incompressible and inviscid. COMSOL Multiphysics software has been used to conduct this study. It is assumed that the flow under consideration is incompressible, turbulent, two-dimensional and unsteady. The flow at the duct's input is pulsating. Since no-slip behaviour is guaranteed by the boundary conditions at the walls, the fluid velocity with respect to the wall is zero. At the channel's output, a boundary condition is implemented to keep the gauge pressure at zero (Pgauge $=$ 0) and the Reynolds number (Re) dictates the mean inflow velocity. The energy equation's residual convergence criteria is set at 10^{-8}, the continuity equation at 10^{-4} and other equations at 10^{-5}. Following the computation of flow and heat transfer variables with the PARDISO solver, the relative tolerance is determined to be 0.005. The duct is subjected to a pulsating velocity boundary condition at the inlet specified as:

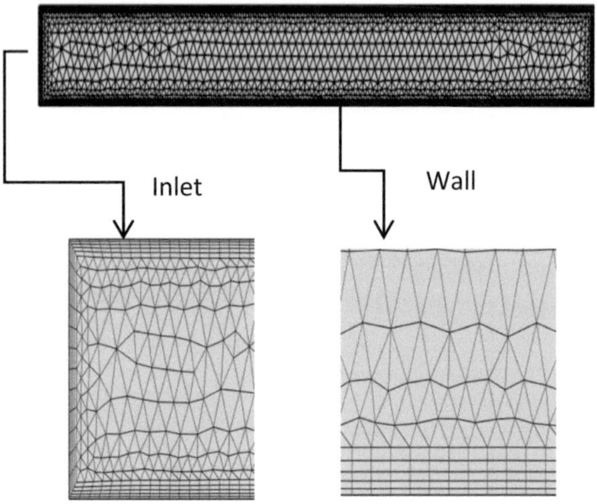

Fig. 2 Mesh domain

$$U = U_0 + A\sin(\omega t)$$

where U_0 is the mean inlet velocity, A is the amplitude of pulsating velocity which is kept constant and ω is the pulsation frequency. The Reynolds number determines the average flow velocity, and the gauge pressure is zero at the channel's exit. For continuity equations, the convergence criterion is set to 10^{-4}, whereas for other equations, it is set to 10^{-5}.

3.1 Grid Independence Study

To ensure that the simulation results are devoid of the computational mesh size, research on grid independence is conducted. Figure 3 plots the mean Nusselt Number of the smooth duct as a function of grid size using an unstructured mesh with a variable number of elements. The variation of average Nusselt Numbers is relatively very small when the mesh element numbers exceed 80,000. Thus to reduce the computation time, a grid containing 80,000 elements is selected for the present study.

4 Data Reduction and Governing Equations

It is assumed that the flow in this study is turbulent, two-dimensional and time dependent. The flow depicted above is governed by the continuity equations, energy equations and Reynolds Averaged Navier–Stokes (RANS) equations. To simulate the turbulent flow, *k*–ω turbulence model is employed. The average Nusselt Number is evaluated by using the heat transfer coefficients as follows [13]:

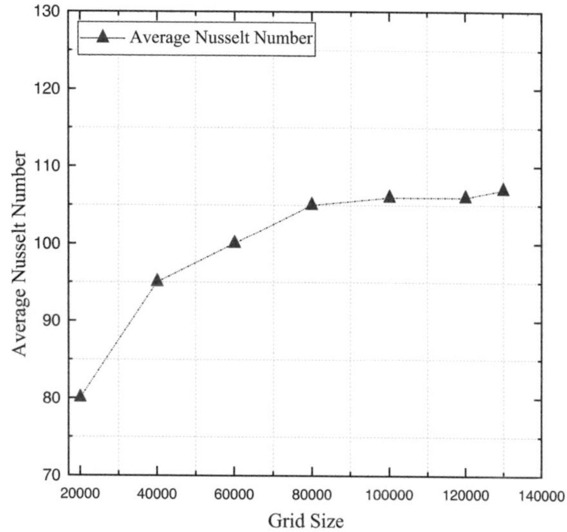

Fig. 3 Grid independence study

$$\overline{\mathrm{Nu}} = \frac{hD_h}{k}$$

In this case, h denotes the coefficient of heat transfer, k represents the thermal conductivity of air and D_h is the channel's hydraulic diameter. The coefficient of heat transfer is computed from the Newton's law of cooling as follows [14]:

$$h = \frac{\dot{q}}{(T_w - T_f)}$$

The mean temperature of the fluid is T_f, the wall temperature is T_w and the heat that was applied is represented by $\dot{q}$.

5 Results and Discussion

5.1 Research Validation

The simulation study is validated by carrying out an experimental analysis. A 100 cm long heated section with a 1000 W/m^2 heat flux applied to the bottom wall and pulsating velocity at the entry make up the experimental setup. Figure 4 presents an experimental and numerical evaluation of the average Relative Nusselt Number variation at various frequencies at a Reynolds Number of 29,000. The Average Nusselt

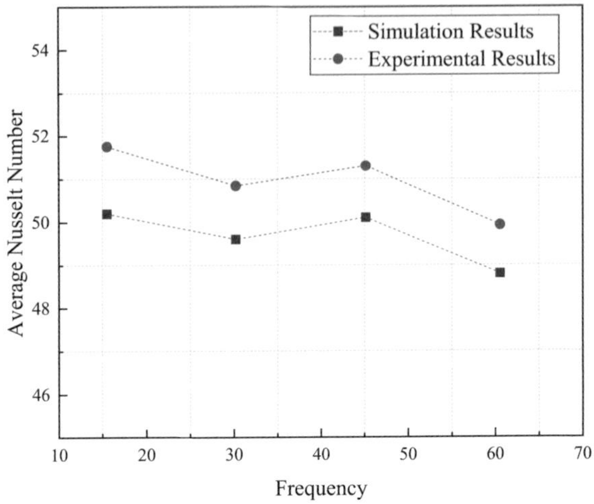

Fig. 4 Validation of the simulation study with the experimental analysis at Re = 12,800

Number variation in experimental analysis follows a similar trend as that of numerical analysis. However, the experimental and numerical results depict a maximum variation of 5.8% in Average Nusselt Number.

5.2 *Heat Transfer Characteristics*

The influence of pulsation frequency on heat transfer augmentation is depicted in Fig. 5.The experimental as well as simulation results indicate that Relative Nusselt Number either increases or decreases with the increase in Reynolds Number and Pulsation Frequency. Pulsating flow conditions can be strategically manipulated to achieve specific heat transfer objectives. By understanding and controlling the impact of pulsations on boundary layer behaviour, turbulence intensity and thermal mixing, we can maximize heat transfer rates, leading to more efficient and effective thermal management in various applications. For a pulsation frequency of 15.5 Hz at a Reynolds Number of 29,000, there is a 19.5% enhancement in heat transfer rate. Figure 6 shows a comparison of experimentally as well as numerically obtained Average Relative Nusselt Numbers at different Reynolds Numbers for different frequencies. From the graph it is clearly evident that the Average Relative Nusselt Numbers obtained in the experimental and numerical analysis follow a similar trend. However, the simulation results are slightly under predicted due to a different mesh size or due to some convective heat losses in the experimental analysis.

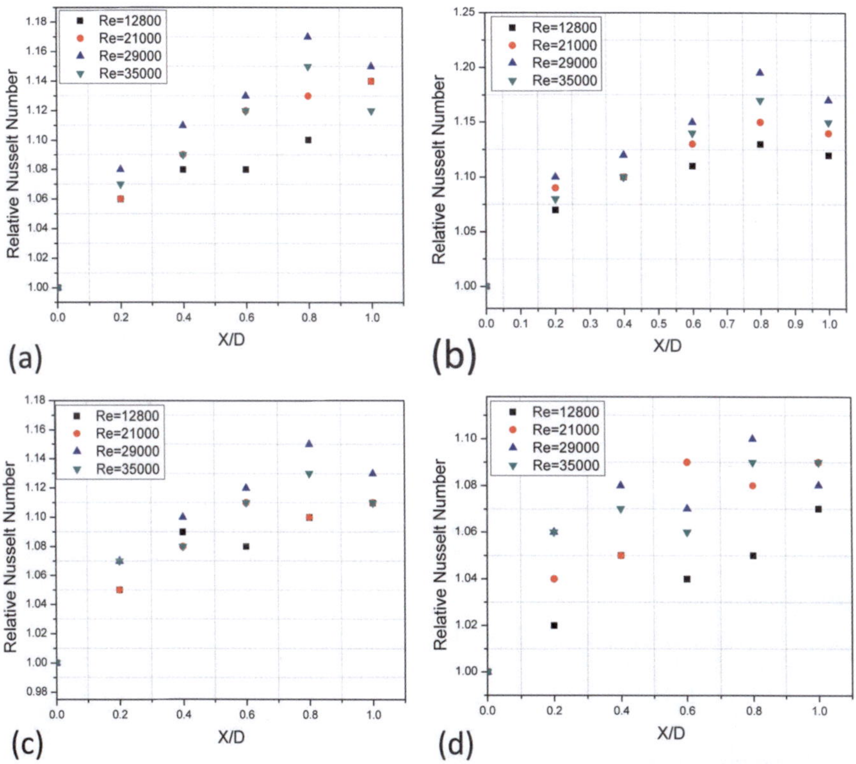

Fig. 5 Span-wise variation of Relative Nusselt Number at different frequencies **a** $f = 15.5$ Hz, **b** $f = 30.2$ Hz, **c** $f = 45.1$ Hz, **d** $f = 60.5$ Hz

6 Conclusions

The present study is focused on investigation pulsating turbulent flow by carrying out a numerical as well as experimental analysis. Some of the prominent conclusions from the present study are:

1. A similar pattern is shown by both the experimental and numerical analyses, which show enhanced heat-transmitting features in a duct due to pulsating flow.
2. Local Nusselt Numbers downstream of the duct initially increases and subsequently decreases near the end of the heated plate.
3. At a frequency of 15.5 Hz and a Reynolds Number of 29,000, a maximum 19.5% increase in heat transfer is noticed.
4. More detailed and systematic experimental studies need to be conducted to explore the influence of various parameters such as pulsation frequency, amplitude and waveform on heat transfer in different duct geometries and investigate the effects of non-uniform pulsating flows, including asymmetric and skewed pulsations, on heat transfer characteristics.

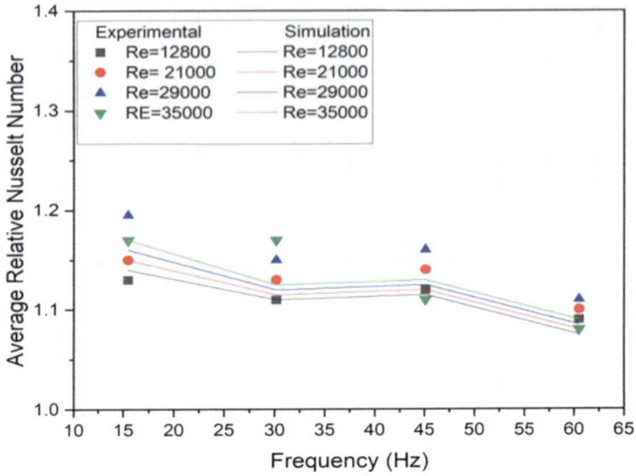

Fig. 6 Comparison between simulation and experimental results

Nomenclature

$\overline{Nu}$	Average Nusselt Number (–)
AR	Aspect ratio (–)
h	Heat transfer coefficient (W/m^2 K)
D_h	Hydraulic diameter (m)
ρ	Density of air (kg/m^3)
ω	Rotor rotational speed (rad/s)
k	Thermal conductivity of air (W/mK)
T_w	Wall temperature (K)
T_f	Fluid temperature (K)

References

1. Kamran M, Qayoum A (2022) Effect of ZnO nanofluids on thermo-hydraulic characteristic of flow in a heated duct. Energy Sources Part A Recover Util Environ Eff 44(4):10681–10693. https://doi.org/10.1080/15567036.2022.2154871
2. Kamran M, Qayoum A (2023) Experimental investigation of natural convection of Fe_3O_4-water nanofluid in a cubic cavity. J Dispers Sci Technol 1–11. https://doi.org/10.1080/01932691.2023.2173607
3. Dilawar M, Qayoum A (2022) Performance study of aluminium oxide based nanorefrigerant in an air-conditioning system. Res Eng Struct Mater 1–16. https://doi.org/10.17515/resm2022.459ma0701
4. Ali B, Qayoum A, Saleem S, Mir FQ (2023) Effect of graphene/hydrofluoroether (HFE-7100) nanofluids on start-up and thermal characteristics of pulsating heat pipe. J Therm Anal Calorim 148(9):3819–3834. https://doi.org/10.1007/s10973-022-11929-w
5. Bhat AY, Qayoum A (2023) Synergistic impact of tube configuration and working fluid on photovoltaic-thermal system performance. Renew Energy 207(September 2022):205–217. https://doi.org/10.1016/j.renene.2023.02.086

6. Yousuf Bhat A, Qayoum A (2023) Performance enhancement of photovoltaic-thermal system using hybrid tubes: an assessment of thermodynamic and thermohydraulic efficiencies. Appl Therm Eng 230(PA):120652. https://doi.org/10.1016/j.applthermaleng.2023.120652
7. Velazquez A, Arias JR, Mendez B (2008) Laminar heat transfer enhancement downstream of a backward facing step by using a pulsating flow. Int J Heat Mass Transf 51(7–8):2075–2089. https://doi.org/10.1016/J.IJHEATMASSTRANSFER.2007.06.009
8. Huang PC, Yang CF (2008) Analysis of pulsating convection from two heat sources mounted with porous blocks 51(25–26):6294–6311. https://doi.org/10.1016/J.IJHEATMASSTRANSFER.2008.04.044
9. Zohir A (2012) Heat transfer characteristics in a heat exchanger for turbulent pulsating water flow with different amplitudes
10. Yakhot A, Arad M, Ben-Dor G (1999) Numerical investigation of a laminar pulsating flow in a rectangular duct. https://doi.org/10.1002/(SICI)1097-0363(19990430)29:8
11. Yoshikawa H, Kai T, Munekata M, Ohba H (2011) Effects of pulsation on separated flow and heat transfer in enlarged channel. J Therm Sci 20(1):70–75. https://doi.org/10.1007/s11630-011-0437-x
12. Mulcahey TI, Pathak MG, Ghiaasiaan SM (2013) The effect of flow pulsation on drag and heat transfer in an array of heated square cylinders 64:105–120. https://doi.org/10.1016/J.IJTHERMALSCI.2012.08.017
13. Sofi AY, Qayoum A (2023) Numerical investigation of thermo-hydraulic performance and irreversibility behaviour in a pulsating turbulent flow ribbed duct. Arab J Sci Eng. https://doi.org/10.1007/s13369-023-07902-w
14. Sofi AY, Qayoum A (2024) Heat transfer and fluid flow characteristics in a two pass duct with variable rib shapes: a CFD analysis. In: Singh KM, Dutta S, Subudhi S, Singh NK (eds) Fluid mechanics and fluid power, vol 3. FMFP 2022. Lecture notes in mechanical engineering. Springer, Singapore. https://doi.org/10.1007/978-981-99-6343-0_35

Numerical Modelling of Flame Characteristics of Radially Slotted Methane Burner

V. Pradeep, S. Raja, Ashish Mittal, Alok Deshpande, Prafull Gandhi, Ketan Parashar, and V. Raghavan

Abstract Domestic gas burners are widely used for household cooking in several countries. With the advent of climate change, there is a need for cleaner combustion. Predominantly, domestic gas burners utilize Liquefied Petroleum Gas (LPG) as a fuel for burning. It is a mixture of several hydrocarbons, with butane and propane being the major constituents. A better alternative for LPG is Natural Gas (NG), which mainly consists of methane, and it releases lower carbon emissions compared to LPG. Due to the differences in chemical characteristics of these fuels, a different burner design is necessary for firing methane. Numerical simulations are helpful in these designs. The present study aims to develop a numerical model that can accurately predict the temperature fields and carbon monoxide (CO) emissions for a multi-slot radial methane burner. The burner has been studied under two configurations: open flame and burner with vessel. Of these, the latter configuration has already been studied in the literature extensively. However, numerical studies on open flames of radially slotted burners are scarce. The results from the present numerical model match quite

V. Pradeep · S. Raja · V. Raghavan (✉)
Department of Mechanical Engineering, IIT Madras, Chennai, India
e-mail: raghavan@iitm.ac.in

V. Pradeep
e-mail: pradeepvaidyanath@gmail.com

S. Raja
e-mail: rajha93@gmail.com

A. Mittal · A. Deshpande · P. Gandhi · K. Parashar
Whirlpool of India, Gurugram, India
e-mail: ashish_mittal@whirlpool.com

A. Deshpande
e-mail: alok_c_deshpande@whirlpool.com

P. Gandhi
e-mail: prafull_gandhi@whirlpool.com

K. Parashar
e-mail: ketan_s_parashar@whirlpool.com

A. Kumar R. et al. (eds.), *Proceedings of Fluid Mechanics and Fluid Power (FMFP) 2023, Vol. 6*, Lecture Notes in Mechanical Engineering,
https://doi.org/10.1007/978-981-96-9600-0_3

well with the experimental data, with an overall accuracy of 80% or more in the temperature and CO predictions.

Keywords Domestic gas burner · Natural gas · Numerical model validation · CO emissions · Open flame configuration

1 Introduction

Domestic gas burners have been in the market for several decades. When it comes to designing a new burner, rigorous experimental testing is required. This involves several iterations of prototyping, manufacturing complex test rigs and using precise measurement equipment. Using a numerical model to simulate the working of a burner eliminates all these costs. However, the numerical model has to be validated against accurate experimental data. The present study focuses on a radially slotted burner, for which past studies are very sparse and little to no information is available on the numerical studies of this design. The challenge lies in modelling the burner under open flame condition. Most of the existing studies model the burner with a vessel on its top. Here, the flame impinges on the vessel surface and curves outwards towards the exhaust. The behaviour of the flame is significantly different under open flame condition and its modelling using a CFD code is non-trivial. The flame tends to curve inwards towards the burner centre, and this causes numerical instability while modelling the flame numerically.

The motivation for the present study stems from the need for a numerical model that can accurately predict the flame behaviour in radially slotted burners under both open flame and burner-with-vessel configurations. The lack of numerical studies of this burner design further reinforces the need for such a numerical model. This model will facilitate efficient simulations of new burner designs in the future by eliminating the need for expensive experimental setups.

2 Literature Review and Objective

Methane burners play a vital role in household heating and cooking applications. Numerical simulations offer cost-effective and efficient means of investigating burner performance, combustion characteristics, and emissions. Methane (CH_4) is considered the fuel of choice in combustion research due to its ease of use, wide availability, and simple, well-understood and tested chemical kinetics mechanism [1]. To efficiently understand the methane combustion in burners and to improve the household burner systems, CFD simulations with at least a short kinetics mechanism are needed [2]. Mostly, simulations by past researchers have been performed with simplified chemical mechanisms, such as the global chemistry or the flamelet approach [3].

However, the proper CO predictions cannot be obtained by these models [4]. Similarly, the formation of other pollutants, such as NOx, cannot be predicted by these models [5]. For modelling turbulent combustion coupled with chemical kinetics, Large Eddy Simulations (LES) are appropriate, but their implementation along with detailed kinetics is not feasible because of the computational cost. Literature reveals that the kinetic mechanisms containing 20–30 species and 80–100 reactions denote an upper limit in terms of mechanism size and are sufficient to model important combustion characteristics of the hydrocarbon fuels [4].

Hosseini and Nobari [6] modelled and simulated a methane-air premixed flame in a cylindrical combustion chamber using the standard k-ε model, finite rate/eddy dissipation model and a 5-step reaction mechanism. They used the discrete ordinate model to account for radiation. The results have been found to be in excellent agreement with experimental outcomes [6]. The standard k-ε model generally provides good results, but its accuracy will be lesser when the flow is restricted by a wall. The k-ω model is better when a wall confinement is present [7]. Regardless of this, the standard k-ε model is widely used due to its simplicity. Juan-Camilo et al. [8] simulated the high-pressure premixed methane-air burner flames. Their study showed that the temperatures predicted by the 2D models are similar to those predicted by the 3D models. However, the flame shape in the 2D model did not exactly match the 3D model. Furthermore, the numerical investigations carried out on domestic gas burners predominantly use a burner with vessel on top configuration [9–11].

Although researches have been carried out to study the detailed chemical kinetics mechanism of methane combustion, simulation of the full 3D domain has not been taken into account in most of the studies due to the limitations of computation cost and time. This study aims to report on a numerical model capable of predicting open flames, to efficiently test new gas burner designs and effectively eliminate the need for expensive experimental setups. This model is validated using in-house experimental data.

3 Experimental and Numerical Methodology

3.1 Experimental Methodology

The schematic of the experimental setup used for the present study is depicted in Fig. 1. It consists of the following components. (i) methane cylinder, (ii) pressure regulator, (iii) inline pressure regulator, (iv) pressure gauge, (v) ball valve, (vi) mass flow controller (Aalborg–DPC), (vii) burner nozzle and (viii) radially slotted burner. All the connections have been made using polyurethane pipe (ID: 10 mm and OD: 14 mm). In order to check the gas leakage, a leak-proof test has been done using N_2 gas. The experiments have been conducted after ensuring there are no leaks at any of the connections made throughout the setup.

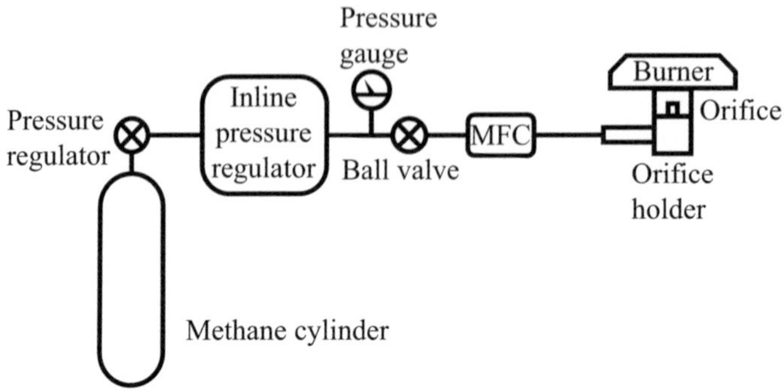

Fig. 1 Schematic of experimental setup

All the dimensions presented in the study have been made non-dimensional with respect to the radius of the burner (*R*). The temperature (*T*) has been made non-dimensional with respect to the ambient temperature (T_o).

In the experiments, temperature and CO fields have been recorded. The vessel used in the experiments has a diameter of 3.64R and a height of 2.73R. Figure 2 highlights the locations where the measurements have been taken around the burner, with a vessel kept over it. These include locations near the flame along the line 0.36R above the base of the burner, radially from a distance of 1.09R to 2.18R from the centre of the burner and near the vessel walls at radial distances of 0.09R, 0.18R and 0.27R each at heights of 0.36R, 0.72R, 1.09R and 1.45R from the base of the vessel. Temperature has been measured using K-type thermocouples fixed near the vicinity of the flame using a typical stand. The thermocouples are connected to an 8-channel data logger to obtain the measurement data. The CO measurements have been obtained using a portable gas analyser. Measurements have been taken for both open flame and burner with vessel configurations.

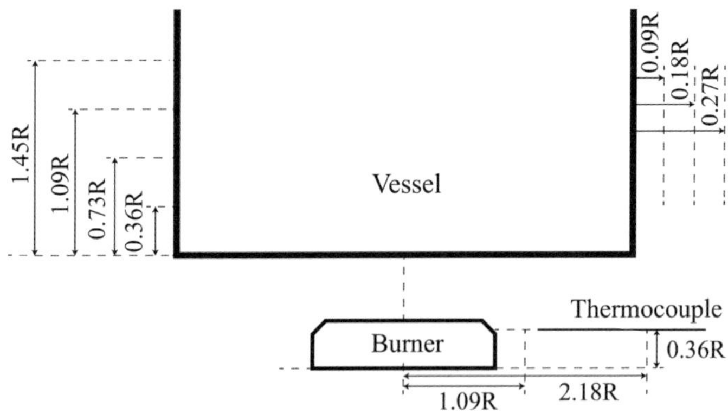

Fig. 2 Measurement locations for burner with vessel condition

The experimental temperature measurement does not account for the convective and radiative losses that happen at the tip of the thermocouple, and the values obtained from the data logger have to be corrected. The correlations suggested by Bradley and Mathews [12] are used to correct the data and account for the losses. Here, the flow velocity at the point of measurement is assumed to be the port velocity since it is difficult to determine it experimentally. This assumption is likely to affect the accuracy of the temperature correction.

3.2 *Numerical Methodology*

The numerical model is developed with the help of Ansys Workbench and the simulations have been carried out using Ansys Fluent with suitable model parameters. The domains used for the numerical models are illustrated in Fig. 3.

Here, four domains are depicted: sector domain, sector domain with an annulus, full three-dimensional (3D) domain with annulus and burner with vessel domain. Annulus refers to pressure inlet boundary condition over the burner head. This is to avoid the instability of flames curving in. The burner used in the present study

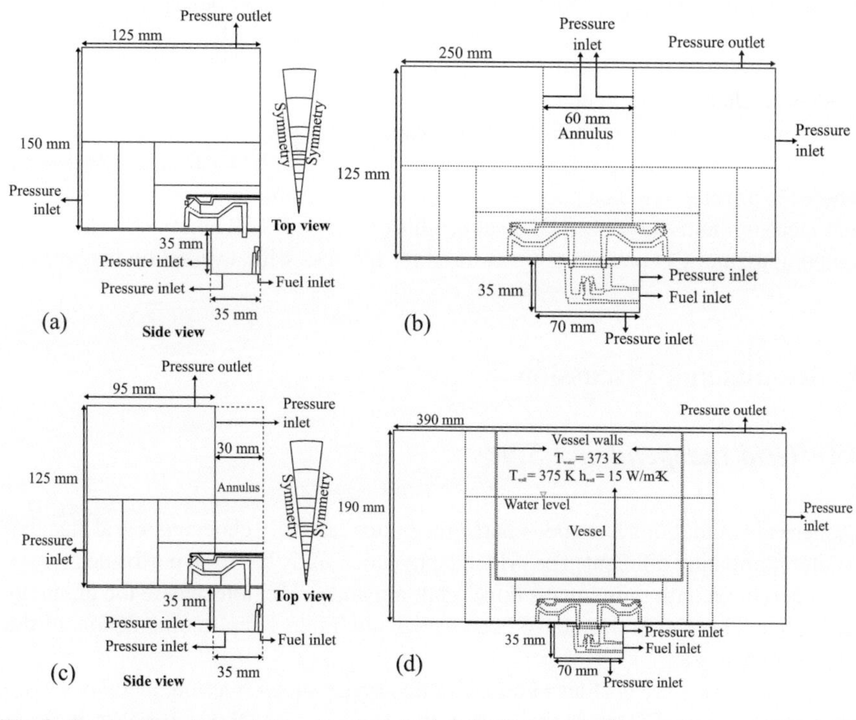

Fig. 3 Computational domain **a** sector domain, **b** full 3D domain with annulus, **c** sector domain with annulus, **d** full 3D domain with vessel

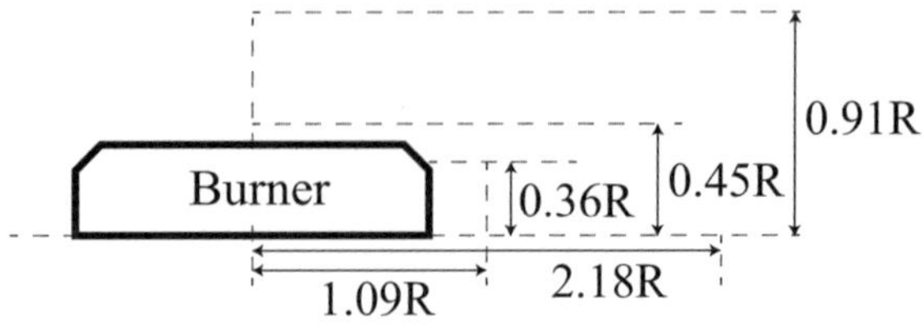

Fig. 4 Measurement locations for open flame condition

is radially slotted with 24 ports. The burner also has two special ports provided to enable ignition and to stabilize the flame in case of extinction. Despite the asymmetry introduced due to the presence of the ignition and the stability ports, the burner geometry can be simplified into an axisymmetric sector case with a single port.

Initially, cold flow simulations are performed. In the fuel inlet, a pressure inlet condition of 1246 Pa is provided. A realizable k-epsilon (k-ε) model with an enhanced wall treatment viscous model is used. Although the nature of the flow in the domain is laminar, the velocity at the exit of the orifice is of the order of 50 m/s. This creates a turbulent flow initially out of the fuel (orifice) inlet. However, the turbulent intensity decreases rapidly, resulting in laminar flames over the burner ports. A species transport model with a finite rate combustion model has been used. Here, the fuel oxidation is represented using a 25 species CHEMKIN mechanism. A full multicomponent and thermal diffusion has been taken into account. Initial trials have been performed using a 4-step methane-air mechanism. However, the literature reveals that this mechanism grossly over-predicts CO emissions around the flame zone [13]. Therefore, the 25 species mechanism has been used in all the simulations.

From the numerical model, the temperature and CO profiles have been obtained for comparison with experimental results. Figure 4 depicts the locations in the domain where the profiles have been obtained in open-flame condition. Here, radial temperature measurements have been taken at heights of 0.36R, 0.45R and 0.91R. Similarly, temperature and CO profiles are measured for the case with a vessel on the burner.

4 Results and Discussion

4.1 Grid Independence Study

Numerical simulations have been performed after carefully choosing a grid size that is computationally efficient. The grid independence study has been performed for the full 3D domain with refinements done predominantly at regions where the gradients are higher. Three grid sizes have been tested and Table 1 reports the details of the mesh for each grid.

Figure 5 shows the profiles of temperature, oxygen mass fraction, methane (CH_4) mass fraction and CO mass fraction at the flame region; the two peaks indicate the flame front. It is observed that grid 2 and grid 3 exhibit similar profiles when

Table 1 Details of grids

Grid	Elements	Orthogonal quality	Aspect ratio
1	841,496	0.1067	31.44
2	973,115	0.1066	31.44
3	1,177,303	0.1044	31.44

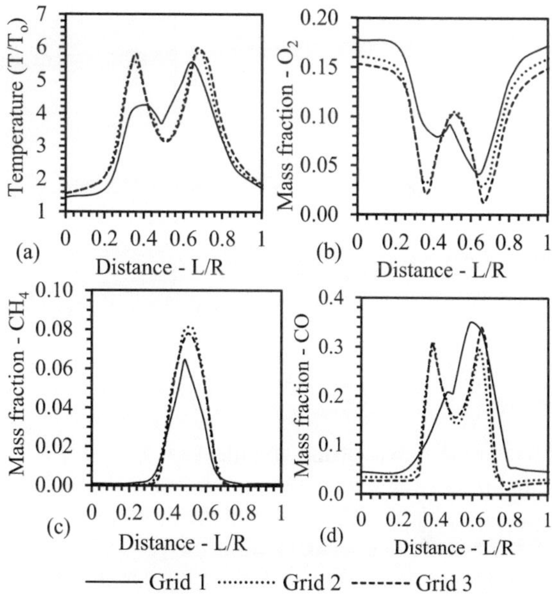

Fig. 5 Profiles of **a** temperature, **b** O_2 mass fraction, **c** CH_4 mass fraction, **d** CO mass fraction

compared to grid 1. Therefore, grid 2 has been considered for the study, considering the accuracy and running time.

4.2 Sector Domain Analysis of Open Flame Condition

Initially, the numerical simulation has been performed using a sector domain without an annulus. It is shown in Fig. 6 that the temperature contours and velocity vectors bend over the cover cap on the top of the burner, displaying an unphysical representation of the flame shape. Numerical results revealed that a strong recirculation zone exists right above the cover cap, causing the bending of the flame. In experiments, flames slightly curve up and the hot gases move vertically upwards.

To solve this numerical instability, a modified domain is used. Here, an annular zone is used coaxially on the top of the cover plate of the burner and a pressure inlet condition is specified on its periphery in the centre of the domain. Flow can either come in or go out of this boundary. A sensitivity study has also been conducted,

Fig. 6 Temperature contour and velocity vectors for open flame condition with sector domain

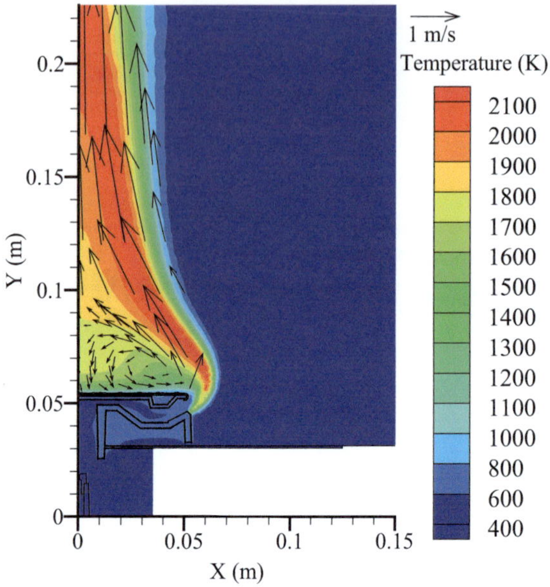

varying the radius of the annulus as 0.45R, 0.55R and 0.63R. Figure 7 shows the radial profiles of temperature predicted at the heights of 0.36R and 0.45R from the base of the burner. The X-axis of the plot denotes the radial distance from the centre of the burner. It is evident from the plot that very little variation exists between the different cases. The domain with 0.55R radius for the annulus shows results similar to the case with 0.45R annular radius. Also, in this case, the vertical and horizontal distance available for secondary air entrainment is the same. Therefore, the domain with 0.55R radius annulus has been chosen as the computational domain for validation and further studies.

After choosing the domain for the open flame study, the numerical model is used to simulate flames under three flow rates, and the predicted temperature contours are compared with the corresponding experimental flame shapes. Figure 8 illustrates the

Fig. 7 Annulus sensitivity study—profiles of temperature at **a** 0.36R height and **b** 0.45R height

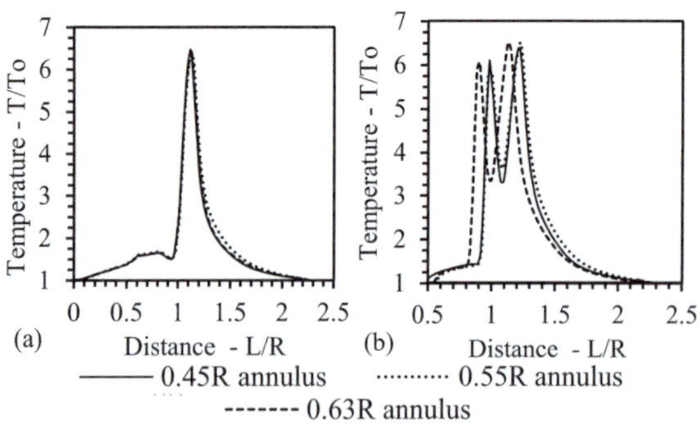

temperature contours along with the velocity vector for 5, 6.7 and 8.3 LPM of flow rates, respectively, along with experimental flame photos. Flame extents between the experimental and numerical cases compare with an accuracy of around 85%.

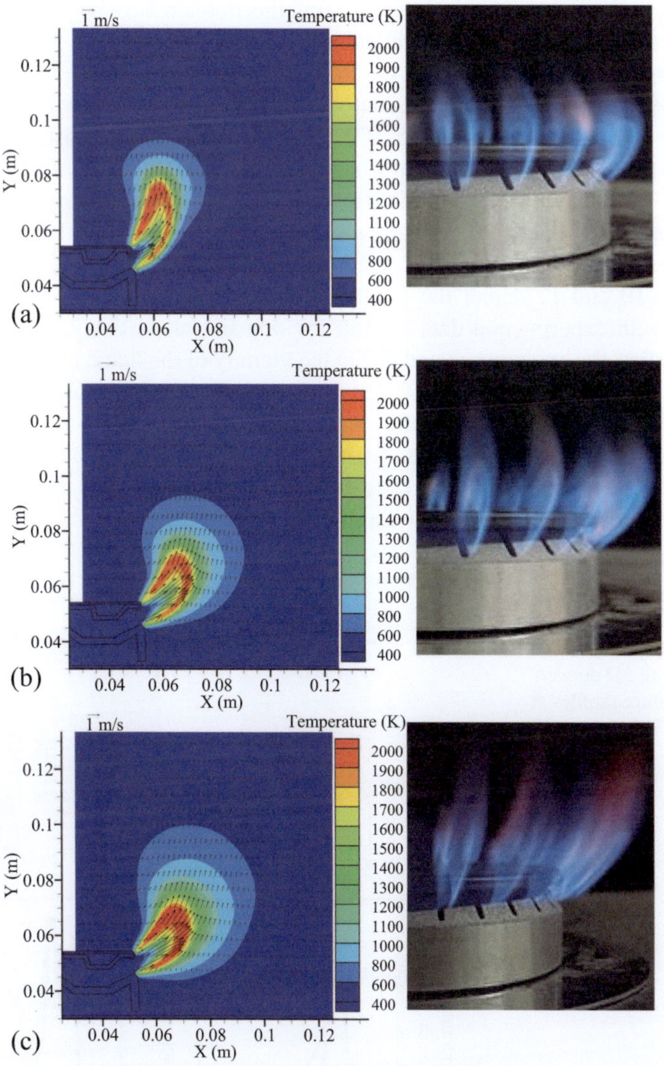

Fig. 8 Flame shape comparison for **a** 5 lpm (3 kW), **b** 6.7 lpm (4.02 kW), **c** 8.3 lpm (4.98 kW) of fuel flow rate

4.3 Full 3D Domain Analysis of Open Flame Condition

The sector domain with annulus has been further developed into a full three-dimensional domain to account for the whole burner geometry, including the ignition slots. Results from the numerical model using this domain have been first compared with the sector domain and then validated with experimental data for temperature and CO profiles.

Figure 9 shows the temperature contour in the full 3D domain with annulus and the profiles of temperature at heights of 0.36R and 0.45R from the base of the burner. In the full 3D domain, measurements have been obtained at four ports to assert the differences that may arise due to the slight asymmetry of the burner due to ignition slots. From Fig. 9, it is seen that the sector domain and full 3D domain produce similar results and the temperature profile trends overlap each other.

Figures 10 and 11 depict the predicted temperature profiles and CO profiles, validated with experimental data for full 3D domain, respectively. The numerical model predicts the temperature and CO in the vicinity of the flame under open flame condition with an accuracy of 81% and 85%, respectively. The gas analyser probe measures the time-averaged CO value. Since the probe is kept near an open flame, the surrounding air dilutes the flue gases. This causes fluctuations in the measurement. The numerical model, being a steady state case, cannot predict the fluctuations, therefore, it results in a minor over-estimation of the CO values.

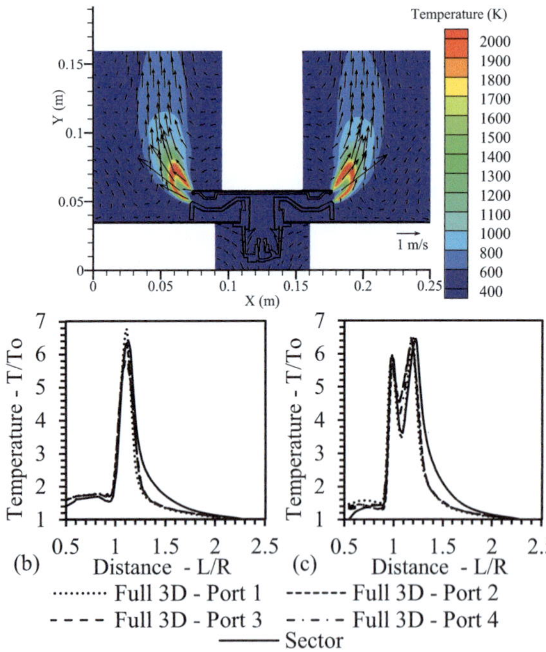

Fig. 9 Comparison of full 3D domain with sector domain **a** temperature contour of full 3D domain and temperature profiles at height, **b** 0.36R and **c** 0.45R

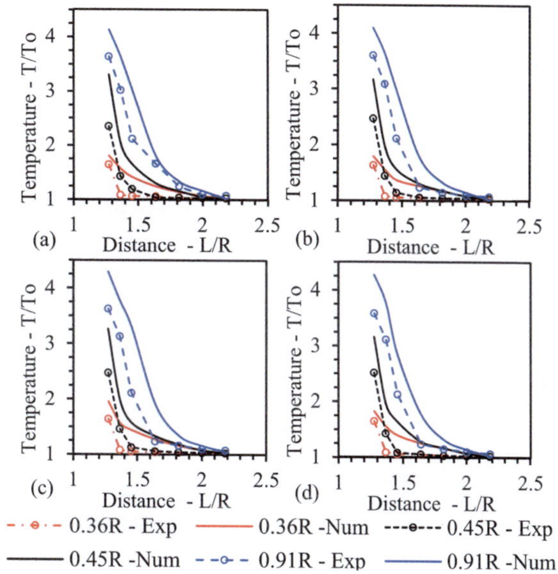

Fig. 10 Experimental validation of temperature profiles of full 3D domain at **a** Port 1, **b** Port 2, **c** Port 3, **d** Port 4

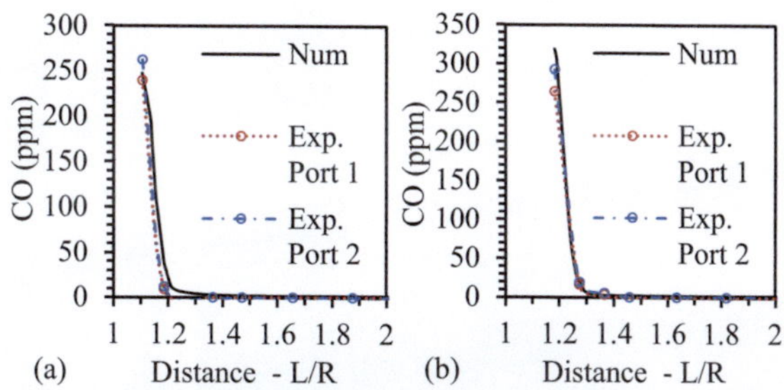

Fig. 11 Experimental validation of CO profile of full 3D domain at height of **a** 0.36R, **b** 0.45R

4.4 *Full 3D Domain Analysis of Burner with Vessel Condition*

Apart from the open flame condition, the numerical model has been validated for the burner with a vessel as well. This condition is well studied in the literature for other burner designs. However, that is not the case for a radially slotted burner. Figure 12 depicts the validation of temperature profiles near the flame as well as near the vessel wall and the CO profile near the flame. Figure 13 compares the temperature contours with the experimental flame shape. The numerical model of burner with vessel condition predicts the temperature and CO with an accuracy of 80% and 86%, respectively.

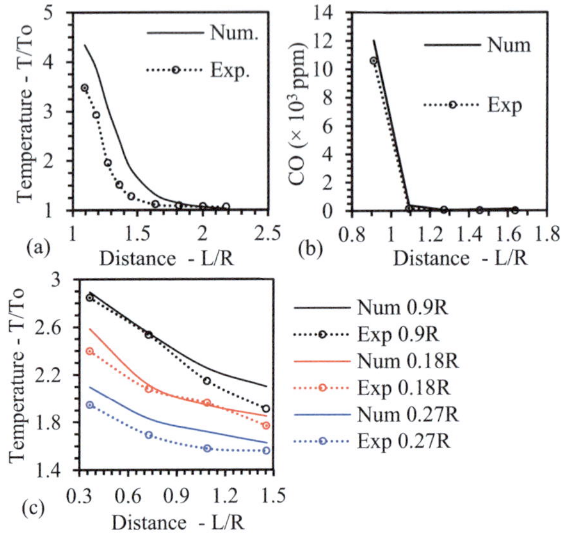

Fig. 12 Experimental validation of profile of full 3D domain with vessel for **a** temperature near flame, **b** CO near flame, **c** temperature near vessel wall

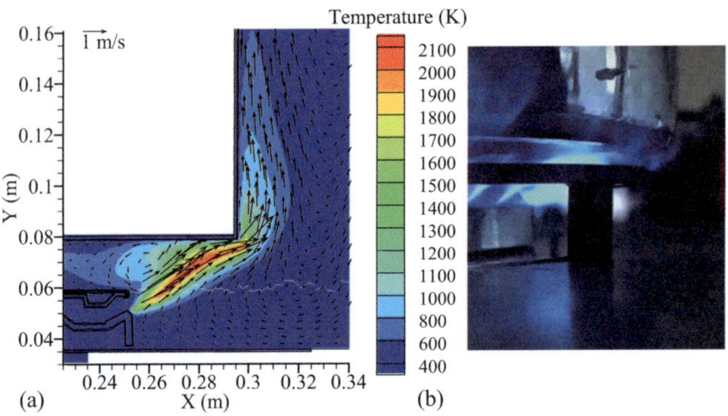

Fig. 13 Flame shape comparison for burner with vessel condition **a** numerical, **b** experimental

5 Conclusions

In the present study, a numerical model has been developed under the portfolio of Ansys FLUENT to predict the flame behaviour in a radially slotted burner fired using methane under open flame condition as well as with a vessel on the top of the burner. A short chemical mechanism having 25 species and 120 reactions has been used to simulate methane oxidation.

Initially, a sector domain considering a single port has been employed in the study. This analysis gave key insights on the numerical instabilities faced in simulating open flame condition. A novel approach has been implemented to the domain, where a central annulus region has been added to eliminate the recirculation over the cover cap that caused an unphysical behaviour. A sensitivity study has been performed using the sector domain with annulus for choosing the radius of the annulus. After this, the sector domain with an annulus concept has been extended to a full 3D domain with a central annulus, and a comparison between the two has been made. Both the sector domain with annulus and full 3D domain with annulus produce similar flame shapes and extents.

The numerical model considering the full 3D domain with annulus has been validated with the experimental data and the model predicted the temperature and CO profiles with an accuracy of 81% and 85%, respectively. Similarly, a numerical model for the burner with a vessel on its top has also been used and validated with corresponding experimental data. Here, the model is found to predict the temperature and CO profiles with an accuracy of 80% and 86%, respectively.

Overall, this study illustrates that the prediction of open flames from radial slot burners is non-trivial due to numerical instability. With a proper boundary condition over the cover plate, the numerical model is able to predict the flame behaviour of a radially slotted burner using methane with more than 80% accuracy as compared to the experimental data. The numerical model parameters are fixed by thorough validation and this model predicts the flame characteristics in both open and with-vessel conditions. The validated numerical model can be used to simulate and test new burner designs and configurations, thereby reducing the cost of the development of experimental rigs for new burner designs. The minor overestimation present in the model may be due to factors such as geometry simplifications, assumptions made in temperature correction, and limitations in measuring the CO under open flame condition. The model accuracy can be improved by using better measurement equipment and higher computational capabilities.

Acknowledgements The authors express their gratitude and thanks to Whirlpool Corporation for providing the radially slotted burners and the funding and aid to conduct this research.

References

1. Glassman I, Yetter RA (2010) Combustion, 4th edn. Elsevier, Amsterdam
2. Lu TF, Law CK (2009) Toward accommodating realistic fuel chemistry in large-scale computations. Prog Energy Combust Sci 35:192–215
3. Fiorina B, Veynante D, Candel S (2015) Modeling combustion chemistry in large eddy simulation of turbulent flames. Flow Turbul Combust 94:3–42
4. Pope SB (2013) Small scales, many species and the manifold challenges of turbulent combustion. Proc Combust Inst 34:1–31
5. Hilbert R, Tap F, El-Rabii H, Thevenin D (2004) Impact of detailed chemistry and transport models on turbulent combustion simulations. Prog Energy Combust Sci 30:61–117
6. Bidi M, Hosseini R, Nobari M (2008) Numerical analysis of methane-air combustion considering radiation effect. Energy Convers Manage 49(12):3634–3647
7. Roy C, Blottner F (2006) Review and assessment of turbulence models for hypersonic flows. Prog Aerosp Sci 42(7–8):469–530

8. Lezcano-Benítez JC, Correa-Restrepo D, Amell-Arrieta AA, Cadavid-Sierra FJ (2011) Numerical simulation of high pressure burner with partially premixed flame. CT&F Ciencia, Tecnología y Futuro 4(4):89–104
9. Das M, Ganguly R, Datta A, Verma MM, Bera AK (2021) Performance improvement of a domestic liquefied petroleum gas cook stove using an extended spill-tray and an annular metal insert. J Therm Sci Eng Appl 13(2):1–10
10. Wang J, Zhang W, Yang T, Yu Y, Liu C, Li B (2022) Numerical and experimental investigation on heat transfer enhancement by adding fins on the pot in a domestic gas stove. Energy 239:122439
11. Wichangarm M, Matthujak A, Sriveerakul T, Sucharitpwatskul S, Phongthanapanich S (2020) Investigation on thermal efficiency of LPG cooking burner using computational fluid dynamics. Energy 203:117849
12. Bradley D, Matthews KJ (1968) Measurement of high gas temperatures with fine wire thermocouples. J Mech Eng Sci 10(4):299–305
13. Laguillo S, Ochoa JS, Ortiz A (2019) Chemical reaction mechanisms assessment for simulation of methane combustion in domestic gas cooking burners. Energy Fuels 33(9):9171–9183

Mean Aerodynamic Flow Field of an Impinging Jet Issuing from a Convergent Nozzle at Different Inlet Pressure

Atri Bandyopadhyay, I. Bharatkrishnan, Tapas Kumar Nag, and Sanjoy Kumar Saha

Abstract The intricate flow physics and broad range of applications of jet impingement on a flat surface are major points of interest for aerodynamics researchers. There are many uses for jet impinging on surfaces, both inclined and flat, from launch vehicles during lift-off to aircraft during vertical takeoff and landing. On the surface where the jets are impinging, there are significant thermal and pressure loads as a result of these impingements. Not only were there high sound levels produced during this impingement, but there was also loss of lift. In order to comprehend the intricate flow characteristics of an impinging jet on a flat surface, numerical flow simulations are performed using the commercial CFD program "ANSYS FLUENT." To determine how different nozzle pressure ratios (NPRs) affect the jet structures, simulations are run for each NPR. The results are thoroughly examined and contrasted with test data from publicly available sources. One can see that the flow phenomena has been well captured. A detailed examination of the flow fields is done in order to identify the critical flow phenomena.

Keywords Lift-plate · Bottom plate · Convergent nozzle · Impinging jet · Mach number · NPR · Lift loss

1 Introduction

Numerous aerospace and related applications involve a high-speed jet impinging on a surface. Jets interact with solid surfaces in many different applications. It is exemplified in the thrust vector control of the nozzle, turbine cooling, jet impingement

A. Bandyopadhyay (✉) · I. Bharatkrishnan · T. K. Nag
Aerospace Engineering, VIT, Bhopal, India
e-mail: bandtintin@gmail.com

S. K. Saha
Aeronautics Entity, Vikram Sarabhai Space Centre, Thiruvananthapuram, Kerala, India

A. Kumar R. et al. (eds.), *Proceedings of Fluid Mechanics and Fluid Power (FMFP) 2023, Vol. 6*, Lecture Notes in Mechanical Engineering,
https://doi.org/10.1007/978-981-96-9600-0_4

on the launch pedestal during aircraft takeoff and landing, and the interaction of aircraft engine exhaust with nearby solid airframes during launch vehicle takeoff. For short takeoff and vertical landing (STOVL) aircraft, especially the Harrier/AV-8 family, these jet flows pose a serious threat during hover mode. In these situations, the flow field created by the impinging high-speed lift jets creates unfavorable local flow conditions, which poses a serious risk of degrading aircraft performance in a number of areas during hover. The highly unstable flow caused by the high-speed jet(s) impinging on the bottom or ground plate and the pressure field created by their natural entrainment are the causes of these unfavorable flow characteristics. The term "ground effects" refers to all of these negative consequences. For example, lift loss causes the airframe to experience low surface pressures, which create a "suckdown" force that opposes lift. The lifting jets' associated flow entrainment is the cause of it. The lift loss typically rises as an aircraft gets closer to the ground and can account for up to 50% of the net lift jet thrust [1].

Unsteady loading leading to fatigue of structural elements adjacent to the nozzle and high-speed jet impingement are also causes for concern due to elevated noise or overall sound-pressure levels. Discrete tones are a defining feature of the noise spectrums, which can exacerbate the issue of sonic fatigue in addition to higher noise levels, particularly when near flight panel frequencies. Furthermore, since the flow creates extremely high shear stresses and wall heat-transfer rates, the impingement of hot, high-speed jets on the bottom plate may result in significant ground erosion. Additionally, during the ingestion of hot gas, the outwash from the hot impinging jets may be drawn into the engine inlets, degrading engine performance and ultimately leading to engine failure. Due to the impact of supersonic jets operating at extremely high temperatures, it is anticipated that the environment for the next generation of supersonic STOVL aircraft will be even more hostile. It follows that from a practical standpoint, research on supersonic impinging jet flows is highly desirable. The intricate nature of the impinging jet flow field, which typically consists of multiple shock and shock/shear layer interactions, subsonic, supersonic, and separated flows, makes this flow exceptionally fascinating from the standpoint of fluid dynamics (Fig. 1). The fluid dynamic and acoustic properties of impinging jet flows have been the subject of study for more than a few decades with Krothapalli et al. [1] being the most recent and Neuwerth [2], Powell [3], Tam and Ahuja [4], Henderson and Powell [5] among others.

The impingement region changes in characteristics based on NPR. When there is a high under expansion, a recirculation zone forms, causing the plate shock to be convex and peak upward (Fig. 1). Slipstreams and tail shocks are also present in triple point structures. After emerging from the triple point, a locally viscous layer known as the slip line spreads downstream and eventually impinges on the landing plate. The velocity differential created by the flow processed by the jet and tail shocks leads to the development of the slip line. At high enough pressures, a stagnation bubble can form when the boundary layer splits apart and expands radially from the surface. The only circumstances in which stagnation bubble flow happens are under-expanded jets, where the degree of interaction between the plate shocks and the plane at a particular plane and location is important.

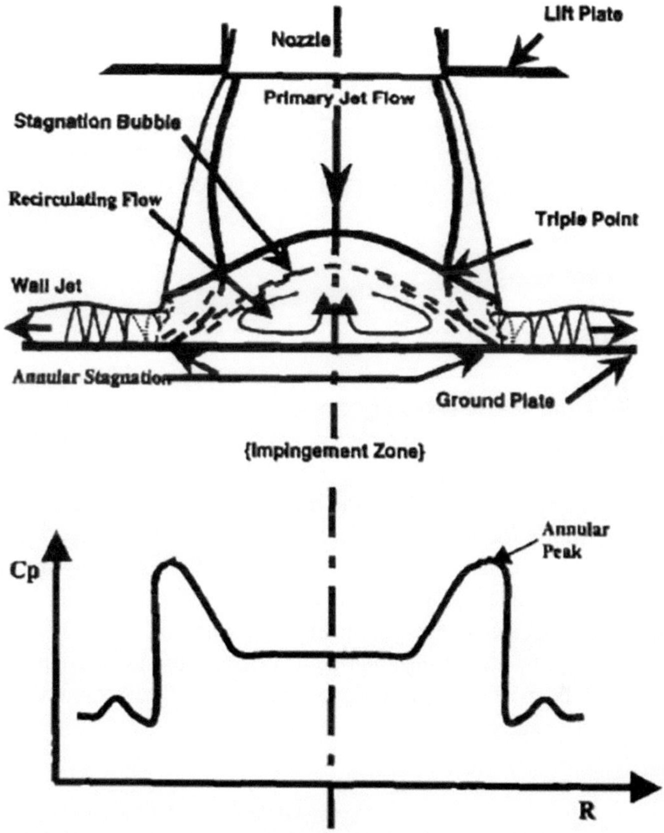

Fig. 1 Schematic of the flow phenomena for normal impinging jet with stagnation bubble [6]

The impinging jet is extremely erratic and oscillatory in nature, which is one of the main findings of these aeroacoustics explorations. The impingement tones, which are caused by a feedback loop, are distinct, high-amplitude acoustic tones that accompany this. Powell [7, 8] was the first to describe this feedback mechanism. While the fluid dynamic and acoustic properties of the flow appear to be closely related, a more thorough understanding of the unsteady properties is outside the purview of this article, since the focus of the current work is on the mean behavior of the impinging jet with variation of NPR as change with respect to impingement distance has been shown in Bandyopadhyay et al. [9].

2 Numerical Simulation

To comprehend the intricate layout of fluid movement triggered by a jet's encounter, computational flow analyses are performed for a convergent nozzle. These studies are conducted on a typical jet impingement situation at varying NPR ratios. The commercial tool ANSYS Fluent [10] is employed for conducting these flow simulations.

2.1 *Geometric Details*

Figure 2 displays the nozzle geometry that is being examined. This nozzle geometry is derived from published research [6], which contains data from wind tunnel experiments. The nozzle has an exit diameter (D) of 25.4 mm and is a converging axisymmetric sonic nozzle. Known as the lift-plate or top plate, this fixed plate has a 10D diameter and is located at the nozzle exit plane. In the location where the jet impinges, the bottom plate—also known as the impinging plate—measures 24D. The impinging plate (bottom plate) that is being considered in this instance is 6.0D away from the nozzle exit plane (h). In Fig. 2, along with the CFD domain, are some additional significant dimensional details of the simulated geometry. Similar geometry has also been used in studying the variation in flow structure with respect to change in nozzle-plate distance in Atri et al. [9].

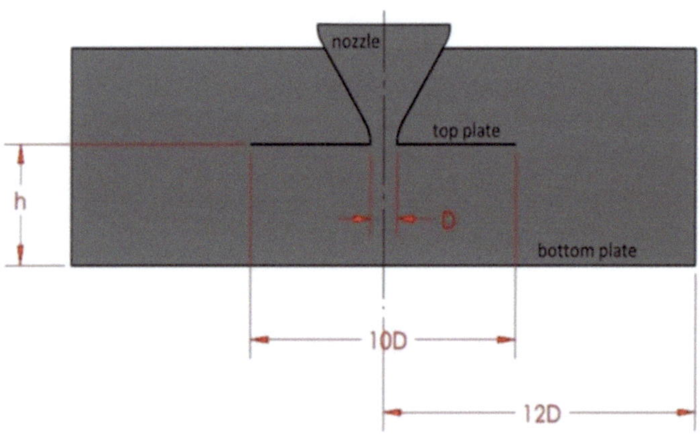

Fig. 2 Dimensional details of the geometry

2.2 Domain and Grid

Understanding the mean flow physics of an impinging jet on a flat plate and how it depends on the nozzle pressure ratio (NPR) is the goal of the current study. The simulation domains take into account a cylindrical shape due to the nozzle's axisymmetry. To minimize the computational expense, a quarter of the entire body is simulated by taking advantage of the body's symmetry. Throughout the simulation domain, grids with various dimensions are employed. For the simulations, the finest grids are used near the jet axis and the coarsest one at the farthest point. Grid independence studies are conducted prior to the grid distribution being finalized. The grid sizes are decided upon based on this study. For a typical case, Fig. 3 displays a typical grid distribution.

As was previously mentioned, the flow is simulated using the commercial CFD program ANSYS. The coupled algorithm solves a coupled system of equations consisting of the momentum and pressure equation in the simulation using a pressure-based coupled solver. An approach based on control volume is used to discretize the governing equation for the conservation of energy, mass, and momentum equations. The steady-state runs used in the RANS-based shear-stress transport (SST) k–ω model, which was based on past research [11]. The simulations continue until the residuals drop to 10^{-4} order and the central line fluid parameter varies by no more than 1%. Identical grid distribution has been utilized for Atri et al. [9].

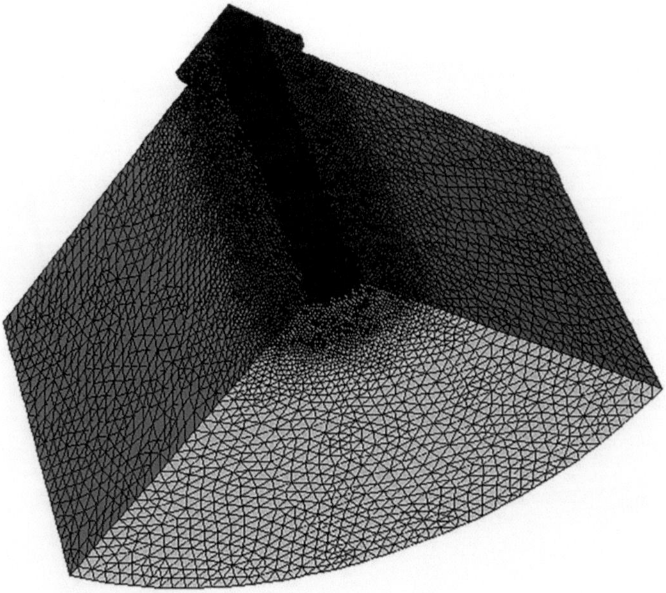

Fig. 3 Grid distribution for $h/D = 6.0$

2.3 Grid Convergence Studies

For this purpose, geometry with $h/D = 6.0$ is chosen. Four different gird distribution is chosen with the aim to resolve the finer details of the mean flow properties. They are

i. Coarse mesh: (4 mm × 10 mm × 20 mm) = 0.12 million
ii. Medium mesh: (2 mm × 5 mm × 10 mm) = 0.74 million
iii. Medium-Fine mesh: (1 mm × 5 mm × 10 mm) = 4.32 million
iv. Fine mesh: (0.5 mm × 3 mm × 10 mm) = 8.71 million

The values in the () indicate the sizes of mesh size used in the domain with the smallest one near the nozzle axis and the largest one far away from the axis.

Flow simulations are carried out at NPR = 2.5 for all the 4 grids mentioned above. The variation of pressure and Mach number along the nozzle axis as well as Mach number along the nozzle exit plane for all 4 grid distributions are monitored and shown in Figs. 4 and 5. It is evident from the figures that the "fine mesh" captures oscillations in the flow field in a better way as expected. Not much difference is noticed for "medium-fine mesh." But the "medium mesh" and the "coarse mesh" capture less oscillations in the flow thereby giving a very indeterminate curve. As mentioned earlier, as the number of grids increases, the variation in flow properties is better captured. Last two highest grids ("fine mesh" and "medium-fine mesh") do not show any further significant improvement. Based on this study, the "medium-fine mesh" is selected for further studies.

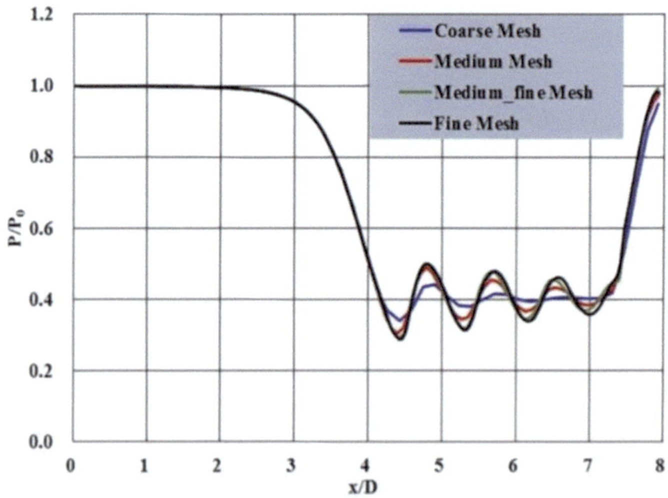

Fig. 4 Variation of P/P0 along the nozzle centralline

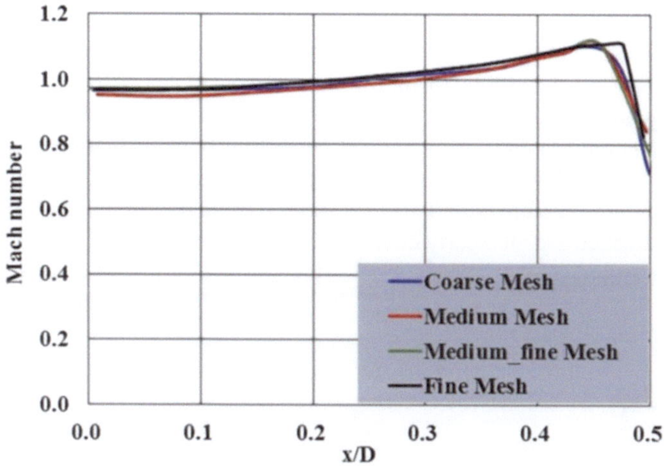

Fig. 5 Variation of mach number at nozzle exit plane

2.4 *Simulation Parameter and Boundary Conditions*

1 bar (100,000 Pa) of ambient pressure and 300 K of ambient temperature were taken into account for this simulation. To cover the various jet profiles, three separate NPR values are simulated. In the simulation, the same NPR values as in the test are taken into account: 2.5, 3.7, and 5. Based on these and NPR values, the nozzle inlet's stagnation pressure (P_0) is fixed at 2.5 bar (over-expanded), 3.7 bar (ideally expanded), and 5.0 bar (under-expanded), in that order. Since air is regarded as an ideal gas in this simulation, its total temperature at the inlet (T_0) is 300 K.

At a distance of 12D from the nozzle center, the far field boundary condition is applied. The simulations use no-slip boundary conditions, which are adiabatic (at $T = 300$ K), because the test report [6] does not specify the precise thermal conditions of the top and bottom plates. Additionally, the nozzle has a no-slip boundary condition. Maximum pressure (P_0) and maximum temperature (T_0) are specified at the nozzle's inlet, where the pressure inlet condition is applied. Applications of symmetry conditions are made at the domain's symmetry planes. Similar simulation parameters have been used in Atri et al. [9].

3 Results and Discussion

As described earlier, a convergent nozzle with nozzle exit diameter of 25.4 mm is considered for numerical simulations. It is operating under pressure ratio (NPR) of 2.50, 3.70, and 5.0, which is much higher than the critical pressure ratio (1.89) for a convergent nozzle to be sonic nozzle. Hence, for the above- mentioned pressure ratios, nozzle will be sonic nozzle and operating as under-expanded jet. NPR = 2.5 (nozzle pressure ratio of 2.5) corresponds to moderate under expansion while NPR = 3.7 and NPR = 5.0 correspond to highly under-expanded jets. When such a jet comes out of the nozzle and impingement takes place at closer nozzle distances, depending on the NPR, complex flow phenomenon consisting both supersonic and subsonic regions develops. Flow structures consisting of a plate shock, annular(tail) shocks, wall jet, constant pressure boundaries, jet shocks, triple points and slip streams, etc., also formed which are all clearly observed in CFD simulations (Fig. 6).

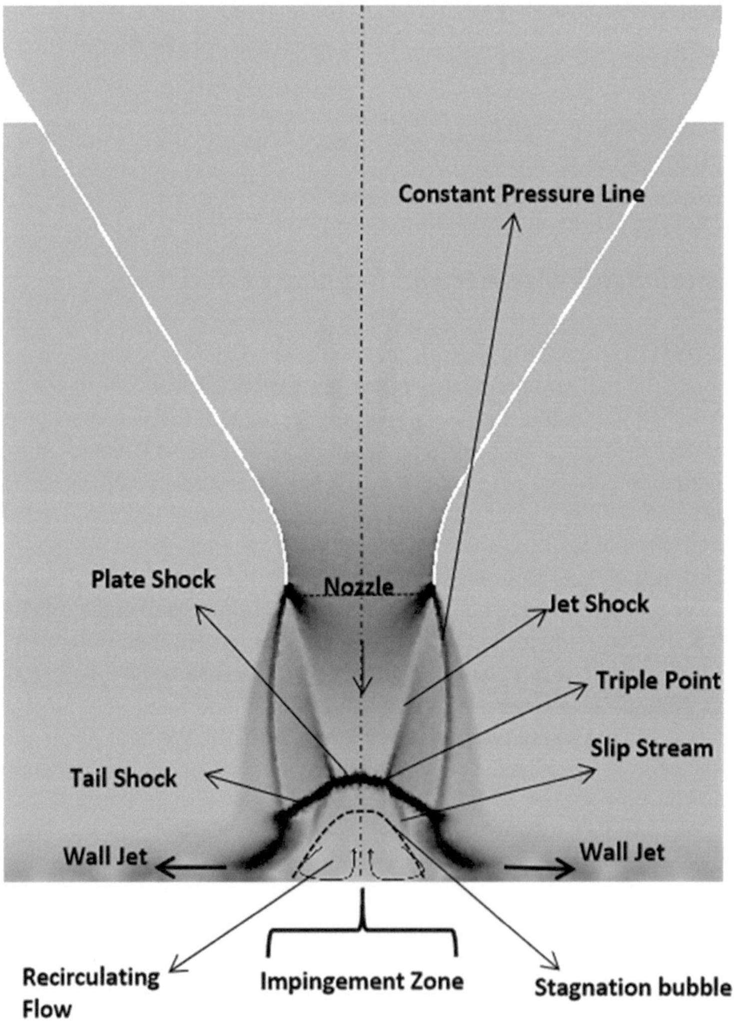

Fig. 6 Flow schematic of a shock layer for $h/D = 2$ and NPR = 5.0

3.1 Validation of the Data

The results of CFD are compared with test data found in the literature [6] before moving forward with the detailed simulations. Below figure presents a comparison between the numerical and literature-available schlieren. Figure 7 displays data at two distinct moments. CFD simulations exhibit a good capture of the flow physics [9].

(a) *h/D* = 2.0 at NPR = 3.7

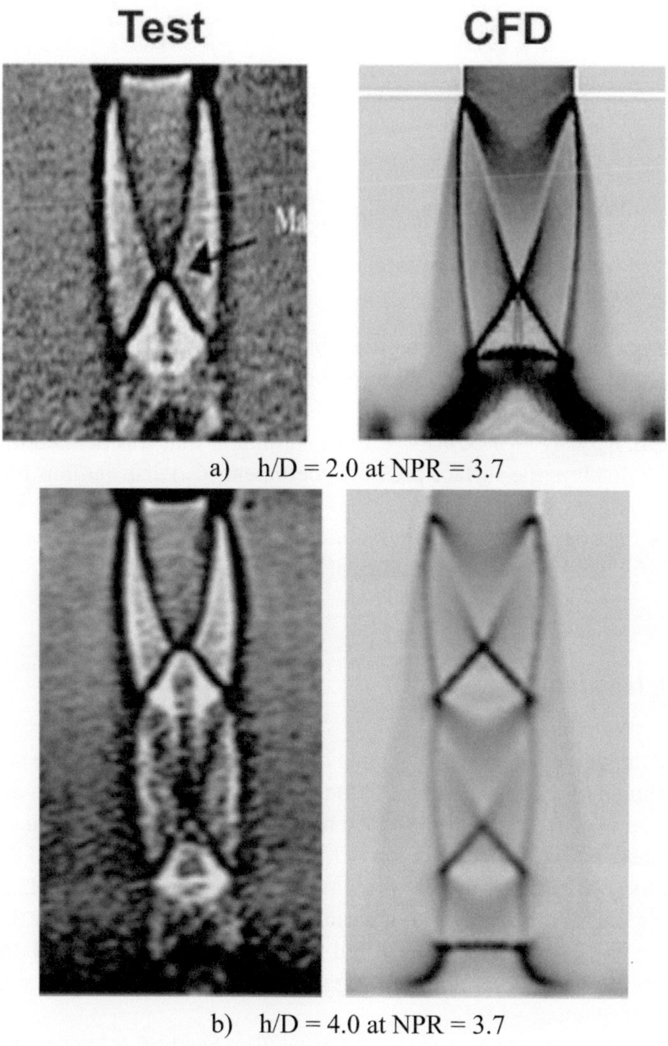

Fig. 7 Comparison of numerical and test schlieren

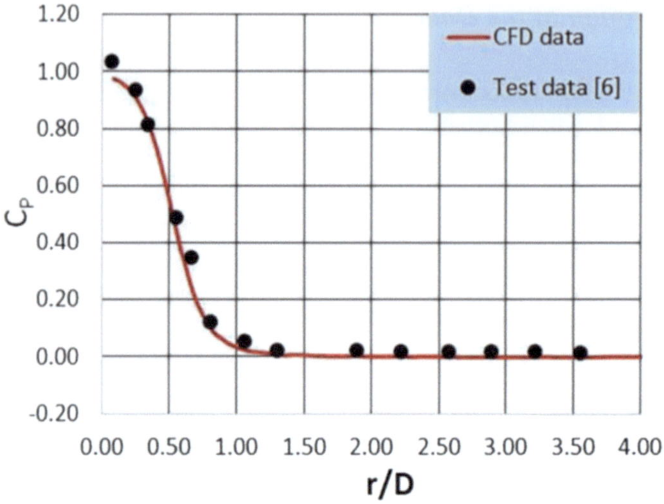

Fig. 8 C_p distribution (no top-plate) at NPR = 2.5

(b) *h/D* = 4.0 at NPR = 3.7

The current simulations are used to estimate the variation of the mean pressure value (C_p), which is then plotted alongside data from the literature along the radial direction of the bottom plate [6]. Here, C_p is non-dimensional surface pressure coefficient where, $C_p = (P - P\infty)/(P0 - P\infty)$, similar to that was defined in literature [6]. Figure 8 displays the surface pressure coefficient (C_p) distributions for the NPR = 2.5 with changes in the radial location (*r*/*D*) at *h/D* 2, 4, and 8. It's evident that the pressure data that were calculated in this way using CFD and the measured data agree well [6]. This bolsters the validity of the data that was produced.

3.2 Mach Palette

The nozzle under consideration is a sonic nozzle, and regardless of the nozzle inlet pressure ratio, $M = 1.0$ is anticipated at the exit plane of the nozzle. It is determined whether there is consistency in the Mach number at the nozzle exit plane for every simulated case. In Fig. 9, the variation in Mach number along the radial direction is displayed. Everywhere but in close proximity to the nozzle wall, the Mach number at the nozzle exit is observed to be the same. Different boundary layer thicknesses at the nozzle exit (which are not captured here) and non-identical grid distribution could be the cause of this variation. As predicted, the nozzle exit has an area average Mach number of about ~1.0 [9].

Since this is a convergent nozzle, the jet will expand after leaving it before it reaches the impinging plate or the bottom. The expansion causes the pressure to

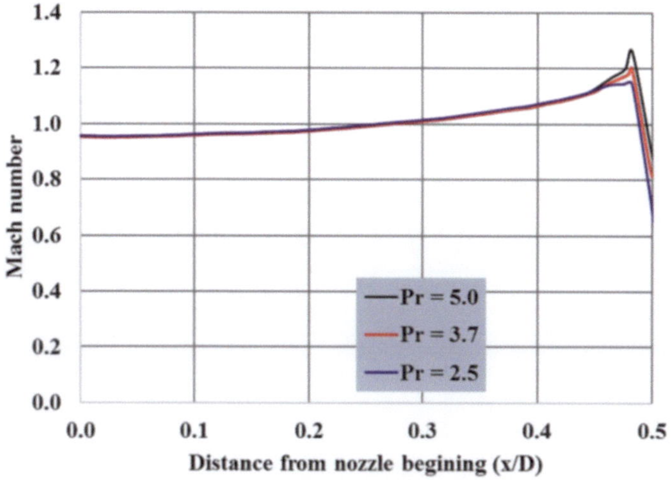

Fig. 9 Mach number distribution along the nozzle exit plane

decrease, which is then followed by shock-induced compression. This expansion and compression process continues till the flow faces the bottom plate. As the distance from nozzle exit increases, the strength of these compression and expansion also decreases. Both the strength of compression and expansion are determined by the NPR values. As the value of NPR increases, the level of compression and expansion also increases. A Mach disk is formed immediately after the nozzle exit for NPR = 5.0 (Fig. 10) and an interception shock is seen for NPR = 2.7.

Mach number distribution along the nozzle center line is extracted for different NPR is shown in Fig. 11. It is noticed that the variation of Mach number is higher for

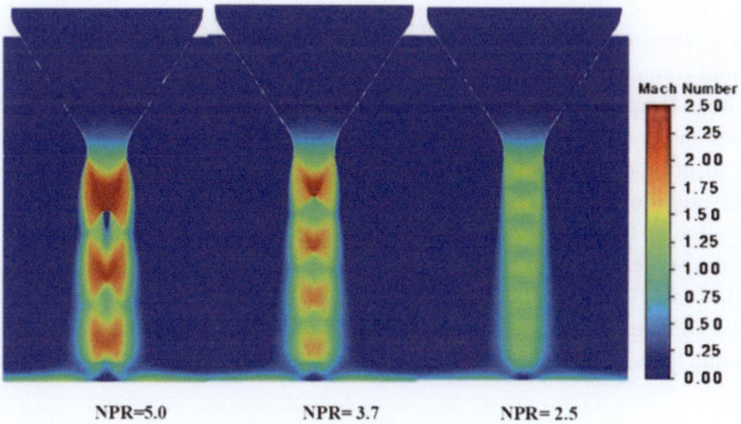

Fig. 10 Mach number palette along the center line

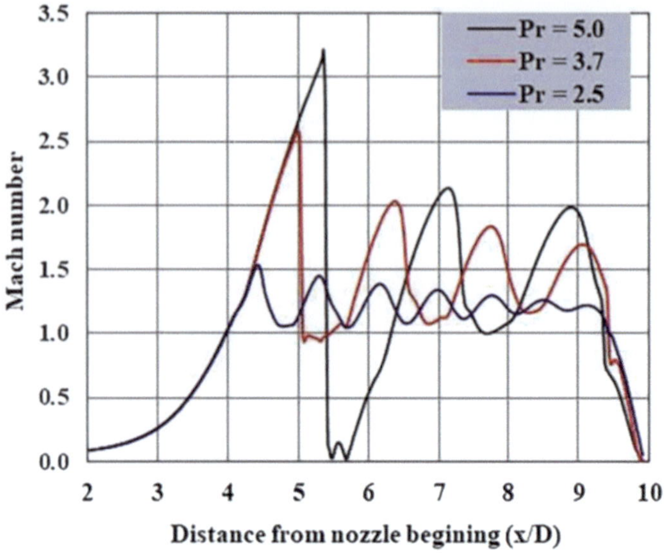

Fig. 11 Mach number distribution along the nozzle center line at $h/D = 6.0$

higher NPR values as expected. Another interesting point to note that the length of shock cell is more for NPR = 5.0 and least for NPR = 2.5. This is also expected, as due to the higher NPR, the jet expands more and take a longer length for the formation of next compression. This is similar to the observation reported in literature [12]. All these gives confidence on the correctness of the results obtained.

After impinging on the plate, flow spread radially outward along the plate. At the impingement point, flow is stagnant. Away from it, based on the surrounding condition (plate shock, Mach disk, etc.), the flow accelerates radially. Hence, the Mach number increases. This flow is called wall jet flow. Along the wall as flow moves, it loses its energy due to interaction with solid surface, viscous action, entrainment of air, and spreading over more area. This will result in a decrease in Mach number of the flow. Because of the two-contradictory effect, the jet Mach number initially increases and reaches to maximum value around $r/D = 1.0$ and later decreases.

The location of peak values and its magnitude depends on the value of NPR. In Fig. 12, variation of jet Mach number along the impinging plate is shown. It is noticed that as value of NPR increases, the magnitude of peak Mach number increases and the location where it occurs shift toward right. The increase in peak value is due to the higher energy of the jet at higher NPR. Based on NPR values, the location where the peak value reached will alter. For NPR $\geq$3.7, the flow field noticed is supersonic. This is due to the increase of energy of the jet with the increase in NPR values.

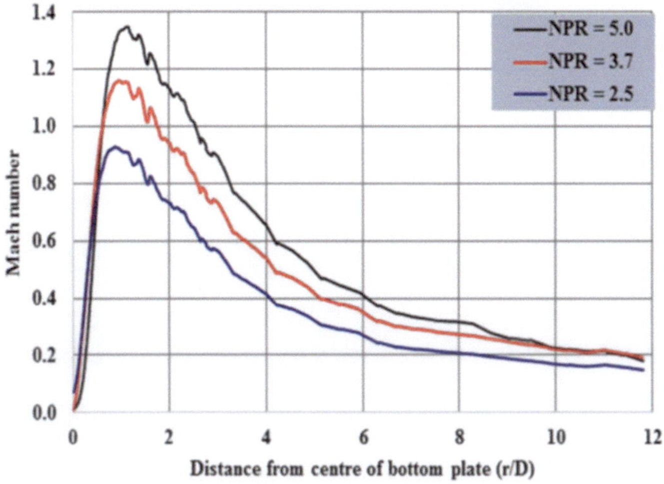

Fig. 12 Mach number variation in radial direction of bottom plate $h/D = 6.0$

3.3 Pressure Palette

Earlier it has been mentioned that based on the NPR values, the shocks are formed once the flow comes out of the nozzle. This formation of shock will lead to higher pressure. Even the number of shock cell is also varying based on the NPR. Earlier it has been mentioned that the shock cell length is maximum for NPR = 5.0. All these are clear from the pressure palette shown in Fig. 13.

Numerical schlieren (i.e., density gradient) obtained for different cases are shown in Fig. 14. Number of shock cells, its length, Mach disk all are clearly visible here. It is noticed that there will be a big separated flow for the case of NPR = 5.0 which

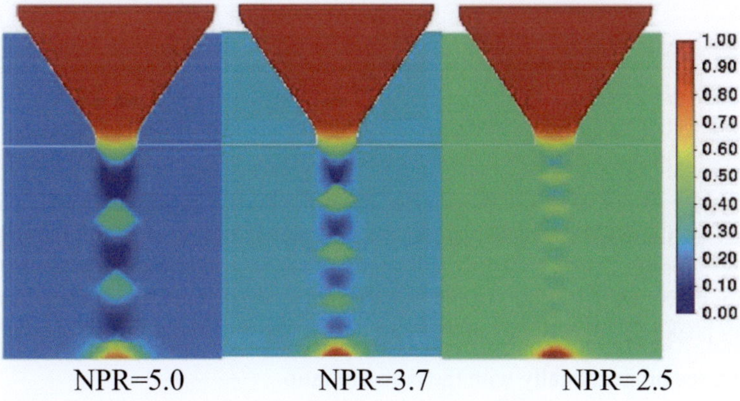

Fig. 13 Pressure palette along the central line

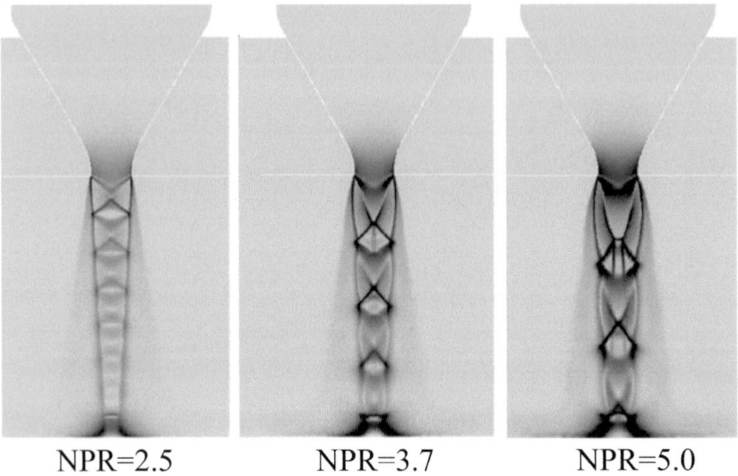

Fig. 14 Numerical schlieren

is absent in other two cases. Plate shock is normal for NPR = 3.7 whereas it is of convex shape at NPR = 5.0. This is due to the differences in approach Mach number to the bottom plate. Which is in line with the literature data [13].

The variation of $P/P0$ along the nozzle axis is clearly depicted in Fig. 15. It shows a fall after $x/D = 3$ indicating the acceleration of the flow in the convergent region. As the flow is further expanded beyond the throat ($h/D = 4$) due to under expansion, there is a further drop in $P/P0$ due to expansion of flow. The amount of expansion depends on the value of NPR. The pressure oscillations are noticed as the flow comes out of the nozzle. This is due to the formation of shock cell, i.e., compression and expansion of flow. It is clear from the figure that, the shock cell lengths will increase with increase in NPR which is as expected. But the number of shock cells is observed to increase with the decrease of NPR.

3.4 Shock Cell

For any under-expanded or over-expanded jet, compression and expansion of flow occurs as the jet comes out of the nozzle. The amount of compression and expansion depends on the NPR and h/D values. Shock cells are regions of alternating high and low- pressure zones that form in supersonic flows, contributing to the complex behavior of the jet. Shock cell length was defined as the distance between subsequent velocity peaks. It is well- known that length of a shock cell in a supersonic jet:

- Increases proportionally with the pressure ratio.
- Escalates in tandem with the exit Mach number.

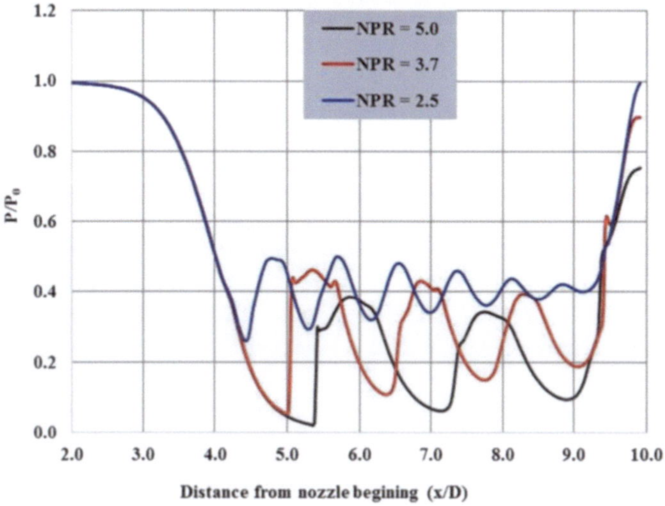

Fig. 15 Pressure (*P*/*P*0) plots along central axis for different NPR

For the present simulations, shock cell length has been found from the Mach number plots. Shock cells are clearly visible through the Mach number or pressure distribution graph. For NPR = 5.0, two shock cells are found. The length of shock cell is about 1.95D for NPR = 5.0 which is the highest among the simulated cases, whereas it is least for NPR = 2.5 (0–0.8D).

Different authors have proposed different formula to calculate the shock cell length for a supersonic jet. The results so obtained from the CFD simulations match well with the experimental and numerical data work Wu et al. [12] and numerical studies by Sanjoy et al. [11]. Variation of length of shock cell with NPR is shown in Fig. 16. It is noticed that as compared to the NPR = 2.5, the length of shock cell is more than double for NPR = 5.0(0.85 versus 1.95D).

3.5 Forces on the Lift and Bottom Plates

There is a drop in pressure at the bottom side of the plate, which is the nozzle exit (top-plate), as a result of surrounding air being entrained with the jet flow. In contrast, the plate's top surface remains unaltered. Thus, the plate experiences a downward force. Depending on the separation between the nozzle and impinging plate, this force's magnitude changes. When the distance increases, it is observed that the force's amplitude diminishes exponentially. However, because there is a greater air suction, this value also rises as NPR does [9].

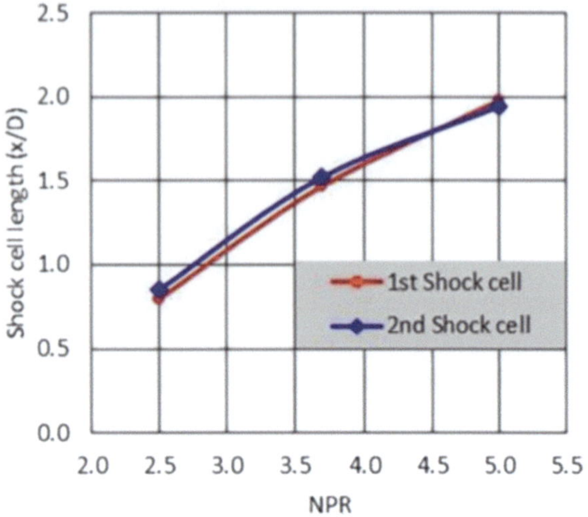

Fig. 16 Variation of shock cell length with NPR

Analogously, the forces acting on the bottom plate are estimated. The value of pressure decreases as the impinging plate's distance grows. The force on the bottom plate will decrease as a result of this.

4 Conclusions

A flat plate is subjected to sonic impinging jet simulations via steady-state flow modeling with ANSYS Fluent. Various jet pressure ratio simulations are run in order to comprehend the variation in flow structures and other parameters concerning NPR. Stagnation bubbles diminish in size as the Mach Disk upstream of it decays with increase in impingement distance. There is no considerable difference between the flow fields with and without lift-plate. Convex Dome like plate shocks which become flat with impingement distance. The shock cell length and the jet's maximum Mach number are observed to increase in tandem with an increase in the NPR value. Lower NPR values result in lower lift loss and lower forces on the bottom plate. The results obtained herewith shows good agreement to the flow fields that were analyzed experimentally prior to this work.

Acknowledgements The authors would like to thank Mr. Anish for his assistance in creating the geometry. We appreciate Dr. Patil, DD AERO, VSSC and Dr. Pankaj Priyadarshi, GD ADSG, VSSC for their assistance with this work.

Abbreviations

D	Nozzle exit diameter (mm)
NPR	Nozzle pressure ratio i.e. ratio of the jet pressure at the nozzle exit plane to free stream pressure
M	Jet flow Mach number
P_0	Chamber pressure (bar)
P	Local static pressure (bar)
$P\infty$	Local free stream pressure (bar)
h	Distance between bottom plate and the nozzle exit plane (mm)
C_p	Pressure co-efficient
x	Distance between plate and the nozzle exit plane (mm)

References

1. Krothapalli A, Rajkuperan E, Alvi F, Lourenco L (1999) Flow field and noise characteristics of a supersonic impinging jet. J Fluid Mech 392:155–181
2. Neuwerth G (1974) Acoustic feedback of subsonic and supersonic free jet which impinges on an obstacle. NASA TT F 12:15719
3. Powell A (1988) The sound-producing oscillations of round underexpanded jets impinging on normal plates. J Acoust Soc Am 83:515–533
4. Tam CKW, Ahuja KK (1990) Theoretical model of discrete tone generation by impinging jets. J Fluid Mech 214:67–87
5. Henderson B, Powell A (1993) Experiments concerning tones produced by an axisymmetric choked jet impinging on flat plates. J Sound Vib 168:307–326
6. Alvi FS, Iyer KG (1999) Mean and unsteady flow eld properties of supersonic impinging jets with lift plates. In: AIAA paper, pp 99–1829
7. Powell A (1953) On edge tones and associated phenomena. Acoustica 3:233–243
8. Powell A (1953) On the mechanism of choked jet noise. Proc Phys Soc Lond B 66:1039–1057
9. Bandyopadhyay A et al (2021) Mean aerodynamic flow field of an impinging jet issuing from a convergent nozzle with varying nozzle to plate distance. https://doi.org/10.21203/rs.3.rs-4653110/v1
10. RI (2022) Ansys fluent theory guide. R1
11. Sanjoy S et al (2022) Numerical prediction of mean flow characteristics of a supersonic jet issuing from a CD nozzle. In: Proceedings of the 9th international and 49th national conference on fluid mechanics and fluid power, December 14–16, 2022, IIT Roorkee, Roorkee
12. Wu J, New TH (2019) Flow characterization of supersonic jets issuing from double-beveled nozzles. J Fluids Eng 141:011202–011203. https://doi.org/10.1016/j.ast.2017.01.003
13. Kalghatgi GT et al (2020) The occurrence of stagnation bubbles in supersonic jet impingement flows. https://doi.org/10.1017/S0001925900007678

Cost-Effective High-Speed Imaging of Droplet Impact with a DSLR Camera

Deependra Kumar, S. P. Santhosh, Abhiram Ramachandran, Adethya S. Nair, Aditya Jain, and M. Kiran Raj

Abstract We developed a cost-effective, high-speed imaging system with a regular digital single-lens reflex camera (DSLR) to explore the impact characteristics of droplets. A stroboscopy-based strategy is followed to vary the timing of the camera flash and thus capture the exact instant of interest using a microcontroller-based circuit, avoiding an expensive high-speed camera. Using this, four different case studies have been examined to demonstrate the capabilities of such off-the-shelf systems to carry out fundamental scientific investigations.

Keywords High-speed imaging · Droplet impact · Liquid pool · Drone propeller · Porous surface

1 Introduction

The study of droplet impact on surfaces has piqued the interest of many researchers due to its relevance in various scientific and industrial applications [1, 2]. Understanding the complex dynamics and behaviour of droplets upon impact is crucial for optimizing processes such as inkjet printing, spray coating and raindrop impact

D. Kumar · M. Kiran Raj (✉)
Department of Applied Mechanics and Biomedical Engineering, Indian Institute of Technology, Madras, Tamil Nadu, India
e-mail: kiran@iitm.ac.in

D. Kumar
e-mail: am22m021@smail.iitm.ac.in

S. P. Santhosh
Department of Electrical Engineering, Indian Institute of Technology, Madras, Tamil Nadu, India
e-mail: ee21b119@smail.iitm.ac.in

A. Ramachandran · A. S. Nair
Department of Mechanical Engineering, Government Engineering College, Thrissur, Kerala, India

A. Jain
Department of Mechanical Engineering, Delhi Technological University, New Delhi, Delhi, India

A. Kumar R. et al. (eds.), *Proceedings of Fluid Mechanics and Fluid Power (FMFP) 2023, Vol. 6*, Lecture Notes in Mechanical Engineering,
https://doi.org/10.1007/978-981-96-9600-0_5

on different surfaces. In recent years, imaging techniques have provided valuable insights into the complex phenomena happening during droplet impact. Worthington showed the first systematic study of droplet splashing in his 1908 book: *A Study of Splashes*. He used high-speed photography to capture images of droplet impact and track the evolution of the splash pattern [3].

The droplet impact on a solid surface can result in the droplet bouncing, spreading, or splashing [1]. The outcome of the impact depends on the impact velocity, the droplet's surface tension and the solid's surface energy. If the impact velocity is low, the droplet will likely bounce. The droplet will likely spread or splash if the impact velocity is high [2]. Numerous research works have explored the topic of droplet impact on solid surfaces. Josserand et al. investigated the dynamics of drop impact on solid substrates and provided detailed insights into the spreading and splashing phenomena [4]. Yarin conducted an in-depth study on drop impact dynamics, covering aspects such as splashing, spreading, receding and bouncing [2]. David et al. focused on the splashing phenomena during high-speed droplet impacts on dry surfaces, shedding light on the underlying mechanisms and dynamics [5]. Liu et al. studied the splashing phenomena during droplet impact and investigated the influence of different parameters [6].

In addition to solid surfaces, droplet impact on porous and liquid pool surfaces is also an area of interest. Several researchers have shown interest in understanding the splashing and pattern formation that occurs when droplets impact porous surfaces. For example, Thijs de Goede et al. showed that droplets splash on surfaces with a high surface roughness. This is because of the formation of air bubbles, which lead to splashing [7]. Rein investigated droplet impact on both solid and liquid surfaces, studying aspects such as splashing, spreading and bouncing [8]. Ray et al. studied oblique drop impact on both deep and shallow liquid pool surfaces [9]. They showed the specific behaviour exerted by the droplets impacting liquid pool surfaces at different oblique angles. Throughout the research, they researched the droplet spreading, crown formation and droplet breakup of oblique droplet impact. Ray et al. explored the different regimes that occur when a droplet impacts a liquid pool. They observed phenomena such as splashing, coalescence, crown formation and bubble formation [10].

The impact of a droplet on the surface of a liquid pool has resulted in crater formation, crown formation and splashing. Crater formation results because of high-impact energy displacing the liquid. Crown formation results because liquid layers splash out of the surface or water. Splashing occurs because of the high-impact energy that causes the liquid to expel out into the air [10].

To investigate droplet impact, high-speed imaging systems have been widely used. However, the cost associated with high-speed cameras often has a barrier to researchers seeking to conduct fundamental scientific investigations. In this study, we propose a cost-effective approach to capture droplet impact using a regular digital single-lens reflex (DSLR) camera. This enables us to capture critical instants during the droplet impact process. Stroboscopy-based high-speed photography, which traces its origins back to the pioneering work of Prof. Harold Edgerton, has been used as an alternative approach. He introduced stroboscopy in the mid-twentieth century, enabling the capture of rapidly occurring events through controlled illumination pulses synchronized with the camera shutter. His innovations opened up a new domain of high-speed imaging and established the foundation for subsequent advancements in the field [11]. By implementing a stroboscopic strategy introduced by Prof. Edgerton and leveraging a microcontroller-based circuit, we can easily control the timing of the camera flash, capturing the exact instant. This allows us to study for fundamental scientific investigations with available equipment while overcoming the limitations of the cost of high-speed cameras. We demonstrate the capabilities of the proposed cost-effective imaging system for exploring droplet impact characteristics through four case studies that showcase its effectiveness and versatility in capturing and analysing droplet impact dynamics.

2 Experimental Methods

2.1 Experimental Setup

The experimental setup (Fig. 1) can be divided into three major parts: The dispensing unit, the sensing unit and the imaging unit. The dispensing unit contains a syringe pump and a syringe containing a dilute solution of food colouring connected to an intravenous (IV) needle placed over the test surface. One can obtain individual droplets of diameter 2.65 ± 0.19 mm by careful calibration. The sensing unit contains a microcontroller (Arduino UNO R3), a laser and a laser detector sensor. The alignment is adjusted so that the laser, laser detector and the tip of the needle lie in a straight line. The imaging unit contains a camera flash (Simpex) connected to the microcontroller via an optocoupler and a DSLR camera (Canon M50) with either a 22 mm or a 45 mm lens. The camera's shutter speed is set to 2.5 s, aperture to F8.0, and ISO to 100. The camera is connected to the computer, running the Canon EOS software, where we set all the necessary parameters and initiate the camera action. The interior of the setup is covered using white cloth for diffusive illumination, and the exterior is covered entirely using a black cloth to block stray light from the outside. Since this will make the setup optically inaccessible, a webcam (HP) is focused on the IV needle to monitor the droplet formation. All the experiments are done at room temperature and pressure conditions. For the imaging process, the syringe pump is operated to release single droplets that are formed on the tip of the IV needle. The shutter button of the camera is activated from the computer. Once the hanging droplet falls and passes through the line of sight between the laser and the laser detector, the flash is triggered after a certain delay so that the image captured by the camera contains the impact of the droplet with the test surface. This certain delay is found experimentally by repeated measurements and is around the value of the time of flight of the droplet.

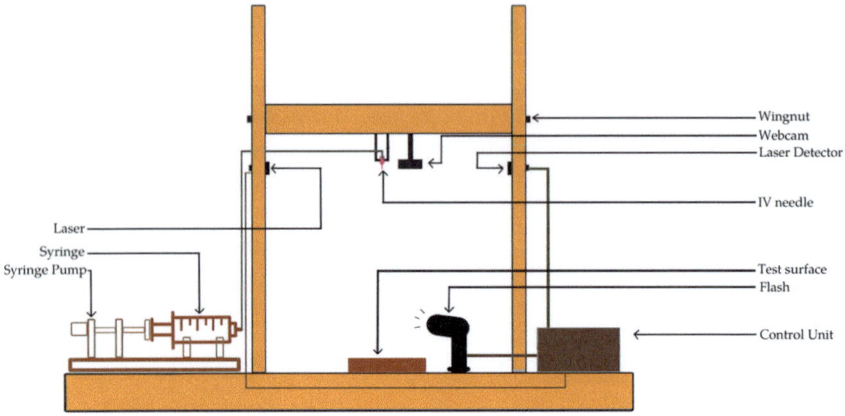

Fig. 1 Experimental setup

2.2 *Operation*

The Arduino code describes a finite-state machine with three states, `IDLE`, `READY` and `DISPENSE`, implemented by switch-case statements in Arduino IDE (Fig. 2). Here, `inp` is a binary variable containing the output of the laser detector. When the liquid droplet blocks the line of sight between the laser and the laser detector, the laser detector cannot detect the laser and `inp` contains the value 1. When there is no droplet between the line of sight between the laser and the laser detector, the laser detector can detect the laser and `inp` contains the value 0. By default, the state machine stays at `IDLE` until a liquid droplet blocks the line of sight between the laser and the laser detector; in that case, it goes to the `READY` state. It stays in the `READY` state until the droplet falls off and the line of sight is re-established; in that case, it goes to the `DISPENSE` state. When the machine is in the `DISPENSE` state, the flash is triggered, after a particular delay in the order of time of flight of the droplet, to ensure that the droplet's impact on the surface is captured.

3 Results and Discussion

Three cases of droplet impact are studied in the present work with the developed experimental setup over (a) a flat porous surface, (b) a deep pool of liquid, (c) a curved drone impeller. For all the cases, only the surfaces were changed, while the dispensing and imaging systems remained the same. Additionally, (d) multi-flash imaging is carried out to investigate the possibility of tracking the droplet trajectory.

3.1 Droplet Impact over a Flat Porous Surface

The impact of a droplet on a porous surface (75 GSM paper) is divided into three key stages (Fig. 3). The first stage is the initial impact when the droplet is about to hit the surface but does not spread where the surface tension of the droplet keeps it still intact. The second stage is the start of spreading, resulting from the inertial forces overcoming the surface tension. The third stage is where the maximum spread occurs when droplets spread out until the surface tension acts at the periphery to stop further extension of the lamella. These results can offer valuable information about the dynamic behaviour of droplet impact on a porous surface.

3.2 Droplet Impact on a Deep Pool of Liquid

Droplet impact on liquid pools is one of the standard problems in fluid mechanics and has been studied elaborately using experimental, theoretical and numerical tools. The analysis focused on the bounce behaviour of droplet impact on a deep-water pool at three different heights (Fig. 4), and thus different Weber numbers. The Weber number is the key non-dimensional number (*We*) given by $We = \frac{\rho u^2 D}{\gamma}$ where ρ is the density, is the surface tension of the liquid, u is the impact velocity and D is the diameter of the droplet. The observed bounce behaviour also changed as the droplet dispensing height increased, resulting from higher Weber numbers. This suggests that droplets with higher impact forces can create smaller droplets that can bounce off from the free surface of the water. Similar results have been shown before, indicating the increased bouncing behaviour in deep pool [2, 4].

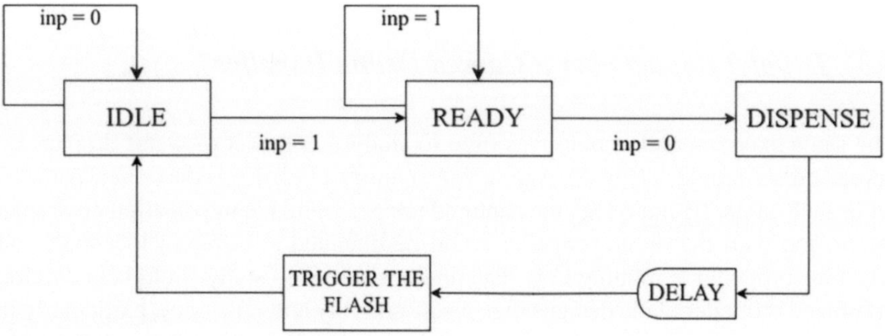

Fig. 2 Experimental setup flow chart

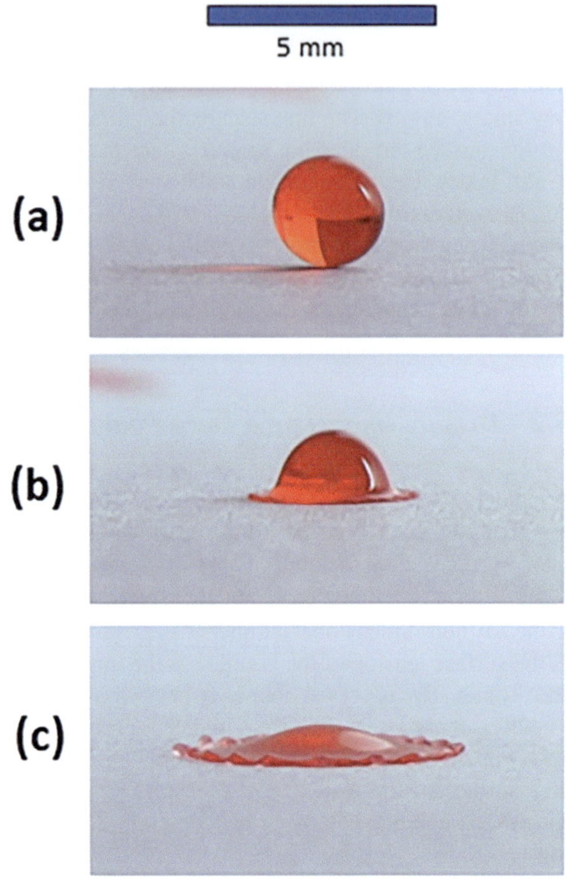

Fig. 3 Image of droplet impact on paper (Weber number = 142.85)

3.3 *Droplet Impact over a Curved Drone Impeller*

The same experimental system was used to study the impact of water droplets on curved drone impellers (Fig. 5). Again, Weber number [12, 13] is the critical parameter in such cases. By analysing the captured images, valuable information on droplet interaction with the drone propeller could be obtained at different locations, and when the propeller is rotating [14]. The findings from this study could help develop optimized propeller blade designs that ensure stability and efficiency, particularly in challenging environmental conditions such as rain [6, 7].

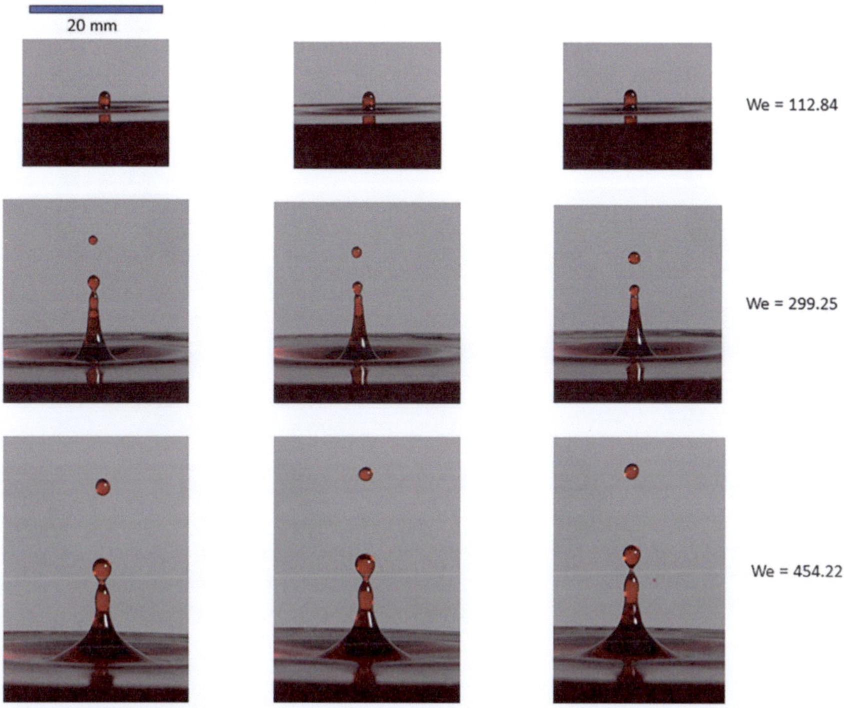

Fig. 4 Image of droplet impact on a deep pool at three different falling heights, and corresponding Weber number **a** 22.7 cm, **b** 46.8 cm and **c** 68.5 cm. [The three images for a particular We are images from three different experiments]

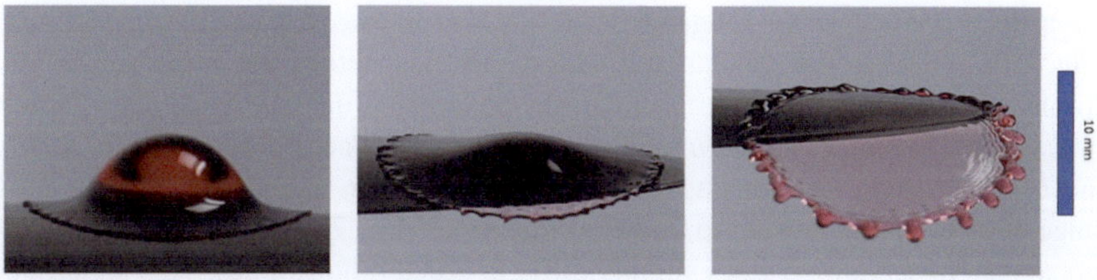

Fig. 5 Droplet impact on a drone propeller (Weber number = 282.13)

3.4 Multi-flash Imaging

As the shutter is open for a long time and only the illuminated instants are recorded, a multi-flashed image will be a superposition of all the events that were illuminated with the flash. We have tested up to seven flashes (Fig. 6), beyond which the images lose clarity. We have used this principle to estimate the acceleration of freely falling droplets.

Fig. 6 Multi-flash image showing the location of the droplets at different time instants but the same time interval between them (enhanced for clarity)

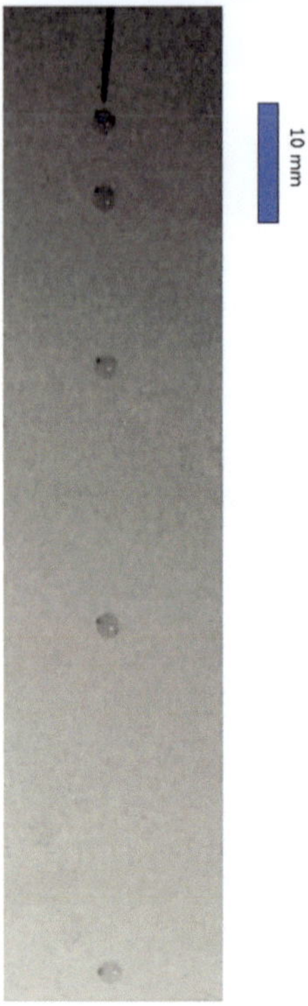

The experimental data observed for the acceleration of water droplets impacting a deep pool of water ranges from 0.00275 mm/msec^2 to 0.01144 mm/msec^2 (Fig. 7). These data were calculated with careful inspection and measurements of droplet displacement during impact. On the other hand, the theoretical calculations are calculated using equation $u^2 = 2gh$, where g is the acceleration due to gravity and h is the droplet dispensing height, which results in an expected acceleration range of 0.00735 mm/msec^2 to 0.00959 mm/msec^2 (Fig. 7). Experimental and theoretical data observation values slightly differ in acceleration values. This underlines the importance of considering the drag effects on the droplet. Further, this can also be attributed to various factors such as experimental errors, variations in droplet properties.

4 Conclusions

We demonstrated that high-speed imaging of the droplet impact could be done with a DSLR-based experimental setup. Although as old as scientific photography, the technique is still relevant to investigate the fundamental aspects of droplet impact

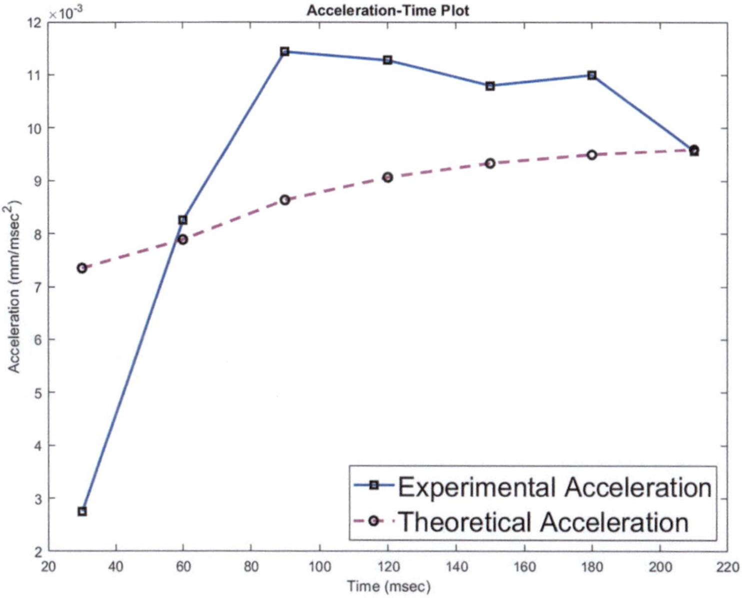

Fig. 7 Acceleration-time plot for a freely falling droplet

with a fraction of the cost of setting up a high-speed camera-based setup. Most of the time, the camera is already available, and only minimal additional procurement is needed. A severe limitation is an inability to capture a series of images, which along with limitations in temporal resolution, makes this method unattractive for some applications. However, we do have a very good spatial resolution. Thus this method is very effective in exploring problems that involve changes in the surface properties, the presence of gradients, etc.

References

1. Liu X, Zhang X, Min J (2019) Spreading of droplets impacting different wettable surfaces at a weber number close to zero. Chem Eng Sci 207:495–503
2. Yarin AL (2005) Drop impact dynamics: splashing, spreading, receding, bouncing.... Ann Rev Fluid Mech 38:159–192
3. Worthington AM (Arthur Mason) (1908) A study of splashes. Longmans, Green, and CO., London
4. Josserand C, Thoroddsen ST (2016) Drop impact on a solid surface. Ann Rev Fluid Mech 48(1):365–391
5. Burzynski DA, Roisman IV, Bansmer SE (2020) On the splashing of high-speed drops impacting a dry surface. J Fluid Mech 892:A2
6. Liu J, Vu H, Yoon S, Jepsen R, Aguilar G (2010) Splashing phenomena during liquid droplet impact. Atomization Sprays 20:297–310

7. de Goede T, de Bruin K, Shahidzadeh N, Bonn D (2021) Droplet splashing on rough surfaces. Phys Rev Fluids 6:043604
8. Rein M (1993) Phenomena of liquid drop impact on solid and liquid surfaces. Fluid Dyn Res 12(2):61
9. Ray B, Biswas G, Sharma A (2012) Oblique drop impact on deep and shallow liquid. Commun Comput Phys 11:1386–1396
10. Ray B, Biswas G, Sharma A (2015) Regimes during liquid drop impact on a liquid pool. J Fluid Mech 768:492–523
11. Turner CE, Jennison MW, Edgerton HE (1941) Public health applications of high-speed photography. Am J Public Health Nations Health 31(4):319–324. PMID: 18015411
12. Almohammadi H, Amirfazli A (2017) Understanding the drop impact on moving hydrophilic and hydrophobic surfaces. Soft Matter 13(10):2040–2053
13. Sykes TC, Fudge BD, Quetzeri-Santiago MA, Castrejón-Pita JR, Castrejón-Pita AA (2022) Droplet splashing on curved substrates. J Colloid Interface Sci 615:227–235
14. Lejeune S, Gilet T, Bourouiba L (2018) Edge effect: liquid sheet and droplets formed by drop impact close to an edge. Phys Rev Fluids 3:083601

Thermocapillary Migration of a Biofluid Droplet

Vishnu Teja Mantripragada and Antarip Poddar

Abstract We present a thin film theory to unveil the interaction between the non-Newtonian effects and the Marangoni flow for a thermally actuated drop on a solid surface. Our numerical simulations with different equilibrium contact angles (θ_e), shear-dependent viscosities (n), and dimensionless thermocapillary strengths (β) reveal a nonlinear influence of the fluid rheology on Marangoni stress and disjoining pressure. For non-Newtonian droplets, we have identified three distinct spreading regimes. The behavior of the Marangoni film regime can be characterized by certain conditions. At lower values of θ_e, higher values of β, and a specific range of n, the film exhibits a distinct linear drop shape. Additionally, for the shear-thickening drops, the thermocapillary time scale occurs early, and the advancing front becomes steeper, while drops with shear-thinning properties maintain a noticeable curvature for an extended period. On the other hand, when θ_e is higher, and (β, n) are lower, the droplets display a consistent shape and move at a uniform speed denoted as (U), which is identified as droplet regime. The droplets of this regime have a complex interaction of n and β, which leads to a considerable rise in U for shear-thinning fluids. Regardless of the regime, the shear-thinning droplet spreading is slower than that of Newtonian droplets, while the shear-thickening droplet spreading is faster in comparison. These findings can be employed in microfluidics to gain control over the spreading of non-isothermal biofluid droplets.

Keywords Biofluids · Microfluidics · Thermocapillary · Thin film

---

V. T. Mantripragada (✉)
Department of Fuel, Minerals and Metallurgical Engineering, Indian Institute of Technology (ISM), Dhanbad, Jharkhand, India
e-mail: vishnuteja@iitism.ac.in

A. Poddar
Department of Mechanical Engineering, Indian Institute of Technology (ISM), Dhanbad, Jharkhand, India
e-mail: antarip@iitism.ac.in

A. Kumar R. et al. (eds.), *Proceedings of Fluid Mechanics and Fluid Power (FMFP) 2023, Vol. 6*, Lecture Notes in Mechanical Engineering,
https://doi.org/10.1007/978-981-96-9600-0_6

1 Introduction

Microbicides delivery to function as anti-HIV agents [1], digital microfluidics [2], treatment of early onset film rupture of tears that cause dry-eye syndrome, etc., are a few examples of the diverse range of bio-microfluidic applications of micro-droplets, where the precise and effective manipulation of micro-droplets is critical.

Thermocapillary actuation of droplets, which involves manipulation of the surface tension of sessile droplets on a substrate with thermal gradient [3], has been acknowledged as an efficient way to facilitate droplet motion and wetting along surfaces. This prompted a number of theoretical and experimental studies over many years [4, 5].

In the analytical study of [4], the lubrication approximation of the Navier–Stokes equations was employed. This approximation takes advantage of the significant difference in length scales between the droplet's spreading direction and height. The study determined that longitudinal cooling of the substrate enhances the spreading of the droplet. Furthermore, heating the substrate along its length affects the isothermal spreading by opposing the capillary driving force, illustrating the significance of heat transfer in droplet manipulation.

Many other routes have been proposed in the literature to tackle the moving-contact line singularity with a decent consideration of the fluid–solid interaction [6]. A contact line singularity model based on the molecular interaction potential can be utilized by incorporating a realistically observed [7] thin precursor film around the drop. This method does not need explicit measurement of the dynamic contact angle and facilitates specifying an independent apparent contact angle. On the other hand, many biological fluids [8] exhibit intricate mechanical responses under the action of external forces, and a constant viscosity assumption is not valid anymore. A common rheological characteristic of these fluids is that the shear stress shows a nonlinear variation with the strain rate.

So far, there has been a lack of investigation regarding the interaction between the non-Newtonian rheology and the thermocapillary driving force on a sessile droplet. The current investigation aims to tackle this issue by conducting numerical simulations and gaining a working insight into the thermocapillary spreading of non-Newtonian droplets. A time-independent, inelastic power-law model for fluid rheology is employed to achieve this. The thin film model-based hydrodynamic analysis reveals the significant impact of non-Newtonian rheology on both the spreading behavior and the morphology of non-isothermal droplets.

2 Mathematical Formulation

Our study examines a 2D complex fluid droplet with viscosity $\widetilde{\mu}$ and density $\widetilde{\rho}$, placed on a solid substrate with varying temperatures. To facilitate the analysis, we have defined a planar Cartesian coordinate system.

The equations that govern the fluid flow ($\widetilde{\boldsymbol{v}}$, $\widetilde{p}$) are the continuity equation and the Cauchy momentum equation

$$\widetilde{\nabla} \cdot \widetilde{\boldsymbol{v}} = 0 \quad \text{and} \quad \widetilde{\rho}\left(\frac{\partial \widetilde{\boldsymbol{v}}}{\partial \widetilde{t}} + \widetilde{\boldsymbol{v}} \cdot \widetilde{\nabla}\widetilde{\boldsymbol{v}}\right) = \widetilde{\nabla} \cdot \widetilde{\mathbf{T}}. \tag{1a}$$

Here, $\widetilde{\mathbf{T}}$ is the stress tensor, given by $\widetilde{\mathbf{T}} = -p\mathbb{I} + \widetilde{\boldsymbol{\tau}}$, and the strain rate tensor is related to the deviatoric stress $\widetilde{\boldsymbol{\tau}}$ as $\widetilde{\boldsymbol{\tau}} = 2\,\widetilde{\mathbf{D}}\,\widetilde{\mu}$. We employ the power-law constitutive relation [9]

$$\widetilde{\mu}\left(\widetilde{\dot{\gamma}}\right) = \mu_0 \left|2\,\widetilde{\mathcal{M}}\,\widetilde{\dot{\gamma}}\right|^{n-1}, \tag{2}$$

where $1/\widetilde{\mathcal{M}}$ is the characteristic deformation rate; μ_0 is the constant viscosity of the Newtonian base fluid. The power-law index is denoted by n where $n = 1$, $n > 1$, and $n < 1$ stand for the Newtonian, shear-thickening, and shear-thinning fluids, respectively. In addition, the strain rate tensor's $\widetilde{\mathbf{D}}$ second invariant, denoted by $\widetilde{\dot{\gamma}}$, is $\widetilde{\dot{\gamma}} = \left[0.5(\widetilde{\mathbf{D}} \colon \widetilde{\mathbf{D}})\right]^{0.5}$, where $\widetilde{\mathbf{D}} = \frac{1}{2}\left[(\widetilde{\nabla}\widetilde{\boldsymbol{v}}) + (\widetilde{\nabla}\widetilde{\boldsymbol{v}})^{\mathrm{T}}\right]$. The power-law model was employed to capture the fundamental physics underlying the stress-dependent rheology of thin films in a wide range of practical scenarios, e.g., thin films breakage [10], vaginal delivery of microbicides [11], etc.

Normal stress and tangential stress equilibrium conditions are applied on the air-liquid interface at $\widetilde{y} = \widetilde{h}(\widetilde{t}, \widetilde{x})$

$$\hat{\boldsymbol{n}} \cdot \widetilde{\mathbf{T}} \cdot \hat{\boldsymbol{n}} = \widetilde{\kappa}\,\widetilde{\gamma} \quad \text{and} \quad \hat{\boldsymbol{n}} \cdot \widetilde{\mathbf{T}} \cdot \hat{\boldsymbol{t}} = \hat{\boldsymbol{t}} \cdot \widetilde{\nabla}\widetilde{\gamma}, \tag{3a}$$

where $\hat{\boldsymbol{n}}$ is the outward unit normal vector at the interface, $\hat{\boldsymbol{t}}$ is the outward unit tangential vector at the interface. The average interface curvature is denoted by $\widetilde{\kappa} = -\widetilde{\nabla} \cdot \hat{\boldsymbol{n}}$. The no-slip and no-penetration conditions are applied at the solid-drop interface ($\widetilde{y} = 0$). The free surface kinematic boundary condition is obtained as

$$\frac{D\widetilde{f}}{D\widetilde{t}} = \frac{\partial \widetilde{f}}{\partial \widetilde{t}} + \widetilde{\boldsymbol{v}} \cdot \widetilde{\nabla} f = 0, \;\; \widetilde{f}(\widetilde{x}, \widetilde{y}, \widetilde{t}) = \widetilde{y} - \widetilde{h}(\widetilde{x}, \widetilde{t}). \tag{4}$$

The interactions between the solid substrate and the droplet at a molecular level are modeled by incorporating a disjoining-conjoining pressure term [12] $\widetilde{\Pi}(\widetilde{h})$, as $\widetilde{\Pi} = \widetilde{\mathcal{K}}\left[\left(\frac{\widetilde{h}_*}{\widetilde{h}}\right)^3 - \left(\frac{\widetilde{h}_*}{\widetilde{h}}\right)^2\right]$, where $\widetilde{h}_*$ is the energetically favored thickness of the molecular film [13]. Thus, the total pressure within the droplet would be

$$\widetilde{p} = -\widetilde{\Pi} - \widetilde{\gamma}\frac{\partial^2 \widetilde{h}}{\partial \widetilde{x}^2}, \tag{5}$$

The first term in the above equation stands for the disjoining pressure $\widetilde{\Pi}(\widetilde{h})$, which arises due to the van der Waals interaction occurring at the interface between the solid and liquid phases. The latter term represents the Laplace pressure, which emerges due to the curvature of the interface. In this model, ahead of the droplet, a thin precursor film is incorporated, experimentally observed to be of thickness ($\sim 10-100\,\text{nm}$) [14], and thereby eliminates the contact line singularity. Additionally, in partially wetting systems, the contact angle at equilibrium is influenced by intermolecular forces, which are accounted for through a specific term $\widetilde{\mathcal{K}}$ in the model, as $\widetilde{\mathcal{K}} = 2\widetilde{\gamma}_0\left(1-\cos(\theta_e)\right)/\widetilde{h}_*$.

The liquid-air interface tension has been considered to be dependent on temperature as [15]

$$\widetilde{\gamma} = \widetilde{\gamma}_0 - \widetilde{\sigma}(\widetilde{T} - \widetilde{T}_0), \tag{6}$$

where $\widetilde{\gamma}_0$ represents the surface tension at $\widetilde{T} = \widetilde{T}_0$ and the positive constant σ denotes the temperature gradient of surface tension. Marangoni stresses are caused at the interface due to these thermally induced gradients in surface tension and can be obtained as $\dfrac{\partial\widetilde{\gamma}}{\partial\widetilde{x}} = \hat{\tau} = \dfrac{\partial\widetilde{\gamma}}{\partial\widetilde{T}}\dfrac{\partial\widetilde{T}}{\partial\widetilde{x}}$. The term representing the inertial forces in the Cauchy momentum equation ($\widetilde{\boldsymbol{v}}\cdot\widetilde{\nabla}\widetilde{\boldsymbol{v}}$ in (1)) becomes negligible under the lubrication approximation. Such a flow regime is dominated by the capillary and viscous effects. By utilizing this approximation, it is possible to obtain a transient differential equation that governs the height of the droplet $\widetilde{h}(\widetilde{x},\widetilde{t})$. This approximation is particularly applicable to small droplet sizes, where the influence of surface tension surpasses that of gravity, thus playing a major role in the deformation and spreading of the droplet [15]. Furthermore, it is valid to assume that the thermal Péclet number is extremely low, implying that the predominant mode of heat transfer within the droplet is conduction [16]. As a result, the temperature profile within the droplet fluid is governed by

$$\frac{\partial^2\widetilde{T}}{\partial\widetilde{y}^2} = 0, \tag{7}$$

The following boundary conditions are applied to solve the energy equation. $\widetilde{T} = \widetilde{T}_w$, at $\widetilde{y} = 0$; and $-\widetilde{k}_l\frac{\partial\widetilde{T}}{\partial\widetilde{y}} = \mathcal{H}_\infty(\widetilde{T} - \widetilde{T}_\infty)$, at $\widetilde{y} = \widetilde{h}$. Here, the liquid's thermal conductivity is denoted by $\widetilde{k}_l$, the ambient temperature is $\widetilde{T}_\infty$, and $\mathcal{H}_\infty$ is the convective heat transfer at the interface. Thus inside the droplet, the temperature profile can be obtained as

$$\widetilde{T} = \widetilde{T}_w - \frac{\mathcal{H}_\infty\widetilde{y}/\widetilde{k}_l(\widetilde{T}_w - \widetilde{T}_\infty)}{1 + \mathcal{H}_\infty\widetilde{h}/\widetilde{k}_l}. \tag{8}$$

It has been found using the lubrication approximation that $\mathcal{H}_\infty\widetilde{h}/\widetilde{k}_l, \mathcal{H}_\infty\widetilde{y}/\widetilde{k}_l \ll 1$ [17], leading to $\widetilde{T} \approx \widetilde{T}_w(\widetilde{x})$. A surface temperature distribution which linear varies as $\dfrac{\partial \widetilde{T}_w}{\partial \widetilde{x}} = \Gamma$. is assumed, which is motivated by silicon-like widely used substrate's high thermal conductivity [18]. As a consequence, we obtain the equation $\hat{\tau} = -\sigma\,\Gamma$ for Marangoni stress, which is greater than 0. Thus, the boundary condition for the tangential stress ((3)) becomes $\widetilde{\tau_{\widetilde{x}\widetilde{y}}}|_{\widetilde{h}} = \hat{\tau}$. The reference values given below are used to non-dimensionalize the variables. (The symbol '~' is being dropped henceforth) $x_c = y_c = a,\ p_c = \gamma_0/a,\ t_c = \mu x_c^4/\gamma_0 h_c^3 = \mu_c a/\gamma_0$, $u_c = x_c/t_c = \gamma_0/\mu_0$, $\mathcal{M}_c = y_c/u_c$, and $\mu_c = \mu_0$. The term $a = \sqrt{\gamma_0/\rho g}$ denotes the capillary length scale.

The kinematic boundary condition uses the no-penetration condition along with the continuity equation and is given by

$$\frac{\partial h}{\partial t} + \frac{\partial Q}{\partial x} = 0, \tag{9}$$

Here, the rate of flow, Q can be expressed as $Q = \int_0^h u\,dy$. The flow velocity's (u) x component can be obtained from the no-slip condition and the relation in (2), as

$$u = \frac{n}{n+1}\frac{\mathcal{M}^{1/n-1}}{p_x}\left[|p_x(y-h)+\beta|^{\frac{1}{n}+1} - |hp_x - \beta|^{\frac{1}{n}+1}\right], \tag{10}$$

where p_x represents the pressure gradient in the axial direction $\partial p/\partial x$. Moreover, β, which is the non-dimensional thermocapillary number, is defined as $\beta = \hat{\tau}/p_c = \hat{\tau}a/\gamma_0$. During the thermocapillary action of thin films on Newtonian fluids, the literature reveals that the two driving forces at the free surface, i.e., a shear flow due to the Marangoni stress $\hat{\tau}$ and a Poiseuille flow because of the pressure gradient p_x linearly superpose to determine the flow velocity [18]. On the contrary, a nonlinear superposition of these two forces can be seen from the streamwise velocity profile in (10), where the net velocity depends on both the power-law index (n) and the thermocapillary number (β). Correspondingly, the flow rate can be obtained as

$$\begin{aligned} Q = \mathcal{M}^{\frac{1}{n}-1}\Bigg[\frac{1}{p_x^2}\frac{n^2}{(n+1)(2n+1)}\frac{1}{p_x^2}\Big\{&\beta^{\frac{1}{n}+2} - |\beta - hp_x|^{\frac{1}{n}+1} \\ &(\beta - hp_x)\Big\} - \frac{h}{p_x}\frac{n}{(n+1)}|\beta - hp_x|^{\frac{1}{n}+1}\Bigg]. \end{aligned} \tag{11}$$

3 Solution Methodology

To solve the governing equation (Eq. 9), numerical simulations were performed using COMSOL Multiphysics® v.6.0. At both the ends of the computational domain for variables h and p, periodic boundary conditions have been applied. In order to realistically represent the molecular film thickness of $\widetilde{h}^* \sim 10$ nm, a grid size in a similar order of magnitude was necessary, resulting in $\Delta x \approx 10^{-6} - 10^{-5}$ to maintain sufficient spatial resolution in the moving-contact line region [16]. However, such fine mesh sizes significantly increase the cost of computation. Hence, following the approach of Gomba and others [16, 19], a value of $h^* = \Delta x = 0.01$ was chosen throughout the simulations. The one dimensional computational domain has a span of 200 to 400 dimensionless length units, as required to capture the essential physics of droplet migration and rupture.

The droplet shape of Gomba et al. [13] under steady state, without a temperature gradient, was chosen as the initial state. Such initial condition ensures the physical consistency of the shape-related parameters of the droplet, such as the drop's maximum surface height ($h_{\max}$), the initial liquid-solid apparent contact angle (θ_e), and the droplet's cross-section area ($\mathcal{A}$). These parameters relate to the equilibrium film thickness around the droplet (h_f). In order to maintain consistency for comparison among outcomes with various power-law indices (n), the same initial conditions were chosen for droplets with varying n.

4 Results and Discussion

Here, we present the findings of our numerical study, focusing on the complex interplay between the shear-dependent fluid rheology and the thermocapillary actuation mechanism. We have selected dimensionless parameters within practical ranges based on published studies [20]. Specifically, we have considered β values ranging from approximately 0.001 to 0.03, and $\mathcal{M}$ of approximately 14. The power-law index (n) is varied between 0.4 and 1.6, which is motivated by the rheological behavior observed in a variety of natural and synthetic polymeric solutions as well as particulate suspensions [11, 21].

4.1 Summary of Spreading Regimes

The various spreading regimes observed as a result of changes in the thermocapillary number β, equilibrium contact angle θ_e, and the power-law index n are illustrated in Fig. 1. These maps serve as a central visualization of the key findings of our work and also highlight the Newtonian spreading regimes reported previously [16]. Results of various Numerical simulations for the different parameter sets (β, n, θ_e) were thoroughly analyzed to classify the dynamical phase regimes. Three distinct droplet spreading regimes were identified, similar to [13], viz., Marangoni film, transition, and the droplet regimes. To delineate these regimes numerically, we consider the relative difference in speeds between the trailing and leading edges of the drop throughout the spreading process. If this relative difference remains below 1% within the entire period of droplet spreading, the drop is placed in the droplet regime. Alternately, if the relative speed difference exceeds the specified threshold and the droplet eventually breaks apart into several parts, it is classified under the transition regime.

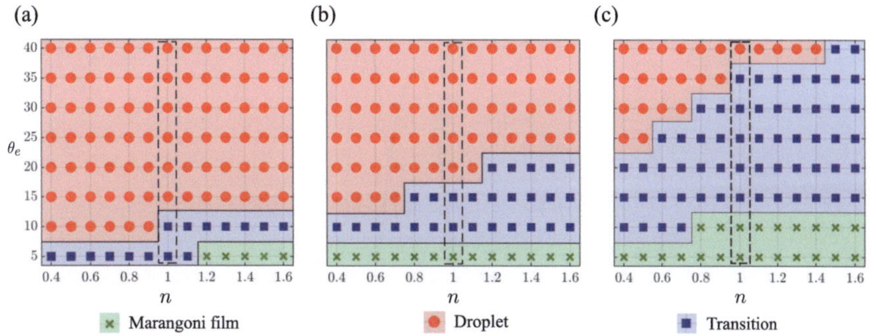

Fig. 1 Droplet spreading regime maps for thermocapillary numbers **a** $\beta = 0.001$, **b** $\beta = 0.007$, and **c** $\beta = 0.03$

Typically, droplets in the transition regime initially exhibit a linear profile, but as they continue to spread, the film disintegrates from the rear and divides into many. The complex and transient behavior, as well as the occurrence of film rupture, make characterizing the transition regime challenging. Lastly, droplets of the Marangoni film regime are characterized by a significant difference between the speeds of both edges. However, they do not dissociate into many drops and maintain a linear profile. The leading edge of the droplet in the Marangoni film regime spreads quicker than the rear (as determined by the specified tolerance), resulting in an increasing droplet width over time. A linear profile is formed, and the trailing edge maintains a characteristic capillary ridge, which becomes less steep as the spreading progresses.

4.2 *Marangoni Films*

For different power-law index n values, the variation of the droplet surface profile along the axial direction is shown in Fig. 2. In comparison, a faster spreading of shear-thickening drops and a slower spreading of shear-thinning drops over the Newtonian drop can be seen here. This observation is counter-intuitive to the general notion of slowing down of contact line due to an increase in viscous resistance by the shear-thickening fluids. Yet, the findings of the isothermal experiments [22] support these observations. It was proposed that the near-contact line region flow accelerates, leading to an enhanced shear-thinning effect. Simultaneously, the escalating shear rate in this zone opposes this acceleration mechanism, thereby displacing the contact line region's corner points faster. As a consequence, the contact angle reduces, which suppresses further spreading, thereby slowing down the contact line motion.

Just like in Marangoni films [23], a capillary ridge forms with a linear bulk shape of the drop, as seen in the figure. Due to the shear-thickening rheology, a steeper front appears at the advancing face of the droplet. While the shear-thickening ($n = 1.4$)

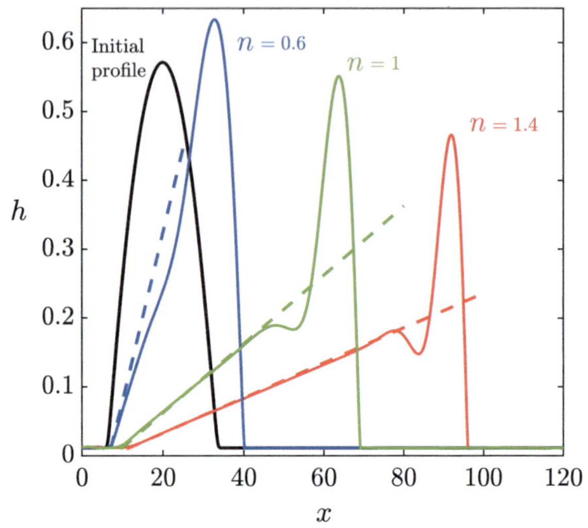

Fig. 2 Comparison of instantaneous droplet profiles for different power-law indices

and the Newtonian ($n = 1$) droplets developed a linear shape, the shear-thinning drop ($n = 0.6$) has only shown a deviation from the original shape.

As can be seen in Fig. 1a–c, high power-law index (n_{cr}) and certain β lead to Marangoni film formation. At the same time, lower θ_e and β values are necessary for the droplet to fall under the Marangoni film zone. For example, for $\beta = 0.001$ and 0.007 the Marangoni zone exists until $\theta_e = 5^0$, but for $\beta = 0.03$ this zone extends till $\theta_e = 10^0$. Suppression of disjoining pressure at large β explains this behavior. Moreover, the shear-dependent viscosity influences the magnitudes of the Marangoni forces and disjoining pressure, which results in the existence of a critical power-law index n_{cr}.

4.3 Uniform Droplet Motion

The effect of curvature becomes significant at high contact angle values, where the shape of the droplets does not resemble a Marangoni film. The enhanced disjoining-conjoining pressure at higher contact angle values becomes comparable to the Marangoni stresses, which manifests the droplet curvature. For different (β) and (n), the droplet profiles are shown in Fig. 3a at $t = 3 \times 10^4$ and $\theta_e = 40^0$.

Within the parametric limits of this zone, the droplet was observed to withhold its shape intact and advance with a fixed velocity U. As the shape is withheld, the droplet's receding and advancing contact lines move with the same velocity. The drop velocity remains uniform, but when compared to a Newtonian fluid, it decreases for shear-thinning fluids and increases for shear-thickening fluids. Furthermore, because of the power-law model's non-linearity, obtaining an analytical expression for drop speed as done with the Newtonian fluids [18, 24] is not feasible.

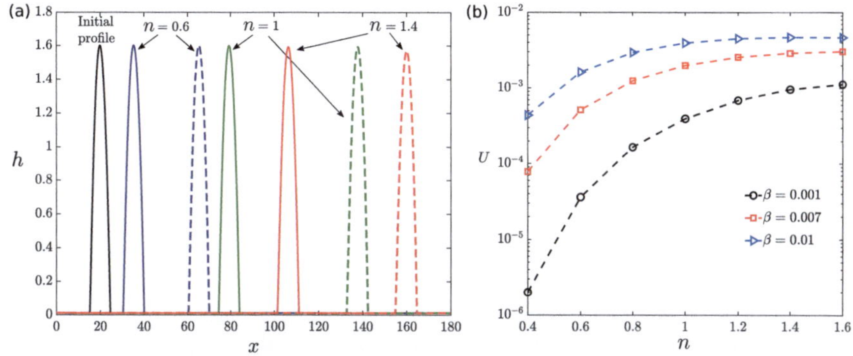

Fig. 3 **a** Axial positions of droplets at $t = 3 \times 10^4$ for different power-law indices. The solid and dotted lines correspond to the cases with $\beta = 0.007$ and $\beta = 0.01$, respectively. **b** Effect of the power-law index on the droplet velocity for different thermocapillary numbers

Figure 3b illustrates the impact of β and n on the migration speed of the droplet. We can observe that the droplet velocity saturates for $n > 1$, but it increases for $n < 1$. Also, the velocity increases for high β values and saturates faster. The effects of the fluid rheology on the Marangoni stress and the disjoining pressure are different for various β, which explains these observations.

4.4 The Transition Zone

Figure 1a–c shows a separation between the Marangoni film and droplet regimes, termed the 'transition' regime, which is characteristic of parent droplets rupturing into several sister droplets. Another peculiar characteristic of this regime is the droplet advancement without rupture but with an irregular shape. In this regime, the magnitudes of the disjoining-conjoining pressure term and the Marangoni stress are comparable, and as a consequence, several sister droplets form due to rupture, which follow the parent droplet. For non-Newtonian droplets, both these mechanisms are modulated by the fluid's shear-dependent rheology in a complex manner. For $\theta_e = 20^0$ and $\beta = 0.007$, the evolution of a shear-thickening droplet in time is in the transitional regime as shown in Fig. 4a, where the formation of multiple smaller droplets due to the rupture of the parent droplet is evident. The shear-thinning droplets also exhibit transition behavior under certain conditions (see Fig. 4b). As observed from these figures, the rupture of the droplets is delayed for shear-thinning fluids at lower β and θ_e values. Notably, for $n = 0.6$, even up to a longer time of $t = 60 \times 10^5$, the drop has an irregular shape and behaves like a single entity.

At a constant (β), this regime begins at a particular angle of contact (θ_{tr}) for Newtonian fluids. However, the critical transition angle of contact θ_{tr} varies with the power-law index n in the case of non-Newtonian droplets. At β as low as $= 0.001$,

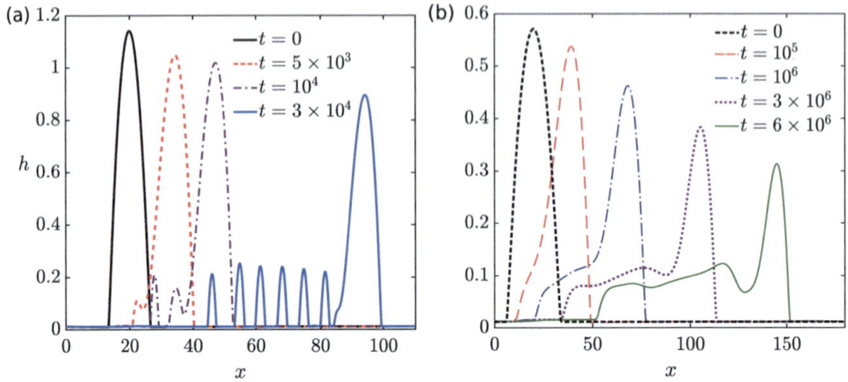

Fig. 4 Shape evolution in the transition regime **a** $\beta = 0.007$, $\theta_e = 20^0$, $n = 1.6$; **b** $\beta = 0.001$, $\theta_e = 5^0$, $n = 0.6$

the transition regime is observed in a limited range of contact angles, i.e., $\theta_e \lesssim 10^0$, as shown in Fig. 1a. On contrary, as shown in Fig. 1b and c, for high β values, broader ranges of θ_e, ranging upto $\theta_e \approx 40^0$ are observed. Furthermore, as revealed from Fig. 1a, the droplet at $\theta_e = 5^0$ does not evolve into Marangoni film but eventually ruptures until $n \lesssim 1.1$. Therefore, in this zone, the effect of n on Marangoni stress becomes less significant compared to its effect on the disjoining pressure. Also, for non-Newtonian fluids with $n > 1.1$, the effects of the shear-thickening viscosity on the Marangoni stresses are so significant that the droplet behaves as a Marangoni film.

5 Conclusions and Remarks

The intricate surface dynamics of a sessile drop with shear-dependent rheology on a differential temperature substrate has been investigated by developing a mathematical model. To address the contact line singularity, the Cauchy momentum equation has been solved using the thin film approximation in conjunction with a disjoining-conjoining pressure model. We discovered that a high power-law index value indicates a stronger proclivity for conversion from the transition regime to the Marangoni film regime. A shear-thickening droplet advances with a steeper front, but a shear-thinning droplet preserves a large curvature for an extended duration. Moreover, the spread of a shear-thinning droplet is slower than a Newtonian droplet. The regime maps show that the drop remains well within the droplet regime by lowering the power-law index. The onset of the droplet regime, unlike in the case of Newtonian droplets, does not begin at a fixed value of θ_e. While certain shear-thinning fluid droplets move with uniform velocity beyond a critical power-law index, the transitional regime kicks in. Moreover, to preserve the droplet shape, the minimum contact angle at equilibrium (θ_{cr}) has been observed to steadily increase with β. Furthermore, the attenuation of U was also highlighted in the $n > 1$ cases. The velocity of the droplet increases overall for large β values. However, it slowly approaches a constant value at an enhanced rate. For shear-thinning fluids, an enhanced rise in the rate of U has been observed with an increase in n. In contrast, with shear-thickening fluids, U saturation has been observed. As can be observed with shear-thinning fluids from Fig. 4b, a delay in the droplet rupture occurs at low β and θ_e values. It has also been observed that droplet rupture is more common in shear-thickening fluids.

To conclude, we discovered the distinct properties of the several spreading regimes resulting from a nonlinear influence of non-Newtonian rheology on the competing effects of disjoining-conjoining pressure and Marangoni stress. It is mentionworthy that one must exercise caution in using the findings of the power-law model at low shear rates. The inadequacies of the power-law model may be addressed by employing Carreau-type models that incorporate finite zero shear rate and infinite shear rate viscosities.

The distinct droplet spreading characteristics that arise due to the fluid's shear-dependent viscosity may be correlated to accurately assess the complex rheology of the fluid. Ultimately, the fundamental insights on the rheological responses can help advance technologies on microfluidic droplet control and can build effective microfluidic systems to help manipulate complex biofluids.

Acknowledgements A. P. acknowledges DST, Govt. of India, for funding the project No: DST(SERB)(346)/2022-2023/940/MECH.

References

1. Hu B, Kieweg SL (2012) The effect of surface tension on the gravity-driven thin film flow of Newtonian and power-law fluids. Comput Fluids 64:83–90
2. Anna SL (2016) Droplets and bubbles in microfluidic devices. Annu Rev Fluid Mech 48:285–309
3. Young NO, Goldstein JS, Block MJ (1959) The motion of bubbles in a vertical temperature gradient. J Fluid Mech 6(3):350–356
4. Ehrhard P, Davis SH (1991) Non-isothermal spreading of liquid drops on horizontal plates. J Fluid Mech 229:365–388
5. Sui Y, Spelt PDM (2015) Non-isothermal droplet spreading/dewetting and its reversal. J Fluid Mech 776:74–95
6. Snoeijer JH, Andreotti B (2013) Moving contact lines: scales, regimes, and dynamical transitions. Annu Rev Fluid Mech 45(1):269–292
7. Starov VM, Kalinin VV, Chen J-D (1994) Spreading of liquid drops over dry surfaces. Adv Colloid Interface Sci 50:187–221
8. Yu T, Malcolm K, Woolfson D, Jones DS, Andrews GP (2011) Vaginal gel drug delivery systems: understanding rheological characteristics and performance. Expert Opin Drug Deliv 8(10):1309–1322
9. Deen WM (1998) Analysis of transport phenomena, vol 2. Oxford University Press, New York
10. Garg V, Kamat PM, Anthony CR, Thete SS, Basaran OA (2017) Self-similar rupture of thin films of power-law fluids on a substrate. J Fluid Mech 826:455–483

11. Kheyfets VO, Kieweg SL (2013) Experimental and numerical models of three-dimensional gravity-driven flow of shear-thinning polymer solutions used in vaginal delivery of microbicides. J Biomech Eng 135(6):061009
12. Mitlin VS, Petviashvili NV (1994) Nonlinear dynamics of dewetting: kinetically stable structures. Phys Lett A 192(5–6):323–326
13. Gomba JM, Homsy GM (2009) Analytical solutions for partially wetting two-dimensional droplets. Langmuir 25(10):5684–5691
14. Popescu MN, Oshanin G, Dietrich S, Cazabat AM (2012) Precursor films in wetting phenomena. J Phys: Condens Matter 24(24):243102
15. Leal LG (2007) Advanced transport phenomena: fluid mechanics and convective transport processes, vol 7. Cambridge University Press
16. Gomba JM, Homsy GM (2010) Regimes of thermocapillary migration of droplets under partial wetting conditions. J Fluid Mech 647:125–142
17. Chaudhury K, Chakraborty S (2015) Spreading of a droplet over a nonisothermal substrate: Multiple scaling regimes. Langmuir 31(14):4169–4175
18. Brzoska JB, Brochard-Wyart F, Rondelez F (1993) Motions of droplets on hydrophobic model surfaces induced by thermal gradients. Langmuir 9(8):2220–2224
19. Mac Intyre J, Gomba JM, Perazzo CA, Correa PG, Sellier M (2018) Thermocapillary migration of droplets under molecular and gravitational forces. J Fluid Mech 847:1–27
20. Chen JZ, Troian SM, Darhuber AA, Wagner S (2005) Effect of contact angle hysteresis on thermocapillary droplet actuation. J Appl Phys 97(1):014906
21. Wang XD, Lee DJ, Peng XF, Lai JY (2007) Spreading dynamics and dynamic contact angle of non-Newtonian fluids. Langmuir 23(15):8042–8047
22. Rafaï S, Bonn D, Boudaoud A (2004) Spreading of non-Newtonian fluids on hydrophilic surfaces. J Fluid Mech 513:77–85
23. Madeline LS, Tsung-Lin H, Wangrun X, Robert DT, Stephen G (2022) Surfactant spreading on a deep subphase: coupling of Marangoni flow and capillary waves. J Colloid Interface Sci 614:511–521
24. Pratap V, Moumen N, Subramanian RS (2008) Thermocapillary motion of a liquid drop on a horizontal solid surface. Langmuir 24(9):5185–5193

Bubble Ascent in the Presence of Vertically Applied Electric Field

Darshan Patel, R. Vishnu, and S. Vengadesan

Abstract In this study, we have examined the phenomenon of a bubble ascent in the presence of a vertically applied electric field. To simulate the problem, a novel electrohydrodynamic solver integrated with the open-source solver interFoam has been utilized. This solver incorporates the solution of the charge convection equation to determine the charge distribution and Coulomb force accurately. The effects of the electric capillary number (Ca_E) and Bond number (Bo) on the bubble's behavior under the imposition of an electric field have been investigated. As the bubble begins to rise, the electrical force exerted at the interface causes the bubble to deform in the vertical direction. It has been observed that higher values of Ca_E result in increased electrical forces, leading to greater deformation of the bubble. A transition from vertical to horizontal deformation is observed as Bo increases from lower to higher values. In the ellipsoidal regime, the electrical force effectively suppresses the bubble's deformation. Moreover, due to the facilitation of bubble motion by the electrical force, the terminal velocity is higher when an electric field is present compared to its absence. An increase in both Ca_E and Bo corresponds to an increase in the terminal velocity.

Keywords Electrohydrodynamics · Bubble dynamics · Leaky dielectric fluids

---

D. Patel · R. Vishnu · S. Vengadesan (✉)
Department of Applied Mechanics and Biomedical Engineering, Indian Institute of Technology, Madras, Tamil Nadu, India
e-mail: vengades@iitm.ac.in

D. Patel
e-mail: am18d003@smail.iitm.ac.in

R. Vishnu
e-mail: am21s015@smail.iitm.ac.in

A. Kumar R. et al. (eds.), *Proceedings of Fluid Mechanics and Fluid Power (FMFP) 2023, Vol. 6*, Lecture Notes in Mechanical Engineering,
https://doi.org/10.1007/978-981-96-9600-0_7

1 Introduction

Bubble rising, a fundamental two-phase flow phenomenon, has garnered considerable attention from researchers over the past several decades due to its industrial relevance, including mixing, aeration, and bubble column reactors. However, understanding bubble ascent under the imposition of external fields, such as electric fields, is relatively limited and requires further investigation. Studying the motion of a single bubble rising in a still liquid under the influence of an electric field has practical implications in various industries, including enhancing heat transfer in microfluidic systems and improving gas separation processes in the oil and gas sector. These insights offer opportunities for more efficient and innovative applications across engineering and scientific domains.

The dynamics of the bubble ascent phenomenon can be attributed to three fundamental forces: buoyancy force, surface tension force, and viscous force [1]. The buoyancy force arises from the density difference between the fluids and serves as the primary driving force for the upward motion of the bubble. The surface tension force, on the other hand, arises from the pressure drop across the interface between the two fluids and minimizes the bubble's surface energy, thereby influencing its shape. Additionally, the viscous force acts as a resistance to the motion of the bubble and contributes to its wobbling behavior [2, 3].

The interplay and relative strength of these forces determine the steady-state shape and terminal velocity of the bubble. To comprehend the relative importance of these forces, various non-dimensional numbers have been introduced, such as the Bond number $\frac{\Delta\rho g D^2}{\gamma}$, Reynolds number $\frac{\rho_c u D}{\mu_c}$, Morton number $\frac{\Delta\rho g \mu_c^4}{\gamma^3 {\rho_c}^2}$, Archimedes number $\frac{{\rho_c}^2 g D^3}{{\mu_c}^2}$, and Galilei number $\frac{d D^3}{{\mu_c}^2}$ [4–6]. These dimensionless parameters facilitate the characterization and understanding of the intricate balance between the buoyancy, surface tension, and viscous forces in the bubble-rising phenomenon.

The Bond number, representing the ratio of buoyancy to surface tension forces, plays a critical role in determining the terminal velocity and shape of a bubble. For lower Bo values, the dominant surface tension force maintains a spherical bubble shape. As the Bond number increases, the bubble shape transitions to an oblate ellipsoidal, dimple ellipsoidal, ellipsoidal cap, and ultimately, a skirted bubble [5, 7]. A topological transformation can also be attributed to the higher values of Bo [8]. Higher Bond numbers correspond to increased terminal velocities. The Reynolds number, on the other hand, quantifies the relative significance of viscosity and inertia [5, 9]. In the ellipsoidal regime, higher Reynolds numbers induce shape and path instabilities [3], while also influencing the terminal velocity. However, Reynolds number has no impact on the steady-state shape of the bubble [10]. The literature explores the effects of additional dimensionless numbers such as Galilei, Morton, and Ohnesorge on bubble dynamics [4, 6, 9, 11].

When a bubble ascends under the imposition of an electric field, it encounters an additional force, apart from hydrodynamic forces, known as the electrical force [12]. Depending on its electrical properties, this force acts at the bubble interface and elongates the bubble either along or perpendicular to the applied field. Consequently, the electrical force influences both the shape and terminal velocity of the bubble. Andalib et al. [13] conducted experimental investigations to examine the impact of uniform and non-uniform electric fields on bubble dynamics, while Rehmat et al. [14] numerically studied the bubble deformation under the imposition of an electric field using the Smoothed Particle Hydrodynamics (SPH) method. The electric capillary number $Ca_E = \frac{\varepsilon_c {E_0}^2 D}{\gamma}$, defined as a ratio of electric to capillary forces, serves as the non-dimensional parameter that quantifies the strength of the applied electric field.

In this study, our objective is to investigate the motion of a bubble under the imposition of a vertically applied electric field. Both fluids are considered leaky dielectric fluids, and Taylor's leaky dielectric model [15, 16] is employed. To accurately calculate the Coulomb force and charge distribution, the charge convection equation is solved. A numerical model is developed using the open-source solver for two-phase flow, interFoam [11]. The volume of fluid method is utilized to capture the interface, and to mitigate the potential for path instability, simulations are conducted in a two-dimensional Cartesian coordinate framework. The remainder of the paper is organized as follows: Sect. 2 presents the problem definition and governing equations, followed by a discussion of the effects of Ca_E and Bo on the motion of the bubble in the results Sect. 3. Finally, our results are summarized in the concluding remarks.

2 Methodology

Initially, a spherical bubble with a diameter D is placed at the center of the channel, positioned at a distance $2D$ from the bottom wall as shown in Fig. 1. The confinement ratio is set to 0.3, and the channel height is chosen as $20D$. To simulate the phenomenon of a bubble rising under a vertically applied electric field, the continuity equation and the Navier–Stokes equation are simultaneously solved. The continuity equation and the Navier–Stokes equation govern the flow dynamics and are expressed as follows:

$$\nabla \cdot \boldsymbol{u} = 0 \tag{1}$$

Navier–Stokes equation:

$$\begin{aligned} \frac{\partial (\rho_m \boldsymbol{u})}{\partial t} + \boldsymbol{u}\nabla \cdot (\rho_m \boldsymbol{u}) = -\nabla p + \boldsymbol{F_{st}} + \boldsymbol{F_e} \\ + \mu_m \left[\nabla \boldsymbol{u} + (\nabla \boldsymbol{u})^T\right] \end{aligned} \tag{2}$$

Here, $\boldsymbol{u}$ is the velocity, $\boldsymbol{F_{st}}$ is the surface tension force, and $\boldsymbol{F_e}$ is the electrical force. α is the volume fraction function (VFF), and it is used to capture the interface. The value of α is 1 in the continuous phase and 0 in the discrete phase. At the interface,

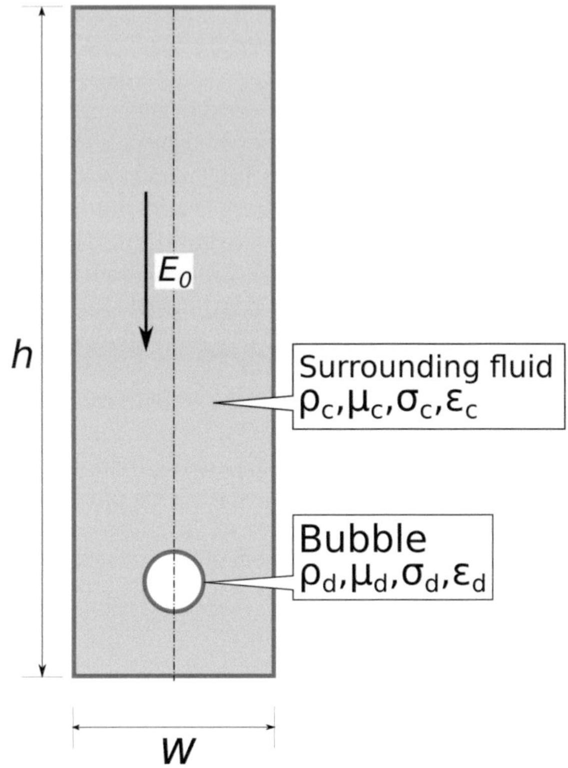

Fig. 1 Schematic diagram of the bubble-rising phenomenon under the imposition of vertically applied electric field

the value of α varies between 0 and 1. A VFF advection equation is solved to capture the interface.

$$\frac{\partial\alpha}{\partial t} + \nabla\cdot(\alpha\boldsymbol{u}) + \nabla\cdot[\boldsymbol{u}_c\alpha(1-\alpha)] = 0 \tag{3}$$

where $\boldsymbol{u}_c$ is the interfacial compressive flux term.

Brackbill's continuous surface tension force method [17] is used to calculate surface tension force.

$$\boldsymbol{F}_{st} = \gamma\kappa\hat{n} \tag{4}$$

where γ = surface tension coefficient, κ = curvature, and $\hat{n}$ = unit normal

$$\kappa = \nabla\cdot\frac{\nabla\alpha}{|\nabla\alpha|}, \hat{n} = \frac{\nabla\alpha}{|\nabla\alpha|} \tag{5}$$

Electrical force is calculated from Maxwell's stress tensor.

$$\boldsymbol{F}_e = \nabla\cdot\bar{\boldsymbol{M}} \tag{6}$$

$$\bar{M} = \varepsilon \boldsymbol{E}\boldsymbol{E}^T - \frac{1}{2}|\boldsymbol{E} \cdot \boldsymbol{E}|\bar{\boldsymbol{I}} \tag{7}$$

where $\bar{M}$ = Maxwell's stress tensor, $\boldsymbol{E}$ = electric field, ε = permittivity, and $\bar{\boldsymbol{I}}$ = identity tensor. The electrical force is calculated as

$$\boldsymbol{F_e} = q\boldsymbol{E} - \frac{1}{2}[\boldsymbol{E} \cdot \boldsymbol{E}]\nabla\varepsilon + \nabla\left[\rho\left(\frac{\partial\varepsilon}{\partial\rho}\right)E^2\right] \tag{8}$$

where q = volume charge density.

Due to the small magnitudes of the dynamic currents within the system, the induced magnetic field can be considered negligible. Consequently, an irrotational electric field is formed, which can be characterized by the gradient of a scalar field known as the voltage potential.

$$\nabla \times \boldsymbol{E} = -\frac{\partial \boldsymbol{B}}{\partial t} = \boldsymbol{0} \tag{9}$$

$$\boldsymbol{E} = -\nabla\phi \tag{10}$$

The voltage potential can be calculated by solving Poisson equation, obtained from Gauss' theorem.

$$\nabla \cdot \boldsymbol{D} = \nabla \cdot (\varepsilon \boldsymbol{E}) = q \tag{11}$$

where q = volume charge density charge distribution can be obtained by solving charge convection equation.

$$\frac{\partial q}{\partial t} + \boldsymbol{u} \cdot \nabla q = -\nabla \cdot (\sigma \boldsymbol{E}) \tag{12}$$

In the single fluid formulation, the electrical properties, such as conductivity and permittivity, are averaged harmonically using the volume fraction function [18].

$$\begin{aligned} \frac{1}{\sigma_m} &= \frac{\alpha}{\sigma_c} + \frac{(1-\alpha)}{\sigma_d} \\ \frac{1}{\varepsilon_m} &= \frac{\alpha}{\varepsilon_c} + \frac{(1-\alpha)}{\varepsilon_d} \end{aligned} \tag{13}$$

2.1 *Validation*

A validation study is conducted prior to the final simulation to assess the capabilities of the solver to capture the interface in the presence of the electric field. The validation results for the EHD drop deformation are presented in Fig. 3, depicting the steady-state deformation parameter (Δ) as a function of Ca_E for the given parameters $(R, S, Ca_E) = (10, 1.37, 0.8)$. In Fig. 2, Δ is defined as $(a_1 - a_2)/(a_1 + a_2)$. Figure 3 illustrates that the present solver exhibits strong agreement with existing literature, specifically Ha and Young's [19] experimental work as well as Jiang et al.'s [20] numerical study, highlighting the consistency between our results and those in the literature.

3 Results and Discussion

In this study, we have conducted simulations to investigate the bubble-rising phenomenon under a vertically applied electric field. Specifically, we have examined the influence of the electric capillary number ($Ca_E = \frac{\varepsilon_c E_0^2 D}{\gamma}$), representing the relative strength of the electric force compared to the surface tension force, and the Bond number ($Bo = \frac{\Delta\rho g D^2}{\gamma}$), which quantifies the relative strength of the buoyancy force to the surface tension force. To ensure consistency in the simulations, we have fixed the property ratios, setting the density ratio at 0.001, the viscosity ratio at 1, the electrical conductivity ratio (0.024), and the permittivity ratio (0.123). The Ohnesorge number has been fixed at 0.1, while the confinement ratio is considered as 0.3. We have investigated the effects of three different values of the Ca_E, namely 0.01, 0.1, and 1.0. Furthermore, we have examined the impact of the Bo using six different values: 0.27, 1, 3, 5, 7, and 9. The results are presented in terms of non-dimensional velocity, which is scaled by the velocity scale factor $\frac{\gamma}{\mu_c}$, and the non-dimensional bubble deformation, which is assessed by defining the aspect ratio (A_r). The aspect ratio is determined as the maximum peripheral distance of the bubble in the horizontal direction to the maximum peripheral distance in the vertical direction. In Fig. 2, the aspect ratio is defined as $A_r = \frac{a_2}{a_1}$.

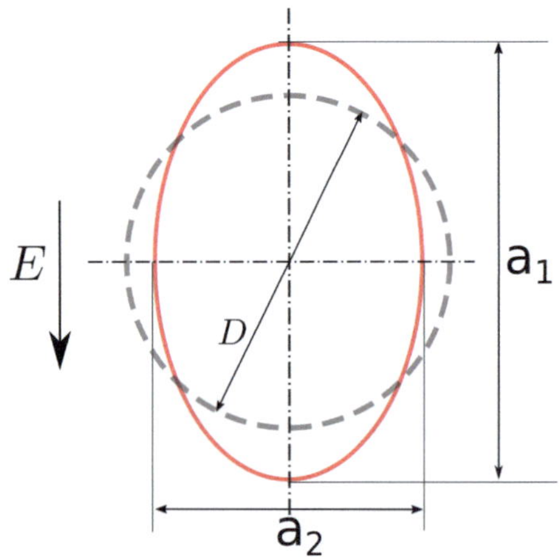

Fig. 2 Definition of dimensionless bubble deformation—aspect ratio (A_r)

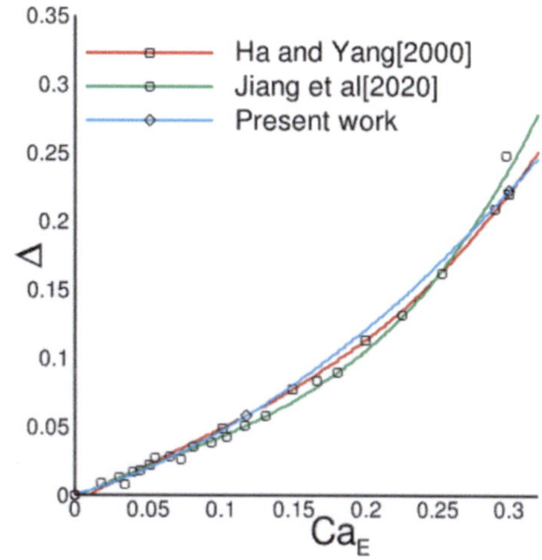

Fig. 3 Deformation parameter Δ against Ca_E for the electrical parameters are $(R, S, \lambda) = (10, 1.37, 0.8)$

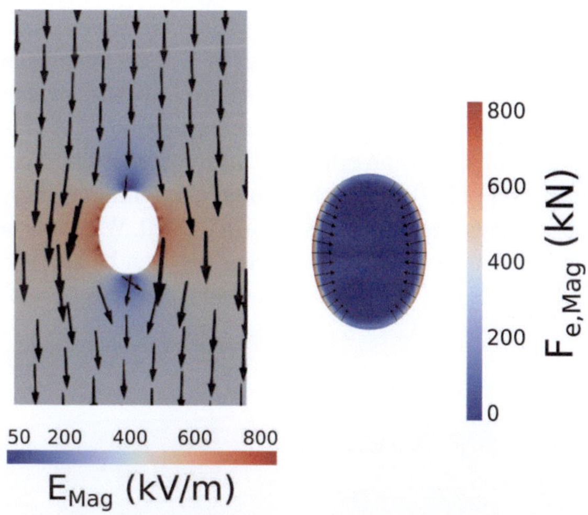

Fig. 4 Distribution of **a** electrical field lines in the vicinity of the bubble and **b** electrical force exerted at the interface

Electrical Force:
Upon the initiation of a bubble ascent under the imposition of an applied electric field, the bubble encounters an electrical force in addition to hydrodynamic forces. The mathematical formulation of this force is elucidated in Sect. 2. The electrical force comprises three primary components: the Coulomb force (CF) resulting from charges accumulated at the interface, the Dielectrophoretic force (DEF) arising from the discontinuity in electrical properties at the interface, and the electrostriction force (EF) attributed to variations in electrical properties with density. In the present study, considering an incompressible flow assumption, the electrostrictive force is neglected. Notably, as charges accumulate at the interface and electrical property variations occur solely at the interface, the electrical force acts exclusively at the

interface. Figure 4 depicts the electrical field lines in the vicinity of the bubble and the electrical force exerted at the interface. It is noteworthy that the electrical force is normal to the interface, with maximum intensity at the bubble's equator. As one moves toward the poles, the electrical force gradually weakens due to the dominance of the dielectrophoretic force over the Coulomb force.

3.1 *Effect of Electric Capillary Number* (Ca_E)

Recalling the definition of electric capillary number (Ca_E), an increase in Ca_E corresponds to a higher strength of the applied electric field. Figure 5 illustrates the distribution of the electric field for Ca_E values of 0.01, 0.1, and 1.0. It is evident from the figure that the electric field distribution is similar for all three configurations. However, the magnitude of the electric field greatly amplifies with an increase in Ca_E. The intensified electric field leads to an increase in both the dielectrophoretic force (DEF) and the Coulomb force (CF), resulting in a higher electrical force. Figure 6 demonstrates the impact of Ca_E on bubble deformation and terminal velocity. As anticipated, an increase in Ca_E leads to a greater electrical force, causing more deformation. Consequently, a decrease in the aspect ratio from unity is observed with increasing Ca_E. The effect of Ca_E on the terminal velocity is negligible when Ca_E increases from 0.01 to 0.1. However, a significant increase in the terminal velocity is observed as Ca_E increases from 0.1 to 1.0. With constant hydrodynamic parameters, this increment in the terminal velocity can be attributed to the prolate shape of the bubble, i.e., the deformation caused by the electrical force.

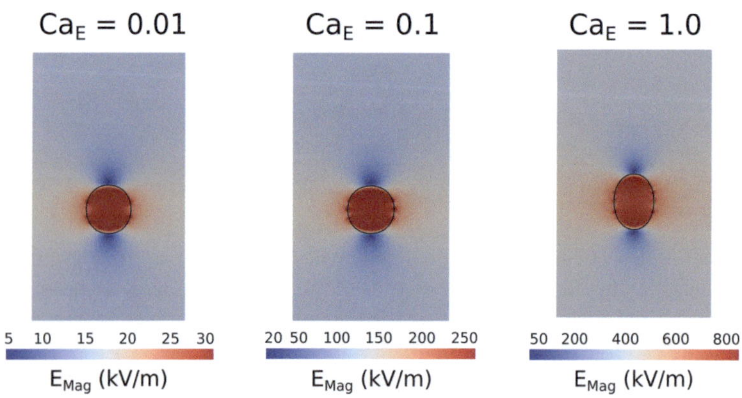

Fig. 5 Effect of Ca_E on the distribution of electric field across the interface

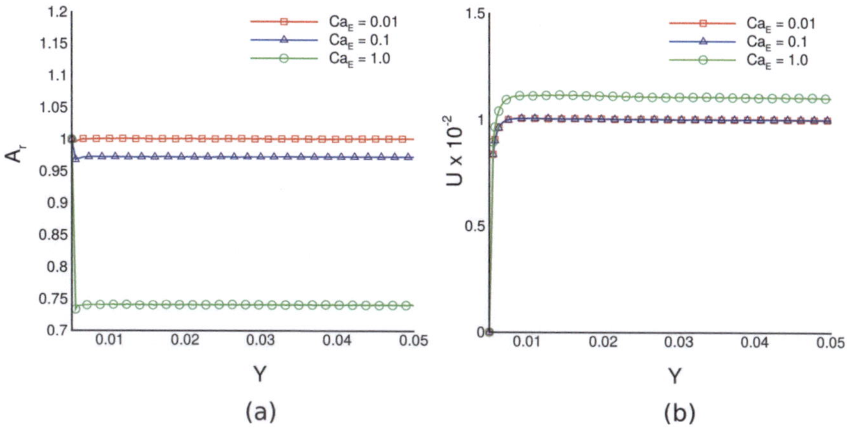

Fig. 6 Effect of Ca_E on **a** bubble deformation, and **b** terminal velocity of the bubble

3.2 *Effect of Bond Number* (*Bo*)

The impact of the Bond number (Bo) on the bubble-rising phenomenon has been firmly established in the absence of an electric field [5, 9]. With lower values of Bo, the surface tension force dominates, maintaining the bubble's spherical shape. As Bo increases, the buoyancy force gradually surpasses the surface tension force, resulting in the deformation of the bubble into an oblate ellipsoidal shape, primarily deforming in the horizontal direction. However, when a bubble begins to rise under the imposition of a vertically applied electric field, the electrical force exerted at the interface causes deformation in the vertical direction. Hence, we have two competing

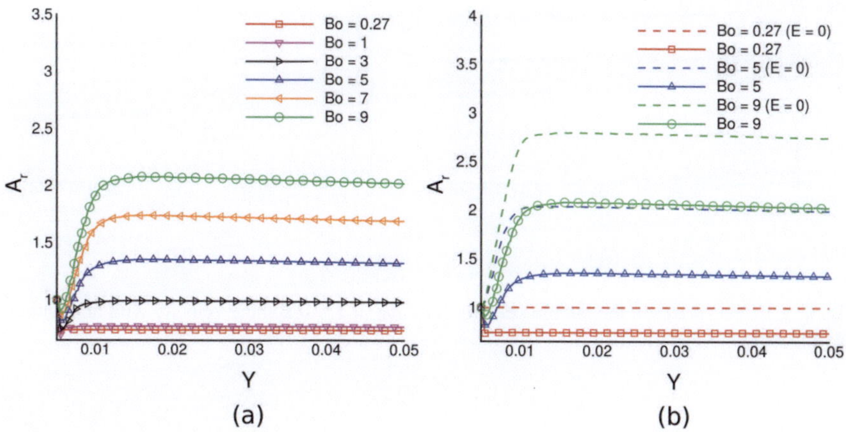

Fig. 7 **a** Effect of Bo on the bubble deformation for $Ca_E = 1.0$ **b** Bubble deformation with and without the imposition of electric field for $Bo = 0.27, 5, 9$, and $Ca_E = 1.0$

forces: one tending to deform the bubble horizontally and the other deforming it vertically. Figure 7b illustrates the effect of Bo on bubble deformation with and without an electric field for Bo values of 0.27, 5, and 9. In the absence of an electric field, for $Bo = 0.27$, the surface tension force retains dominance, resulting in the bubble maintaining a spherical shape. However, when subjected to an electric field, the electrical force deforms the bubble primarily in the vertical direction. Furthermore, for $Bo =$ 5 and 9, the bubble acquires an oblate shape, primarily deforming horizontally. As the electrical force acts to deform the bubble vertically, it counteracts the horizontal deformation for $Bo = 5$ and 9. Consequently, less deformation is observed under the imposition of an electric field compared to the absence of an electric field. Figure 7a illustrates the impact of Bo under the imposition of an electric field. As Bo increases from 0.27 to 9.0, the bubble undergoes a transition from vertical deformation to horizontal deformation, leading to an increase in the aspect ratio (A_r).

An increase in Bo results in a higher terminal velocity for the bubble, while the presence of a vertically applied electric field further facilitates the motion of the bubble. Consequently, when a bubble rises under the imposition of a vertically applied electric field, its terminal velocity surpasses that of a bubble rising without an electric field. Figure 8b visually demonstrates this effect for three different Bo values: 0.27, 5, and 9. Figure 8a depicts the influence of Bo on the terminal velocity of the bubble. Similarly to the bubble ascent in the absence of an electric field, an increase in velocity is observed with an increase in Bo for the case of a bubble ascent under the imposition of an applied electric field.

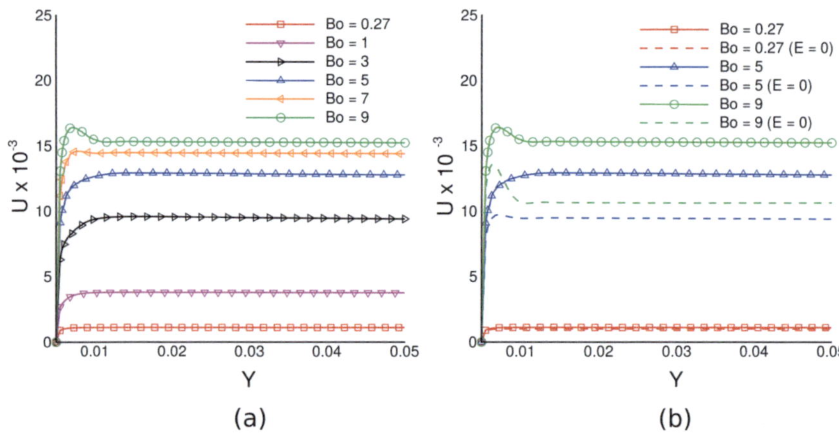

Fig. 8 **a** Effect of Bo on the bubble velocity for $Ca_E = 1.0$ **b** Bubble velocity with and without the imposition of electric field for $Bo = 0.27, 5, 9$, and $Ca_E = 1.0$

4 Conclusions

In the present study, we have conducted simulations to investigate the bubble-rising phenomenon under the imposition of a vertically applied electric field. The influence of the capillary electric number (Ca_E) and the Bond number (Bo) on the dynamics of the phenomenon has been studied. An increase in Ca_E results in a greater electrical force exerted at the interface, leading to increased deformation and terminal velocity of the bubble. As Bo increases from a lower value to a higher value, a transition is observed from vertical deformation to horizontal deformation. The aspect ratio increases with an increase in Bo, but under the imposition of an electric field, the electrical force deforms the bubble primarily in the vertical direction, resulting in a suppression of the aspect ratio with increasing Bo. The electrical force facilitates the motion of the bubble, causing an increase in the terminal velocity with increasing Bo. Additionally, the terminal velocity is higher when the bubble rises under the imposition of an electric field compared to the absence of an electric field.

Nomenclature

A_r	Aspect ratio
B	Magnetic field (T)
Bo	Bond number
Ca_E	Electrical capillary number
D	Initial bubble diameter (m)
E	Electric field (V/m)
F_{st}	Surface tension force (N)
F_e	Electrical force (N)
J	Current density (A/m^2)
q	volume charge density (C/m^3)
Oh	Ohensorge number
Re	Reynolds number
VoF	Volume of fluid
α	volume fraction
γ	Surface tension coefficient (N/m)
κ	curvature (1/m)
μ	Dynamic viscosity (Pa s)
ϕ	Voltage potential (V)
ρ	Density (kg/m^3)
σ	Electrical conductivity (S/m)
ε	Permittivity (F/m)

References

1. Roland C, Grace J, Weber M (1978) Bubbles, drops, and particles
2. de Vries AWG, Biesheuvel A, van Wijngaarden L (2002) Notes on the path and wake of a gas bubble rising in pure water. Int J Multiph Flow 28(11):1823–1835
3. Duineveld PC (1995) The rise velocity and shape of bubbles in pure water at high Reynolds number. J Fluid Mech 292:325–332
4. Kumar P, Vanka SP (2015) Effects of confinement on bubble dynamics in a square duct. Int J Multiph Flow 77(June):32–47
5. Norman CE, Miksis MJ (2005) Dynamics of a gas bubble rising in an inclined channel at finite Reynolds number. Phys Fluids 17(2):1–13
6. Tripathi MK, Sahu KC, Govindarajan R (2015) Dynamics of an initially spherical bubble rising in quiescent liquid. Nat Commun 6:1–9
7. Bhaga D, Weber ME (1981) Bubbles in viscous liquids: shapes, wakes and velocities. J Fluid Mech 105:61–85
8. Chen L, Garimella SV, Reizes JA, Leonardi E (1999) The development of a bubble rising in a viscous liquid. J Fluid Mech 387:61–96
9. Patel T, Patel D, Thakkar N, Lakdawala A (2019) A numerical study on bubble dynamics in sinusoidal channels. Phys Fluids 31(5)
10. Zukoski EE (1966) Influence of viscosity, surface tension, and inclination angle on motion of long bubbles in closed tubes. J Fluid Mech 25(4):821–837
11. DeBisschop KM, Miksis MJ, Eckmann DM (2002) Bubble rising in an inclined channel. Phys Fluids 14(1):93–106
12. Saville DA (1997) Electrohydrodynamics: the Taylor-Melcher leaky dielectric model. Annu Rev Fluid Mech 29(1962):27–64
13. Andalib S, Hokmabad BV, Esmaeilzadeh E (2013) Study of a single coarse bubble behavior in the presence of D.C. electric field. Colloids Surfaces A 436:604–617
14. Rahmat A, Tofighi N, Yildiz M (2016) Numerical simulation of the electrohydrodynamic effects on bubble rising using the SPH method. Int J Heat Fluid Flow 62:313–323
15. Melcher JR, Taylor GI (1969) Electrohydrodynamics: a review of the role of interfacial shear stresses. Annu Rev Fluid Mech 1(1):111–146
16. Taylor GI (1964) Disintegration of water drops in an electric field. Proc R Soc Lond. Series A 280(1382):383–397
17. Brackbill JU, Kothe DB, Zemach C (1992) A continuum method for modeling surface tension. J Comput Phys 100(2):335–354
18. Tomar G, Gerlach D, Biswas G, Alleborn N, Sharma A, Durst F, Welch SWJ, Delgado A (2007) Two-phase electrohydrodynamic simulations using a volume-of-fluid approach. J Comput Phys 227(2):1267–1285
19. Ha J-W, Yang S-M (2000) Deformation and breakup of Newtonian and non-Newtonian conducting drops in an electric field. J Fluid Mech 405:131–156
20. Jiang Z, Gan Y, Luo Y (2020) Effect of viscosity ratio on the dynamic response of droplet deformation under a steady electric field. Phys Fluids 32(5)

Numerical Simulation of Drag Cum Lift Horizontal Axis Wind Turbine

Nandan Kumar, Bakhshi Mehul, Rakesh Aggarwal, Abhilash Anand, and Pritanshu Ranjan

Abstract The current research focuses on static computational fluid dynamics simulations of our patented Horizontal Axis Wind Turbine design. The innovative concept concentrates on combining lift and drag forces, thereby eliminating the issue of reverse drag when the blade panel is moving in opposition to the wind. The design focuses on converting reverse drag into an effective lift force. The investigation used the finite volume-based software Ansys Fluent at various blade positions. The simulations were performed using the pressure-based solver and the realizable $k - \epsilon$ turbulence model. The simulations were conducted for a quarter of the rotation, as the blades acquired the same initial position after each quarter by continuously adjusting their angle of attack using advanced controls. The simulation results demonstrate a significant resistance by the drag force and an upward perpendicular force by the lift, with both forces contributing to energy production.

Keywords Wind turbine · Ansys fluent · Lift · Drag

N. Kumar · R. Aggarwal (✉) · A. Anand
TDR Foundation, New Delhi, Delhi, India
e-mail: rakesh@comcon.co.in

N. Kumar
e-mail: nandan@svurg.com

A. Anand
e-mail: abhilash@svurg.com

B. Mehul · P. Ranjan
Department of Mechanical Engineering, K.K. Birla Goa Campus, BITS Pilani, Sancoale, Goa, India
e-mail: f20190670@goa.bits-pilani.ac.in

P. Ranjan
e-mail: pritanshur@goa.bits-pilani.ac.in

A. Kumar R. et al. (eds.), *Proceedings of Fluid Mechanics and Fluid Power (FMFP) 2023, Vol. 6*, Lecture Notes in Mechanical Engineering,
https://doi.org/10.1007/978-981-96-9600-0_8

1 Introduction

Wind turbines are prominent renewable energy devices that offer sustainable alternatives to conventional fossil fuel-based power generation methods. The fundamental function of wind turbines is to convert the kinetic energy of the wind into electrical power. There have been many technological advancements in recent years that have made wind turbines more efficient and affordable.

Currently, one of the most successful wind turbines is the Horizontal Axis Wind Turbine (HAWT), which has an approximate 60% theoretical efficiency [5]. Modern HAWTs are primarily 3-blade turbines powered by computers and motors [3]. However, because the drag force is generated perpendicular to the rotation, it cannot impact or contribute to the lift force generated by the blades. In the past, Vertical Axis Wind Turbines (VAWTs) have been developed, which typically operate close to the earth [5]. The two most common types of VAWTs are Savnoius and Darrieus [4]. In certain VAWTs, the reverse drag is minimized in reverse half rotation by utilizing blades with a cylindrical shape. However, these turbines were inefficient, rendering them inappropriate for industrial applications. The analysis of all previous research reveals a paucity of drag cum lift-based wind turbine design technology.

This investigation aims to conduct a numerical analysis of our patented wind turbine named Gurv [2], which integrates lift and drag forces in a super-efficient design. The paper begins by describing the geometry, computational domain, and methodology and concludes with the results obtained through steady-state simulations of different turbine blade positions. The paper discusses the lift and drag forces obtained through the computational simulations, the pressure distribution, and our future work in transient simulations.

2 Computational Methodology

The wind turbine technology combines lift and drag forces to convert wind energy into electricity. Unlike traditional Horizontal Axis Wind Turbines (HAWTs) that rely solely on lift force and Vertical Axis Wind Turbines (VAWTs) that rely on drag force, our turbine utilizes a simple rectangular blade profile and variable angle of attack as shown in Fig. 1a to optimize energy conversion. With the turbine rotating in one direction and the blades rotating in the opposite direction, this technology harnesses both forces, as depicted in Fig. 1b, enhancing efficiency and power generation.

2.1 Geometry and Boundary Conditions

The turbine geometry consists of four blades, with dynamic control to optimize torque from forward drag and prevent reverse drag, transforming it into usable lift force.

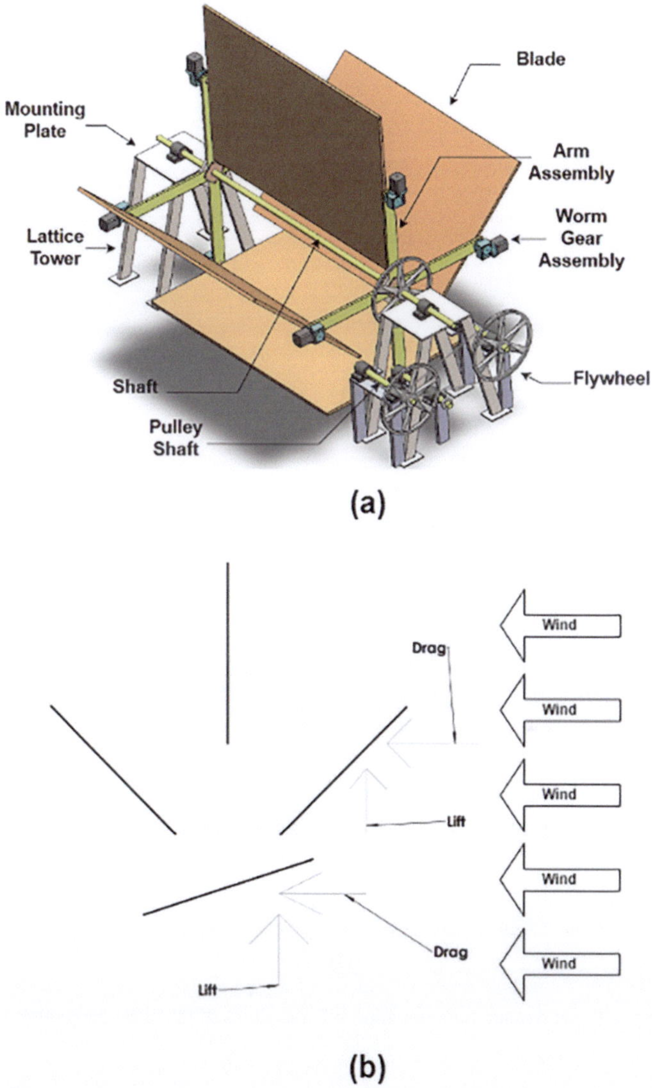

Fig. 1 **a** Wind turbine with gear-controlled dynamics **b** Lift and drag forces acting on the turbine blades

Considering the computational cost and the fact that the blades are the only turbine sections in direct contact with the wind to produce usable energy, only the blades were evaluated in the computational simulations. The dimensions of the external domain and blades are depicted in Fig. 2. The wind speed of 12.5 m/s was taken at the inlet, and the no-slip condition was applied to the walls.

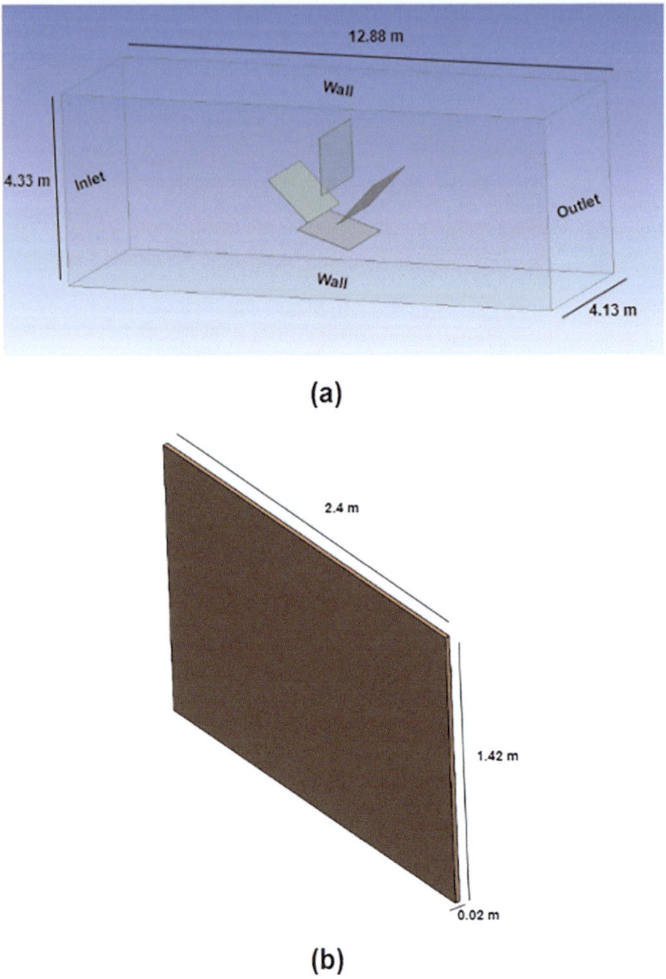

Fig. 2 **a** Computational domain for simulation **b** Dimensions of the blade geometry

2.2 Meshing Details

Tetrahedral meshes with approximately 1.3 million elements were chosen to provide a framework for performing calculations and obtaining results at specific points within the system. An inflation layer has been applied around the blade, comprising five layers with a height of 0.001 units. This inflation layer is designed to accurately capture the flow behaviour near the blade surface, allowing for a more comprehensive understanding of boundary layer phenomena and their impact on the system's performance (Fig. 3).

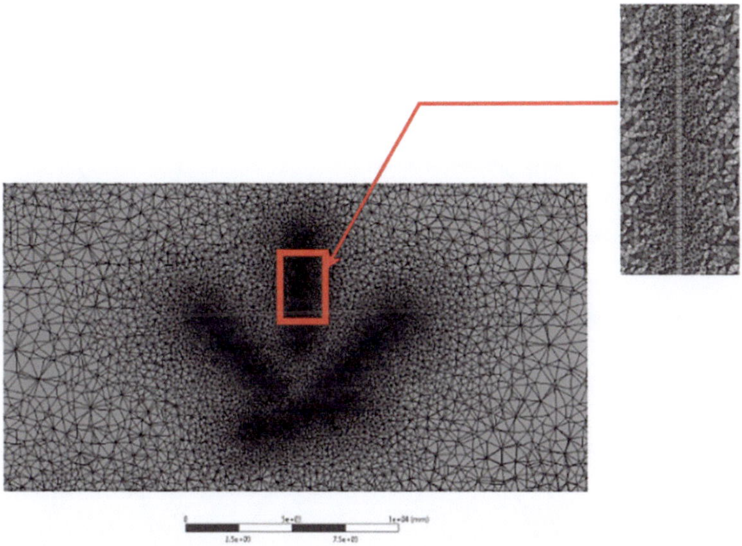

Fig. 3 Meshing of the computational domain with refinement near the blades

2.3 *Governing Equations*

The flow field simulation was conducted using the pressure-based solver and realizable $k - \epsilon$ turbulence model [1]. In addition to the turbulent model, conservation of mass (Eqn. 1) and momentum (Eqn. 2) were the two main governing equations solved in our simulation.

$$\frac{d\rho}{dt} + \nabla.(\vec{v}\rho) = 0 \tag{1}$$

$$\frac{d\rho\vec{v}}{dt} + \nabla.(\vec{v}\vec{v}\rho) = -\nabla p + \nabla.(\tau) + \rho\vec{g} + \vec{F} \tag{2}$$

The stress tensor τ is given by

$$\tau = \mu[(\nabla\vec{(v)} + \nabla\vec{v}^T) - \frac{2}{3}\nabla.\vec{(v)}I]$$

3 Results and Discussion

As shown in Fig. 4, the turbine's blades rotate along the central axis and axis passing through its centre to combine the lift and drag forces. The blades rotate at a specified RPM about its central axis to reach the initial position of the adjacent blade after

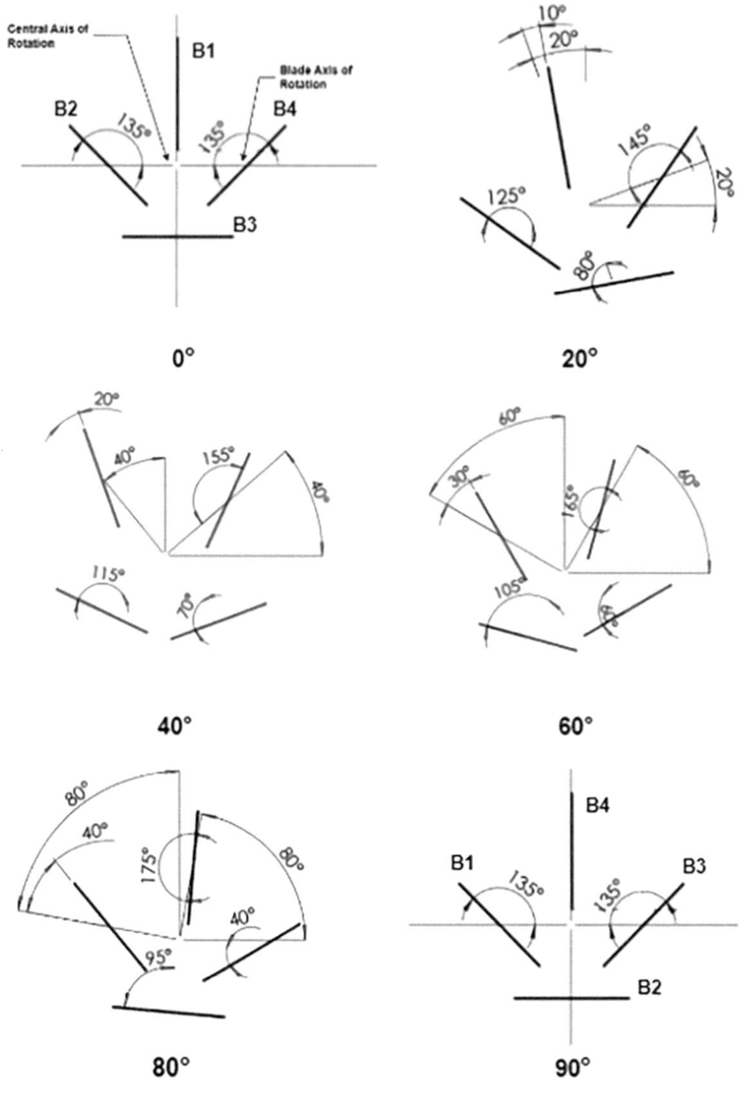

Fig. 4 Different positions of blades (B1, B2, B3, and B4) at different angles of rotation of the turbine

each quarter rotation of the turbine, as illustrated in Fig. 4, 90°. The static simulations were conducted at five different turbine rotation angles (0°, 20°, 40°, 60°, and 80°), as depicted in Fig. 4.

Table 1 Comparison of lift forces for five different positions of the turbine at 0° and 72° angle of attack of the lower blade

Rotation of turbine about the central axis (Degree)	Case 1 (N)	Case 2 (N)
0	−88.53	270.08
20	−130.02	279.10
40	−321.04	189.39
60	−321.04	192.53
80	−272.05	177.34

3.1 Calculation of Lift and Drag Forces

The lift and drag forces were computed at each of the five blade positions. At all five positions, we observed negative values for the lift force (Table 1). After a thorough inspection, it was determined that the lower blade (blade 3) is not contributing to the overall lift force of the turbine. Thus, an effort was made to increase the angle of attack from 0° (Case 1) at the initial position of the turbine, resulting in improved outcomes. As depicted in Fig. 5, we observed a significant contribution from blade 3 to lift force at an attack angle of 72° (Case 2).

We proceed with analysing the simulation results for case 2 at five different angles of rotation (0°, 20°, 40°, 60°, 80°) about the central axis, as the revised configuration of blade 3 produces better results. The blades in case 2 behave in the same manner, acquiring the position of the adjacent blade after each quarter rotation. Figures 6 and 7 depict the lift and drag forces exerted by the turbine with variable characteristics at various angles.

The turbine exhibits positive lift values at all angles, indicating an upward force perpendicular to the flow direction. An overall decrease in the value of lift force can be observed between the initial and final position of the wind turbine.

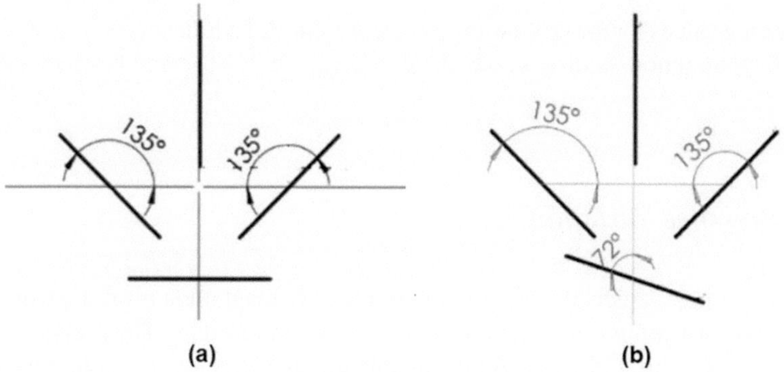

Fig. 5 Initial position of turbine blades **a** Case 1 **b** Case 2

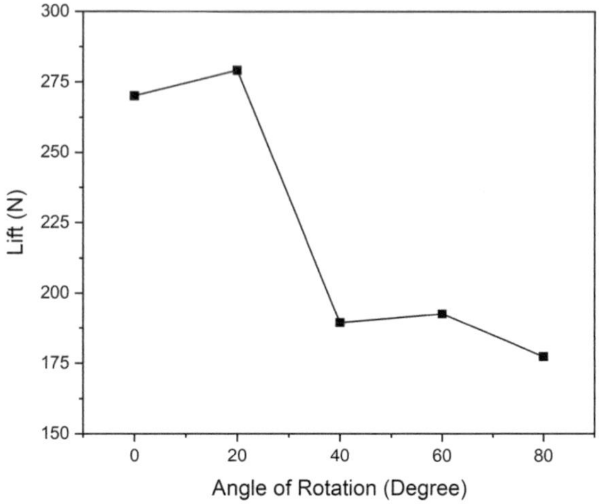

Fig. 6 Variation in turbine lift force at different rotation angles

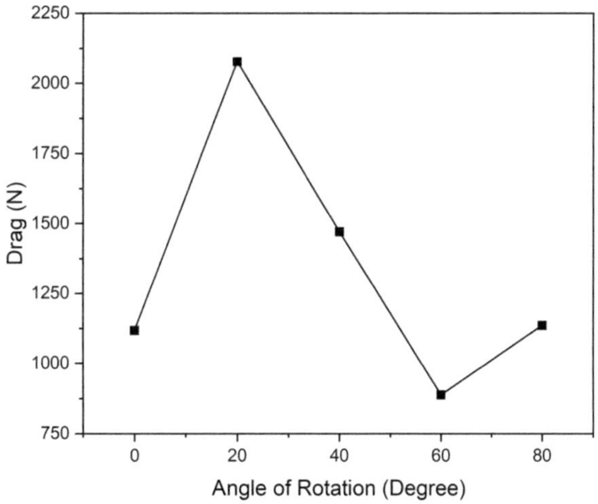

Fig. 7 Variation in turbine drag force at different rotation angles

The drag values for the turbine are high, representing significant resistance parallel to the flow direction. A drop in the value of drag can be observed between 20° and 60°.

3.2 Pressure Variation

The contour plot depicts the distribution of pressure magnitude resulting from CFD simulations at various angles of attack, as illustrated in Fig. 8. The pressure acting on the wind turbine's blades exhibits notable disparities across different quadrants. Specifically, the collective pressure exerted on the blades in the second and third

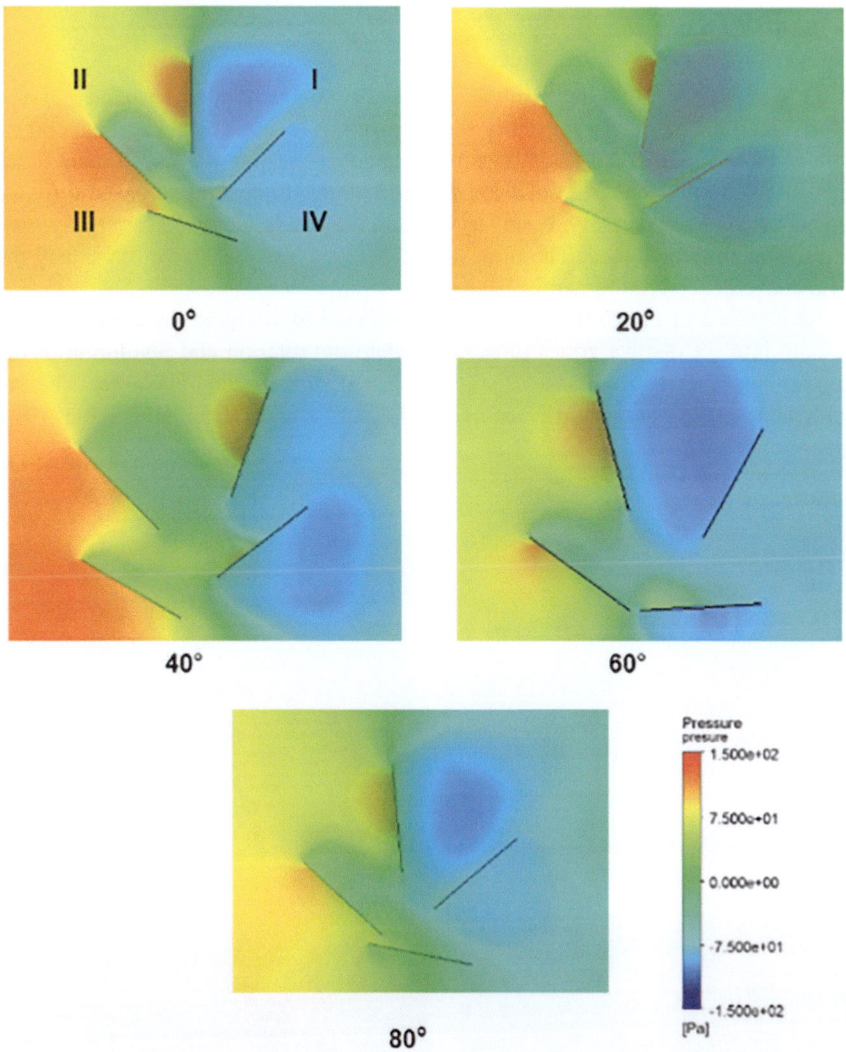

Fig. 8 Pressure contours in different quadrants and at different positions of the wind turbine

quadrants surpasses that experienced by the blades in the first and fourth quadrants. This discrepancy is pivotal in facilitating the clockwise rotation of the main shaft. The rotation of the shaft is further influenced by the lift force generated by the blade in the third quadrant. According to Bernoulli's equation, this lift force arises from the pressure disparity on either side of the blade. By capitalizing on this pressure differential, the blade imparts a rotational force to the shaft, causing it to rotate clockwise.

3.3 Transient Simulation

After obtaining encouraging results from steady-state simulations, we have moved on to transient simulations for our turbine in an effort to attain even better outcomes. Due to the computational limitations of the student version of Ansys Fluent, transient simulations are carried out using the open-source finite volume-based software OpenFOAM. The geometry, boundary conditions, and model used for steady-state simulation are maintained. The time step of 1e-3 sec was considered to satisfy the CFL convergence requirements.

As depicted in Fig. 9, the meshing is conducted by dividing the computational domain into six distinct zones to accommodate the rotation and revolution of the turbine blades. Each zone was separately meshed with tetrahedral elements before being combined. The meshing information for each zone is detailed in Table 2. The

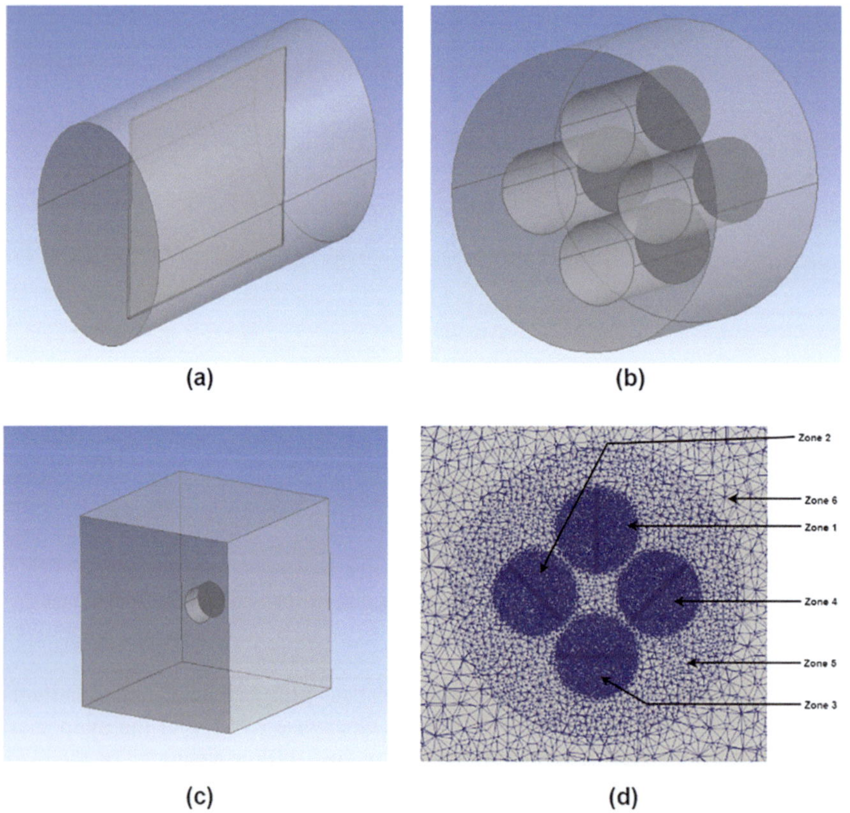

Fig. 9 **a** Blade and zone 1 (zone 2, 3, and 4 have the same domain) **b** Zone 5 **c** Zone 6 **d** Merged meshes of all the zones

Table 2 Mesh element size of each zone

Zone	Element size (mm)
Blades	20
1	50
2	50
3	50
4	50
5	150
6	500

Table 3 Torque values for transient simulation of the turbine

Torque	Value (Nm)
Average	695.59
Minimum	444.55
Maximum	1454.39

rotational velocities for zones 1, 2, 3, 4, and 5 were determined so that after each quarter rotation, the blades acquire the position of their adjacent blades.

Currently, we are aiming to achieve the best possible results from transient simulations. The average, minimum, and maximum torque values obtained to date are displayed in Table 3.

4 Conclusions

The performance of our patented wind turbine was analysed using computational simulations. The steady-state simulations were performed for five different turbine positions (0°, 20°, 40°, 60°, 80°) and different blade configurations by rotating the blades about their own central axis. From the findings of our study, the following conclusion can be drawn:

(1) An effective lift was achieved by adjustment of the angle of attack of a single turbine blade.
(2) At all positions of the turbine, positive lift force acting upward in the perpendicular direction of airflow was attained due to pressure differences on either side of the blades. The produced lift force aids in the rotation of the wind turbine.
(3) High values of drag force, representing substantial resistance against the airflow, were obtained, thereby supporting the rotation with the turbine and minimizing force in the reverse direction, thus leading to greater energy production.

We began our numerical investigation with steady-state simulations, which yielded promising results at a reduced computational expense. Now, we are advancing to the transient simulations, which may yield results that are closer to the original physical scenario.

Nomenclature

p Static pressure [Pa]

μ Molecular viscosity [Pa.s]

I Unit tensor –

References

1. ANSYS Inc (2015) ANSYS Fluent Theory Guide 12.0
2. Aggarwal R (2020) A drag cum lift based wind turbine system having adjustable blades. http://www.google.it/patents/WO2021148859A1, 1 2020, Google Patents, WO2021148859A1
3. Eid AM, Abdel-Salam M, Abdel-Rahman MT (2006) Vertical axis wind turbine modelling and performance with axial flux permanent magnet synchronous generator for battery charging applications. In: The eleventh international middle east power systems conference, December 19–21, El-Minia, Egypt, pp 162–166
4. Kumar PM, Sivalingam K, Narasimalu S, Lim TC, Ramakrishna S, Wei H (2019) A review on the evolution of Darrieus vertical axis wind turbine: small wind turbines. J Power Energy Eng 7(4):27–44
5. Saad MM, Asmuin N (2014) Comparison of horizontal axis wind turbines and vertical axis wind turbines. IOSR J Eng (IOSRJEN) 4(8):27–30

Experimental Investigation of Droplet Impact on Spherical Particles

K. Niju Mohammed, Abhijith Vijayan, P. S. Tide, Franklin R. John, and Ranjith S. Kumar

Abstract Extensive studies have been conducted on the droplet impact behaviour on symmetric surfaces like spheres and flat surfaces. The literatures regarding the impact of hydrophobically coated symmetric surfaces and super heated symmetric curves with diameter ratios below 1. In this study, we experimentally examined how water droplets intract when they come into contact with solid spherical targets. Studies of droplet impact and post-impact outcomes have been carried out on hydrophilic, hydrophobic and super heated surfaces. The thermo hydrodynamic behaviour of millimeter-scale water droplets impacting on spherical profiles is established through experiments. Mono dispersed water droplets are impinged on surface which is the Leidenfrost temperature maintained 180 °C. High-speed imaging techniques are employed to visualise droplet evolution, and Weber numbers range from 5 to 170. In order to study impact phenomena in side and top views, high-speed imaging techniques are used. Using geometrical parameters, such as the length of spreading and thickness of the liquid layer near the north pole of the target surface, the morphological effects of this collision process are quantitatively analysed. Observations reveal that as spherical target size decreases, covering increases, yet the thickness of the liquid coating near the target surface's north pole decreases. The liquid layer's thickness was plotted over time at the North Pole target, and it was discovered that it matched theoretic predictions from earlier investigations. Different morphologies

K. Niju Mohammed (✉) · A. Vijayan (✉) · P. S. Tide · F. R. John
Department of Mechanical Engineering, School of Engineering, Cochin University of Science and Technology, Cochin, Kerala, India
e-mail: nijumk@cet.ac.in

A. Vijayan
e-mail: abhijithv1752@gmail.com

K. Niju Mohammed · R. S. Kumar
Department of Mechanical Engineering, College of Engineering Trivandrum, Trivandrum, Kerala, India
e-mail: ranjith@cet.ac.in

A. Kumar R. et al. (eds.), *Proceedings of Fluid Mechanics and Fluid Power (FMFP) 2023, Vol. 6*, Lecture Notes in Mechanical Engineering,
https://doi.org/10.1007/978-981-96-9600-0_9

are seen when droplets interact with hydrophilic, hydrophobic and super heated surfaces. This investigation might help us better understand how fluids with intricate forms and geometries behave when they are wetted, sprayed or coated.

Keywords Droplet impact · Spread factor · Rim thickness · Residence time · Spherical surface

Nomenculture

We	Weber number
Re	Reynolds number
T_S	Surface temperature
T_L	Leiden frost temperature
τ	Non-dimensional residence time
$\frac{S_l}{D_d}$	Non-dimensional spreading length
$\frac{h_r}{D_d}$	Non-dimensional rim height or rim thickness

1 Introduction

Drop impact happens when a liquid drop lands on a surface, either solid or liquid. A liquid droplet will typically spread over a dry solid surface when it collides with it. However, if the impact is strong enough to force the droplet to spread more widely than it would have otherwise due to its constantly lowering contact angle, it will retract. The size, speed, surface tension, viscosity, roughness and contact angle between the drop and the surface all have a significant role in the precise outcome of an impact.

The droplet impact on solid surface have broad spectrum of applications. Interaction between impacting a drop of liquid and a surface is relevant in nature, industrial applications and agricultural processes. Typical phenomena include rain and mist deposition, spray painting, ink jet printing, fuel injection, spray cooling, cooling towers and pesticide spraying [1–3]. In addition to flat surfaces the droplet impact on curved surfaces are investigated. It is possible for icing to happen on electrical lines or fan blades in the power industry, which poses a serious risk to the equipment's safety and functionality [4]. It is obvious that electrical cables or fan blades have curved surfaces, especially towards their leading edges. In spray cooling, chemical coating and other applications, the droplet impacting process on a curved surface usually appears in a constrained area if the surface is rough or features micro structures [5–7]. Studies on the droplet impacting dynamics on a spherical surface, which is a typical and regular curved surface, may be helpful in understanding the mechanism of droplet impact on the curved surface. A number of variables, including liquid characteristics, surface shape, droplet size and impact velocity, affect the outcome of the droplet impact. Spray coating, crop spraying, painting and printing are a few examples of processes where droplets collide on surfaces at room temperature. The re-flooding of nuclear fuel rods, fuel droplet impact on combustion chamber walls, spray cooling of metals and electronic equipment, and fire suppression are some examples of drops impacting on hot surfaces [8–10].

The most significant experiment was performed by Liu et al. [11] to investigate the bouncing of liquid droplets on super hydrophobic cylindrical surfaces maintained at room temperature.

Transverse rim instability analysis was used by Huang et al. [12]to study the coronal fingering of a water droplet upon impact on a super hydrophobic surface. In the range of Weber numbers from 60 to 160, they theoretically calculated the number of spires that would form during recoil for a given droplet size and impact velocity, and then compared their results to experimental data. Bakshi et al. [13] report on a droplet's effect on spherical surfaces. The initial drop deformation phase, the phase where inertia is dominant and the phase where viscosity is dominant are the three temporal stages of film dynamics that have been documented. When water droplets strike a solid spherical target, Khurana et al. [14] investigation looks at how they behave.

At various impact velocities, Kim and Kim [15] performed drop impact studies on copper fibres of various diameters. They identified catching, single drop falling and splitting as the three effects of drop impact on a fibre. Water droplets impinging on cylindrical surfaces with diameters corresponding to those of the droplets were examined by Khurana et al. [16]. Through experimentation, the effects of the impact We, diameter ratio and surface wettability were investigated.

Numerous studies were conducted to comprehend the intricate dynamics of a single droplet impact on solid surfaces after Worthington's seminal experiment [17]. Water droplet impingement studies were performed by Wachter et al. [7] on a heated gold surface at 400 °C, and they discovered that the droplets break up into smaller ones after rebounding from the surface at 30 We and above. According to Hatta et al. [18] investigation into the dynamics of water droplet collisions on a heated surface above the Leiden frost temperature. Additionally, the maximum spreading factor and angle of incidence of drops were discovered to have an impact on the heat transfer process [19–23]. To determine the rate of evaporation, concomitant drop levitation and subsequent recoil cycles, the impact We and surface super heat are important. [24–26].

With systematic studies examining the fluid dynamics of droplet impact on such surfaces, flat stiff surfaces are the primary focus of the majority of research on droplet impact behaviour. On spherical surfaces, however, there has not been much research on droplet impact studies. There has been only limited work done on the droplet impingement heat transfer properties on spherical surfaces. There is a dearth of studies on spherical surfaces maintained at Leidenfrost temperature with diameter ratios lower than unity.

2 Methodology

Deionized water is utilised and dispensed through a burette, while droplets of various sizes are generated using a glass syringe. Stainless steel spheres with diameters of 2.5, 3.5 and 7 mm are employed. The droplets descend freely through the needle tip under the force of gravity. To ensure mono disperse droplets, a flow control valve is employed in relation to droplet size solely determined by the needle diameter. The droplets with diameters 1.9, 2.3, 2.9 and 3.5 mm are measured by dispensing a fixed liquid volume using a dispenser and verified during the post-processing of captured images. The droplets are permitted to impact from desired heights for imparting velocities between 0.3 and 2 m/s. Several trials are conducted to ensure accurate droplet impact at the north pole of the spheres. Each case is repeated twice on three different sample targets to ensure repeatability. A mixture of thermal paste and araldite is used to adhere the steel ball to the flat surface, providing a temperature range tolerance of 420 °C. Applying Rust-Oleum Never Wet Rain Repellent Spray to the surface will create a hydrophobic coating. Back lighting is provided by an LED lamp with a diffuser plate, and shadow images of the droplets are captured. The experiments are carried out in a lab environment with controlled temperature and atmospheric pressure. The ambient temperature and relative humidity were maintained at 27 °C and 55% throughout the trials under "inertial force" conditions. The temperature of the test surfaces (T_S), which ranges between 180 and 190 °C, is controlled by a temperature controller, which is mounted on a heater. A high-speed camera is used to take shadow pictures of the droplet impacts as soon as the surface temperature exceeds the required preset value under steady state conditions. The trials are carried out in succession for both spherical and flat profiles. The droplets' shadows are photographed with an LED light that has a diffusion plate. Using free, open-source software called ImageJ, post-processing of the photos is done to quantify variables like spread factor and the thickness of the liquid sheet near the north pole. Using software called Text Comparer, image discrepancies are determined (Fig. 1).

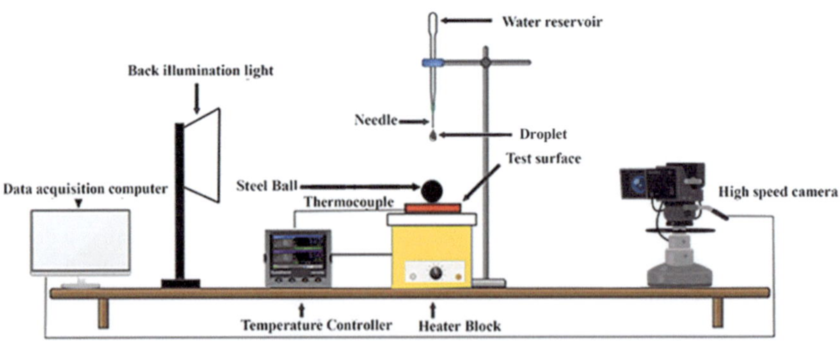

Fig. 1 Experimental setup

3 Results and Discussion

We may track changes in the dynamics of the film flow by charting the variations of the non-dimensional liquid film thickness and the spread factor against the non-dimensional residence time or time step. These variations are crucial markers for comprehending how liquid films behave under impact.

$$\tau = \frac{t}{\sqrt{\frac{\rho D_d^3}{\sigma}}}$$

$$\mathrm{We} = \frac{\rho U^2 D_\mathrm{d}}{\sigma}$$

$$\mathrm{Re} = \frac{\rho v D_\mathrm{d}}{\mu}$$

$$\text{Non-dimensional Spreading length} = \frac{S_l}{D_\mathrm{d}}$$

$$\text{Non-dimensional rim height} = \frac{h_\mathrm{r}}{D_\mathrm{d}}$$

$$\text{Diameter ratio } D^* = \frac{D_\mathrm{d}}{D_\mathrm{p}}$$

where t is the evolution's time, D_d is the droplet's diameter after impact, D_p is the particle's diameter, ρ is the liquid's density, σ is its surface tension, μ is its dynamic viscosity, S_l is its spreading length and h_{rim} is its rim height or rim thickness.

The outcomes of the droplet impact on hydrophilic (a–e) ,hydrophobic (f–j) and super heated (k–o) surfaces are illustrated in Fig. 2 for $D^* = 0.414$, We = 39.47. A droplet has a tendency to spread out on a hydrophilic spherical surface when it makes contact with it. This happens as a result of the droplet's low contact angle with the surface, which is typically less than 90°. On impact with the hydrophilic surface the liquid wets the surface, and the droplet flattens and forms a thin film as a result of the adhesive forces between the liquid and the solid. Rapid wetting is encouraged by hydrophilic surfaces, meaning the droplet immediately contacts the surface and spreads out. A droplet makes a large contact angle with a hydrophobic spherical surface when it strikes it, usually greater than 90°. Because of this, the droplet usually maintains its roughly spherical shape upon impact and does not greatly spread out. The liquid beads up and bounces off the surface due to the low wettability of the surface rather than saturating it. An air cushion is trapped between the droplet and the sphere due to the surface's anti wetting characteristic. The Leidenfrost effect is a phenomena that happens when a droplet strikes a super heated sphere. When a liquid droplet contacts a surface that is much hotter than the liquid's boiling point, the Leidenfrost effect happens. This occurs due to the intense heat of the super heated

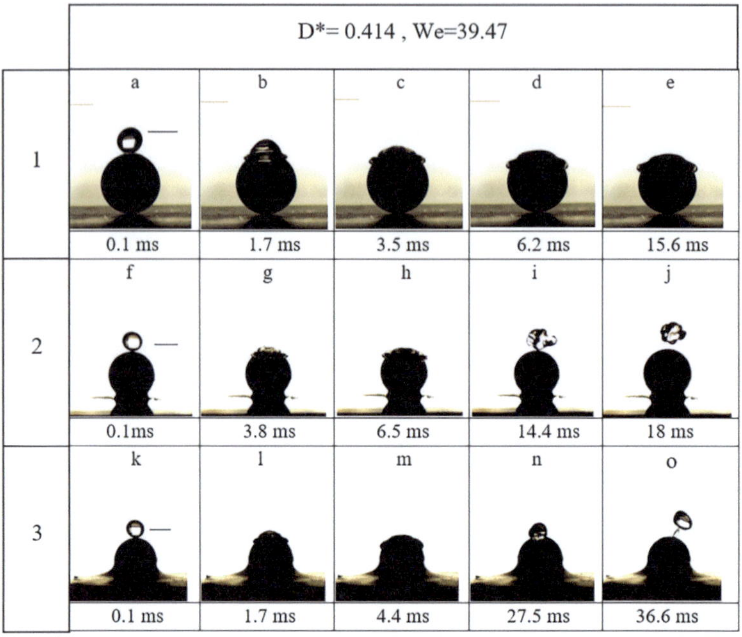

Fig. 2 Illustrate the post-impact collision of droplets over (**a–e**) hydrophilic, (**f–j**) hydrophobic and (**k–o**) super heated structures at $D^* = 0.414$, We = 39.47. Scale bar = 3.5 mm

sphere and the bottom of the droplet quickly evaporates upon collision. A thin vapour layer is created between the droplet and the sphere. This vapour layer serves as an insulating cushion, keeping the droplet from coming into direct contact with the hot surface. As a result, the droplet is levitated above the surface and the rate of heat transfer between them is decreased. The heat transfer rate between the super heated sphere and the droplet is significantly decreased by the Leidenfrost effect. As a result, compared to droplets contacting non-super heated surfaces, the droplet takes a longer time to evaporate.

On impingement, the droplet spreads on the super heated surface and formation of vapour layer occurs. The drop is retracted back to its original shape due to surface tension forces acting on it. The fluid's radial momentum is then transformed to vertical momentum, which causes the droplet to rise upwards after a certain amount of residence time.

Figure 3 explains how the non-dimensional spreading length varies depending on the surfaces being studied, including hydrophilic, hydrophobic and super heated surfaces. For all three surfaces, the spreading length increases over time. Due to capillary forces and the low contact angle on hydrophilic surfaces, the droplet's spreading length rises quickly. Compared to super heated and hydrophobic surfaces the hydrophilic surfaces have higher capillary forces resulting in higher spreading length.

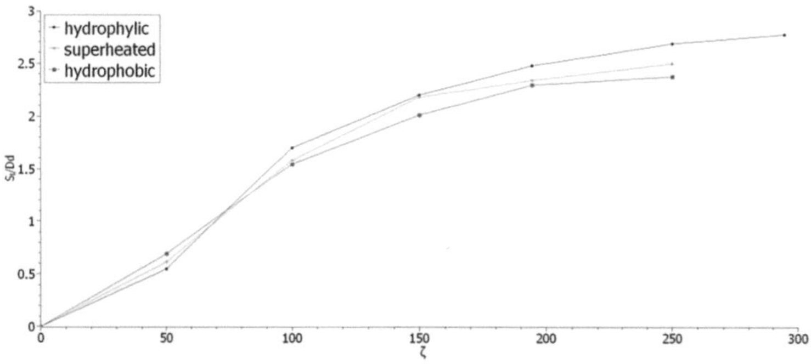

Fig. 3 Variation of non-dimensional spreading length w.r.t residence time for hydrophilic, hydrophobic and super heated surfaces at $D^* = 0.414$, We = 39.47

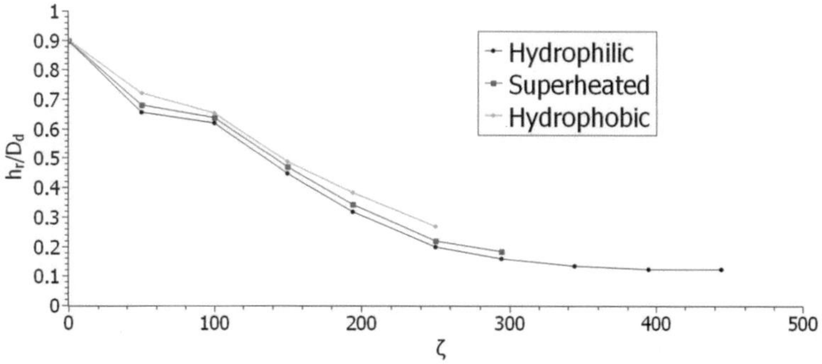

Fig. 4 Variation of non-dimensional rim height w.r.t residence time for hydrophilic, hydrophobic and super heated surfaces at $D^* = 0.414$, We = 39.47

Figure 4 explains how the non-dimensional rim height varies depending on the surfaces being studied, including hydrophilic, hydrophobic and super heated surfaces. The resultant trend is directly in opposition to what we saw in the non-dimensional spreading length. The relationship between residence time and rim height is inverse. As time goes on, it is seen that the rim thickness decreases until it is the smallest possible value for all three surfaces. The minimum and maximum rim thicknesses are displayed for hydrophilic and hydrophobic surfaces, respectively. Variation of non-dimensional max spreading length w.r.t residence time for hydrophilic, hydrophobic and super heated surfaces are illustrated in Fig. 5. For all We, spheres having hydrophilic nature, the spread factor is roughly uniform throughout the early stages of spreading due to lower impact velocity. The high We spheres on the other hand show increasing spreading when the curves diverge. The reason for this divergence is that when We increases, the inertial forces grow along with it and finally outweigh the viscous force. As a result, the impact velocity rises, causing the droplet to completely engulf the sphere. Larger targets generate more viscous shear. During the early inertial phase of spreading, the viscous forces and the target's curvature

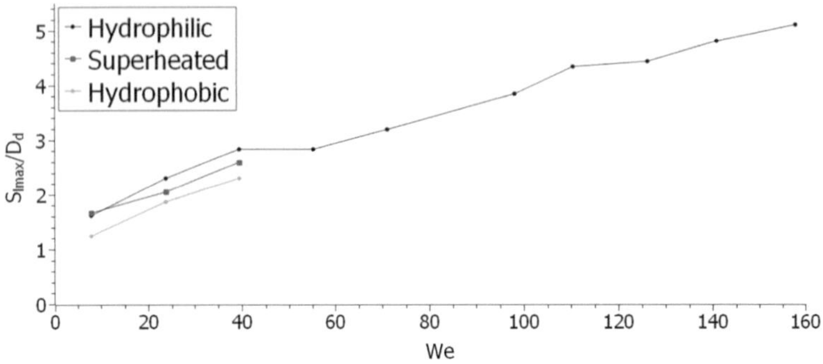

Fig. 5 Variation of non-dimensional max spreading length w.r.t residence time for hydrophilic, hydrophobic and super heated surfaces at $D^* = 0.414$

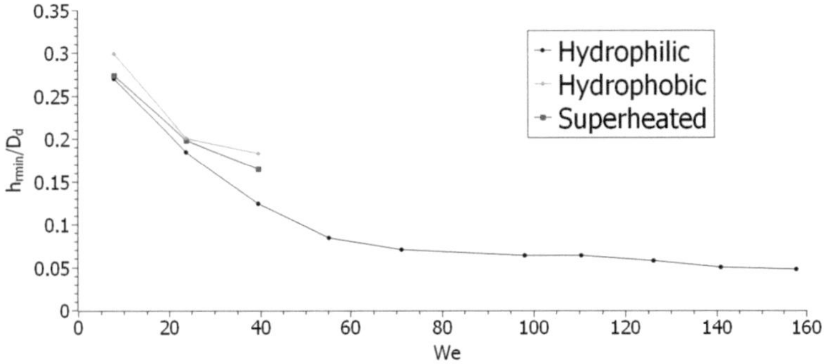

Fig. 6 Variation of non-dimensional min rim height w.r.t residence time for hydrophilic, hydrophobic and super heated surfaces at $D^* = 0.414$

prevent the film from flowing. The huge curvatures of the spheres in the latter phases favour gravity-induced spreading, leading to a discernible increase in the spread factor for smaller targets. Viscous resistance of target surfaces is proportional to their surface area, which is significantly lower for smaller targets and promotes larger spreading. However, the spread factor is essentially similar in the early phases for extremely hot and hydrophobic surfaces. The droplet entirely covers the sphere as the We is increased, increasing the impact velocity over the spheres. Early breaking of the droplets occurs as a result of the hydrophobic and extremely hot surfaces' non-wetting characteristics. In comparison with the vapour layer that forms over the super heated surface, the air cushion that covers the hydrophobic surface offers more resistance. Because of this, super heated surfaces have a longer maximum spreading length than hydrophobic surfaces. The trend is the opposite when it comes to minimum rim height. As seen in Fig. 6, the height of the north pole's rim inevitably decreases. Due to their largest spreading length, hydrophilic surfaces have the lowest rim height, followed by super heated and hydrophobic surfaces.

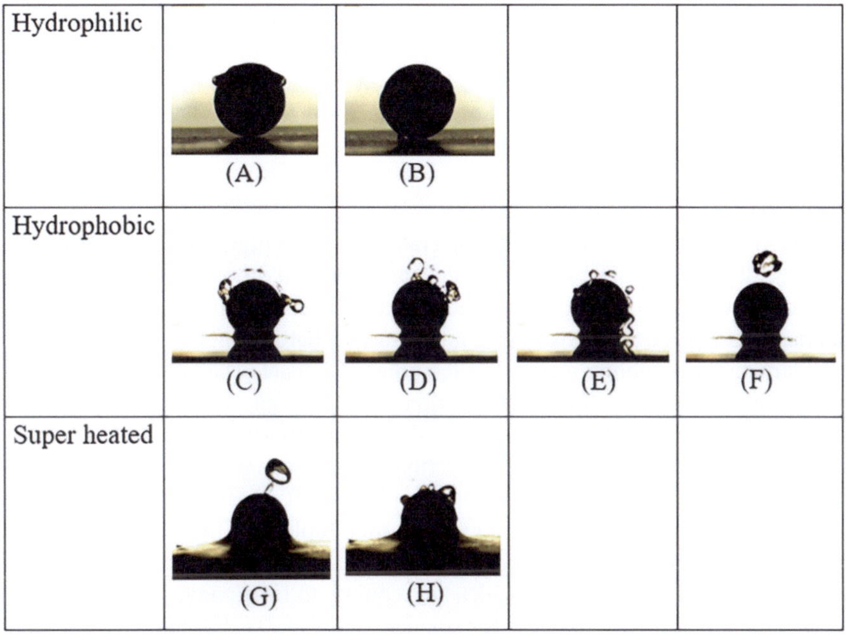

Fig. 7 Morphologies of droplet impact on hydrophilic, hydrophobic and super heated surfaces. **a** Partially covered, **b** fully covered, **c** and **h** breakup after covering, **d** breakup after bouncing, **e** Corona splash, **f** and **g** bouncing

Major morphologies that occur when a droplet impact over a spherical surface in this experiment involves spreading, fully covering, breakup,and corona splash Fig. 7. The behaviour of the droplet as it makes contact and spreads out across the surface following impact is referred to as the "spreading phenomenon" in droplet impact over a spherical surface. A key factor in the spreading phenomena is the contact angle between the liquid-air interface and the spherical surface of the droplet. The contact angle is often smaller than 90° on a hydrophilic surface, which promotes more spreading. On the other side, the contact angle is greater on a hydrophobic surface, and the droplet may exhibit minor spreading or partial rebound. The phrase "fully covering phenomenon" describes a unique behaviour in which a droplet entirely covers the surface of a sphere upon impact, generating a thin, continuous liquid film that stretches across the whole surface. Another name for this occurrence is "complete wetting".

The droplet's kinetic energy at impact must be high enough to overcome surface tension forces and cause the fully covering phenomena, which wets the whole surface of the sphere. When a droplet impacts a sphere, the term "breakup" refers to the droplet's fragmentation or disintegration into smaller droplets. The interaction of several forces and instabilities may cause a droplet to break up when it strikes a spherical surface with enough force. Droplet size, impact velocity, surface characteristics and the dynamics of the liquid flow are some of the variables that have an impact on the breakdown process. Higher Weber numbers increase breakdown because the inertial forces outweigh the surface tension forces. The integrity of the droplet is often maintained by the liquid's surface tension. However, the cohesive forces could not be strong enough to keep the droplet from disintegrating at high impact velocities or low surface tension.

Low viscosity may encourage disintegration whereas high viscosity tends to stabilise the droplet. Droplet breakdown and partial or total droplet rebound can sometimes coexist. Corona splash, also called the crown splash, is an amazing phenomena that happens when a droplet strikes a liquid surface hard enough to form a thin, circular structure resembling a crown at the point of impact. The typical symptom of this phenomena, which is the radial ejection of small droplets, is high-speed droplet impact events. The droplet spreads out on the liquid surface and forms a central hollow as a result of impact. Two main morphologies are seen in hydrophilic spheres as in Fig. 8. Lower Weber numbers cause the droplet impact across the sphere to partially cover the sphere. As we continue to impact the droplets by increasing the height, they attain enough kinetic energy to completely surround the spheres. Additionally, when D^* value falls, greater velocities are required to fully engulf the droplet over the sphere. Arrest of spread occur for smaller values of $D*$. For higher D^* a rim is formed at the north pole by the liquid film as it runs over the target. Further lowering of D^* results in a dramatic stop of the spreading characteristics. The hydrodynamics of droplet impact on flat surfaces and impact on large spheres are very similar. High We cases cause the lamella to expand to greater lengths, which results in the formation of a thin wetting layer at the sphere's north pole.

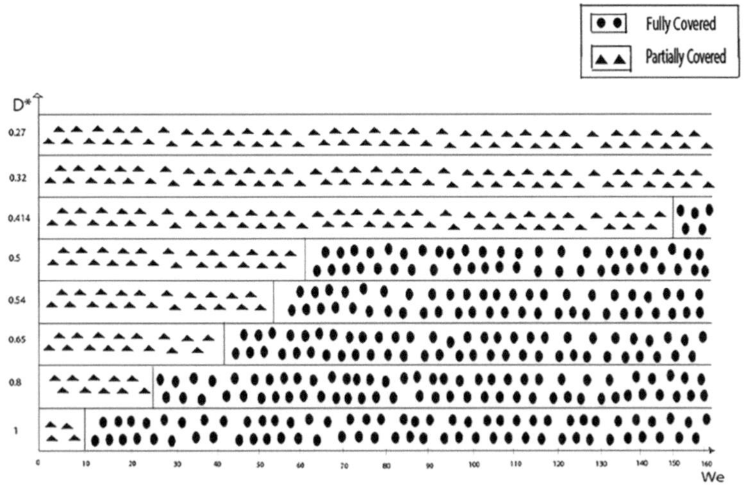

Fig. 8 Regime map of hydrophilic spheres having varying D^*

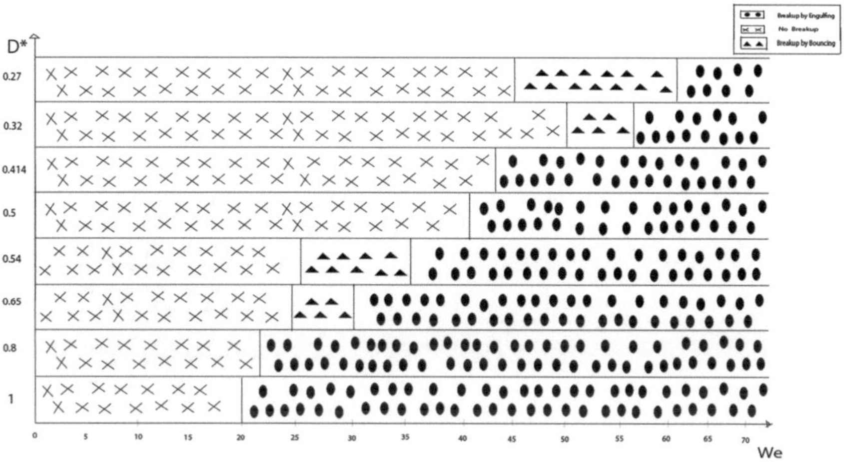

Fig. 9 Regime map of hydrophobic spheres having varying D^*

When a droplet impacts hydrophobic spherical surfaces Fig. 9, there are two different forms of breakage that are seen. Breakup after engulfing happens at higher Weber numbers at 3.5 and 2.9 mm drop diameters. However, the initial droplet break happens after bouncing for droplet diameters of 2.3 and 1.9 mm. When the spherical surface diameter is greater than the droplet diameter at an intermediate We, breakage after bouncing is typically observed. There is not enough energy in the impact droplet to completely envelop the spherical particle. This causes a breakup after bouncing. The velocity of impact rises as we continue to grow. The droplet then breaks apart after engulfing the spherical particle. The droplet touches the spherical surface at a lower Weber number and begins to spread. The Leidenfrost effect causes evaporation to begin, which then causes a vapour layer to form between the solid and liquid interface. The north pole's shortest spreading length results in a relatively greater rim height due to the low impact velocity. The creation of the vapour layer occurs on a tiny surface area. As a result, the droplets did not fragment. The droplets collided with each other more quickly when we raised the We number Fig. 10. After striking the spherical particle, the droplets try to engulf the surface. The droplet's spreading length increases, and it covers a larger area. This causes a layer of vapour to build over the sphere. Finally the heat causes the completely engulfed layer to begin splitting.

4 Conclusions

The post-impact hydrodynamics of water droplets on spherical objects of various sizes is experimentally studied. In this paper, we discuss droplet spreading and lamella generation on spherical substrates under various impact conditions. To fully

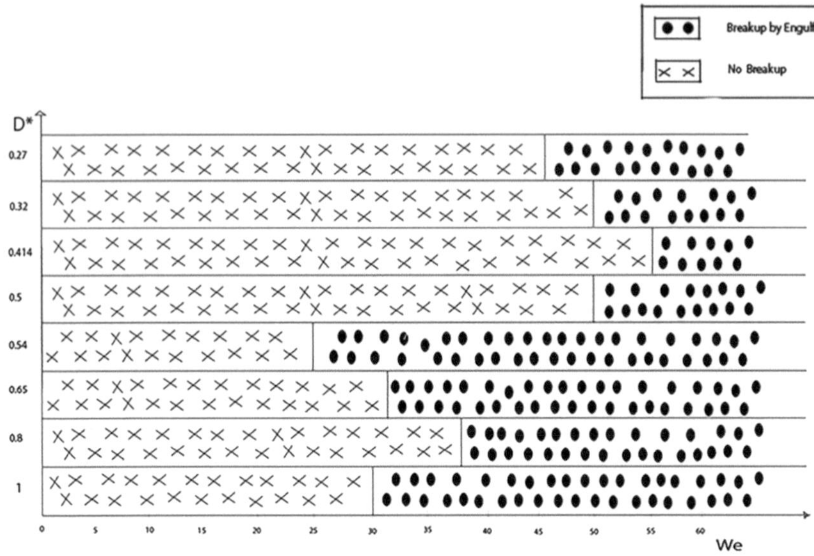

Fig. 10 Regime map of super heated spheres having varying D^*

assess these activities, high-speed photography is used. The spread factor, maximum spreading factor, minimum rim thickness ratio and rim thickness ratio at the north pole are geometric factors that are identified and evaluated. It has been noted that the droplet spreading pattern, recoiling and rebound vary for hydrophilic, hydrophobic and super heated surfaces. For surfaces that are hydrophilic, there are mainly two morphologies , Fully and partial engulfment. The wetting behaviour of the spheres is the cause of this. At lower Weber values, bouncing occurs for super heated surfaces. Then, jetting is observed. The droplets break by swallowing the spherical after a partial rebound. For hydrophobic surfaces, the morphology is similar. However, breakup after engulfing happens at higher Weber numbers at 3.5 and 2.9 mm drop diameters. However, the initial droplet break happens by bouncing for droplet diameters of 2.3 and 1.9 mm. When the Weber number is raised further, droplet breaks after enveloping. At smaller Weber numbers, the droplet lacks enough kinetic energy to encircle the sphere. The air cushion that has developed between the droplet and the hydrophobic sphere is what causes it to be propelled upward as a result. However, as the Weber number increases, the high speed of the droplets impacts the surface and overcomes the resistance of the air cushion, engulfing the sphere and shattering the drop. Hydrophilic surfaces display more spreading when compared to super heated, hydrophobic and hydrophilic surfaces. The air cushion over the hydrophobic surface provides higher resistance to spreading than the vapour layer formed over the super heated surface [27].

Acknowledgements The authors would like to acknowledge the Department of Mechanical Engineering College Of Engineering Trivandrum for providing access to the Microfluidics laboratory facilities and resources, which were essential for conducting the experiments.

References

1. Crooks R, Cooper-White J, Boger DV (2001) The role of dynamic surface tension and elasticity on the dynamics of drop impact. Chem Eng Sci 56(19):5575–5592
2. Park H, Carr WW, Zhu J, Morris JF (2003) Single drop impaction on a solid surface. AIChE J 49(10):2461–2471
3. Roux DCD, Cooper-White JJ (2004) Dynamics of water spreading on a glass surface. J Colloid Interface Sci 277(2):424–436
4. Yarin AL (2006) Drop impact dynamics: splashing, spreading, receding, bouncing. Annu Rev Fluid Mech 38:159–192
5. Ueda T, Enomoto T, Kanetsuki M (1979) Heat transfer characteristics and dynamic behavior of saturated droplets impinging on a heated vertical surface. Bulletin JSME 22(167):724–732
6. van Dam DB, Le Clerc C (2004) Experimental study of the impact of an ink-jet printed droplet on a solid substrate. Phys Fluids 16(9):3403–3414
7. Wachters LHJ, Westerling NAJ (1966) The heat transfer from a hot wall to impinging water drops in the spheroidal state. Chem Eng Sci 21(11):1047–1056
8. Fujimoto H, Hatta N (1996) Deformation and rebounding processes of a water droplet impinging on a flat surface above Leidenfrost temperature
9. Strotos G, Gavaises M, Theodorakakos A, Bergeles G (2008) Numerical investigation of the cooling effectiveness of a droplet impinging on a heated surface. Int J Heat Mass Transf 51(19–20):4728–4742
10. Zhao Z, Poulikakos D, Fukai J (1996) Heat transfer and fluid dynamics during the collision of a liquid droplet on a substrate–I. Modeling Int J Heat Mass Transf 39(13):2771–2789
11. Liu Y, Andrew M, Li J, Yeomans JM, Wang Z (2015) Symmetry breaking in drop bouncing on curved surfaces. Nat Commun 6(1):10034
12. Huang X, Wan KT, Taslim ME (2018) Axisymmetric rim instability of water droplet impact on a super-hydrophobic surface. Phys Fluids 30(9)
13. Bakshi S, Roisman IV, Tropea C (2007) Investigations on the impact of a drop onto a small spherical target. Phys Fluids 19(3)
14. Khojasteh D, Bordbar A, Kamali R, Marengo M (2017) Curvature effect on droplet impacting onto hydrophobic and superhydrophobic spheres. Int J Comput Fluid Dyn 31(6–8):310–323
15. Kim SG, Kim W (2016) Drop impact on a fiber. Phys Fluids 28(4)
16. Khurana G, Sahoo N, Dhar P (2019) Post-collision hydrodynamics of droplets on cylindrical bodies of variant convexity and wettability. Phys Fluids 31(2)
17. Worthington AM (1877) XXVIII. On the forms assumed by drops of liquids falling vertically on a horizontal plate. Proc R Soc Lond 25(171–178):261–272
18. Hatta N, Fujimoto H, Takuda H, Kinoshita K, Takahashi O (1995) Collision dynamics of a water droplet impinging on a rigid surface above the Leidenfrost temperature. ISIJ Int 35(1):50–55
19. Bertola V (2009) An experimental study of bouncing Leidenfrost drops: comparison between Newtonian and viscoelastic liquids. Int J Heat Mass Transf 52(7–8):1786–1793
20. Chen R-H, Chiu S-L, Lin T-H (2007) Resident time of a compound drop impinging on a hot surface. Appl Therm Eng 27(11–12):2079–2085
21. Cossali GE, Marengo M, Santini M (2008) Thermally induced secondary drop atomisation by single drop impact onto heated surfaces. Int J Heat Fluid Flow 29(1):167–177
22. Cossali GE, Marengo M, Santini M (2005) Secondary atomisation produced by single drop vertical impacts onto heated surfaces. Exp Therm Fluid Sci 29(8):937–946

23. Kandlikar SG, Steinke ME, Singh A (2001) Effects of weber number and surface temperature on the boiling and spreading characteristics of impinging water droplets. In: Proceedings of NHTC 1
24. Chow CK, Attinger D (2003) Visualization and measurements of microdroplet impact dynamics on a curved substrate. Fluids Eng Div Summer Meet 36975:405–413
25. Hardalupas Y, Taylor AMKP, Wilkins JH (1999) Experimental investigation of sub-millimetre droplet impingement on to spherical surfaces. Int J Heat Fluid Flow 20(5):477–485
26. Liang G, Guo Y, Yang Y, Guo S, Shen S (2013) Special phenomena from a single liquid drop impact on wetted cylindrical surfaces. Exp Thermal Fluid Sci 51:18–27
27. Gaitonde DV (2015) Progress in shock wave/boundary layer interactions. Prog Aerosp Sci 72:80–99

A Phase Field Model for Viscosity Induced Droplet Splitting Phenomenon at a T-Junction Microchannel

Joy Mandal and Sandip Sarkar

Abstract A 2D simulation is carried out using dimensionless phase field formulation to delineate the fate of a droplet at a T-junction microchannel for different viscosity ratios. The model is validated by experimental findings and scaling law from the existing literatures with good agreement. We captured three regimes (for fixed droplet length) namely, breakup with tunnels, breakup with discontinuous occlusion, and no-breakup. Furthermore, the variation in viscosity ratio revealed the existence of regimes that were not found in the literature, for the investigated droplet length, for different Capillary number. Next, we show new breakup characteristics as the viscosity ratio is varied. These classifications are mainly based on the relative dominance of surface tension represented in the Capillary number and viscosity ratio. Finally, we present a phase diagram to illustrate the boundary transitions of different regimes.

Keywords Droplet breakup · Viscosity ratio · Phase field · Microchannel · T-junction

1 Introduction

Microfluidic flows observed in nature and human-made devices are of multiphase origin. Two-phase flows in confined geometries elicited the attention of numerous researchers due to the precise control over the fluid particle trajectories that can be achieved along with the handling of the small discrete entities (droplets)

J. Mandal (✉) · S. Sarkar (✉)
SERB Sponsored Microfluidics Laboratory, Department of Mechanical Engineering, Jadavpur University, Kolkata, India
e-mail: joymandal2014@gmail.com

S. Sarkar
e-mail: sandipsarkar.mech@jadavpuruniversity.in

A. Kumar R. et al. (eds.), *Proceedings of Fluid Mechanics and Fluid Power (FMFP) 2023, Vol. 6*, Lecture Notes in Mechanical Engineering,
https://doi.org/10.1007/978-981-96-9600-0_10

with a high degree of control and reproducibility [1, 2]. This is otherwise difficult in macrochannels, leading to a wide range of applications in emulsification and engineering [3].

Shia-Yen et al. [4] summarized droplet breakup methods as passive fission and active fission. Unlike active fission, passive fission has attracted the attention of many researchers because it can split droplets at precise locations into controlled volumes by designing the microchannel geometry without any external power. Among the various microchannel geometries employed for passive droplet fission, the T-junction represents one of the most fundamental building blocks in microfluidic devices and has therefore been extensively investigated. Link et al. [5] pioneered the introduction of T-junctions to passively split droplets into miniature sizes, following which substantial research efforts have been devoted toward elucidating the mechanism of droplet breakup in a microfluidic T-junction. As reported by Jullien et al. [6], when droplets pass through a T-junction, there exist three distinct typical flow patterns, i.e., non-breakup, breakup with tunnels and breakup with permanent obstruction. It was found that the flow pattern of the passing droplet depends on the Capillary number(Ca) and the dimensionless initial droplet length.

Viscosity ratio (μ_r) of the droplet to the surrounding medium plays an important role in the droplet dynamics at a T-junction bifurcation. Menech et al. [7] applied a phase field method to model the flow and breakup of droplets in a T-shaped junction. They found that the critical capillary number for droplet breakup depends on the viscosity ratio between continuous and disperse phases with a trend that is analogous to that observed for individual isolated droplets in hyperbolic flow. Yet, the literatures had mainly emphasized on the effect of varying the droplet length as a function of Ca to study the droplet breakup at a bifurcation. This has indeed, revealed existence of various breakup regimes but however, not much attention has been given to μ_r. Many researchers have studied the regime map for droplet splitting in Ca-droplet length plane for a few μ_r [8]. But, we believe that an entire spectrum of μ_r should be investigated to unravel the breakup dynamics and to examine how the viscous forces compete against the surface tension.

In this article, we will present a numerical model based on the phase field theory coupled with the Navier–Stokes equations to simulate droplet dynamics in microchannel. A phase field method is mass-conserving and calculation of surface tension can be done with confidence. It is also referred to as a diffuse-interface method since the interface is assumed to be diffuse with constant thickness [9]. The phase field method is somewhat like the Lattice Boltzmann method since they both are based on the postulate of the free energy of the system.

The remainder of this paper is organized as follows: In Sect. 2, we have described the physical system and formulated the model mathematically. Next, the numerical verification is presented, followed by the discussion on the reported results in Sect. 3. Finally, Sect. 4 summarizes the research findings.

2 Methodology

We solve the continuity, the Cahn–Hilliard, and Navier–Stokes equations to obtain the pressure and velocity field of the two-liquid flow system (Fig. 1), which are expressed as

$$\nabla.u = 0 \tag{1}$$

$$\frac{\partial(\rho u)}{\partial t} + \nabla.(\rho u u) = -\nabla p + \nabla.[\mu(\nabla u + \nabla u^T)] + G\nabla\phi \tag{2}$$

$$\frac{\partial\phi}{\partial t} + u.\nabla\phi = \nabla.(M_\phi \nabla G) \tag{3}$$

Here, $G\nabla\phi$ is the phase field-dependent interfacial tension force. The physical properties of the droplet fluid and surrounding medium considered in the simulation are density (ρ) and viscosity (μ) as shown in Fig. 1.

2.1 *Normalization of the Governing Equations*

In the present section, the aforementioned governing equations are non-dimensionalized to attain a profound understanding. The width (l) of the microchannel is considered as the characteristic length scale of the flow. The following set of dimensionless variables is defined for normalization of the governing equations:

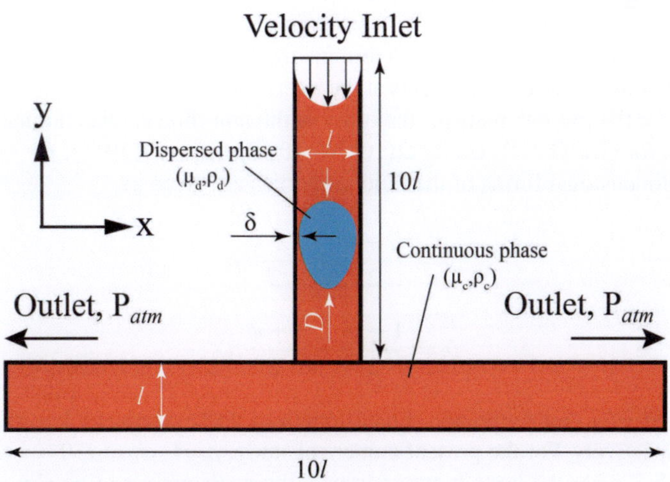

Fig. 1 Schematic of the microfluidic setup (not to scale)

$$U = \frac{u}{u_{avg}}, T = \frac{tu_{avg}}{l}, P = \frac{pl}{\mu_c u_{avg}}, \bar{\mu} = \frac{\mu}{\mu_c}, \bar{\rho} = \frac{\rho}{\rho_c}, \bar{\nabla} = l\nabla$$

Using the above set of non-dimensionalized variables, the governing transport equations, i.e., Cahn–Hilliard–Navier–Stokes equations along with the continuity reduce to

$$\bar{\nabla}.U = 0 \tag{4}$$

$$\begin{aligned} \bar{\rho}\frac{DU}{DT} &= \frac{1}{Re}[-\bar{\nabla}P + \bar{\nabla}.\bar{\mu}(\bar{\nabla}U + \bar{\nabla}U^T)] \\ &+ \frac{1}{ReCaCn}\frac{3}{2\sqrt{2}}[(\phi^3 - \phi) - Cn^2\bar{\nabla}^2\phi]\bar{\nabla}\phi \end{aligned} \tag{5}$$

$$\frac{\partial\phi}{\partial T} + U.\bar{\nabla}\phi = \bar{\nabla}.[\frac{3}{2\sqrt{2}}\frac{1}{Pe}\bar{\nabla}((\phi^3 - \phi) - Cn^2\bar{\nabla}^2\phi)] \tag{6}$$

Thus, the present problem of two-liquid flow system is characterized by the following set of dimensionless parameters:

$$Re = \frac{\rho_c u_{avg} l}{\mu_c}, Ca = \frac{\mu_c u_{avg}}{\sigma}, Cn = \frac{\epsilon}{l}, Pe = \frac{\epsilon l u_{avg}}{M_\phi \sigma}$$

The Reynolds number, Re, interprets the relative significance of the inertia force to the viscous force; the Cahn number, Cn, is the ratio of interface thickness to the characteristics length; the capillary number, Ca, is the ratio of the viscous force to the interfacial force; the interface Péclet number, Pe, is the ratio of the advection of the phase field parameter (ϕ) to its diffusion. This interface Péclet number, specific to phase-field modeling, is defined to determine the relative strength of advection over the diffusion of the fluid components at the interfacial region. It is worth mentioning here that, for the present analysis, the order of the non-dimensional numbers is considered as, $Re \sim O(10^{-1})$, $Ca \sim O(10^{-1} - 10^{-3})$, $Cn \sim O(10^{-2})$, $Pe \sim O(10^{1})$. The non-dimensional forms of the fluid properties are given as:

$$\bar{\rho} = \frac{1-\phi}{2} + \frac{1+\phi}{2}\rho_r \tag{7}$$

$$\bar{\mu} = \frac{1-\phi}{2} + \frac{1+\phi}{2}\mu_r \tag{8}$$

Note that $\bar{\rho}$, $\bar{\mu}$ denote the non-dimensionalized phase-averaged density, and viscosity, respectively. For the present numerical study: ρ_r=1, $\mu_r \sim (10^{-3} - 10^3)$, and $D/l = 2.1$. As per the boundary conditions, the walls are set to no slip with contact angle π. A parabolic velocity profile is imposed at the inlet for the continuous phase while atmospheric pressure boundary condition is specified at the outlets.

Figure 2a provides the geometry details used for the simulation while Fig. 2b shows the meshing for $Cn = 0.02$. The inset of Fig. 2b represents an enlarged view of the mesh comprising triangular elements. The unsteady governing transport equations together with the boundary conditions in the present study are solved by using the finite element framework of COMSOL Multiphysics.

2.2 Validation

To validate the mathematical formulation for the two-phase system in micro confinement, we perform a set of simulations in a T-shaped junction for different flow rate ratio(Q^*) and Ca. We compare the simulation results for droplet generation with the scaling model proposed by Garstecki et al. [10] and also with the experimental data by Christopher et al. [11]. The variation of the droplet length with the flow rate ratio can be seen in Fig. 3. It can be observed that the cluster of points obtained from our model lies in the vicinity of both the scaling law and the experimental findings. Furthermore, the deviation from reported literature data and model results are well within 5%.

3 Results and Discussion

To ensure that the model results are not susceptible to solver/modeling uncertainties, a set of simulations with different Cn numbers have been performed; $0.05 \leq Cn \leq 0.01$ keeping Ca=0.096, μ_r=0.02, and Re=0.3 fixed, see Fig. 4. The simulation results

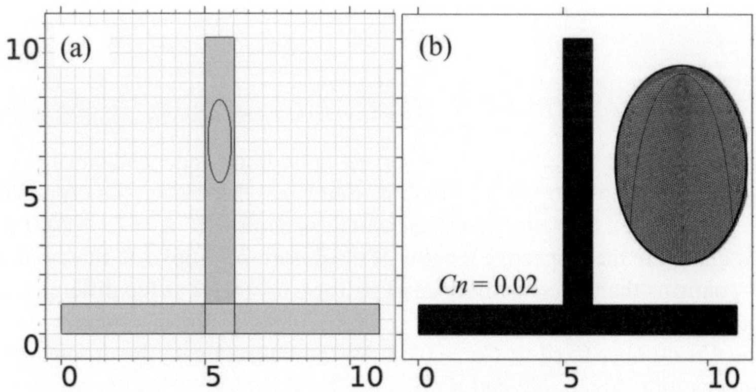

Fig. 2 Details of the geometry and meshing as carried out in COMSOL

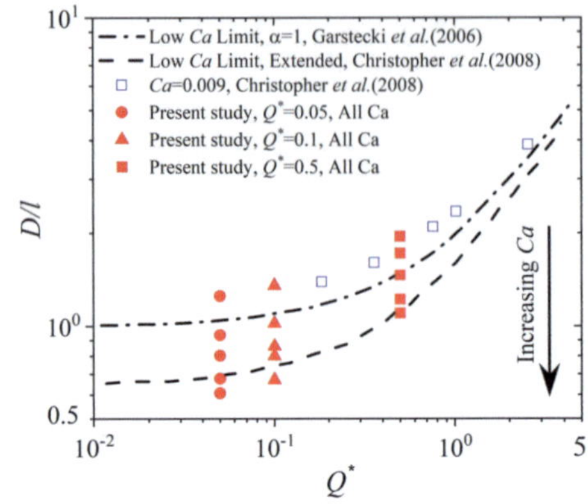

Fig. 3 Comparison of measured droplet lengths corresponding to the scaling law of Garstecki et al. [10] and experiments by Christopher et al. [11] with the present model. Q^* denotes the flow rate ratio

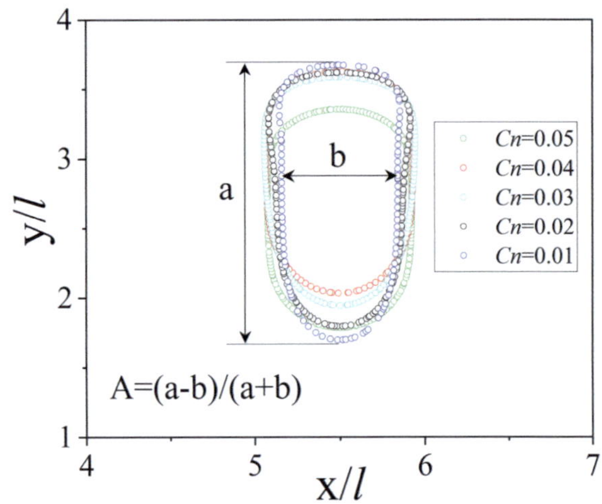

Fig. 4 Droplet deformation captured for various Cahn numbers; $Re = 0.3$, $Ca = 0.096$, $\mu_r = 0.02$, and $D/l = 2.1$

with $Cn = 0.01$ and $Cn = 0.02$ are observed to be fairly independent of the interface width in terms of droplet deformation. See Fig. 4. Moreover, the droplet deformation is quantified by defining a dimensionless parameter $A = (a - b)/(a + b)$, where a and b are the respective lengths of the deformed droplet as shown in Fig. 4. Table 1 confirms the grid sensitivity analysis for $Cn = 0.02$ as the difference in the computed quantity (A) is within 1%. Therefore, in the following computations, we keep the Cn number fixed to $Cn = 0.02$. We now investigate the droplet splitting mechanisms for different Ca and μ_r.

Table 1 Droplet deformation parameter evaluated for various Cahn numbers; $Re = 0.3$, $Ca = 0.096$, $\mu_r = 0.02$, and $D/l = 2.1$

Sl. No.	Cn	A	Error(%)
1	0.05	0.194	–
2	0.04	0.221	13.91
3	0.03	0.210	04.97
4	0.02	0.205	04.28
5	0.01	0.203	00.10

3.1 Breakup with Tunnel (TB)

The tunnel breakup regime (TB) at the T-junction is illustrated in Fig. 5a. As the droplet enters the T-junction, it deforms, leading to the formation of gaps on both sides. Simultaneously, higher Ca results in dominant viscous stresses acting on the droplet interface, eventually leading to droplet splitting. A circular neck can be observed to from in the center of the T-junction that stretches with time during the breakup process. This scenario can be attributed to classical Rayleigh-Plateau instability, through which a cylindrical liquid thread can decrease its total surface area by breaking when its length exceeds its circumference. A tunnel between the daughter droplets and the upper wall is found to exist during the entire splitting phenomenon. As a result, the pressure never accumulates upstream of the T-junction. This breakup regime (for $D/l = 2.1$) has not been found in an analogous study conducted by Chen et al. [8] for three different viscosity ratios. They could only capture the DOB regime associated with $D/l = 2.1$ probably because of their smaller ranges of Ca and μ_r.

3.2 Breakup with Discontinuous Occlusion (DOB)

In this regime, the droplet obstructs the path leading to the two downstream channels and blocks the two horizontal channels, as shown in Fig. 5b. As the flow progresses, a transition from obstruction to the appearance of a gap occurs, which can be described as a tunnel opening and accumulated pressure released through the gaps. A small tunnel between the daughter droplets and the upper wall is found to exist during the splitting phenomenon. This splitting phenomenon is hence termed a discontinuous occlusion breakup regime (DOB).

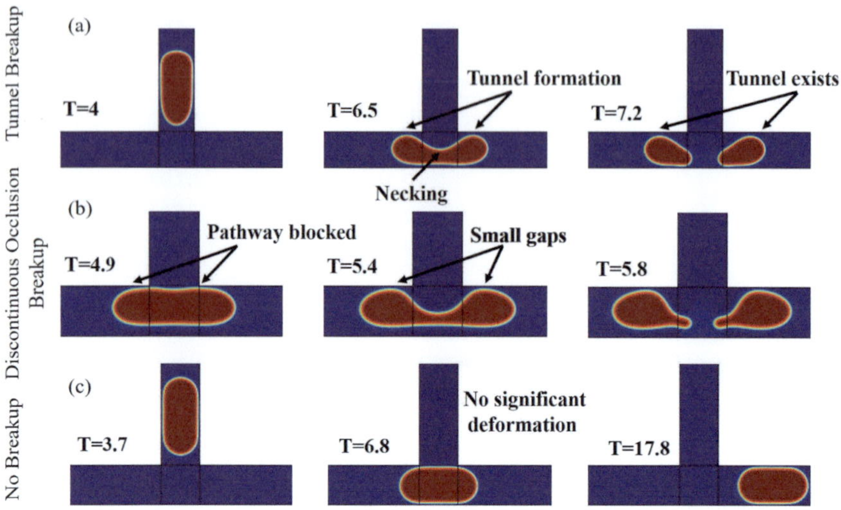

Fig. 5 Regimes of droplet splitting captured **a** Breakup with tunnels, $Ca = 0.07$; **b** Breakup with discontinuous occlusion, $Ca = 0.25$; **c** No-breakup, $Ca = 0.01$. $Re = 0.3$, $\mu_r = 0.02$, and $D/l = 2.1$

3.3 No-Breakup Regime (NB)

We have also found no-breakup (NB) regime where the splitting ceases to occur, and the droplet goes to either of the two downstream channels at very low Ca. Figure 5c shows that when the droplet fully enters the bifurcation, negligible deformation occurs at the T-junction. Moreover, the necking cannot be observed, and the droplet takes either of the two downstream channels. This phenomenon can be attributed to the low value of Ca resulting in lower squeezing force from the upstream continuous phase, and the viscous stress cannot struggle out the constraint of surface tension as the droplet retains its shape. This regime for such a large-sized droplet was not found by Link et al. [5] probably, for their choice of D/l range.

3.4 Phase Regime Map

Lastly, we represent these regimes by capillary number and viscosity ratio for the two-liquid system via. a phase diagram in Fig. 6. It can be observed that at low Ca, the NB regime is found throughout the entire spectrum of μ_r. Such a low value of Ca prevents the droplet breakup as the surface tension force becomes dominant. On the other hand, as Ca increases, viscous effects start to dominate the surface tension, consequently leading to droplet splitting owing to necking and deformation (TB regime). This happens even at very high Ca, but as the droplet becomes more

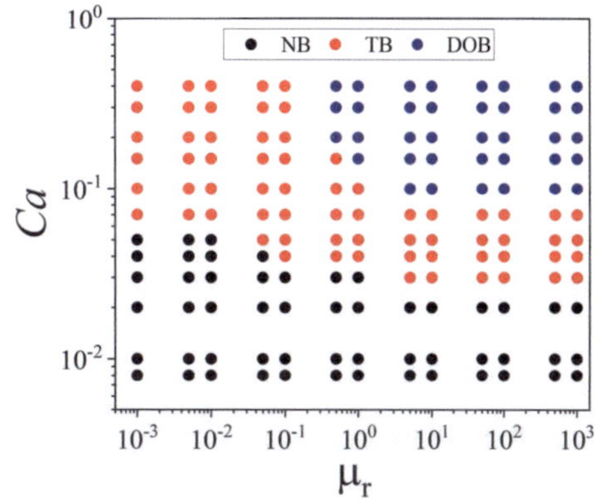

Fig. 6 Phase diagram between the two-liquid phases; $Re = 0.3$, $D/l = 2.1$

viscous (increasing μ_r), it enhances stretching and deformation of the droplet at the bifurcation and results in the DOB regime. Subsequently, the droplet splits up into daughter droplets via. necking after initially blocking the entire pathway. The fact that droplet breakup in the T-junction occurs more easily as the viscosity ratio is increased [7] can be verified in Fig. 6. The numerical results for the transitions among the droplet breakup regimes are found to be consistent with the findings of Chen et al. [8] within their parameter spaces.

4 Conclusions

We have investigated the breakup and non-breakup dynamics of a droplet (length fixed) at a T-junction microchannel using a dimensionless phase field model. We have conducted the study for wide ranges of Ca and μ_r in the quest for new findings. The results were found to be consistent with the experimental data and empirical relations from existing literature. Two different breakup patterns have been captured: breakup with tunnels and discontinuous occlusion, followed by the no-breakup regime. We present a phase diagram in the $Ca - \mu_r$ plane to understand the regime transition phenomenon. Two new findings are reported; (a) the NB regime is found to exist for such a large-sized droplet at very low Ca, (b) the TB regime is also found to exist in the Ca-μ_r plane in the same phase map. The relationship between viscosity and surface tension can be well apprehended from Ca in the phase regime map. At low $Ca \sim 10^{-2}$, surface tension dominated region is observed where the deformation ceases to exist (NB regime) as viscous stresses are negligible. On the other hand, as the Ca increases, shear stresses start to prevail surface tension and viscous dominated region is observed. This results in stretching and splitting of the droplet.

Nomenclature

u velocity vector [m/s]
t time [s]
p pressure [Pa]
G chemical potential
M_ϕ Interfacial mobility factor [m-s/kg]
D Droplet length [mm]
l channel width [mm]

Greek Symbols

ϕ phase field parameter
ρ density [kg/m^3]
μ viscosity [Pa-s]
σ surface tension coefficient [N/m]
ϵ parameter controlling interface thickness [m]

Subscripts

c continuous phase property
d dispersed phase property
r ratio of dispersed to continuous phase
avg average flow velocity

Acknowledgements J.M acknowledges AICTE to provide financial support for doing Ph.D. work under the AICTE-ADF scheme at Jadavpur University.

References

1. Mandal J, Sarkar S, Sen S (2021) A deterministic model for bubble propagation through simple and cascaded loops of microchannels in power-law fluids. Phys Fluids 33(7)
2. Mandal J, Sarkar S (2023) Morphology and kinematics of a train of power-law droplets in a corrugated microchannel. Chem Eng Sci 274:118691
3. Song H, Chen DL, Ismagilov RF (2006) Reactions in droplets in microfluidic channels. Angew Chem Int Ed 45(44):7336–7356
4. Teh S-Y, Lin R, Hung L-H, Lee AP (2008) Droplet microfluidics. Lab Chip 8(2):198–220
5. Link DR, Anna SL, Weitz DA, Stone HA (2004) Geometrically mediated breakup of drops in microfluidic devices. Phys Rev Lett 92(5):054503
6. Jullien M-C, Tsang Mui Ching M-J, Cohen C, Menetrier L, Tabeling P (2009) Droplet breakup in microfluidic t-junctions at small capillary numbers. Phys Fluids 21(7)
7. De Menech M (2006) Modeling of droplet breakup in a microfluidic t-shaped junction with a phase-field model. Phys Rev E 73(3):031505

8. Chen B, Li G, Wang W, Wang P (2015) 3d numerical simulation of droplet passive breakup in a micro-channel t-junction using the volume-of-fluid method. Appl Therm Eng 88:94–101
9. Badalassi VE, Ceniceros HD, Banerjee S (2003) Computation of multiphase systems with phase field models. J Comput Phys 190(2):371–397
10. Garstecki P, Fuerstman MJ, Stone HA, Whitesides GM (2006) Formation of droplets and bubbles in a microfluidic t-junction—scaling and mechanism of break-up. Lab Chip 6(3):437–446
11. Christopher GF, Noharuddin NN, Taylor JA, Anna SL (2008) Experimental observations of the squeezing-to-dripping transition in t-shaped microfluidic junctions. Phys Rev E 78(3):036317

Binary Droplet Impact on Square-Micropillared Hydrophobic Substrate

Mukesh Kumar Yadav and Nagesh D. Patil

Abstract The binary water droplet impact dynamics on square-micropillared hydrophobic surface is investigated experimentally. A primary droplet is first impacted on the substrate and allowed to achieve static condition, which is observed as in the Wenzel state. Thereafter, the secondary droplet is impacted as vertical head-on collision with the primary droplet. The subsequent dynamics behavior is studied considering the effect of the impact velocity (0.22 and 0.5 m/s) and the pitch of the pillars. The maskless photolithography is used to fabricate the square-micropillared surfaces of different pitches ranging from 30 to 76 μm, on which the equilibrium contact angles of water droplets fall within the range of 129° to 139°. High-speed photography and image analysis are performed to present the binary droplet impact dynamics qualitatively and quantitatively, respectively. In quantitative analysis, the temporal variation of the interfacial contact length between the droplets is analyzed. The results show that at lower impact velocity, the secondary droplet bounces off on lower-pitch substrates, while it merges with the primary droplet on higher-pitch substrates. Based on the variation of the impact velocities and pitch of the pillars, three different regimes, namely, no merging/bouncing, late merging, and gentle merging are investigated.

Keywords Binary droplet impact · Micropillared substrate · No merging · Late merging · Gentle merging

M. K. Yadav
Department of Mechanical Engineering, IIT Bhilai, Bhilai, India
e-mail: mukeshkd@iitbhilai.ac.in

N. D. Patil (✉)
Department of Mechanical Engineering and Department of Bioscience and Biomedical Engineering, Indian Institute of Technology Bhilai, Bhilai, India
e-mail: nageshpatil@iitbhilai.ac.in

A. Kumar R. et al. (eds.), *Proceedings of Fluid Mechanics and Fluid Power (FMFP) 2023, Vol. 6*, Lecture Notes in Mechanical Engineering,
https://doi.org/10.1007/978-981-96-9600-0_11

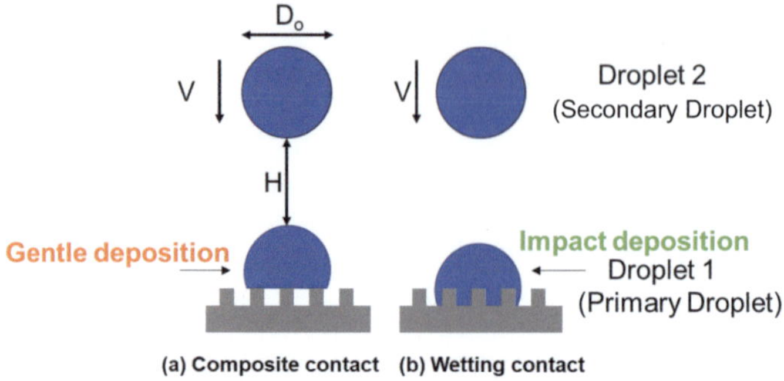

Fig. 1 Schematic diagram representing the initial conditions at the boundary of the primary droplet (droplet 1)

1 Introduction

Understanding the binary droplet or droplet-on-droplet impact on different wettability surfaces is very crucial for numerous technical applications such as industrial mixing [1], surface cleaning [1], pesticide spray coating [2], and raindrop impact [3]. In literature, the present understanding of binary droplet impact dynamics is still evolving, especially when compared to that of the understanding of single droplet dynamics. Considering this, the current study focuses on the impact dynamics of binary droplets.

Further, in the case of binary droplet impact, their impact dynamics greatly depend on the placement of the primary droplet. The primary droplet (droplet1) can be placed on the substrate in two ways, namely, gentle deposition ($We_{droplet1} < 1$) and impact deposition ($We_{droplet1} > 1$) as shown in Fig. 1 (as reported in Ref. [4]). In literature, there is a scarcity of studies that consider the binary droplet impact under the condition of impact deposition case. While there are several studies available, which consider the binary droplet impact under the condition of gentle deposition case on different wettability substrates such as hydrophobic [5, 6], coated superhydrophobic [7–9], and hydrophilic substrates[5–7], as discussed below in literature review section.

2 Literature Review and Objective

Wakefield et al. [9] investigated a binary droplet impact with different velocities having Weber numbers (We) in the range of 10–270 on Teflon ($\theta_{eq} \sim 90°$). Their main emphasis was on investigating the impact of We on the spreading of the droplet and energy budget analysis. They observed that at higher Weber numbers, a greater amount of kinetic energy is dissipated, as the liquid volume deforms through the processes such as crown formation, movement of the contact line (CL), and spreading. Wang et al. [8] experimentally and theoretically studied the binary droplet impact on the polydimethylsiloxane (PDMS) hydrophobic substrate (θ_{eq} ~ 101.7°) and these head-on collisions between droplets exhibited four distinct responses: complete rebound, partial rebound with conglutination, coalescence, and coalescence combined with conglutination. Jaiswal and Khandekar [10] investigated the binary droplet impact on carbon-soot-based superhydrophobic surfaces (θ_{eq} ~ 152°) both experimentally and numerically. For different We (1.05, 8.24, and 14.48), they observed three impact regimes late merging, gentle merging, and no merging (droplet bounce-off without merging).

Damak and Varanasi [11] conducted an experiment involving binary droplet collision on a superhydrophobic surface (θ_{eq} ~ 165°). They performed a scaling analysis that involves the expansion and retraction dynamics of the droplets, considering the We and the droplet diameter. Gunjan et al. [12] reported the recoil phenomena of water droplets over the course of vertical inline coalescence on superhydrophobic substrate. Somwanshi et al. [13] studied the merging/coalescence of droplet with lower Weber number (We < 1) that were vertically aligned initially and impacted on the superhydrophobic ($\theta_{eq} = 150°$) surfaces and they reached a conclusion that the disparity in surface energy serves as the motivating factor for the separation or lift-off of the merged droplet following coalescence.

Recently, Ramírez-Soto et al. [14] experimentally studied the impact of two different liquid droplets (oil and water droplet with contact angle ($\theta_{eq} > 150°$) onto a superamphiphobic surface. The study investigated different outcomes of the impact, which were found to depend on the Weber number (We) and the offset between the impacting and sessile droplets. Additionally, the conversion of energy during such impacts was analyzed using direct numerical simulation (DNS). Jaiswal and Khandekar [4] studied the significance of the initial boundary condition (i.e., impact and gentle deposition of primary droplet) during the binary droplet impact dynamics behavior on flat superhydrophobic substrates. In both cases, the difference was substantial. For former case, the binary droplet combined mass spread, retract, and bounced-off. While for latter case, the pinning of moving contact line during the retracting phase of combined mass was observed.

Based on the above presented literature, to the best of our knowledge, it was found that the binary droplet impact dynamics behavior for impact deposition condition (refer Fig. 1b) is scarcely studied. Moreover, the binary droplet impact dynamics on textured substrates is scarce. Only Jaiswal and Khandekar [4] studied the impact deposition condition for such binary droplet impact dynamics but they considered impact dynamics on flat superhydrophobic substrates. Thus, in view of this, in the present work, we investigate the binary droplet impact dynamics considering the impact deposition condition on micropillared substrates. In this study, objective is to study the binary droplets head-on collision dynamic behavior under the effects of pitch of the pillars as well as impact velocity of the droplet. We consider six values of pitch of the pillars (30, 34, 40, 47, 62, and 76 μm) and two impact velocities (0.22 and 0.5 m/s).

3 Materials and Methods

3.1 Fabrication of Micropillared Surfaces

Fabrication of the micropillared surface has been carried out using maskless photolithography (Micro Writer ML®3 Pro). The square micropillars were fabricated on silicon wafer (size: 2-inch, P type, orientation) using SU8 2100 epoxy photoresist polymer. The various steps followed during the fabrication process are presented in Fig. 2, which are discussed as follows. First, the silicon (Si) wafer was thoroughly cleaned using acetone, IPA, and DI water, in succession, to remove the absorbed or suspended impurities from the surface. It is then followed by drying via hot plate at a temperature of 100 °C for 10 min. To improve the adhesion of SU8 on silicon wafer, plasma cleaning was performed for 5 min prior to coating. Next, spin coating of the SU8 photoresist was carried out at a speed of 500 rpm for 5 s and 6000 rpm for following 60 s with an acceleration time of 20 s. Then, this coated sample was heated at 65 °C for 5 min followed by an additional heating at 95 °C for 10 min. Once the sample cooled down to ambient conditions, it was placed inside the lithography machine. The desired geometry was created in the software which consisted of six micropillar arrays of size 10 mm × 10 mm. The wafer was subjected to UV radiation exposure having an intensity and wavelength of 90 mJ/cm^2 and 365 nm, respectively. The pixel resolution of irradiating UV beam was kept as 5 μm corresponding to which the time for lithography completion was around 30 min. In the next step, the samples were post-heated at 65 °C for 2 min followed by an additional heating at 95 °C for 6 min. Thereafter, the samples were immersed in SU8 developer for 5–6 min and subsequently the samples were cleaned using isopropanol (IPA). Finally, the sample was hard baked at 120 °C for a duration of 10 min. Further, field emission scanning electron microscopy (Carl Zeiss Inc., GEMINI SEM 500 KMAT) characterization at 2000X magnification was performed as shown in Fig. 3, to visualize the arrangement of the micropillars on coated Si wafer. For all the surfaces, the measured height (h) is 28 ± 2 μm and width (a) is 20 ± 2 μm. Using the Advance Drop Shape Analyzer (Kruss, DSA25), the contact angle measurement was carried out by depositing a 2.57 μL volume droplet on the Si wafer. The measured equilibrium contact angles (θ_{eq}) of water droplets fall within the range of 129°–139° for different pitches (P) ranging from 30 to 76 μm. These observations align with the findings documented by Patil et al. [15] for SU8 micropillared hydrophobic surfaces.

3.2 Experimental Setup

The representation diagram of experimental arrangement has been shown in the Fig. 4, comprised of a high-speed camera incorporating a lens (FASTCAM Mini

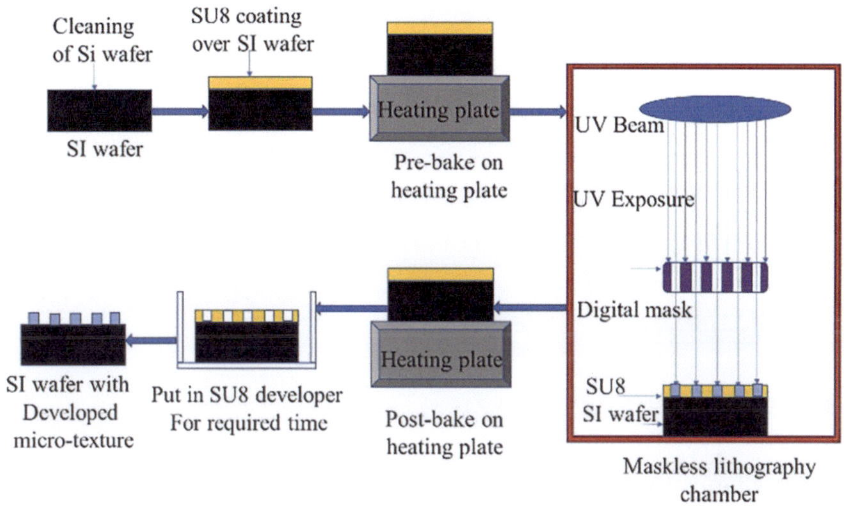

Fig. 2 Schematic diagram for the development of micropillared surfaces through maskless lithography[16]

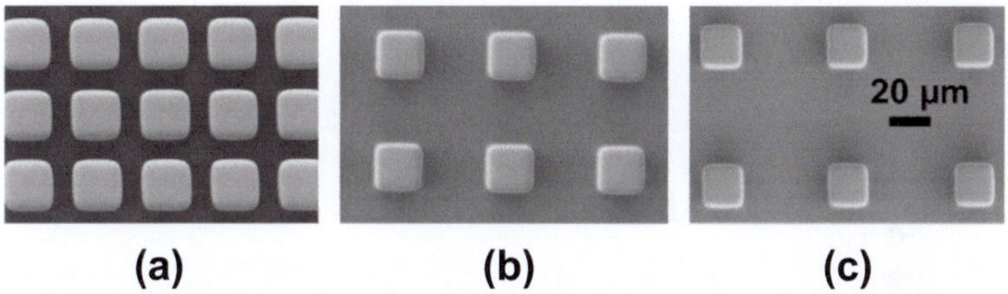

Fig. 3 Sample FESEM images of square micropillars of **a** 30, **b** 47, and **c** 62 μm pitches

UX50, Photron) for the side visualization of impacting droplets and screw-controlled leveling platform (i.e., base).

Droplets of deionized water, measuring approximately 1.7 ± 0.01 mm in diameter, are produced by a syringe pump equipped with a 31-gauge needle syringe. The droplet impact velocity (V) ranging from 0.22 to 0.5 m/s is managed with manipulation of syringe height with respect to substrate. The uncertainty in experimental measurement associated with this variation is ±0.01 m/s. The surfaces with micropillars undergo a thorough cleaning process using isopropanol, ensuring meticulous removal of any contaminants. Subsequently, they are left to air-dry completely. This experimental setup is similar to the one described in the references [15]. For the experiment, a white LED light serves as backlight source, offering the required illuminations. The high-speed camera captures images at a resolution of 768 × 768 pixels, with a recording speed of 3200 frames per second. After recording the videos, the data was post-processed using the ImageJ software, specifically for extracting

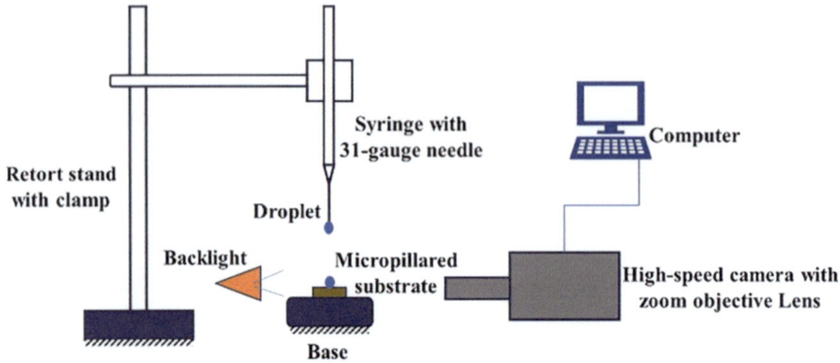

Fig. 4 Schematic diagram of experimental setup

relevant information from the images. This enabled the determination of various droplet related parameters such as droplet diameter, contact angle, wetted diameter, volume, and height by analyzing the pixel data obtained from the instantaneous images. Throughout the experiments, the ambient temperature was kept at 22 ± 3 °C, while the relative humidity was kept at 35 ± 5%. To ensure the reliability and reproducibility of the results, each experiment was conducted a minimum of three times.

4 Results and Discussion

4.1 Effect of Pitch of the Pillars

In this section, the water droplets dynamics impacting on micropillared substrate in primary (Fig. 5) and binary (Fig. 6) mode have been presented. The transient evolution of the impacting droplets captured using high-speed visualization has been analyzed. In the first part of our analysis, Fig. 5 displays the side depicting images of the primary water droplets impacting at 0.22 m/s velocity for different time instants. Based on our findings, it is observed that corresponding to the pitch values of 30, 34, 40, 47, and 62 μm, non-bouncing regime prevailed for all these cases. However, a bouncing regime is found to exist for the droplet impacting on the micropillared substrate having pitch of 76 μm. These observations are found to agree well with the findings given in the literature [15], which validates the present single droplet impact dynamics on the fabricated substrates.

Now, the experimental findings pertaining to binary droplet (i.e., droplet-on-droplet) impact have been demonstrated. Here, it is to be noted that the secondary droplet is impacted on the already existing impact-deposited primary sessile droplet

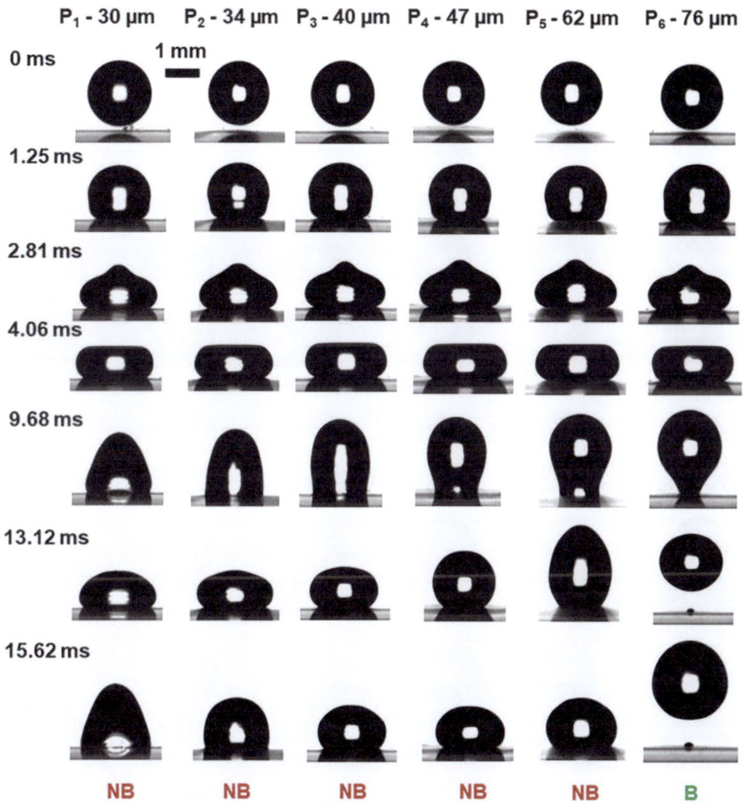

Fig. 5 Visualization of the instantaneous interfaces through high-speed photography, during primary water droplet impact on SU8 micropillared substrate with a velocity of 0.22 m/s (NB—non-bouncing, B—bouncing)

as shown earlier in Fig. 1. Further, as per the experimental observation, the impact-deposited primary droplet undergoes Wenzel transition on the micropillared surface [15]. Consequently, the droplet fluid penetrates the space between the pillars leading to the pinning of the CL. It has been observed that the primary droplet's wetted diameter (D_{wetted}) existing in the Wenzel state is lower on micropillared substrates having a smaller pitch and is higher on micropillared substrates having a larger pitch and this wetted diameter changes during impact of secondary droplets for 40, 47, 62, and 76 μm pitch while almost constant for 30 and 34 μm pitch (see dashed lines as shown in Fig. 7). This is primarily due to the increased resistance offered by the micropillars toward spreading in the case of a surface with a lower-pitch value. Since it will have higher pillar density causing increased friction. Thus, restricting the CL motion and giving rise to a lower-wetted diameter.

In the case of impact-deposited droplets on micropillared surfaces, there is a distinction regarding the equilibrium contact angle at different pitch values. For the surfaces with a higher pitch, a reduction in the contact angle is manifested and

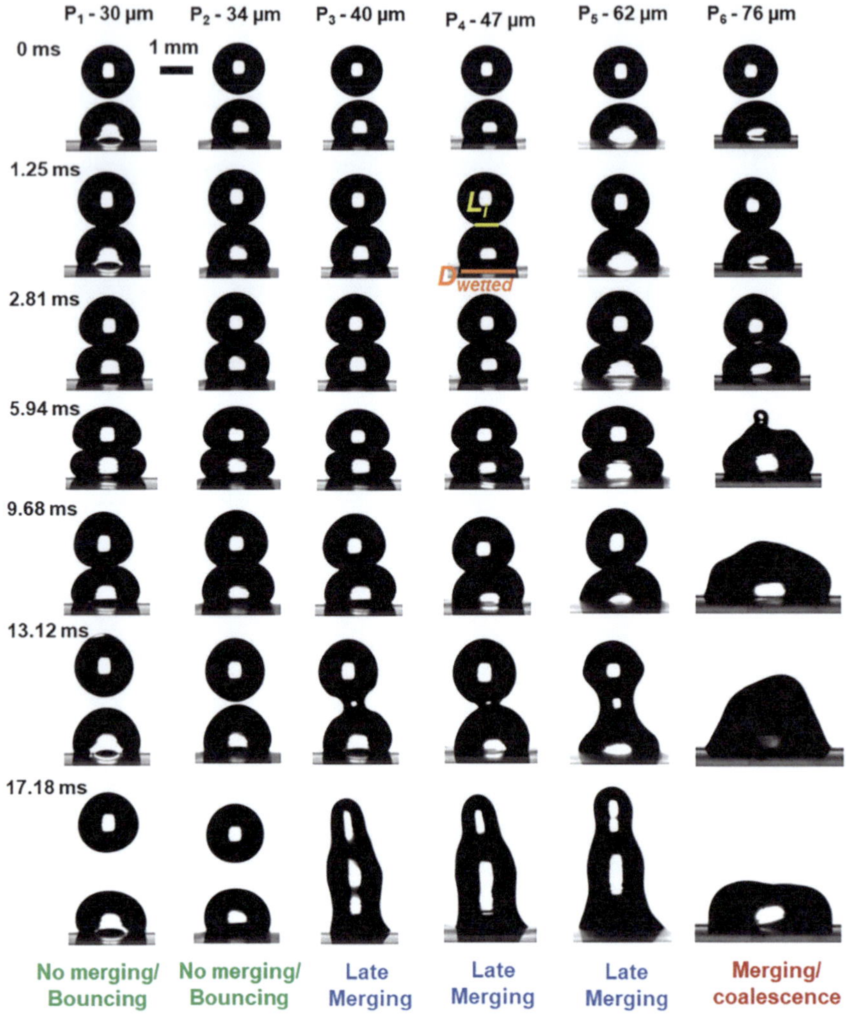

Fig. 6 Visualization of the instantaneous interfaces by high-speed photography, during binary water droplet impact on SU8 micropillared surfaces with a velocity of 0.22 m/s

vice-versa for lower pitch. So, for the case of secondary droplet impact on primary sessile droplets on surfaces with low pitch values (30 and 34 μm), bounce-off regime (i.e., no merging of secondary droplet with primary droplet) is prevailed. As per the observations, it is found that the dynamic variation of the Laplacian pressure caused due to local changes in the interfacial curvature leads to higher surface energy [17]. For the case of micropillared surfaces with other pitch values (40, 47, 62, and 76 μm), merging of the secondary droplet with primary droplet is existed in all cases. However, there lies a difference in terms of their merging times which have been classified as late merging (for 40, 47, and 62 μm) and gentle merging (for 76 μm) regimes.

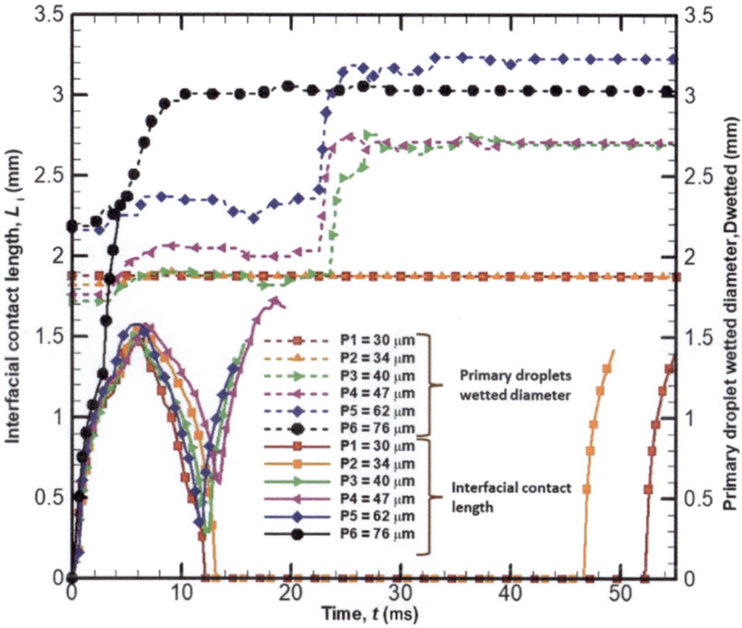

Fig. 7 Temporal variation of apparent interfacial contact length of two droplets for 30, 34, 40, 47, 62, and 76 μm pitch substrates

Interestingly, despite the existence of almost identical initial droplet contact angle values for the surfaces with 34 and 40 μm pitch, a distinctive bouncing and non-bouncing behavior, respectively, is prevailed, which demarcates the two cases. This is caused due to the strong pinning of the CL in the case of the surface with 34 μm pitch leading to higher contact angle of primary droplet as compared to that of the 40 μm pitch. Further, this can be noted that, even after secondary droplet impacts on the primary droplet's apex region, the primary droplet's wetted diameter on lower-pitch surface is almost pinned. While on higher-pitch surface, the primary droplet's wetted diameter spreads further or CL moves further when the secondary droplet head-on collision occurs. In short, here we observe that for lower-pitch surface, the primary droplet acts like an elastic body with fixed ends in comparison to that the higher-pitch surface case. In addition to these findings, it can be also hypothesized that when the secondary droplet interacts the primary droplet, the higher curvature of the interface at primary droplet's apex region do exist for the case of lower-pitch surface in comparison to that of the higher-pitch surface. Because of this, the secondary droplet head-on collision consequences in the formation of a depression near the apex area of the primary droplet, which may cause air entrapment [18] there and an increase in the dynamic pressure of the entrapped air between the apex of the primary droplet and the bottom surface of the secondary. This air acts as a cushion and the secondary droplet bounces-off without merging with the primary droplet, while in case of higher-pitch surface, the curvature of the interface at primary droplet's apex region may not withhold such aircushion and thereby late/gentle merging is observed.

Therefore, when the secondary droplets impact the primary droplet on lower-pitch surfaces with an impact velocity of 0.22 m/s, the secondary droplet bounces-off due to the elastic behavior of primary droplet or we can say that the entrapped air between the droplets is not removed completely so the secondary droplet is not able to puncture the interface of primary droplet. Further, as per this observation, the droplet-on-droplet impact on micropillared surfaces for different pitch values is categorized to exhibit no merging, late merging, and gentle merging behavior. The secondary droplet shows late merging for 40, 47, and 62 μm, and gentle merging for 76 μm pitch values.

Further, we also study the binary droplet impact dynamics behavior on all these micropillared substrates with an impact velocity of 0.5 m/s. It shows no merging/bouncing for 30–40 μm pitch, late merging for 47 μm pitch, and gentle merging 62–76 μm pitch. For the sake of brevity, we have not presented instantaneous variation of their impact dynamics.

4.2 *Interfacial Contact Length Between Two Droplets*

To quantify the binary droplet impact dynamics, we define an apparent interfacial contact length (as marked in Fig. 6).

Figure 7 shows the variation of interfacial contact length between two droplets with respect to time. In figure, dashed lines and solid lines show the time-based variation of primary droplet's wetted diameter and the apparent interfacial contact length between the two droplets, respectively, for surfaces of different pitch (30, 34, 40, 47, 62, and 76 μm). The apparent interfacial contact length increases to L_i ~ 1.5 mm till t ~ 6 ms for all surfaces (except 76 μm). Thereafter, from t ~ 6 to 12 ms, for 30 and 34 μm surfaces, the interfacial contact length decreases till the value of L_i becomes zero, which indicates that bounce-off takes place; for 40, 47, 62 μm surfaces, the interfacial contact length decreases till the value of L_i reaches ~0.3 mm and after that suddenly it increases, which indicates that late merging is occurred, while for 76 μm surface, the interfacial contact length increases monotonically from $t = 0$ to 6 ms, which indicates that the gentle merging is occurred.

5 Conclusions

The study presents the binary droplet impact dynamics behavior for impact deposition condition on the square-micropillared substrates, which is scarce in literature. We vary the pitch of pillars and impact velocity to understand the dynamic behavior. The qualitative and quantitative analysis show that three different regimes are occurred. The droplet-on-droplet impact dynamics is substantially influenced by the initial boundary condition of the primary droplet deposition on the substrate. At 0.22 m/s impact velocity, the secondary droplet, bounces-off on lower-pitch surfaces 30 and 34 μm, lately merges on intermediate pitch surfaces 40, 47, and 62 μm, and gently merges on highest pitch surfaces 76 μm. This was attributed mainly due to the pinned CL and higher curvature of primary droplet on lower-pitch surfaces in comparison to that of the higher value pitch surfaces. These fundamental findings can be helpful in effectively minimizing the requirement of pesticides and paints in spray pesticides and spray-painting applications, respectively.

Acknowledgements N.D.P. gratefully acknowledges the financial support provided by the SPARC (Scheme for Promotion of Academic and Research Collaboration) scheme from the MoE (Ministry of Education), Government of India, by Grant number SPARC/2019-2020/P2502/SL, (Title of Project- Droplet Impact Dynamics on Substrates with Spatial Wettability Gradient: Experimental and Numerical Study). N.D.P. also acknowledges the financial support from the Science and Engineering Research Board (SERB), Department of Science and Technology, Government of India, New Delhi, by the grant number SRG/2020/001947, for developing the experimental setup used in the present work. The authors are thankful to the Central Instrumentation Facility (CIF), IIT Bhilai, for providing access to the scanning electron microscope (SEM) for imaging of textured substrates.

Nomenclature

V	Velocity of binary droplet (m/s)
D_o	Diameter of droplet (mm)
H	Height between droplets (m)
L_i	Interfacial contact length between the droplets (mm)
P	Pitch of the pillars (μm)
a	Width of the pillars (μm)
h	Height of the pillars (μm)
t	Time (ms)
θ_{eq}	Equilibrium contact angle (°)
We	Weber number
D_{wetted}	Primary droplet wetted diameter (mm)

References

1. Rabbi R, Kiyama A, Allen JS, Truscott T (2022) Droplet lift-off from hydrophobic surfaces from impact with soft-hydrogel spheres. Commun Phys 5(1):1–7. https://doi.org/10.1038/s42005-022-01114-8
2. Bergeron V, Bonn D, Martin JY, Vovelle L (2000) Controlling droplet deposition with polymer additives. Nature 405(6788):772–775. https://doi.org/10.1038/35015525
3. Singh HH, Khare N (2021) KPFM study of flexible ferroelectric polymer/water interface for understanding the working principle of liquid–solid triboelectric nanogenerator. Adv Mater Interf 8(12):1–7. https://doi.org/10.1002/admi.202100032

4. Jaiswal AK, Khandekar S (2021) Dynamics of a droplet impacting a sessile droplet on a superhydrophobic surface: role of boundary conditions during droplet placement. J Flow Vis Image Process 28:4
5. Graham PJ, Farhangi MM, Dolatabadi A (2012) Dynamics of droplet coalescence in response to increasing hydrophobicity. Phys Fluids 24(11):513. https://doi.org/10.1063/1.4767513
6. Batzdorf S et al (2017) Heat transfer during simultaneous impact of two drops onto a hot solid substrate. Int J Heat Mass Transf 113:898–907. https://doi.org/10.1016/j.ijheatmasstransfer.2017.05.091
7. Kapur N, Gaskell PH (2007) Morphology and dynamics of droplet coalescence on a surface. Phys Rev E Stat Nonlinear Soft Matter Phys 75(5):25–28. https://doi.org/10.1103/PhysRevE.75.056315
8. Wang FC, Feng JT, Zhao YP (2008) The head-on colliding process of binary liquid droplets at low velocity: high-speed photography experiments and modeling. J Colloid Interf Sci 326(1):196–200. https://doi.org/10.1016/j.jcis.2008.07.002
9. Wakefield J, Tilger CF, Oehlschlaeger MA (2016) The interaction of falling and sessile drops on a hydrophobic surface. Exp Therm Fluid Sci 79:36–43. https://doi.org/10.1016/j.expthermflusci.2016.06.022
10. Jaiswal AK, Khandekar S (2021) Drop-on-drop impact dynamics on a superhydrophobic surface. Langmuir 37(43):12629–12642. https://doi.org/10.1021/acs.langmuir.1c01779
11. Damak M, Varanasi K (2018) Expansion and retraction dynamics in drop-on-drop impacts on nonwetting surfaces. Phys Rev Fluids 3(9):1–12. https://doi.org/10.1103/PhysRevFluids.3.093602
12. Gunjan MR, Somwanshi P, Agrawal A, Khandekar S, Muralidhar K (2015) Recoil_of_drops_during_coalescence 3(2):203–220
13. Somwanshi PM, Muralidhar K, Khandekar S (2020) Coalescence of vertically aligned drops over a superhydrophobic surface. Phys Fluids 32(5):20–54. https://doi.org/10.1063/5.0007419
14. Ramírez-Soto O, Sanjay V, Lohse D, Pham JT, Vollmer D (2023) Lifting a sessile oil drop from a superamphiphobic surface with an impacting one. Sci Adv 6(34):eaba4330. https://doi.org/10.1126/sciadv.aba4330
15. Patil ND, Bhardwaj R, Sharma A (2016) Droplet impact dynamics on micropillared hydrophobic surfaces. Exp Therm Fluid Sci 74:195–206. https://doi.org/10.1016/j.expthermflusci.2015.12.006
16. Yadav MK, Patil ND (2024) Water droplet impact dynamics comparison on solid and hollow square micropillared substrates. Phys Fluids 36(3). https://doi.org/10.1063/5.0191814
17. Gennes P-G, Brochard-Wyart F, Quéré D, Reisinger A, Widom B (2004) Capillarity and wetting phenomena: drops, bubbles, pearls, waves. Phys Today Phys Today 57:66–67. https://doi.org/10.1063/1.1878340
18. Sykes TC et al (2020) Surface jets and internal mixing during the coalescence of impacting and sessile droplets. Phys Rev Fluids 5(2):1–21. https://doi.org/10.1103/PhysRevFluids.5.023602

Fluid-Structure Interaction of a Collapsible Tube Under the Effect of Buoyancy

R. Aashish, P. Deepu, Kamlesh Kumar, and Subrata Kumar

Abstract Flow through collapsible tubes holds significant physiological importance, which can be widely seen in our body, especially arteries, veins (thin-walled and highly adaptive blood vessels), vocal cords (behaves like an airline), urethra (acts as thin-walled elastic tube), pulmonary capillaries, etc., and finds widespread application in various industries as well. In this study, we focus on the dynamics of a thin collapsible tube subjected to internal airflow and its interaction with the surrounding fluid medium. The tube response exhibits varied oscillatory behavior across a range of Reynolds numbers (Re), showing bifurcating patterns. The free water surface also shows similar oscillatory features in response to the disturbances induced by the oscillations of the tube. Our research contributes valuable insights into the dynamics of a collapsible tube and its interaction with fluid flow, which could have far-reaching implications in physiological research and diverse industrial sectors such as petrochemical, pharmaceutical, nuclear engineering.

Keywords Dynamics · Self-excited oscillation · Coupling · Free surface

R. Aashish (✉) · P. Deepu · K. Kumar · S. Kumar
Department of Mechanical Engineering, Indian Institute of Technology Patna, Patna, Bihar, India
e-mail: aashish_2221me19@iitp.ac.in

P. Deepu
e-mail: deepu@iitp.ac.in

K. Kumar
e-mail: kamlesh_1821me03@iitp.ac.in

S. Kumar
e-mail: subrata@iitp.ac.in

A. Kumar R. et al. (eds.), *Proceedings of Fluid Mechanics and Fluid Power (FMFP) 2023, Vol. 6*, Lecture Notes in Mechanical Engineering,
https://doi.org/10.1007/978-981-96-9600-0_12

Nomenclature

Q	Air flow rate (lpm)
Re	Reynolds number
ρ	Density of air (kg/m^3)
$\bar{V}$	Average velocity air flow (m/s)
D	Diameter of rigid tube (m)
μ	Viscosity of air (Pa-s)
KE	Average kinetic energy of air flow (J)
TP	Transmural pressure (Pa)

1 Introduction

The development of modern non-linear dynamics including chaotic theory has widened the scope of the study of collapsible tube dynamics. Due to this advancement, the self-excited oscillations of a collapsible tube conveying a fluid have become a topic of great importance in the area of research because of its relevance to biomedical and engineering applications as well as its fascinating and rich dynamical behavior [1–3]. Flow through collapsible tubes is of great importance in both physiological contexts within the body and across various industrial applications. Physiologically, these tubes are prevalent in crucial anatomical structures such as arteries, veins, vocal cords, urethra, and pulmonary capillaries, playing vital roles in biological processes. The dynamic behavior of the collapsible tube shows relevance to medical conditions like asthma wheezing, snoring in obstructive sleep apnea, and Korotkoff sounds in sphygmomanometry. In the industrial sector, flow-induced oscillations in elastic tubes were investigated to enhance the membrane flux [4, 5] and for micro-filtration processes [6].

Even after the significantly large number of studies so far been done on this topic but still, a comprehensive study of the fluid-structure interaction mechanism underlying these oscillations remains elusive. Most of the studies so far have been made by choosing the combination of external and internal fluids in such a way that the effect of buoyancy on the oscillations can be neglected. Some of the researchers used water as both external and internal fluid [7], while others used air as both [8]. In the present study, however, we are deliberately incorporating the effects of the buoyancy on the tube oscillations by choosing different external (water) and internal (compressed air) fluids to offer valuable insights into the behavior of the collapsible tube under the effect of buoyant force. In addition, we seek to explore the coupling between tube oscillations and the corresponding response of the free surface to it.

2 Experimental Setup

Figure 1 illustrates a detailed schematic of our experimental setup. We conduct a series of experiments over a range of Reynolds numbers (Re) between 0 and 2104 on a Starling resistor, measuring 300 mm in length, 115 mm in width, and 120 mm in height. It consists of an open rectangular channel made from a transparent acrylic sheet, filled with water to a depth of 95 mm above the base. The two ends of the channel are provided by a pair of rigid tubes of 33.5 mm diameter. These tubes are connected internally by a flexible tube, a commercially available condom (about 70 μm thick), to facilitate compressed air flow. The rate of airflow (Q) traversing the conduit is measured using a flow meter in liter per minute (lpm), positioned near the upstream end of the tube. We calculate Re using the relation $\mathrm{Re} = \rho \bar{V} D/\mu$, where ρ and μ represent the density and dynamic viscosity of the air at room temperature, respectively, $\bar{V} = 4Q/\pi D^2$ is the average velocity through the rigid tube, while D is its diameter. In addition, a valve is also installed on the flow meter, which allows us to regulate the airflow. This configuration ensures the controlled and measured airflow through the tube, and hence, we can calculate the Re, which acts as the controlling parameter in our experiments. The system response, encompassing both the tube behavior and the corresponding free surface oscillation, is captured using a high-speed camera, MIRO LAB 3a10, which operates at a frame rate of 100 frames per second (fps) with a full resolution of 1280 × 1280 pixels. The acquired images are analyzed using an image processing technique in MATLAB to understand the dynamics of the collapsible tube and its interaction with the free surface oscillation, which will be further elucidated in the subsequent section.

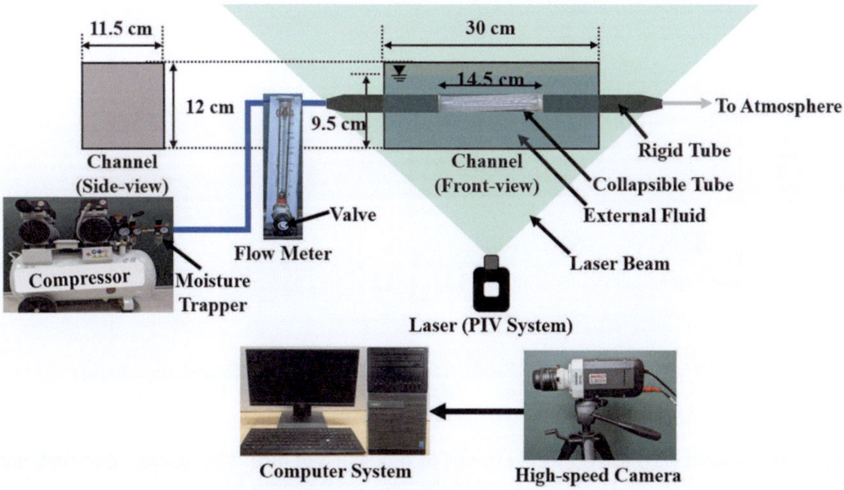

Fig. 1 A detailed schematic of our experimental setup

3 Results and Discussion

The response of the flexible tube (Y_{tube}) is quantified by measuring its mid-point distance from the centerline, as shown in the left top inset of the bifurcation diagram (Fig. 2). This diagram depicts a comprehensive representation of the different dynamics observed in our study by carrying out a bifurcation analysis. This figure plots the peak values of Y_{tube} against the Reynolds number (Re) on the primary y-axis and Y_{fs} on the secondary y-axis over time. The tube dynamics are distinguished into three zones namely, steady collapsed (SC), oscillatory regime (OR), and steady distended (SD). At lower Re values (Re $\leq$ 715), the tube undergoes a significant negative transmural pressure (TP), which denotes pressure difference ($P_{in} - P_{out}$) across its walls. This causes the tube to collapse or narrow, forming a neck corresponding to the minimum cross-section area. The tube collapses entirely when there is no flow through it.

However, the pressure inside the tube rises when no fluid flows through it, particularly at the upstream end, leading to an increase in TP. On the other hand, the pressure drops toward the downstream end due to viscous losses. As a result, the tube undergoes expansion from its upstream end, causing the formation of the neck at its downstream end. This state of the tube is known as the neck at downstream (ND), which is categorized within the SC zone and is marked by a single point in the bifurcation diagram. In addition, when the flow through the tube begins, its volume increases, leading to the action of buoyancy forces on it. This results in the bending of the tube (hogging), causing the tube to oscillate around a curved axis, unlike the horizontal oscillations observed without buoyancy effects [8].

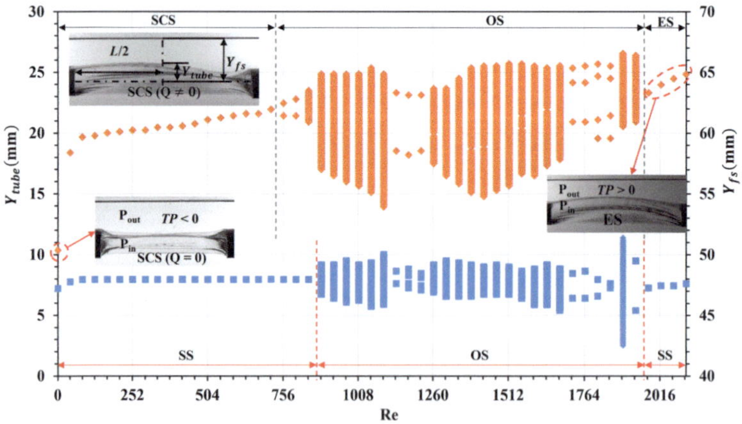

Fig. 2 Bifurcation diagram: the local extrema of oscillation of tube, Y_{tube} (orange diamond) and of free surface oscillation, Y_{fs} (blue square) in the time series against Re

As Re surpasses a critical value of approximately 757, the tube transitions from SC to OR. Previous studies noted self-sustained oscillations starting at a critical Re of around 100, but those experiments used a relatively thicker wall tube (with a thickness-to-radius ratio of 0.37 [9], while this ratio is 0.004 in the current study) in an air-pressurized channel. The average kinetic energy (KE) of the flow inside the tube increases at higher Re. The KE of the pressurized airflow is converted into pressure energy near the downstream neck. This phenomenon leads to increased pressure at the upstream end of the tube, exceeding the opposing pressure at the neck and inducing a negative TP throughout the tube length as the neck opens, releasing the surged pressure. This allows the neck to form at the upstream end, leading to a configuration known as the neck at upstream (NU). This initiates a cycle where the neck oscillates along the tube, resulting in self-sustained tube oscillations. The characteristic behavior of the OR mostly depends on the nature of the movement of the neck position.

We employ the image processing technique in MATLAB to get the time-series data of Y_{tube}. Following this, we perform the spectral analysis using the Fast Fourier Transform (FFT) algorithm to get the frequency spectra and the phase-plane trajectory. This analysis allows us to further classify the OR based on distinct dynamical behaviors, such as limit cycle oscillations, weak and strong chaotic oscillations, and period-doubling bifurcations.

As the flow rate is further increased beyond another threshold value of Re, the internal pressure rises sufficiently, leading to a positive TP throughout the tube length. As a result, the tube fully expands or widens. This state of the tube is referred to as the steady distended (SD) zone, depicted by a point in the bifurcation diagram.

The corresponding response of the free surface, denoted as Y_{fs}, is similarly quantified by measuring its mid-point displacement from the same centerline (refer to left top inset of Fig. 2). The response of free surface response is categorized into two states: steady state (SS) and oscillatory regime (OR). To distinguish between different dynamics in the OR, a similar image processing technique is employed, and the various dynamics so obtained are presented on the same bifurcation diagram on the secondary y-axis (Fig. 2) as mentioned earlier. The free surface remains unresponsive corresponding to the SC and SD configuration of the tube, as expected. This condition is referred to as the steady state (SS), illustrated by a single dot in the bifurcation diagram. Near the Hopf bifurcation point of the tube response, although the tube undergoes self-excited oscillation, its amplitude is so low that the disturbances created by its oscillation could not reach the free surface, resulting in the SS of the free surface. This causes a relatively delayed onset of the oscillation of the free surface as compared to that of the tube.

The oscillatory response of the tube is predominantly chaotic, occasionally showing period-1 and period-2 oscillation. A very small amplitude period-1 oscillation occurs in the Re ranging between 757 and 799, while in the range of 1136–1220, it exhibits significantly higher amplitude, which is depicted by a pair of points in the bifurcation diagram. The periodic oscillation of the tube is a result of the periodic motion of the neck position (ND to NU and vice-versa) along the tube length. In addition, period-2 oscillation is also observed in the Re range of 1725–1851, as illustrated

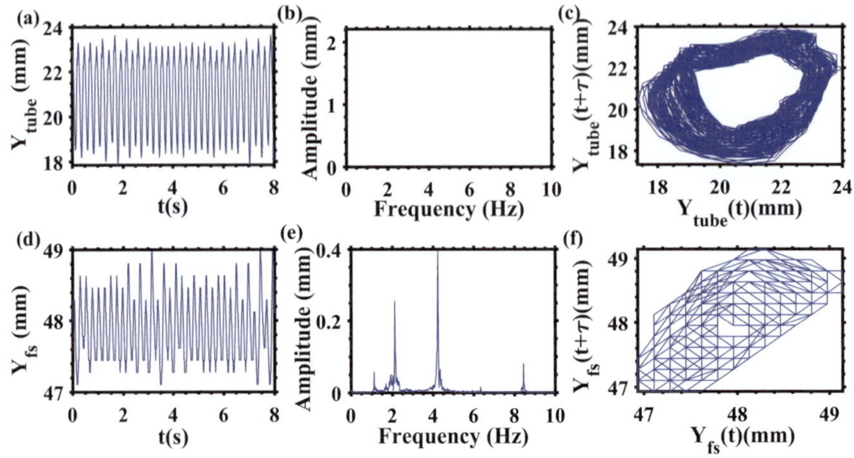

Fig. 3 Time series (**a**, **d**), frequency spectra (**b**, **e**), and phase-plane trajectory (**c**, **f**) corresponding to the periodic oscillations of the tube (top) and the free surface (bottom) at Re = 1178

in Fig. 2 by either three points or four points. During this case, the neck undergoes two distinct movements: one with small displacement between ND to NU near the mid-section of the tube) and the other with larger displacement, extending the NU beyond the mid-section within a single period, which repeats over time. The time-series data, frequency spectra, and the phase-plot path for the period-1 oscillation of the tube at Re = 1178 are shown in Fig. 3. The time-series plot (Fig. 3a) shows a consistent periodic variation in amplitude, indicating the periodic tube response. The associated frequency spectra (Fig. 3b) exhibit a sharp peak around 4.1 Hz, and the phase-plane path (Fig. 3c) forms a closed loop, further supporting the period-1 oscillation. It is important to highlight that to avoid the potential errors associated with the numerical differentiation techniques, we employ the delay coordinate method [10] for plotting phase-plane trajectories. This method involves arbitrarily choosing a delay parameter, denoted as τ, ensuring its topological character remains consistent within a certain range of the selected value. The free surface response at Re = 1178 also shows low-amplitude periodic oscillation, as seen in its time-series plot (Fig. 3d). Its respective frequency spectra Fig. 3e exhibit two major peaks: one corresponds to the forced frequency, equal to the tube oscillation frequency, and the other represents the sub-harmonic frequency, indicating a robust interplay between the system responses. The small amplitude of the free surface oscillation results in the phase-plane trajectory (Fig. 3f) not forming a regular closed loop but rather a filled, erratically stepped area.

On the other hand, the tube performs a strong chaotic oscillation (SCO) within the Re range approximately from 841 to 1094 and 1893 to 1935, while weak chaotic oscillation (WCO) occurs nearly between 1262 and 1682. In Fig. 4, we present the

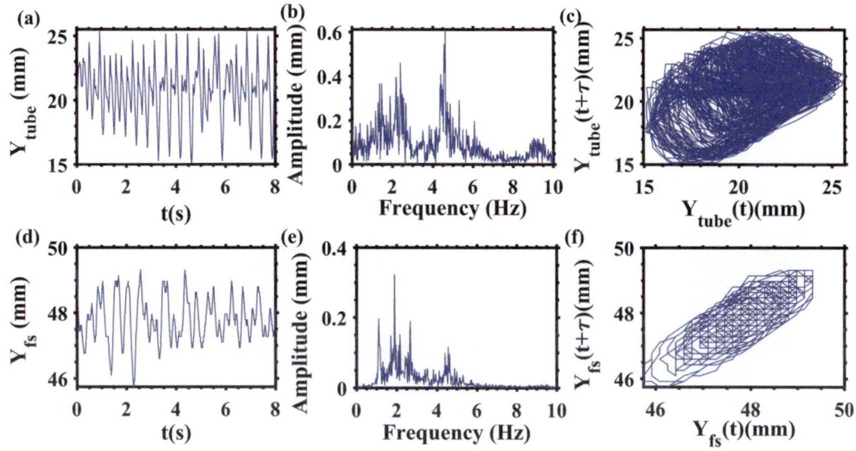

Fig. 4 Time series (**a**, **d**), frequency spectra (**b**, **e**), and phase-plane trajectory (**c**, **f**) corresponding to the chaotic oscillations of the tube (top) and the free surface (bottom) at Re = 1052

time-series plot, frequency spectra, and the phase-plot trajectory for a chaotic behavior of the tube oscillation at Re = 1052. The time-series plot (Fig. 4a) shows an irregular and non-repeating variation in amplitude over time, while the frequency spectra (Fig. 4b) exhibits significant amplitude across a range of frequencies, confirming the chaotic nature of the tube response. The phase-plane path (Fig. 4c) shows a filled and dense zone, further supporting the chaotic behavior of the tube oscillation. The corresponding free surface response also demonstrates a chaotic response, evident from its aperiodic time-series plot (Fig. 4d). The frequency spectra (Fig. 4e) and the phase-plane trajectory (Fig. 4f) show a similar nature as the tube response. The chaos in the oscillatory behavior is possibly caused by the internal pressure gradient along the tube, where viscous forces cause a pressure drop from the inlet to the outlet. This gradient may lead to flow separation at the downstream neck, which in turn could trigger vortex shedding. The interaction of these unsteady forces with the inherent periodic oscillations of the tube might result in an irregular, non-repetitive movement pattern between ND and NU configurations. Thus, the tube performs chaotic oscillation which is represented by a large number of points (theoretically infinite but limited by the time domain) in the bifurcation diagram (Fig. 2).

It is interesting to note that for chaotic responses, the dominating free surface frequency no longer coincides with the peak frequency of the tube oscillation. This observation indicates that the system responses show a perfect coupling during periodic oscillations, while they do not exhibit such coupling during chaotic oscillations. Further analysis of this system behavior is presented in Fig. 5, plotting their respective frequency spectra against the range of Re. As seen in Fig. 5b, the free surface predominantly oscillates at a frequency close to its fundamental or second harmonic frequency [11] at relatively lower Re. On the other hand, at higher Re, the frequency response of the free surface includes frequencies that match both the dominant tube

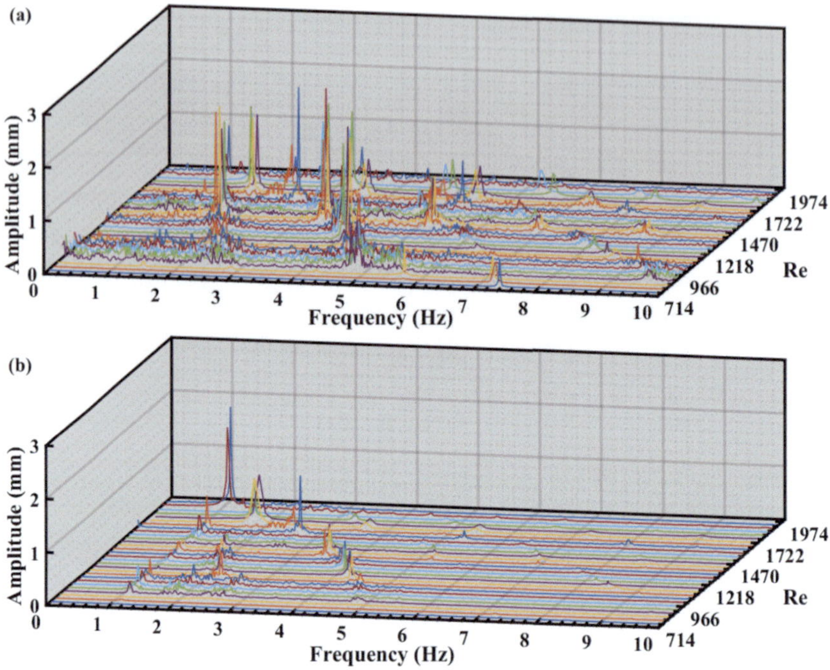

Fig. 5 Frequency spectra of **a** tube oscillations, and **b** free surface corresponding to the oscillatory regime

frequency and its inherent frequency. Additionally, Fig. 5a shows a significant shift in the dominant tube frequency from higher to lower values with increasing Re. The observed trend can be understood by considering the inlet pressure and flow kinetic energy (KE), which are lower at reduced Re. As the flow nears the downstream neck, a quicker shift from ND to NU configuration occurs due to the diminished inlet pressure, leading to a rapid onset of negative transmural pressure (TP) and, consequently, a higher oscillation frequency. On the flip side, at elevated (Re), the greater inlet pressure results in a more gradual flow adjustment, slowing the ND to NU transition and thus decreasing the oscillation frequency. Additionally, a notable increase in the amplitude of the free surface oscillation is observed when the forcing frequency is close to the fundamental free surface frequency due to resonating effects.

4 Conclusions

Our study investigates the dynamics of a collapsible tube subjected to compressed air flowing through it under the influence of the buoyant force. Initially, at lower

Reynolds numbers (Re), the tube maintains a steady collapsed (SC) state. As Re crosses a critical threshold, the tube transitions into an oscillatory regime (OR), exhibiting a spectrum of behaviors. This includes the occasional occurrence of periodic-1 oscillations of varying amplitude and period-doubling bifurcation, along with chaotic oscillations of different strengths, observed for a majority of Re. Further, at lower flow rates, the dominant frequency of the free surface oscillation is either aligned to the fundamental frequency or second harmonic frequency, indicating no coupling with the tube oscillation. On the other hand, its frequency spectra at higher Re primarily include two frequencies, one equal to the dominant tube frequency and the other equal to its natural frequency, observed in periodic or weak chaotic cases. Hence, the synchronization between the free surface and the tube oscillation becomes more pronounced at the higher Re for such cases. An increase in flow rate at high Re augments the tube volume, enhancing the buoyant force exerted on it, which in turn causes the tube to bulge out and move closer to the free surface. This proximity facilitates the transmission of the oscillatory disturbances from the tube to the free surface, leading to a high amplitude oscillation of the free surface. However, in the presence of chaotic or aperiodic tube responses, this synchronization is notably subdued. Furthermore, when the tube oscillation frequency approaches the natural frequencies (fundamental or second harmonic) of the free surface oscillation, a pronounced periodic oscillation with substantial amplitude is observed on the free surface because of resonance.

Acknowledgements D.P. gratefully acknowledges the financial support received from the Department of Science and Technology, Govt. of India under the DST-FIST Program (Grant No. SR/FST/ET-II/2018/240(C)) for the acquisition of PIV (for flow field measurement) system.

References

1. Hai-Jun W, Jia L-B, Yin X-Z (2015) Experiments on self-excited oscillation in a thin-walled collapsible tube. Acta Mech Sin 31:817–826
2. Anderson P, Fels S, Green S (2016) Experimental characterization of the oscillatory behavior of a quasi-two-dimensional collapsible channel. J Fluids Struct 66:254–268
3. Podoprosvetova A, Zayko J, Yushutin V, Vedeneev V (2021) Experimental study of the flow regime effect on the stability of collapsible tubes conveying fluid. Phys Fluids 33(6):064104
4. Stairmand JW, Bellhouse BJ, Jamal Z, Lewis RW, Urban JP, Entwistle CC (1986) Separation of plasma from whole blood by membrane filtration in oscillatory flows. Life Support Syst J Eur Soc Artif Organs 4(3):193–204
5. Rodgers VGJ, Sparks RE (1993) Effects of solution properties on polarization redevelopment and flux in pressure pulsed ultrafiltration. J Membr Sci 78(1–2):163–180
6. Bertram CD, Hoogland MR, Li H, Odell RA, Fane AG (1993) Flux enhancement in crossflow microfiltration using a collapsible-tube pulsation generator. J Membr Sci 84(3):279–292
7. Kounanis K, Mathioulakis DS (1999) Experimental flow study within a self oscillating collapsible tube. J Fluids Struct 13(1):61–73
8. Kumar K, Prabhakaran D (2022) Dynamics of a collapsible tube with internal constriction. Phys Fluids 34(12):121905

9. Bertram CD, Castles RJ (1999) Flow limitation in uniform thick-walled collapsible tubes. J Fluids Struct 13(3):399–418
10. Takens F (2006) Detecting strange attractors in turbulence. In: Dynamical systems and turbulence, Warwick 1980: proceedings of a symposium held at the University of Warwick 1979/80. Springer, pp 366–381
11. Kundu PK, Cohen IM, Dowling DR, Capecelatro J (2015) Fluid mechanics. Academic Press

Numerical Study of the Flow Characteristics of Wings with Low Aspect Ratio at Low Reynolds Number

Dattatraya S. Kulkarni, B. N. Rajani, and Vinayak Kulkarni

Abstract The in-house incompressible flow code 3D-PURLES has been used to simulate the effect of wing aspect ratio on the aerodynamic and transition characteristics at $Re = 6 \times 10^4$ based on the root chord (C) of the wing. The SD7003 rectangular wing non-dimensionalized with respect to the root chord and having different semi aspect ratio (sAR) of 0.25, 0.5 and 1 are considered for this study. The wing aspect ratio is found to significantly affect both the transition phenomenon and the aerodynamic characteristics. The transition onset for wing with lowest sAR is found to be delayed and having a longer transition zone. The aerodynamic characteristics of this wing in terms of the stall angle, maximum lift coefficient and drag coefficient are found to be superior making it a suitable candidate for MAV applications operating at this low Reynolds number regime. The influence of Reynolds number in the low Re regime on the performance of the SD7003 rectangular wing with 0.25 a semi aspect ratio is also discussed in this paper.

Keywords Incompressible flow · Pressure-based finite volume method · $\gamma - Re_\theta$ SST transition model · Aerodynamic performance · Transition onset

1 Introduction

The fascinating and complex physics involved in the low Reynolds number flow has lead to a continuous research to understand and predict these complex features. This low Reynolds number flow finds application in flying objects found both in nature

---

D. S. Kulkarni · B. N. Rajani
CTFD Division, CSIR-NAL Bengaluru, Bengaluru, Karnataka, India
e-mail: rajani@nal.res.in

D. S. Kulkarni (✉) · V. Kulkarni
Department of Mechanical Engineering, IIT Guwahati, Guwahati, Assam, India
e-mail: kulkarni@nal.res.in

V. Kulkarni
e-mail: vinayak@iitg.ac.in

A. Kumar R. et al. (eds.), *Proceedings of Fluid Mechanics and Fluid Power (FMFP) 2023, Vol. 6*, Lecture Notes in Mechanical Engineering,
https://doi.org/10.1007/978-981-96-9600-0_13

(insects, birds and bats) and man-made aerial vehicles. The man-made aerial vehicles like MAVs, UAVs and NAVs are generally characterized by small size which operate in the lower Reynolds number flight regime ($10^4 - 10^5$). The small size and low Reynolds number pose a major challenge to aerodynamicist especially to the CFD community. The flow at these Reynolds number are dominated by strong viscous effects and undergoes flow transition, separation and reattachment. Further at these Reynolds number, the separation induces the flow transition leading the formation of laminar separation bubble (LSB). The key geometric feature of this class of vehicle is the thin cambered wing section with low aspect ratio. This thin wings with low aspect ratio may pose difficulties in generating a good quality grid to simulate the flow. Several researchers have carried out experiments [1–4] and numerical study [5–7] in order to understand the effects of wing span and wing section camber on aerodynamic characteristics at low Reynolds number. There are well tested approaches to predict fully laminar and fully turbulent flows in CFD which are extensively used by the designers. However, on the other hand, due to the lack in proper understanding of the physics related to flow transition, there is no single reliable and computationally affordable tool to simulate these flows. The simpler empirical based approach to handle transition lacks generality and cannot be applied to complex practical problems. While methods like direct numerical simulation (DNS) and large eddy simulation (LES) are capable of predicting flow transition are computationally too expensive to be used for design purpose. However, the Reynolds Averaged Navier Stokes (RANS)-based approach is the most viable tool for the designer as it gives reasonably accurate mean flow quantities at a lower computational cost. The recent approach to model flow transition in the RANS framework is to use additional transport equations which include the effects of transition. In this direction, Menter et al. [8, 9] proposed a correlation based transition model using local variables which is based on Menter's SST model [10] and solves additional two transport equations, one for intermittency (γ) and another for momentum thickness based Reynolds number (Re_θ) for finding the transition onset. The present work aims at understanding the effect of aspect ratio on the aerodynamic and transition characteristic of the non-dimensionalized SD7003 rectangular wing at $Re = 6 \times 10^4$ based on the wing root chord (C). The influence of Reynolds number in the low Re regime on the performance of the non-dimensionalized SD7003 rectangular wing is also studied. The low Re SD7003 airfoil whose thickness and camber is 8.5% and 1.4% is selected for this study as its major characteristic is laminar separation followed by transition which is observed for range of Reynolds number and angles of attack [11–14]. The present transitional flow simulation is carried out using the Menter's $\gamma - Re_\theta$ SST model available in the in-house incompressible finite volume multi-block flow solution code [15] 3D-PURLES (**3D P**ressure based **U**nsteady **RANS LES**).

2 Methodology

The Reynolds averaged Navier Stokes equations for unsteady incompressible flow in the coordinate-free form:

Mass conservation:

$$\nabla.\rho U = 0 \tag{1}$$

Momentum conservation:

$$D_t \rho U = -\nabla P + \nabla.((\mu + \mu_t)(\nabla U + \nabla^t U)) \tag{2}$$

where μ and ρ are fluid viscosity and density, p and U are the pressure and velocity vector, respectively. The eddy viscosity μ_t is evaluated through turbulence/transition models. The $\gamma - Re_\theta$ SST transition model implemented in 3D-PURLES uses the constants, damping function, etc., as given by Menter [9] except for the two correlation functions viz. $Re_{\theta C}$ and F_{length} which are adopted from Malan et al. [16]. However, the $Re_{\theta C}$ correlation has been suitably modified as given below based on our validation study carried out for the flat plate T3A test case [15].

$$Re_{\theta C} = min\left(0.665\tilde{R}e_{\theta t} + 66.5, \tilde{R}e_{\theta t}\right) \tag{3}$$

3 Results and Discussion

The γ-Re_θ SST transition model available in the in-house code 3D-PURLES has been used to study the effect of the wing aspect ratio and effect of Reynolds number in the low Re regime on the aerodynamic and transition performance of the non-dimensionalized SD7003 rectangular wing which are discussed in the following sub-sections.

3.1 *Effect of AR on the Wing Flow Characteristics*

The effect of the three different aspect ratio (AR) is analysed upto the stall angle of attack (α) at $Re = 6 \times 10^4$. The simulations have been carried out for half the wing by specifying a symmetry boundary condition. This analysis uses a H-H grid topology having a grid size of 590 × 93 × 178, 590 × 109 × 178 and 590 × 129 × 178 for the wing with semi aspect ratio (sAR) of 0.25, 0.5 and 1, respectively. The grids are stretched appropriately towards the wing surface in order to obtain the near wall y^+ to be less than 1. This grid size was decided based on our earlier study carried out for MAV wings [17]. In this study, the transition onset location is obtained based on the Reynolds shear stress threshold criterion as suggested by Ol et al. [12]. In this criteria, the onset location is determined whenever the normalized Reynolds shear stress exceeds more than 0.001 and a clear visible rise is observed. The Reynolds shear stress contours on the wing surface and volume streamlines obtained at $\alpha = 8°$ for the three different wings analysed are shown in Fig. 1. The transition onset location (X_T/C) obtained based on the above criteria is shown for the wing root and tip sections. These figures clearly shows that the transition onset along the wing span is almost constant from the root to the region close to the tip section. The contours at the tip section are observed to be quite different which may be due to the formation of the tip vortices. The transition onset location (X_T/C) at the tip is observed to be much delayed (occurring downstream) compared to the root section. It is observed that the level of the turbulent intensity (quantified by the value of Reynolds stress) is much less at wing tip as compared to the root section for all the three wings. The wing with lowest aspect ratio is found to be more laminar compared to the other two wings. In other words, the transition onset moves upstream leading a lesser laminar region with increase in the wing aspect ratio. It is further observed that the length of transition zone is also found reduce (i.e. the flow become turbulent faster which is indicated by higher value of Reynolds stress contour) with increase in sAR. The influence of the wing span on the transition onset location at different angles of attack is shown in Fig. 2. The wing with lowest aspect ratio ($sAR = 0.25$) is found to have a delayed transition onset (longer laminar region) at all the angles of attack as compared to other two wings. The volume streamlines obtained at $\alpha = 8°$ for the three different wings analysed are shown in Fig. 1 The streamlines shows the formation of the tip vortex for all the three wings. These figures clearly depict that the formation of the tip vortices in terms of its strength and onset location is not greatly influenced by the wing aspect ratio.

The effect of the wing aspect ratio on the aerodynamic characteristics is shown in Fig. 3. The influence of the wing span on the aerodynamic coefficients is observed to be quite insignificant for $\alpha \leq 0°$. The difference especially in the lift coefficient obtained by the three wings are found to be enhanced with angle of attack. The lift coefficient (C_L) curve for the wing having the lowest sAR (Fig. 3a) is observed to be non-linear especially at the lower angles of attack and this non-linearity disappears with increase in aspect ratio. The wing with the lowest aspect ratio though has lesser C_L is observed to have a soft and delayed stall angle around $\alpha = 42°$. On the other hand, for the wing with higher aspect ratio an early and sharp stall is observed with the stall angle reducing with increase in aspect ratio. The stall around $\alpha = 22°$ and $\alpha = 16°$ are obtained for the wing with aspect ratio 0.5 and 1, respectively. It is further observed that the maximum C_L reduces with increase in the wing aspect ratio but the C_L before the stall angle is highest for the wing having highest aspect ratio. The variation of drag coefficient C_D (Fig. 3b) for the three wing with different sAR are almost identical with the wing with lowest aspect ratio having slightly lower drag especially beyond $\alpha = 4°$. This lower drag obtained by this wing may be attributed to reason that the flow over this wing has a longer laminar region compared to the other two wings. The lift to drag ratio L/D (Fig. 3c) is observed to reduce with the reduction in the wing aspect ratio, however, the maximum L/D is found to occurs at the same angle of attack ($\alpha = 8°$) for all the three wings. The wing with the lowest

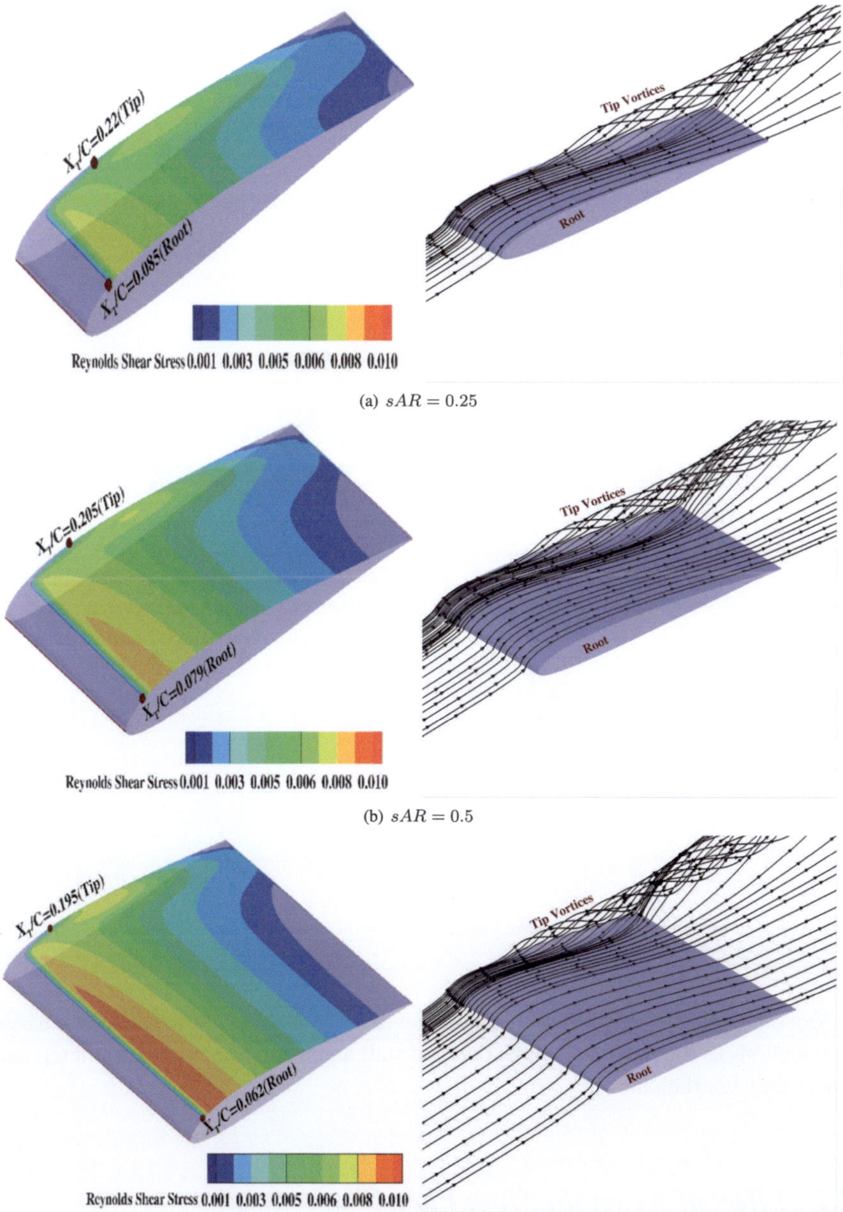

(a) $sAR = 0.25$

(b) $sAR = 0.5$

(c) $sAR = 1$

Fig. 1 Reynolds stress contours and volume streamlines for different wing spans for $Re = 6 \times 10^4$ and $\alpha = 8°$ (left figure: surface contours, right figure: volume streamlines)

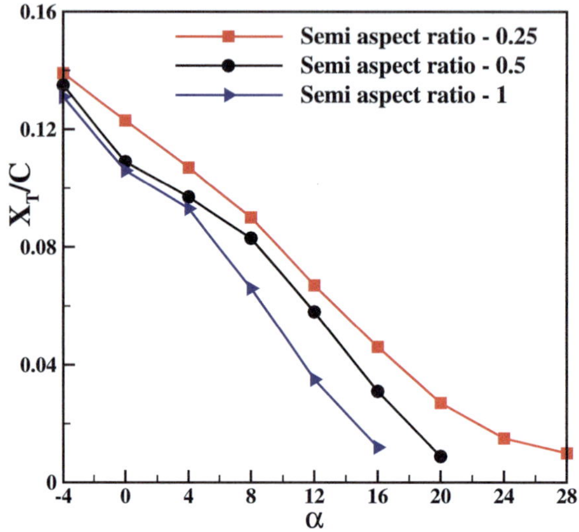

Fig. 2 Effect of wing span on the transition onset location at $Re = 6 \times 10^4$

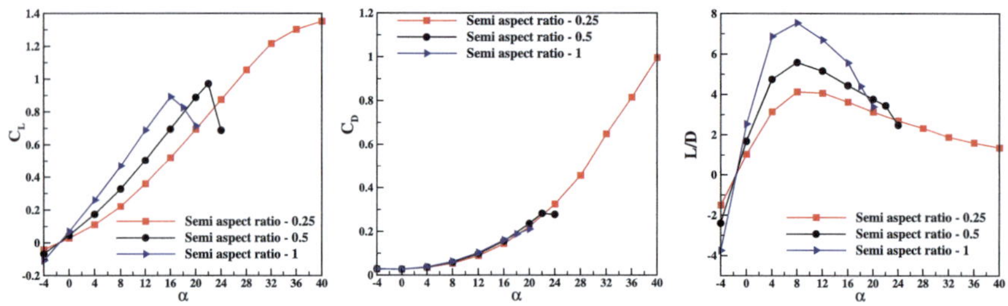

(a) Coefficient Lift (b) Coefficient Drag (c) Lift to Drag Ratio

Fig. 3 Effect of wing span on the aerodynamic coefficient at $Re = 6 \times 10^4$

aspect ratio ($sAR = 0.25$) though having lower L/D ratio may be preferred for MAV applications as this wing has much delayed stall angle and higher maximum C_L and lower drag coefficient.

3.2 Effect of Re on the Wing Flow Characteristics

The influence of Reynolds number in the low Re regime on the performance of the SD7003 rectangular wing with 0.25 a semi aspect ratio are studied for three Reynolds number (4×10^4, 6×10^4 and 1.5×10^5) using the γ-Re_θ SST model. The aerodynamic coefficients obtained at the three Reynolds number are shown in Fig. 4. It is evident form this figure that increasing the Reynolds number from

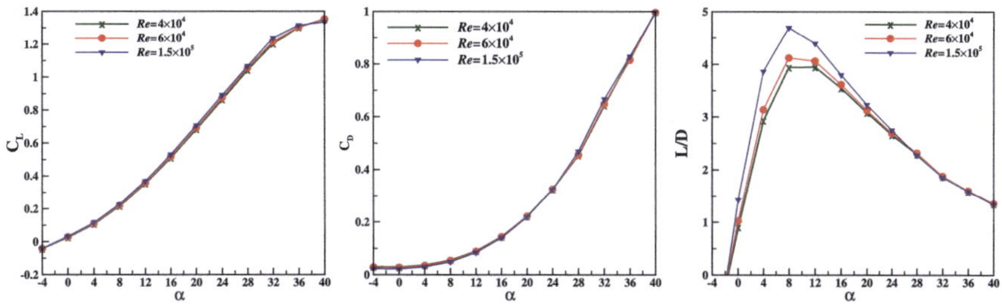

(a) Coefficient Lift (b) Coefficient Drag (c) Lift to Drag Ratio

Fig. 4 Effect of Reynolds number on the aerodynamic coefficient for $sAR = 0.25$

4×10^4 to 1.5×10^5 has not significantly influence the variation of the lift and drag coefficients for the SD7003 rectangular wing with $sAR = 0.25$. However, a close examination of these aerodynamic coefficients reveal that increasing the Reynolds number has produced higher lift coefficient and lower drag coefficient. This higher C_L and lower C_D at higher Re has resulted in the higher lift to drag ratio. In order to examine the influence of varying the Reynolds number on the transition onset and flow structure, the Reynolds stress contours on the wing surface and volume streamlines obtained for the three Reynolds number is shown for $\alpha = 8°$ in Fig. 5. The contours of the Reynolds stress on the wing surface clearly indicate that the level of the Reynolds stress which depict the turbulence intensity reduces with increase in Re. This feature of reduction in the turbulence level with increase in Re was also observed in the experimental study on the 2D and 3D large scale wavy wall carried out by Hamid et al. [18]. The transition onset location at the wing root section and tip section is observed to move upstream as the Reynolds number is increased. The computed volume streamlines clearly reflect the formation of the wing tip vortices at all the three Reynolds numbers.

4 Conclusions

The $\gamma - Re_\theta$ SST transition model has been successfully used to simulate the effect of wing aspect ratio at $Re = 6 \times 10^4$ and influence of Reynolds number for semi aspect ratio 0.25. The SD7003 rectangular wing with three semi aspect ratio (sAR) viz. 0.25, 0.5 and 1 were found to significantly influence both the transition and aerodynamic characteristics. The wing with the lowest aspect ratio ($sAR = 0.25$) may be considered as a suitable candidate for MAV application at this low Reynolds number because of its better aerodynamic and transition performance. The simulations indicated that aerodynamic, transition and flow characteristics of wing with $sAR = 0.25$ was not greatly influenced when the Reynolds number was increased from $Re = 4 \times 10^4$ to 1.5×10^5. However, the aerodynamic performance in terms of the lift to drag ratio was found to slightly better at $Re = 1.5 \times 10^5$. The turbulence

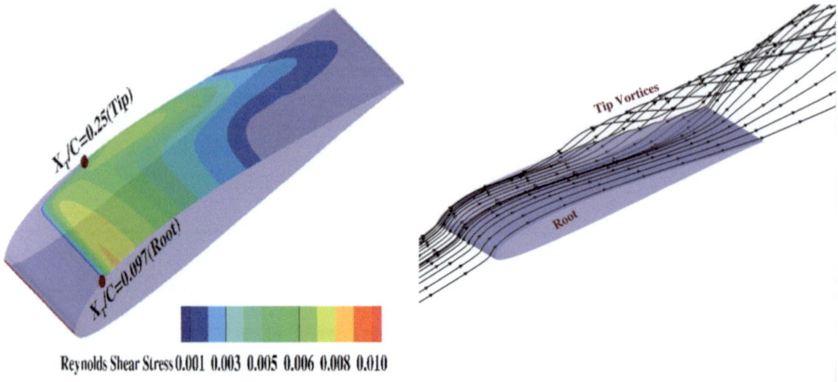

(a) $Re = 4 \times 10^4$

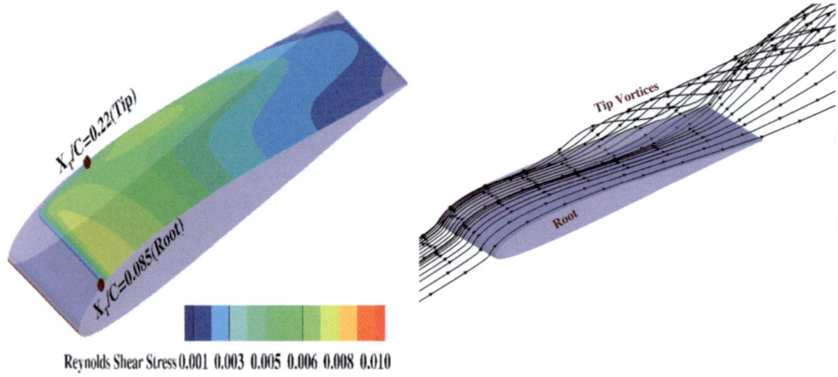

(b) $Re = 6 \times 10^4$

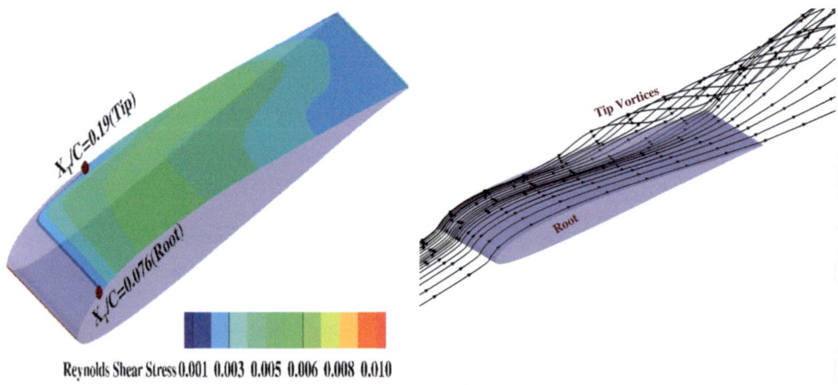

(c) $Re = 1.5 \times 10^5$

(Left figure : Surface contours, Right figure : Volume streamlines)

Fig. 5 Reynolds stress contours and volume streamlines at different Reynolds number for $sAR = 0.25$ and $\alpha = 8°$ (left figure: surface contours, right figure: volume streamlines) level post transition was found to reduce with increase in Reynolds number but the the transition onset occurred little earlier.

Nomenclature

SD	Selig-Donovan
CFD	Computational Fluid Dynamics
MAV	Micro Air Vehicle
NAV	Nano Air Vehicle
UAV	Unmanned Air Vehicle
Re	Reynolds number
sAR	Semi aspect ratio
α	Angle of attack
C_L	Lift coefficient
C_D	Drag coefficient
L/D	Lift to drag ratio
X_T/C	Transition onset location
F_{length}	Transition length correlation
$Re_{\theta c}$	Momentum thickness based Re correlation

Acknowledgements The authors wish to acknowledge the Director CSIR-NAL and Head CTFD Division, CSIR-NAL for their encouragement and support to carry out this work.

References

1. Torres GE, Mueller TJ (2004) Low-aspect-ratio wing aerodynamics at low Reynolds numbers. AIAA J 42:865–873
2. Pelletier A, Mueller Thomas J (2000) Low Reynolds number aerodynamics of low-aspect-ratio thin, flat, cambered-plate wings. J Aircraft 37:825–832
3. Reid MR, Kozak J (2006) Thin/cambered/reflexed airfoil development for micro air vehicle applications at Reynolds numbers of 60,000 to 100,000. In: AIAA atmospheric flight mechanics conference and exhibit, Keystone, Colorado, number 2006-6832
4. Zuo LX, Wang JJ (2010) Planform and flexibility on lift characteristics for flow over low-aspect ratio wings. J Aircr 23(1):1537–1542
5. Null W, Shkarayev S (2004) Effect of camber on the aerodynamics of adaptive-wing micro air vehicles. J Aircr 42(6):1537–1542
6. Chen ZJ, Qin N (2013) Planform effects for low-Reynolds-number thin wings with positive and reflex cambers. J Aircr 50(3):952–964
7. Kulkarni DS, Rajani BN, Kulkarni V (2021) Transitional flow past low Re SD airfoils with varying camber. In: 48th national conference on fluid mehcnaics and fluid power. BITS-Pilani, Rajasthan, India, December 2021

8. Menter FR, Langtry R, Huang PG, Volker S (2006) Transition modelling for general purpose CFD code. Flow Turbul Combust 77:277–303
9. Langtry R, Menter FR (2009) Correlation based transition modeling for unstructured parallelized computational fluid dynamics codes. AIAA J 47:2894–2906
10. Menter FR (1994) Two-equation eddy-viscosity turbulence models for engineering applications. AIAA J 32(8):1598–1605
11. Galbraith MC, Visbal MR (2008) Implicit large-eddy simulation of low Reynolds number flow past the SD7003 airfoil. In: 46th AIAA aerospace sciences meeting and exhibit, Reno, Nevada, number 2008-225
12. Ol MV, McAuliffe BR, Hanff ES, Scholz U, Kalher C (2005) Comparison of laminar separation bubble measurements on a low Reynolds number airfoil in three facilities. In: 35th AIAA fluid dynamics conference and exhibit, Toronto, Ontario Canada, number 2005-5149
13. Radespiel R, Graage K, Brodersen O (1991) Transition predictions using Reynolds-averaged Navier-Stokes and linear stability analysis methods. AIAA Paper 91-1641
14. Kulkarni DS, Rajani BN, Kulkarni V (2019) Numerical simulation of transitional flow past SD7003 airfoil over a range of Reynolds number. In: 21st annual AeSI CFD symposium. NAL, Bangalore, India, August 2019
15. Manu YC, Rajesh A, Subrahmanya MB, Kulkarni DS, Rajani BN (2015) Simulations using transition models within the framework of RANS. In: Sengupta TK et al (eds) Advances in computation. Modeling and Control of Transitional and Turbulent Flows. World Scientific, Singapore
16. Malan P, Suluksna K, Juntasaro E (2009) Calibrating the $\gamma - Re_\theta$ transition model for commercial CFD. In: 47th AIAA aerospace sciences meeting, Orlando, Florida, number 2009-1142
17. Subrahmanya MB, Rajani BN (2018) Transition prediction for flow over a MAV using the correlation based model. In: 8th symposium on applied aerodynamics and design of aerospace vehicles
18. Hamed AM, Kamdar A, Castillo L, Chamorro LP (2015) Turbulent boundary layer over 2D and 3D large-scale wavy walls. Phys Fluids 27:1–14

Time-Resolved Imaging of Droplet Collision in Interacting Sprays

Surya Ghosh and Srikrishna Sahu

Abstract The current study focuses on droplet collisions in interacting sprays, generated from two swirl injectors. We employ the high-speed shadowgraph technique to capture the time-resolved images at various downstream positions within the interaction zone. The images are recorded at 5000–20,000 frames per second (fps). Closer to the injector exit, the droplet collisions are more frequent, so the images are captured at a higher frame rate, enabling us to track the colliding droplets in the instantaneous images. The non-dimensional parameters, such as the collision Weber number (We) and the impact factor (*B*), are calculated based on the instantaneous images of the colliding droplets. Also, at each downstream location, these parameters are evaluated based on the characteristic droplet size and velocity. It is observed that the parameters based on mean value are in good agreement with the average of the instantaneous values. The We–*B* regime map depicting the outcome of binary droplet collision indicated that in the current interacting sprays, the collision outcome is mostly stretching separation, which is also demonstrated in the experimental images. It is found that the mean Weber number decreases, and the impact parameter increases downstream of the injector exit in the interaction zone. This suggests the reduction in droplet collision probability. Both droplet size and axial velocity distribution are measured in the current study. The results reveal that in the near nozzle locations, where droplet collision is dominant in the spray interaction zone, the characteristic droplet size reduces, and the mean axial velocity increases as compared to the single injector.

Keywords Droplet collision · Stretching separation · Collision angle · SMD · Axial velocity · Weber number · Impact factor

S. Ghosh (✉) · S. Sahu
Department of Mechanical Engineering, IIT Madras, Chennai, India
e-mail: suryaghosh55@gmail.com

S. Sahu
e-mail: ssahu@iitm.ac.in

A. Kumar R. et al. (eds.), *Proceedings of Fluid Mechanics and Fluid Power (FMFP) 2023, Vol. 6*, Lecture Notes in Mechanical Engineering,
https://doi.org/10.1007/978-981-96-9600-0_14

1 Introduction

The widespread occurrence of droplet collisions in both natural processes and numerous industrial applications demands a detailed investigation of such phenomenon. Droplet collision plays a crucial role in various fields, including atmospheric studies, combustion engines, spray drying processes, etc. Several studies were conducted to predict the outcome of the droplet collisions. Accurate modeling of the collision behavior plays an essential role in various applications. Experiments were conducted to understand the different collision outcomes, and the corresponding regimes are mainly coalescence, bouncing, reflexive separation, and stretching separation [1]. In this regard, a study was reported on the collision outcome by varying the size ratio of colliding water droplets [2]. Subsequently, a model that accounts for the stretching separation boundary was implemented, explicitly incorporating viscosity's influence through a two-parameter approach [3]. The study by Quin and Law [4] highlights that the differences in collision behavior between water and hydrocarbon droplets result from the difference in their viscosity and surface tensions. A comprehensive model was developed that considers viscosity and size ratio effects in initiating reflexive separation [5]. Moreover, while the previous studies report similar liquid droplets, there have been extensive works on collisions between droplets of different liquids, e.g., experiments with diesel-diesel droplets, diesel-ethanol droplets [6], and ethanol-ethanol droplets collision [7]. The study on droplet-particle collisions results in mainly two collision regimes, such as agglomeration and stretching separation [8].

However, most of the previous studies were based on binary droplet collision, and it is essential to study droplet collision in sprays. For instance, in a practical scenario, a combustion chamber utilizes multi-element injectors for power generation. Similarly, multiple injectors are employed in cooling tower units. The interaction between the sprays occurs due to the aerodynamic forces resulting from the swirling or accelerated airflow. There are very few works of literature available in this direction. Gavaises et al. [9] conducted a study on the effect of droplet collision on interacting sprays from two hollow cone nozzles subjected to cross-flow air [9]. Also, the study incorporated a spray model in the CFD code to evaluate droplet collisions. Most studies on droplet collisions for interacting sprays are based on numerical calculations, which are deterministic or stochastic models, where the pair of parcels is studied. With the stochastic droplet collision model, Lain and Sommerfeld [10] investigated the droplet collision on a hollow-cone spray.

In the present investigation, we aim to study droplet collisions in two interacting sprays near the injector using the high-speed shadow imaging technique. The droplet size and velocity are measured at various downstream locations, mainly in the interaction zone. Additionally, we calculated the crucial non-dimensional parameters, such as the collision Weber number and impact factor, by tracking the colliding droplets at different instances. Comparisons were made between the average values obtained from these instantaneous images and the mean Weber number and mean impact factor calculated near the injector exit. To generate the interacting sprays, two gas-centered swirl co-axial injectors are employed where the swirling liquid sheet interacts with high-speed air and creates hollow-cone sprays. The droplet measurements are obtained mainly in the interaction zone. Also, the droplet size and axial velocity distribution are obtained in single injectors at the same locations.

The paper is structured as follows: Sect. 2 includes the experimental arrangement and optical techniques used for the study, followed by a brief description of the image processing algorithm. Section 3 describes the results obtained from the droplet collisions for the interacting sprays. The evolution of the variation of the non-dimensional parameters related to droplet collision has been presented in this section. Also, a modification in droplet size and velocity has been reported. Moreover, a summary and conclusions of the present study can be found in Sect. 4.

2 Materials and Methods

2.1 Experimental Setup

This study used two gas-centered swirl co-axial injectors to generate interacting sprays. These injectors were integrated into a specially designed manifold. Figure 1 illustrates the experimental setup employed for this investigation. The experiment was conducted under atmospheric pressure and room temperature. The working fluids used for this study were air and water. A pump was used to deliver the water into the manifold chamber, from where it proceeded through tangential holes with a diameter of 1 mm in the injectors. For air supply, separate provisions were made, with each injector receiving air through an opening with a diameter of 10.4 mm. Volume flow rates of the working fluids were measured using rotameters. The water rotameter had a range of 0.5–5 lpm with $\pm$ 2% accuracy, while the air rotameter had a range of 100–1000 lpm with $\pm$ 2% accuracy. The sprays from both injectors interact at the outer region of the spray axes. The measurement window is focused mainly on this interaction zone.

2.2 Operating Condition

The mass flow rate of the air ($\dot{m}_g$) was 14.5 g/s, and the mass flow rate of the liquid water ($\dot{m}_l$) was kept at 16.6 g/s. The liquid velocity and the air velocity are 0.6 m/s and 110 m/s, respectively, for this study. Accordingly, the momentum-flux ratio for the current study is 51.7.

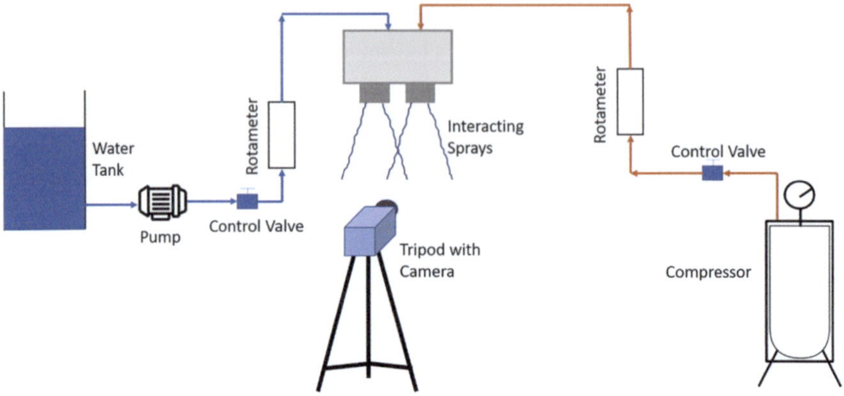

Fig. 1 Schematic of the experimental setup for the spray-interaction study

2.3 Data Acquisition and Visualization Technique

To visualize the collision dynamics of droplets and their resulting outcomes, high-speed shadowgraph images are acquired at a frame rate of 5000–20,000 fps using a high-speed camera (FASTCAM Photron 12bit, NOVA). The camera's exposure time was set to 3 μs, and the measurements were taken at a location of $z = 30$ mm, 50 mm, 70 mm, and 90 mm downstream of the injector exit. Close to the injector exit (at 30 mm), the images were captured at 20,000 fps at the interaction zone. As this location is more probable for droplet collision, capturing the images at a higher frame rate helps in tracking and visualizing the colliding droplets. At each location, the Sauter Mean Diameter (SMD) of droplet size distribution and average velocity are calculated with MATLAB in-house codes. For droplet size "canny" filter was used with a proper threshold for detection of the droplet edges and "imfill" function was used to fill the holes. Droplet centroids and diameters are calculated by "regionprops" in-build function. The Particle Tracking Velocimetry Algorithm (PTV) was used to calculate the velocity of the droplets. To obtain the droplet velocity, from the time-resolved images, the droplets are tracked across a pair of consecutive images. A search window is assigned to find the displacement of the identified droplets from two successive images. If the disparity in droplet size is within 5%, the droplets are considered corresponding pairs. With knowledge of the centroid positions of the droplets from the MATLAB function "regionprops" and the inter-frame time, the droplet velocity was determined by evaluating their displacements between both images. The above process was applied to all the identified droplets.

The colliding droplets at close to the injector exit are tracked from the time-resolved recorded images, and collision outcomes are also visualized from the images. MATLAB software was used with the above-mentioned algorithm to calculate colliding droplets' size and velocity. Further, the important parameters, such as collision Weber number (We) and impact factor (B), which govern the droplet collision, were calculated as sketched in Fig. 2. Also, these non-dimensional parameters

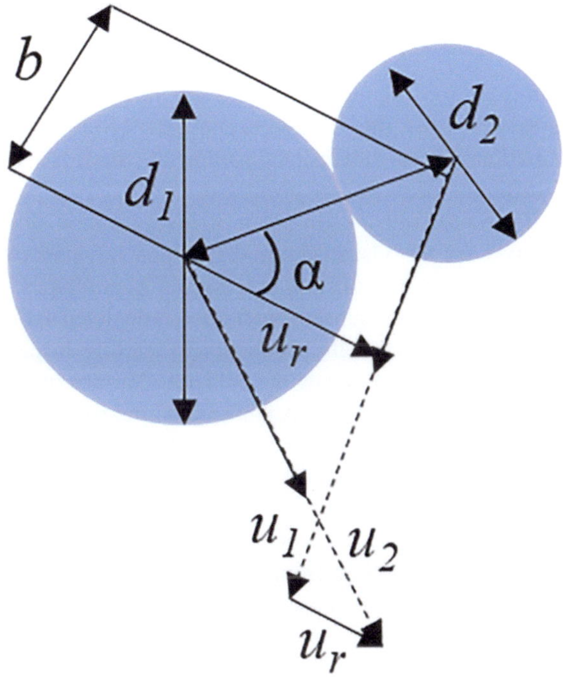

Fig. 2 Geometric parameters of droplet collision

are calculated at each measurement location based on mean droplet size and velocity. The collision Weber number (We) is calculated as,

$$We = \frac{\rho u_r^2 d}{\sigma}$$

where d is the average diameter ($(d_1 + d_2)/2$) of the colliding droplets, and u_r is the relative velocity between two colliding droplets. The relative velocity equals to $u_r = \left(u_1^2 + u_2^2 - 2u_1u_2\cos\alpha\right)^{1/2}$. The velocities of the colliding droplets are u_1 and u_2, respectively, and α is the collision angle. The density (ρ) and surface tension (σ) of the water are 998.5 kg/m^3 and 72.5 mN/m, respectively. The non-dimensional parameter, the impact factor (B) is defined as,

$$B = \frac{2b}{d_1 + d_2}$$

where b represents the perpendicular distance from the center of one colliding droplet to the relative velocity vector originating from the center of the other droplet, and the diameter of the colliding droplets are d_1 and $d_{2,}$ respectively. In this study, the collision outcomes of the droplets are investigated by plotting the We–B regime map [11]. This comprehensive approach provides valuable insights into the collision process and its associated outcomes.

3 Results and Discussion

In this section, the regime map for droplet collisions based on the analysis of instantaneous images, the discussion about the effect of droplet collision, as well as the variations of non-dimensional parameters, such as Weber number and impact factor, at different locations, are presented.

Time-resolved imaging at a 30 mm location downstream of the injector exit allowed for the visual identification and tracking of colliding droplets, enabling observation of the collision process and its outcomes. An example of such colliding droplets, tracked in subsequent frames, is depicted in Fig. 3. The instantaneous images provided evidence of the production of satellite droplets through a stretching separation mechanism following the collision. Various other droplet collision events were identified from different instantaneous images. The collision Weber number (We) and impact factor (*B*) were calculated for these colliding droplets using the specified equations. Subsequently, the values of We and *B* were plotted in the We–*B* map (Fig. 4). The above figure also indicates different regimes of the outcome of the binary droplet collision process [11]. The analysis of the regime map revealed that a majority of the tracked droplets fall into the stretching separation regime, while only a few collisions resulted in droplet coalescence and bouncing regimes. The accuracy of the regime map was verified against experimental observations. In the event of droplet interactions for the intersecting sprays, multiple collisions between the droplets have been recorded. Qualitatively, more than 90% of the collisions ended with producing satellite droplets from the parent droplets via the stretching-separation mechanism. Tracking all collision events in the spray from the images to find relevant parameters is challenging. Rather, we have calculated We and *B* based on the mean size and velocity. It can be observed that the We–*B* pair, calculated using the mean values, fairly represents the collision outcome since the majority of data points are concentrated within the stretching-separation region.

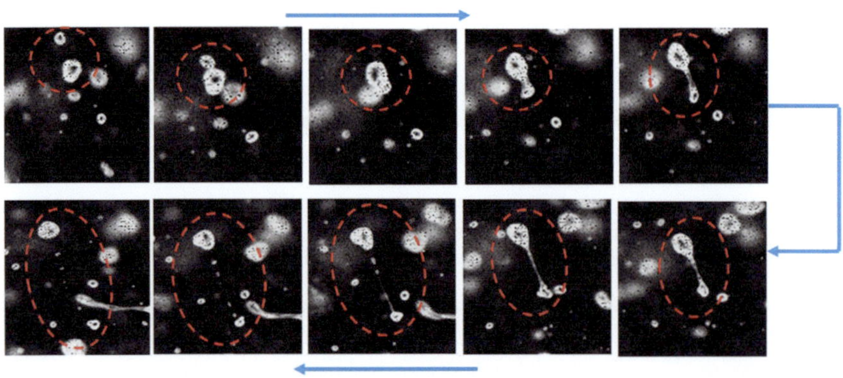

Fig. 3 Time series of instantaneous images depicting droplet collision leading to generation of new droplets via stretching-separation mechanism

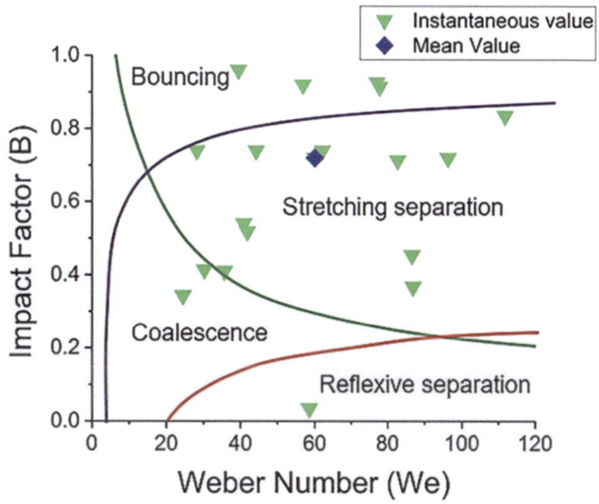

Fig. 4 Regime map for outcome of droplet collision in We–B plot with respective boundary lines between the different regimes

To further investigate the impact of droplet collisions on spray characteristics, the droplet size and axial velocity were computed for both individual and interacting sprays. The histograms of droplet size and axial velocity distribution at the interaction zone are presented in Figs. 5a and b. Notably, the peak in the droplet size distribution shifted toward smaller droplet sizes in the case of interacting sprays. Consequently, SMD reduced from 345 μm ± 6% for individual sprays to 321.5 ± 6% μm for interacting sprays. This reduction in characteristic droplet size was attributed to droplet collisions, as evidenced by the observation of smaller droplet generation in the instantaneous images. Similarly, the axial velocity distribution showed a shift in the peak toward higher velocity magnitudes for interacting sprays, where droplet collisions occurred more frequently. The mean axial velocity increased from 3.6 m/s ± 4% for individual sprays to 4.5 ± 4% m/s for interacting sprays. This phenomenon was attributed to the quick adjustment of the smaller droplets to the high-speed airflow, leading to increased acceleration and higher droplet velocities. In summary, the study demonstrated that droplet collisions in interacting sprays have a significant influence on spray characteristics at the interaction zone.

From the instantaneous measurement, the average droplet size and velocity are calculated close to the injector exit in the interaction zone, and the same is measured considering the characteristic droplet size and mean velocity.

It is found that both We and B values are similar in both cases, which result in the same location within the regime map, specifically falling under the stretching separation regime. So, the We and B values, considering characteristic droplet size and mean velocity, indicate that they are representative of the outcome of droplet collisions in the interaction zone. A similar analysis was carried out at various downstream locations, and the results are presented in Figs. 6a and b. As the measurement location moves further downstream from the injector exit, the We values for interacting sprays exhibit a gradual reduction. This decline can be attributed to the smaller droplet size and the diminishing difference in relative velocity between the colliding

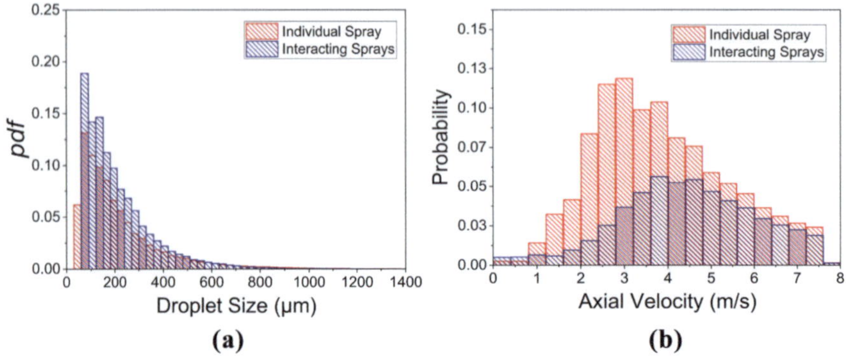

Fig. 5 **a** Droplet size and **b** axial velocity distribution for individual spray and interacting sprays

droplets. Conversely, the impact factor shows an increasing trend as we move downstream, approaching a value close to 1. These trends collectively suggest that the likelihood of droplet collisions diminishes significantly at locations farther away from the injector exit. Furthermore, the We–*B* plot indicates a transition from the stretching separation regime to the bouncing regime as we progress downstream. This shift implies that the reduction in droplet size due to droplet collision becomes less likely as the distance from the injector exit increases.

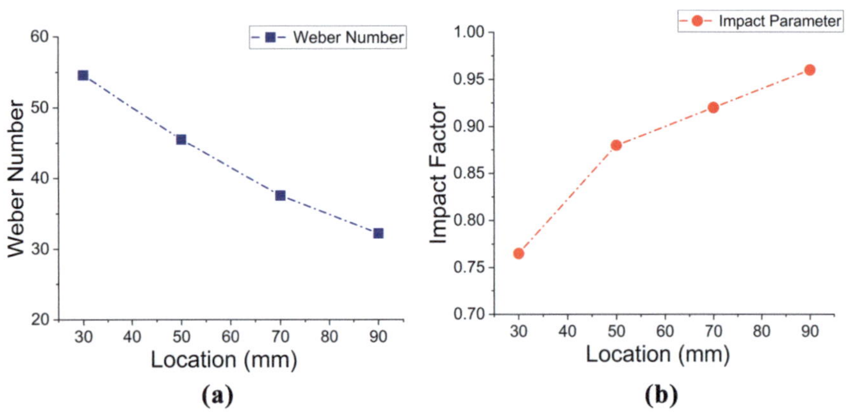

Fig. 6 Evolution of **a** collision Weber number and **b** impact factor for interacting sprays at different downstream locations

4 Conclusions

An experimental investigation on droplet collision process in interactive sprays generated from gas-centered swirl injectors is reported in this work. The results provide some insights into the collision process and its influence on spray characteristics. In the near nozzle location, the outcome of droplet collision on the We–B regime map indicated the dominance of stretching separation in the interacting sprays, while droplet coalescence and bouncing events were rare. The analysis of droplet size and axial velocity distributions revealed that droplet collisions in interacting sprays led to a reduction in droplet size and an increase in mean axial velocity. Downstream of the injector We reduces, while B increases toward one, implying reduction in the collision probability. Such understanding on the modifications in spray characteristics at the spray interaction zone is crucial for optimizing the spray-based applications. Taken together, these findings underscore the substantial influence of droplet collisions on spray characteristics.

Nomenclature

SMD	Sauter Mean Diameter (–)
g/s	Gram per sec (–)
lpm	Liter per minute (–)
α	Collision angle (°)
We	Weber number (–)
B	Impact factor (–)
d	Diameter of droplet (μm)

References

1. Orme M (1997) Experiments on droplet collisions, bounce, coalescence and disruption. Prog Energy Combust Sci 23(1):65–79
2. Ashgriz N, Poo J (1990) Coalescence and separation in binary collisions of liquid drops. J Fluid Mech 221:183–204
3. Jiang Y, Umemura A, Law C (1992) An experimental investigation on the collision behaviour of hydrocarbon droplets. J Fluid Mech 234:171–190
4. Qian J, Law C (1997) Regimes of coalescence and separation in droplet collision. J Fluid Mech 331:59–80
5. Tang C, Zhang P, Law CK (2012) Bouncing, coalescence, and separation in head-on collision of unequal-size droplets. Phys Fluids 24(2):022101
6. Chen R-H (2007) Diesel-diesel and diesel-ethanol drop collisions. Appl Thermal Eng 27:604–610
7. Estrade J-P, Carentz H, Lavergne G, Biscos Y (1999) Experimental investigation of dynamic binary collision of ethanol droplets—a model for droplet coalescence and bouncing. Int J Heat Fluid Flow 20:486–491

8. Pawar SK et al (2016) An experimental study of droplet-particle collisions. Powder Technol 300:157–163
9. Gavaises TL, Theodorakakos A, Bergerles G, Brenn G (1996) Evaluation of the effect of droplet collisions on spray mixing. Proc Inst Mechanical Engineers 210:465–465
10. Lain S, Sommerfeld M (2020) Influence of droplet collision modeling in Euler/Lagrange calculations of spray evolution. Int J Multiph Flow 132:103392
11. Sommerfeld M, Pasternak L (2019) Advances in modelling of binary droplet collision outcomes in sprays: a review of available knowledge. Int J Multiph Flow 117:182–205

Flow Structures and Heat Flux Parameterizations in Dry Convective Atmospheric Boundary Layers

Bela Lodh, Niranjan S. Ghaisas, Vishal Dixit, K. Raghuvamshi, and Kaushal Gujrathi

Abstract Large-eddy simulations (LES) of the unstably stratified dry convective atmospheric boundary layer are reported. The flow is driven by heating from the bottom and a geostrophic wind is imposed on the top. LES of this configuration are performed using a high-fidelity code with a state-of-the-art subrid-scale model. Results of a previously studied configuration are reproduced and excellent agreement is obtained with previously published results. For a fixed geostrophic wind speed and varying heat flux at the bottom wall, the flow structures show distinct patterns in their planform organization. The flow structures transition from convective rolls at low surface heating to convective cells for large surface heat fluxes. The eddy-diffusivity mass-flux (EDMF) parameterization, which has been developed for zero geostrophic wind, is evaluated using the LES data. We find that the EDMF parameterization needs a correction to be able to quantitatively match the LES results. Quadrant analysis is performed based on the vertical heat flux and its results suggest that the corrections required to the EDMF parameterization may be correlated with a portion of the cold updraft and warm downdraft events.

Keywords Atmospheric boundary layer · Large-eddy simulation · Dry convection · Parameterization

B. Lodh
Department of Earth and Environmental Sciences, Indian Institute of Science Education and Research Bhopal, Madhya Pradesh, India

N. S. Ghaisas · K. Raghuvamshi · K. Gujrathi
Department of Mechanical and Aersospace Engineering, Indian Institute of Technology Hyderabad, Telengana, India

N. S. Ghaisas (✉)
Department of Climate Change, Indian Institute of Technology Hyderabad, Telengana, India
e-mail: nghaisas@mae.iith.ac.in

V. Dixit
Interdisciplinary Program in Climate Studies, Indian Institute of Technology Bombay, Maharashtra, India

A. Kumar R. et al. (eds.), *Proceedings of Fluid Mechanics and Fluid Power (FMFP) 2023, Vol. 6*, Lecture Notes in Mechanical Engineering,
https://doi.org/10.1007/978-981-96-9600-0_15

1 Introduction

The atmospheric boundary layer extends from the Earth's surface up to a height of roughly 1–2 km and accounts for all of the momentum and thermal exchanges between land and air. When the surface is warmer than the ambient air, it results in a net heat flux to the air from the ground leading to convective updrafts, or motions of rising air parcels due to the effects of buoyancy. A canonical convective boundary layer is a configuration which is heated from below by a heat flux q_w and forced horizontally due to pressure differences between different regions. The horizontal forcing can be expressed as a geostrophic wind, U_g, imposed at the top of the boundary layer. A number of previous studies have focused on the dry convective boundary layers [2, 4, 7] where phase changes, e.g., of water vapor are neglected.

Large-scale weather and climate models (either regional or global) require parameterizations of the processes happening within the boundary layer. The traditional approach to parameterizing small-scale processes is to use down-gradient closures with a specified profile for the eddy-diffusivity. A relatively recent modification to these approaches was proposed by Siebesma et al. [5] wherein a mass-flux term was added to the usual eddy-diffusivity term to represent 'non-local' mixing effects. This combined eddy-diffusivity mass-flux (EDMF) parameterization was proposed and evaluated for dry convective boundary layers in their study. The parameterization has also been extended to moist convective boundary layers [8]. In most cases, these studies include representation of moist processes with an aim to reproduce the real-time moist convective boundary layer while assuming that the structural form of EDMF originally designed using dry CBL is applicable. Recent studies show that even *dry* CBL shows a rich variety in 're-organization' of flow structure (patterns in planform) when forced with a combination of geostrophic wind shear and surface heat fluxes [4]. In conclusion, whether EDMF can capture changes in heat and momentum transport, in response to changes in the flow structures, remains to be evaluated.

The aim of this paper is to characterize flow structures and the EDMF heat flux parameterization in dry convective atmospheric boundary layers. Radiation effects are neglected throughout this paper because the focus is on dry convective boundary layer. Radiative cooling, although important in general, may not play a significant role in air devoid of water vapor. Similar omission of radiative effects is found in several previous studies [2, 4, 7], including the work of Siebesma et al. [5], which is the immediate motivation for our work. High-fidelity large-eddy simulations (LES) are carried out of this configuration for a range of U_g and q_w forcings. The flow structures in the planform are analyzed followed by an evaluation of the EDMF parameterization. A quadrant analysis is performed on the convective heat flux to attempt, diagnose, and correct for the discrepancies between the LES results and the EDMF parameterization.

The numerical methodology is described in Sect. 2. Results and conclusions are presented in Sects. 3 and 4, respectively.

2 Methodology

2.1 LES Methodology

Large-eddy simulations (LES) are performed of the dry convective boundary layer in a three-dimensional rectangular box of dimensions $L_x \times L_y \times L_z$. Here, x and y are the horizontal directions and z is the vertical direction. The flow is driven by an imposed geostrophic wind, U_g, and a surface heat flux, q_w. The governing equations are the incompressible-Boussinesq, LES-filtered Navier-Stokes equations. These include the equations governing conservation of mass, momentum (velocities u_i or equivalently u, v, w) and potential temperature (θ). The momentum equation includes body-force terms due to buoyancy and Coriolis forces corresponding to a latitude of 43.3° N. The two horizontal directions are periodic. No-penetration ($w = 0$), slip velocity ($u = U_g$, $v = 0$) conditions are imposed at the top of the domain. The bottom surface is a no-penetration ($w = 0$), slip wall ($u, v \neq 0$) with the slip velocities obtained from an imposed wall shear stress. For the thermal boundary conditions, a constant lapse rate of $d\theta/dz = 0.003$ K/m is imposed at the top wall while the bottom surface temperature is given by a wall model. Standard wall models for the shear stress and the heat flux based on the Monin-Obhukhov Similarity Theory are used.

The in-house Fortran code, 'PadeOps-igrid' [6], is utilized for the simulations. The code employs Fourier-spectral discretization in x and y, 6th-order staggered compact finite-differences in z, and a 3rd-order Runge-Kutta method for time advancement. The Anisotropic Minimum Dissipation [3] subgrid-scale model is used for turbulence closure. This code has been validated for unstratified and stably-stratified atmospheric boundary layers, as well as for flows over wind farms. Here, we extend its validation to the unstable or convective boundary-layer regime.

2.2 Cases Simulated

Two sets of simulations are performed. The first set is forced by $U_g = 10$ m/s and $q_w = 0.24$ K-m/s with a domain size of $5 \times 5 \times 2$ km^3 to match the setup of Abkar and Moin [1]. These simulations are carried out on grids ranging from 48^3 to 96^3. The second set of simulations uses a fixed $U_g = 9$ m/s and wall heat flux varying from 0.0075 K-m/s to 0.36 K-m/s. The domain size is $12 \times 12 \times 2$ km^3 and 160^3 grid points are employed for each simulation. Each simulation is initialized with a capping inversion at 1 km and with quiescent flow conditions. The simulations are conducted for roughly 20 large-eddy turnover times, where one turnover time is z_*/w_*. Statistical averaging is performed over x, y and over time after attainment of the quasi-stationary state (roughly 12–15 turnover times). Here, z_* is the boundary-layer height [4] and w_* is the overturning velocity. Both these quantities are obtained a-posteriori, z_* being the height where the buoyancy flux is minimum

Table 1 Summary of LES cases studied

Ug (m/s)	q_w (K-m/s)	Grid	z_* (m)	$\|L\|$ (m)	$-z_*/L$	u_* (m/s)	w_* (m/s)	w_*/u_*
10	0.24	48^3	1021	59.6	17.1	0.570	2.00	3.51
10	0.24	72^3	1014	58.9	17.2	0.568	1.99	3.50
10	0.24	96^3	1032 (1035)	59.2 (57.9)	17.4 (17.9)	0.569 (0.573)	2.00 (2.00)	3.51 (3.49)
9	0.36	160^3	1381	39.8	34.7	0.57	2.53	4.42
9	0.18	160^3	1131	62.6	18.1	0.53	1.88	2.56
9	0.06	160^3	1043	147.2	7.09	0.49	1.27	2.61
9	0.03	160^3	1043	253.1	4.12	0.46	1.01	2.18
9	0.015	160^3	1031	456.9	2.26	0.446	0.795	1.78
9	0.0075	160^3	1044	807.1	1.29	0.43	0.63	1.48

and $w_* = (g\theta_0 q_w/z_*)^{1/3}$, where $g = 9.81$ m/s^2 and $\theta_0 = 301.78$ K. Key parameters of the simulations are summarized in Table 1. The Obhukhov length (L) and friction velocity (u_*) are given by standard relations [4].

3 Results

A widely studied convective boundary-layer setup is simulated first, with $U_g = 10$ m/s and $q_w = 0.24$ K-m/s. LES of the same configuration have been reported by [1, 2, 7]. The mean velocities obtained from our LES are compared to the results of [1] in Fig. 1a. The total momentum flux, given by $\langle u'w' \rangle^{tot} = \langle u'w' \rangle + \tau_{13}^{sgs} + \tau_{13}^{wm}$, is plotted in Fig. 1b. Here, the superscripts 'tot,' 'sgs,' and 'wm' refer to total, subgrid-scale, and wall-model, respectively. The buoyancy flux, $\langle w'\theta' \rangle^{tot}$ (g/θ_0) with contributions from the resolved, subgrid-scales and wall-model portions, is compared in Fig. 1c. It is clear that all the vertical profiles are in excellent agreement with the previous reference results. Table 1 shows a further quantification of the results in

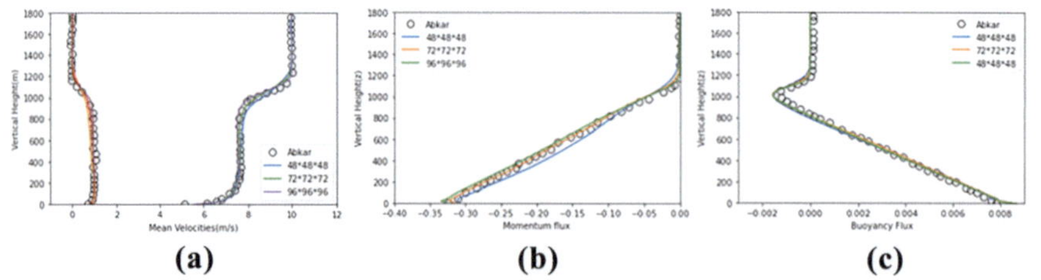

Fig. 1 Comparison of the convective boundary-layer simulation results with previously published results [1]. **a** Mean velocities in m/s, **b** momentum flux in m^2/s^2 and **c** buoyancy flux in m^2/s^3

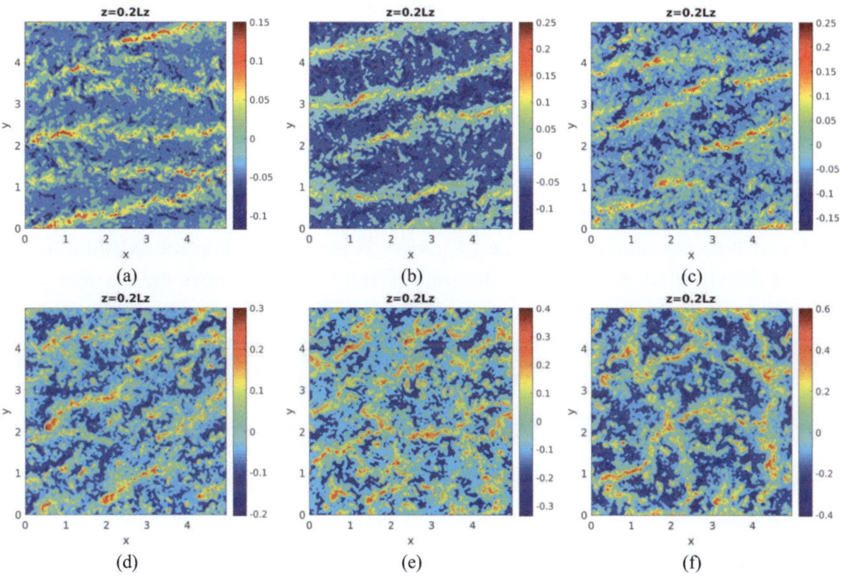

Fig. 2 Flow structures for $U_g = 9$ m/s and varying q_w visualized using instantaneous contours of w/U_{ref} at $z = 0.2L_z$ for q_w values of **a** 0.0075 K-m/s, **b** 0.015 K-m/s, **c** 0.03 K-m/s, **d** 0.06 K-m/s, **e** 0.18 K-m/s and **f** 0.36 K-m/s

terms of the boundary-layer height, Obhukhov length, friction velocity and vertical velocity scale. Quantities in the brackets in the third row are the values reported by [1]. Our LES results show very small variations with the grid size and agree very well with the reference results.

The second set of simulations is analyzed next, with the geostrophic wind fixed at $U_g = 9$ m/s and varying the surface heat flux. The effect of varying the surface heat flux keeping the geostrophic wind unchanged is seen in Fig. 2. Contours of the instantaneous vertical velocity, w'/U_{ref} ($U_{ref} = 10$ m/s), at a height of 400 m from the surface are shown. For smaller surface heating, the updrafts are organized as convective rolls. For the two largest q_w values, the updrafts resemble convective cells, similar to what is observed in classical Rayleigh-Benard convection. A gradual transition is observed from rolls to cells at intermediate values of q_w. The same conclusions can be drawn from contours of θ' but are not shown here for brevity. This transition from rolls to cells has been observed previously by [4].

Modeling the heat flux in terms of resolved quantities, the surface heat flux and the geostrophic forcing is important for closure in large-scale models. The eddy-diffusivity mass-flux (EDMF) parameterization proposed by [5] is evaluated using our LES data in Fig. 3. The EDMF parameterization gives the heat flux as

$$\left\langle w'\theta' \right\rangle_{EDMF} = -K \frac{d \left\langle \theta \right\rangle}{dz} + a_u w_u \left(\theta_u - \left\langle \theta \right\rangle \right), \tag{1}$$

where K is an eddy-diffusivity, and the quantities with subscripts 'u' are averages over updrafts, or regions where $w > 0$. The right-hand side is calculated directly from a few snapshots of the LES data and averaged over time and horizontal directions. Figure 3 shows that the resolved heat flux obtained from the LES is predicted qualitatively correctly by the original EDMF model. However, there are large over-predictions closer to the ground and under-predictions closer to the top of the boundary layer. The original EDMF formulation was proposed and tested for purely thermal-driven boundary layers, i.e., $U_g = 0$. It is possible that the introduction of shear in the simulations of purely thermal-driven flows enhances the vertical transport of hear flux which is not accounted for in the EDMF. Hence, Our results suggest that a modification to the original EDMF formulation is necessary to obtain a quantitatively accurate model of the heat flux for convective boundary layers that are also driven by shear ($U_g \neq 0$).

The correction required to the EDMF parameterization is evaluated as

$$\langle w'\theta' \rangle_{corr} = \langle w'\theta' \rangle - \langle w'\theta' \rangle_{EDMF} \tag{2}$$

and plotted in Fig. 4 for different cases after normalizing by the respective q_w. For the three higher q_w values, the required corrections fall on approximately the same curve. For $q_w \leq 0.03$ K-m/s, roughly the same profile is followed till $z/z_* \leq 0.5$ but the required corrections diverge above that height. The divergence can be traced to the fact that the EDMF parameterization magnitude becomes excessively large close to the top of the boundary-layer height, which is in turn due to a small mismatch in the prediction of z_*. Excluding these three cases for the present, the corrections required for the higher surface heating values are averaged and a cubic fit to the average is shown in Fig. 4. Incorporating this correction in the EDMF model leads to a significant improvement in the prediction of the heat flux, as seen in Fig. 3. The improvement is particularly significant for $z/z_* \leq 0.6$. Closer to the boundary-layer top, there continues to be significant mismatch between the LES and the predicted values.

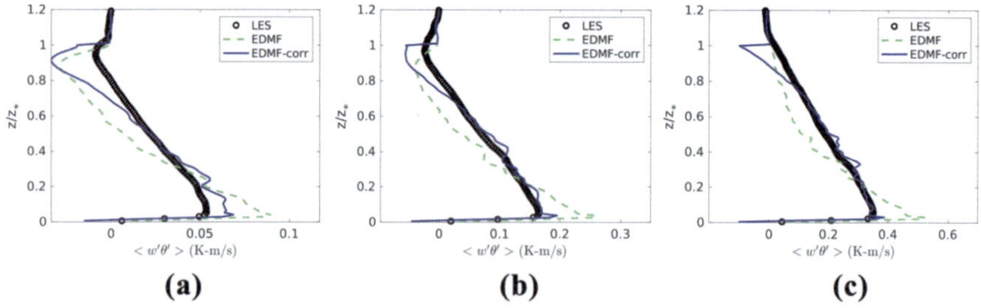

Fig. 3 LES results compared to EDMF and corrected EDMF parameterizations for q_w equal to **a** 0.06 K-m/s, **b** 0.18 K-m/s and **c** 0.36 K-m/s

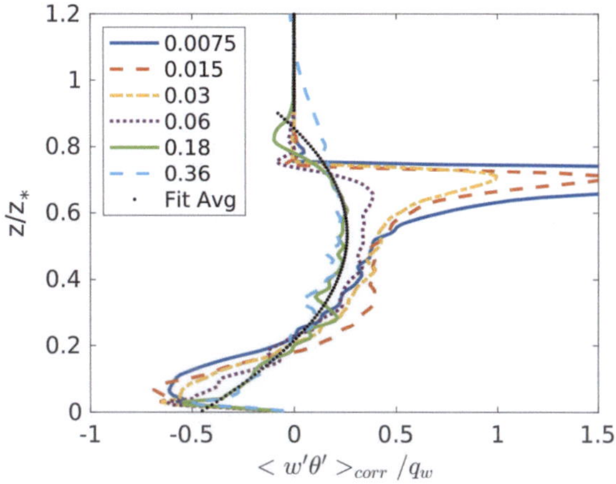

Fig. 4 Correction to the EDMF model. Legend shows the wall heat fluxes in K-m/s. 'Fit Avg' denotes cubic fit to the average of the last three curves

To provide a physical basis for the corrections, we perform a quadrant analysis of the heat flux. The quadrants are identified by plotting w' on the vertical axis and θ' on the horizontal axis. Thus, the four quadrants are Q1: ($\theta' > 0$, $w' > 0$), Q2: ($\theta' < 0$, $w' > 0$), Q3: ($\theta' < 0$, $w' < 0$), and Q4: ($\theta' > 0$, $w' < 0$). The mass-flux portion of the EDMF parameterization accounts explicitly for Q1 (warm updraft) and Q3 (cold downdraft) events. Thus, it is clear that the Q2 (cold updraft) and Q4 (warm downdraft) events need to be modeled by the eddy-diffusivity portion and the correction. The fraction of events that can be classified as Q2 and Q4 are shown in Fig. 5a and c, respectively. The corresponding contribution to the heat flux (e.g. $|\langle < w'\theta' \rangle >_{Q2} |/| \langle < w'\theta' \rangle |$ for Q2) are shown in Fig. 5b and d, respectively, for Q2 and Q4. There is a striking similarity between the Q2 and Q4 events and contributions across different cases. The Q2 events/contributions are almost identical while the Q4 results show a clear dependence on the surface heating. Our current efforts are targeted at determining a relation between the Q2/Q4 contributions and the composite contribution of the eddy-diffusivity and corrections. This will eventually enable the development of a modified EDMF parameterization that is applicable to shear-driven convective boundary layers.

4 Conclusions

We have conducted large-eddy simulations of dry convective or unstable atmospheric boundary layers. The flow is forced by the geostrophic wind on the top and by a surface heat flux from the bottom. The relative strengths of these two forcings determine the flow structures and mean flow properties such as friction velocity, overturning velocity scale, Obhukhov length and the boundary-layer height. Our results are validated by comparing to previously published reference results for $U_g = 10$ m/s and $q_w = 0.24$ K-m/s. Excellent agreement is obtained between our results

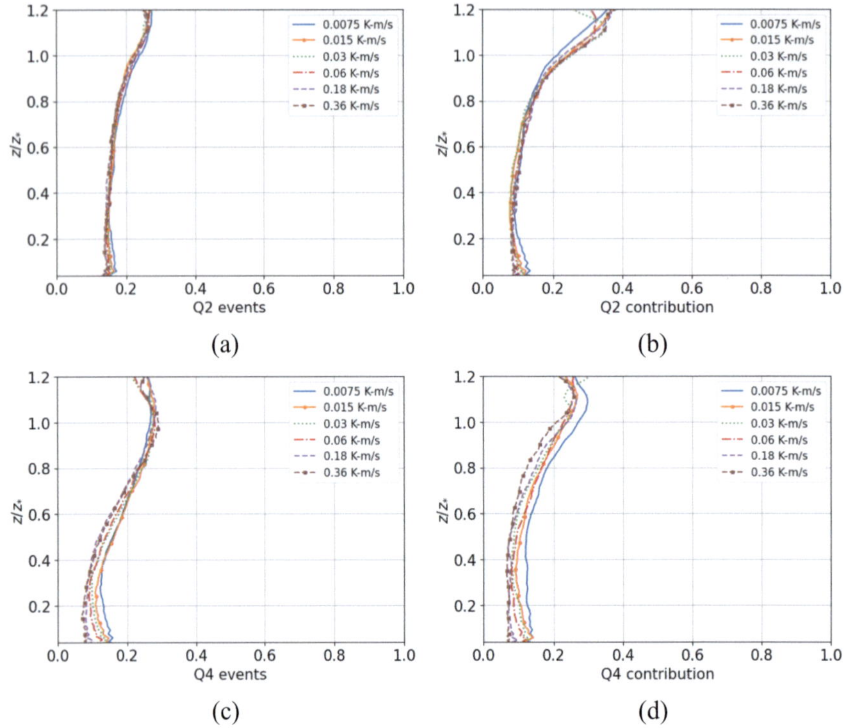

Fig. 5 Quadrant analysis based on w' and θ'. **a** Fraction of Q2 events, **b** contribution to $\langle w'\theta' \rangle$ of Q2 events, **c** fraction of Q4 events and **d** contribution to $\langle w'\theta' \rangle$ of Q4 events

and the reference results for vertical profiles as well as averaged quantities. For a fixed $U_g = 9$ m/s and varying surface heatings, the flow structures show a transition from rolls (for lower q_w) to cells (for higher q_w), consistent with previously reported results.

The eddy-diffusivity mass-flux (EDMF) parameterization is found to give qualitatively similar profiles of the heat flux. However, there are significant quantitative differences. These differences between the actual heat flux and the heat flux predicted by EDMF are quantified and are shown to be similar across a range of q_w values. Finally, the flow events are classified into four quadrants based on the perturbations of the vertical velocity and the potential temperature around their respective horizontal means. The events in the second and fourth quadrants are not represented by the mass-flux portion of the EDMF model. Results of this quadrant analysis will be used in the future to develop an improved EDMF parameterization, which will be suitable for convective boundary layers that are driven by a vertical shear in addition to the surface heating.

Nomenclature

u_* Friction velocity $[L/T]$
w_* Vertical overturning velocity $[L/T]$
z_0 Aerodynamic roughness $[L]$
z_* Boundary-layer height $[L]$
L Obhukhov length $[L]$
U_g Geostrophic wind speed $[L/T]$
q_w Surface heat flux $[KL/T]$
κ von-Karman constant $[-]$

Acknowledgements NSG thanks the SERB Core Research Grant (CRG/2022/000675) for enabling access to computational resources on the ParamSeva cluster at IIT Hyderabad.

References

1. Abkar M, Moin P (2017) Large-eddy simulation of thermally stratified atmospheric boundary-layer flow using a minimum dissipation model. Bound-Layer Meteorol 165:405–419
2. Moeng C-H, Sullivan P (1994) A comparison of shear and buoyancy-driven planetary boundary layer flows. J Atmos Sci 51:999–1022
3. Rozema W, Bae HJ, Moin P, Verstappen R (2015) Minimum-dissipation models for large-eddy simulation. Phys Fluids 27:085107
4. Salesky S, Chamecki M, Bou-Zeid E (2017) On the nature of the transition between roll and cellular organization in the convective boundary layer. Bound-Layer Meteorol 163:41–68
5. Siebesma A, Soares P, Teixeira J (2007) A combined eddy-diffusivity mass-flux approach for the convective boundary layer. J Atmos Sci 64:1230–1248
6. Subramaniam A, Ghate A, Ghaisas NS, Howland MF et al (2021) PadeOps GitHub repository, Available: https://github.com/FPAL-Stanford-University/PadOps/tree/igridSGS. Last Accessed: 5 Aug 2021
7. Sullivan P, Williams J, Moeng C-H (1994) A subgrid-scale model for large-eddy simulation of planetary boundary-layer flows. J Atmos Sci 51:999–1022
8. Tan Z, Kaul CM, Pressel KG, Cohen Y, Schneider T, Teixeira J (2018) An extended eddy-diffusivity mass-flux scheme for unified representation of subgrid-scale turbulence and convection. J Adv Modeling Earth Syst 10(3):770–800

Experimental Investigation of Hydrodynamic Instabilities Due to Air Flow Over a Liquid Film in a Horizontal Channel

Sudipta Saha, SK. Hasan Raja, Biswanath Chakraborty, Aranyak Chakravarty, Sourav Sarkar, Achintya Mukhopadhyay, and Swarnendu Sen

Abstract Hydrodynamic instabilities generated due to interaction of air flow over liquid films are a phenomenon of substantial importance that is dealt in various natural and industrial processes. One of the major instances where this can be observed is our respiratory tract, where underlying a liquid layer (mucus) air flow. Such interaction can lead to instability and can results in formation of droplets and ligaments of different sizes and shape. An experimental investigation is carried out to understand the phenomenon and distribution of droplets and ligaments in a channel which is subjected to air flow over an underlying liquid layer. The analysis is carried out for different air flow rate ranging from 5 to 10 lpm considering a constant liquid feed rate of 1 ml/min. The effect of liquid viscosity is also studied by considering two liquids—water and glycerol—having different viscosities. Co-current flow of air and liquid is considered with focus on understanding the underlying physics behind the destabilization of the liquid film under different flow conditions. The extent of destabilization of the liquid film is observed to be dependent on the magnitude of the air flow rate. Viscosity is also observed to play an important role in determining the flow pattern and destabilization of liquid film. Formation of ligaments is observed to be more probable for lower air flow rates, while droplets are formed for higher air flow rates. Size of the droplets formed is observed to be in the sub-millimeter range.

Keywords Kelvin–helmholtz instability · Droplets · Liquid film · Fragmentation · Stratified flow

S. Saha · A. Chakravarty (✉)
School of Nuclear Studies and Application, Jadavpur University, Kolkata, India
e-mail: aranyak.chakravarty@jadavpuruniversity.in

SK. Hasan Raja · S. Sarkar · A. Mukhopadhyay · S. Sen
Department of Mechanical Engineering, Jadavpur University, Kolkata, India

B. Chakraborty
Department of Mechanical Engineering, NIT Durgapur, Durgapur, India

A. Kumar R. et al. (eds.), *Proceedings of Fluid Mechanics and Fluid Power (FMFP) 2023, Vol. 6*, Lecture Notes in Mechanical Engineering,
https://doi.org/10.1007/978-981-96-9600-0_16

1 Introduction

Liquid films are encountered in several natural phenomenon as well as in various technological applications. Understanding the dynamic and velocity-dependent interactions between an air and liquid film is of great importance in fundamental scientific problems, which are of interest to many industrial processes. Such processes include micro—and nanofluidic engineering [1, 2] as well as biomedical engineering [3, 4]. A typical example of liquid film is the mucus layer lining the inner surface of respiratory tract in the human body.

The Kelvin–Helmholtz (KH) instability [5, 6] is one of the three most commonly used interfacial hydrodynamic instabilities and, in the context of inertial confinement fusion, frequently follows the acceleration driven Rayleigh–Taylor (RT) instability [7, 8]. It was experimentally investigated the parallel flow of two immiscible liquids in a Hele-Shaw cell and reported that the instability threshold in the liquid-liquid interface is controlled by inertial force, while the wave velocity and its growth rate are controlled by the viscosity [9].

When air flows through over a thickly lined mucus layer in airways, the interaction between air and liquid layer develops an interfacial shear. Thus, the liquid layer can be propelled in the direction of air flow when the velocity or flow rate of air becomes sufficiently high to develop a shear force which can overcome the viscous resistance of mucus layer [10]. This destabilizes the air-liquid film interface and may advance the rupture of the film interface under certain circumstances. The rupture of film due to the instability may results in formation of droplets and ligaments of several shapes and sizes. These droplets are transported along the airways in the direction of air flow and play an important role in progression of respiratory diseases as well as pulmonary drug delivery [11]. To properly predict the phenomenon, the nature of liquid film shearing that effects the interaction must be studied. Such process depends acutely on the relative velocity of the moving air and liquid film. At different velocities (flow rates), the hydrodynamic instability between air and liquid film in a channel provides a major knowledge to the interfacial deformation liquid film and air. Also, the physiological properties of the liquid in the liquid films, i.e., viscosity, density, surface tension, etc., play a pivotal role in determining the interaction phenomenon.

2 Objective

Only limited experimental and computational works have been carried out to understand the interaction between two fluids in a channel [10, 12–18]. The objective of our present study is to experimentally investigate the hydrodynamic instability due to interaction between air flow and liquid film in a horizontal channel. The effects of liquid viscosity and relative air-liquid flow rates are primarily investigated in this study. The results are reported in terms of the evolution of air-liquid interface over time as well as the size of droplets/ligaments formed due to the hydrodynamic interaction between air and the liquid film. The aspect ratio of the channel is kept fixed for all the investigations reported here. The impact of channel aspect ratio will be studied in the future.

3 Experimental Methodology

3.1 Materials

A horizontal transparent channel of rectangular cross section (4 mm × 6 mm) is considered. The channel is subdivided into two different sections—a constricted section and an expanded section. The constricted section is the present focus of our study. The length of constricted section is 5.8 cm. Two different liquids, i.e., distilled water and glycerol, are used. A syringe pump (New Era NE-1000) is used to deliver liquids into the channel through a separate inlet at a constant flow rate. The dispensing accuracy of the syringe pump is between − 1 and + 1%. A mass flow controller (ALICAT) ranging from 0 to 10 l per mins is used for maintaining constant flow rate of air throughout the channel. The uncertainty of the flow of the controller is ±0.32% reading or ±0.02% full scale, whichever is greater. Rhodamine is additionally used as a dye to enable easier visual demarcation of the air-liquid interfaces. A magnetic stirrer is used to mix the liquids uniformly. A reciprocating compressor for air supply is used. A LED light source with regulatable intensity and a high-speed camera (The Photron FASTCAM mini UX100) with lens of 70 mm focal length (Tamron SP 24-70 mm F/2.8 Di VC USD model: A007N) are used for visualization. The images are captured at 12500 fps using Photron Fastcam Viewer (PFV). The schematic of experimental setup is shown in Fig. 1.

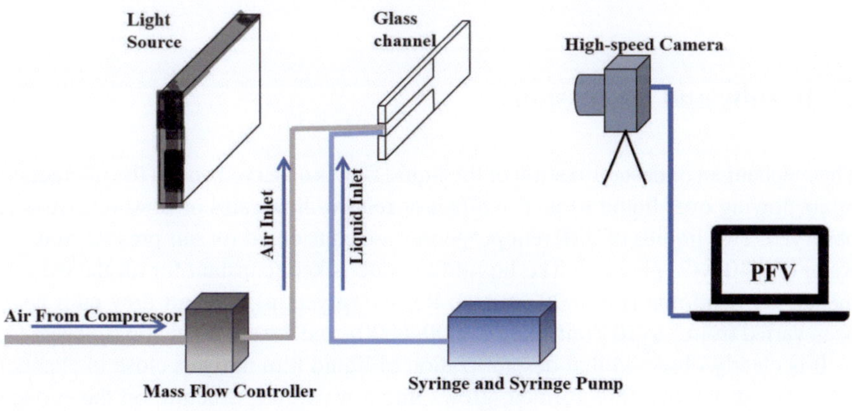

Fig. 1 Schematic diagram of the present experimental setup

Table 1 Properties of air and liquids considered in our experiments

Material	Density (kg/m^3)	Dynamic viscosity (kg/ms)	Surface tension (N/m)
Air	1.225	1.81×10^{-5}	–
Water	997.05	0.00087	Air–water—0.072
Glycerol (40 wt%)	1096.1	0.00314	Air–glycerol—0.069

3.2 *Methods*

Experiments were performed using two different liquids injected into the channel—distilled water and glycerol (40 wt%). Compressed air is injected into the channel at different flow rates varying from 5 to 10 lpm. The properties of the fluids are listed in Table 1. The flow rate for both the liquids is kept constant at 1 ml/min for all the cases and liquids are supplied with the help of syringe pump. Rhodamine is mixed with the flowing liquids, and the mixing is done with the help of magnetic stirrer for at least an hour at a rotation of 1000 RPM to identify the liquid in the channel without significant change in water properties. Air is supplied in the channel underlying the liquid layer of fluids through a reciprocating air compressor, and the flow rate is controlled with the help of mass flow controller. An initial flow rate and thickness of liquid layer are already maintained before supplying the air. A suitable length scale and time scale are attached with the setup to get the real time values. The phenomenon is captured with the help of high-speed camera, and the results are being post processed with the help of in-house MATLAB algorithm. This provides us information regarding droplet size distribution for all the cases. A specific region of domain in the expanded section of the channel is considered for the analysis. With the help of appropriate scale factor and a binary conversion threshold, the boundaries of the droplets are identified, and an equivalent circle is determined based on boundary length. The diameter of the circle is considered to be diameter of droplet. The algorithm is described in Gangopadhyay et al. [16].

4 Results and Discussion

The mechanism of destabilization of the liquid film is observed due to the interaction of air flowing over liquid film. The effect of relative flow rates or flow velocities is observed. Two liquids of different properties are considered for our present study—water and 40 wt% glycerol. The liquid flow rate is kept constant for all the experiments, i.e., 1 ml/min ($Re_l \sim 20$ (water); $Re_l \sim 6$ (glycerol)), and air flow rates have been varied from 5 to 10 l/min ($Re_a \sim 3000$–6000), and the phenomenon is observed.

It is clearly observed that destabilization of liquid film initiates close to channel inlet. The instability, thus formed, grows into a wave-like structure on the surface of the liquid film. The temporal evolution of the liquid film instability is illustrated in Fig. 2 using captured images. It is observed that the phenomenon is primarily initiated due to the existing relative velocity between the two phases. Higher air velocity drags the liquid film in the direction of air flow which destabilizes the surface leading to onset of Kelvin–Helmholtz instability on the film surface. This instability grows in the direction of air flow and causes stripping of ligaments and droplets from the surface of the liquid film. The droplets or ligaments, thus formed, are carried downstream along with the air flow. This mechanism persists until the instability is evolved along the liquid film. The propagation of the instability forces liquid from the film to spill over outside the constricted portion and leads to the liquid film getting depleted (see Fig. 2). It is also observed that the ligaments undergo a flapping motion—in the expanded section of the channel—due to the local air flow in that region as shown in Figs. 2 and 3.

Effect of Relative Flow Rate

The relative flow rates of the injected liquid and air are observed to have a major contribution in determining the liquid film destabilization. Figure 3 illustrates the distribution of phases within the channel for $Q_a = 6$ and 10 l/min for a constant liquid flow rate $Q_l = 1$ ml/min. It can be seen that while Kelvin–Helmholtz instability is established for both flow rates cases, the extent of destabilization of the liquid film is dependent on the magnitude of the air flow rate.

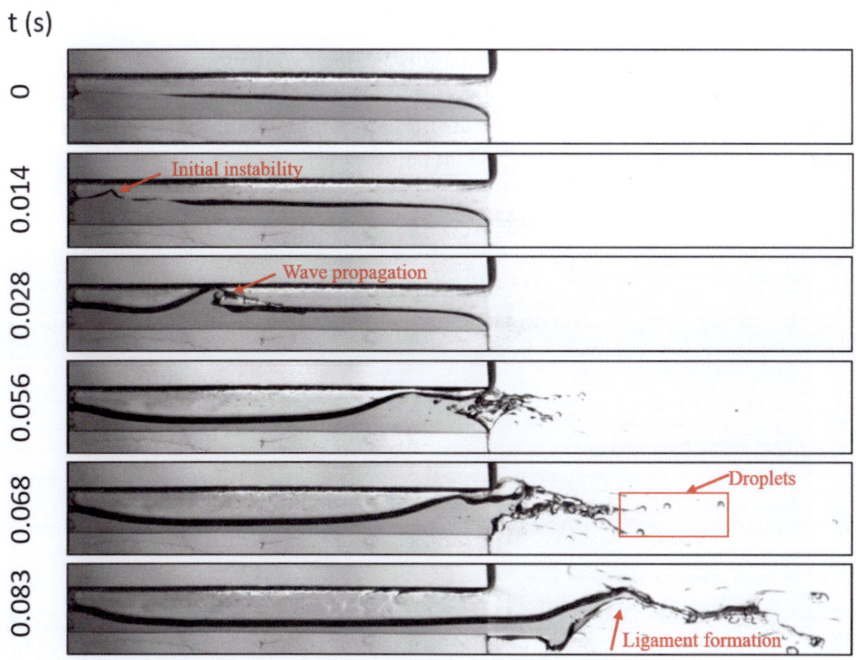

Fig. 2 Development of instability in liquid (water) film and formation of droplets and ligaments due to breakup of film at $Q_a = 6$ l/min and $Q_l = 1$ ml/min

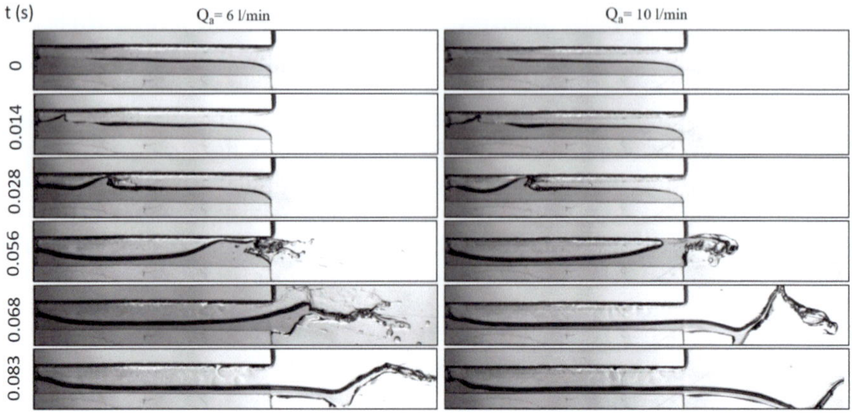

Fig. 3 Comparison of liquid (water) film destabilization with change in air flow-rate $Q_l = 1$ ml/min

Effect of Viscosity

The effect of variation in viscosity is also observed in our present study. Two liquids are considered, i.e., water and 40 wt% glycerol with the latter being more viscous than water (see Table 1). Experiments are carried out for Q_a (5–10 lpm) and Q_l (1 ml/min). Figure 4 compares the observed phenomena for a representative flow rate of air of 7 l/min.

It is clearly evident that the K–H instability is much more pronounced for glycerol (more viscous) as compared to water (less viscous). Multiple surface waves are observed to be formed in case of glycerol. The wave length of the surface waves is also observed to be much smaller (as compared to water).

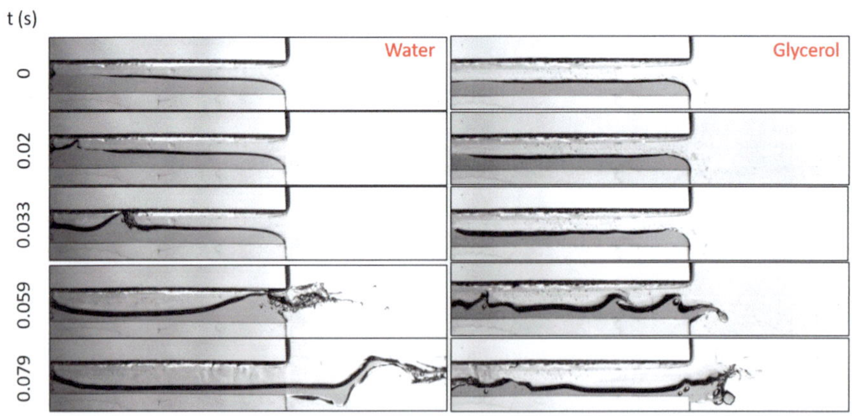

Fig. 4 Comparison of liquid film destabilization with change in viscosity ($Q_l = 1$ ml/min, $Q_a = 7$ l/min)

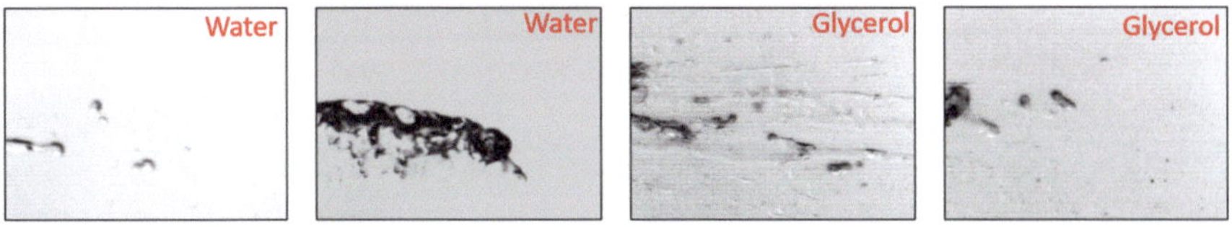

Fig. 5 Representative images showing various droplets/ligaments shapes and sizes as observed experimentally

Also, the flow of glycerol solution is much slower than water as solution if glycerol is more viscous and glycerol has strong interparticle force of attraction whereas water has weaker interparticle force of attraction. The flow pattern for the water air interaction in the channel remains stratified, while for the same configuration and flow parameters, the flow pattern changes from stratified to stratified wavy for the flow of air glycerol. So, it is evident that viscosity has a role in determining the flow pattern and destabilization of liquid film. For the case of water, the stability of liquid film achieved much earlier than that of glycerol. The evidence is shown in Fig. 4.

Droplet Shapes and Sizes

Figure 5 shows the different shapes and sizes of droplets and ligaments that are experimentally observed to be formed as a result of film rupture. The probability distribution of droplet size is shown in Fig. 6. It is evident that the droplets are mostly in sub-millimeter range (mostly 200–600 μm). However, few droplets are observed having diameter more than 1 mm and also there can be seen ligaments for all the cases whose diameters ~ 2 mm. The probability of formation of ligaments is observed to be more for the cases where air flow rate is lower. This is due to lower drag as the shear is not large enough to fragment the ligaments into droplets. However, it is observed to be varying between the two liquids studied.

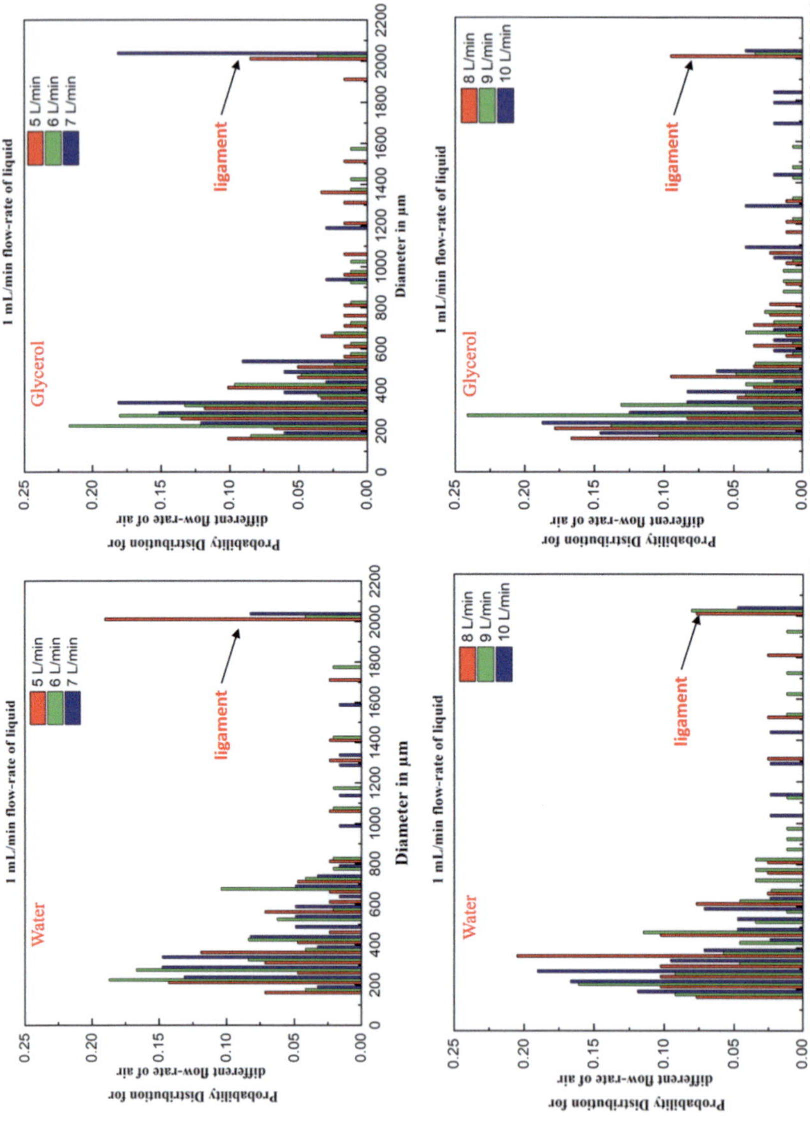

Fig. 6 Probability distribution of droplet size obtained for both the cases for different air flow rates

5 Conclusion

The following major observations were made from the experimental analysis:

a. Kelvin–Helmholtz instability is the dominant hydrodynamic phenomena that destabilizes the air–liquid interface within the channel.
b. The shearing action, due to the Kelvin–Helmholtz instability, leads to formation of ligaments/droplets beyond a certain condition. No ligament/droplet formation was observed when the air flow rate remained below 5 l/min in the current configuration.
c. Increase in relative flow rates of the fluid enhances surface wave propagation as well as droplet formation. Smaller droplets were formed at higher air flow rates, and the droplets are majorly in sub-millimeter range (200–400 μm). Ligaments ($\geq$ 2 mm) are observed to be formed for lower air flow rates for both the liquids (water and glycerol).
d. The K–H instability is observed to get pronounced with increase in viscosity of the liquid film. Multiple surface waves are observed with smaller wavelengths. However, no significant change is observed in the sizes of the droplets/ligaments.

Acknowledgements The authors gratefully acknowledge the funding support provided by the SERB, DST, Govt. of India.

References

1. Whitesides GM (2006) The origins and the future of microfluidics. Nature 442:368–373
2. Prakash M, Gershenfeld N (2007) Microfluidic Bubble Logic. Science 315:832–835
3. Taylor CA, Draney MT (2004) Experimental and computational methods in cardiovascular fluid mechanics. Annu Rev Fluid Mech 36:197–231
4. Chen KJ et al (2014) Hyperthermia-mediated local drug delivery by a bubble-generating liposomal system for tumor-specific chemotherapy. ACS Nano 8:5105–5115
5. Kelvin L (1871) Hydrokinetic solutions and observations. Phil Mag 42(281):362–377
6. von Helmholtz H (1868) On discontinuous movements of fluids. Phil Mag 36(244):337–346
7. Rayleigh L (1882) Investigation of the character of the equilibrium of an incompressible heavy fluid of variable density. Proc Lond Math Soc s1-14(1):170–177
8. Taylor GI (1950) The instability of liquid surfaces when accelerated in a direction perpendicular to their planes I. Proc Math Phys Eng Sci 201(1065):192–196
9. Gondret P, Rabaud M (1997) Shear instability of two-fluid parallel flow in a hele–shaw cell. Phys Fluids 9(11):3267–3274
10. Kim CS, Greene MA, Sankaran S, Sackner MA (1986) Mucus transport in the airways by two-phase gas-liquid flow mechanism: continuous flow model. J Appl Physiol 60(3):908–917
11. Chakravarty A, Panchagnula MV, Mohan A, Patankar NA (2022) Pulmonary drug delivery and retention: a computational study to identify plausible parameters based on a coupled airway-mucus flow model. PLoS Comput Biol 18(6):e1010143
12. Cherlo SKR, Kariveti S, Pushpavanam S (2010) Experimental and numerical investigations of two-phase (liquid– liquid) flow behavior in rectangular microchannels. Ind Eng Chem Res 49(2):893–899

13. Balasubramanian P, Kandlikar SG (2005) Experimental study of flow patterns, pressure drop, and flow instabilities in parallel rectangular minichannels. Heat Transf Eng 26(3):20–27
14. Nezihovski Y, Gelfgat A, Ullmann A, Brauner N (2022) Experimental measurements versus linear stability analysis for primary instability of stratified two-phase flows in a square rectangular duct. Int J Multiphase Flow 153:104061
15. Rajendran RR, Banerjee A (2021) Effect of non-Newtonian dynamics on the clearance of mucus from bifurcating lung airway models. J Biomech Eng 143(2):021011
16. Gangopadhyay T, Ravikrishna RV, Mukhopadhyay A (2016) ILASS-Asia 18th Annual Conference on liquid atomization and spray systems, India
17. Saha S, Sarkar S, Chakravarty A, Mukhopadhyay A (2022) A computational analysis of hydrodynamic instability during co-current airflow over a thin liquid film. In: ILASS-Asia 22nd Annual Conference on Liquid Atomization and Spray Systems
18. Yao C, Zhao Y, Zheng J, Zhang Q, Chen G (2020) The effect of liquid viscosity and modeling of mass transfer in gas–liquid slug flow in a rectangular microchannel. AICHE J 66(5):e16934

Numerical Simulation of Shear-Induced Particle Migration of Concentrated Suspension in a Networked-Bifurcating Channels

Rajnish Kumar Pandey, Rohit Kumar, and Anugrah Singh

Abstract Suspensions are made with solid rigid particles in viscous fluids often used in many industries as well as biological processes and natural phenomena. In some phenomena, separating the particles from the bulk suspension is critical. Suspension flow even in simple geometry has complex nature. In the current work, numerical investigation of the shear-induced particle migration of concentrated suspension in a network of bifurcating channels is performed. A numerical simulation of monodisperse suspension in a two-dimensional (2D) network of bifurcating channels was performed. The velocity profile of suspension has a marked difference from the velocity profile of the pure fluid having same effective viscosity. Moreover, the concentration field shows inhomogeneous particle distribution resulting from shear-induced migration. There is unequal partitioning of particles in the daughter branches, and the partitioning of particles does not follow the partitioning of the bulk suspension.

Keywords Numerical simulation · Particle fractionation · Monodispersed suspension · Particle migration · Fluid flow

1 Introduction

The significance of rheology has been increasing very rapidly in demand. Suspension of solid particles in viscous fluid are commonly encountered in petroleum industry, food processing, paints, and cosmetics, etc. In biological system common examples are milk and blood. Moreover, in nature it can also be observed in the form of mud and sediments in rivers. In solid-in-liquid suspension, the dispersed phase is

R. K. Pandey · R. Kumar (✉) · A. Singh
Department of Chemical Engineering, Indian Institute of Technology Guwahati, Guwahati, India
e-mail: rohit1492kumar@gmail.com

R. Kumar
Department of Mechanical Engineering, Indian Institute of Technology Guwahati, Guwahati, India

A. Kumar R. et al. (eds.), *Proceedings of Fluid Mechanics and Fluid Power (FMFP) 2023, Vol. 6*, Lecture Notes in Mechanical Engineering,
https://doi.org/10.1007/978-981-96-9600-0_17

finely distributed in the continuous dispersion medium. In the present work, we will emphasize the suspensions more precisely, addressed as the dispersion of the solid particle in a continuous liquid phase. The forces in suspension are classified into two categories: hydrodynamic forces and non-hydrodynamic forces. Hydrodynamic forces are occurred basically by the viscous force and inertial forces. The action of these forces in a flowing suspension gives rise to many interesting phenomena such as shear-induced particle migration.

Firstly, Gadala Maria and Acrivos [1] saw a decrease in viscosity during the inhomogeneous shearing of concentrated suspension in a cylindrical Couette device. They could not explain the reason behind the viscous decrease when shearing occurred. After this, Leighton and Acrivos [2] explained the decrease in viscosity upon shearing. Abbott et al. [3] investigated the particle migration in Couette flow. They found that particle migration is directly proportional to the particle diameter, shear rate, and independent of the suspending fluid viscosity. Using NMR imaging, Fang and Phan-Thien [4] observed shear-induced particle migration in concentrated suspensions, undergoing flow between the rotating eccentric cylinders. Shear-induced migration in pressure-driven flow in circular tubes was studied by Hampton et al. [5]. Diffusive flux model (DFM) was used based on the scaling arguments of Leighton and Acrivos [6], developed by Phillips et al. [7]. According to this, the shear-induced particle migration fluxes responsible for the particle migration were due to spatially varying collision frequency (Nc) and viscosity ($N\eta$). The diffusive flux model successfully explains the shear-induced particle migration phenomena in rectilinear geometry, like channels and pipe flows. After this, the model was improved by Krishnan et al. [8], which includes the flux due to the radius of curvature (Nr) into the total migration flux (Nt), proposed by Phillips et al. [7].

Reddy and Singh [9] conducted numerical simulations for bidisperse suspensions based on a model proposed by Kanehl and Stark [10] in a T-shaped converging and diverging channels. Moreover, the results are compared against the obtained results with monodisperse suspensions. They concluded that in a bidispersed system the large particle dominates near the center leading to the depletion of small particles. In this work, we have restricted ourselves to a suspension that is non-colloidal, non-Brownian, and the flow is non-inertial.

2 Modeling and Validation

In the present work, the diffusive flux model is used to perform the simulations with monodispersed suspension. The two essential aspects of this model are: (a) assuming the whole suspension to be a generalized Newtonian fluid with effective viscosity (η) as a function of particle volume fraction (φ), and (b) the particle conservation equation is solved for the distribution of solid particles [7].

The steady-state continuity and momentum equations for the suspension are given as

$$\nabla.U = 0 \tag{1}$$

$$\rho\nabla.UU = -\nabla P + \nabla.\tau \tag{2}$$

where U is the bulk suspension velocity, P is the pressure, ρ is the density of the suspension, and τ is the stress tensor which is given as

$$\tau = -2\eta(\varphi)E \tag{3}$$

where $\eta(\varphi)$ is the effective viscosity of the suspension and is given by the Kreiger [11] correlation as

$$\eta(\phi) = \eta_0(1 - \frac{\phi}{\phi_c})^{-1.82} \tag{4}$$

In the Eq. (3), E is the strain rate tensor and is given by the mathematical expression:

$$E = \frac{1}{2}(\nabla U + \nabla U^T) \tag{5}$$

In the present model, one more additional transport equation is solved to predict the particle migration. The steady-state particle conservation equation is given by [9]

$$\nabla.U\varphi = -\nabla.(N_t) \tag{6}$$

In the above equation, N_t represents the total migration flux, which includes the net effect of spatially varying interaction frequency (N_c), viscosity (N_η), Brownian diffusion (N_b) and curvature induced flux (N_r).

$$N_t = N_c + N_\eta + N_b + N_r \tag{7}$$

2.1 Geometry

The schematic diagram of the 2D network of bifurcating channels is shown in Fig. 1. The direction along the flow in the inlet branch of channel geometry is taken as X axis and across the channel is considered as Y axis. The inlet branch is also known as the parent branch, with a width (0.0018 m). The computational geometry contains two inlet branches and three outlet branches, which bifurcate at the angle of 60° into the daughter branches. Every parent branch bifurcates into two daughter branches, in which one daughter branch is connected to the main branch. The width of oblique bifurcating daughter branch is 0.0009 m. The bifurcation of the main branch takes place three times. The minimum length required for a fully developed flow is calculated for different sections. The particle radius taken for simulations is 37.5 µm. The length of each straight section of the inlet branch (L) as well as the bifurcating daughter branch is 0.3615 m. However, the length of the last outlet section of the middle channel (L_1) is 0.0495 m. Other geometry details and simulation parameters are mentioned in Table 1.

The analysis is done for laminar and steady flow at low Reynolds numbers. Fine grids is used near the near the wall of solid–fluid domain and grid independent test were carried out. For grid independence solution, we have performed two simulations (for $\varphi = 0.4$) with 270,000 and 150,000 grids. For both the grids, the velocity and concentration profiles follow similar trend, indicating that the simulations are grid independent as shown in Fig. 2 (Table 2).

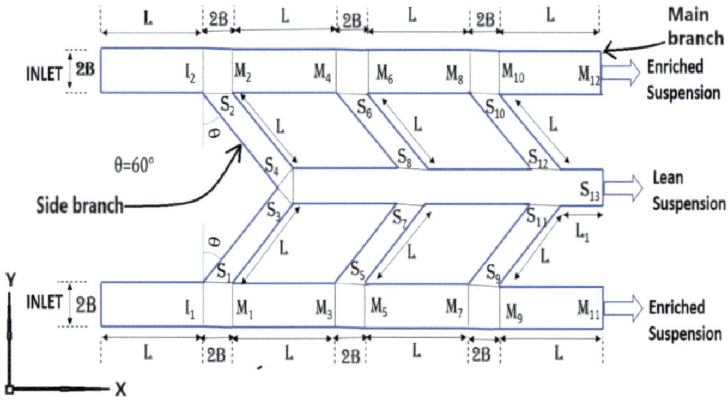

Fig. 1 Schematic diagram of network of bifurcating channel and locations where the concentration and the velocity profile are analyzed

Table 1 Geometry details and simulation parameter

S. No.	Flow parameter	Symbol	Value
1	Suspended fluid viscosity	η_0 (Pa S)	0.48
2	Suspended fluid density	ρ_f (kg/m^{-3})	1180
3	Particle radius	a (µm)	37.5
4	Width of the channel	$H = 2B$ (m)	0.0018
5	Length of each branch	L, L_1 (m)	0.3615, 0.0495
6	Ratio of channel half width to particle radius	B/a	24
7	Maximum particle volume fraction	Φ_c	0.68
8	Average inlet particle volume fraction	Φ	0.3, 0.35, 0.4

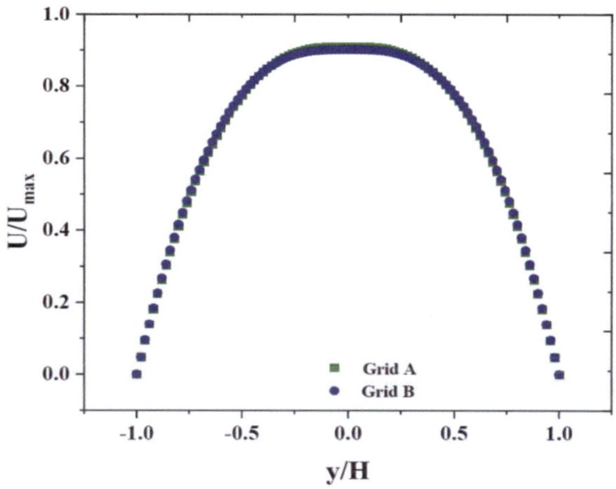

Fig. 2 Velocity profiles in the inlet section at location l_1

Table 2 Simulation parameters and geometry details in the rectangular conduit for the validation of model

1	Parameter	Symbol	Value
2	Particle radius	a [μm]	50
3	Ratio of channel width to particle radius	H/a	18
4	Fluid viscosity	η_0 [Pa.s]	0.48
5	Fluid density	ρ [Kg.m^{-3}]	1190

2.2 *Model Validation*

In Fig. 3, the simulation results using the diffusive flux model are compared with the experimental and analytical data for the case of suspension flow in a straight rectangular channel. The diffusive constants are same that were used in the studies by Phillips et al. [7]. The values of the diffusive constants are: $K_c = 0.41$ and $K\eta = 0.62$. Due to shear-induced migration, particles migrate from a high shear rate at the wall to a low shear rate region at the center. Due to this, the particle concentration is higher at the center than at the wall. The simulation results agreed with the experimental solution except for the walls as shown in Fig. 3. In shear-induced particle migration, most experiments reported difficulty in collecting reliable data in the wall region.

3 Results and Discussion

The effect of particle phase on the velocity field is studied by comparing the suspension velocity with that of pure fluid having the same effective viscosity (Fig. 3). The velocity magnitude is normalized with the fully developed pure fluid velocity ($U_{\max}$).

Fig. 3 Comparison of suspension velocity profile with experimental, analytical and same viscosity Newtonian fluid 0.4

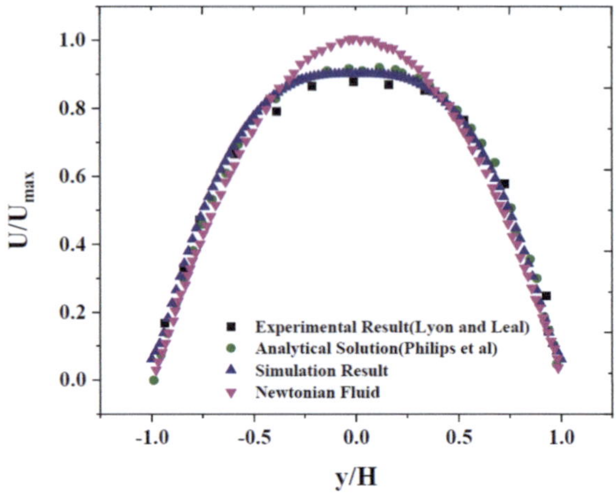

The purpose of the normalization of the velocity field is to show the comparison. The velocity profile of the suspension flowing in the network of bifurcating channels was observed to be blunted.

The velocity profiles at various locations in the inlet branch, main branch, and side branch are shown in Fig. 4. The difference in velocity profiles can be clearly noticed, and we observe that due to less width the profiles in the side branch are more blunted. Deceleration in velocity is higher for higher concentration. The full length of the network of bifurcating channels (for the clarity of flow partitioning) is not shown here. The corresponding particle volume fractions at these locations are shown in Fig. 5. The flow section along the inlet branch is called main branch or parent branch. The branch of bifurcation is side branch which connects the main branch with side daughter branch. The suspension flow in the parent branch leads to the migration of particles from the wall side to the center. This means particles are moving from the high shear rate region to the low shear rate region. The inlet velocity was taken ($U = 0.002$ m/s). It is observed that before bifurcation (location I_1, I_2) fully developed concentration profile persists. In the side branch, lower concentration is observed. Moreover, the main branch linked daughter branch contains high-concentration suspension. Again, the main branch bifurcates (Location M_3, M_4) into the side branch that connects with daughter branch. Similar phenomena happen in these branches. The higher concentration stream flows in the main branch linked daughter branch and the lean suspension flows in the side branch and daughter branch. Once again, the main branch linked side branch bifurcates (location M_{12}, M_{13}) into two branches, and enriched suspension flows in the main branch and lean suspension flows in the side branch. At location S_{13}, the flows of side branches are getting mixed. Similarly, at location S_{14}, the flows of side branches are getting mixed. Similar phenomena occur at location S_{15}. Using these data, after the velocity and concentration analysis at every bifurcation, the particle partitioning and fluid partitioning at every bifurcation can be calculated. Similarly, the overall separation efficiency in the network of bifurcating channels can be calculated.

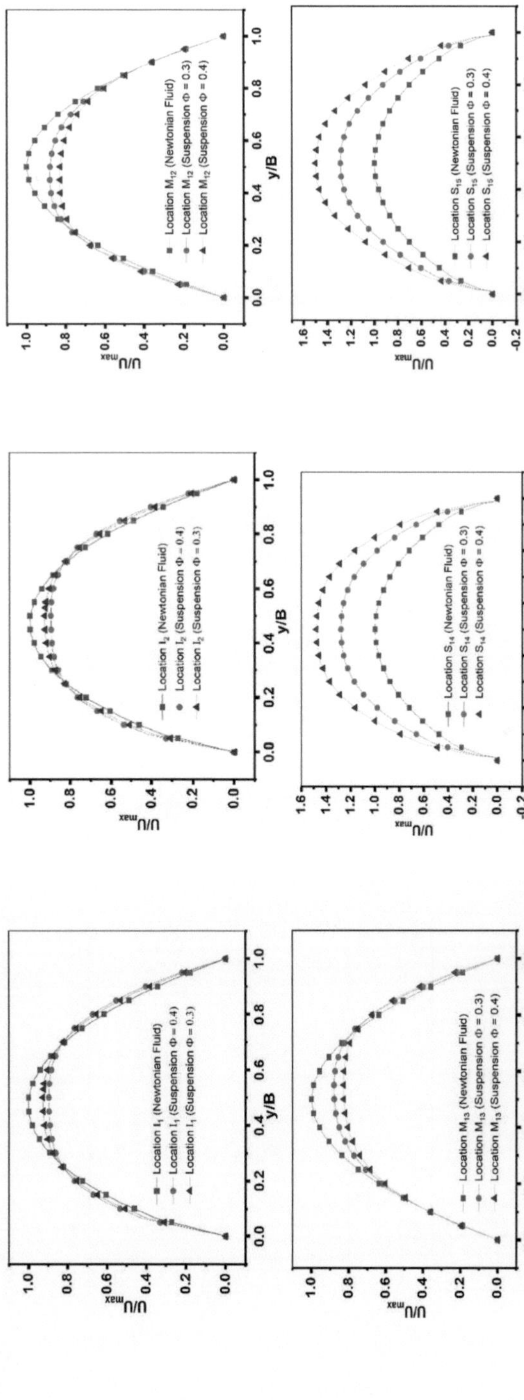

Fig. 4 Comparison of velocity profile of suspension with Newtonian fluid at every bifurcation

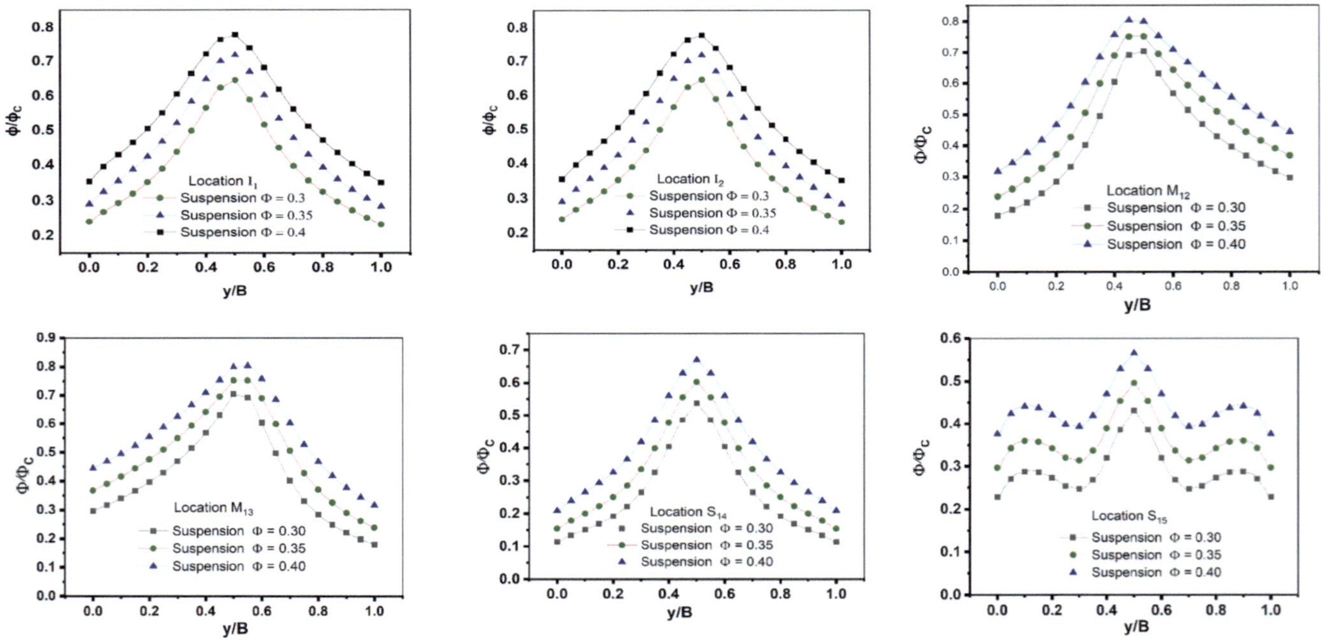

Fig. 5 Comparison of particle volume fraction at every bifurcation

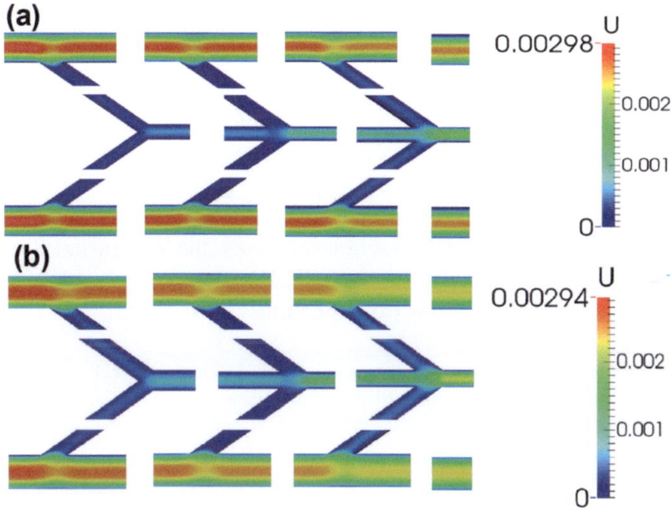

Fig. 6 Comparison of velocity contour of **a** Newtonian fluid with **b** the suspension fluid

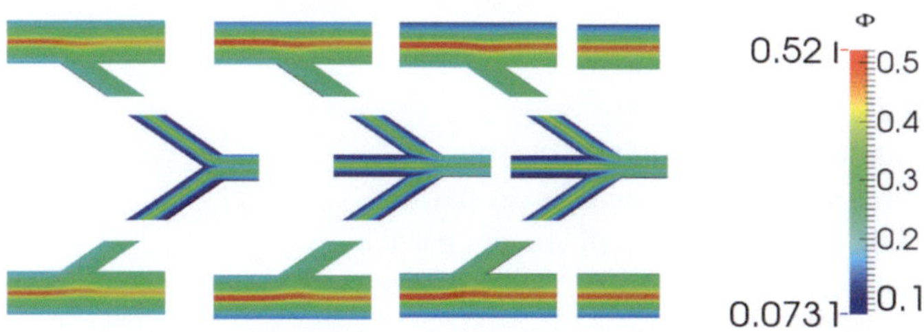

Fig. 7 Particle concentration contour plot by using diffusive flux model

The magnitude of the velocity contour of pure fluid is shown in Fig. 6. Using DFM, the corresponding contour for suspension with a 30% particle volume fraction is shown in Fig. 7.

4 Conclusions

The primary purpose of this work is to study the shear-induced migration effect in the network of bifurcating channels and velocity and particle distribution at every bifurcation of the channel. Moreover, the present study is useful in determining the separation efficiency and to predict the number of bifurcation networks required to achieve the desired efficiency. This investigation analyzes how much separation takes place at every bifurcation and the overall separation in the enriched and lean sections. In the present study, the particle considered was non-colloidal, monodispersed, and neutrally buoyant in the Stokes regime. The shear-induced migration of particles gives rise to non-uniform distribution, which influences many aspects of suspension flow in a network of bifurcation. Non-uniform distribution significantly affects every branch's velocity and concentration profile, particle partitioning, and flow features. The suspension fluid velocity profile significantly differs from the pure fluid velocity profile. The particles do not follow the suspension partitioning at the bifurcation points. As the particle fraction increases, the suspension fluid velocity profile becomes blunter. Our study can be utilized to design a network of bifurcation channels for separation of particles from the fluid.

Acknowledgements We would like to express our gratitude for the project's financing from the Science and Engineering Research Board (SERB), India. Project no. CRG/2021/005121.

Nomenclature

a	Radius of the particle [m^2]
a_L	Radius of larger particle
a_S	Radius of smaller particle
Φ	Particle volume fraction
Φ_c	Maximum particle volume fraction
φ_L	Particle volume fraction of larger particle
φ_S	Particle volume fraction of smaller particle
η_{eff}	Suspension viscosity
η_0	Carrier fluid viscosity
R_{ep}	Reynolds number
P_e	Peclet number
γ	Shear rate
K	Boltzmann's constant
P	Pressure field
K_c	Empirical parameter of the collision flux
$K\eta$	Empirical parameter of the viscosity flux

References

1. Gadala-Maria F, Acrivos A (1980) Shear-induced structure in a concentrated suspension of solid spheres. J Rheol 24(6):799–814
2. Leighton D, Acrivos A (1987) The shear-induced migration of particles in concentrated suspensions. J Fluid Mech 181:415–439
3. Abbott JR, Tetlow N, Graham AL, Altobelli SA, Fukushima E, Mondy LA, Stephens TS (1991) Experimental observations of particle migration in concentrated suspensions: couette flow. J Rheol 35(5):773–795

4. Fang Z, Phan-Thien N (1995) Numerical simulation of particle migration in concentrated suspensions by a finite volume method. J Nonnewton Fluid Mech 58(1):67–81
5. Hampton RE, Mammoli AA, Graham AL, Tetlow N, Altobelli SA (1997) Migration of particles undergoing pressure-driven flow in a circular conduit. J Rheol 41(3):621–640
6. Leighton D, Acrivos A (1987) Measurement of shear-induced self-diffusion in concentrated suspensions of spheres. J Fluid Mech 177:109–131
7. Phillips RJ, Armstrong RC, Brown RA, Graham AL, Abbott JR (1992) A constitutive equation for concentrated suspensions that accounts for shear-induced particle migration. Phys Fluids A 4(1):30–40
8. Krishnan GP, Beimfohr S, Leighton DT (1996) Shear-induced radial segregation in bidisperse suspensions. J Fluid Mech 321:371–393
9. Reddy MM, Singh A (2019) Shear-induced particle migration and size segregation in bidisperse suspension flowing through symmetric T-shaped channel. Phys Fluids 31(5)
10. Kanehl P, Stark H (2015) Hydrodynamic segregation in a bidisperse colloidal suspension in microchannel flow: a theoretical study. J Chem Phys 142(21)
11. Krieger IM (1972) Rheology of monodisperse latices. Adv Coll Interface Sci 3(2):111–136
12. Lyon MK, Leal LG (1998) An experimental study of the motion of concentrated suspensions in two-dimensional channel flow. Part 1. Monodisperse systems. J Fluid Mech 363:25–56

Influence of Slip Models and Wall Surface Roughness on Micronozzle Flow Behavior

S. Kishore, Pulkit Pandey, and S. R. Shine

Abstract This research delves into different velocity-slip and temperature jump models applicable near the inner wall of micronozzles during numerical simulations using *N–S* equations. It also investigates how wall surface roughness influences flow parameters. The study compares the Maxwell first-order slip and Langmuir–Maxwell slip models to determine their accuracy in predicting flow behavior. Moreover, it examines the impact of wall surface roughness on flow characteristics, including pressure, temperature, Mach number, and mass flow rate. The study reveals intriguing micronozzle sub-choking phenomena, leading to unexpected flow restrictions. Understanding these phenomena is crucial for the efficient and dependable design of micronozzles. The research findings contribute to advancing microfluidic and aerospace technologies, optimizing micronozzle performance across various applications.

Keywords CFD · OpenFOAM · Navier stokes (NS)

Nomenclature

H_t/W	Nozzle throat diameter (μm)
ΔP	Pressure difference (KPa)
$\frac{X}{L^*}$	Non dimensional length –
Ma	Mach number –
Kn	Modified Knudsen no –
λ	Mean free path (m)
n	Number density (m^{-3})
d	Diameter of air molecule (m)

S. Kishore (✉) · P. Pandey · S. R. Shine
Department of Aerospace Engineering, Indian Institute of Space Science and Technology, Thiruvananthapuram, Kerala, India
e-mail: kishore.s1811@gmail.com

S. R. Shine
e-mail: shine@iist.ac.in

A. Kumar R. et al. (eds.), *Proceedings of Fluid Mechanics and Fluid Power (FMFP) 2023, Vol. 6*, Lecture Notes in Mechanical Engineering,
https://doi.org/10.1007/978-981-96-9600-0_18

1 Introduction

In the context of satellite maneuvering in orbits, minimizing torques on a satellite's gyroscope in a drag-free environment is crucial. Typically, this is achieved by employing micro thrusters that generate thrust in the 1–5 mN range. To design efficient micro thrusters, it is essential to understand the flow behavior of micronozzles. Such knowledge aids in optimizing the attitude control system of satellites [3].

In micronozzles, the Reynolds number (Re) may exhibit a substantial decrease, spanning several orders of magnitude (Re $\sim 10^1$–10^3), while the surface-to-volume ratios experience significant amplification. This results in a distinct flow pattern marked by supersonic movement within an environment where significant viscous forces are noticeable. In this specific situation, the effectiveness of the nozzle is diminished because increased viscous losses occur within the subsonic layers adjacent to the wall. Previously, there have been many studies done on micronozzles. Rothe [14] has investigated the flow characteristics like density, pressure, temperature, etc. using electron beam studies on conical micronozzles with throat diameter varying from 2.5 to 5.5 mm, with ambient backpressure of 1 atm. It was studied in the Re range of 55–550. Bayt [2] and Choudhary et al. [5] conducted experimental studies, establishing that as the Reynolds number (Re) decreased below 1500, both specific impulse and nozzle efficiency experienced a rapid decline. They observed fully viscous flows at a Re of approximately 1000, leading to a decrease in velocity and an increase in wall boundary layer thickness in the divergent section. Reed [13] demonstrated a direct relationship between nozzle dimensions, throat Reynolds number, and thrust, impulse, and nozzle discharge coefficient. Liu et al. [9] observed that as the Reynolds number (Re) increased, the efficiency and coefficient of discharge also rose. At the same time, the influence of rarefaction effects lessened with larger throat dimensions. Louisos and Hitt [10] investigated the impact of Re by variation of nozzle throat depth, leading to a higher performance at high Re to the inverse scaling relationship between depth of the nozzle and the relative magnitude of losses created due to the viscous forces, which displayed a linear correlation. Torre et al. [17] observed a significant performance loss of up to 10% due to viscous effects in micronozzles with a thrust level of 1 mN. They reported that the boundary layer covered 30–50% of the outlet area for micronozzles ranging from 1 to 0.1 mN. The research conducted by Gomez and Groll [6] indicated that by decreasing the length of the diffuser and raising the Reynolds number, it was possible to alleviate the effects caused by the viscous boundary layer. Manu et al. [16] have investigated the geometry effects on the flow characteristics of a 2D microscale nozzle by varying the divergence angle of the nozzle, flow in the nozzle has been analyzed. The results showed that continuum approach is valid in the convergent section of the nozzle and the DSMC approach can be used in the divergent section of the nozzle. Banuti et al. [1] has studied the flow characteristics of monopropellant microscale nozzles. In the study, it has been found that an assumption of a dominant isentropic core with moderate viscosity corrections is not suitable for the design aspects of the micronozzle. It has also been concluded from the study that the length from the nozzle throat to the exit plane of the nozzle is a much more critical parameter compared to the expansion ratio. Singh and Kumar et al. [15] have studied about the effect of the varying area ratio of micronozzle and the effect of semi-divergence angle in the divergent section of the nozzle. In that study, different flow parameters like mass flow rate, Mach number, and temperature have been compared for both Maxwell first-order slip and no-slip boundary conditions. Li et al. [8] have studied various slip models for micronozzles using H_2O_2 and Argon gases as propellants. These models have been validated with DSMC data. This study shows that for $Kn < 0.1$, the Maxwell 1st order slip model is effective, and for higher Kn regimes other slip models like the Langmuir–Maxwell model, which also accounts for the fraction of gas molecules adsorbed on the wall surface, seem effective. Yao et al. [19] have investigated about the phenomenon of sub-choking in micronozzles and it has been attributed to the increase in surface-to-volume ratio of a micronozzle compared with the conventional nozzles. Craig White et al. have studied in detail about Langmuir–Maxwell and Langmuir–Smoluchowski jump conditions for thermal flow simulations in hypersonic aerodynamics.

Ngalande et al. [11] conducted initial studies and found that wall surface effects had no significant impact for Reynolds numbers (Re) greater than 100, but they observed a 10% reduction of mass flow and thrust when $Re < 100$. Yan et al. [18] discovered that roughness in the divergent section of the nozzle could lead to supersonic multi-waves, leading to performance degradation and fluctuations in flow characteristics. According to Torre et al. [17], the quantity and magnitude of these shocks were contingent upon the amplitude and direction of the surface roughness. These shocks resulted in an overall reduction in velocity and stagnation pressure. Cai et al. [4] attempted to establish a connection between the micro milling's axial depth of cut, the roughness of internal surfaces, and the exit velocity of nozzles featuring a throat diameter (d_t) of 1 mm. The loss in velocity was approximately 5% when the critical depth of cut was 16% of d_t. Giovannini and Abhari (2015) utilized different manufacturing techniques to produce aero spike nozzles with varying surface roughness levels (milling with CNC machine for an average roughness of 0.49 μm; EDM for 0.023 μm; and RP for 14.157 μm). The velocity variation was less than 2%, and the expansion of the kinetic boundary layer demonstrated no notable impact. Cai et al. [4] observed that for certain values of surface roughness, the variation in the performance of micronozzles was less significant compared to large-scale nozzles. With the above literature survey, it has been found that there is a gap in the study of micronozzles where there has been no detailed analysis of Maxwell first-order slip boundary condition and the surface roughness parameter and its impacts on the sub-choking of micronozzles.

2 Methodology

The CFD simulations are performed using the rhoCentralFoam solver of the open-source software OpenFOAM (refer to the OpenFOAM version). The model of compressible flows with various densities is the main emphasis of the rhoCentralFoam

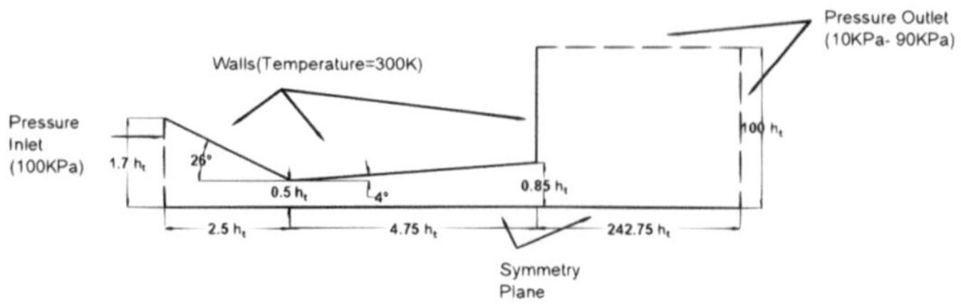

Fig. 1 Schematic of micronozzle [7]

solver. The compressible Navier–Stokes equations are solved using the finite volume method and the central differencing methodology. Density is the primary variable from which other flow parameters are produced in the solver's density-based method, which enables reliable prediction of flows with significant density fluctuations.

A. **Geometry and Grid Size**

The CFD simulations employ a two-dimensional micro nozzle, as shown in Fig. 1. In this study, it also becomes necessary to capture the flow characteristics in the far-field downstream of the nozzle exit because the flow is driven by the backpressure conditions. So a suitable large domain is used downstream of the nozzle exit plane. The domain size is determined based on a domain independence study. The domain size is chosen in such a manner that there is no backflow in the domain. For effective meshing and precise results, the nozzle is separated into four blocks, each with a thickness of one millimeter. To choose a suitable grid size that adequately reflects the flow features, a grid independence analysis is carried out. To find out how the diverging angle affects the flow properties, changes in the Mach number and Knudsen number are analyzed. The right mesh size is chosen after conducting grid independence research. A coarser mesh of 200 cells in Domain 1 and 1200 cells overall is initially employed. The number of cells in Domain 1 and Domain 2 are increased, resulting in a finer mesh. Grid independence is determined by pressure tolerance. For the current simulation, a mesh structure with 3600 cells in Domain 1 and 9000 cells in Domain 2 is selected based on the study.

B. **Governing Equations and Boundary Conditions**

The problem is modeled using the Compressible Navier–Stokes equations and the energy equation. Using the initial values of pressure, temperature, dynamic viscosity, and specific heat at constant pressure, the solver discretizes the equations and finds solutions.

Boundary Conditions

At the walls of the nozzle, a case study with two different slip non-equilibrium boundary conditions has been used. Those two boundary conditions are as follows.

Maxwell 1st Order Slip

$$u_s = u_w + C_1\lambda\frac{\partial u_x}{\partial n} \tag{1}$$

$$C_1 = \frac{2-\sigma_U}{\sigma_U} \tag{2}$$

$$T_s = \frac{2-\sigma_T}{\sigma_T}\frac{2\gamma}{(\gamma+1)\mathrm{Pr}}\lambda\frac{\partial T}{\partial n} + T_w \tag{3}$$

where $C_1$is the slip coefficient, σ_U is the accommodation coefficient which accounts for the surface roughness parameter, and σ_T is the temperature accommodation coefficient.

Langmuir–Maxwell Slip

$$u_s = u_w + \frac{1}{1-\alpha}\lambda\frac{\partial u_x}{\partial n} \tag{4}$$

$$T_s = T_w + \frac{1}{1-\alpha}\frac{2\gamma}{(\gamma+1)\mathrm{Pr}}\lambda\frac{\partial T}{\partial n} \tag{5}$$

where α accounts for the fraction of molecules adsorbed on the surface of the wall which directly governs the effusion of gas molecules that account for the friction offered by the wall surface.

3 Results

In our study, we extensively simulated micro nozzles to examine how changes in Mach number and mass flow rate were affected by various boundary conditions. Additionally, we carefully considered the potential effects of surface roughness coming from manufacturing procedures on nozzle performance. Designing effective and optimized micro nozzles for diverse purposes, especially in satellite attitude control and orbital maneuvering, requires an understanding of these effects.

Throughout our research, we systematically altered the surface roughness of the micronozzle to comprehend its effects on flow behavior. Surface roughness is an inherent factor in real-world applications and can significantly impact nozzle performance. By studying these variations, we gained valuable insights into designing micronozzles that perform efficiently and effectively under diverse conditions.

The impact of micro nozzle size on the Knudsen number, which is crucial in defining the flow properties, was also investigated. We were able to analyze the variations in the Knudsen number by adjusting the micro nozzle's scale and showing how the size of the nozzle affects its flow regime and overall performance.

By combining the investigations into boundary conditions, surface roughness, and nozzle size, our study contributes to a comprehensive understanding of micro nozzle behavior.

A. **Validation**

To ensure accuracy, numerical mass flow rates have been found with the simulated data for the nozzle with 1 μm thickness. To validate it with the experimental mass flow rate data of Rothe nozzle [7] with thickness of 120 μm, numerical mass flow rate for thickness 120 μm has been found using the simulated data of 1 μ thickness nozzle. It is shown in Fig. 2. It is found that the numerical values match well with the experimental values, within an error of ± 5%. It is possible to explain this disparity by the existence of transition flow conditions in some areas where the Knudsen number is within that range. Figure 2 provides a visual illustration of the results and the differences in mass flow rate at various back pressures. The plot offers insightful information about the system's interaction between back pressure and mass flow rate. To validate the use of the Maxwell first-order slip boundary condition, modified Kn number [12] has been computed for various back pressure conditions using the following approach.

$$\lambda = \frac{1}{n\pi d^2} \tag{6}$$

$$\mathrm{Kn} = \frac{\lambda \nabla X}{X} \tag{7}$$

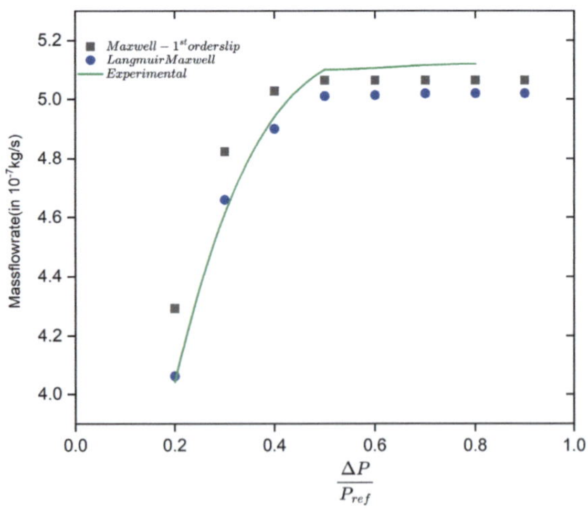

Fig. 2 Comparison of experimental flow rates with numerical flow rates

Fig. 3 Distribution of center line mach number in micro nozzles with different back pressure for $h_t = 20$ μm

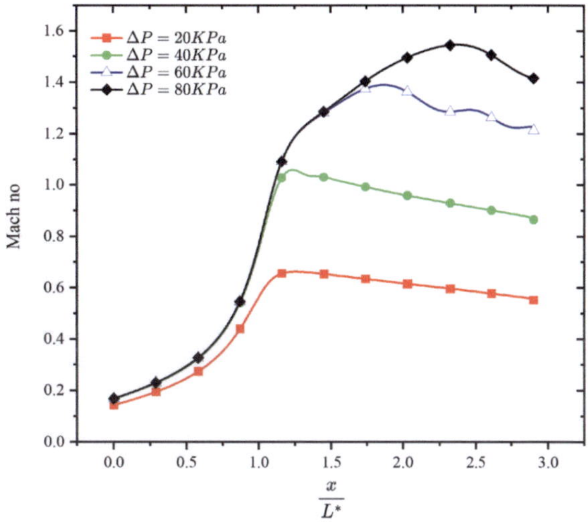

where X can be any flow property like density, pressure, temperature. Based on previous literature surveys it is found that $\frac{\nabla\rho}{\rho}$ has been accurately depicting the Knudsen nos. So in the current investigation same approach has been used for computing Kn. These Kn contours are depicted in Fig. 8.

The variation in mass flow rate while using various boundary conditions is shown in Fig. 2. When the Knudsen number is in the slip flow regime ($0.001 < \text{Kn} \leq 0.1$) and transition flow regime ($0.1 < \text{Kn} \leq 10$), the Maxwell's first-order boundary condition and Langmuir–Maxwell's boundary condition have been added to validate their applicability. When the results are compared to experimental data, the error is within $\pm$ 5 %. This error may result from the micronozzle simulation using perfect computational models, indicating the need for further improvement in computational methods. Along the center line axis of the nozzle, the Mach number has been plotted with respect to the non-dimensional distance along the nozzle in Fig. 3. It can be seen that these numerical values well predict the actual physical phenomena.

In Fig. 4, the Mach number has been plotted along the center line axis of the nozzle for different scales of throat width. It has been seen that as the size of the throat increases, surface forces are becoming dominant compared to other forces, thus there is increased dissipation leading to the displacement of the sonic Mach number downstream of the throat.

In the figure, it is evident that the mass flow rate stabilizes at approximately 60 kPa when using the Maxwell first-order slip boundary condition, while it reaches around 50 kPa with the Langmuir–Maxwell boundary condition.

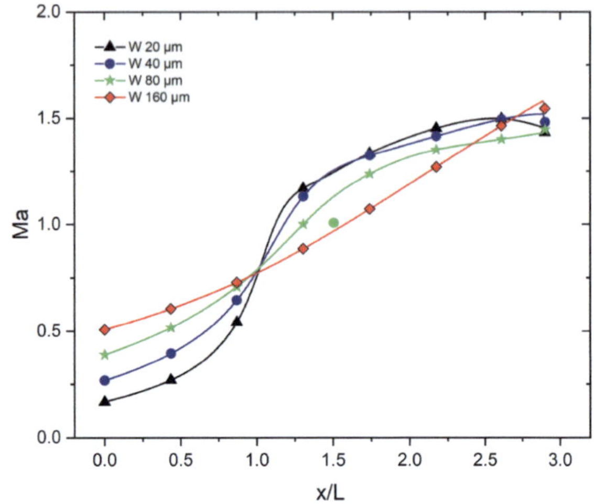

Fig. 4 Distribution of center line mach number in micro nozzles with different throat width (W)

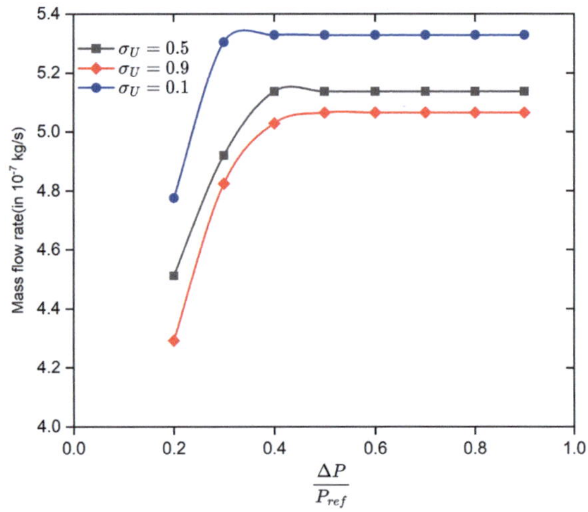

Fig. 5 Comparison of surface roughness mass flow rate

Figure 5 clearly shows how surface roughness affects the mass flow rate and the sub-choking phenomenon. Higher mass flow rates are demonstrated by the micro nozzle's smoother surfaces, which also postpone the onset of sub-choking because the boundary layer growth is not that dominate. In contrast, the mass flow rate drops as the micro nozzle surface grows rougher, and sub-choking occurs earlier in the flow phase.

The significant impact of surface smoothness on the Mach number inside the micronozzle is illustrated in Figs. 6 and 7. Micronozzle performance is directly impacted by nozzle roughness, as smoother micronozzles have lower temperatures and subsequently higher Mach numbers than rougher ones. In Figs. 6 and 7, the centerline Mach numbers and temperature of two micronozzles with different surface

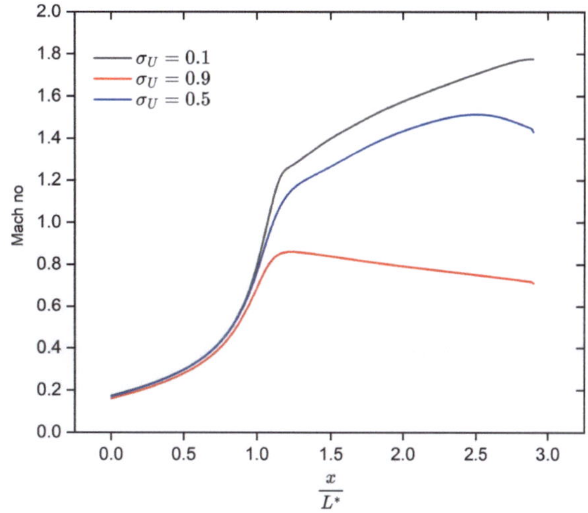

Fig. 6 Mach number comparison on surface roughness

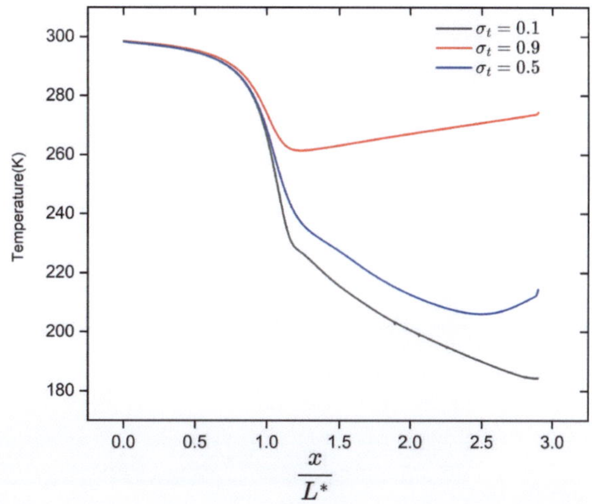

Fig. 7 Temperature comparison on surface roughness

conditions—one with a smoother surface ($\sigma = 0.1$) and the other with a rougher surface ($\sigma = 0.9$)—are contrasted. This comparison enables us to comprehend how surface roughness affects the micronozzle's overall performance.

In Fig. 8, knudsen contours are being plotted. From Fig. 8, it is clear that the Knudsen number continues to exist in the continuum domain upstream of the throat. However, the Knudsen number enters the slip regime downstream of the neck. This finding emphasizes how the scale of the nozzle also affects the change from continuum to slip flow regimes.

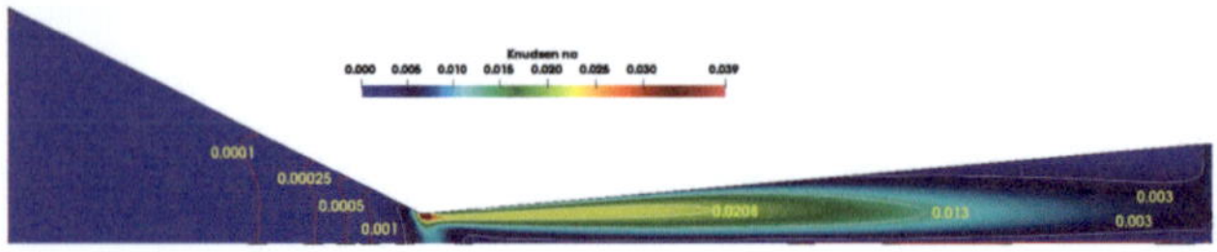

(a) W=7μm

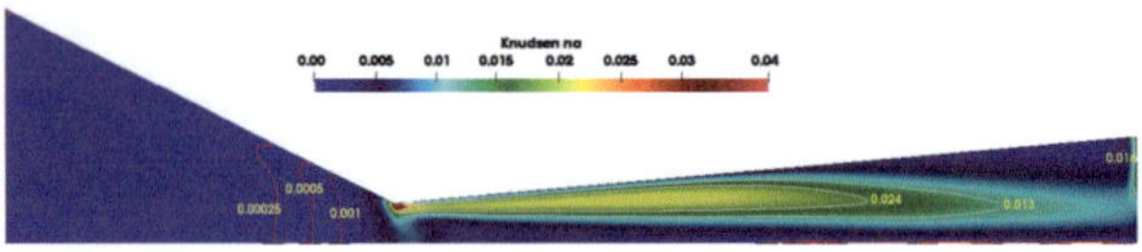

(b) W=10μm

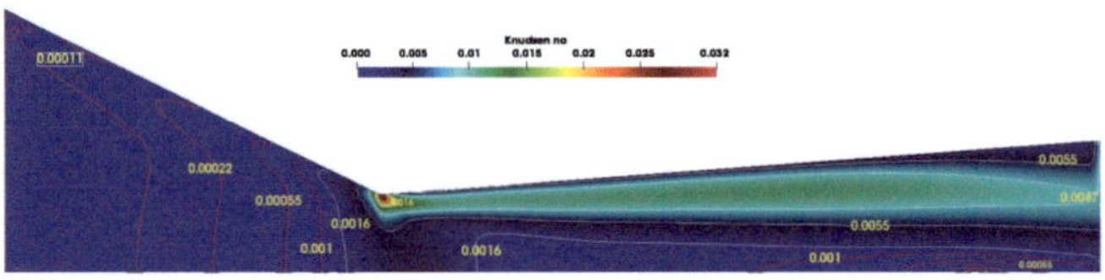

(c) W=20μm

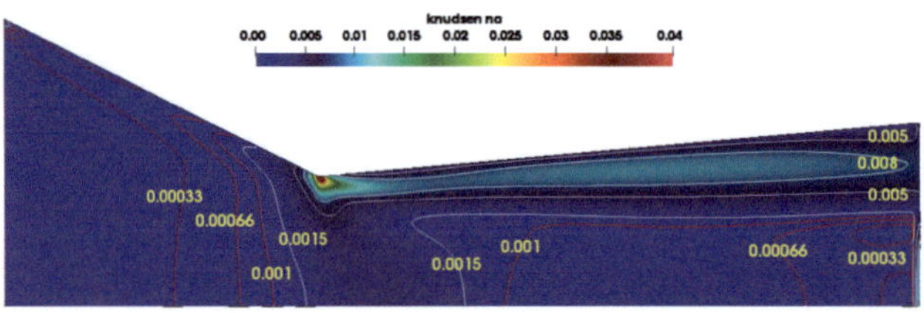

(d) W=40μm

Fig. 8 Knudsen number contour in micro nozzles with different throat width (W)

4 Conclusions

In this investigation, we explored the impact of various first-order slip boundary conditions on the ability to predict flow properties in micronozzles. The validation of the computational predictions with experimental data showed how these boundary conditions, mainly when the Knudsen number is in the slip regime and transition regime, may accurately and consistently predict the flow behavior in micronozzles. Simulations revealed that greater surface roughness increased the mass flow rate due to the reduced boundary layer thickness near the nozzle wall. It is observed that smoother nozzles experienced sub-choking at lower pressure differences compared to rougher ones. This valuable insight aids in understanding the performance characteristics of micronozzles and optimizing their design. In summary, this study contributes to a deeper comprehension of micronozzle behavior by considering the influence of boundary conditions and surface roughness. The validated boundary conditions provide a robust foundation for future simulations, and examining surface roughness highlights crucial design factors for enhancing satellite attitude control and maneuvering performance.

References

1. Banuti DT, Grabe M, Hannemann K (2019) Flow characteristics of monopropellant micro-scale planar nozzles. Aerosp Sci Technol 86:341–350
2. Bayt RL (1999) Fabrication, and testing of a mems-based micropropulsion system. Technical report, Massachusetts Institute of Technology, FDRL TR 99-1, Ph.D. Thesis
3. Boyd lD, Jafry Y, Beukel JV (1994) Particle simulations of helium microthruster flows. J Spacecraft Rockets 31(2):271–277
4. Cai Y, Liu Z, Song Q, Shi Z, Wan Y (2015) Fluid mechanics of internal flow with friction and cutting strategies for micronozzles. Int J Mech Sci 100:41–49
5. Choudhuri A, Baird B, Gollahalli S, Schneider S (2001) Effects of geometry and ambient pressure on micronozzle flow. In: 37th Joint propulsion conference and exhibit, p 3331
6. Gomez J, Groll R (2016) Pressure drop and thrust predictions for transonic micronozzle flows. Phys Fluids 28(2):022008
7. Hao P-F, Ding Y-T, Yao Z-H, He F, Zhu K-Q (2005) Size effect on gas flow in micro nozzles. J Micromech Microeng 15(11):2069
8. Li X, Yuan J, Ren X, Cai G (2022) Simulation applicability verification of various slip models in micro-nozzle. Acta Astronaut 192:68–76
9. Liu M, Zhang X, Zhang G, Chen Y (2006) Study on micronozzle flow and propulsion performance using DSMC and continuum methods. Acta Mech Sin 22(5):409–416
10. Louisos WF, Hitt DL (2012) Viscous effects on performance of three-dimensional supersonic micronozzles. J Spacecr Rocket 49(1):51–58
11. Ngalande C, Lilly T, Killingsworth M, Gimelshein S, Ketsdever A (2006) Nozzle plume impingement on spacecraft surfaces: effects of surface roughness. J Spacecr Rocket 43(5):1013–1018
12. Rabani R, Heidarinejad G, Harting J, Shirani E (2019) Effect of temperature difference between channel walls on the heat transfer characteristics of nanoscale-confined gas. Int J Therm Sci 137:13–25
13. Reed B (2003) Decomposing solid micropropulsion nozzle performance issue, AIAA 2003–0672. In: 41st Aerospace sciences meeting & exhibit. Reno, NV
14. Rothe DE (1971) Electron-beam studies of viscous flow in supersonic nozzles. AIAA J 9(5):804–811
15. Singh SK and Arun Kumar R (2022) A parametric study on the fluid dynamics and performance characteristic of micronozzle flows. J Fluids Eng 144(3):031208
16. Sukesan MK, Shine SR (2021) Geometry effects on flow characteristics of micro-scale planar nozzles. J Micromech Microeng 31(12):125001

17. La Torre F, Kenjeres S, Kleijn CR, Moerel J-LPA (2010) Effects of wavy surface roughness on the performance of micronozzles. J Propul Power 26(4):655–662
18. Yan H, Zhang W-M, Peng Z-K, Meng G (2015) Effect of three-dimensional surface topography on gas flow in rough micronozzles. J Fluids Eng 137(5)
19. Yao Z-H, He F, Ding Y-T, Shen M-Y, Wang X-F (2004) Low-speed gas flow subchoking phenomenon in a long-constant-area microchannel. AIAA J 42(8):1517–1521

Investigation of Frequency Response of Bladeless Cylindrical Wind Turbine

Abhishek Pednekar and Shailesh R. Nikam

Abstract Flow over bluff body (cylindrical body) results in flow-induced vibration due to flow separation and alternate shedding of vortices. This flow-induced vibration/oscillations can be used to harness wind energy by converting it into electric energy using electromagnetic technique where frequency of oscillations play important role. The goal of the present study is to experimentally investigate the vortex-induced vibration for designing a bladeless wind turbine (BWT). Cylinder is taken as BWT. Experiments are performed in wind tunnel of 410 × 410 mm test section. Measurement of oscillations of the BWT is done by a three axis accelerometer. The Vortex shedding frequency is measured by the Honeywell Differential pressure. The test results which relate the natural frequency the Oscillating frequency and the Vortex shedding frequency of model with flow speed. It is found that the Oscillations closed to natural frequency can have maximum effect for energy produced. The increase in flow velocity increases the frequency of vortex shedding, i.e., the Strouhal Frequency for stationary cylinder, in other case when cylinder is elastic the forcing frequency gets locked in with natural frequency and a "Lock in" phenomenon is observed.

Keywords Bluff body · Bladeless wind turbine (BWT) · Vortex-induced vibration · Natural frequency · Lock in

1 Introduction

Wind turbines are used to harness wind energy. Conventional wind turbine uses rotor with the blade to harness wind energy. Various factors which define wind energy harnessing efficiency are terrain features on desirable location, wind velocity, direction of wind, air density, swept area of wind turbine, turbine blade profile,

A. Pednekar · S. R. Nikam (✉)
Somaiya Vidyavihar University, K J Somaiya College of Engineering, Mumbai, India
e-mail: shailesh.n@somaiya.edu

A. Pednekar
e-mail: pednekar.av@somaiya.edu

A. Kumar R. et al. (eds.), *Proceedings of Fluid Mechanics and Fluid Power (FMFP) 2023, Vol. 6*, Lecture Notes in Mechanical Engineering,
https://doi.org/10.1007/978-981-96-9600-0_19

tip speed ratio of rotor, and orientation of axis of turbine. Thus there are various factors which need to take into account for effective utilization of wind energy using wind turbine. Also there are environmental issues and economical aspects associated with wind turbine installation. Environmental issues associated are physical restrictions, noise levels, tower design constraints, disturbance of local ecology, effects on radio communication and television reception, zoning restriction, and impact on bird life. Economic issues associated are cost for analysis and design, fabrication, assembly and operational testing, transportation, final testing, and installation. Considering these operational, environmental, and economic issues associated with conventional wind turbines, new technology should be used to harness wind energy. Present work addresses the issues associated with conventional wind turbine technology by using bladeless wind turbine (BWT) which works on the principle of vortex-induced vibration/oscillations (VIV). Flow over bluff body (cylindrical body) results in flow-induced vibration which occurs due to flow separation and alternate shedding of vortices from separated shear layer in the base region of body. This phenomenon causes noise, vibration, and system drag which are detrimental especially in aerospace application. However, this flow-induced vibration/oscillations can be used to harness wind energy. This flow-induced vibration energy will be converted into electric energy using electromagnetic or piezoelectric technique. Objective of the present work is to optimize geometry of cylindrical bluff body and flow conditions to get required oscillations from cylindrical body. Investigation is done through experiment in wind tunnel.

2 Literature Review and Objective

Considering increase in the global energy demand and need to use renewable source of energy many researchers are working toward various technology to harness wind energy. Mechanical energy embedded in vibrating medium can be converted to electrical energy [1]. The analysis predicts that the power produced is proportional to the cube of the frequency of vibration. To have a larger scale energy production, research in the field of producing vibration/oscillation in the body using wind energy has been a topic of interest. Computational, theoretical, and experimental methodology is being used by many researchers for investigation. Meninger et al. [2] proposed to convert ambient mechanical vibration into electrical energy for use in powering autonomous low-power electronic systems. The energy is transuded through the use of a variable capacitor. Salvador et al. [3] have done experimental study on vortex-induced vibration on cylinder as a bladeless wind turbine, using arc and disk type pendulum to produce electricity. The vortex-induced vibration of cylinder is used to move the pendulum below as shown in Fig. 1. David Jesus Yanez Villarreal 2018 [4] stated that the aeroelastic effects can produce oscillating movement in the body if its natural resonance frequency and the vortex shading's wake frequency are similar. The vibrations induced in a body by vortices are known as VIV phenomena and electric energy conversion is done as shown in Fig. 2.

Fig. 1 Arc-shaped pendulum [3]

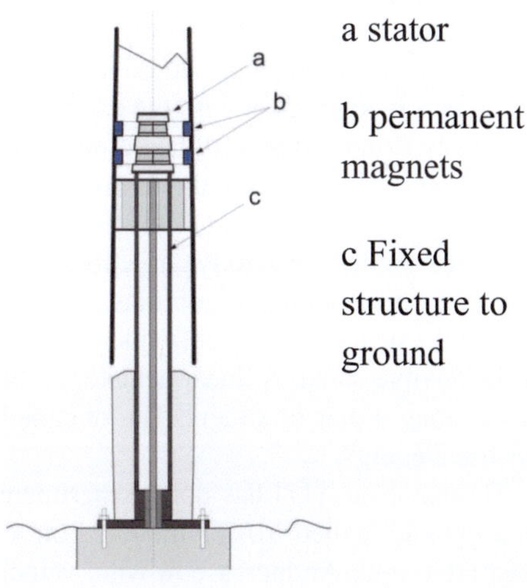

Fig. 2 Arrangement of the alternator and tuning system [4]

Chizfahm et al. [5] studied the dynamic modeling of four configurations of vortex-induced vibrations of a bladeless wind turbine (BWT). The effects of the wind speed on the induced lift force, turbine deflection, and generated power of four BWTs are investigated. It is verified that the amplitude of the vibrations of the BWT increases significantly when the vortex shedding is synchronized with the structural oscillations. The dynamic model is validated through comparison with a 3D CFD-FEM numerical simulation.

Gang et al. [6] experimentally investigated the performance of a circular cylinder-based piezoelectric wind energy harvester with two small-size rod-shaped attachments on the main circular cylinder. Three different cross-sectional shapes: circular, triangular, and square were used. The force measurement results indicated that the circular cylinder with triangular rods at the circumferential location of 60° had larger transverse force coefficients than the other two rod-type cases.

Ordonez and Duke [7] presented a technical analysis of the VIV in the residential sector of Puerto Cortés; a power generation and demand relation were studied. A real-time analysis was done and the behavior of a Vortex Nano at a height of 12 m was studied using the model. Average wind study for 5 ms^{-1}. The Power generation at different speeds was considered and found that maximum generation is when speed is around 7 ms^{-1}.

Aballe et al. [8] used a spring mechanism to produce a linear generator with spring mechanism for vortex bladeless wind turbine to somehow increase the generated voltage. This research offers a new method of gathering energy by focusing on the design of the generator that can produce higher voltage through the use of spring mechanism and serves as a benchmark to future studies of bladeless wind turbines and focuses on the design of the generator that can be beneficial to the industry and future power grids. An experimental analysis of low-speed vortex-induced vibration on cylinder placed horizontally was done by Farsi et al. [9]. Cylinder mounted on four springs is subjected to steady flow of the wind and a fixed cylinder is mounted in front of it and the amplified effect is studied. Also, numerical analysis was done which was in agreement with the experimental results.

Magdy Roman et al. [10] based on numerical simulation presented a novel semi-active approach for tuning the turbine response to widen the effective power generation region. When subjected to wind flow, the mast alternately sheds vortices and generates oscillating aerodynamic force and in turn produces vortex-induced vibration of the flexibly supported mast. A part of the proposed tuning technique is to change the structural natural frequency. This is done by changing the effective length of the flexible stand. A linear actuator is used to move the supporting sleeve up and down along a pair of sliders. The obtained mechanical vibration is converted into electrical energy.

Thomai et al. [11] has done experimental analysis of vortex-induced vibration on a cylinder called MAST mounted on a spring. Four different configurations of mast are tested. An increase in wind velocity led to a terrible change in the amplitude of vibration. For the various mast cross sections analysis is done through time domain signals and the corresponding Fast Fourier Transform (FFT) for various flow Reynolds's Numbers (Re). Amplitude level variation of vibration signal at different wind speeds is verified to identify the performance of good mast cross-section. They found out that the uniform tapered rectangular section is better to harvest more energy.

Gautam et al. [12] used the principle of vortex-induced vibration in mast body which can move in any direction depending on the direction of the wind. Different possible arrangements of the field with respect to the coil to harness electrical power are discussed in detail, and an efficient arrangement was proposed for maximum voltage generation. They studied a model in Ansys by arranging bar magnets around the coil in layers and checked the performance. The layered configuration has proved to be promising.

Francis et al. [13] have made Ansys models and CFD analysis of glass fiber rod and carbon fiber rod of different length on the bladeless wind turbines, and they found out that the glass fiber rod is most suited in terms of maximum oscillation.

Bhardwaj and Teja [14] proposed different electromagnetic field configurations for bladeless wind turbines. They have suggested replacing permanent magnets in the electromechanical system due to cost and limitation on the flexibility of changing the field magnitude and direction. In this paper, an electromagnetic field arrangement suitable for bladeless wind turbines is proposed. Various field coil configurations are presented and analyzed using Ansys. The results are compared, and the best possible field coil configuration is proposed. As the wind speed and direction are not uniform during the time of a day so a field with high flux change for slightest vibration and flux change in all directions was considered.

Many researchers around the world initiated work on utilizing vortex-induced vibration to harness wind energy in recent years. Work in this area is at initial stage. Very little experimental study is done. From the literature, it is observed that comprehensive study on effect of various parameters affecting vortex-induced vibration/oscillation is missing. Various geometric parameters of cylindrical body will be verified experimentally at different velocities to get required oscillation. Oscillations will be recorded using accelerometer and data acquisition system.

3 Materials and Methods

Experiments are conducted in open jet wind tunnel with 410 × 410 mm test section. Required velocity of flow is maintained using variable frequency drive connected with the fan motor of tunnel. The cylinder model is placed very close to wind tunnel at the center of the cross section of the wind tunnel. potential core of the open jet wind tunnel is established through Pitot tube measurement in y direction so that model can be kept in region of constant velocity. Figure 3 shows region of potential core of tunnel (representative).

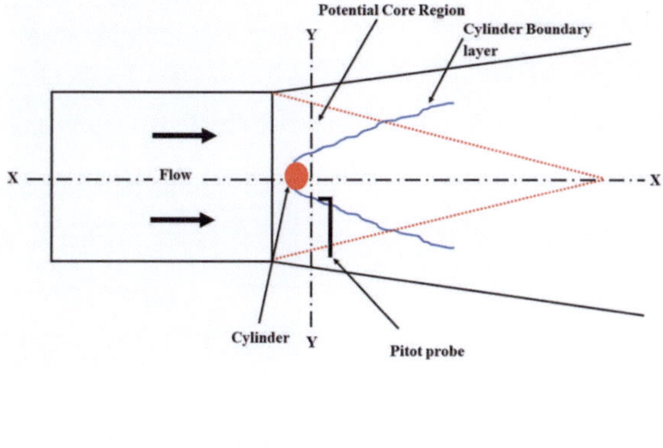

Fig. 3 Flow domain of open jet wind tunnel

Accelerometer with maximum frequency range 1 kHz in X and Y direction and frequency range of 550 Hz in Z direction is used to measure frequency of oscillations. The Vortex shedding frequency is measured by the Honeywell Differential pressure sensor of range 2.5 inch of water. The schematic of experimental setup, photograph of the setup and the coordinate system is shown in Figs. 4, 5, and 6, respectively.

Figure 7 shows the glass fiber model of 57 mm outer diameter and 400 mm length mounted on the vertical leaf spring (hacksaw blade) having width 12 mm and thickness 0.62 mm. Mechanism is developed for changing the height of the leaf spring and also for placing the mass at different locations on the cylinder so that natural frequency can be varied. The hacksaw blade on which cylinder is mounted is fixed in a vice and is used as a leaf spring so that the motion or oscillations are orthogonal to the flow. With this arrangement the height of the spring can be adjusted. The accelerometer is placed on the top of the cylinder.

Vortex shedding frequency is given by nondimensional quantity; Strouhal number $\mathbf{St} = \frac{\mathbf{fsD}}{\mathbf{U}}$

For flow over circular cylinder with Reynolds number, $40 < \mathrm{Re} < 5 \times 10^5$, it has been confirmed from literature that Strouhal number is 0.2 which remains constant

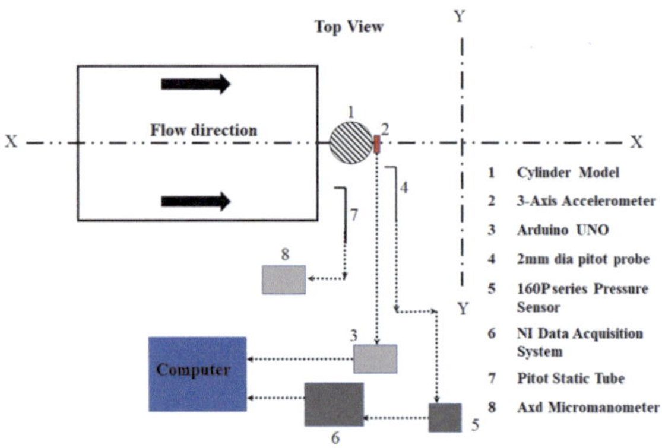

Fig. 4 Schematic representation of experimental setup

Fig. 5 Photograph of experimental setup

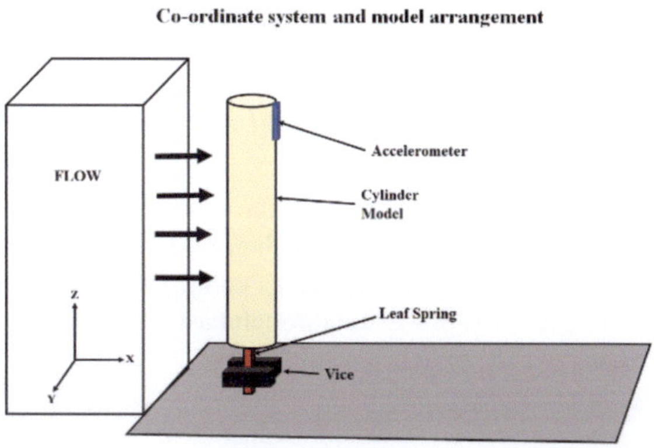

Fig. 6 Coordinate system

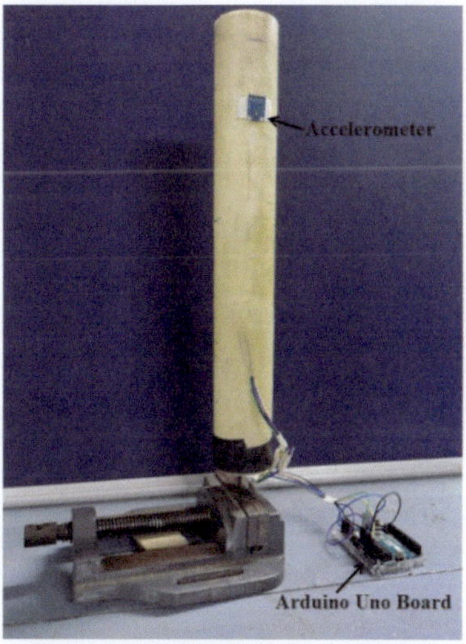

Fig. 7 Glass fiber model with accelerometer and Arduino

for a given cylindrical two-dimensional body. From the vibration dynamics theory the natural frequency of the system can be obtained using equation a $f_n = \frac{1}{2\pi}\sqrt{\frac{k}{m}}$.

4 Results and Discussion

4.1 Estimation of Natural Frequency of Cylinder (f_n)

Natural frequency is estimated through experiment by fixing the leaf spring in vice with the cylinder mounted on it, then giving a displacement to the model and recording the acceleration by accelerometer connected with Arduino board to computer. Time-domain result is processed through Fast Fourier Transform (FFT) to get dominating frequency. The obtained results are as shown in Fig. 8 for height of spring 5 mm and 40 mm (minimum and maximum height results are shown here).

Cylindrical model as shown in Fig. 7 is held in a vice, the height of the spring is adjusted so different natural frequencies can be obtained. The model is fixed at a particular height of 5 mm of spring and weight of 10gm is fixed to cylinder at locations from top at **Z = 0H, 0.24H, 0.48H, 0.71H, 1H**, adding mass changes the natural frequency thus height and mass are two important factors in the design of the model. The mass added at different locations creates nodes and the nodes affect the amplitude of vibration so at top the mass causes the amplitude of vibration

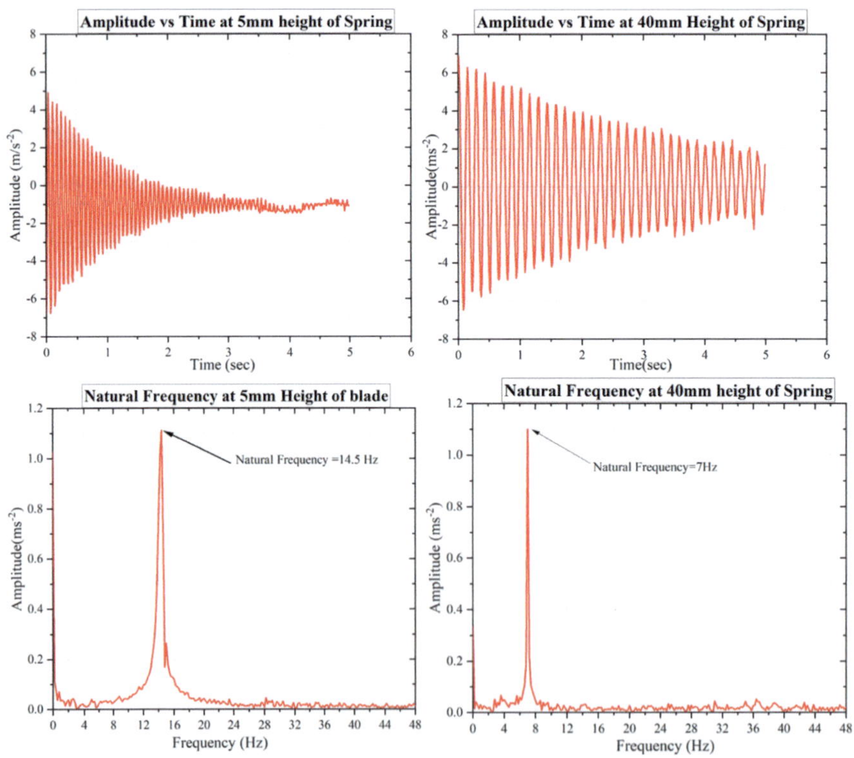

Fig. 8 Natural frequency by experimental method

Table 1 Variation of natural frequency

Height variation		Mass variation along height	
H	f_n	Z/H	f_n
5	14.5	0.0	11.4
10	12.6	0.2	12.2
20	10.4	0.5	13.4
40	7.4	0.7	14.2
60	5.6	1.0	14.8

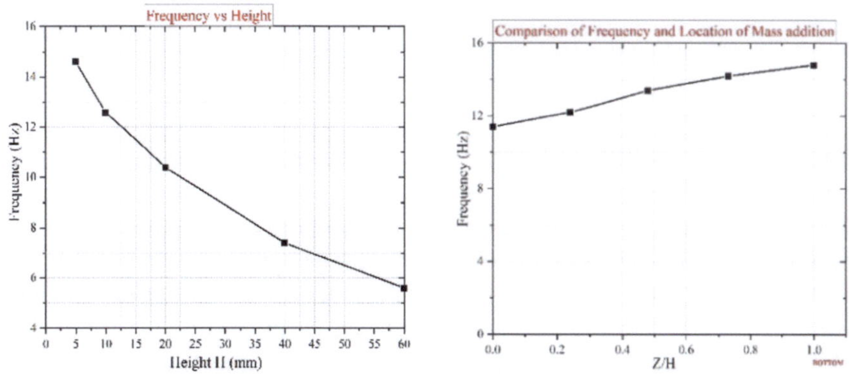

Fig. 9 Variation of natural frequency with height and mass

to decrease so low natural frequency. Variation of natural frequency with height at which the model is kept is shown in Table 1. The natural frequency is maximum at low height and the frequency decreases with increasing height. The effect of addition of mass at different height on natural frequency is also given in Table 1. The natural frequency is lowest when mass is on top location. The more mass at top affects the oscillations which arise at the top or it causes a damping effect to the oscillation which results in reduced natural frequency. Figure 9 shows the variation of height and mass on natural frequency.

4.2 *Estimation Oscillation Frequency of Cylinder* (f_o)

Oscillation frequency is determined by operating wind tunnel at different velocities and taking the accelerometer readings mounted on top of the cylinder. Here, just the frequency was adjusted to lower value by modifying the blade just to see if the oscillations increase with speed or is a Lock-in. Time domain readings are processed to get frequency of oscillations. Comparison of natural frequency and oscillating frequency at flow velocity 1.7 and 4.2 m/s is shown in Fig. 10

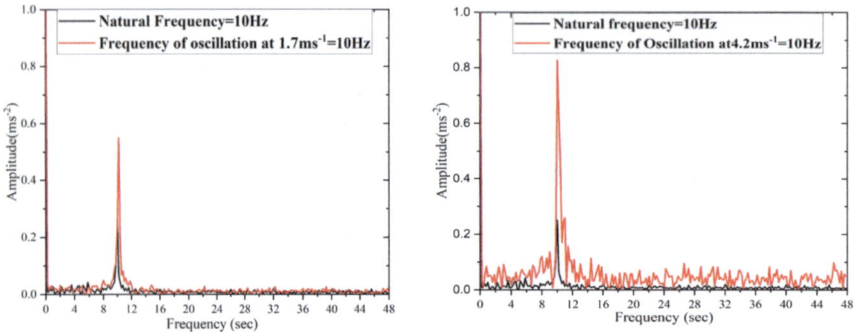

Fig. 10 Comparison of natural frequency and oscillating frequency at 4.2 ms^{-1}

It has been observed that frequency of oscillation does not change with the increase in flow velocity which is due to Lock-in phenomenon. Lock-in or synchronization is often perceived as the regime where the frequency of oscillation for cylinder (f_o), as well as the vortex formation frequency (f_v), are close to the natural frequency (f_n). The force of the vortex is not strong enough to force the spring to oscillate beyond its natural frequency; thus, the spring tends to oppose the motion. So cylinder oscillating frequency remains constant when the flow speed is increased.

4.3 Estimation of Vortex Shedding Frequency (f_v)

Vortex shedding frequency is determined by using pressure Sensor measurement in Y direction. At different flow conditions pressure readings are taken in Y-direction to identify wake region by traversing the pitot probe from center of cylinder to the point where the pressure obtained is equal to that in the flow and also to ensure that the reading is taken outside the wake region. From Fig. 11 it is well understood that at $Y = 1.4$D and $X = 3$D can be fixed for the pressure measurements to obtain vortex-induced vibrations. This position indicates no pressure fluctuations and away from wake region. Pressure probe is traversed in the Z direction along the height of cylinder (from top to bottom) at Z = 1D, 3D, 5D, and 6D.

Measurements are taken for different flow velocity and height (spring) is kept as 5 mm. This gives natural frequency of 14.5 Hz which was the same as discussed in Fig. 8 and as shown in Fig. 12. Figure 12 shows comparison of natural and flow oscillating frequency for different Z/D locations for flow speed of 4.2 m/s. It has been observed that: At top Z = 1D the frequency of flow is dominant as the model is cantilever so the oscillation at top is more and also the other frequency which is Natural frequency is present but less in amplitude. At $Z = 3$D and $Z = 5$D along the downward direction of cylinder the oscillations seem to decrease due to the flow so the pitot readings. At $Z = 6$D both the natural frequency and flow frequency captured by the pitot probe are same in amplitude as the cylinder at bottom is same

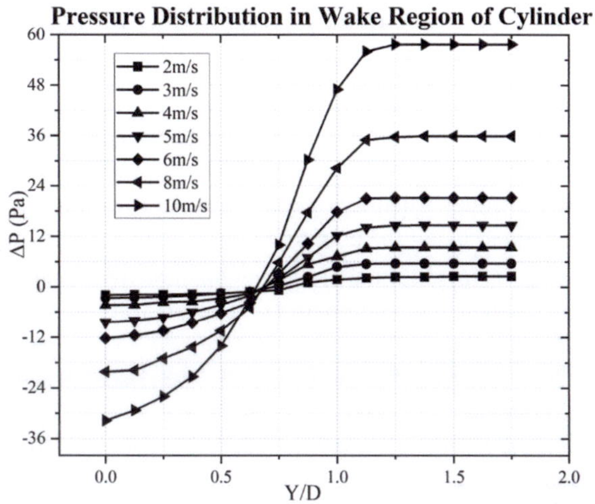

Fig. 11 Pressure distribution behind the cylinder along Y-axis

as a stationary cylinder so it gives the frequency same as the Strouhal frequency. The phenomenon is the Lock In as the forcing frequency is not strong enough to oscillate the cylinder beyond its natural frequency.

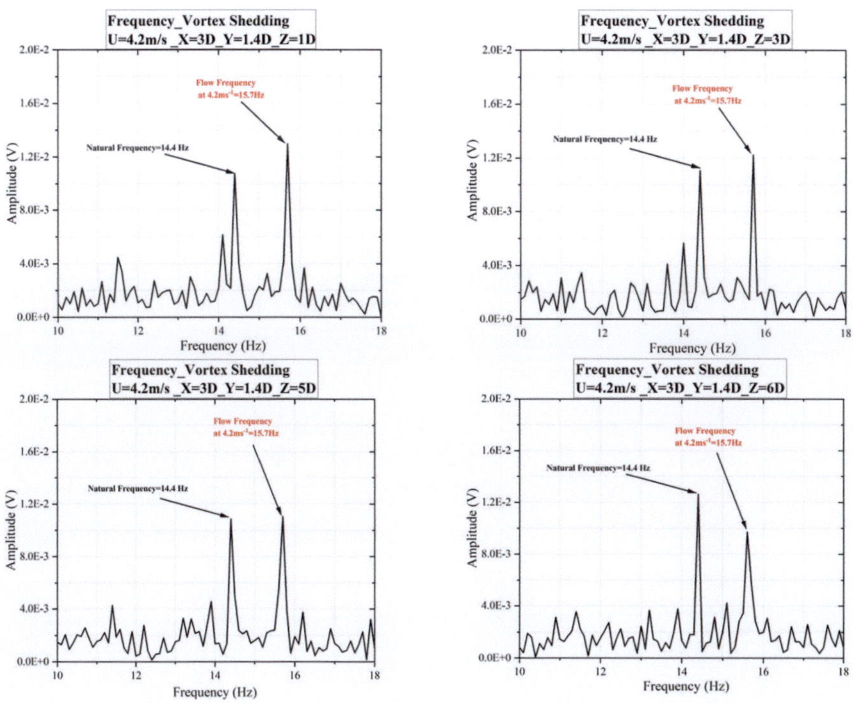

Fig. 12 Comparison of natural and flow frequency at flow velocity 4.2 m/s at different locations

4.4 Comparison of Different Frequency Response

The modified model with 10 Hz natural frequency is taken in this case for the conditions used for Fig. 10 and with varying speed the oscillation frequency f_o is experimentally determined The Strouhal Frequency, f_s^* is determined with the same cylinder when it is fixed and plotted with reduced velocity. Reduced velocity is obtained using the equation $\mathrm{Ur} = \frac{U}{foD}$ as given in literature [15]. Frequency ratio $f^* = f_o/f_n$ (oscillation frequency/natural frequency) and $f = f_s^*/f_n$ (experimental flow frequency of stationary cylinder/natural frequency). These values are compiled in Table 2.

It is observed that the $f^* = f_o/f_n$, i.e., oscillating frequency to natural frequency ratio remains 1 which is also seen in literature [15] is a Lock in condition. Variation of all the frequencies with reduced velocity is given in Fig. 13. It gives information about optimal condition of flow speed.

Table 2 Frequency ratios at different velocity

U	D (mm)	f_n	f_s (St = 0.2)	f_s^* (experimental)	f_o	f^*	f	U_R
1.7	57	10	6	8	10.2	1	0.8	3.0
2.8	57	10	10	12	10.2	1	0.12	4.9
3.4	57	10	12	16	10.2	1	0.16	6.0
4.3	57	10	15	18	10.2	1	0.18	7.5
5.1	57	10	18	22	10.2	1	0.22	9.0

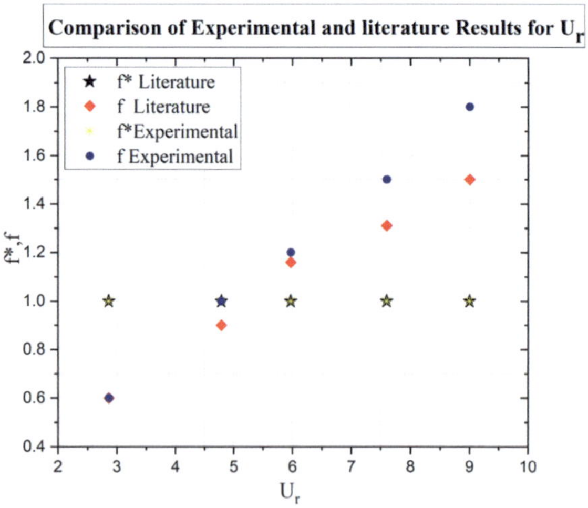

Fig. 13 Comparison of frequency ratio with reduced velocity

5 Conclusion

- The selection of material of the model is very important in terms of strength and weight as it will directly affect the natural frequency which is our main governing factor. In this case the glass fiber material was proved to be best in terms of weight and strength.
- The model was placed on a leaf spring made from a Hacksaw blade so its stiffness was known and as it is fixed in a vice with changing height of blade the natural frequency could be varied so at lowest height we get a high value of natural frequency.
- The mass added at different locations creates nodes and the nodes affect the amplitude of vibration so at top the mass causes the amplitude of vibration to decrease so low natural frequency.
- The phenomenon of Lock in is found in all cases of natural frequency as the forcing frequency of the flow is not strong enough to oscillate the model beyond its natural frequency. There is a damping effect of blade.

Acknowledgements Authors are thankful to Dr. S. D. Sharma for providing inputs at various stages of this study. Authors are also thankful to Prof. Vineeth Nair for allowing using wind tunnel facility at IIT Bombay.

Nomenclature

BWT	Bladeless Wind Turbine
D	Diameter of cylinder (mm)
ρ	Density (kg/m^3)
μ	Dynamics viscosity (kg/m-s)
f_n	Natural Frequency
f_o	Oscillating Frequency
f_v	Vortex shedding Frequency for oscillating cylinder
f_s	Strouhal Frequency of stationary cylinder
H	Cylinder height
I	Mass moment of Inertia
k	Spring Stiffness (N/m)
L	Characteristics Length (m)
m	Mass (gm)
Re	Reynolds Number
St	Strouhal Number
U	Flow Speed (m/s)
Ur	Reduced velocity
VIV	Vortex Induced Vibration

References

1. Williams CB, Yates RB (1995) Analysis of a micro-electric generator for microsystems. In: Proceedings of the international solid-state sensors and actuators conference—TRANSDUCERS '95, pp 369–372. https://doi.org/10.1109/SENSOR.1995.717207
2. Meninger S, Mur-Miranda JO, Amirtharajah R, Chandrakasan A, Lang JH (2001) Vibration-to-electric energy conversion. IEEE Trans Very Large Scale Integrat (VLSI) Syst 9(1):64–76. https://doi.org/10.1109/92.920820
3. Salvador CS, Teresa JA, Martinez JM, Bacasnot MC, Orilla KV, Cabana RJ, Ladaran WI (2017) Design and construction of arc shaped and disc shaped pendulum for vortex bladeless wind generator. In: 2017 25th International conference on systems engineering (ICSEng), pp 363–369
4. Yáñez Villarreal DJ (2018) VIV resonant wind generators
5. Chizfahm A, Azadi YE, Eghtesad M (2018) Dynamic modeling of vortex induced vibration wind turbines. Renew Energy 121(C):632–643. https://EconPapers.repec.org/RePEc:eee:renene:v:121:y:2018:i:c:p:632-643
6. Hu G, Tse KT, Wei M, Naseer R, Abdelkefi A, Kwok KCS (2018) Experimental investigation on the efficiency of circular cylinder-based wind energy harvester with different rod-shaped attachments. Appl Energy 226:682–689. ISSN 0306–2619. https://doi.org/10.1016/j.apenergy.2018.06.056
7. Ordoñez O, Duke AR (2021) Wind resource assessment: analysis of the vortex bladeless characteristics in Puerto Cortés, Honduras. In: IOP conference series: earth and environmental science. 11th International conference on future environment and energy 28–30 January 2021, Tokyo, Japan, vol. 801
8. Aballe ABD, Cruz KYC, Rosa VJHD, Magwili GV, Ostia CF (2020) Development of a linear generator with spring mechanism for vortex bladeless wind turbine. In: 2020 IEEE 12th international conference on humanoid, nanotechnology, information technology, communication and control, environment, and management (HNICEM), pp 1–6. https://doi.org/10.1109/HNICEM51456.2020.9399999
9. Farsi M, Shariatzadeh MJ, Bijarchi MA et al (2022) Low-speed wind energy harvesting from a vibrating cylinder and an obstacle cylinder by flow-induced vibration effect. Int J Environ Sci Technol 19:1261–1272. https://doi.org/10.1007/s13762-021-03241-1
10. Roman M, Sobh R, Sedrak M, Ali M (2021) Power performance enhancement of vortex-induced vibration wind turbines using a semi-active control approach. Energy Sources Part A: Recovery Utilization Environ Effects.https://doi.org/10.1080/15567036.2021.2006830
11. Thomai MP, Kharsati L, Rama Samy N, Sivamani S, Venkatesan H (2019) Experimental analysis of vortex induced vibration in the bladeless small wind turbine. In: Proceedings of the ASME 2019 gas turbine India conference. Chennai, Tamil Nadu, India. December 5–6, V002T06A009.ASME. https://doi.org/10.1115/GTINDIA2019-2484
12. Gautam A, Srinivas SS, Ravi Teja AV (2020) Efficient electro-mechanical conversion system in bladeless wind turbines. In: 2020 IEEE 29th international symposium on industrial electronics (ISIE), pp 1009–1014. https://doi.org/10.1109/ISIE45063.2020.9152509
13. Francis S, Umesh V, Shivakumar S (2021) Design and analysis of vortex bladeless wind turbine. Mater Today: Proc 47(Part 16). https://doi.org/10.1016/j.matpr.2021.03.469
14. Bhardwaj V, Ravi Teja AV (2021) Electromagnetic field configurations for bladeless wind turbines. In: 2021 IEEE 30th international symposium on industrial electronics (ISIE), pp 1–5. https://doi.org/10.1109/ISIE45552.2021.9576306
15. Feng CC (1968) The measurement of vortex induced effects in flow past stationary and oscillating circular and D-section cylinders

Numerical Simulation of Linear Compressor Cascade at Transonic Flow

S. Ragupathy and Abhijit Kushari

Abstract In this paper, numerical simulation was carried out on the stall Flutter in compressor blades. The main objective of this present simulation is to identify the Flutter behaviours of oscillating blades influenced by the stationary blades in the linear cascade setup. In this, there are two geometry configurations: one is a DCA profile with a flat bottom surface, and the other is a double symmetrical shape. As it involved transonic flow conditions, the ninth standard configuration (STC-9) airfoil has been chosen for the present study. The cascade consists of 5 blades, set up in rows linearly with a stagger of 60 and a solidity of 0.75. Initially, the steady-state simulation was done and validated with the original paper. Since the current CFD setup agreed with the experiment results, the time-dependent analysis was carried out on two incidence angles (0, 8) with the variation of reduced frequency (0.25,0.5,0.75,1). The torsional oscillation is given to blades (+ 1, − 1) which are adjacent to the centre blade. All the unsteady measurements were taken on the stationary centre blade. In this, the energy method was adopted to quantify the work done per cycle and aerodynamic parameters. The double symmetry profile leads to instability in all frequencies, but the flat blade is in the stable zone at a low incidence angle.

Keywords Turbomachinery · Aeroelasticity · Reduced frequency · Transonic flow · Aerodynamic damping

---

S. Ragupathy (✉) · A. Kushari
Department of Aerospace Engineering, IIT Kanpur, Kanpur, India
e-mail: ragupathys20@iitk.ac.in

A. Kushari
e-mail: akushari@iitk.ac.in

A. Kumar R. et al. (eds.), *Proceedings of Fluid Mechanics and Fluid Power (FMFP) 2023, Vol. 6*, Lecture Notes in Mechanical Engineering,
https://doi.org/10.1007/978-981-96-9600-0_20

1 Introduction

Today, most commercial aircraft are equipped with turbojets and turbofan engines to meet the requirements of lightweight, compact, and efficient fan and compressor units. So, to get higher aerodynamic and propulsion efficiency, these are the demands that lead to the choice of axial flow compressors.

Most commercial aircraft are currently flown in transonic flight conditions. Significantly, the transonic Mach number is encountered by the axial flow compressor's fan and first-stage compressor blades. As a result, the relative flow over the rotor blade tip section became supersonic locally. And a passage shock forms on the blade's surface [1]. Since these blades are lightweight and, it finally leads to the dynamic instabilities called 'Flutter'. It is caused by the energy exchange of positive feedback between the body and fluid flow.

Initially, the system oscillates at a certain amplitude and total/net damping level (sum of aerodynamic and structural damping), reached either zero or negative, and the amplitude of the vibration increases continuously at each step of the cycle. Finally, it enters the Flutter zone.

There are two types: classical and stall Flutter. First, the coupling of aerodynamics and structure is necessary to cause the Flutter. But the latter is nonlinear, decoupled of both modes. If the flow periodically separates over the blade, either partially or completely, it will manifest a different phenomenon than the classical Flutter. This flow separation, which occurs at a high angle of attack/incidence angle, especially at part speed conditions in turbomachinery, is the source of nonlinear aerodynamics [16]. Due to shock and boundary layer separation within the passage, flow separation and shock oscillations are the primary causes of transonic stall Flutter [2, 10].

Some theoretical methods have been developed [16], and it shows a good match with the experimental results for fully stalled conditions. And found that critical reduced frequency for stalled cases is much higher. The primary purpose of this simulation is to study the stall Flutter in low and high-incidence angles when the passage changes by modifying the geometry. So, the chosen STC-9 configuration DCA airfoil and geometry changes by keeping one flat bottom profile and the other double symmetric profile [5]. The influenced coefficient method was used to calculate unsteady load acts on the blade, and data show good agreement with both linearized cascade theory and experimental results. Fundamentally, torsion and bending mode oscillation lead to Flutter. But compared to the flexural mode, the torsional vibration mode is more vigorous and easily leads to instability regions. Phase lag and lag of pressure coefficients with the blade motion also quantified and emphasized the double symmetric profile leads to instabilities at both incidences, but the flat bottom surface stays in Flutter-free zone.

2 Literature Review and Objective

The development of axial flow compressor studies started in the 1950s by the NACA at LEWIS research centre in the USA [11]. The axial flow compressor's performance improves as the stage pressure ratio and thrust-weight ratio are increased. To improve the T/W ratio, either increase thrust or decrease weight. Currently, the axial flow fan and compressor have reached a point where increasing aerodynamic efficiency is difficult. However, designers are still striving to improve further. Even a small improvement in efficiency has a significant impact on the fuel and operational costs. Rather than increasing efficiency, reduce stage weight by keeping the blades light, slender, and thin without sacrificing efficiency [3].

However, there are significant mechanical and aeroelastic issues that arise when the weight and size of the compressor blades are reduced. The Collar was the first to explain that aeroelasticity is a result of the interaction of the forces of inertia, aerodynamic force, and structural dynamics [9, 10, 12]. A few notable vibration problems were also brought up. When the J85-21 engine's production began, some vibration issues appeared at the fan blade assembly stage, even though many of the engines had already passed the qualification test in flying condition. The worst failure on a Cathy Pacific flight was at RB211-524, which was in flight conditions at 21,000 feet [14].

The AGARD reports for aeroelasticity issues in turbomachinery were published in two volumes in the years 1987 and 1988. For the purpose of studying turbomachinery aeroelastic behaviours, the first report went into great detail about unsteady turbomachinery aerodynamics. In addition, structural dynamics and various analysis techniques used on isolated blades and blade assemblies were covered in the second report. Unfortunately, none of the volumes covered the experimental data and Flutter susceptibility that were used to confirm theoretical findings. However, due to the high cost of creating specialized wind tunnels, such experiments are very difficult to conduct in real turbomachines [8].

To quantify the aeroelastic phenomenon requires the application of unsteady aerodynamics. Calculating the aerodynamic damping parameters and predicting the pressure loads acting on the blades due to variations in the flow stream is advantageous. When the variation is periodically detected by the blades during each revolution, forcing functions result. Shock and boundary layer separation within the passage make even simple configurations, such as a flat plate with zero stagger, computationally complicated [2, 14]. Over time, the bending Flutter with small mean deflection has diminished [15]. Mostly, flexural Flutter occurs at low reduced frequencies (based on the chord) between 0.27 and 0.33, which are acceptable levels. However, torsional Flutter appears at 0.8 to 1.6, which has a significant impact on the stability parameters.

Several aeroelasticity experiments are carried out and documented in the report [13]. However, there is no clear understanding of the flow phenomenon at stall Flutter in transonic flow conditions. The primary goal of this simulation is to investigate attached and separated flow in high subsonic Mach numbers of 0.8 and reduced frequency 0.25, 0.5, 0.75,1 in two different geometries. One is a DCA airfoil with a flat bottom surface, and the other is a double symmetrical surface, as shown in Figure [1]. The current simulation used the ninth standard configuration (STC-9) and five airfoils in a cascade, with the blade adjacent to the centre blade oscillating at different frequencies. The stationary centre blade provides steady and unsteady data. STC-9 data was compared to steady data at Mach 0.8 for validation. It has demonstrated good agreement with current data and numerical data [13]. With that assurance, an unsteady simulation was performed and emphasized in this report.

In this study, the energy method will be used to calculate the aerodynamic damping parameters. It's based on the sum of the work done by the aerodynamic force generated over the blade on one cycle of period [4, 12]. A framework has been developed to determine the work done per cycle and aerodynamic damping values. It shows that changing the passage shape by modifying the blades (double symmetry) leads to instability in all the incidence angles.

3 CFD Model

The time-dependent and frequency response numerical simulation was carried out in this present simulation. So, the geometrical configuration used in this analysis which is included in the 9th standard configuration detail explained by Verdon [13], and due to unsteadiness and frequency involved, removed flow variation in radial direction by doing the 2-dimensional simulation (Fig. 1).

From the rotating reference, chosen single periodic sector is to keep linear cascade analysis. The cascade consists of 5 blades with a spacing of 34.7 mm and a chord 46.3 mm (Fig. 2).

The cascade was designed at Ansys SpaceClaim, and the mesh was generated at ICEM CFD. As its transonic condition, the flow behaves as a complex and nonlinear phenomenon. The grid independence study is carried out to finally get 80,000 mesh elements on the blade passage. So, to capture the local behaviour like flow separation

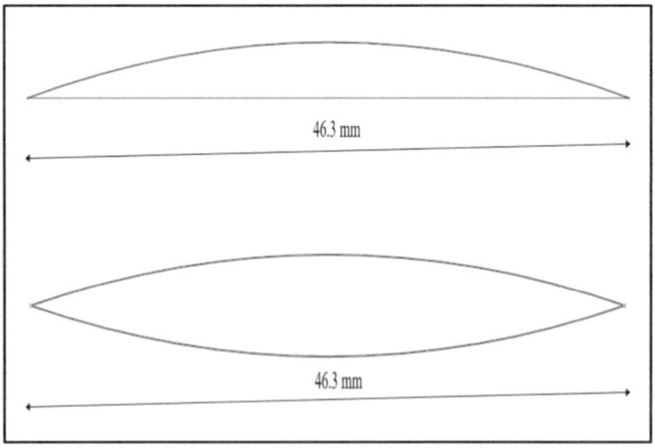

Fig. 1 Schematic of the flat bottom and double symmetric profile

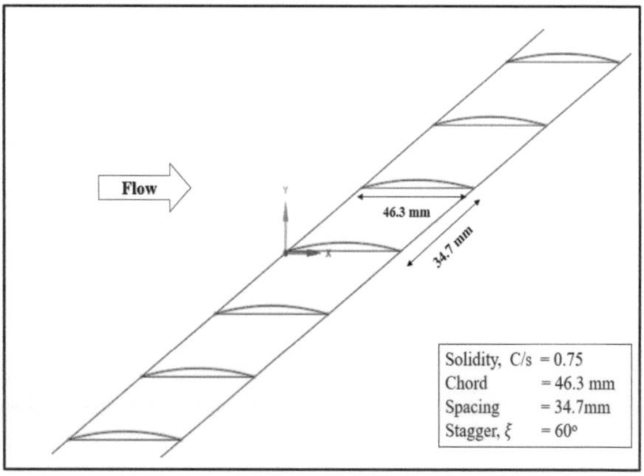

Fig. 2 Schematic of the linear cascade setup

over airfoil due to high unsteadiness and oscillation, used a fine H-type grid with the inflation layer. Due to mesh distortion on high frequency at leading and trailing edges, it has carefully maintained Y + value 1 at all test conditions.

4 CFD Analysis

4.1 Steady-State Analysis

Initially, the time-dependent analysis was carried out on the stationary blades and with the boundary conditions of the inlet at 1.54 bar and 300 K, shown in Fig. 3. To validate the CFD setup, the Mach contour from the steady simulation was compared with the Verdon paper, which aligned at a stagger of 60 deg and incidence of 0 deg (Figs. 4 and 5).

The steady-state simulation has been done for the cases of Mach 0.8 and incidence angle (0, 8 deg), and the comparison of the Cp plot in two geometric configurations is shown in the above figure. In this, shows that at the zero incidence, the flow accelerates highly in the flat bottom configuration, but in contrast happens in the symmetric configurations [5, 6] (Fig. 6).

4.2 Unsteady Analysis

The torsional vibration is provided at the blades (− 1, 1), which are adjacent to the centre blade. And the unsteady pressure distribution on the centre blade at different reduced frequencies was quantified. Initially, steady-state analysis was carried out

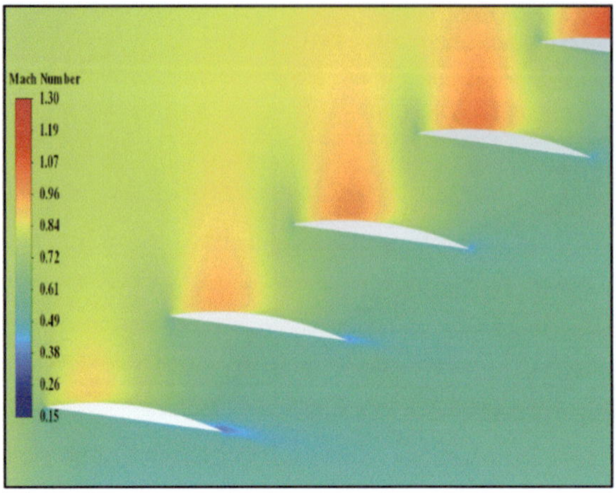

Fig. 3 Mach contour of $M = 0.8$ in flat bottom profile

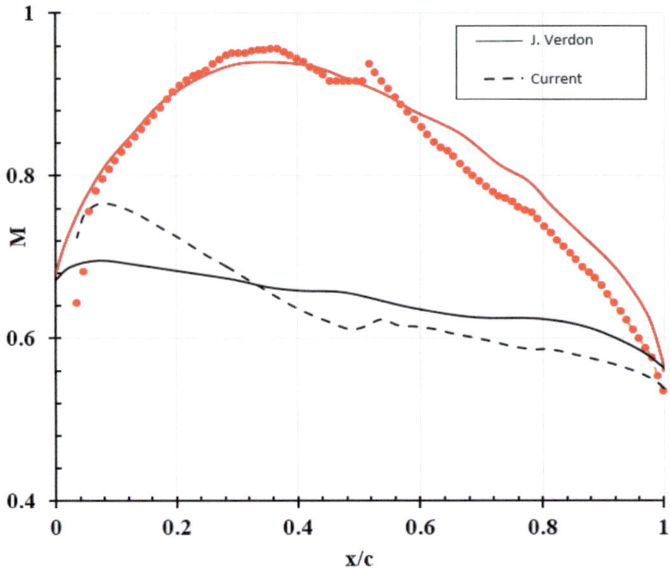

Fig. 4 Mach number distribution over suction (red) and pressure side (black) with variation of frequency

at Mach 0.8 and incidence angle (Angle of Attack) 0 and 8 deg. Then, give the steady-state results as input for the unsteady analysis.

There is a total of 16 cases carried out in the unsteady simulation (For the both profiles, varying the frequency of k=0.25, 0.5, 0.75, 0.10). Due to the blade vibrating at very high-speed oscillations, the blade amplitude is restricted up to 2 deg. And the unsteady pressure distribution has been calculated by taken the first harmonic pressure fluctuation from the Fourier transform in the frequency domain. It was

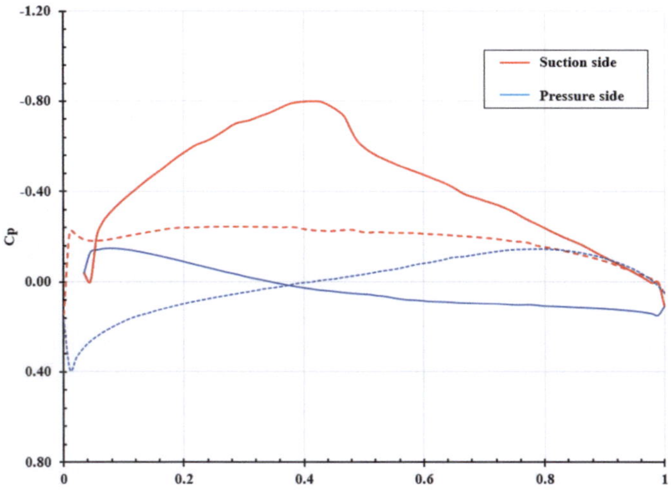

Fig. 5 Cp distribution in flat (—) and symmetry profile (-----) at $i = 0$ deg

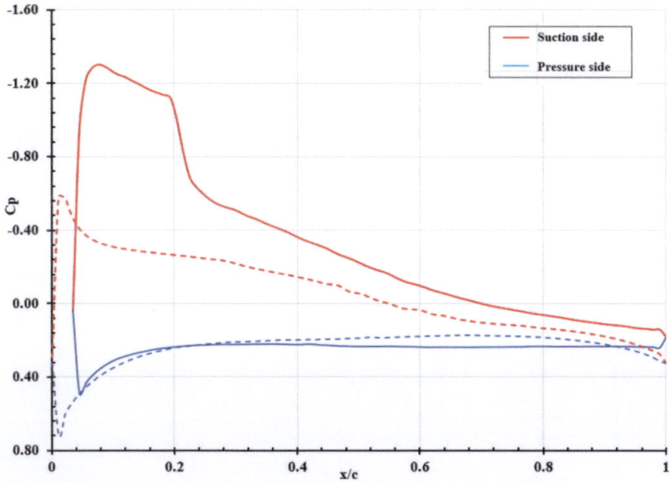

Fig. 6 Cp distribution in flat (—) and symmetry profile (-----) at I $= 8$ deg

further normalized by the oscillation amplitude A, shown in the below equation.

$$Cp, A\left(\frac{x}{c}, t\right) = \frac{P\left(\frac{x}{c}, t\right)}{\frac{1}{2}\rho v^2 A} \tag{1}$$

$$\Delta Cp = Cp_u - Cp_l \tag{2}$$

Figures 7, 8, 9, and 10 show the unsteady pressure coefficient and phase difference with respect to the chord of Mach 0.8. There are shows the comparison of incidence angle ($i = 0, 9$) and different reduced frequency. The Cp plot demonstrates that the pressure variation over the upper surface at low AOA is consistent across all the frequencies. However, there weren't much discrepancies between their values in double symmetry blades when it comes to frequency changes.

However, the flat bottom blades experience considerable pressure fluctuations on the suction side during operation. When the blades are subjected to high angles of attack (AOA), significant irregularities emerge at the leading edges. These irregularities are less pronounced at lower AOAs, where instead, the trailing edges exhibit increased fluctuations. The variations in pressure along the chord, which are induced by flow separation, contribute to different aerodynamic behaviours. In some regions, these pressure differences can enhance stability, whereas in other areas, they can lead to instabilities.

The pressure fluctuation in the phase difference plot shows that it has in phase manner in the forward region of blades, which leads to the local instabilities and, but rearward blades have an opposite phase, which leads to local stability. As the adjacent blades are vibrated at a prescribed different frequency on the first mode shape, the incidence angle at blade 0 changes and the unsteady aerodynamic loads continuously varies. And calculate the work performed by the blade using the variation in the unstable aerodynamic force at each step increment in blade displacement (Fig. 11).

The aerodynamic damping ratio is further computed using the oscillating blade's work at each step and the average vibrational kinetic energy at the end of each cycle.

$$\delta = \frac{\Delta W/2\pi}{W}; \tag{3}$$

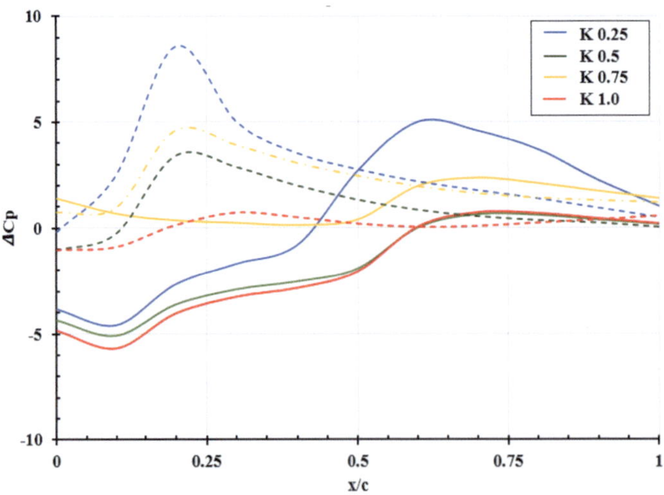

Fig. 7 Comparison of Cp at low and high AOA with variation of frequency in flat bottom profile

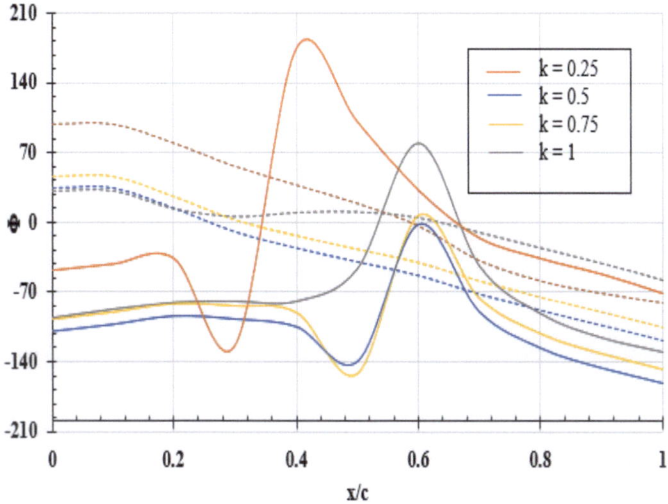

Fig. 8 Comparison of phase difference at low and high AOA with the variation of frequency in flat bottom profile

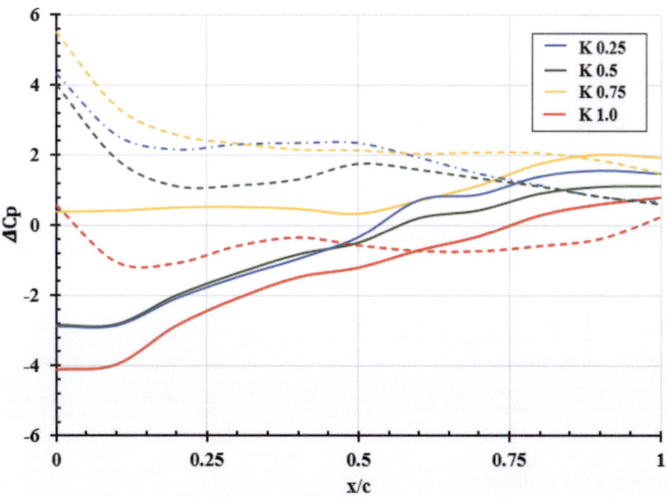

Fig. 9 Comparison of Cp at low and high AOA with the variation of frequency in symmetric bottom profile

$$W = KE = \frac{1}{2}m(a\omega)^\wedge 2 \tag{4}$$

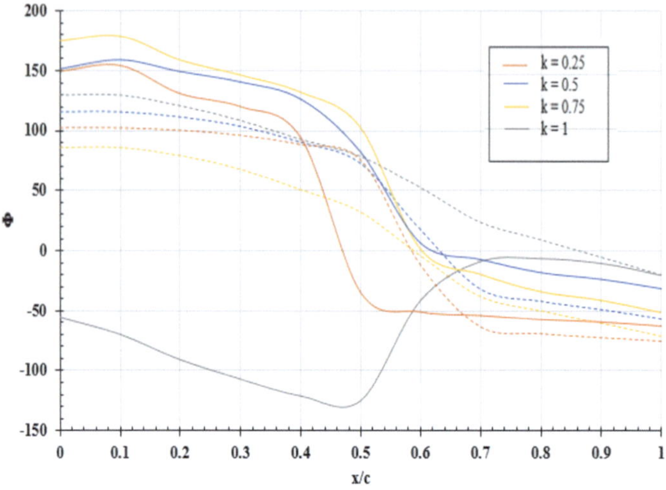

Fig. 10 Comparison of phase difference at low and high AOA with variation of frequency in symmetric bottom profile

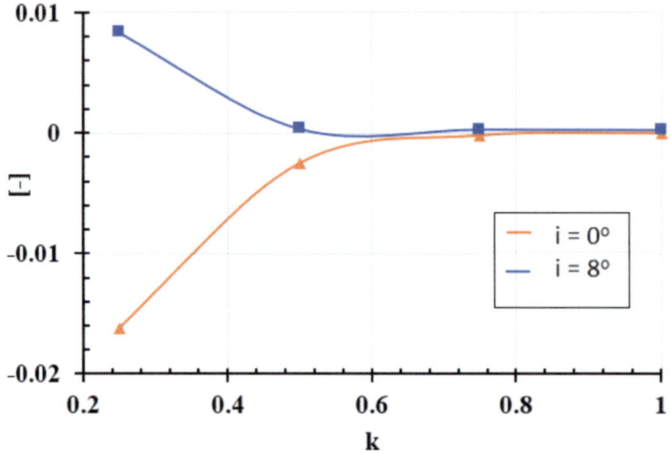

Fig. 11 Damping ratio at flat bottom

The blade absorbs energy from the fluid when the damping ratio is negative, making it prone to Flutter. The energy is transferred from the blade to the fluid during each cycle period if there is positive damping.

Therefore, the oscillation amplitude will decrease in every stage. The above damping plot shows that in flat bottom blades, it indicates that the blades lead to failure at a low reduced frequency, but damping increases with increase the frequency. In high AOA, it happened in reversed manner, the increasing frequency leads to decrease of damping of blades. Figure 12 shows that double-symmetry blades lead to unstable

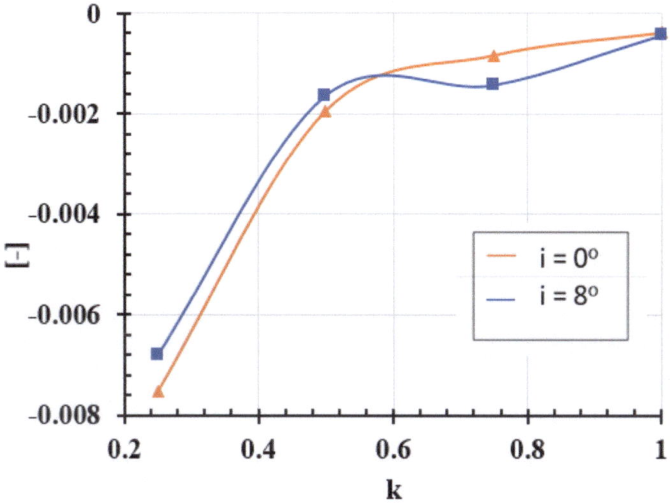

Fig. 12 Damping ratio at the double symmetric bottom

at low reduced frequency, but increases frequency tends to increase stability in both incidence angles.

5 Conclusions

The oscillating blade in a linear cascade setup was simulated using computational methods. The main goal of this simulation was to determine marginal Flutter stability in two sets of geometries. The CFD technique developed for steady-state analysis at $M = 0.8$ initially shows a good comparison between the original paper and current simulation data. Based on prior knowledge, the unsteady pressure coefficient and phase lead were calculated using time-dependent analysis. A total of 16 cases were simulated, each with a different frequency and incidence angle.

To calculate the aerodynamic damping and Flutter stability in oscillating blades, the energy method and work per cycle are used. The interaction between structural and aerodynamic modes has been neglected to adopt classical methods. The phase angle plot demonstrates that the pressure variation over the centre blade in the high-incidence angle is greater than the low-incidence blade. As the frequency increases in the low AOA blades, the damping also increases, and high AOA decreases in the flat bottom surface. But at double symmetry surface, both incidences follow the same pattern to increase damping with the increase of the reduced frequencies.

Nomenclature

M	Mach number [–]
i	Incidence angle (°)
Cp	Coefficient of pressure [–]
δ	Damping ratio [–]
A	Oscillation amplitude (°)
ρ	Density of air [kg/m^3]
ω	Angular frequency [rad/s]
k	Reduced frequency [–]
ϕ	Phase angle (°)
ΔW	Work done per cycleJ
KE	Kinetic energyJ

References

1. Armstrong EK (1960) Some practical aspects of compressor blade vibrations. The Aeronaut J 117–130
2. Aotsuka M, Watanabe T, Machina Y (2003) Role of shock and boundary layer separation on unsteady aerodynamics characteristics of oscillating transonic cascade. In: ASME Turbo Expo, Atlanta, Georgia, USA
3. Benser W (1974) Transonic compressor technology advancements. In: Pennsylvania state university fluid mechanics, acoustics, and design of turbomachinery, USA
4. Buffum D, Fleeter S (1989) Experimental investigation of transonic oscillating cascade aerodynamics. In AIAA aerospace science meeting, Reno, USA
5. Buffum D (1995) Blade row interaction effects of flutter and forced response. J Propuls Power 11(2)
6. Buffum D, Capece V, King A, EL-Aini Y (1998) Oscillating cascade aerodynamics at large mean incidence. J Turbomachinery 173(1):122–130
7. Forsching H (1994) Aeroelastic stability of cascade in turbomachinery. In: Progress in aerospace science
8. Gallus H, Broichhausen K, Henne J (2015) Experimental unsteady shock boundary layer interaction at single blades and in linear cascade. In: Turbo expo: power for land, sea and air
9. Gan J, Im H, Zha G (2017) Stall flutter simulation of a transonic axial compressor stage using a fully coupled sluid structure interaction. In: AIAA aerospace science meeting
10. Kerrebrock J (1981) Flow in transonic compressor. AIAA 19(1)
11. Marshall J, Imregun M (1996) A review of aeroelasticity methods with emphasis on turbomachinery application. J Fluids Struct 10(3):237–267
12. Platzer M, Carta O (1987) Unsteady turbomachinery aerodynamics. AGARD, USA
13. Srinivasan A (1997) Flutter and resonant vibration characteristics of engine blades. J Eng Gas Turbine Power 119(4):742–775
14. Whitehead D (1959) Vibration of cascade blades treated by actuator disk methods. In: Proceeding of the institution of mechanical engineers
15. Yashima S, Tanaka H (1978) Torsional flutter in stalled cascade. J Eng Gas Turbine Power 100(2):317–325

16. Bolcs A, Sari O (1984) Unsteady flow measurements in a turbine cascade installed transonic flow. Cambridge England, Switzerland

Design and Development of a Model Rocket with Active Stabilization

B. V. Praneeth, Mohammed Zaidur Rahman, S. Harsha Vardhana, and Jayahar Sivasubramanian

Abstract This work is to present the design and development of a model rocket equipped with active stabilization, aiming to improve flight stability, trajectory accuracy, and overall safety. The traditional model rocketry community has long relied on passive stabilization methods, such as fins, to ensure stability during flight. However, as model rockets become increasingly sophisticated and ambitious in their design and performance goals, the need for more advanced stabilization systems has emerged. The study begins with a review of existing passive and active stabilization techniques, including their advantages and limitations. It then focuses on the development and integration of an active stabilization system utilizing modern sensor technology, actuators, and an on-board microcontroller. This active stabilization system actively adjusts the rocket's orientation during flight, counteracting disturbances and enhancing stability. In addition to evaluating performance, the safety aspect is meticulously examined. The paper discusses the potential risks associated with active stabilization systems and outlines safety protocols adopted during the research and development process. The findings from this research provide valuable insights for the model rocketry community and inspire further investigations into advanced stabilization techniques for model rockets. This work contributes to the advancement of model rocketry technology, fostering safer and more efficient designs for future high-performance model rockets and possibly extending the knowledge to larger-scale rockets and space vehicles.

Keywords Model rocketry · Solid rocket motor · Avionics · Gimbal

B. V. Praneeth · M. Z. Rahman · S. Harsha Vardhana · J. Sivasubramanian (✉)
Department of Aerospace Engineering, M S Ramaiah University of Applied Sciences, Bengaluru, India
e-mail: jayahar.ase.et@msruas.ac.in

A. Kumar R. et al. (eds.), *Proceedings of Fluid Mechanics and Fluid Power (FMFP) 2023, Vol. 6*, Lecture Notes in Mechanical Engineering,
https://doi.org/10.1007/978-981-96-9600-0_21

1 Introduction

In recent years, the field of model rocketry has experienced a remarkable evolution, driven by advancements in materials, electronics, and propulsion systems [1]. As enthusiasts push the boundaries of what is achievable in this hobby, the need for improved stability and control during flight has become increasingly apparent. Traditionally, model rockets have relied on passive stabilization methods, such as fins and stable airframe design, to maintain their orientation and ensure a safe trajectory. However, as rocket designs become more complex and ambitious, these conventional methods have shown limitations in providing the level of stability required for optimal performance and safety.

This research paper delves into the design and development of a model rocket equipped with an active stabilization system. The active stabilization system represents a paradigm shift in model rocketry, where modern sensor technology, actuators, and microcontrollers work in synergy to actively adjust the rocket's orientation during flight [9]. By continuously counteracting external disturbances, the active stabilization system aims to enhance the rocket's stability, improve trajectory accuracy, and increase overall flight safety.

Model rockets are a class of small rockets designed to be of low weight (1500 g maximum) and can reach altitudes of 200 to 2000 feet. Model rocketry is done for the purpose of research and experimentation, and majorly enjoyed as a hobby. Most of the small model rocket motors are single-use engines, with lightweight molded clay nozzles and cardboard bodies, ranging in impulse class from A to G. Model rockets generally use commercially manufactured black-powder motors [3]. A model motor's class is determined by its impulse. Motors are divided into classes from A to G. Black powder rocket motors are only commonly produced up to Class F. Each class's upper limit is double the upper limit of the previous class.

There are two types of static stability, longitudinal stability and directional stability. Longitudinal stability is the stability of rocket in the pitching plane []. Rockets or indeed any structure traveling in free fall are inherently unstable in flight, in that any external disturbance or internal misalignment of the thrust vector with the center of mass will cause the vehicle to deviate from its intended path, and a restoring force must be provided by the vehicle itself. To overcome this problem, conventional rockets and missiles have used active stabilization, where electromechanical control systems to detect such deviations from the predetermined path and correct for them. A wide range of these systems has been developed. One such method is the gimbaled thrust system [2]. In a gimbaled thrust system, the nozzle of the rocket can be swivel from side to side. As the nozzle is moved, the direction of the thrust is changed relative to the center of gravity of the rocket. Using this method, the stability condition is achieved. The center of gravity (Cg) is ahead of the center of pressure (Cp). Rocket will be more stable if the distance between Cg and Cp is larger.

Further MATLAB® Simulink was used to simulate thrust vector control (TVC) mechanism and to obtain proportional integral derivative (PID) gains for the flight computer code for the actual launch. The control algorithm for the actuators is based out of PID. The pitch and yaw data from one of the sensors of the flight computer was used to control TVC mechanism. The mechanism consists of a gimbal, which was 3D printed using polylactic acid (PLA) and two actuators or servos. The TVC mechanism is discussed in detail in Sect. 3.

In order to recover the rocket after the launch piston base parachute deployment mechanism has been adapted. A ground test of parachute deployment was done in the process. The deployment mechanism is discussed in Sect. 3. Using Autodesk® Fusion 360 CAD models were created. For the flight computer Autodesk® Eagle was used to create a 2-layer PCB. In Sect. 3, the experimental setup and methodology used to evaluate the model rocket's performance are explained. This section also discusses the safety protocols adopted to mitigate potential risks associated with active stabilization. The results and analysis of the conducted tests are presented in Sect. 4, comparing the performance of the active stabilization system against traditional passive stabilization methods.

The research concludes with a discussion of the implications of the findings, the limitations of the study, and future directions for further research and development in the field of model rocketry with active stabilization. Ultimately, this research aims to contribute valuable insights into the application of advanced stabilization techniques, driving the hobby of model rocketry towards safer, more reliable, and high-performance rockets.

2 Literature Review and Objective

A crucial component of rocket design and operation is rocket's stability. It speaks to a rocket's capacity to keep the appropriate orientation and trajectory while in flight. Without adequate stability, a rocket may move unevenly or even swerve off track, which might endanger the safety of the crew or cause mission failure. Rocket stability is governed by a number of principles. Such as, Center of Gravity, Center of Pressure, Static Margin, and Thrust Vectoring.

The conditions for a stable rocket are that the center of pressure must be located below the center of gravity. There is a simple test that can be done on a model rocket to determine the stability. Tie a string around the body tube at the location of the center of gravity. The Cg and Cp for a rocket is illustrated in Fig. 1.

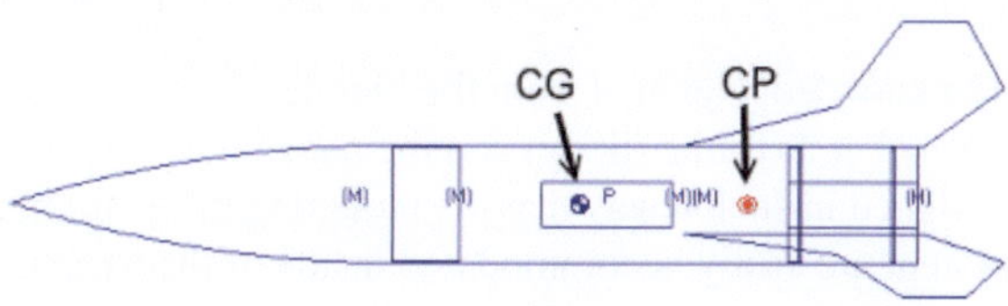

Fig. 1 Placement of centre of pressure (Cp) and centre of gravity (Cg)

2.1 *Passive Stabilization Methods*

The simplest of all passive controls is a stick. Fire arrows were simple rockets mounted on the ends of sticks that kept the center of pressure behind the center of mass. Fire arrows were inaccurate in spite of this. Air had to be flowing past the rocket before the center of pressure could take effect. While still on the ground and immobile, the arrow might lurch and fire the wrong way. The accuracy of fire arrows was improved considerably years later by mounting them in a trough aimed in the proper direction. The trough guided the arrow until it was moving fast enough to become stable on its own.

2.2 *Active Stabilization Methods*

Canards and tilting fins seem very similar to one another; the only major distinction is where they are located on the rocket. At the front are canards, while at the back are tilting fins. The fins and canards tilt like rudders while in flight to deflect airflow and steer the rocket in a different direction. Unintentional directional changes are detected by motion sensors on the rocket, and they can be corrected by slightly tilting the fins and canards. These two gadgets have an advantage due to their size and weight. Compared to big fins, they are lighter and smaller and create less drag.

Fins and canards can be completely eliminated by other active control techniques. By adjusting the angle at which the exhaust gas leaves the rocket engine, course corrections can be done while the rocket is in flight. Exhaust direction can be changed using a variety of methods. Vanes are tiny fin-like components installed inside the rocket engine's exhaust. The exhaust is deflected by tilting the vanes, and the rocket responds by pointing the other way as a result of action-reaction.

Gimbaling the nozzle is a further technique for altering the direction of the exhaust. An exhaust gas-passing nozzle that can wobble is known as a gimbaled nozzle. The rocket adjusts by altering course in response to tilting the engine nozzle in the right direction. The thrust vectoring mechanism is shown in Fig. 2.

3 Materials and Methods

The knowledge gained from the literature review to custom design and manufacture the rocket airframe (body) and the parachute recovery system. The rocket body was designed in such a way that the mounting of the flight computer and also the gimbal system are easily accommodated inside of the rocket.

Fig. 2 Thrust vectoring

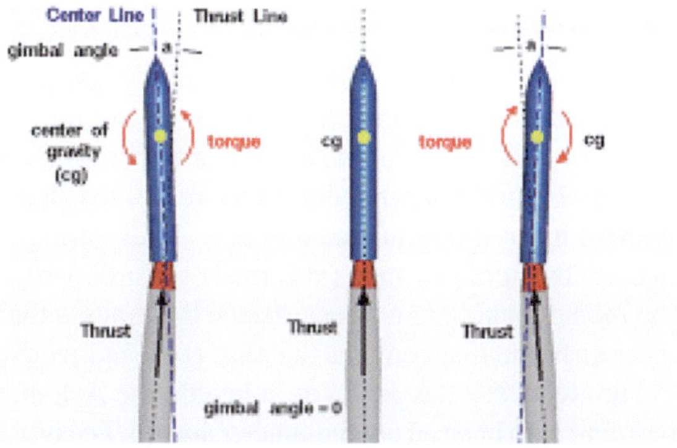

3.1 Rocket Motor

For a rocket weighing 1.28 kg and having a defined apogee at 200 ft, an F type solid rocket motor was required. Since commercial-off-the-shelf F type solid rocket motors are not available in India and are significantly expensive to ship, a home-made motor was the only option. The fuel that we chose was a sugar-based rocket propellant. Model rocket propellants include sugar-based propellants, also known as rocket candy. Its main fuel is sugar, and its main oxidizer is low energy nitrate. In some circumstances, the fuel, the oxidizer, and the additives could be the propellant's constituents. Using sugar as fuel in a rocket propellant is considered as an excellent active component; however, it is limited with its inconsistent characteristics between batches of propellant due to its persistent caramelization and excessively high-pressure index. In our case, the fuel formulation was 35% sugar and 65% potassium nitrate (KNO3), with no added additives/catalysts. This mixture is popularly known as KNSU. This formulation was chosen for its optimum propellant performance. The rocket motor is shown in the Fig. 3.

Fig. 3 KNSU rocket motor

3.2 *Airframe*

The airframe of the rocket is made out of cardboard. Because it is lightweight and simple to deal with, cardboard was an appropriate option. The avionics bay and recovery bay were separated from the rest of the airframe based on calculation. Each segment has a different function and stores various components, but when joined together by a coupler tube (also made of cardboard), they form the entire airframe. The model rocket has a base airframe that houses the motor and the avionics and an upper airframe that contains the nose cone and recovery mechanism.

The Recovery Bay is 0.3 m in length, the Avionics and Propulsion Bay is 0.5 m long, and both have an outer diameter and thickness of 78 mm and 3 mm, respectively. The coupler tube has the same thickness and has an outside diameter of 75 mm. In order to balance the needs of attaining flight to the specified height and providing the room needed within the rocket for the other subsystems, i.e., the motor and electronics bay, a diameter size of 75 mm was found to be adequate. The lower airframe may accommodate a larger motor tube with a 75 mm diameter, boosting GodSpeed's to maximum altitude. The CAD model of the rocket was created using Autodesk® Fusion 360 and is shown in the Figs. 4 and 5.

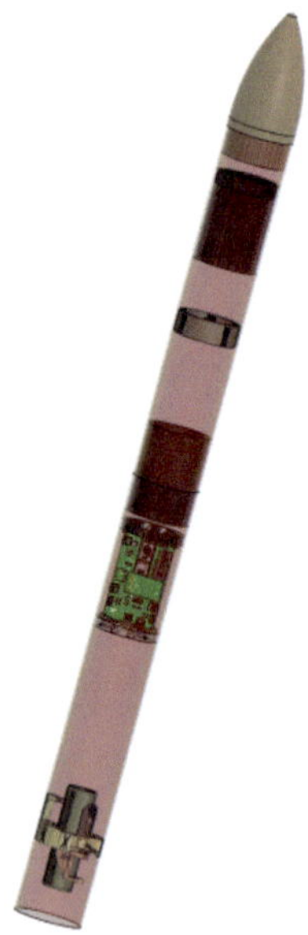

Fig. 4 CAD model of GodSpeed rocket

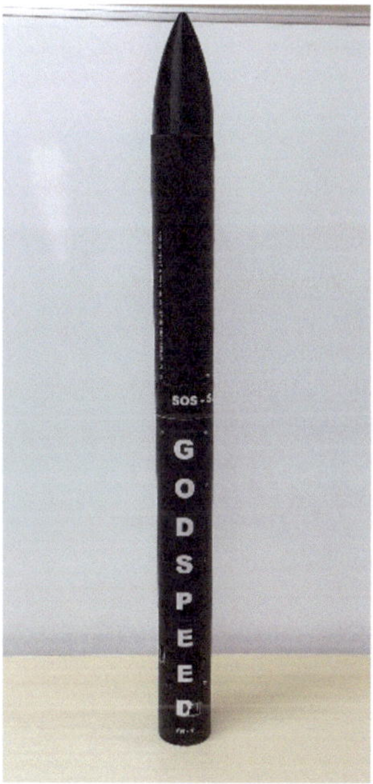

Fig. 5 CAD model of GodSpeed rocket

3.3 Recovery System

The recovery system of a model rocket is an essential component that ensures a safe descent and landing after the rocket's flight. It is designed to slow down the rocket's descent and minimize the impact forces upon landing. The recovery system's design considers the rocket's weight, size, and desired descent rate. The size and type of parachute is chosen to provide a controlled descent that avoids excessive drifting or high-speed descents. Proper stability during descent is crucial to ensure the recovery system functions effectively, avoiding tumbling or instability. The type of recovery device used in our rocket was an ejection charge. An ejection charge is a small pyrotechnic device that produces gas or smoke when ignited. It is typically placed inside the rocket's body and is activated by an electronic or mechanical timer. The ejection charge creates pressure that deploys the recovery system, pushing it out of the rocket's body. We used an electric match containing red phosphorous for the purpose. The recovery system is shown in the Fig. 6.

3.4 Gimbal Mechanism

The gimbal mechanism shown in the Fig. 7 aides the rocket to stabilize along X and Y-axis, which represents pitch and yaw axis of a rocket, respectively. 3D printing technology helped to develop light weight mechanism. The gimbal is driven by 2

Fig. 6 Parachute ejection system

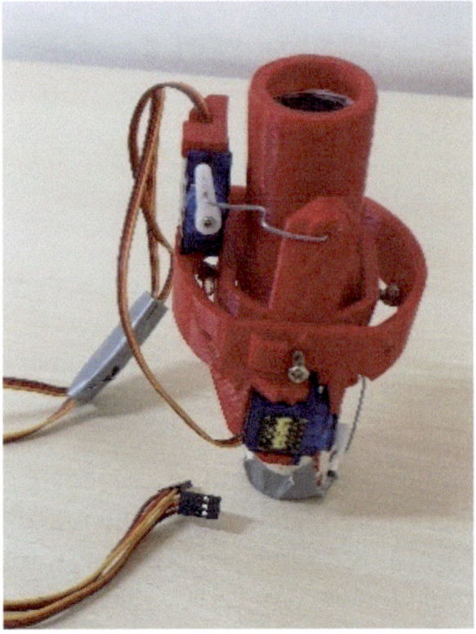

Fig. 7 Thrust vectoring gimbal

SG90 servos, which can control each axis individually. The mechanism can move $\pm$ 15 degree in X and Y-axis. The inner part of the gimbal will hold the solid rocket motor.

3.5 *Flight Computer Algorithm*

The control algorithm involves various stages such as:

1. Launch Detection: The flight computer will detect the launch by looking at the acceleration data in Y-axis. If Acceleration along Z-axis > 2 the flight computer will turn on the data logging and stabilization system.

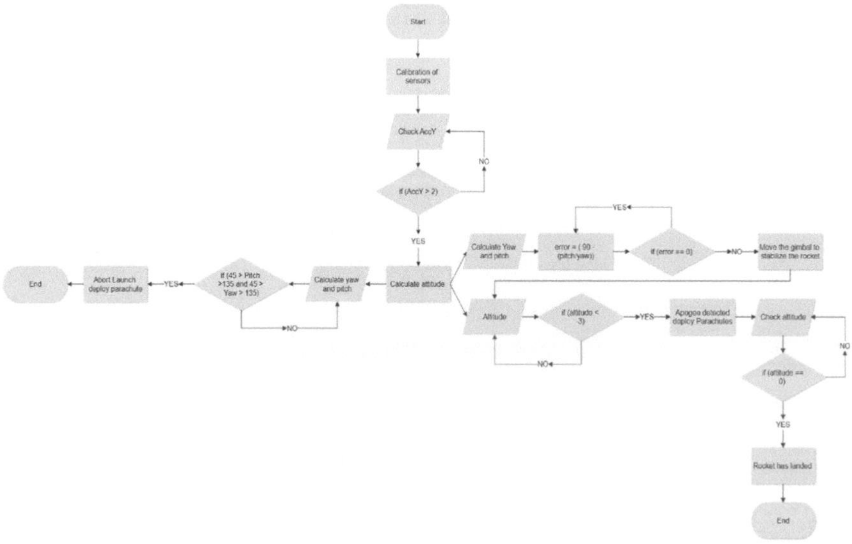

Fig. 8 Flowchart for flight computer algorithm

2. Apogee Detection: The flight computer will look at the barometric sensor and finds the altitude. If the current altitude keeps on decreasing flight computer will deploy parachutes from the recovery bay.
3. Launch Abort: Launch abort is a sequence where the flight computer will detect the change in angle from the sensors right after the launch. If change in angle is greater than 45 deg, flight computer will deploy the parachutes and aborts the launch.
4. Motor Burn Out: Motor burn out is the sequence where the flight computer will detect the acceleration in Z-axis. If acceleration is less than 0. The GSFTI will turn off the servos or stability controller.

The flowchart for the algorithm is shown in the Fig. 8.

3.6 *Avionics*

The Avionics system of the rocket will control its position and record all the data from onboard sensors to SD card. Arduino Nano is used as a processor in GodSpeed Telecommunication Interpreter (GSFTI) flight computer. GSFTI uses a MPU6050 an inertial measurement unit (IMU) sensor for position control and BMP-180 a barometric pressure sensor for altitude, pressure and temperature data. The IMU is 3-axis accelerometer and 3-axis gyroscope. The pitch and yaw data from the sensor is then used to control the position of the rocket. PCB design for the flight computer was done using Autodesk® Eagle. The Flight computer houses couple of power MOSFET

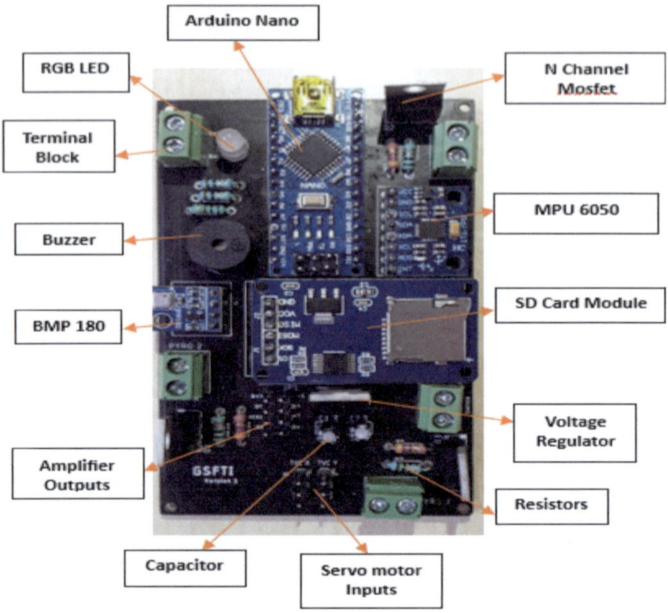

Fig. 9 Avionics-GodSpeed flight telecommunication interpreter

outputs for igniting electric matches for deploying parachute and also uses a light emitting diode (LED) and a buzzer to know the state of the rocket. GSFTI can be powered by 12 V battery. GSFTI is shown in the Fig. 9.

3.7 Control System Design

Matlab Simulink was used to the design the control model to obtain flight trajectories and to tune the PID values. The Simulink model is shown below. 3DOF block is used, which uses thrust data from the plot and returns orientation or position of the rocket. The control system uses PID controller to stabilize the rocket. The PID values are tuned for the given rocket mass and are shown in the Table 1, moment of inertia (MOI), moment arm, and thrust curve of the motor. Mass moment of inertia can be calculated using Eq. 1. Simulink inputs are shown in the Table 1. Pre Flight tuning of the rocket was done to obtain the PID values from the Simulink model Fig. 10. Figure 11 represents the experiment carried out to get the cycle time (C_t). The C_t is value is obtained by swivel rocket around its Cg and recording the time of peak to peak. The PID values obtained from Simulink are used in control algorithm of flight computer, which will rotate the servos without any overshoot (Table 2).

Table 1 Simulink inputs

Parameter	Value
Mass of the rocket (kg)—M	1.235 kg
Cycle time—(C_t)	1.4 s
Cg (center of gravity)	45.5 cm
Length from Cg to rope—COM	35.5 cm
Length of rope from table to Vehicle (m)—RL	63 cm
Mass Moment of Inertia—mMOI	0.12032 $kg-m^2$

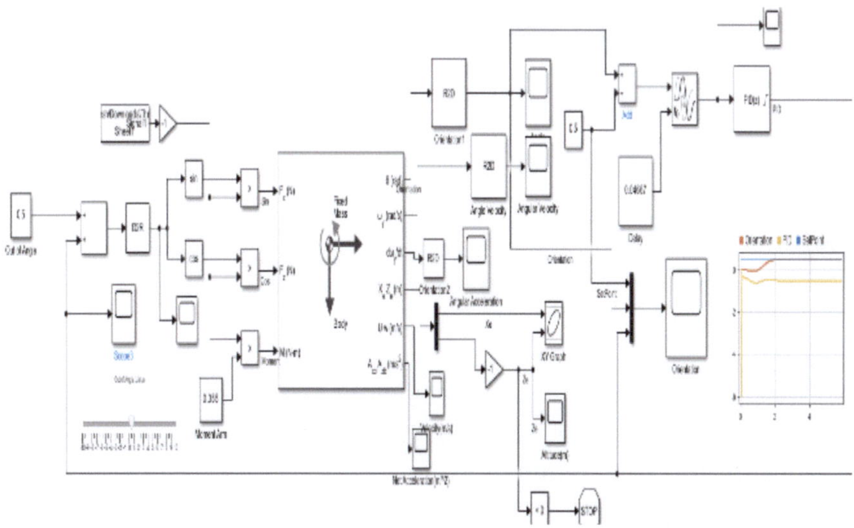

Fig. 10 Simulink model for TVC simulation

Fig. 11 Pre-flight rocket tuning

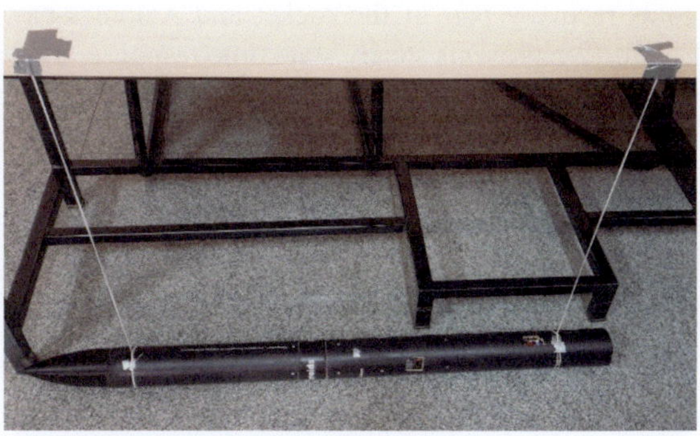

Table 2 PID gains for control algorithm

PID gains	Values
Proportional (P)	0.65
Integral (I)	0.7
Derivative (D)	0.2

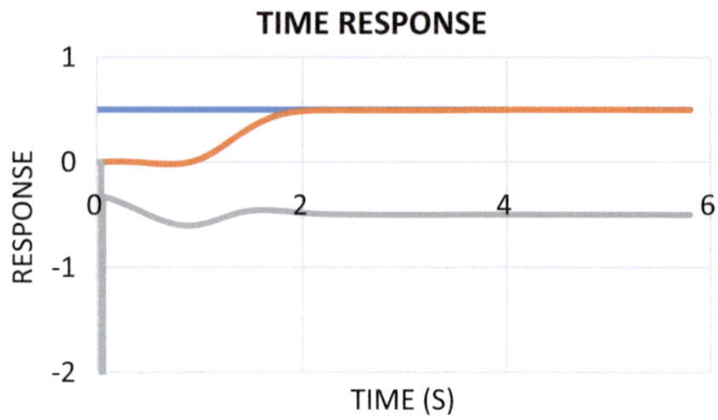

Fig. 12 Steady state time response for actuators

$$mMOI = \frac{\left(M \times g \times Ct^2 \times COM^2\right)}{\left(4 \times \pi^2 \times RL\right)} \quad (1)$$

4 Results and Discussion

The GodSpeed total weight was 1.235 kg. Figure 13 shows blow-up view of GodSpeed which shows all the systems. It has two sections Recovery-bay which holds recovery part of the rocket and Avionics-bay which carried rockets flight. Figure 12 represents steady state time response. The Blue line represents change in angle and orange line shows time taken by the rocket to reach that angle (Fig. 14).

5 Conclusions

The project aimed to improve the stability and control of the model rocket during flight, ultimately enhancing its performance and safety. The gimbal system offers a practical and efficient solution to the inherent instability of rockets, providing a stable and controlled flight experience. Throughout the project, extensive research was conducted on the design and development of model rocketry, stability principles, and active stabilization methods. The team paid close attention to detail, ensuring proper alignment, balance, and secure attachment of the components to guarantee optimal performance and safety. Various tests were performed, measuring the rocket

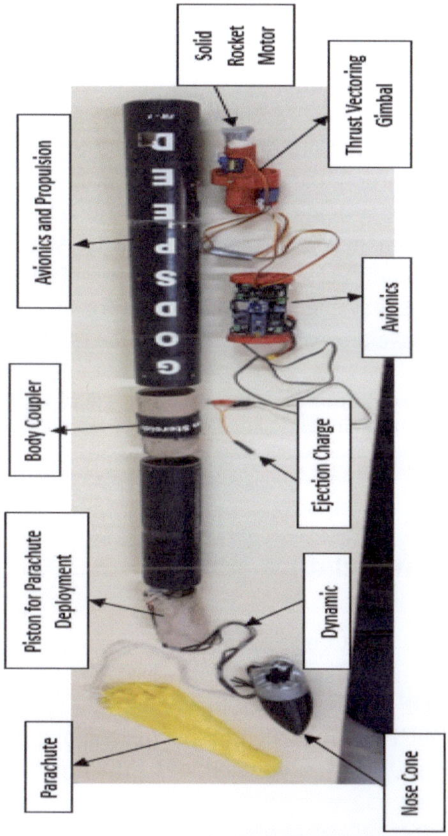

Fig. 13 Blow-up view of GodSpeed rocket

Fig. 14 GodSpeed launch

motor's static thrust production, the unfurling of the parachute recovery system, among other tests. The data collected during these tests provided valuable insights into the performance of the rocket and the gimbal system and allowed for iterative improvements and adjustments. We also acquired hands-on experience in fabrication, testing, and project management, honing our technical and collaborative skills. In conclusion, the project has successfully achieved its objectives of creating a model rocket system that incorporates an active stabilization mechanism using a gimbal system. In conclusion, the project has successfully achieved its objectives of creating a model rocket system that incorporates an active stabilization mechanism using a gimbal system. We are grateful for the opportunity to have worked on this project and we are excited to see what the future holds.

Acknowledgements The authors are grateful to M S Ramaiah University of Applied Sciences for the support in development of the test bench. Also, a special thanks to Dr. Jayahar Sivasubramanian for the support to our project anytime.

Nomenclature

Cp	Centre of Pressure [–]
Cg	Centre of Mass [–]
DOF	Degrees of Freedom [–]
GSFTI	GodSpeed Flight Telecommunication Interpreter [–]
mMOI	Mass Moment of Inertia [kg–m^2]

References

1. Stine GH, Stine B (2007) In: Handbook of model rocketry, 7th edn
2. Ünal A, Yaman K, Okur E, Adli M (2018) Design and implementation of a thrust vector control (TVC) test system. J Polytechnic 10
3. Campbell TA, Seufert ST, Reis RC, Brewer JC, Tomiozzo RL, Whelan CE, Okutsu M (2016) Model rocket projects for aerospace engineering course: simulation of flight trajectories. In: Proceedings of the AIAA aerospace science meeting, AIAA Science and technology form and exposition (SciTech)
4. Sutton G, Biblarz O (2017) Rocket propulsion elements. Wiley & Sons Inc., Hoboken
5. Mandell GK (1973) Topics in advanced model rocketry. MIT Press, Massachusetts
6. Du W (2010) Dynamic modeling and ascent flight control of Ares-I crew launch vehicle, Ph.D. Thesis, Aerospace Engineering Department, Iowa State University
7. O'Connor C, Mehiel EA (2007) Horizon: a system modeling and simulation framework for systems engineering utility analysis, Master's thesis, California Polytechnic State University, San Luis Obispo
8. Perkins J, Greenberg A, Sharp J, Cassard D, Massey B (2003) Free software and high-power rocketry: The Portland State aerospace society, pp 245–258
9. Guerrero VA, Barranco A, Conde D (2018) Active control stabilization of high power rocket, mechanical engineering senior theses. pp 81

Numerical Investigation of Fluid Flow and Heat Transfer in the Microchannel Channel Heat Sink with Different Geometric Fins

Rohit Kumar, Nabadeep Choudhury, and Manmohan Pandey

Abstract Modern electronic components and their compact dimensions have led to the miniaturization of the systems and at the same time the development of high-performance devices. The demand for heat flux dissipation also increased with the reducing the size of these devices and reached up to 1000 W/cm^2. An effective cooling technique is required for effective high heat flux removal from these devices for their safe and efficient operation. A microchannel heat sink (MCHS) can remove large heat flux from electronic components and high-performance computing devices. Heat transfer and flow behavior in microchannel heat sinks were examined. This work investigates numerical measurement of three-dimensional conjugate heat transfer and fluid flow at MCHS. Two different geometric structures of microfins arranged in MCHS are studied. For case 1, rectangular straight fins are considered, and for case 2, square pin fins are considered. The analysis is performed at different operating input parameters by varying Reynolds numbers and heat fluxes. Case 1 shows a better thermal conduction than case 2, which is improved by increasing the Reynolds number. Case 1 has better cooling and lower wall heating temperature than case 2. Hydraulic resistance for case 1 was higher at a low Reynolds number, which became slightly lower than case 2 on increasing Reynolds number.

Keywords Microchannel heat sink · Fluid flow · Heat transfer · Microfins · Computational fluid dynamic

R. Kumar (✉) · M. Pandey
Department of Mechanical Engineering, IIT Guwahati, Guwahati, India
e-mail: rohit1492kumar@gmail.com

N. Choudhury
Department of Mechanical Engineering, National Institute of Technology, Silchar, India

A. Kumar R. et al. (eds.), *Proceedings of Fluid Mechanics and Fluid Power (FMFP) 2023, Vol. 6*, Lecture Notes in Mechanical Engineering,
https://doi.org/10.1007/978-981-96-9600-0_22

1 Introduction

Miniature thermal management refers to the techniques and technologies used to manage heat in small-scale electronic devices and systems. As electronic components become increasingly powerful, efficient heat management becomes crucial to prevent overheating and extend the lifespan of the devices. The rising demand for accelerated computational speed and device compactness raises the issue of effective thermal management. Water cooling is coming out as an effective solution for thermal management. Water cooling is one of the popular methods employed in miniature thermal management. It involves using a liquid, typically water, to dissolve heat from electronic components. Water has a high heat capacity and thermal conductivity, making it an actual medium for absorbing and removing heat away from flashpoints. Although it is practically difficult to create a cooling system that completely resolves all of a heat-producing source's thermal management issues, an ideal, nearly feasible cooling system can be formed with several physics and principles. The configuration and shape of the channels has been shown to have a significant impact on the behavior of the cooling water and the overall heat transfer [1, 2]. In conduit diameter analysis, the rectangular design showed superior thermal performance and lowest thermal resistance contrasted to other channel profiles [3]. Corrugated microchannels with porous fins have been introduced to reduce shock and thermal resistance [4]. Porosity has a positive effect on water flow, reducing shock and thermal resistance [5].

It is necessary to reduce the thermal impedance of the cooler, especially the convection thermal impedance, which plays an important role in the development of heat in microchannels [6]. Additionally, temperature uniformity must be maintained by proper design of the inlet to maintain a smooth flow. These experiments include geometric modifications to achieve high performance [7]. Despite numerous studies conducted on microchannel heat sinks, there is a critical need for continuous research to address the escalating cooling challenges faced by miniature electronic devices. It is important to improve the thermal conduction rate and overall thermal performance [8]. This study focuses on a rectangular parallel microchannel and a square-size multiple fin configuration. The manipulation of fin numbers introduces an open-type microchannel configuration. A parametric analysis determines the most suitable heat sink geometry that maximizes heat transfer performance.

2 Numerical Modeling

The viscous laminar flow system is selected for the examination. Select liquid water and copper as materials. The conjugate heat transfer problem is investigated by applying constant heat flux from the bottom wall of the miniature heat sink. The constant physical properties of solid are considered while the temperature-dependent properties of fluid are taken. The intake conditions are taken as constant inlet velocity while the outlet pressure is maintained at 1 atmospheric pressure. The no slip condition is assumed.

2.1 Geometry and Setup

The heat sink was of square top view with 25 mm of side length. Rectangular fins are 500 μm in width and 25 mm in length while facing the inlet axially. The height of the fins, as explained, was 3.2 mm. The lower wall of the radiator was 3 mm thick. A heat source of 50 W/cm^2 was applied over 25 mm × 25 mm—mass flow inlet of 1.2977 Lpm (0.0186 kg/s, Re = 200, laminar). Variation of the temperature of the heater surface under laminar conditions is analyzed. The straight-fin microchannel heat sink is called Case 1, and the multiple square-fin microchannel heat sink is called case 2 (Fig. 1).

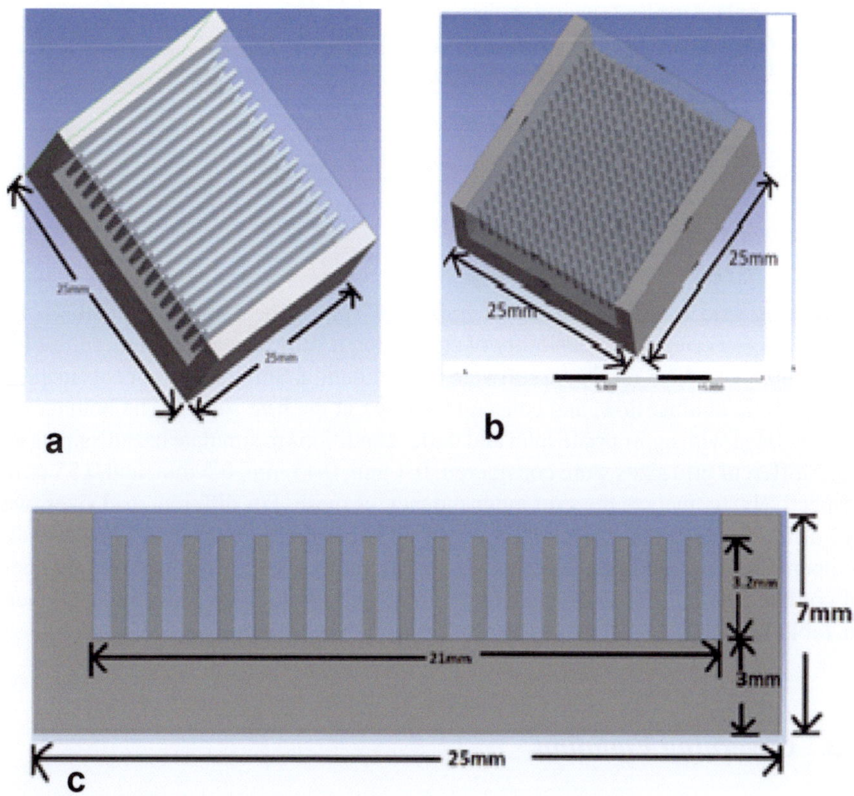

Fig. 1 **a** Isometric view case-1, **b** isometric view case-2, **c** side view

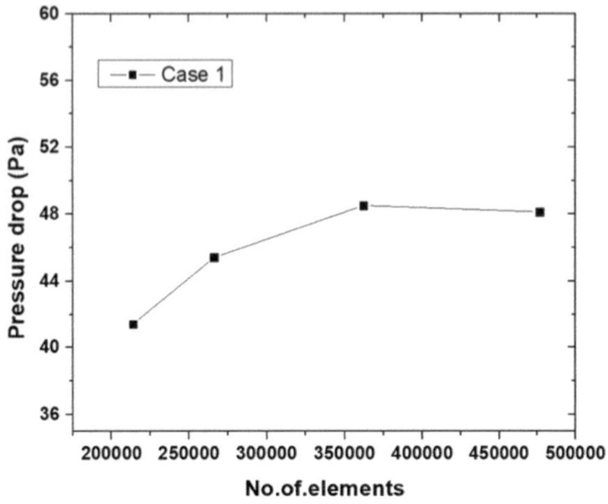

Fig. 2 Pressure drop versus number of elements for case-1

Table 1 Details of mesh selection in channel design

Sl. No.	Channel design	Minimum mesh size (mm)	Number of elements	Number of nodes
1	Case-1	0.45	362,381	65,547
2	Case-2	0.45	357,821	63,881

2.2 Grid Independence Test

Measuring grid independence is an important step in heat sink's simulation. Its purpose is to evaluate the sensibility of simulation results to changes in mesh resolution or size. Use ANSYS Fluent software for simulation. Simulation concepts include steady-state, laminar flow, and constant heat flux at the heat source. The boundaries were defined with appropriate inlet and outlet conditions to simulate heat dissipation. Four different grid sizes were considered, 0.4 mm, 0.45 mm, 0.5 mm, and 0.55 mm, respectively, to analyze the grid independency of mesh. For different grid sizes, we get different elements' values along with different pressure drop and heater wall temperature values. Figure 2 shows the variation of pressure drop with the number of components in Case 1. The details of optimized mesh of both geometry is given in Table 1.

2.3 Governing Equation

The equations for solving tri-dimensional conjugate thermal conduction and fluid flow are as follows (Table 2):

Table 2 Detailed plan for grid independence test for case 1

Sl. No.	Minimum element size (mm)	Number of element	Number of nodes	Pressure drops (Pa)	Error (%)
1	0.55	213,846	39,084	39.4259	4.611
2	0.50	266,275	48,488	39.3987	4.677
3	0.45	362,381	65,547	39.8941	3.478
4	0.40	476,606	86,217	41.3319	0

Mass balance equation (continuity equation):

$$\nabla \cdot \left(\rho_l \vec{V}\right) = 0 \tag{1}$$

Conservation of momentum (Navier–Stokes equation):

$$\rho_l \vec{V} \cdot \nabla \vec{V} = -\nabla p + \nabla \cdot \mu_l \left[\left(\nabla \vec{V} + \nabla \vec{V}^t \right) - 2/3 \nabla \cdot \vec{V} \right] + \rho_l \vec{g} \tag{2}$$

Energy equation:

Energy balance equation for fluid flow:

$$\rho_l c_{p,l} \vec{V} \cdot \nabla T = \nabla \cdot (k_l \nabla T) \tag{3}$$

Energy balance equation for solid:

$$k_s \nabla^2 T = 0 \tag{4}$$

The temperature-dependent variation of the thermophysical properties of the fluid is incorporated into Fluent to solve the governing equations [8]. The dependencies used are as follows:

$$\rho(T) = 765.33 + 1.8142T - 0.0035T^2 \tag{5}$$

$$\begin{aligned} c_p(T) = {} & 28070 - 281.7T + 1.25T^2 - \left(2.48 \times 10^{-3}\right)T^3 \\ & + \left(1.857 \times 10^{-6}\right)T^4 \end{aligned} \tag{6}$$

$$k(T) = -0.5752 + \left(6.397 \times 10^{-3}\right)T - \left(8.151 \times 10^{-6}\right)T^2 \tag{7}$$

$$\mu(T) = (9.67 \times 10^{-2}) - \left(8.207 \times 10^{-4}\right)T + (2.344 \times 10^{-6})T^2 - (2.244 \times 10^{-9})T^3 \tag{8}$$

3 Model Validation and Data Reduction

The numerical results of model are compared with the experimental data of Qu and Mudawar [9] which is found to be in acceptable range with MAE of 19%. Figure 3 shows the validation of model by comparing the pressure drop data on varying Reynolds number.

Basic data: Process the results by analyzing the numbers and calculating the necessary parameters. The convective heat transfer rate is determined accordingly:

$$q = h(T_w - T_m) \tag{9}$$

T_w is defined as the channel wall temperature and T_m is expressed as the average bulk fluid temperature.

$$h = \frac{q}{(T_w - T_m)} \tag{10}$$

$$T_m = \frac{\iint u \cdot TdA}{\iint u \cdot dA} \tag{11}$$

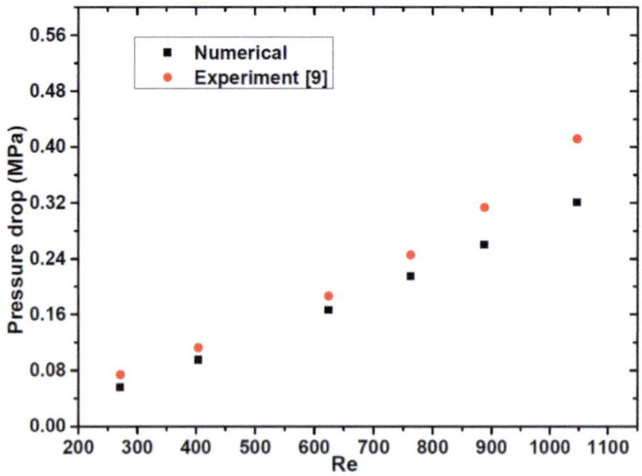

Fig. 3 Comparison of pressure drop numbers with experimental data [9]

4 Results and Discussion

The maximum and minimum heater wall temperatures of 365.368 K and 336.464 K were recorded for case 1, and the maximum and minimum heater wall temperatures of 372 K and 344.8 K were observed for case 2. The hydraulic diameter of the rectangular channel is 0.00672 m. The pressure drop between the inlet and outlet of case 1 is 17.9353 Pa (Re = 200), and the pressure drop between the inlet and outlet of case 2 is 15.5028 Pa (Re = 200).The variation in heater wall temperature, pressure drop is analyzed with the Reynolds number variation.

The temperature distribution must be estimated to identify changing properties and determine local heat if present in the heat sink. To understand this, the corresponding bottom wall temperature profiles for Re = 200 and $q = 50$ W/cm^2 are shown in Fig. 3a and b, where the fin height is $H_f = 3.2$ mm. Fluid flow is in the direction of rectangular fins axially from up to down. It should be noted that the blue to red color on the temperature contour map represents temperature from low to high. The maximum heater wall temperature is 365.385 K for Re = 200, the minimum heater wall temperature is noted as 336.464 K for Re = 800 for case 1, maximum heating wall temperature is 377.949 K for Re = 200, and the minimum heater wall temperature is 344.8 K for Re = 800 for case 2. In addition, since the fin is made of copper material, its thermal conductivity is higher than the cooler, causing a temperature gradient and a higher temperature than the cooler, as shown in Fig. 3a and b.

The velocity contour analysis revealed important insights into the airflow patterns within the heat sink. Figure 4c–f illustrates the velocity distribution across the heat sink surface, obtained through computational fluid dynamics simulations. The contours indicate variations in airflow speed, with higher velocities depicted by regions of closer contour lines. Upon analyzing the velocity contours, several key observations were made. Firstly, the highest velocities were observed near the leading edge of the heat sink, where the airflow enters. This shows that the air gets good results and has a good effect on heat transfer from the radiator base to the environment.

In contrast, the lowest velocity was observed at the trailing edge; This indicates the potential for heat accumulation in this region. The minimum inlet velocity is 0.029762 for (Re = 200), and the maximum is 0.119048 for (Re = 800). We observe the velocity contour for the midplane and upper plane in the solid–fluid domain. Overall, the velocity contour analysis highlighted the importance of optimizing airflow within the heat sink. The results emphasize the need to carefully consider fin design, inlet and outlet configurations, and overall heat sink geometry to enhance convective heat transfer and minimize thermal resistance.

The trend in pressure drop variation with Re is shown in Fig. 5. Figure 6 revealed that an increase in the Reynolds number increases the Nusselt number value. Figure 7 revealed that increasing Reynolds numbers decreases the heater wall temperature.

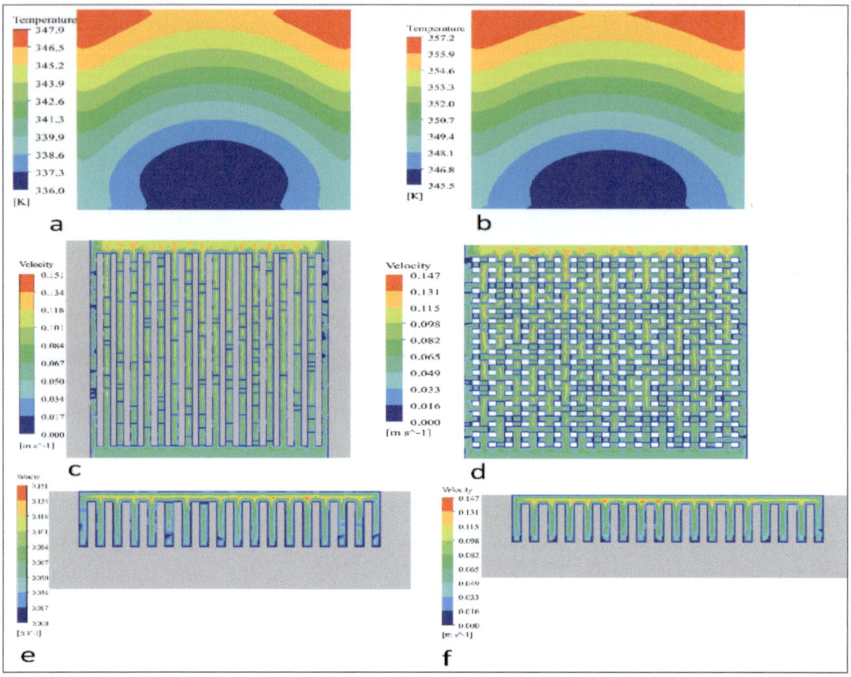

Fig. 4 **a** Temperature profile of case 1, **b** Temperature profile of case 2, **c** Velocity profile of upper plane of case 1, **d** Velocity profile of upper plane of case 2, **e** Velocity profile of mid plane of case 1, **f** Velocity profile of mid plane of case 2

Fig. 5 Pressure drop variation with Reynolds number

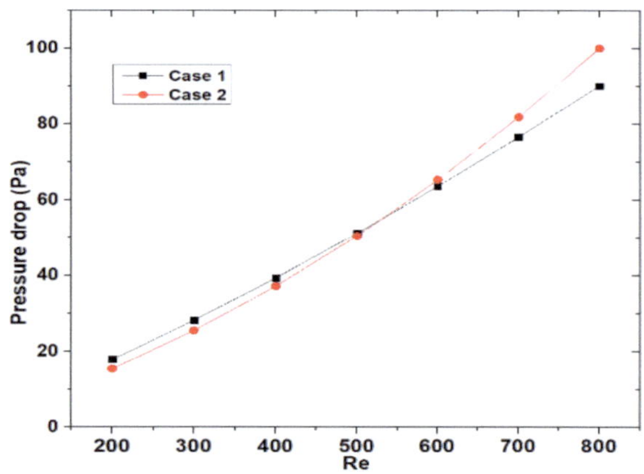

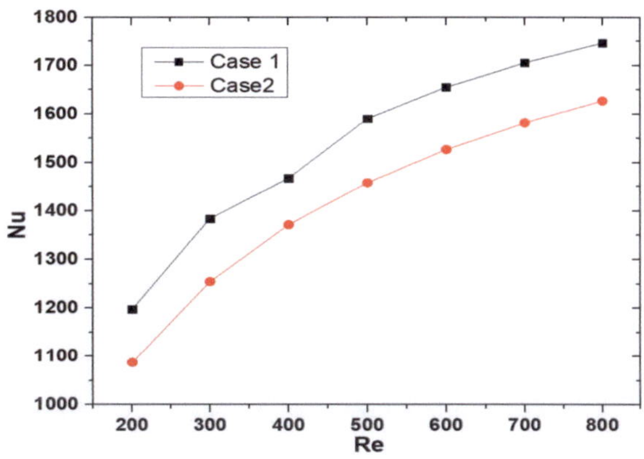

Fig. 6 Nusselt number variation with Reynolds number

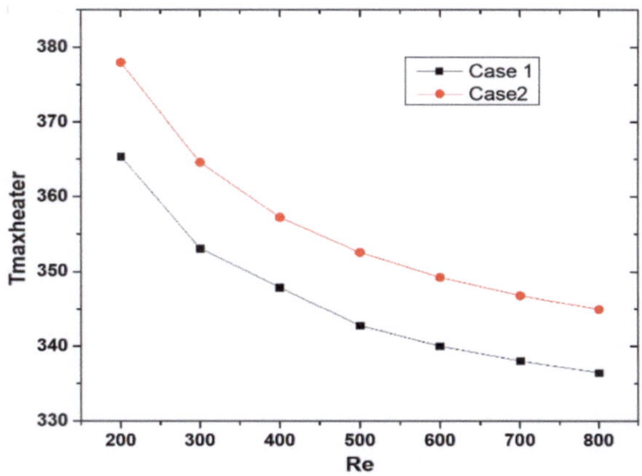

Fig. 7 Maximum heater wall temperature versus Reynolds number

5 Conclusions

Three-dimensional numerical analysis of MCHS was performed for two different microfin geometries. The Reynolds number varied from 200 to 800, and hot air was applied through the bottom wall of MCHS. The analysis of the work can be summarized below:

- The MCHS with straight fins arranged parallel in MCHS, considered case 1, shows better thermal performance than square microfins arranged inline, named case 2.
- The pressure drops in case 1 were higher at a low Reynolds number which became inferior at a high Reynolds number compared to case 2.
- The pressure drop increases with increasing Reynolds number, but the rise in flow resistance was more stepper in case 2.
- Case 1 has better cooling ability and maintains a lower heater wall temperature than case 2. Also, the maximum temperature in case 1 was low.

- In both cases the Nusselt number increases with increasing flow Reynolds number, but for case 1 the value is always higher.
- The heater wall temperature decreases with increased Reynolds number of flow because of the change in efficiency, but the wall temperature for case 1 was consistently lower compared to case 2.

Nomenclature

A	Cross-sectional area of inlet (m^2)
AR	Aspect ratio (–)
D_h	Hydraulic diameter (m)
h	Heat transfer co-efficient (W/m^2 K)
Nu	Nusselt number
q	Heat flux (W/m^2)
Re	Reynolds number
T	Temperature (K)
v	Velocity inlet (m/s)
ρ	Density of air (kg/m^3)
μ	Viscosity (kg/m s)

References

1. Bhandari P, Prajapati YK (2021) Thermal performance of open microchannel heat sink with variable pin fin height. Int J Therm Sci 159:106609
2. Pedram M, Nazarian S (2006) Thermal modeling, analysis, and management in VLSI circuits: principles and methods. Proc IEEE 94(8):1487–1501
3. Visaria M, Mudawar I, Pourpoint T, Kumar S (2010) Study of heat transfer and kinetics parameters influencing the design of heat exchangers for hydrogen storage in high-pressure metal hydrides. Int J Heat Mass Transf 53:2229–2239
4. Mudawar I, Lee J (2009) Low-temperature two-phase microchannel cooling for high-heat-flux thermal management of defense electronics. IEEE Trans Compon Packag Technol 32(2):453–465
5. Wang L, Wu W, Li X, Way N (2013) Numerical and experimental investigation of mixing characteristics in the constructal tree-shaped microchannel. Int J Heat Mass Transf 67:1014–1023
6. Kumar N, Singh P, Redhewal AK, Bhandari P (2015) A review on nanofluids applications for heat transfer in micro-channels. Proc Eng 127:1197–1202
7. Zhao J, Huang S, Gong L, Huang Z (2016) Numerical study and optimizing on micro square pin-fin heat sink for electronic cooling. Appl Therm Eng 93:1347–1359
8. Chiu HC, Hsieh RH, Wang K, Jang JH, Yu CR (2017) The heat transfer characteristics of liquid cooling heat sink with micro pin fins. Int Commun Heat Mass Transf 86:174–180

9. Qu W, Mudawar I (2002) Experimental and numerical study of pressure drop and heat transfer in a single-phase micro-channel heat sink. Int J Heat Mass Transf 45(12):2549–2565
10. Wang H, Chen Z, Gao J (2016) Influence of geometric parameters on flow and heat transfer performance of micro-channel heat sinks. Appl Therm Eng 107:870–879

Large Eddy Simulation of the Flow Over Two-Dimensional Smooth Hills with Increasing Slopes

Ankita Maity, Jay A. Patel, and Niranjan S. Ghaisas

Abstract The study of atmospheric boundary layer (ABL) flow over smooth hills is an important topic of research with several engineering applications. We study the ABL flow over smooth two-dimensional hills using large-eddy simulations. The numerical framework is based on a high-order Cartesian solver along with an immersed boundary method to represent the non-flat bottom boundary due to the hills. The framework is validated by comparing to previously reported wind-tunnel experiments. We present mean velocities and turbulence intensities of the streamwise velocity for hills of different slopes. Our results show that increasing the hill slope leads to significantly modified flow features as compared to smooth hills. Particularly, flow reversal and separation are not observed for the least steep hill studied here and increasingly large regions of recirculation with large turbulence intensities are observed for steep hills.

Keywords Large eddy simulations · Immersed boundary method · Smooth hill · Hill wakes · Atmospheric boundary layer

Nomenclature

$\overline{U}$	Upstream friction velocity [L/T^2]
TI	Turbulence intensity [–]
h_{ref}	40 mm, height of Hill 2 [L]
z_0	Aerodynamic surface roughness [L]
z_h	Hill profile [L]

A. Maity (✉) · J. A. Patel · N. S. Ghaisas
Department of Mechanical Engineering, Indian Institute of Technology Hyderabad, Kandi, Telengana, India
e-mail: remove@gmail.com

N. S. Ghaisas
Department of Climate Change, Indian Institute of Technology Hyderabad, Kandi, Telengana, India

A. Kumar R. et al. (eds.), *Proceedings of Fluid Mechanics and Fluid Power (FMFP) 2023, Vol. 6*, Lecture Notes in Mechanical Engineering,
https://doi.org/10.1007/978-981-96-9600-0_23

L	Half width of the hill	[L]
u_*	Friction velocity	[L/T^2]
U_∞	Velocity outside boundary layer	[L/T^2]
Z_{ib}	Height of immersed boundary	[L]

1 Introduction

Studying the characteristics of wind flowing over complex terrain has received increasing attention in recent decades. Turbulent boundary layer flow over complex terrain is marked by features such as flow separation and reattachment. These flow features are affected by the shape and size of the terrain. A thorough understanding of these is crucial for several applications such as in designing wind farms [1, 2], assessment of aerodynamic loading over mechanical structures [3], and fire propagation in forest areas [4].

Flows over two-dimensional (2D) and three-dimensional (3D) smooth hills have been examined extensively both experimentally and numerically [5–9]. In particular, the wind-tunnel experiments of Cao and Tamura [5] on a model hill of height 40 mm is a good experimental database that can be used for validating numerical simulations. Liu et al. [9] studied numerically the effects of hill shapes and sizes on the flow features. A significant effect on the mean velocity, speedup, second-order turbulent statistics and flow structures of the hill shapes and sizes are described in their work. Our aim in this paper is to conduct a similar comparison of the flow features for hills of varying slopes and further elucidate the flow features. We obtain hills of increasing slopes by changing the hill height without changing the length of the base.

The height of the hill indeed plays a substantial role within the atmospheric boundary layer [9]. A relevant investigation conducted by Wu and Porté-Agel [10], shares similarities with our work. They aimed to explore turbulence statistics when the velocity kept constant at the hub height of a wind turbine. Their work signifies the identifying turbine wakes in particular scenarios. In contrast, by investigating the flow over a 2D hill one can identify the regions over a hill with higher losses before installing the wind farm. As well as the distinguishing two or more different site locations while setting up wind farms to optimize the wind farm output. Unlike the focus on turbine wake dynamics, our research centers on the distinct turbulence or wakes generated by the hills with different slopes. This will help to optimize wind energy in wind farms over hills. The surface roughness is kept constant since the slope effect are the main focus of investigation. The surface roughness considered is similar to approximate the grass land over actual hill [11].

In the present study, three-dimensional large-eddy simulations (LES) of the flow over smooth 2D hills are reported. We consider hills of different heights each with a cosine-squared profile. Flow fields such as mean streamwise velocity field and turbulence intensity are systematically examined. The effects of changes in the hill slopes on the hill wakes are studied.

2 Methodology

2.1 Overview

We employ a high-order accurate LES code, PadeOps-igrid [12], as our primary computational tool. Incompressible Navier–Stokes equations are solved on a uniform Cartesian grid. The horizontal derivatives in the x and y directions are calculated using the Fourier-spectral method while vertical (z) derivatives are calculated using a 6th-order compact finite-difference scheme. A 3rd-order Runge-Kutta method is used for time advancement. The state-of-the-art anisotropic minimum dissipation model (AMD) is used for subgrid-scale modeling.

The streamwise and spanwise directions have periodic boundary conditions. To simulate realistic inflow conditions to the hill, the concurrent-precursor simulation method is used as shown in Fig. 1. In both the domains, the top boundary is an impregnable free-slip wall. At the bottom end, a second-order accurate immersed boundary method (IBM) coupled with a wall model (WM) [13] (IBWM) is used. The IBWM treatment is used to represent a flat surface in the precursor domain, and a flat as well as a 2D hill surface in the main domain. The fully-developed boundary layer flow in the precursor domain is used as the target flow field for forcing the flow in the main domain. This is achieved by adding a body force in the momentum equations within the fringe region of the main domain [14].

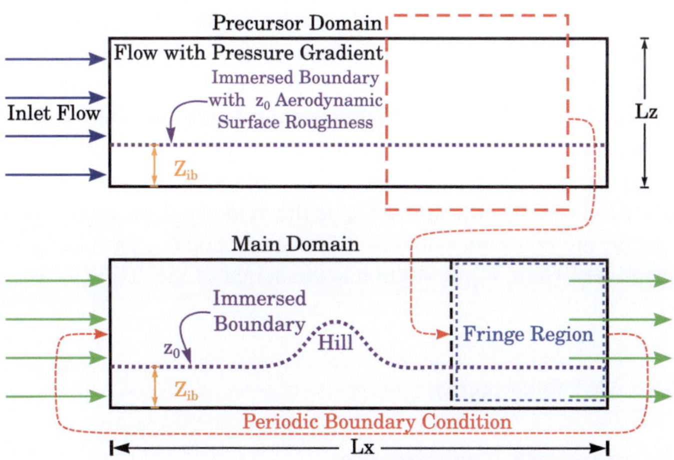

Fig. 1 Schematic of the precursor and main domains in the concurrent-precursor simulation methodology and the geometries simulated using the immersed boundary method

Table 1 Description of cases simulated

Case name	Hill 1	Hill 2	Hill 3
Grid resolution	Grid 3	Grid 1 Grid 2 Grid 3	Grid 3
Hill height h (in mm)	20 (0.5 h_{ref})	40 (1.0 h_{ref})	80 (2.0 h_{ref})
Max u_* over a hill	1.6641	1.8037	2.8691
Min u_* over a hill	0.4532	0.1054	0.1208
Max TI	0.1335	0.1931	0.2413

2.2 Description of Cases Simulated

The simulation domain is a Cartesian box of size $(Lx \times Ly \times Lz)$, with $Lx = 1.2$ m, $Ly = 0.24$ m, and $Lz = 0.3$ m [8]. The immersed domain is placed at a height of $Z_{\text{ib}} = 0.06$ m making the effective height $Lz = 0.24$m. The "precursor" domain has a flat surface (represented as an immersed boundary) displaced by a distance Z_{ib} from the lower end of the domain and the main domain consists of a two-dimensional cosine-squared hill. Both domains have the same aerodynamic surface roughness $z_0 = 0.4 \times 10^{-5}$ m.

The hill profile is $z_h(x) = h\cos^2(x\pi/2L)$ for $x \in (-L, L)$ and 0 otherwise [8]. Here, $L = 0.1$ m and h are the half-width and the height of the 2D hill, respectively. The crest of the 2D hill is centered at a distance of $x = Lx/4$ [8] from the left end of the main domain. We simulate three different cases with hill heights h of 20, 40, and 80 mm. Details regarding these cases are provided in Table 1. For validation purposes, the results of case "Hill 2" are compared with the wind-tunnel experimental results of Cao and Tamura [5]. To match the experimental conditions, the friction velocity is $u^*/U_\infty = 0.03$ [8], where U_∞ is the velocity at the top of the boundary layer. All results are non-dimensionalized using U_∞ and h_{ref} as velocity and length scales, respectively. Here, $h_{\text{ref}} = 40$ mm is the height of the "Hill 2" case.

3 Results and Discussion

3.1 Validation with Experiments

The results of case Hill 2 are compared first to the experimental results of Cao and Tamura [5]. Figure 2 shows LES profiles and experimental data at nine distinct locations spanning from $x/h_{\text{ref}} = -2.5$ to 7.5. Simulations results on grids of different sizes are shown, namely Grid 1 ($128 \times 64 \times 80$), Grid 2 ($192 \times 80 \times 120$), Grid 3 ($256 \times 128 \times 160$). The LES results for the mean streamwise velocity (Fig. 2a) and

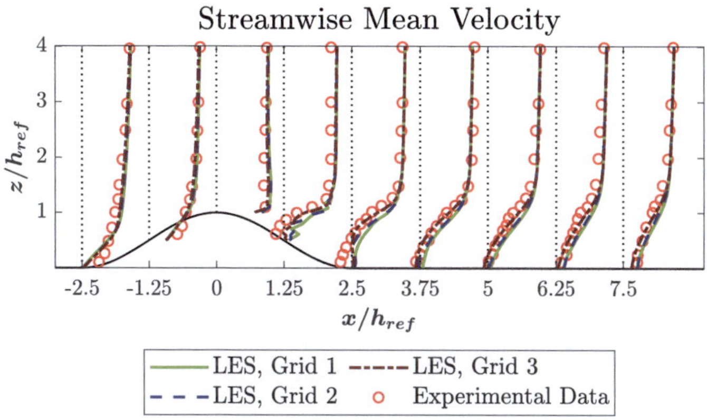

(a)

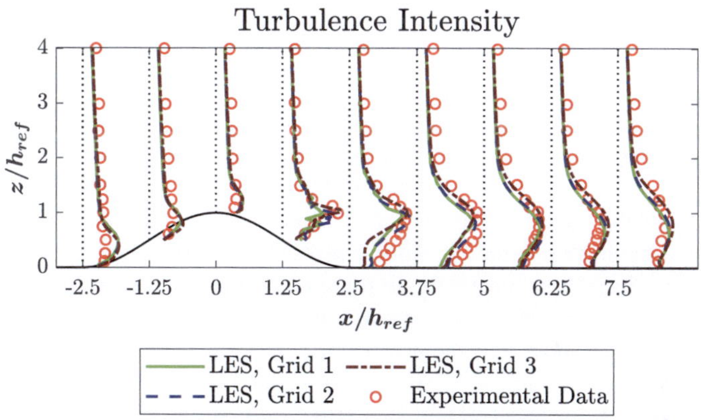

(b)

Fig. 2 Comparison of **a** mean streamwise velocity and **b** turbulence intensity from the current LES results on three different grids to the experimental results of Cao and Tamura [5]

the turbulence intensity (Fig. 2b) are not very sensitive to the grid size and show convergence as the grid is refined. Furthermore, the LES results are in good agreement with the experimental results, especially for finest grid.

3.2 Effect of Hill Slope

The effect of increasing the hill slope is studied in this section. We attain this by changing the height h of the hill while keeping its base length fixed at L. Figure 3 shows the mean velocity profiles ($\overline{U}$) at nine locations for all three hills. For

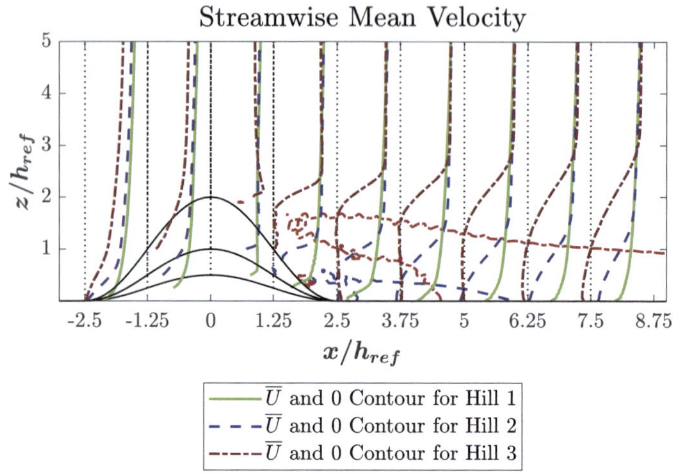

Fig. 3 Mean streamwise velocity ($\overline{U}$) profiles at different x locations and 0-contour lines for three hills

each case, the contour line where the magnitude is 0 is also shown. These "0-contour" lines demarcate the regions of flow reversal in each case. Figure 4 shows the contours of the mean streamwise velocity. Similar plots of the turbulence intensity (TI) are shown in Figs. 5 and 6, respectively. In these plots, the location of the maximum of TI at each x location is also shown for all hills.

Figures 3 and 4a show that there is no flow reversal for the smallest hill height ($h = 20$ mm). With an increase in the hill slope, a recirculation region emerges and increases in size. A comparison between Hill 2 and Hill 3 shows that the recirculation region appears closer to the hill crest in Hill 3. Furthermore, the height of the recirculation region (the height of the upper 0-contour line in Fig. 4b) in Hill 2 is approximately half the height of the hill, whereas in Hill 3 (upper 0-contour line in Fig. 4c) it is almost the same height as the hill itself.

The trace of the value of $\overline{U}$ at h_{ref} is also depicted as a contour line in Fig. 4. Hill 1, which is the least steep, has a longer region around its crest with a higher wind speed compared to Hill 2 and Hill 3. Moreover, it can be concluded that in the wake of a hill, this trace follows at least at the height of h_{ref}. For Hill 3, it is at $2h_{\text{ref}}$, almost the same as the hill's height.

The maximum and minimum values of friction velocity (u_*) in the ABL can provide information about the level of turbulence in the surface layer of the atmosphere. As tabulated in Table 1, the maximum value of u_* increases as the slope increases. The increment is 8.39% from Hill 1 to Hill 2, but 59.1% from Hill 2 to Hill 3. Thus the differences in the near-wall flow are very large between Hill 2 and Hill 3 as compared to between Hill 1 and Hill2. The minimum value has not deviated much from the value 1.0 for Hill 1, indicating relatively lower turbulence at the surface level. The minimum value stands for the least turbulence present at the bottom most part. Surprisingly, the minimum value is lower for Hill 2 compared to Hill 3. The

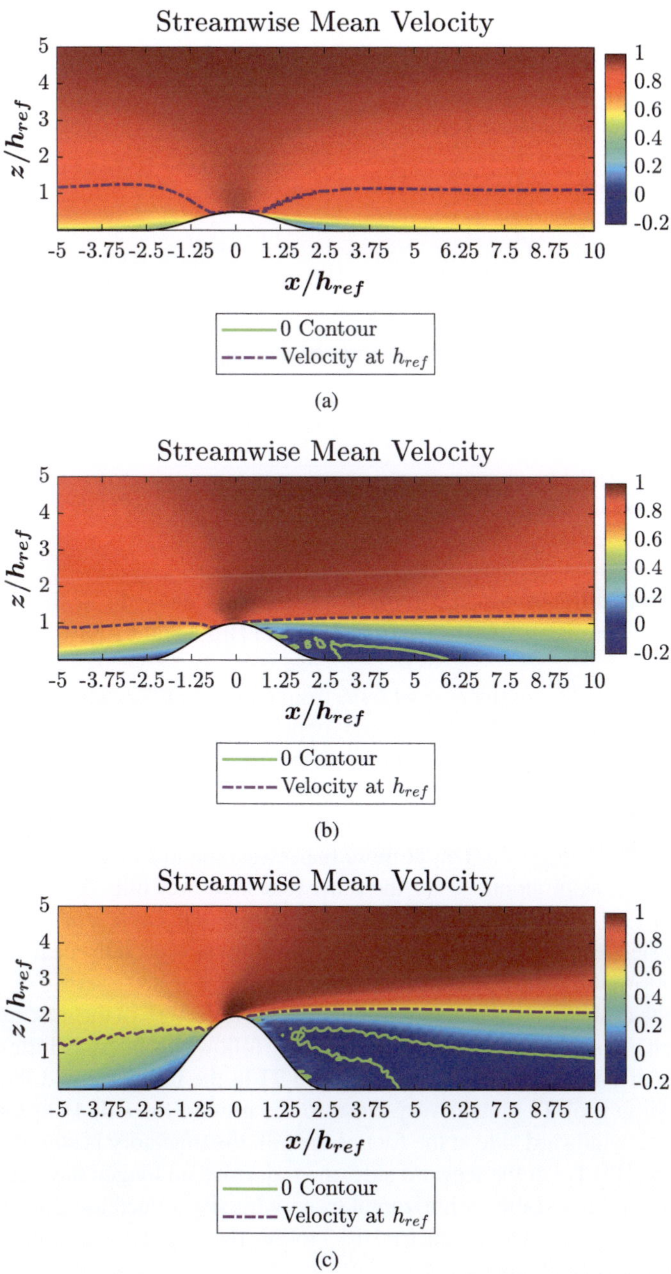

Fig. 4 Contours of mean streamwise velocity, $\overline{U}$, 0-contour lines and contour lines corresponding to the value at h_{ref} for **a** Hill 1, **b** Hill 2, and **c** Hill 3

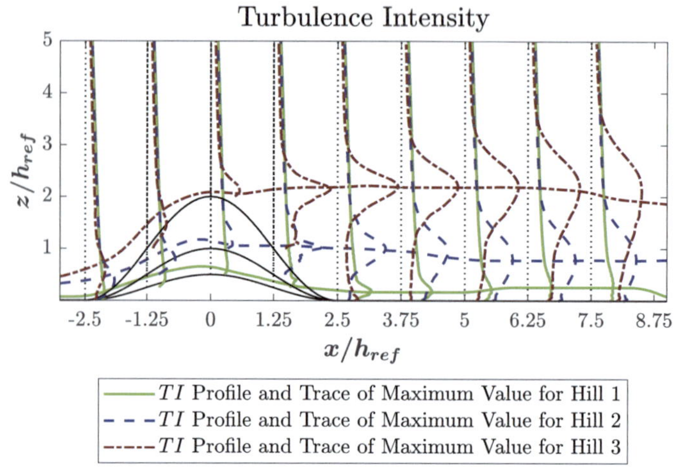

Fig. 5 Profiles of the turbulence intensity, TI, and the trace of maximum TI for three hills

observations of difference in maximum and minimum value indicate that it increases rapidly, with increments of 40.25% from Hill 1 to Hill 2 and 61.53% from Hill 2 to Hill 3.

The maximum TI increment is 44.64% from Hill 1 to Hill 2, while the increment from Hill 2 to Hill 3 is 24.97%. This trend is opposite to that of other typical quantities, where the percentage increment from Hill 2 to Hill 3 is higher.

The flow features upstream of the hills are impacted by the presence of the hills. This is clearly observed for the steepest hill (Fig. 4c) which induces a slowdown in the flow even at $z/h_{\text{ref}} = 5$, i.e., at more than twice the hill height. The slowdown upstream is not as pronounced for the other two shallower hills. This trend is also apparent in Fig. 3, particularly upstream of the hill (e.g., $x/h_{\text{ref}} = -2.5, -1.25$), where the steepest hill exhibits the highest slowdown, especially near the bottom surface, followed by Hill 2 and Hill 1.

The black dashed line in Fig. 6a–c represents the trajectory of the maximum value of TI, which is also shown in Fig. 5. The trajectory follows the hill with little deviation in the case of Hill 1. The region of enhanced TI in the hill wake and the location of the maximum of TI remain very close to the ground for Hill 1. In the case of Hill 2, on the windward side at the foot of the hill, the trajectory is slightly elevated compared to Hill 1. On the leeward side, the trajectory no longer stays close to the ground surface but exhibits a horizontal trend with slight decrease downstream of $x/h_{\text{ref}} = 5$. The peak TI locations for Hill 3 follow the same trend as Hill 2, but with significantly higher TI values.

The dashed-dotted line shows the trajectory of the TI value at h_{ref}. For Hill 1 this trace has an almost unchanged height at h_{ref}. For Hill 2 and Hill 3, this trace increases to roughly $2h_{\text{ref}}$ and $3.5h_{\text{ref}}$, respectively, indicating a large region in the hill wake with enhanced TI values.

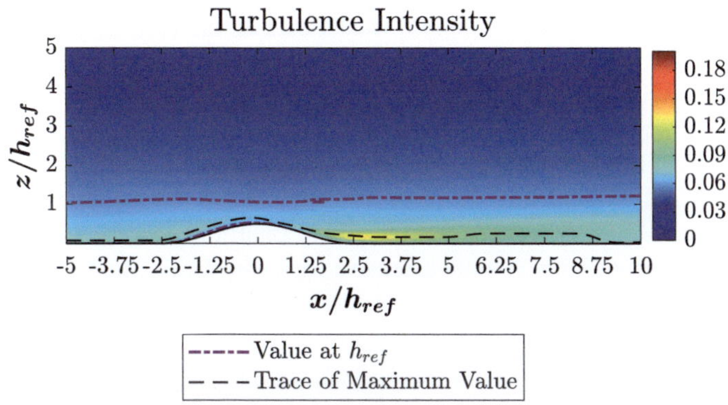

(a)

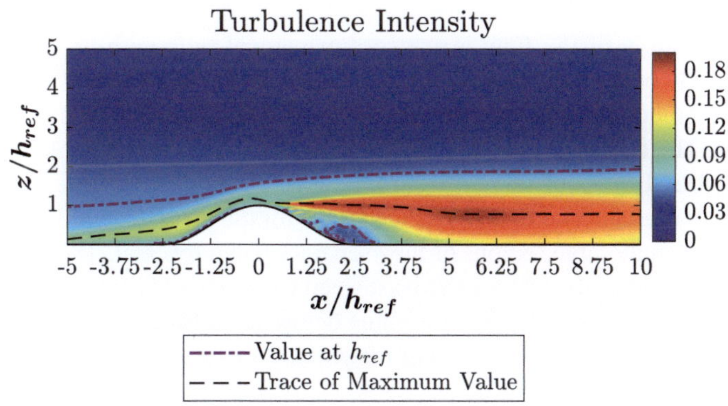

(b)

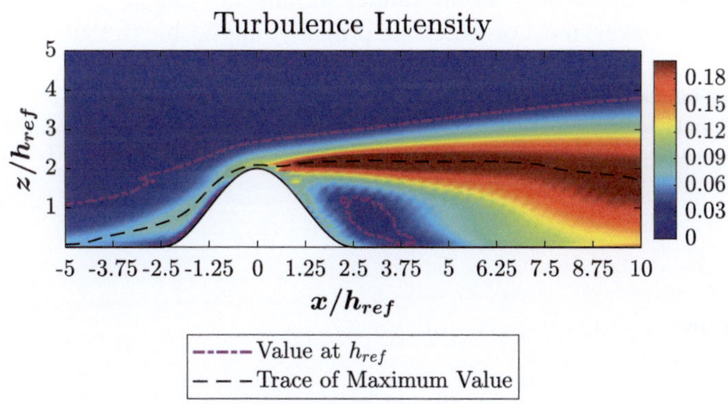

(c)

Fig. 6 Contours of TI, maximum TI contour lines and trace of the TI value at h_{ref} **a** for Hill 1, **b** Hill 2 and **c** Hill 3

The increase in the width of the hill wakes with an increase in hill height is also accompanied by an increase of the maximum values of TI. As the hill slope increases, the maximum width of the hill wake is achieved slightly further downstream from the hill. For example, for Hill 1, the maximum wake width occurs at approximately $x/h_{\text{ref}} = 2$. For Hills 2 and 3, the maximum wake widths are located at approximately $x/h_{\text{ref}} = 5$ and 9, respectively. The low-pressure created immediately behind a hill results in a 'dead water' region where the TI is small due to very small velocities. As the low-pressure region is more extensive for steeper hills, the widest part of the wake occurs farther downstream from the hill and extends over a greater distance.

4 Conclusions

A numerical study of the flow over two-dimensional hills with varying slopes is reported here. The study has been carried out using an in-house code with high-order numerical method and a recently-developed coupled immersed boundary-wall model framework. The framework is validated by comparing to previously reported experimental results. Effects of increment in hill slope on the flow features are studied. The flow remains attached to the ground surface for the least steep hill. Steeper hills exhibit increasingly large regions of flow separation and reversal. The maximum TI follows the profile of the hill for the least steep case, unlike steeper hills, where TI jumps at the hill then remains at same value before it starts decaying very far downstream. The thickness of the wake and magnitudes of turbulence intensity increase with increasing hill slope.

The results presented in this paper suggest that modeling of hill wakes is more challenging for steeper hills. In the future, we aim to evaluate reduced/analytical models for hill wakes, most of which are primarily applicable to shallow hills.

Acknowledgements NSG thanks the the National Supercomputing Mission (Grant DST/NSM/R&D_HPC_Applications/2021/28) and the SERB Core Research Grant (CRG/2022/006735) for providing computational resources on ParamBrahma (IISER Pune) and ParamSeva (IIT Hyderabad) clusters.

References

1. Shamsoddin S, Porté-Agel F (2018) Wind turbine wakes over hills. J Fluid Mech 855:671–702
2. Politis ES, Prospathopoulos J, Cabezon D, Hansen KS, Chaviaropoulos PK, Barthelmie RJ (2012) Modeling wake effects in large wind farms in complex terrain: the problem, the methods and the issues. Wind Energy 15(1):161–182
3. Davenport AG, King JPC (1990) The influence of topography on the dynamic wind loading of long span bridges. J Wind Eng Ind Aerodyn 36:1373–1382
4. Loureiro JBR, Alho ATP, Freire APS (2008) The numerical computation of near-wall turbulent flow over a steep hill. J Wind Eng Ind Aerodyn 96(5):540–561

5. Cao S, Tamura T (2006) Experimental study on roughness effects on turbulent boundary layer flow over a two-dimensional steep hill. J Wind Eng Ind Aerodyn 94(1):1–19
6. Ishihara T, Hibi K, Oikawa S (1999) A wind tunnel study of turbulent flow over a three-dimensional steep hill. J Wind Eng Ind Aerodyn 83(1–3):95–107
7. Ishihara T, Fujino Y, Hibi K (2001) A wind tunnel study of separated flow over a two-dimensional ridge and a circular hill. J Wind Eng 89:573–576
8. Liu L, Stevens RJAM (2020) Wall modeled immersed boundary method for high Reynolds number flow over complex terrain. Comput Fluids 208:104604
9. Liu Z, Hu Y, Wang W (2019) Large eddy simulations of the flow fields over simplified hills with different roughness conditions, slopes, and hill shapes: a systematical study. Energies 12(18):3413
10. Wu Y, Porté-Agel F (2012) Atmospheric turbulence effects on wind-turbine wakes: an LES study. Energies 5(12):5340–5362
11. Stull RB (1988) An introduction to boundary layer meteorology, vol 13. Springer, Berlin
12. Subramaniam A, Ghate A, Ghaisas NS, Howland MF et al (2021) PadeOps GitHub repository. https://github.com/FPAL-Stanford-University/PadOps/tree/igridSGS. Accessed 5 Aug 2021
13. Patel JA, Maity A, Ghaisas NS (2023) A coupled immersed boundary method and wall modelling framework for high-Reynolds number flows over complex terrain. In: Submitted to 14th Asian computational fluid dynamics conference-2023
14. Stevens RJAM, Graham J, Meneveau C (2014) A concurrent precursor inflow method for large eddy simulations and applications to finite length wind farms. Renew Energy 68:46–50

Optimizing Sorting of Micro-sized Bio-Cells in Symmetric Serpentine Microchannel Using Machine Learning

Sayan Karmakar, Md. Safwan Mondal, Anish Pal, and Sourav Sarkar

Abstract Precise sorting of target cells is vital for the progression of cellular research and the enhancement of medical diagnostics. Inertial microfluidics is a recent development in technology providing a very promising approach for high throughput, label-free particle separation. This study presents an extensive CFD numerical investigation on particle migration and segregation within a symmetric serpentine microchannel. The research applies the Eulerian framework to resolve fluid dynamics and the Lagrangian approach to trace particle trajectories, investigating the influence of flow Reynolds number and serpentine channel loops on sorting efficacy. A numerical set of CFD simulations that were executed over 200 randomly set-up data configurations was used for constructing a robust data-driven model. Based on the dataset, a machine learning model was developed to forecast inflow parameters representative of specific sorting efficiencies. In particular, 92% prediction accuracy for the Channel Reynolds Number was achieved in validation. Nevertheless, predicting the number of loops that will ensure optimal sorting performance is still quite a challenge.

Keywords Inertial microfluidics · Particle sorting · Separation efficiency · Machine learning · Focusing

---

S. Karmakar
Department of Civil Engineering, Jadavpur University, Kolkata, India
e-mail: sayank.civil.ug@jadavpuruniversity.in

Md. S. Mondal · A. Pal
Department of Mechanical Engineering, University of Illinois-Chicago, Chicago, IL, USA
e-mail: mmondal@uic.edu

A. Pal
e-mail: apal8@uic.edu

S. Sarkar (✉)
Department of Mechanical Engineering, Jadavpur University, Kolkata, India
e-mail: souravsarkar.mech@jadavpuruniversity.in

A. Kumar R. et al. (eds.), *Proceedings of Fluid Mechanics and Fluid Power (FMFP) 2023, Vol. 6*, Lecture Notes in Mechanical Engineering,
https://doi.org/10.1007/978-981-96-9600-0_24

1 Introduction

In the realm of biomedical research, understanding, and analyzing cells have proven crucial for detecting and diagnosing various diseases. For instance, anomalous cells, such as circulating tumor cells (CTCs), can indicate cancer progression and metastasis. Additionally, separating circulating fetal cells (CFCs) allows for non-invasive prenatal diagnosis, while the isolation of stem cells enables effective regenerative medicine. Consequently, cell separation techniques have become essential in biotechnology, cancer research, regenerative medicine, and targeted drug delivery [1].

Particle separation techniques can be categorized as active or passive. Active techniques rely on external force fields, while passive techniques utilize channel geometry and hydrodynamic forces. Traditional sorting methods like centrifugation and fluorescent-activated cell sorting have limitations, including high costs, the need for reagents, and potential damage to cells [2]. Inertial microfluidics, however, has emerged as a promising label-free particle sorting technique that overcomes these limitations. It offers faster sample processing, lower costs, and higher precision. Among the microchannel structures used in inertial microfluidics, the serpentine channel stands out as a viable choice due to its small footprint, ease of parallelization, and potential for high throughput. It has been seen from the works of Zhang et al. [3] that serpentine microchannel can be used as an efficient passive particle sorter, in which parameters like channel Reynolds number (Re_L), particle diameter ratio (n), channel aspect ratio (AR), and number of serpentine loops (N) dictates the separation efficiency of particles. This happens due to the variation of the forces like inertial lift, dean drag, and centrifugal forces acting on the particles [4].

In this paper, the authors produced a numerical model of a symmetric serpentine microchannel particle sorter. The performance of the microchannel with respect to efficiency (η) and throughput has been investigated by subjecting to variation of the characteristic flow parameters by utilizing a machine learning model. It is difficult to obtain the exact combinations of the mentioned flow parameters to efficiently separate user-specific particle combinations, as it requires numerous trial simulations. Thus, a machine learning model is implemented to overcome this computational expense. CFD simulation results corresponding to randomly generated 200 datapoints have been used to train and develop a data-centric ML model, to predict the required channel Reynolds number and the minimum number of loops based on the required sorting efficiency and throughput for a specific diameter ratio.

2 Problem Formulation and Modeling

2.1 Geometry and Boundary Conditions

This paper presents an inertial focusing in a two-dimensional microfluidic channel, initially having AR = 0.25 and 30 loops, as shown in Fig. 1. It has a 2D width of 200 μm, and is considered to have a depth of 50 μm in 3D. The outlet is bifurcated for a separate collection of particles. Figure 1a depicts the working of the serpentine microchannel, and a zoomed-in plot of the inlet and outlet is shown in Fig. 1b and c. Later the computational domain of the microchannel was altered as the number of loops was varied, keeping other dimensions unaltered, in order to study the effect of variation of the number of loops on particle focusing and separation efficiency.

For accurate modeling of the fluid–particle interaction, the particle release time was selected which enabled each particle to be treated as the instantaneous position of one or more particles. All the particles were released into the flow at $t = 1$ s, so that the steady flow regime is achieved before the particle release. Along with this, the particle diameter ratio defined as $n = \frac{d_{\text{large}}}{d_{\text{small}}}$, (ratio of the diameter of the largest and smallest particle), is also crucial in reflecting the particle behavior which in turn affects the flow in the confined microchannel. For every simulation, only 2 types of particles are present at a time with 1000 particles each. In this study a wide range of particle diameter ratios is considered, varying from 1.10 to 2.00, where 7.5 μm is considered as the base diameter. As shown in Fig. 1b and c, at the microchannel walls, no-slip and particle bounce boundary conditions were applied, and then the calculated flow field was used to trace the particles. At the outlets, zero pressure boundary conditions were imposed.

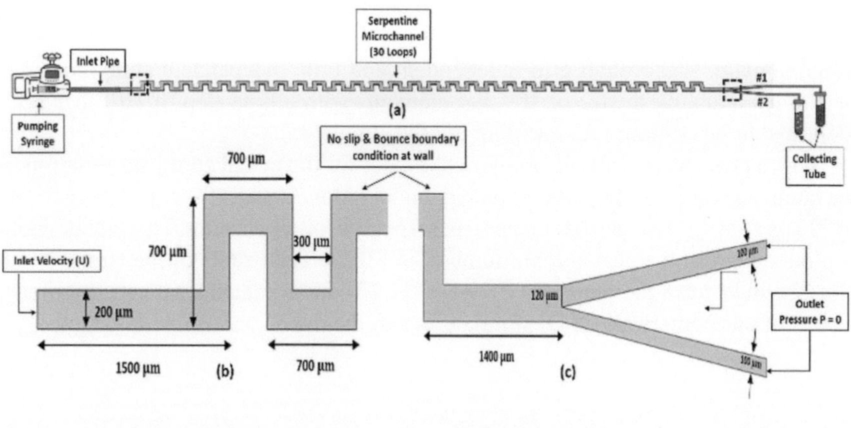

Fig. 1 **a** Schematic of working of a serpentine microchannel particle sorter. **b** Microchannel inlet. **c** Bifurcated microchannel outlet with Boundary conditions

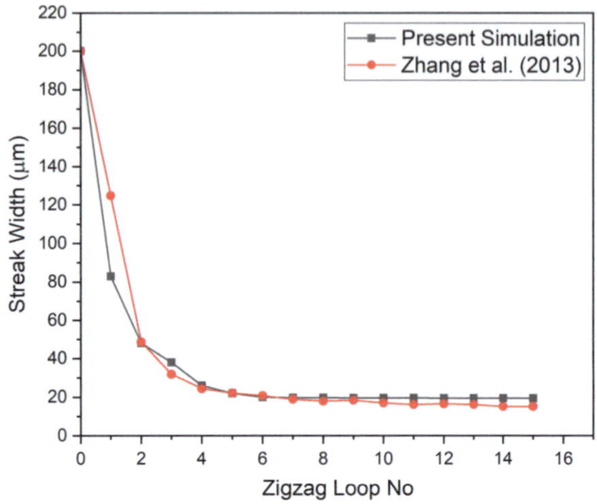

Fig. 2 Streak width versus Loop number plot for particle motion within each loop of the serpentine microchannel, from experimental results of Zhang et al.

2.2 Numerical Modeling

In order to model the particles within the fluid in the microchannel discrete phase modeling has been utilized. An Eulerian–Lagrangian framework has been used. Due to the small dimensions of a microchannel, relatively low velocity, and consequently low relative Reynolds Number (Re_L), the flow field was solved using a laminar flow module in COMSOL [5] by an Eulerian approach. Particles injected at the inlet, which mimic the bio-cells of varying sizes, are solved in the Lagrangian frame. The fluid is considered to be a Newtonian fluid with an average density $\rho = 994\ \text{kg/m}^3$ and dynamic viscosity $\mu = 0.0004$ Pa s [6]. The injected particles are treated as solid particles with a fixed shape, size, and density of $1110\ \text{kg/m}^3$ [7].

An unstructured triangular mesh is used for discretization of the computational domain. A thorough grid-independent and time-independent study has been performed and a mesh size of 109,394 elements with a time step of 10^{-4} has been identified to be optimum for carrying out the simulation.

Before proceeding with the numerical model for the actual study, a validation of the laminar fluid flow and particle tracing model is done against the experimental data of Zhang et al. [3] for inertial focusing in serpentine microchannel. This streak width is plotted against the zigzag loop number, in Fig. 2. The results of the streak width plot obtained from the paper by Zhang et al. [3] are compared with the predicted results of our numerical (CFD) simulation, and the match was found to be good.

2.3 Machine Learning Modeling

Given the high computational load and extensive time consumption involved in the numerical modeling to generate simulation results, in this study, we leverage machine

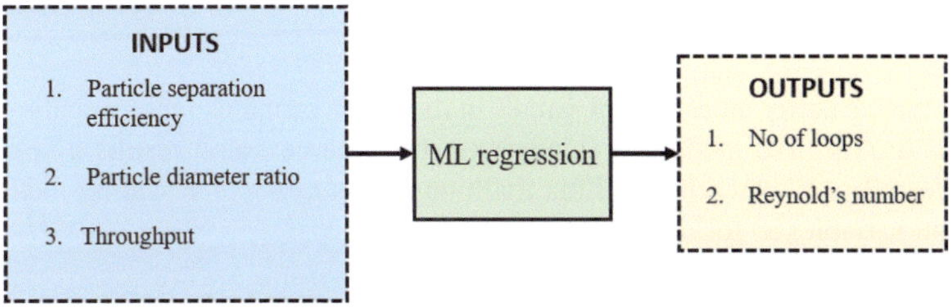

Fig. 3 Machine learning model with input and output features

learning (ML)-based techniques to model the relationship between the parameters of the symmetric serpentine microchannel particle sorter. ML approaches are widely recognized in the scientific community for their high accuracy, achieved through their capability to capture and model highly nonlinear behaviors [8–10]. Unlike traditional models developed from theoretical or established physical reasoning, ML models are "data-driven," built exclusively on observed data [11]. For the particle sorter, we have tried to model the correlation between particle separation efficiency, throughput, flow characteristics (Reynolds number), number of loops, and particle diameter ratio through five different machine learning-based regression techniques. The particle separation efficiency, throughput, and particle diameter ratio were considered as the inputs whereas the Reynolds number of the flow and the number of loops were considered as the outputs to the ML models (see Fig. 3). From the developed numerical model (see Sect. 2.2), 200 data points were synthesized for the ML techniques. A larger subset (70%) of those data was used for the training of the ML algorithms (training data) and the rest data was kept unseen to the trained algorithms and used for the validation of the algorithms (testing data).

To check the stability of the trained models, 50 different random combinations of the testing-training data were tested on the ML models and standard performance metrics were calculated for the assessment of the trained ML models.

2.4 Machine Learning Methodology

2.4.1 ML Models

In this work, we have considered three parameters as inputs and two outputs to predict what makes the problem a "multi input multi output" (MIMO) regression problem. As a benchmark, we started with the linear regression model, which produced a poor accuracy score suggesting a strong nonlinear relationship between the inputs and outputs. Hence, we utilized five other regression techniques [12]; K-nearest neighbors (KNN), support vector regression (SVR), random forest regression (RF), bagging regression (BR), gradient boosting regression (GBR) that can handle the nonlinear pattern between the multiple inputs and outputs.

2.4.2 Feature Scaling

Due to having different ranges of values in the input variables feature scaling technique is used in ML models to standardize the variables, as it results a faster and stable learning. We incorporated the traditional *StandardScaler* scaling method on the input variables as follows:

$$z_{ji} = \frac{x_{ij} - \mu_i}{\sigma_i}$$

Here, z_{ji} is the normalized value of jth observation of feature i, x_{ij} is value of jth observation of feature i, μ_i is the mean value of feature i, σ_i is the standard deviation of feature i. First, scaling was done in the training dataset, then the same scaling parameters were transformed into the testing dataset to prevent snooping/leakage effects [13].

2.4.3 Model Evaluation Metrics

We used three widely accepted metrics, mean absolute error (MAE), root mean squared error (RMSE), and coefficient of determination (R^2 score) to measure the performance of the ML models. Essentially, MAE and RMSE are utilized to capture the difference between the true target values y_i and predicted target values $\hat{y}_i$ over n data points and R^2 score is a measure to determine the goodness-of-fit of the regression models; basically, it represents how much variation in the dataset is explained by the regression prediction model.

$$\text{MAE} = \frac{1}{n}\sum_{j=1}^{n}|y_i - \hat{y}_i|$$

$$\text{RMSE} = \sqrt{\frac{1}{n}\sum_{i=1}^{n}(y_i - \hat{y}_i)^2}$$

$$R^2 \ \text{score} = 1 - \frac{\sum_{i=1}^{n}(y_i - \hat{y}_i)^2}{\sum_{i=1}^{n}(y_i - \overline{y})^2}$$

Overall, MAE, RMSE, and R^2 scores are intuitive metrics that helped us to evaluate the performance of the ML models.

3 Results and Discussion

3.1 Separation and Focusing Mechanism

During the propagation of particles within a serpentine microchannel they mainly experience drag force, lift forces, and centrifugal force. The phenomenon of migration of microparticles within a channel under the influence of these competing forces, into a single focused streak is called focusing. Depending on particle size, they get focused in differential equilibrium positions and can be separated by appropriate outlet bifurcation. Figure 4a shows the change in particle separation efficiency with flow rate (Q), which in turn depends on Channel Reynolds Number. Figure 4b depicts the variation of separation efficiency with the number of loops, for the particle diameter ratio 2, at an optimized flow rate of 250 µL/min.

It is evident from the Fig. 4a that, for a particular diameter ratio of $n = 2$, the separation efficiency steadily increases with the flow rate (Q), up to a certain limit. This can be accounted due to the fact that with the increase in the flow rate, the inertial forces also increases, which in turn increases the lift force on the particles, thus facilitating inertial migration within the microchannel.

Again from Fig. 4b, an effective steady separation efficiency can only be observed when the number of loops is 6 or more. This can be explained in the context of centrifugal force acting on the particles at each bend of the serpentine channel. Thus, with the increase in loops, the particle gets more and more deflected, which leads to focusing and eventually separation.

From the above data it is seen that in the case of particular diameter ratio 2, where 7.5 µm is the base diameter, we need an optimized flow rate of 250 µL/min and a minimum of 6 looped serpentine microchannels, for efficient particle sorting. Thus, it is computationally very tiresome to determine these optimized flow conditions for a particular particle size combination, as it involves a large number of trials.

Hence, the effects of channel Reynolds number (Re_L), particle diameter ratio, and the number of serpentine loops on separation efficiency and throughput, have been analyzed using a ML model. The channel Reynolds number is varied from 50 to 200, the number of serpentine loops varied from 2 to 20 and particle diameter ratio varied from 1.10 to 2.00. The machine learning methodology has been elaborated in Sect. 2.4, and the corresponding results are discussed in the subsequent sections.

3.2 Machine Learning Results

As previously mentioned, we carried out evaluations of 50 different train-test data splits across the five machine learning models. From these 50 splits, mean values of the MAE and RMSE error metrics for the two output variables—Reynolds number and the number of loops—are tabulated in Tables 1 and 2. Notably, ensemble learning techniques such as gradient boosting regression (GBR), bagging regression (BR),

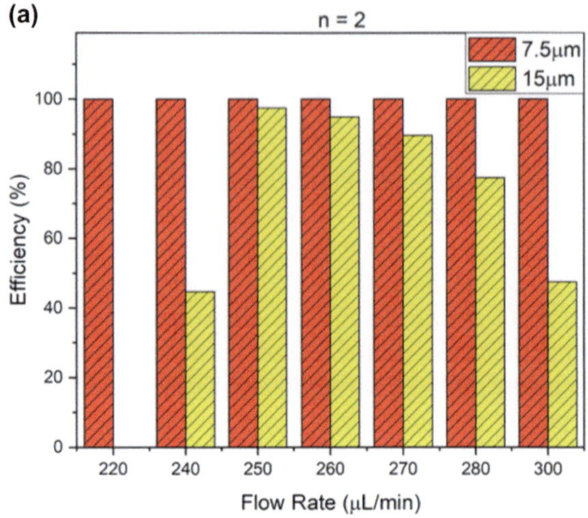

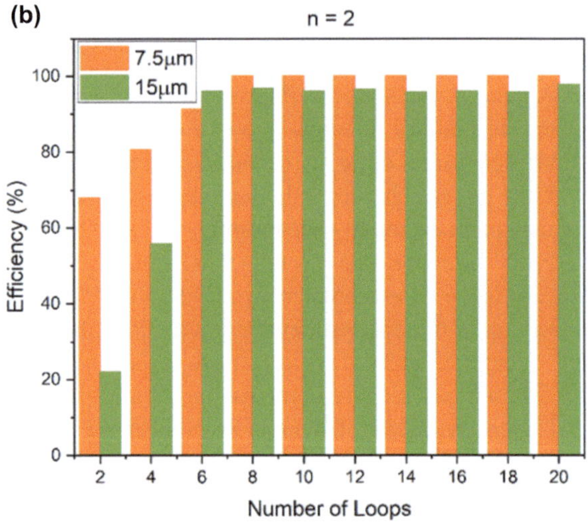

Fig. 4 **a** Plot for flow rate (Q) versus separation efficiency. **b** Plot for number of loops versus separation efficiency

and random forest (RF) demonstrated superlative performance in comparison to the K-nearest Neighbors (KNN) and support vector regression (SVR) methods. This superior performance can be attributed to the dataset size; when the dataset is relatively small, conventional regression methods tend to overfit, thus generating higher errors on the test dataset. On the contrary, ensemble learning approaches have the potential to enhance model generalization and mitigate overfitting. They achieve this by creating a variety of training subsets, ensuring the model learns from assorted perspectives, and avoiding overfitting to the original dataset [14].

Additionally, we evaluated the coefficient of determination (R^2 score) for both the training and testing datasets corresponding to the two output variables. Figure 6

Table 1 Performance error metrics on Re target variable

ML model	MAE		RMSE	
	Train data	Test data	Train data	Test data
KNN	12.96	16.57	17.49	21.85
RF	3.63	9.40	5.10	12.94
BR	5.97	9.97	8.11	13.33
GBR	6.12	10.03	7.44	12.80
SVR	29.45	29.80	34.97	35.21

Table 2 Performance error metrics on the number of loops target variable

ML model	MAE		RMSE	
	Train data	Test data	Train data	Test data
KNN	2.46	3.09	3.21	3.93
RF	1.19	3.00	1.60	3.77
BR	1.93	3.17	2.37	3.84
GBR	1.56	2.79	1.98	3.60
SVR	2.95	3.35	3.69	4.01

clearly indicates that the machine learning models are more adept at fitting Reynold's number (Re_L) than the "Number of loops" output variable. The Re_L output exhibits less variation (i.e., more stable) in the R^2 score values compared to the "Number of loops," which can be ascribed to the continuous nature of the Re_L compared to the discrete integer datatype of the "Number of loops" variable. Among all the techniques, the gradient boosting regression (GBR) outperforms others with a mean R^2 score of 0.98 (training) and 0.92 (testing) for Re_L, and 0.86 (training) and 0.51 (testing) for the "Number of loops." Figure 5 demonstrates the actual versus predicted plot for the output variables using GBR, reinforcing that the trained model tends to correspond closely with the best-fit line (Fig. 6).

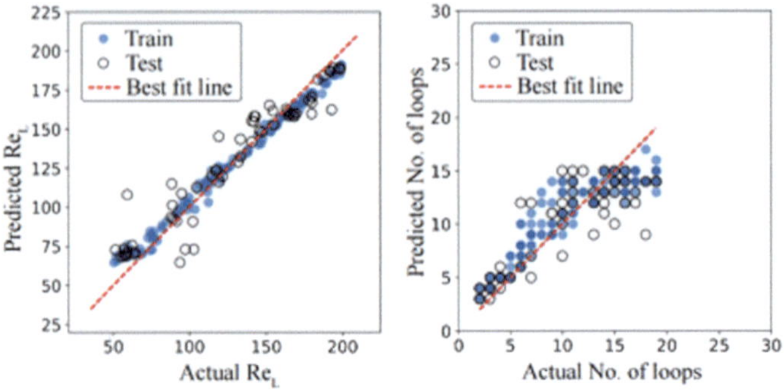

Fig. 5 Actual versus predicted Reynolds number and number of loops, using GBR

Fig. 6 R^2 score comparison for Reynolds number and number of loops

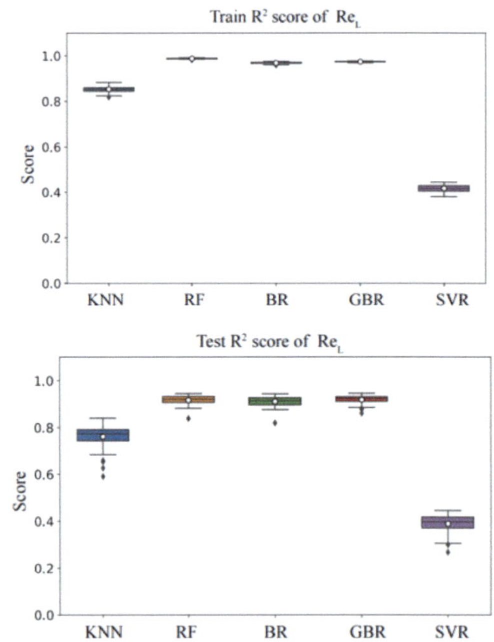

4 Conclusion

In this research endeavor, we investigated the sorting efficiency and throughput of a symmetric serpentine microchannel particle sorter subjecting the channel to various flow parameters and henceforth applied a data-driven machine learning approach to model the correlation between the flow parameters and both efficiency and throughput. Our comparative analysis of five distinct non-parametric machine learning regression techniques revealed ensemble-based regression methods as the most effective in predicting the ideal Reynolds number and the number of loops for

Fig. 6 (continued)

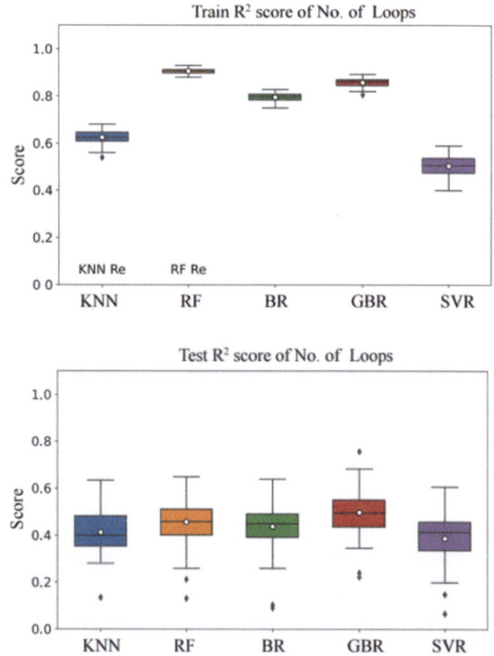

achieving a targeted particle separation efficiency. In particular, using the gradient boosting regression method, we achieved a mean R^2 score accuracy of 98% in training and 92% in testing for predicting Reynolds number, and a mean R^2 score accuracy of 86% in training and 51% in testing for predicting the number of loops. The relatively lower prediction accuracy for the number of loops, attributable to its discrete datatype, will be addressed with more suitable machine learning methodologies in future stages of this work.

Nomenclature

Re_L	Channel Reynolds number (–)
AR	Aspect ratio (–)
n	Number of loops (–)
η	Particle separation efficiency (–)
ρ, ρ_f	Liquid density (kg/m^3)
ρ_p	Particle density (kg/m^3)
μ	Dynamic viscosity of the fluid (kg/m s)
U_m	Channel inlet velocity (m/s)
a	Particle diameter (m)
D_h	Hydraulic channel diameter (m)

R_p	Particle Reynolds number (–)
f_c	Lift coefficient (–)
r	Radius of curvature (m)
v_{fr}	Radial fluid velocity (m/s)
v_{pr}	Radial particle velocity (m/s)
v_{pt}	Tangential particle velocity (m/s)
F_L	Inertial lift force (N)
F_D	Drag force (N)
F_{Cent}	Centrifugal force (N)

References

1. Yu W, Liu R, Zhou Y, Gao H (2020) Size-tunable strategies for a tumor targeted drug delivery system. ACS Cent Sci 6(2):100–116. https://doi.org/10.1021/acscentsci.9b01139
2. Pereira H, Schulze PSC, Schüler LM, Santos T, Barreira L, Varela J (2018) Fluorescence activated cell-sorting principles and applications in microalgal biotechnology. Algal Res 30(December 2017):113–120. https://doi.org/10.1016/j.algal.2017.12.013
3. Zhang J, Li W, Li M, Alici G, Nguyen NT (2014) Particle inertial focusing and its mechanism in a serpentine microchannel. Microfluid Nanofluidics 17(2):305–316. https://doi.org/10.1007/s10404-013-1306-6
4. Zhang J, Yan S, Sluyter R, Li W, Alici G, Nguyen NT (2014) Inertial particle separation by differential equilibrium positions in a symmetrical serpentine micro-channel. Sci Rep 4(ii):1–9. https://doi.org/10.1038/srep04527
5. COMSOL Multiphysics (2015) The COMSOL Multiphysics Reference Manual. Manual, pp 1–1336. www.comsol.com/blogs
6. Nader E et al (2019) Blood rheology: key parameters, impact on blood flow, role in sickle cell disease and effects of exercise. Front Physiol 10(Oct):1–14. https://doi.org/10.3389/fphys.2019.01329
7. Norouzi N, Bhakta HC, Grover WH (2017) Sorting cells by their density. PLoS ONE 12(7):1–16. https://doi.org/10.1371/journal.pone.0180520
8. Lee D, Derrible S (2020) Predicting residential water demand with machine-based statistical learning. J Water Resour Plan Manag 146(1):04019067
9. Asif K, Zhang L, Derrible S, Indacochea JE, Ozevin D, Ziebart B (2022) Machine learning model to predict welding quality using air-coupled acoustic emission and weld inputs. J Intell Manuf 1–15
10. Dayeen FR, Sharma AS, Derrible S (2020) A text mining analysis of the climate change literature in industrial ecology. J Ind Ecol 24(2):276–284
11. Jafari Gukeh M, Moitra S, Ibrahim AN, Derrible S, Megaridis CM (2021) Machine learning prediction of TiO_2-coating wettability tuned via UV exposure. ACS Appl Mater Interfaces 13(38):46171–46179
12. James G, Witten D, Hastie T, Tibshirani R (2013) An introduction to statistical learning, vol 112. Springer, New York, NY, p 18
13. Kaufman S, Rosset S, Perlich C, Stitelman O (2012) Leakage in data mining: formulation, detection, and avoidance. ACM Trans Knowl Discov Data (TKDD) 6(4):1–21
14. Brownlee J (2018) Better deep learning: train faster, reduce overfitting, and make better predictions. Mach Learn Mastery

Comparative Evaluation of S-CO_2 with Antifreeze Heat Transfer Fluids as an Application in Solar Flat Plate Collectors

Wasim Ashraf and M. Ramgopal

Abstract In developing countries like India, where access to reliable electricity is limited, solar water heaters that run on the sun's free energy are an appealing alternative. Here, the effectiveness of a solar water heater is evaluated in the cold climatic region of Ladakh, India. The high altitude, low humidity, and ample sunshine in Ladakh make it an excellent choice for solar water heating systems. Due to very cold climatic conditions, it is a big challenge to prevent water freezing in solar water heating systems (SWHS). Hence, in this study, the analysis of a flat plate collector (FPC) based SWHS at ISO 9806:2017 conditions under three sky models is done. The effectiveness of an FPC is compared by considering supercritical carbon dioxide (s-CO_2) and antifreeze heat transfer fluids (HTF) like ethylene glycol water mixtures (EGW 40%). It was observed that using s-CO_2 in FPC provides non-freezing, higher useful heat gain resulting in higher efficiency, lower viscosity resulting in higher Reynolds number, lower pressure drop and pumping power requirement compared to EGW mixtures. s-CO_2-based solar collectors comprise high working pressures, phase-changing possibilities with surrounding temperature, and significant property fluctuations near its critical point, making SWHS design challenging. Therefore, these topics need thorough research.

Keywords Solar water heating system · Analytical analysis · Supercritical-CO_2 · Ethylene glycol · Antifreeze HTF

W. Ashraf (✉)
School of Energy Science and Engineering, IIT Kharagpur, Kharagpur, India
e-mail: wasimashraf@iitkgp.ac.in

M. Ramgopal
Department of Mechanical Engineering, IIT Kharagpur, Kharagpur, India
e-mail: ramg@mech.iitkgp.ac.in

A. Kumar R. et al. (eds.), *Proceedings of Fluid Mechanics and Fluid Power (FMFP) 2023, Vol. 6*, Lecture Notes in Mechanical Engineering,
https://doi.org/10.1007/978-981-96-9600-0_25

1 Introduction

Today, there is a rising demand for energy all over the globe as a result of continued economic growth, urbanization, and population expansion [1]. Despite their widespread availability, widespread technical support, and simplicity of use, traditional energy resources pollute the environment by emitting hazardous gases and radiation. As a result, energy resource limitations and worries about global warming are fast increasing. Over the course of the last several decades, a variety of various sources of renewable energy for power production have been under consideration [2–4]. Solar power is relatively clean because of how it is used. Solar-powered systems and improvements in their efficiencies have gathered interest. Solar energy is a viable alternative to its conventional equivalents as a result of the advancements in technology that have been made and the widespread availability of solar energy. It is now the most dependable renewable source of power. But the low intensity prevents its direct usage for generating electricity. There is a wide range of technology that may be used for developing efficient concentrating systems, it can power anything from household appliances to factories. Solar power may be converted directly into electricity by photovoltaic systems and then stored in batteries [3].

In the past several years, solar energy has made a discernible difference in India's energy landscape. Applications powered by solar energy have helped millions of people in rural India with their cooking, water heating, lighting, and other energy requirements in a sustainable way. Solar energy has a lot of untapped potential in India. The intake of solar energy at Indian landmass is approximately 5000 trillion kilowatt-hours (kWh) annually, or 4–7 kWh per sq. m. each day. Capturing solar energy efficiently offers enormous development in India. When it comes to the reliability of energy supply, solar is the safest option because of its abundance. Theoretically, the energy requirements of the country might be satisfied by collecting and implementing to use just a little percentage of the total solar energy that strikes at the landmass of the country [2, 5].

Another method by which solar energy may be harvested and used is the utilization of sensible heating of fluid that circulates constantly through tubes in the solar collectors. The FPC is the least expensive, most convenient, and most efficient to install and operate [6]. This collector may use both direct and dispersed solar irradiation, unlike concentrated collectors. FPCs are now prevalent in residences, schools, hospitals, medical institutions, dairies, educational institutions, agricultural and industrial regions, etc. FPCs have an absorber plate, riser tubes, glazing, thermal insulation, and casing. There are a number of characteristics, each of which has an operational impact on the solar collector's overall thermal performance. These include the collecting plate location, the coating on top of the collector plate, the substance used for coating, the material used for glass, the distance between the riser tubes with regard to the diameter, the flow rate, the intensity of the incoming radiation, and the thickness of the insulation on the bottom and the sides [6]. The FPC is enclosed in a metal case with a protective glass top. The top side of the FPC has absorber tubes that are linked with a metallic absorber plate. The rear side of the FPC has thick

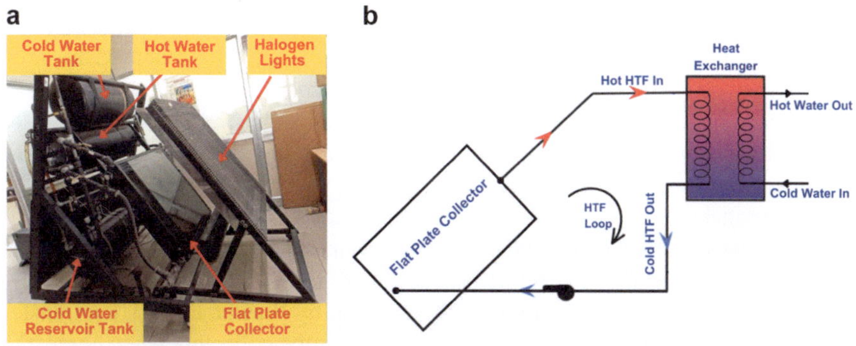

Fig. 1 **a** Photograph of experimental test-rig, **b** A simplified diagram of an indirect SWHS

insulation. The another type of solar collector, evacuated tube collector consists of a double-walled glass tube that is evacuated so that it may prevent convection heat losses.

Solar energy can also be utilized to heat water by using heat pipes. Heat pipes are used to store hot water produced through the integrated collector storage system (ICS). According to Singh et al. [7], the ICS is an entity that performs the function of an absorber and stores the hot water at the same time. Two distinct kinds of SWH systems may be distinguished by the manner in which heat is transferred: direct and indirect. In a system known as a direct system, the collector is where the water is heated. In indirect heating systems, water for residential use is heated by first passing a heat transfer fluid via a collector, where it is heated, and then through a heat exchanger. The mode of operation of these systems may be either active or passive, depending on the means through which the HTF is circulated. A pump is employed in an active system to continuously circulate the working fluid, while in passive systems, the working fluid is allowed to flow by itself rather than being pumped. It takes advantage of the thermo-syphon effect, which is when fluid flows because of differences in density [8].

In this study, an indirect SWHS is considered for the analysis under cold climate conditions. Fig. 1a shows the test-rig setup where experiments were performed considering water as a working fluid and Fig. 1b shown below is a simplified diagram of an indirect SWHS.

2 Literature Review

Researchers from all over the world have made steady and rigorous progress in engineering and technology that has helped FPSC advance forward. Here are some short summaries of what many experts have done that has helped to improve the performance of FPSCs. Fara et al. [9] used exergetic modeling to figure out the most effective way for an FPC to work. The optimum efficiency can be reached by taking into account four factors: fluid flow rate, exergy efficiency, heat absorption area, and heat storage volume. Alvarez et al. [10] did experiments and used the finite element method (FEM) to simulate a modified solar collector with a curved channel and a larger surface area open to a stream that conducts heat. The authors came to the conclusion that a fluid-transporting duct with a serpentine shape works better than one with a parallel shape. Kaichun et al. [11] examined the cross-section of the various thermal collector designs by using both numerical analysis and experimental testing. It was observed that elliptical cross-sectional tubes had a better rate of heat transfer than circular tube collectors because of the larger contact area provided by the elliptical tubes. According to the research done by Razi et al. [12], the transfer coefficient was significantly increased for flattened tube collectors when compared to circular tubes under the same flow circumstances. Rosli et al. [13] conducted an in-depth theoretical investigation of the round, square, and rectangular tube shapes. They claimed that the performance of heat transmission in rectangular cross-sections was much better than that of their round and square counterparts. Solar energy stimulated by improving absorption material optical clarity. More glass covers decreased heat loss. The correct heat transfer fluid may boost these collectors' performance and efficiency. Heat transfer fluid thermal characteristics affect energy absorption, heat transmission, and system efficiency. Water-based systems have shown thermal efficiency and excellent heat transmission in several studies [4, 6]. In frigid climates, water freezes, limiting its use. Employing different nanofluids as HTF, like Al_2O_3-water, SiO_2-water, CuO-water, MgO-water, and TiO_2-water nanofluids [14, 15], may improve the efficiency with which water delivers heat. A nano-fluid is a liquid in which particulates are usually between 1 and 100 nm in dimension. Nanofluids were employed to boost the HTF's thermal properties, diffusivity, and rate of convective heat transfer. Genc et al. [16] performed theoretical analysis of a SWH considering variable mass flow rates (0.01–0.05 kg/s) using Al_2O_3-water nanofluids. At a higher mass flow rate of 0.06 kg/s, the heat efficiency was found to be about 83%.

Apart from the various nanofluids as a possible HTF, researchers have shown an immense interest in the use of carbon dioxide (CO_2) as a potential alternative heat transfer fluid in solar FPC due to its unusual features and environmental benefits. There are a number of advantages to using CO_2 as a heat transfer fluid. It can absorb and transport heat efficiently inside the collector because of its excellent capacity for heat transfer. CO_2 also has a very low boiling point (− 78.5 °C) at atmospheric pressure, making it useful across a wide range of temperatures. In addition to being a safer and greener option, CO_2 has the added benefits of being non-flammable, non-hazardous, and easily accessible [17, 18]. Even though there have been a lot of studies on FPC using solar energy [7–18], there hasn't been much research that uses computations. Simulation-based research on solar FPC became more and more important for advancing the research forward. Still, using s-CO_2 as the heat transfer fluid could make these systems even more effective. Researchers think that s-CO_2 may give better performance, but a review of the literature shows that there hasn't been much actual or theoretical study on s-CO_2-based solar water heaters. So, the goal of this study is to make a model for figuring out how well a solar FPC works by using s-CO_2 as the main working fluid in a closed loop of an indirect solar water heating system and its performance comparison with antifreeze HTF like ethylene glycol–water mixture under the cold climatic conditions of Ladakh, India.

3 Methodology

3.1 Selection of Working Site: Ladakh, India

The abundant sunlight, high altitude, and low humidity levels at Ladakh [19] contribute to better performance for the solar energy systems. Figure 2a and b shows calculated yearly and hourly solar irradiation on June 21st (summer solstice), at Ladakh, India.

Further, the remote location and the limited access to conventional energy sources make it an ideal candidate for energy independence through solar power. In order to satisfy its water heating demands, Ladakh may become less reliant on petroleum and other fossil fuels by adopting the SWHS. Additionally, SWHS offer long-term economic benefits, as the abundant sunlight ensures a high energy yield, resulting in significant energy savings and reduced utility bills over time. These systems are effective in harsh winters, providing hot water even in freezing conditions, reducing dependence on traditional heating methods, and saving energy costs.

An FPC is used here for capturing solar energy in an experimental test setup for a SWHS. In addition, the halogen lights serving as an artificial solar flux generator, the test rig also has a hot water storage tank, flow control valves, and a data collecting unit. Table 1 gives thorough information about the technical features of the solar FPC.

This work develops a mathematical model to estimate the flow behavior, the outlet temperature, and useful heat gain per tube which is connected below to the absorber plate. The mass flow in every riser tube is equal to the average mass flow rate in the header tube divided by the total number of riser tubes.

In this study, s-CO_2 is recommended for usage as a fluid for transferring heat in solar collectors for indirect SWHS so that it is possible to gain the advantages of the enhanced heat transmission capabilities of CO_2 at or near its critical point temperature. It has been found that the properties variations of s-CO_2 as compared to the water are very large. So, in case of s-CO_2 it is very essential to capture the property variations and hence discretization technique is applied here. The variation of specific heat with operating temperature along the riser tubes is shown in Fig. 3.

In the present study, the performance of the s-CO_2 based SWHS is compared with antifreeze HTF (ethylene glycol–water mixtures). The thermophysical properties of ethylene glycol water mixtures at 30 °C are shown in the Table 2.

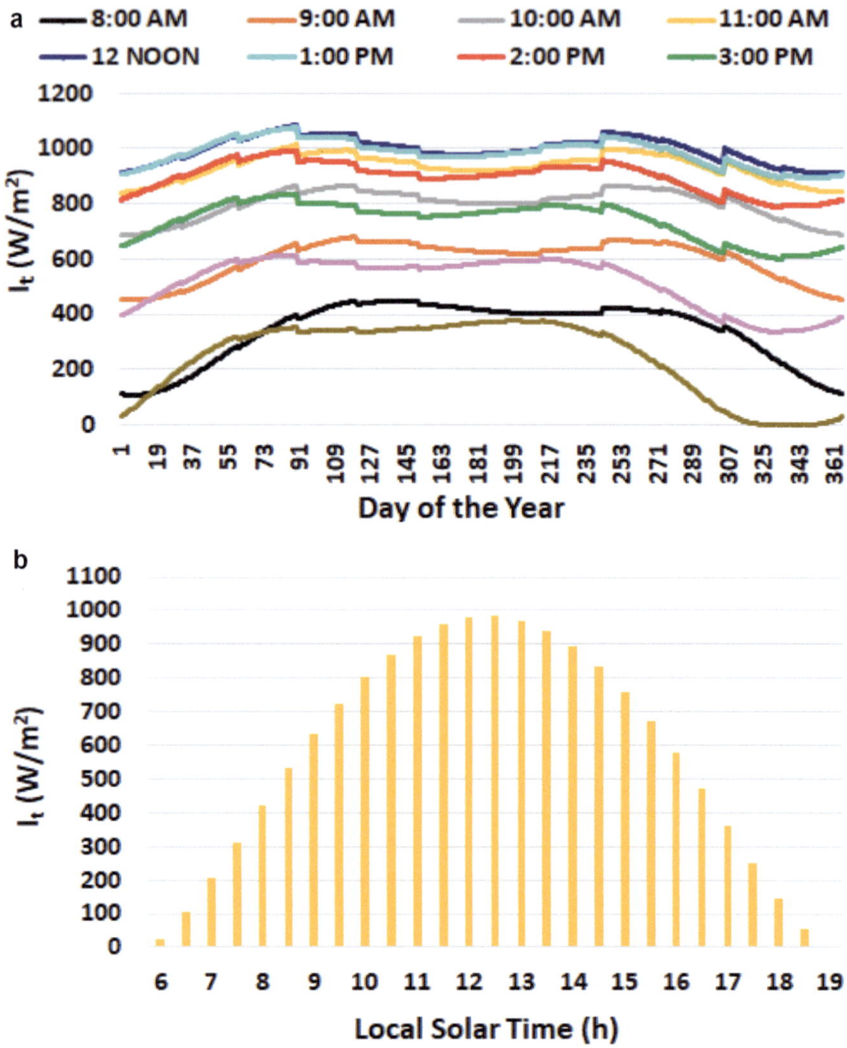

Fig. 2 **a** Yearly solar irradiation. **b** Hourly solar irradiation

3.2 Governing Equations

To make the analysis easier, it has been presumed that both fluid circulation and heat conduction are at a steady state, the top surface of the collector receives uniform amount of solar irradiation, and heat is dissipated in the same environment from both the top and bottom surfaces of the collector.

For mathematical modeling, the specifications of the various components of the FPC which are glazing, absorber plate, riser tube spacing, rise tube inner diameter,

Table 1 Technical features of the collector [5]

S. N.	Parameters	Value
1	Absorber plate area	1.7 m^2
2	Riser tubes inner diameter	8 mm
3	Riser tubes outer diameter	9 mm
4	Thermal conductivity (absorber plate)	240 W/mK
5	Thermal conductivity (back insulation)	0.04 W/mK
6	Thickness (absorber plate)	0.5 mm
7	Number of riser tubes	10
8	Heat transfer fluids	Water

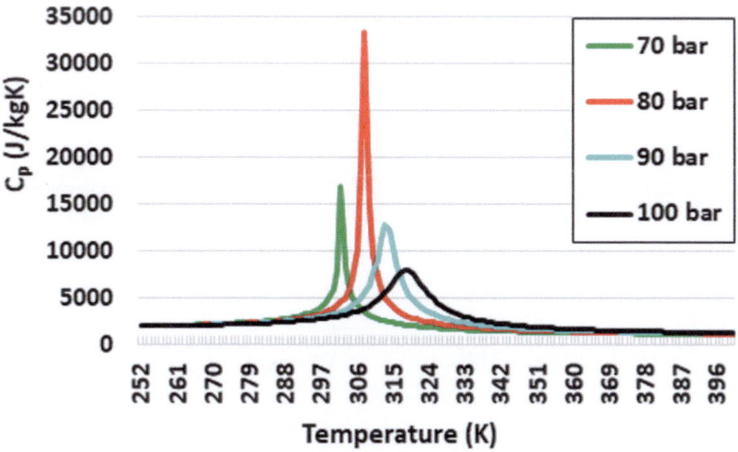

Fig. 3 Variation of specific heat of s-CO_2

Table 2 Thermo-physical properties of EGW [20]

HTF	Freezing point (°C)	C_p (J/kgK)	ρ (kg/m^3)	μ (mPa.s)
Water	0	4180	995.6	0.799
EGW (20:80)	− 7.9	3815	1029.7	1.65
EGW (40:60)	− 23.5	3468	1059.6	2.96

and the riser tube thickness all match those of the testing apparatus. A small number of preliminary experiments provided the basic information for the modeling, including the ambient temperature, inlet water temperature, insolation at the FPC top surface, and wind speed. With the help of energy, continuity, and momentum equations, we can figure out the outlet temperature, useful heat-gain as well as the temperature of the absorber plate. The average flow rate in the riser tubes determines the degree of warmth of the water coming out of the collector and the temperature of the absorber plate. It is observed that the flow with low Reynolds number produces homogeneous temperature distribution in riser tubes [21].

The gain of useful heat is calculated by [5]:

$$Q_u = \dot{m}C_p\left(T_{f,\text{out}} - T_{f,\text{in}}\right) \tag{1}$$

Here, heat capacity, rate of mass flow of the fluid, exit temperature of the fluid, and fluid inlet temperature are denoted by C_p, $\dot{m}$, $T_{f,\text{out}}$, and $T_{f,\text{in}}$, respectively. In order to calculate how much useful heat a working fluid may absorb from the absorber plate, we can use the Eq. (1). The heat loss coefficient and collector optical efficiency both have an impact on thermal efficiency, which is not accounted for in this estimation. The following equations are used to take into account thermal and optical losses.

$$Q_u = F_r A_p\{S - U_l(T_{f,\text{in}} - T_a)\} \tag{2}$$

$$S = \eta_o I_t \tag{3}$$

where F_r is the heat removal factor and S is the measure of solar energy which the collector absorbs per m^2 of surface area. η_o is the optical efficiency of the collector.

To get the heat removal factor, we use the formula: [22]

$$F_r = \frac{\dot{m}}{A_p U_l}\left[1 - \exp\left(-\frac{F' U_l A_p}{\dot{m}C_p}\right)\right] \tag{4}$$

The total heat loss rate from top surface is calculated as: [22]

$$\dot{Q}_{t,l} = U_t A_p\left(T_p - T_a\right) \tag{5}$$

Here, U_t is the convective heat loss coefficient for top surface and can be calculated as [22]:

$$U_t = \left[\frac{1}{\frac{C}{T_p}\left[\frac{T_p - T_a}{1+f}\right]^e} + \frac{1}{h_w}\right]^{-1} + \left[\frac{\sigma\left(T_p + T_a\right)\left(T_p^2 + T_a^2\right)}{\frac{1}{\varepsilon_p + 0.00591 h_w} + \frac{1+f+0.133\varepsilon_p}{\varepsilon_g} - 1}\right] \tag{6}$$

Here, the temperatures of the surrounding air, the collecting surface, and the absorber plate are denoted by T_a, T_c, and T_p, respectively.

The thermal efficiency of the FPC is obtained by: [5]

$$\eta_{\text{th}} = \frac{\dot{m}C_p\left[\left(T_{f,\text{in}} - T_a - \frac{S}{U_l}\right)\left(\exp\left(-\frac{F\prime U_l A_p}{\dot{m}C_p}\right) - 1\right]}{A_p I_T} \tag{7}$$

The Reynolds number is obtained by: [18]

$$R_e = \frac{\rho V_{\text{avg}} D_i}{\mu} \tag{8}$$

The Nusselt number for CO_2 is evaluated by Pitla et al. [23],

$$Nu = Nu_b \left(\frac{\mu_b}{\mu_w}\right)^{0.11} \left(\frac{k_w}{k_b}\right)^{0.33} \left(\frac{h_w - h_b}{C_{pb}(T_w - T_b)}\right)^{0.35} = \frac{\alpha_t d}{k_b} \tag{9}$$

where, Nu_b is the Nusselt number, which is obtained at bulk temperature of the working fluid [23]

$$\text{Nu}_b = \frac{(f/8)\text{RePr}}{12.7\sqrt{f/8}\left(\text{Pr}^{2/3} - 1\right) + 1.07} \tag{10}$$

The pressure drop in the riser tubes and its variation along the length of the riser tubes was observed and it was calculated by considering discretization of the tubes and the same can be calculated as: [17]

$$p_f^i - p_f^{i+1} = \frac{G_f^2}{2\rho_f}\left[f\frac{\Delta L}{d} + \left(\frac{\rho_f^i}{\rho_f^{i+1}} - \frac{\rho_f^{i+1}}{\rho_f^i}\right)\right] \tag{11}$$

where, ρ_f is the fluid density and f is the friction factor. The friction factor for laminar and turbulent fluid flow can be obtained as: [5]

$$f = 64/\text{Re} \tag{12}$$

$$f = [0.79\ln(\text{Re}) - 1.64]^{-2} \tag{13}$$

where, Re is the Reynolds number calculated as per Eq. (8).

3.3 Grid Independence Analysis

Grid independence analysis may determine the extent to which a simulation's results rely upon the design. Researchers can find the best grid condition by comparing

Table 3 ISO 9806:2017 standard conditions

S. N.	Parameter	Blue sky (W/m^2)	Hazy sky (W/m^2)	Grey sky (W/m^2)
1	I_b	850	440	0
2	I_d	150	260	400

findings from different grid sizes. This work uses grid independence analysis on very small riser pipe divisions $N = 10, 20, 30, 35$, and 50 to capture characteristic fluctuations of s-CO_2 near its critical point. After 35 riser tube segments, analytical findings show very small changes. Hence, $N = 35$ segments were chosen for system analysis because the thermal effectiveness was 85.94% at 35 segments and 85.95% at 50 segments [5].

4 Results and Discussion

Studies conducted for solar FPC using water as a HTF were used to verify the preciseness of the created MATLAB algorithm. There is a significant qualitative and quantitative consistency between the findings of analytical analysis and experiments. The energy efficiency of the FPC varies with the collector performance factor, fluid inlet temperature, and mass flow rates [5].

4.1 Analysis of FPC at ISO 9806:2017 Conditions

The recently introduced Standard Reporting Conditions (SRC) are included in ISO 9806:2017 [24] so that collector outputs may be reported consistently regardless of testing methodology or collector technology. "Blue Sky," "Hazy Sky," and "Grey Sky" are the three types of weather conditions recognized by the SRC (Table 3). Certain irradiation levels are specified for each of these states. A fluid flow and thermal analysis is conducted to learn how solar FPC performs when subjected to three distinct sky models.

4.2 Performance Under Cold Climate Conditions

The performance analysis of the solar FPC is done for Ladakh region at a very low ambient condition of – 20 °C and therefore the EGW (40%) mixture [20] was considered as an antifreeze solution for comparison with s-CO_2 and the results are presented below. From Fig. 4 it can be concluded that the thermal efficiency of the

FPC with s-CO_2 is higher for all sky models as compared to EGW (40%) but the outlet temperature is slightly higher with EGW as shown in the Fig. 5.

In the case of s-CO_2, the pressure drop is much lower leads to reduced pump power consumption (Fig. 6) due to its lower density.

A higher Reynolds number can lead to increased convective heat transfer rates, which can be advantageous for faster heating of the system. The comparison of the Reynolds number is shown in Fig. 7. The s-CO_2 offers a very high Reynolds number, which ultimately results in higher thermal efficiency due to higher useful heat gain for the FPC.

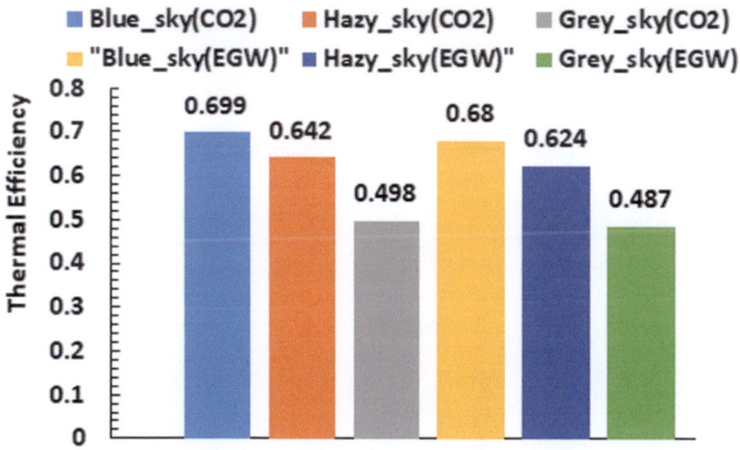

Fig. 4 Thermal efficiency

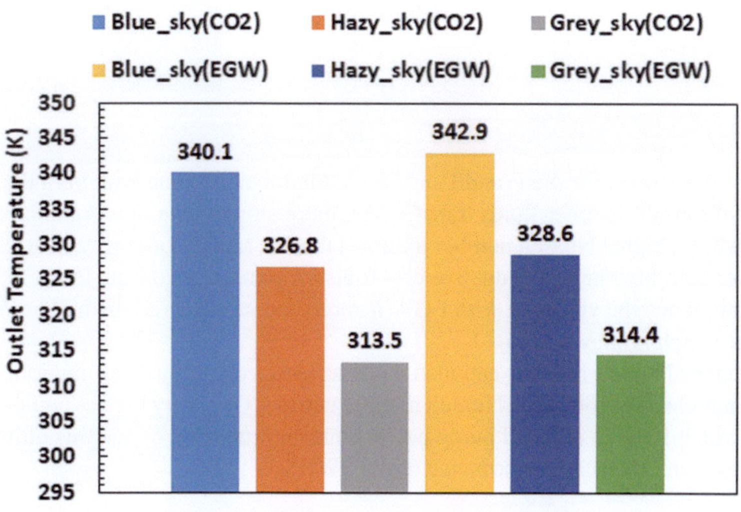

Fig. 5 Outlet temperature

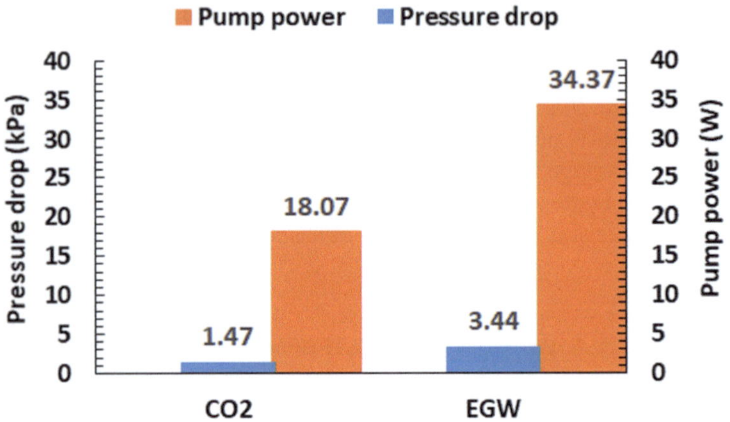

Fig. 6 Pressure drop and pump power

Fig. 7 Reynolds number

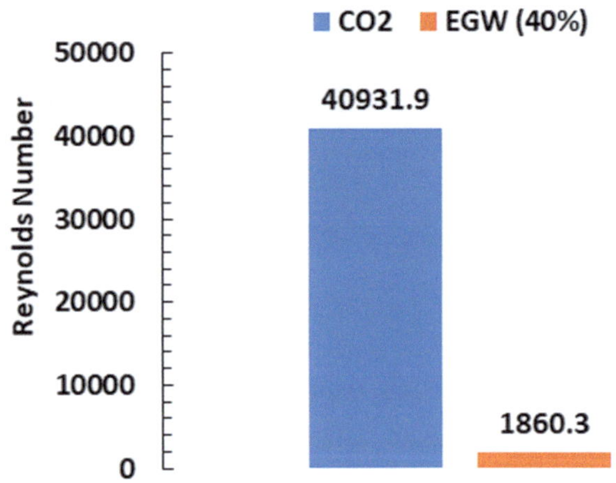

The comparison for the Prandtl and Nusselt numbers is done with both the HTF. A higher Prandtl number fluids transfer heat more slowly than low Prandtl number fluids while a higher Nusselt number indicates that the rate of convective heat transfer is higher than the rate of conductive heat transfer. Here, the working temperature is very low, hence the viscosity with EGW is more and so observed the higher Prandtl number which is shown in Fig. 8.

A higher Nusselt number can also leads to reduced energy consumption. From Fig. 8, it is clear that obtained Nusselt number with s-CO_2 is very high as compared to EGW which leads to reduced pump power consumption (Fig. 6) and this ultimately provides better thermal performance of the SWHS.

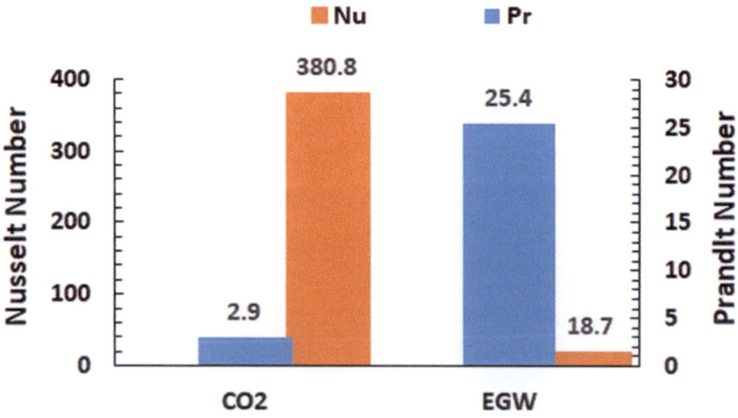

Fig. 8 Nusselt and Prandtl number

5 Conclusion

An elementary computational model is developed to recommend design and operation configurations and look forward to the outcomes of a solar FPC under the standards of ISO 9806:2017 and the cold environment of Ladakh, India. Experimental findings using water as the HTF verify the developed model. The theoretical findings are in qualitative and quantitative agreement with the experimental ones. It was observed that the variations in the thermo-physical properties of s-CO_2 are very large hence it is important that the fluid properties should not be considered as constant with operating temperature and pressure to get more accurate results. Hence, the grid-independent analysis is done and it is seen that more than 35 elements of the riser tube do not affect the improvement of the results. Therefore, discretization is done by considering 35 elements of the riser tube with four operating conditions. The performance of both the fluids (s-CO_2 and EGW) are compared at various working conditions. The theoretical results show that the blue sky model with s-CO_2 provides superior thermal performance compared to EGW under similar operating conditions. In the case of s-CO_2, the pressure drop is much lower (2.38 times) as compared to EGW, hence the power required for the pump to circulate the HTF is very lower (2.34 times) with s-CO_2 compared to EGW. The thermal efficiency is higher with s-CO_2 as compared to ethylene glycol water mixtures. Apart from this fluid flow analysis with s-CO_2 offers much better performance as compared to EGW mixtures.

In addition, studies may be conducted to investigate how factors like heat transfer fluids (nanofluids), absorber plate material, dimension, riser tube gaps, and radius affect the functionality of the solar FPC.

Nomenclature

A_p	Absorber plate surface area, [m^2]
D_i	Tube inner diameter [mm]
I_T	Solar irradiance [W/m^2]
$\dot{m}$	Mass flow rate [kg/s]
Nu	Nusselt number
Pr	Prandtl number
$\dot{Q}u$	Useful heat gain [W]
U_l	Net heat loss coefficient [W/m^2K]
β	Tilt angle of the collector [deg]
υ	Kinematic viscosity of fluid [m^2s^{-1}]
η_{th}	Thermal efficiency

References

1. World energy outlook (2019)
2. The ministry of new and renewable energy, India. https://mnre.gov.in/solar/current-status
3. The central public health and environmental engineering organisation. http://cpheeo.gov.in
4. Zhang R, Geoffrey QP, Ni M, Wong JKW (2018) Techno-economic feasibility of solar water heating system: overview and meta-analysis. Sustain Energy Technol Assess 30:164–173
5. Ashraf W, Ramgopal M, Reddy VM (2022) Performance analysis of an s-CO_2 based solar flat plate collector, advances in clean energy and sustainability Springer, Proceedings of ICAER
6. Pandey KM, Chaurasiya R (2017) A review on analysis and development of solar flat plate collector. Renew Sustain Energy Rev 67:641–650
7. Souliotis M, Singh R, Papaefthimiou S, Lazarus IJ, Andriosopoulos K (2016) Integrated collector storage solar water heaters: survey and recent developments. Energy Syst 7:49–72
8. Kalogirou S (2009) Thermal performance, economic and environmental life cycle analysis of thermosiphon solar water heaters. Solar Energy 39–48
9. Fara I, Luminosu L (2005) Determination of the optimal operation mode of a flat solar collector by exergetic analysis. Energy 30:731–747
10. Alvarez A, Cabeza O (2010) Experimental and numerical investigation of a flat-plate solar collector. Energy 3707–3716
11. Lia K, Lib T, Taoa H (2015) Numerical investigation of flow and heat transfer performance of solar water heater with elliptical collector tube. Energy Procedia 70:285–292
12. Razi P, Behabadi MA, Saeedinia M (2011) Pressure drop and thermal characteristics of CuO–base oil nanofluid laminar flow in flattened tubes under constant heat flux. Int Commun Heat Mass Transfer 38:964–971
13. Rosli MAM, Ping YJ, Misha S (2018) Simulation study of computational fluid dynamics on photovoltaic thermal water collector with different designs of absorber tube. J Adv Res Fluid Mech Therm Sci 52(1):12–22
14. Hawwash AA, Ali K, Rahman A, Nada SA (2018) Numerical investigation and experimental verification of performance enhancement of flat plate solar collector using nanofluids. Appl Therm Eng 130:363–374
15. Said Z, Saidur R, Rahim NA (2016) Energy and exergy analysis of a flat plate solar collector using different sizes of aluminium oxide based nanofluid. J Cleaner Production 133:518–530

16. Genc M, Ezan AM, Turgut MA (2018) Thermal performance of a nanofluids based flat plate solar collector. Appl Therm Eng 395–407
17. Sarkar J (2013) Performance of a flat plate solar thermal collector using supercritical carbon dioxide as heat transfer fluid. Int J Sustain Energy 32(6):531–543
18. Biradar MK, Parmar DN, Yadav AK (2022) CFD and exergy analysis of subcritical/supercritical CO_2 based naturally circulated solar thermal collector. J Renew Energy 189:865–880
19. https://www.accuweather.com
20. https://www.engineeringtoolbox.com/ethyleneglycol-d146.html
21. Sarwar J, Khan MR, Rehan M, Asim M, Kazim AH (2020) Performance analysis of a flat plate collector to achieve a fixed outlet temperature under semi-arid climatic conditions. Sol Energy 207:503–516
22. Duffie JA, Beckman WA (2006) Solar engineering of thermal processes. Wiley third ed., New York etc
23. Pitla SS, Robinson DM, Groll EA, Ramadhyani S (1998) Heat transfer from supercritical carbon dioxide in tube flow: a critical review. HVAC & R Res 4(3):281–301
24. https://www.iso.org/standard/67978.html

Investigation of Flow Past a Tandem Flat Plates Airfoil with Modified Trailing Edges

Aryasomayajula Sri Prabha Harshini, A. Kaivalya Pranathi, and Shailesh Kr Jha

Abstract In present study, the aerodynamic characteristics of tandem flat plate airfoils with modified trailing edge have been investigated by using commercial software ANSYS fluent 21. We found that the lift and drag coefficient of the first plate is higher than the second case for both 1c and 4c tandem airfoil. Also we observed that in 1c tandem case when the leading edge vortex (LEV) of the first plate reaches the second plate at that moment leading edge vortex start generated at the leading edge (LE) of the second plate. While for the 4c tandem it was noticed that as the trailing edge vortex (TEV) is about to shed at that moment the LEV is start forming at the LE of the second plate.

Keywords Vortex size · Lift · Drag 1c and 4c tandem

1 Introduction

The study of 2-D flow past a multiple airfoil is a most complex problem. Due to immense application in engineering field the flow past a tandem body attracts large section of the researcher. To understand the basic flow physics over the flat plate it has been extensively explored by several researchers but the detailed systematic numerical investigation of the flow past a tandem flat plate airfoil with modification along the trailing edge has received less attention from the research community. A profound understanding of flow phenomena such as vortex shedding behind the tandem wings is essential in the design of airfoils since it finds immense application determining dynamic loading on neighboring structures in fans, wind turbines, propellers, etc. and is a highly challenging field of research. In the present problem attention has been paid to investigate the effect on fluid flow due to modification along the trailing edge of the tandem flat plate airfoil under various operating conditions. Some of the relevant literatures on the flat plate airfoils are given below: Jha

A. S. P. Harshini · A. K. Pranathi · S. K. Jha (✉)
National Institute of Technology Jamshedpur, Jharkhand, India
e-mail: shailesh.me@nitjsr.ac.in

A. Kumar R. et al. (eds.), *Proceedings of Fluid Mechanics and Fluid Power (FMFP) 2023, Vol. 6*, Lecture Notes in Mechanical Engineering,
https://doi.org/10.1007/978-981-96-9600-0_26

et al. [1] numerically investigated the flow features behind a circular trailing edge for different Reynolds numbers and angle of attack. They found that, maximum lift is obtained when leading edge vortex attains maximum size and minimum lift when trailing edge vortex attains maximum size. They identified that in the near wake a single primary vortex shedding is formed and the vortices convect alternately, but in far wake secondary oscillations are predominant over the primary vortex shedding and the vortices of opposite sign move in pairs. Lam and Wei [2] numerically, investigated the shedding phenomena from an inclined flat plate for the range of AOA from 20° to 45°. In their study the found two series of vortices from the LE and trailing edges (TE) of the plate at a Reynolds number of 2×10^4, thus forming a vortex street behind the plate. They predicted the production mechanism of lift and drag and its connection with vortex shedding. Han et al. [3] analyzed the unsteady aerodynamics of tandem configuration flying near the ground. They computed the wake shapes of both single flat plate and flat plates in tandem configuration; by comparison they found that the unsteady ground effect has considerable impact on the performance of wings in tandem configuration. Jones et al. [4] experimentally investigated the aerodynamic performance of two wing configuration at a Re (Reynolds number) of 100,000. They found that the aerodynamic performance of two wing configuration is reduced as compared to that of single wing for small incidence angles. They also found that for the same, the aerodynamic performance increases and the stall is reduced at post stall angles of attack. It was noticed that the performance of two wing configuration heavily depends on the strength of inter wing flow and interaction between separated shear layers. Erfan et al. [5] studied a tandem flapping wing mechanism for a bio mimetic air vehicle. They designed a mechanism capable of maintaining good consistency and aerodynamic performance such that it can be used in future bio mimetic micro vehicle designs. They found that the tandem wings generate 50% higher lift and improve stability which will allow hovering.

1.1 Objectives of the Present Study

Flow phenomena over the tandem flat plate, airfoils were studied by the several researchers but the detailed comparative investigations on the flow phenomena effect due to the modification in flat plate trailing edge is almost untouched, which shapes the objective of the present work. Also post stall study of tandem airfoil has not paid much attention. Thus, in the present study detailed investigation of tandem flat plate airfoils with modified edges in post stall condition are carried out in order to understand the complex physics such as vortex shedding from leading and trailing edges, effects on aerodynamics characteristics, for 30° AOA (in post stall condition) and at a Reynolds numbers of 1.82×10^5. The current Reynolds number chosen based on Lam and Wei [2] paper.

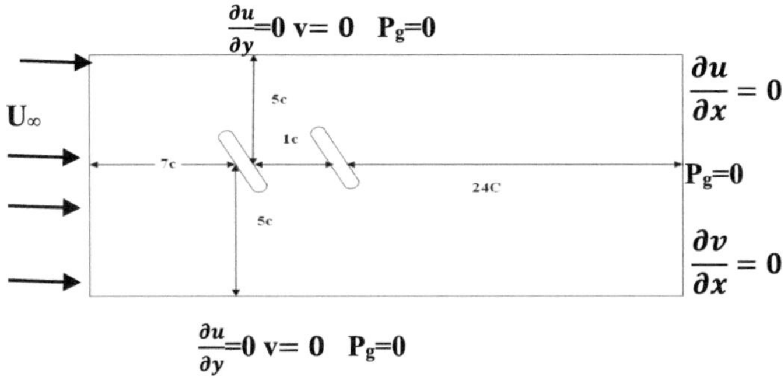

Fig. 1 Domain and boundary condition

2 Methodology

2.1 Computational Domain, Boundary Condition Grid, and Grid Sensitivity Test

The physical geometry, computational domain, and boundary conditions for the present problem are shown in Fig. 1. The computational domain was kept as − 7c and 24c & 1c between two plates (where c = chord length of plate) in the stream wise direction and ± 5c [2] normal to the stream wise direction. The structured grid with varying mesh size was constructed using ANSYS CFD. Near the flat plate aerofoil fine mesh were used in order to capture the flow effects as they are significant near the plate and thus to resolve the boundary layer properly and while in the region where flow effects are absent (or insignificant) grids are made coarse. The computations were carried out using pressure based implicit solver. We have taken constant density fluid for present study. The two equation SST K-ω turbulence model with standard values of model constants was used for present computation. SST K-ω model is fairly shown to give good results for problems involving wall bounded flows as well as boundary layers subjected to unfavorable pressure gradients with reduced computational time. Based on the grid sensitivity test we have taken 100,391 grid for our present problem solution.

The boundary layer near the flat plate airfoil is resolved by limiting the maximum wall y + value within about 35–110.

2.2 Governing Equations

The 2-D incompressible flow equations for the flow past a flat plate airfoil is given below.

Mass conservation equation:

The equation for conservation of mass can be written as follows:

$$\frac{\partial u}{\partial x} + \frac{\partial v}{\partial y} = 0 \tag{1}$$

where, ***u***, ***v*** are velocity along ***x*** and ***y*** axis respectively.

Momentum Conservation Equation:

(a) Along x- axis

$$\frac{\partial u}{\partial t} + \frac{\partial \left(u^2\right)}{\partial x} + \frac{\partial (uv)}{\partial y} = -\frac{1}{\rho}\frac{\partial P}{\partial x} + \upsilon\left(\frac{\partial^2 u}{\partial x^2} + \frac{\partial^2 u}{\partial y^2}\right) \tag{2}$$

(b) Along y-axis

$$\frac{\partial v}{\partial t} + \frac{\partial (uv)}{\partial x} + \frac{\partial \left(v^2\right)}{\partial y} = -\frac{1}{\rho}\frac{\partial P}{\partial x} + \upsilon\left(\frac{\partial^2 v}{\partial x^2} + \frac{\partial^2 v}{\partial y^2}\right) \tag{3}$$

where, ***u***, ***v*** are velocity along ***x*** and ***y*** axis respectively, *P* is the pressure, ρ is the fluid density which is constant for the present work since flow is considered as incompressible and ***υ*** is the kinematic viscosity of the fluid.

2.3 Validation

The drag coefficients obtained from the present computation shown in Table 1 at 30° angle of attack matched well with those observed by Lam and Wei [2] also c_d variation with time showed good agreement (within 10%) with the one obtained by Lam and Wei [2] (Fig. 2).

Table 1 Comparison of maximum drag coefficient obtained from present simulation for 30° angle of attack with those predicted by Lam and Wei [2], at the same parametric conditions

Maximum drag coefficient calculated at same boundary condition as that of Lam and Wei at 45° angle of attack	Maximum drag coefficient obtained for Lam and Wei [2] for 45° angle of attack	% deviation between Lam and Wei [2] and present simulation
1.86	2.02	8.2016

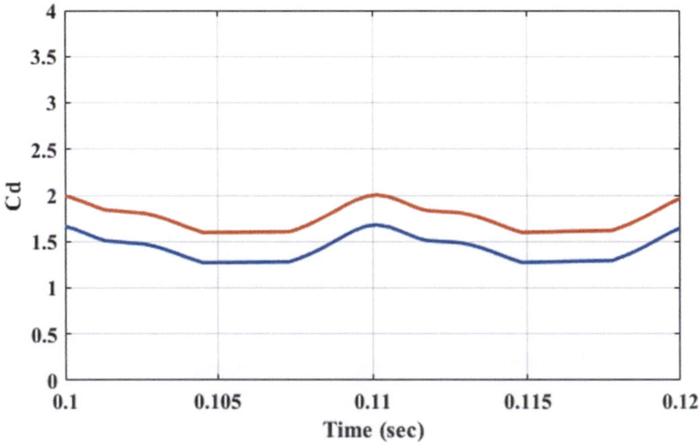

Fig. 2 Comparison drag coefficient versus time (sec) flat plate of current work with that of Lam and Wei [2]

3 Result and Discussion

3.1 Vortex Shedding

The flow begins to separate once it impinges the LE of the flat plate (Fig. 3) and a clockwise vortex forms from the separated shear layer, which grows in the stream wise direction and reaches the trailing edge of the plate (Fig. 3). The separated shear layer reaching the trailing edge creates higher suction pressure (Fig. 3) near the surface and sheds from the surface and moves along with the flow. The presence of suction pressure causes the formation of anticlockwise vortex from the tip of the TE, which grows toward the leading edge. The anticlockwise vortex from the TE sheds from the surface in the clockwise direction and moves along with the flow after attaining maximum size. As the vortex from the LE reaches to the second plate separated shear layer is formed near the LE of the second plate which further result in formation of clockwise vortex from the LE as it grows, a suction is created near the TE which leads to formation of TEV, at the same time the vortex from the TE of the flat plate of the first plate reaches to the second plate kept at 1c distance apart. The pattern of vortex formation is almost same for both the plate. Also it move along the plate in same way for both plate (i.e., clockwise from the LE and anticlockwise from TE). The mechanism of vortex shedding pattern is same for all the tendam arrangement (1c and 4c) studied, as mentioned above.

It is observed that the leading as well as the trailing edge shedding structures merge at a certain distance downstream of the TE.

A very interesting observation is noticed in the pattern of vortex shedding for the two cases we studied, i.e., 1c tandem and 4c tandem flat plate airfoil. In case of 1c tandem flat plate airfoil the vortex formation on the second plate is started as the

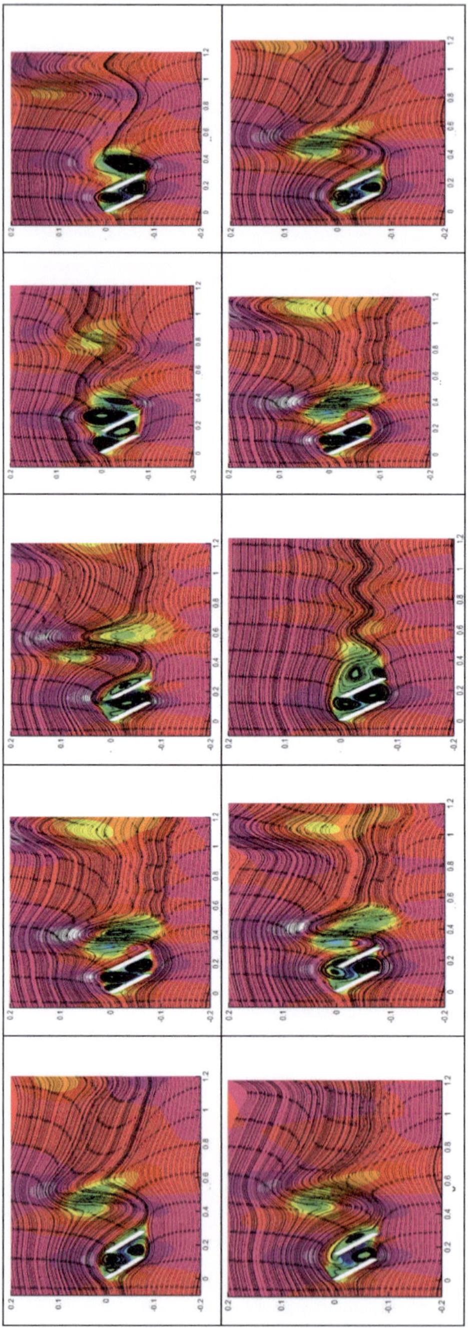

Fig. 3 Streamlines for tandem flat plate airfoils with rounded trailing edge **a** 1c tandem and **b** 4c tandem

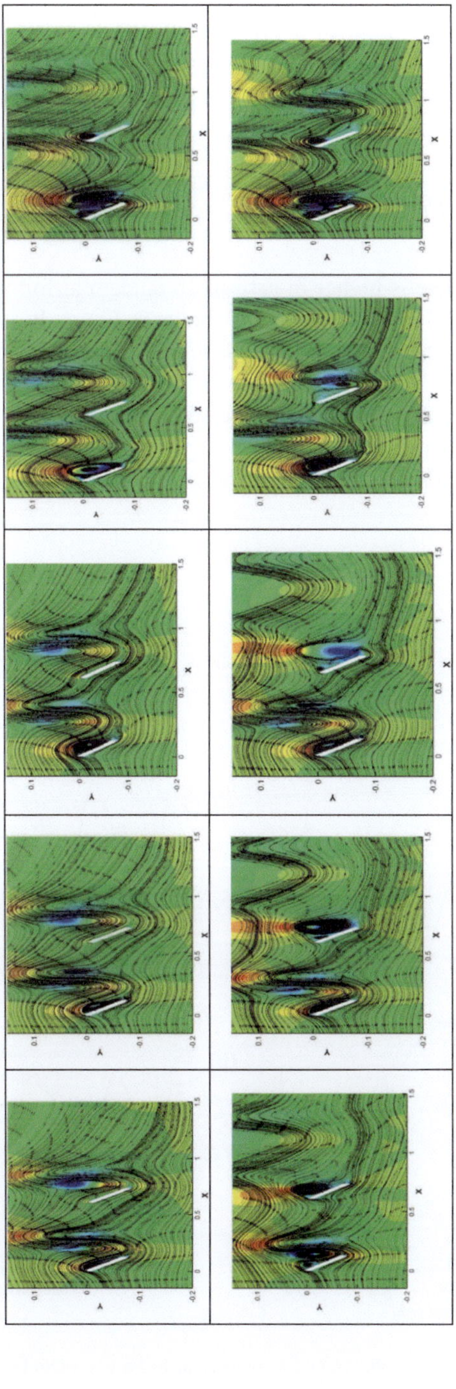

Fig. 3 (continued)

LEV from the first plate is about to shed while in case of 4c tandem airfoil the vortex formation from the leading edge of the second plate start as the TEV of the first plate is about to shed. Also for 1c tandem flat plate the leading vortex gets stressed and could not fully grow due to the presences of the second plate at 1c distance from it. While in case of 4c tandem vortex from the first plate has fully grown.

3.2 *Lift and Drag Fluctuation*

The lift of the second plate is higher in case of 1c tandem airfoil while lift of first plate is higher in case of 4c tandem airfoil as shown in Fig. 4. It may be due to the vortex shed from the first plate may leads to develop high pressure difference about the second plate in case of 1c tandem. While in case of 4c tandem airfoil CL (lift coefficient) of the first plate is higher than the second plate reason for this may be due to low pressure difference around the second plate. The drag of 4c tandem airfoil is observed to be higher than 1c tandem airfoil for both the plates and is shown in Fig. 5.

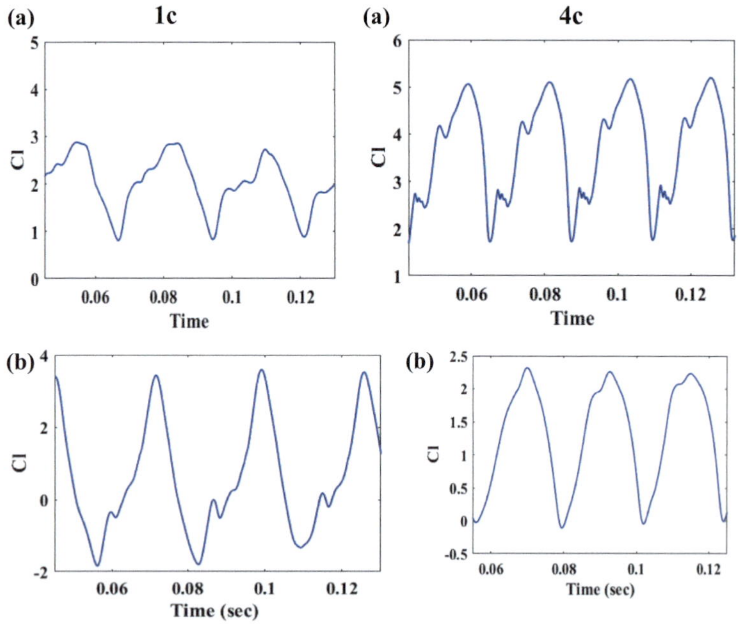

Fig. 4 Cl versus time (sec) for 1c tandem and 4c tandem airfoil **a** first plate **b** second plate

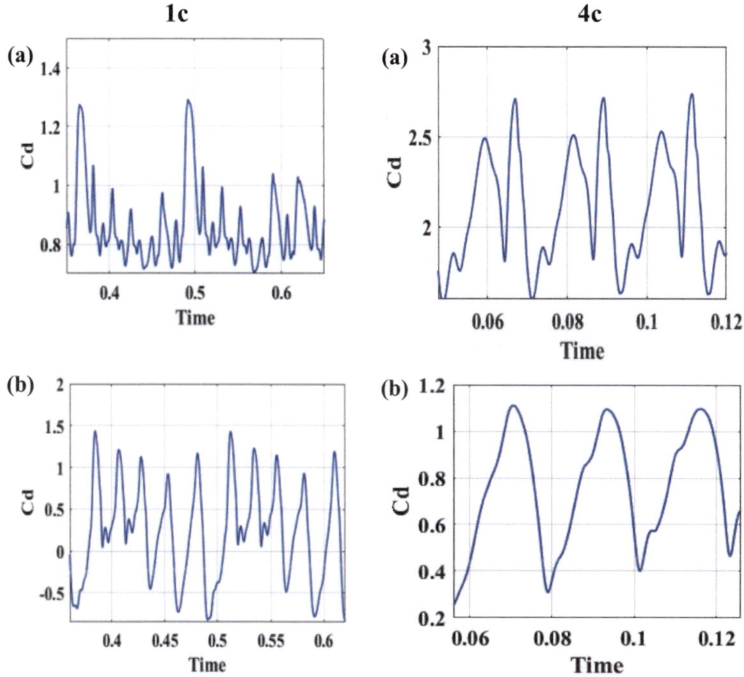

Fig. 5 Cd versus time (sec) for 1c tandem and 4c tandem airfoil **a** first plate **b** second plate

4 Conclusion

Numerical studied are carried out to get indepth understanding of the effect of tandem position of the plate on its aerodynamics characteristics. It is observed that the vortex formation at the LE of the second plate of 1c tandem starts as the LED vortex of the first flat plate foils reches the second plate. In case of 4c tandem the LEV of the second plate start forming as the TEV of the first plate is about to shed. Also the lift of the first plate of 4c tandem is higher than that of second plate at the same time drag of second plate of 4c tandem is higher than the first plate. Also for the 1c tandem the lift of second plate is observed to be higher than the first plate. The streamlines show that the pattern of shed vortices from the LE and TE are almost same for both the tandem (i.e., 1c and 4c) studied. Further, it reveals the merging of LE and TE vortices at a some distance downstream of the TE. It indicates that the speed of the shed vortex from the TE is higher than the LE, since the LEV sheds prior to the TEV.

Acknowledgements The authors gratefully acknowledge that the current work is supported by Department of Mechanical Engineering NIT Jamshedpur.

References

1. Jha SK, Narayanan S, Kumaraswamidhas LA (2019) Investigations of flow phenomena behind a flat plate with circular trailing edge. J Braz Soc Mech Sci Eng 41(5):1–17
2. Lam KM, Wei CT (2010) Numerical simulation of vortex shedding from an inclined flat plate. Eng Appl Comput Fluid Mech 4(4):569–579
3. Han C, Yoon Y, Cho J (2002) Unsteady aerodynamic analysis of tandem flat plates in ground effect. J Aircr 39(6):1028–1034
4. Jones R, Cleaver DJ, Gursul I (2015) Aerodynamics of biplane and tandem wings at low Reynolds numbers. Exp Fluids 56:1–25
5. Salami E, Montazer E, Ward TA, Nik Ghazali NN, AnjumBadruddin I (2022) Aerodynamic performance of a dragonfly-inspired tandem wing system for a biomimetic micro air vehicle. Front Bioeng Biotechnol 10:787220

Numerical Simulation of a Damped-Driven MNLS Equation to Study Turbulence Behavior

Praveen Kumar, R. Uma, and R. P. Sharma

Abstract This article introduces a numerical model for examining wave turbulence in fluids. The model equation is derived by considering energy conservation and the standard fluid equations for compressible flow. The nonlinearity in the equation arises from the temperature increase caused by the high amplitude of the acoustic wave. The modified nonlinear Schrodinger equation (MNLS) is derived and solved using numerical methods. Artificial damping is added to the MNLS equation, and turbulence generation is investigated through a numerical simulation. The numerical simulation utilizes a pseudo-spectral method for spatial domains and a finite difference method for temporal evolution. The results demonstrate a periodic pattern resembling the Fermi-Pasta-Ulam (FPU) recurrence for the NLS equation. At the same time, turbulence generation disrupts the FPU recurrence in the case of the MNLS and damped MNLS equation. The energy spectrum in the inertial range follows a Kolmogorov-Zakharov (KZ) scaling of approximately $k^{-1.2}$ in the inertial range and steeper than k^{-3} beyond the inertial range.

Keywords Wave turbulence · Damping · Nonlinear Schrodinger equation

Nomenclature

c_0	Speed of sound at T_0	(cm/s)
e	Energy per unit mass	(ergs/g)
n	Number density of particle in fluid	–
k_B	Boltzmann constant	($cm^2 gm/s^2 K$)
R	Universal gas constant	(ergs/mol K)
T	Absolute temperature	(K)

P. Kumar (✉) · R. Uma · R. P. Sharma
Department of Energy Science and Engineering, Indian Institute of Technology Delhi, New Delhi, India
e-mail: praveen.mech2011@gmail.com

A. Kumar R. et al. (eds.), *Proceedings of Fluid Mechanics and Fluid Power (FMFP) 2023, Vol. 6*, Lecture Notes in Mechanical Engineering,
https://doi.org/10.1007/978-981-96-9600-0_27

δT	Temperature change	(K)
T_0	Initial temperature	(K)
ρ_0	Background density	(gm/cm^3)
k	Wave number	(cm^{-1})
ψ_0	Initial amplitude of the wave	–
ω	Angular frequency	(Hz)
α	Perturbation wavenumber	(cm^{-1})
β	Relative magnitude of the perturbation	–

1 Introduction

Turbulence, a category of nonlinear motion, has received an extensive amount of attention in the field of fluid dynamics. However, in dispersive media, turbulence is associated with the formation of complex interacting nonlinear waves. In the1960 s, Zakharov has done a primary investigation on weak turbulence known as wave turbulence (WT) in dispersive media [1]. One of the significant achievements in WT theory, discovered by V. E. Zakharov in 1965, is the identification of a new type of solution to kinetic equations known as Kolmogorov-Zakharov (KZ) spectra have been observed in various experiments conducted in turbulent wave systems [2]. These solutions exhibit a constant energy flux across different scales. In addition, turbulence can also refer to the complex spatial and temporal behavior of chaotic, random waves within physical systems. The dynamics described above can be adequately described by employing integrable nonlinear partial differential equations, such as the Korteweg-de Vries (K-dV) equation or the one-dimensional nonlinear Schrödinger (NLS) equation. The field of integrable turbulence, which Zakharov initiated, has become a vibrant research area supported by theoretical (primarily numerical) developments and experimental observations. Unlike in standard wave turbulence, integrable systems lack resonances, leading to fundamentally different underlying mechanisms. Consequently, the study of integrable turbulence requires distinct theoretical approaches.

The phenomenon known as the Kolmogorov-Zakharov cascade becomes evident when the spectrum or scales of nonlinear random waves can be divided into three distinct regions. These regions include a pumping spectral range where external energy is supplied, typically at larger scales, a dissipation range occurring at smaller scales, and an intermediate inertial range where no dissipation occurs [1, 2]. The research conducted by Le Reun et al. [3] illustrates the presence of the Kolmogorov-Zakharov spectrum in numerical simulations of inertial wave turbulence within domains that are both substantially anisotropic and three-dimensional. In their investigation, Zakharov et al. [4] discovered that when examining the defocusing nonlinearity in a one-dimensional wave turbulence model, the observed outcome displays a k^{-1} scaling. Meanwhile, Rumpf et al. [5] study on radiating nonlinear solitary wave pulses, focusing on the nonlinearity of a different nature, demonstrates that the energy spectrum closely follows a power law resembling the KZ spectra of k^{-1} for the case of defocusing nonlinearity and is steeper (approximately $k^{-1.25}$) than the KZ spectra of wave turbulence. Lamb [6] demonstrated that by using α to represent the amplitude of the wave, ν to describe the kinematic viscosity of the fluid, and k to represent the wave number of the decaying wave, a relationship $d\alpha/dt = -2\nu k^2\alpha$, leads to the classical law for viscous decay of waves of amplitude $\alpha \sim exp(-2\nu k^2 t)$. Kraichnan [7] anticipated using analytical methods, the energy spectra scaling are $k^{-5/3}$ and k^{-3} for $k < k_f$ and $k > k_f$, respectively. Kraichnan formulas have been discussed in many experimental and numerical simulations [8–10]. Recently, Gupta et al. [11], in their investigation on energy spectra in the inertial-dissipation region for 2D turbulence, show that the energy spectra are steeper than k^{-3} for $k > k_f$.

This current article utilizes a damped one-dimensional MNLS equation to study turbulence generation by numerical simulation. In the previous work [12], we have numerically solved the MNLS equation for defocusing nonlinearity ($\sigma = -1$) where the source of nonlinearity is due to the rise in temperature. However, this article concentrates on the study of turbulence due to defocusing nonlinearity with damping effect. We introduced artificial damping to simulate viscosity at small scales as $d/dt \rightarrow (d/dt + \gamma)$, where γ is damping. It is worth noting that several papers employ the same type of artificial damping for dissipation (one can refer [13–15]). The source of nonlinearity is the rise in temperature due to the acoustic wave's high amplitude. Accounting for temperature change, the evolution of the localized structure in space and time is analyzed. The FPU recurrence relation is also tested to reveal how the FPU recurrence [16, 17] is broken to develop the system's nonlinearity. The energy spectrum is KZ spectra in the inertial range and steeper than k^{-3} beyond the inertial range. Finally, the results are discussed in the context of broader physical wave turbulence scenarios.

The paper is organized as follows: Sect. 2 describes the methodology and energy transfer in the MNLS equation. Section 2.1 presents the setup for the numerical simulations and validation conducted for the model equation. The results and discussion are provided in Sect. 3, and finally, the conclusions drawn from the findings are summarized in Sect. 4.

2 Methodology

The equations for an inviscid and compressible fluid, along with all necessary assumptions, are outlined in the following section. It does not consider any external forces acting on the system, and the only factor contributing to the unsteady flow is the presence of sound waves.

$$\frac{\partial \rho}{\partial t} + \nabla \cdot (\rho \mathbf{u}) = 0 \tag{1}$$

$$\rho \frac{D\mathbf{u}}{Dt} + \nabla p = 0 \tag{2}$$

$$\rho \frac{De}{Dt} + p\nabla \cdot \mathbf{u} = 0 \tag{3}$$

$$\frac{\partial^2 p}{\partial t^2} - c^2 \nabla^2 p = 0. \tag{4}$$

Equations (1–4) represent the continuity equation, momentum conservation equation, energy conservation equation, and acoustic wave equation, respectively. Here, p, $\mathbf{u}$, and ρ are the pressure field, velocity field, and density of the fluid, respectively. The speed of sound $c = \sqrt{\gamma RT}$ $(T = T_0 + \delta T)$. A linearized pressure-density relationship is required, which depends upon the thermodynamical properties of the fluid and is given as $p = c^2 \rho$. The total state consists of a background quantity plus unsteady and non-uniform perturbations: $\rho = \rho_0 + \rho_1$, $p = p_0 + p_1$, and $u = u_0 + u_1$. The quantities with the subscript "0" are the background values, while those with the subscript "1" are the perturbed values. These fluctuations are significantly smaller compared to the average density. Utilizing these values Eq. (4) one can get

$$\frac{\partial^2 \rho_1}{\partial t^2} - c_0^2 \left(1 + \frac{\delta T}{T}\right) \frac{\partial^2 \rho_1}{\partial x^2} = 0. \tag{5}$$

The modified temperature (δT) can be calculated using the energy conservation equation (Eq. 4). Now, let's consider that the fields fluctuate in a harmonic manner

$$\rho_1(x, t) = \rho' \mathrm{e}^{\iota(kx - \omega t)}, \tag{6a}$$

$$u_1(x, t) = u' \mathrm{e}^{\iota(kx - \omega t)}. \tag{6b}$$

Here, ρ' and u' are the density perturbation amplitude and velocity perturbation amplitude (a slowly varying function of x and t) Utilizing Eq. (6a) into Eq. (5) and neglecting the terms proportional to $(\delta T/T)\, \partial^2 \rho'/\partial x^2$ or $(\delta T/T)\, \partial \rho'/\partial x$ since ρ' is slowly varying in space (and multiplied here by the small perturbation parameter δT)

$$-2i\omega \frac{\partial \rho'}{\partial t} - c_0^2 \frac{\partial^2 \rho'}{\partial x^2} - 2ik_x c_0^2 \frac{\partial \rho'}{\partial x} + c_0^2 k_x^2 \frac{\delta T}{T_0} \rho' = 0. \tag{7}$$

We write Eq. (7) by neglecting the term $\left(\partial^2 \rho'/\partial t^2\right)$ in the slowly varying envelope approximation, where $\left(\partial^2 \rho'/\partial t^2 << \omega \partial \rho'/\partial t\right)$. The following relation, written as a perturbed quantity, was obtained from Eq. (2)

$$\rho_0 \left(u_1 \frac{\partial u_1}{\partial x}\right) = -\nabla p_{\text{non}}. \tag{8}$$

where p_{non} is nonlinear pressure due to nonlinearity in the system. From Eq. (8), one can easily get

$$p_{\text{non}} = -\frac{1}{2}\rho_0 |u_1|^2. \tag{9}$$

Here, $|u_1|^2 = |u_1 \times u_1^*|$ and u_1^* is complex conjugate of u_1. Now energy equation (Eq. 3) can be written in the form of perturbed quantity using relations: $T = T_0 + \delta T$ and $e = 3nk_B T/2\rho_0$ as

$$\left\{\frac{3}{2}nk_B u_1 \frac{d}{dx}(\delta T)\right\} = -p_{\text{non}} \nabla u_1. \tag{10}$$

Here, δT is nonlinear temperature. Now combining these relations, $u' = \omega\rho'/\rho_0 k$, (which is obtained from continuity Eq. 1), Eqs. (9), and (10), the temperature change can be expressed as

$$\delta T = \frac{\omega^2 |\rho'|^2}{6nk_B k^2 \rho_0}. \tag{11}$$

Substituting Eq. (11) into Eq. (7), one can get the dynamical MNLS equation.

$$\iota \frac{2\omega}{k^2 c_0^2} \frac{\partial \rho'}{\partial t} + \frac{1}{k^2} \frac{\partial^2 \rho'}{\partial x^2} + \iota \frac{2}{k} \frac{\partial \rho'}{\partial x} - \frac{\omega^2}{6nk_B T_0 \rho_0 k^2} |\rho'|^2 \rho' = 0. \tag{12}$$

Now normalizing Eq. (12) gives

$$\iota \frac{\partial \psi}{\partial \tilde{t}} + \frac{\partial^2 \psi}{\partial \tilde{x}^2} + \iota c_1 \frac{\partial \psi}{\partial \tilde{x}} - |\psi|^2 \psi = 0. \tag{13}$$

Equation (13) is the normalized one-dimensional (1D) MNLS equation with the normalized quantities: ψ, $\tilde{t}$, $\tilde{x}$, and a constant $c_1 (= 2.0)$. The normalization parameters are: $t_n = 2\omega/k^2 c_0^2$, $x_n = 1/k$, and $\rho_n = (6nk_B T_0 \rho_0 k^2/\omega^2)^{1/2}$. Adding artificial damping (γ) into Eq.(13) we can easily get

$$\iota \left(\frac{\partial \psi}{\partial t} + \gamma \psi\right) + \frac{\partial^2 \psi}{\partial x^2} + \iota c_1 \frac{\partial \psi}{\partial x} - |\psi|^2 \psi = 0 \tag{14}$$

Equation 14 is the model equation called the damped MNLS equation. Here, $\psi(x, t)$ is a complex function in space and time with real constant $\gamma \geq 0$ representing the dissipation rate [14]. We have neglected the tilde symbol for simplicity. The equation reduces to the MNLS equation as $\gamma = 0$ and to the NLS equation as both constants γ & c_1 are zero.

2.1 Nonlinear Energy Transfer in MNLS Equation

In Fourier space, Eq. 14 can be written as

$$\iota \left(\frac{\partial}{\partial t} + \gamma \right) \hat{\psi}(k,t) = \left\{ k^2 + 2k - N(k,t) \right\} \hat{\psi}(k,t) \tag{15}$$

Here, $\hat{\psi}(k,t)$ and $N(k,t)$ are the Fourier transform of $\psi(x,t)$ and nonlinear term respectively. Multiplying Eq. 15 with $\hat{\psi}^*(k,t)$ and then adding to its complex conjugate yields

$$T(k,t) = (-2k - k^2)E(k,t) \tag{16}$$

The energy spectrum corresponding to $\hat{\psi}(k)$ is defined as the squared magnitude of the Fourier transforms $E(k,t) = |\hat{\psi}(k,t)|^2/2$. Here, $T(k,t)$ is nonlinear energy transfer rate [18]

$$T(k,t) = -\sum Re\{\hat{\psi}(k_1,t)\hat{\psi}(k_2,t)\hat{\psi}(k_3,t)\hat{\psi}^*(k,t)\} \tag{17}$$

The overall energy in real space, according to Parseval's theorem, can be written as the sum of the energies associated with particular modes, as indicated by the equation: $E = \sum_k < E(k,t) >$. If $T(k,t) = 0$, we will get a constant flux condition that shows Kolmogorov spectra $k^{-5/3}$. However, when RHS of Eq.(16) is not zero, for small wavenumber ($k < 1$), leading to a decrease in flux [19] and a deviation from the Kolmogorov $5/3^{rd}$ law. As a result, for $\gamma = 0$, we will get shallower spectra (see Fig. 4), and for weak damped system ($\gamma = 0.1$), we will get spectra steeper than k^{-3} for $k > 1$ beyond the inertial range.

3 Simulation Setup

To study the turbulence generation Eqs. (14) is numerically solved in a computational domain to obtain the evolution of the density perturbation. The size of the computational domain $2\pi/\alpha$. Here, α is normalized by x_n^{-1} with 1024 grid points and step size is $\Delta t = 10^{-5}$. The current study utilized a pseudo-spectral scheme for spatial integration, the finite difference method for time integration, and the predictor-corrector approach [20]. We will look at the equation with the initial condition $\psi(x,0) = \psi_0\{1 + \beta cos(\alpha x)\}$ and periodic boundary conditions in this article (Fig. 1).

Fig. 1 Schematic diagram illustrating the periodic boundary condition (PBC) of the numerical model. Here $\psi_1 = \psi_{n+1}, \psi_{-1} = \psi_{n-1}$ $(n = 1, 2, 3...)$

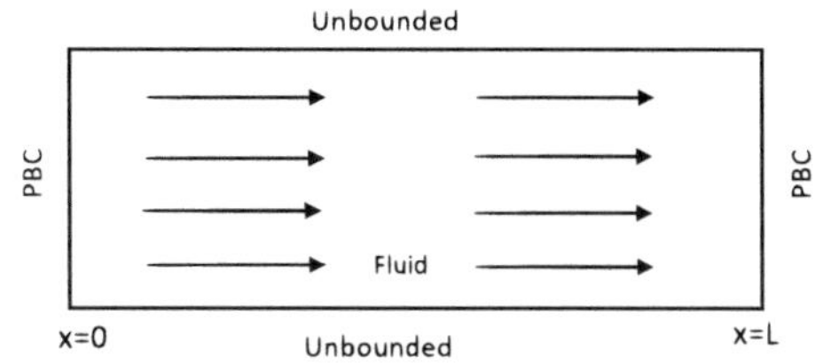

For the numerical simulation, the employed parameters are: $T_0 = 300K$, $c_0 \approx 3.44 \times 10^4$cm/s, $\omega = 6.28 \times 10^3$ rad/s, $k = 0.1826$ cm^{-1}, $x_n = 5.474$ cm, $t_n = 0.318$ ms, $\rho_n = 4.58 \times 10^{-13}$ g/cm^3, $\rho_0 = 1.0$, $\psi_0 = 1$, and $\beta = 0.1$. The different values of α have been taken in the simulation, which has been mentioned in the descriptions of the corresponding figures.

3.1 Validation

After testing the method to the desired accuracy, we obtained results anticipated by this equation on localized structures and turbulence generation. We found the spectral index in the same range as expected from the existing data in the works of literature. The well-known NLS equation is used to check the code accuracy with the help of the consistency of the conserved quantity (N) (i.e., the density of particles), $N = \sum_k |\psi_k|^2$). The invariant quantity (N) was conserved throughout the simulation up to the accuracy of the order of 10^{-5}. Following that, the technique was modified to solve a nonlinear dimensionless equation. The testing of the invariant density of particles with the variation of the grid points and the time step is given in Tables 1 and 2. Furthermore, we tried to validate our results based on the literature [19] available on nonlinear energy transfer rate.

Table 1 At the grid points 1024

Time step	10^{-4}	10^{-5}	10^{-6}	10^{-7}
N	1.00498	1.00500	1.005000	1.005000

Table 2 At time step 10^{-5}

Grid points	256	512	1024	2048
N	1.00499	1.00499	1.005000	1.005000

4 Results and Discussion

The primary objective of this study is to examine the turbulence generation in a fluid caused by a high amplitude of acoustic waves along with damping. The high amplitude of the sound wave serves as the source of nonlinearity for turbulence production. To investigate the Fermi-Pasta-Ulam (FPU) recurrence, the formation of localized structures, and the scaling behavior of turbulence at different length scales, the numerical solution of Eq. (14) has been computed. The FPU recurrence refers to the system returning to its initial state after a specific amount of evolution, which has been previously documented in the literature [21, 22] for NLS equation-driven systems. Figure 2a illustrates the occurrence of the Fermi-Pasta-Ulam (FPU) recurrence as observed in the NLS equation through the implemented code. The energy exchange process initiates with power in the $k = 0$ mode, which then propagates to higher modes, regrouping back into the $k = 0$ mode, where the FPU recurrence occurs. As depicted in Fig. 2a, energy is transferred to higher modes as the wave propagates.

On the other hand, Fig. 2b showcases the breakdown of the FPU recurrence in the case of the damped-driven MNLS system. The normalized density for $k = 0$, $k = \alpha$, and $k = 2\alpha$ is calculated to examine the energy exchange between different Fourier modes in the current nonlinear system governed by Eq. (14). Figure 2b depicts this energy exchange characterization. It is evident from the graph that the FPU recurrence is disrupted, indicating the development of nonlinearity in the system.

The study also examined the localization of the nonlinear wave in both the NLS and MNLS cases, with the perturbation wavenumber (α) playing a significant role. Figure 3 depicts the evolution of density in the spatiotemporal domain for perturbation wave number $\alpha = 0.3$ in the NLS, MNLS, and damped-driven cases, respectively. In the NLS case, the structures exhibit a periodic pattern. However, in the MNLS and damped-driven MNLS cases (as shown in Fig. 3b, c), these recurring patterns are disrupted, and localization becomes evident as we progress in the spatial domain, resulting in the generation of random structures. The complex localization and the formation of chaotic systems observed in Fig. 3 indicate the potential existence of turbulence in nonlinear waves. This type of turbulence is called "wave turbulence," as sound waves play a significant role in dissipating energy in this phenomenon. The ensemble-averaged energy spectrum shown in Fig. 4 presents twelve spectra between $t = 70$ and $t = 92$ for $\alpha = 0.01$, with the red line and green line representing the reference scaling. The energy spectra $E(k, t)$ exhibit a notable variation, indicating a $k^{-1.2}$ scaling for $\gamma = 0$ case at small wave numbers ($k < 1$), which corresponds to the nonlinear regime known as the KZ spectra. The density spectrum closely approximates the KZ scaling ($\sim k^{-1}$) in the inertial sub-range. While for damped case ($\gamma = 0.1$), the spectrum shows k^{-5} scaling in the dissipation range for $k > 1$.

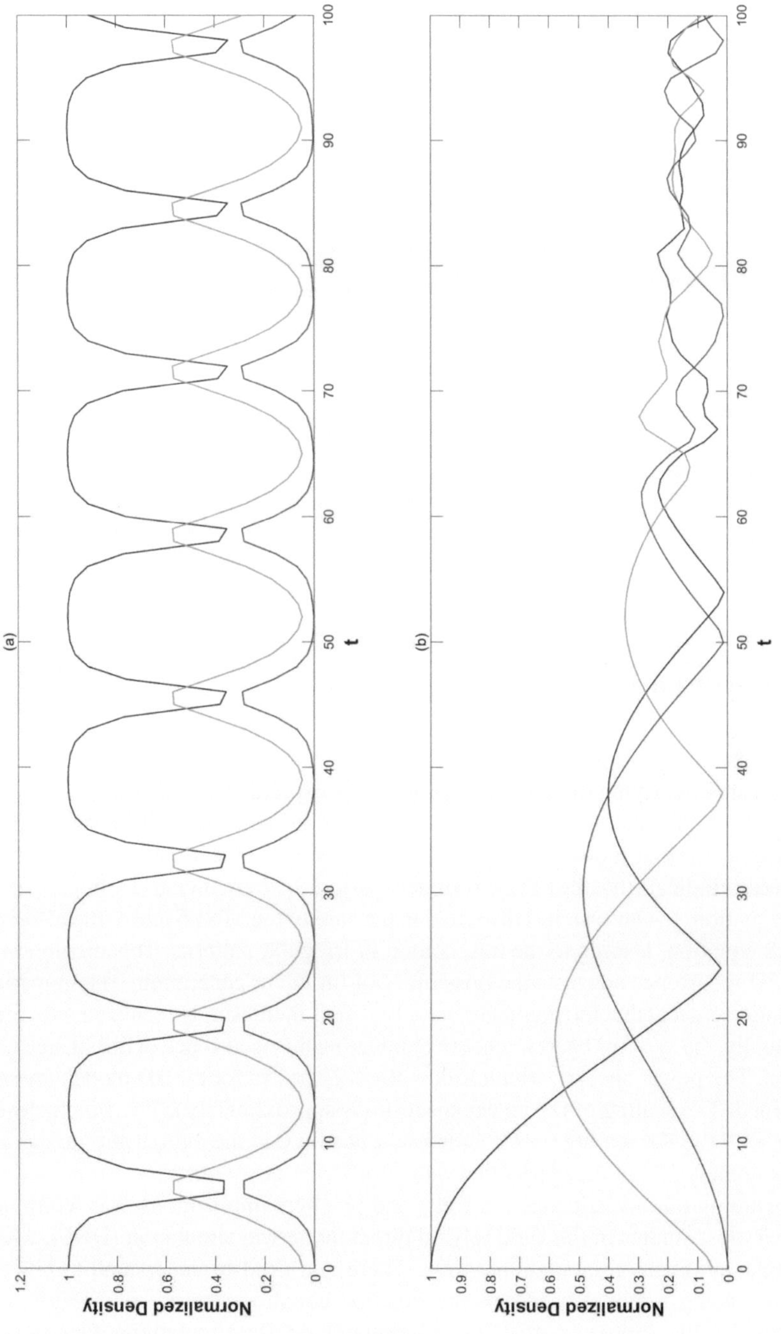

Fig. 2 Time evolution of normalized density for three initial Fourier modes, i.e., $k = 0$ (blue curve), $k = \alpha$ (green curve) and $k = 2\alpha$ (red curve) with t **a** FPU recurrence for NLS equation (periodic) and **b** present system showing recurrence breakdown in case of damped-driven MNLS

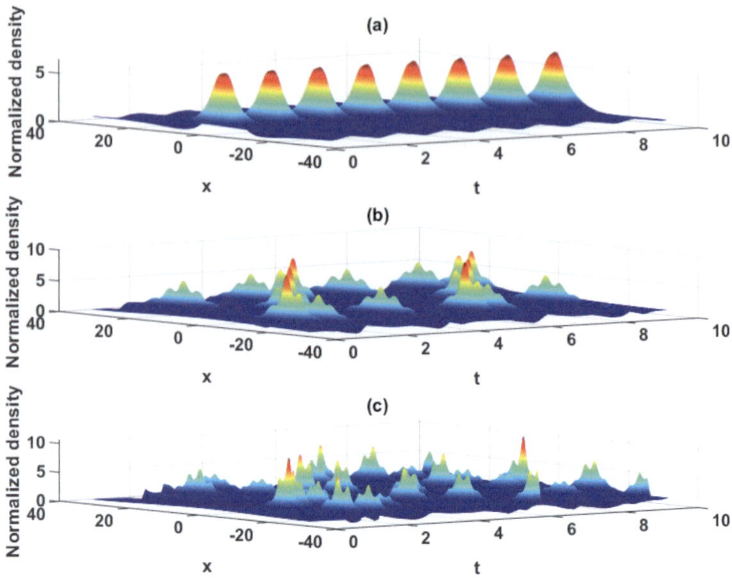

Fig. 3 The variation of normalized density for **a** NLS model ($\gamma = 0\& c_1 = 0$) equation, **b** MNLS case ($\gamma = 0$) and **c** damped MNLS case, for $\alpha = 0.3$ in the spatiotemporal domain

5 Conclusions

In conclusion, the primary objective of this study is to gain insight into the physical mechanism underlying the formation of localized structures and investigate the generation of turbulence caused by temperature variations along with artificial damping. A one-dimensional model equation (MNLS) is developed and numerically solved. The results indicate that the FPU recurrence (periodic pattern) occurs in the case of the NLS equation. In contrast, it is disrupted in the case of the MNLS and damped-driven MNLS equation, leading to the observation of irregular patterns. This disruption of the FPU recurrence suggests the possibility of turbulent generation. The numerical simulations reveal that the nonlinear wave becomes increasingly localized over time. Eventually, the systems become chaotic, indicating the occurrence of turbulence generation. The power spectra, which follow the KZ spectra for the 1D model, are also examined. The scaling of the power spectrum is approximately $k^{-1.2}$, which closely aligns with the KZ scaling ($\sim k^{-1}$) for $k < 1$ at $\gamma = 0$ in the inertial sub-ranges and approximately k^{-5} in the dissipation range for $k > 1$ at $\gamma = 0.1$.

Regarding our investigation's novelty and practical implications, it is worth noting that most studies in this field rely on direct numerical simulation (DNS), which requires significant computational time. There is a need to understand turbulence behavior using a model based on conservation equations (mass, momentum, and energy), as this approach offers faster computation for simulations. Our present work represents a small step in that direction. It is the first attempt, to the best of our

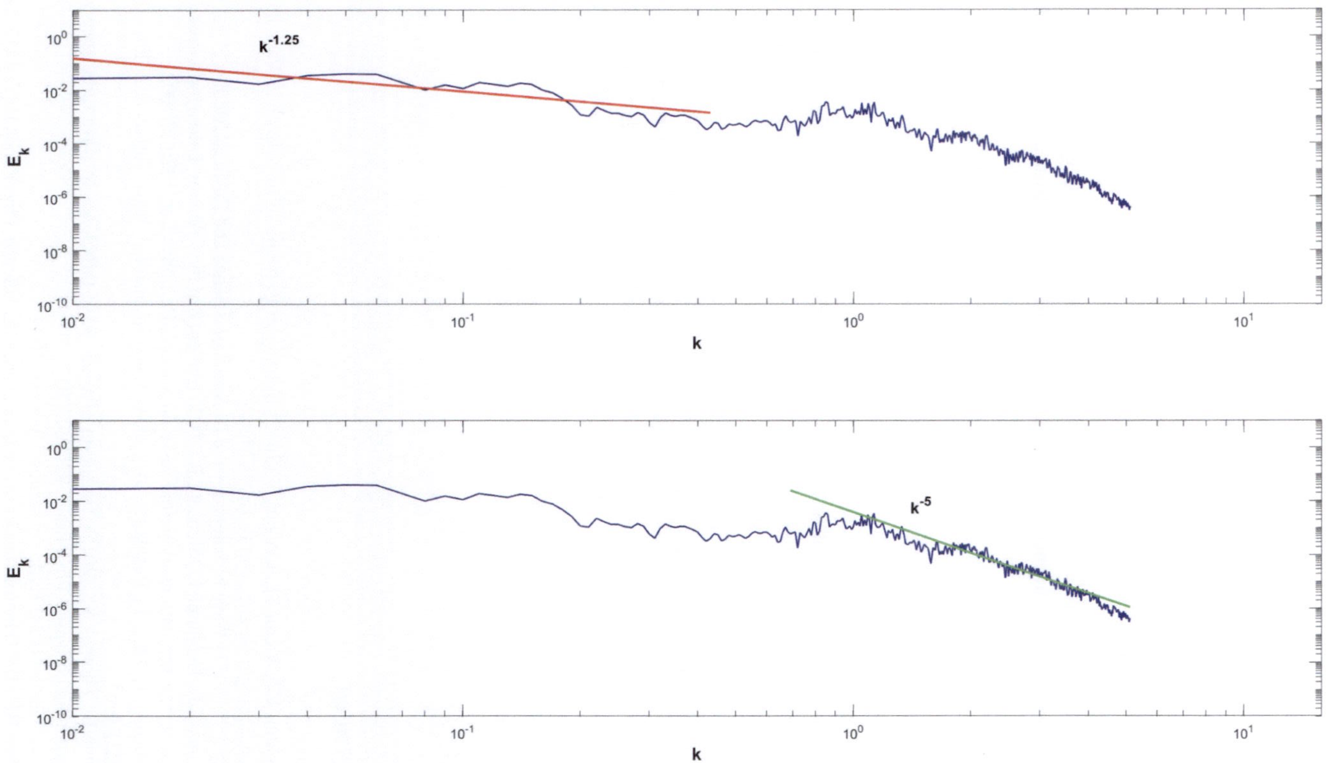

Fig. 4 Ensemble average power spectrum (E_k vs k) between the normalized time, $t = 80$ to $t = 92$ for $\gamma = 0$ (top) and $\gamma = 0.1$ (bottom) at $\alpha = 0.01$

knowledge, to explore the behavior of turbulence caused by temperature rise due to high-amplitude acoustic waves using the damped MNLS equation. This effort contributes to advancing the understanding of turbulence in fluid dynamics. By observing the disruption of localized structures, we gain insights into the underlying physics of turbulent behavior.

In spite of the outcomes presented in this article, the damped MNLS equation has some limitations too. The one spatial dimension in which the 1D MNLS equation explains waves limits its applicability to certain systems and may cause it to ignore the dynamics of waves in systems with multiple dimensions. Dissipation occurs in many physical systems for a variety of reasons, including damping, shear, and more. It might be necessary to modify the MNLS equation in the future or utilize whole different equations in order to account for these considerations. The MNLS equation makes certain assumptions, such as minor damping, minor spatial fluctuations, and weak nonlinearities. To get around these limitations, more comprehensive models could be utilized, such as higher dimensional generalizations of the MNLS equation or numerical simulations that take shear, damping, and dissipation into account.

Expanding our future research to 2D and 3D models may yield more substantial results with practical applications in real-life scenarios. For instance, the study of turbulence in space reentry vehicle problems, where friction typically plays a role, could benefit from understanding the potential temperature rise caused by high-amplitude acoustic waves and their impact on turbulence. Such turbulence could affect the vehicle's trajectory, leading to deviations from the intended destination. Therefore, comprehending the dynamics of these phenomena would be highly beneficial. Additionally, applications in boundary layer flows, ocean waves, and underwater bodies can be explored. Future research on wave turbulence in higher dimensional MNLS equations would be particularly intriguing.

Acknowledgements The authors acknowledge the hospitality of the Indian Institute of Technology Delhi, India, where this work was initiated.

References

1. Vladimir EZ, Victor SL, Gregory F (2012) Kolmogorov spectra of turbulence i: Wave turbulence. Springer Science and Business Media
2. Sergey N (2011) Wave turbulence, vol 825. Springer Science and Business Media
3. Thomas LR, Benjamin F, Michael LB (2021) Evidence of the zakharov-kolmogorov spectrum in numerical simulations of inertial wave turbulence. Europhys Lett 132(6):64002
4. Zakharov V, Dias F, Pushkarev A (2004) One-dimensional wave turbulence. Phys Rep 398(1):1–65
5. Benno R, Thomas YS (2015) Transition of weak wave turbulence to wave turbulence with intermittent collapses. Phys Rev E 92(2):022927
6. Horace Lamb (1924) Hydrodynamics. University Press
7. Robert HK (1967) Inertial ranges in two-dimensional turbulence. Phys Fluids 10(7):1417–1423
8. Frisch U, Sulem P-L (1984) Numerical simulation of the inverse cascade in two-dimensional turbulence. Phys Fluids 27(8):1921–1923

9. Jérôme P, Patrick T (1997) Experimental observation of the two-dimensional inverse energy cascade. Phys Rev Lett 79(21):4162
10. Yong LX, Charles-Henri B, Hamid K (2011) Direct numerical simulations of 2d channel flows in the presence of polymers. Europhys Lett 95(6):64003
11. Akanksha G, Rohith J, Anando GC, Shubhadeep S, Ravi S, Mahendra KV (2019) Energy and enstrophy spectra and fluxes for the inertial-dissipation range of two-dimensional turbulence. Phys Rev E 100(5):053101
12. Praveen Kumar P, Nandal R, Uma R, Sharma RP (2023) Numerical simulation of modified nonlinear Schrodinger equation and turbulence generation. Phys Fluids 35(7):075138
13. Alexander ID, Alexander OK, Vladimir EZ (2004) Weak turbulent Kolmogorov spectrum for surface gravity waves. Phys Rev Lett 92(13):134501
14. Guangyu W, Liu Y, Dick K, Yue P (2006) A note on stabilizing the benjamin-feir instability. J Fluid Mech 556:45–54
15. Vladimir EZ, Aleksandr OK, Andrei NP, Aleksandr ID (2005) Mesoscopic wave turbulence. J Exp Theor Phys Lett 82:487–491
16. Kimmoun O, Hsu HC, Branger H, Li MS, Chen YY, Kharif C, Onorato M, Edmund J, Kelleher J, Bertrand Kibler R, Akhmediev N et al (2016) Modulation instability and phase-shifted fermi-pasta-ulam recurrence. Sci Rep 6(1):1–9
17. Gaetan VS, Emplit P, Haelterman M (2001) Experimental demonstration of the fermi-pasta-ulam recurrence in a modulationally unstable optical wave. Phys Rev Lett 87(3):033902
18. Mahendra KV, Ramgopal A, Pradeep Kumar Y, Sanjay P (2023) Nonlinear energy dissipation and transfers in coarsening systems. Phys Rev E 107(3):034207
19. Mahendra KV (2021) Variable energy flux in turbulence. J Phys A: Math Theoretic 55(1):013002
20. Gazdag J (1976) Time-differencing schemes and transform methods. J Comput Phys 20(2):196–207
21. Hie TM (1990) Homoclinic crossings and pattern selection. Phys Rev Lett 64(4):412
22. Henry CY, Warren EF (1978) Relationship between benjamin–feir instability and recurrence in the nonlinear schrödinger equation. Phys Fluids 21(8):1275–1278

The Effect of Microfin Orientation in Fluid Flow and Heat Transfer in Microchannel Heat Sink

Rohit Kumar, Abhipsa Das, and Manmohan Pandey

Abstract Miniaturization of electric components aces the performance by rapidly increasing transistors on microchips with increased power and, consequently, higher heat fluxes. The discrepancy lies in the fact that increasing heat flux would lead to an exponential rise in temperature by generating thermal increase or heating issues proving to be a significant disadvantage in miniaturization. To reduce the effect of overheating, thermal management on a miniature scale is an essential measure. An initiative to contribute to the potential storage is deliberated throughout the paper. Three-dimensional modeling of the micro-channel heat sink with microfins incorporated has been simulated and explored. The microchannel heat sink with microfins having two different orientation cases is examined. The rectangular microfins of the same cross-sectional area are taken but vary the orientation along the flow. The microfins arrangement with length by width (l/b) 0.5 is taken as Case 1, and the l/b ratio of 2 is named Case 2. The thermal performance of the Case 1 is much better than Case 2, with enhancement in heat transfer by four times. The pressure drop penalty was higher in Case 1, but the augmentation in thermal performance was much more prominent.

Keywords Microchannel heat sink · Conjugate heat transfer · Numerical simulation · Heat transfer enhancement · Fluid flow

1 Introduction

Tuckermann and Pease initially introduced microchannel heat sinks in 1981 [1]. Microchannel heat exchangers can achieve high heat transfer coefficients and low to moderate pressure drops generating heat flux at 1000–11,000 W/cm^2. For most

R. Kumar (✉) · M. Pandey
Department of Mechanical Engineering, IIT Guwahati, Guwahati, India

A. Das
Department of Chemical Engineering, NIT Agartala, Agartala, India

A. Kumar R. et al. (eds.), *Proceedings of Fluid Mechanics and Fluid Power (FMFP) 2023, Vol. 6*, Lecture Notes in Mechanical Engineering,
https://doi.org/10.1007/978-981-96-9600-0_28

applications through conventional channels, it gets inconvenient for the system to cross a heat flux limit of 100 W/cm^2. High heat flux cooling is mandatory in many applications, such as power electronics, plasma-facing components, high heat-load optical components, laser diode arrays, medical devices, and power electronics in hybrid devices. Another trivial application of high heat flux cooling may occur in electric vehicles (EVs) flourishing technology depending upon specific requirements and vehicle design [2]. High heat flux incorporated in microchannel heat exchangers due to high surface-to-volume ratio. Owning to the benefits of microchannels, the significant challenges are the outcomes of fabrications, filtering of the working fluid, pressure drop changes, and pumping power requirements.

The future scope of further enhancement of single-phase heat transfer performance is expected to be described by the conventional equations, as mentioned in a survey article by Kandlikar et al. [3], which gave many comparative studies to conclude that instabilities, high-pressure drop, and complex flow behavior are avoidable considerations in single phase flow, providing accuracy in thermal performance determination. Rosa et al. [4] found that the scaling effect with cross-section geometry and measurement uncertainties may create discrepancies if not considered carefully, while the scaling effect may significantly influence and is a vital consideration in single-phase flow and heat transfer characteristics in the microchannel. Also, single-phase liquid flow is advantageous over gas flow in microchannel heat sinks as rarefaction effects can impact it, and conventional hydrodynamics and heat transfer equations are primarily applicable for liquid flow.

Wu et al. [5] revealed that interrupted micro-fins and engineered fluid with additives are promising enhancement techniques for single-phase flow. A microchannel heat sink (MCHS) with D_h and L identified as hydraulic diameter and channel length was modeled by Kumar et al. [6] to augment the thermal performance of single-phase flow in the microchannel heat sink (MCHS). Using computational fluid dynamics (CFD), investigation of the thermal-fluid flow features, including fluid flow characteristics and heat transfer characteristics, in microchannels of various sizes with or without microfin is possible channels with micro fin exhibited significantly higher-pressure drops compared to channels without microfin, regardless of their size in single phase flow within microchannels. A numerical investigation was conducted to assess the impact of fin height on heat transfer in a microchannel heat sink. The study by Prajapati et al. [7] focused on analyzing the effect of various fin heights on the heat transfer coefficient. Different Reynolds numbers (100, 200, 300, 400) and heat fluxes (50 and 100 W/cm^2) were considered in the analysis. Among the seven fin heights examined, a fin height of 0.8 mm demonstrated the most favorable results regarding the average Nusselt number. The simulation exercise was performed using ANSYS FLUENT software, replicating the numerical model of the experiment. The analysis focused on the Case with a fin height of 0.8 mm, a Reynolds number 300, and consistent dimensions. Temperature variations along the flow direction were recorded and compared to a novel design with uniform surface temperature and a triangular fin microchannel. The Nusselt number was employed as a measure of convective heat transfer. The effect of fin dimension on the performance of micro pin fin heat sinks has been extensively researched. Compared to microchannel heat sinks, micro pin fin heat sinks offer advantages such as increased convective area, secondary fluid flow, vortex production, and thinner boundary layers.

Bhandari et al. [8] evaluated in a review article that the design for an effective micro pin fin heat sink should have considerations that include the shape of the pin fin, with sharp pointed shapes demonstrating better heat transfer performance compared to streamlined shapes.

2 Geometry and Numerical Model

The two Cases of miniature channels with varying fin arrangements and fixed flow direction have been presented. Figure 1(top left) represents a 3-dimensional square channel (10mm × 10mm) modeled as a solid material, encompassing rectangular configuration microfins. The microfins have dimensions of 0.9 mm × 3.75 mm (width). The erected fins are positioned at 0.725–0.5 mm from the edge of the plane surface of the square channel. The ANSYS Design Modeler is used to construct the 3-dimensional channel geometry. Using the commercial program ANSYS Fluent, it is possible to numerically explore the conjugate heat transfer and fluid flow in multiple directions. Figure 1(top right) represents the top view with a detailed illustration of the total height (H) having 5.75 mm, and the region for fluid flow with descriptive fins of height (h) is 3.75 mm. The model has 54 micro-fins for analyzing fluid behavior. Figure 1(bottom left and right) demonstrates the flow direction for each Case 1 and Case 2.

The present investigation uses computational fluid dynamics to navigate the thermal-fluid flow features, fluid flow characteristics, and heat transfer characteristics of single-phase fluid flow in microchannels under the commercial software ANSYS FLUENT. The microchannels display two-pin fin arrangements of 0.9–0.45 mm for a particular flow direction. Water has high latent heat and commonly used in literature so far, to analyse the heat transfer and fluid flow characteristics of liquid flow through microchannel heat sink in this work. Water is selected as liquid-fluid with temperature dependent property. Therefore, conventional fluid flow theory is applicable. The SIMPLE numerical method is activated under the pressure-velocity coupling factor. The second-order upwind scheme is employed to solve the equations in this study. The convergence criteria for the solution are set as residuals of 6×10^{-4} for continuity, 10–4 for momentum equations, and 10–6 for the energy equation. Copper is regarded as the solid material in this model because of its high thermal conductivity. A constant heat flux is applied to the bottom surface of the wall, while a constant inlet flow velocity is imposed on the fluid.

To simplify the simulation of heat flow, several reasonable assumptions made in this study includes [9, 10]:

1. The fluid being Newtonian and incompressible.
2. Steady-state single-phase and laminar flow at 293 K.

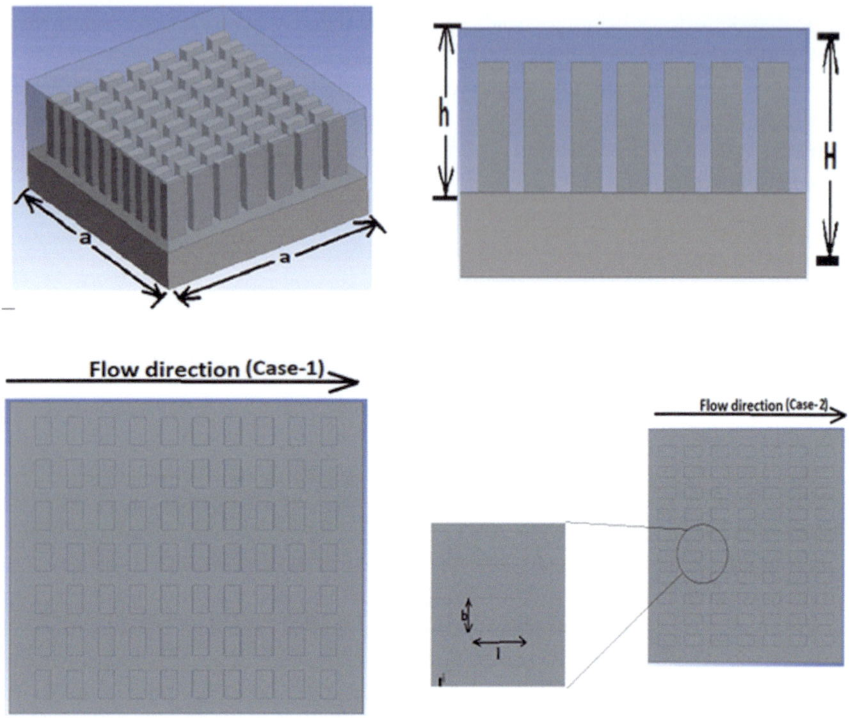

Fig. 1 Model representation of isometric (top left) and top view (top right) of case1 and case 2 according to flow direction

3. Constant thermo-physical properties of the solid; piecewise-linear variation of thermo-physical properties of the fluid with temperature, considering temperature-dependent viscosity.
4. Consideration of steady-state heat transfer.
5. Neglecting thermal radiation effects.
6. Assuming negligible influence from viscous dissipation, gravity forces, volume forces, and other body forces.

By considering the assumptions, the general governing continuity equation, momentum equation, and energy equation for the solid as well as fluid domain in Case of 3-dimensional conjugate heat transfer are derived as [6]:

The conservation of mass (Continuity equation):

$$\nabla.(\rho \boldsymbol{V}) = 0 \tag{1}$$

The conservation of momentum (Navier-Stokes equation):

$$\rho(\boldsymbol{V}.\nabla \boldsymbol{V}) = -\nabla p + \nabla.(\mu \nabla \boldsymbol{V}) \tag{2}$$

Energy equation:
The conservation of energy for fluid flow:

$$\rho c_p(\boldsymbol{V}.\nabla T) = \nabla\left(k_f \nabla^2 T\right) \tag{3}$$

The conservation of energy for solid domain, where the velocity vector $\boldsymbol{V}$ is equal to zero:

$$k_s \nabla^2 T = 0 \tag{4}$$

In the derived equations, "K_f" and "K_s" denotes the thermal conductivity of fluid and solid domain in J/s-m-K, including velocity vector $\boldsymbol{V}$ in m/s, temperature T in K, pressure in Pa, fluid density ρ in kg/m^3, dynamic viscosity μ in Pa-s, specific heat c_p of water in J/kg-K represents the thermal conductivity of the solid.

Fluent incorporates temperature-dependent thermophysical properties of the fluid as variables in the solution of the governing Eq. (10). The corelations used are given as:

$$\mu(T) = \left(9.67 \times 10^{-2}\right) - \left(8.207 \times 10^{-4}\right)T + \left(2.344 \times 10^{-6}\right)T^2 - \left(2.244 \times 10^{-9}\right)T^3 \tag{5}$$

$$\rho(T) = 756.33 + 1.8142T - 0.0035T^2 \tag{6}$$

$$k(T) = -0.5752 + \left(6.397 \times 10^{-3}\right)T - \left(8.151 \times 10^{-6}\right)T^2 \tag{7}$$

$$c_P(T) = 28070 - 281.7T + 1.25T^2 - \left(2.48 \times 10^{-3}\right)T^3 + \left(1.857 \times 10^{-6}\right)T^4 \tag{8}$$

In this numerical study, the following boundary conditions were implemented: at the inlet of the microchannel, a uniform velocity boundary condition has been applied with an inlet temperature of $T_i = 293$ K. At the outlet of the microchannel, the average static pressure boundary condition is adopted, assuming a relative pressure of zero. The model considered two domains: the solid domain and the fluid domain. The side walls of the microchannel set as symmetry surfaces. The contact surfaces between the solid and liquid interfaces were defined as coupling faces. Remaining wall surfaces were assigned adiabatic and no-slip boundary conditions.

Grid Independency Test

In this specific study, unstructured tetrahedral grid elements are utilized to discretize the entire computational domain of the proposed design channels. A highly refined meshing technique is explicitly employed at the interfaces between the fluid and solid regions and along the fin wall. Boundary conditions are applied to the computational domain during the numerical analysis, followed by meshing. This meticulous meshing approach aims to capture intricate details of fluid flow and heat transfer interactions near the wall, enhancing accuracy in modeling these phenomena. It is worth noting that increasing the mesh counts leads to higher accuracy in the results, regardless of the running time during the analysis. The minimum element size of the grid is taken at 0.3 mm, the maximum at 0.4 mm, and the optimized mesh is obtained at 0.35 mm. An accurate result with finer mesh produced is shown in Fig. 2 with isometric and front view of the orientation.

Data Reduction

The Nusselt number is a dimensionless parameter used to evaluate the effectiveness of convective heat transfer relative to conductive heat transfer. It represents the ratio of convective heat transfer to conductive heat transfer. The Nusselt number has been generalised for Laminar flow in this context (Fig. 3).

$$\mathrm{Nu} = \frac{hD_h}{k_f} \tag{9}$$

where, h is heat transfer coefficient, D_h is hydraulic diameter, and K_f is conductivity of the fluid.

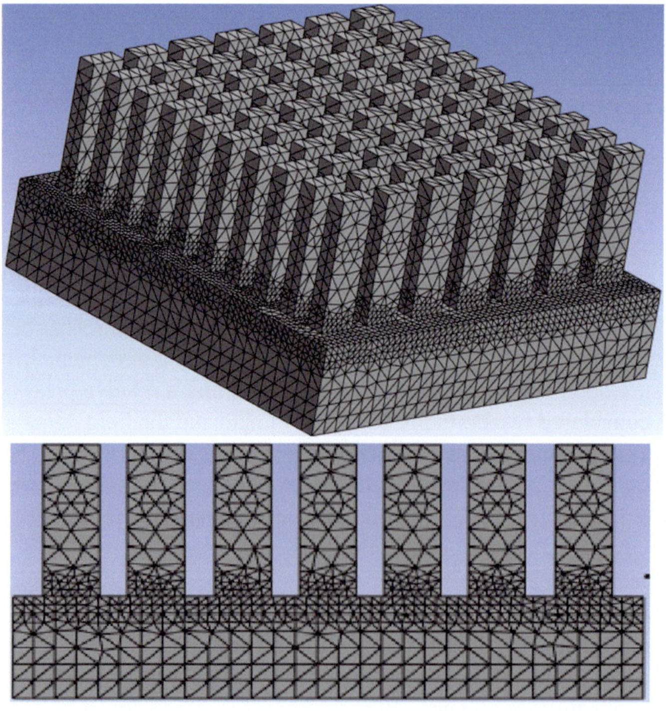

Fig. 2 Meshing of MCHs

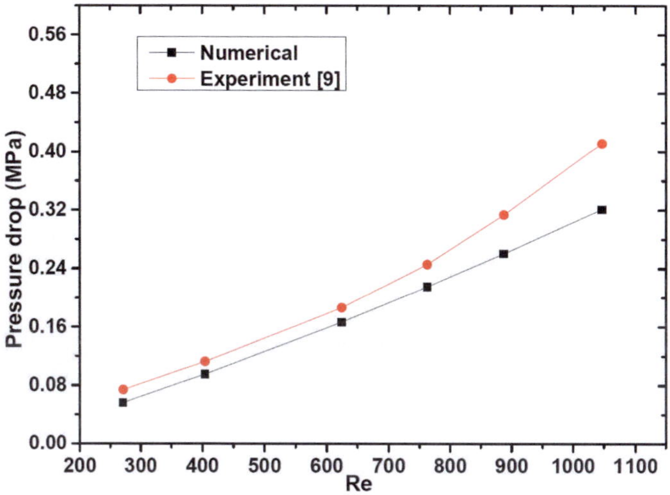

Fig. 3 Validation of model with experimental data [11]

$$h = \frac{Q_{\text{eff}}}{t_w - t_m} \tag{10}$$

$$Q_{\text{eff}} = \frac{qA_{\text{hs}}}{A_{\text{Cs}}} \tag{11}$$

A_{hs} and A_{Cs} denote the surface area used for heating and convective processes, respectively. q represents the supplied heat flux, while T_w signifies the average temperature of the solid domain on the convective surface and T_m specifies the mean temperature of fluid domain. The other equations required for data reduction are taken as [6]:

The Reynolds number (Re) are defined as:

$$R_e = (\rho U D_h)/\mu \tag{12}$$

where ρ represents density with the unit of kg/m^3, U represents velocity with unit of m/s, and μ is the dynamic viscosity of the fluid with unit of Pa-s. D_h represents hydraulic diameter at the microchannel inlet and is defined as:

$$D_h = 2\text{WH}/(W + H) \tag{13}$$

With width (W) of the microchannel in mm and H as total height of the microchannel in mm.

Δp is the pressure drop that occurs between inlet and outlet of microchannel and expressed as: Δp=p_{in}−p_{out}.

A_s is contact surface area of the fluid and microchannel in mm^2 and is expressed as:

$$A_s = (W + 2H)\,L \tag{14}$$

3 Results and Discussion

The present study revolves around numerical simulation of 3-dimensional microchannel heat sink with varying fin arrangements. The working fluid is assumed to be a Newtonian fluid that is viscous in nature and promotes flow resistance during fluid flow in channel decreasing the pressure drop. Two Cases of proposed channel design with pin fins having different fin length along the flow direction are examined. The numerical result of both the Cases are compared with plane channel to assess the enhancement in heat transfer and thermal performance, using thermal performance factor (TPF) which adopted in context to plane channel geometry as in [12]:

$$\mathrm{TPF} = \frac{\mathrm{Nu}/\mathrm{Nu}_0}{\left(\Delta_P/\Delta_{P_0}\right)^{\frac{1}{3}}} \tag{15}$$

where, Nu_0 and Δ_{P_0} is the Nusselt number and pressure drop of plain miniature channel.

The general trend observed for the numerical analysis is that the maximum heater wall temperature decreases with increase in Reynold number (Re), while with increasing heat flux $T_{\max}$ (heater wall) increases for both the Cases. Figure 4 (top left) and (top centre) represents temperature contour for Case 1 and Case 2 at inlet Reynold number 400 and heat flux 300 KW/m^2. Figure 4 (top centre) has lower maximum heater wall temperature due to decrease in fin area the inlet the flow resistance decreases which lowers the heat transfer coefficient and hence, the thermal performance reduces. The average fluid wall temperature is 295.1 K and channel wall temperature is 301.7 K, for this Case which is lower in comparison to Case 1 with broader fin area at the inlet. The maximum heater wall temperature for Case 1 is 314.6–312.6 K for Case 2. A progressive trend has been followed in the maximum heater wall temperature as the effective heat flux increases and hence, an unwavering thermal performance is noticed for Case 1 compared to Case 2. The velocity contour is demonstrated in Fig. 4 (top right and bottom left) for both Case 1 and Case 2 with cross section taken at middle position of fin height and at the top surface of fin height. The maximum flow velocity achieved in Case 1 is highest compared to Case 2 for of middle and upper fin position as shown in Fig. 4 (bottom centre and bottom right) taken to explore the velocity contour. Since the region of high velocity area is large in Case 1, the pressure drop trend for the same mass flow rate is described as: Case 1 > Case 2. Also, the heat transfer coefficient for convection in Case 1 is dominant than in Case 2. The effect of Reynolds number on fluid flow behaviour for both the Cases has been illustrated in Fig. 5a decreases in Re maximum heater temperature of wall decreases for both the Cases showing a steep decrease in slope. The variations in Nu due to the effect of Re increases while for change in heat flux Nu remains almost constant with negligible variations. Figure 5b shows high variations for Case 1 and minimal variations for Case 2. Both Case 1 and Case 2 for Fig. 5c shows an exponential rise in pressure drop with Re, but for wall heat flux with 100 KW/m^2 increase from initial value the pressure drop remains invariant. Hence, Case 1 represents better fluid flow characteristic. Figure 5d denotes heat transfer coefficient for change in Re where Case 2 has minimal increase in HTC compared to Case 1 denoting that Case 1 has better thermal performance as well as higher fluid resistance is provided by the same.

4 Conclusions

The present study addresses the critical issue of thermal management in miniaturized electronic devices. The findings emphasize the significance of optimizing pin fin dimensions, shapes, and arrangements in microchannel heat sinks to achieve high heat transfer coefficients and reduce pressure drops. The study reveals that a particular design (Case 1) with a broader fin area at the inlet exhibits superior thermal performance and fluid flow characteristics compared to another design (Case 2). The pressure drops for Case 1 is twice in comparison to Case 2. The heat transfer coefficient and Nusselt number are five to eight times greater than Case 2. The pressure drops and the Nusselt number increases with the Reynolds number. There is a marginal difference in heater wall temperature for both Cases, but Case 2 has a lower temperature than Case 1.

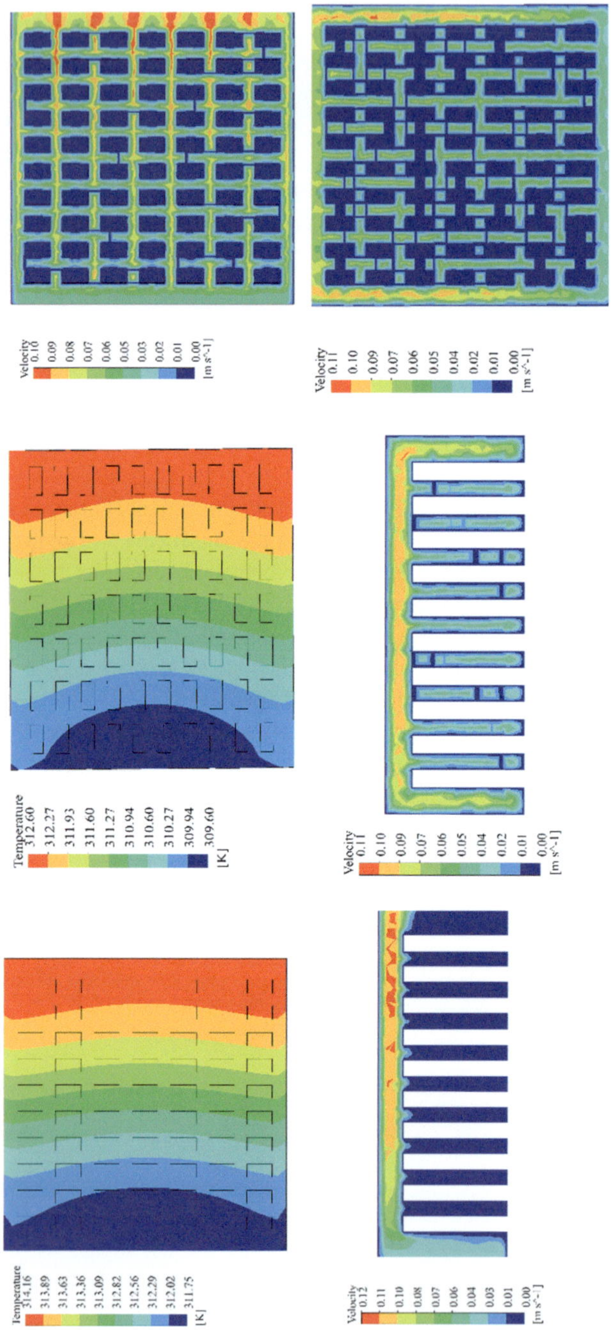

Fig. 4 Temperature and velocity contour

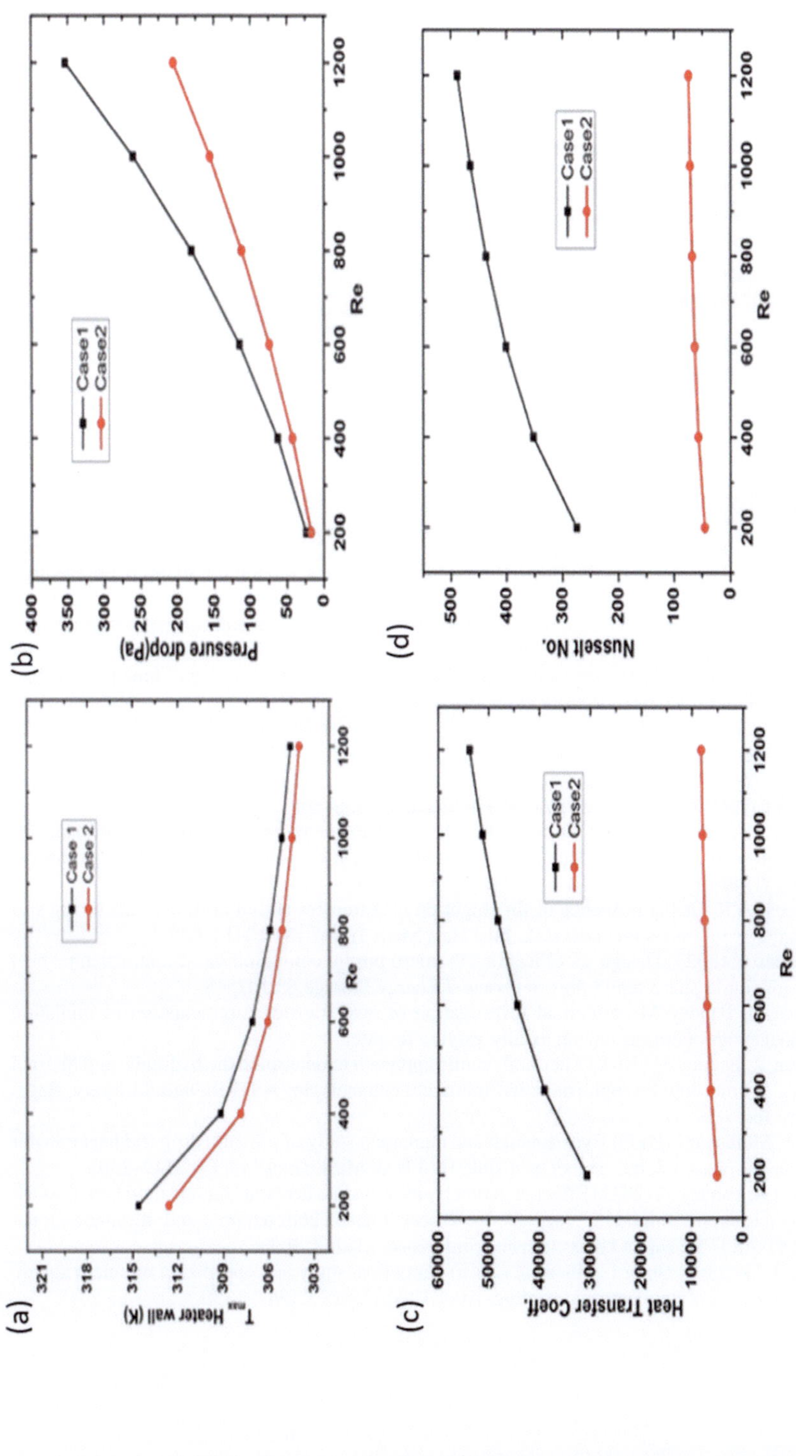

Fig. 5 Variation of reynold number (Re) with **a** pressure drop, **b** nusselt number, **c** heat transfer coefficient, **d** T_{heater} wall

Nomenclature

A	Cross-sectional area of Inlet (m^2)
AR	Aspect ratio
D_h	Hydraulic diameter (m)
h	Heat transfer co-efficient (W/m^2k)
Nu	Nusselt number
q	Heat flux (W/m^2)
Re	Reynolds number
T	Temperature (K)
v	Velocity inlet (m/s)
ρ	Density of air (kg/m^3)
μ	Viscosity (kg/m s)

References

1. Tuckerman DB, Pease RFW (1981) High-performance heat sinking for VLSI. IEEE Electron Device Lett 2(5):126–129
2. Ohadi M, Choo K, Dessiatoun S, Cetegen E (2013) Next generation microchannel heat exchangers. Springer, New York, pp 33–65
3. Kandlikar SG (2012) History, advances, and challenges in liquid flow and flow boiling heat transfer in microchannels: a critical review, 034001
4. Rosa P, Karayiannis TG, Collins MW (2009) Single-phase heat transfer in microchannels: the importance of scaling effects. Appl Therm Eng 29(17–18):3447–3468
5. Wu JM, Zhao JY, Tseng KJ (2014) Parametric study on the performance of double-layered microchannels heat sink. Energy Convers Manag 80:550–560
6. Kumar SR, Singh S (2022) Numerical analysis for augmentation of thermal performance of single-phase flow in microchannel heat sink of different sizes with or without micro-inserts. Fluids 7:149
7. Prajapati YK (2019) Influence of fin height on heat transfer and fluid flow characteristics of rectangular microchannel heat sink. Int J Heat Mass Transf 137:1041–1052
8. Bhandari P (2023) Design modifications in micro pin fin configuration of microchannel heat sink for single phase liquid flow: a review. J Energy Storage 66:107548
9. Kumar R, Pandey M Numerical investigation of heat transfer Augmentation in miniature channels with microfins having axially varying fin pitch
10. Kumar R, Pandey M (2022) Thermodynamic approach to determine the hydraulic and thermal losses in fluid flow through triangular miniature channels. In: ASTFE Digital Library. Begel House Inc
11. Qu W, Mudawar I (2002) Experimental and numerical study of pressure drop and heat transfer in a single-phase micro-channel heat sink. Int J Heat Mass Transf 45(12):2549–2565
12. Kumar R, Pandey M (2021) Effect of geometry in conjugate heat transfer and fluid flow through triangular copper miniature channels. In: Proceedings of 26th national and 4th international ISHMT-ASTFE heat and mass transfer conference, Madras, India
13. Zhao J, Huang S, Gong L, Huang Z (2016) Numerical study and optimizing on micro square pin-fin heat sink for electronic cooling. Appl Therm Eng 93:1347–1359

Numerical Simulation of High-Velocity Mixing Inside a Miniature Jet Pump

Vimal Kotak, Samiran Sengupta, Sugilal Gopalakrishnan, and Sujay Bhattacharya

Abstract With increase in computational power using modern computers, advanced numerical techniques like Large Eddy Simulation (LES) have emerged as useful tools for solving industrial problems to get more accurate solutions and for flow visualization for getting better scientific insights. In this paper, Embedded Large Eddy Simulation (ELES) technique was applied to a miniature jet pump to predict the mixing behaviour in the mixing chamber and diffuser using LES. Other parts of the jet pump were modelled through conventional Reynolds Average Navier–Stokes with standard k–ε turbulent model (RANS-SKE). Comparison of this detailed simulation results with that using RANS-SKE in the whole computational domain was done. It was observed that both the techniques could predict overall performance in terms of jet pump efficiency well in agreement with experimental results. However, considerable differences were observed with respect to jet behaviour, length of the developing region, etc. Moreover, the miniature size of the jet pump considerably affected the flow regimes formed inside the mixing chamber. Turbulent coherent structures were well visualized in the mixing region by simulation using LES.

Keywords CFD · LES · Flow visualization · Mixing chamber · Miniature jet pump · Turbulence

V. Kotak (✉) · S. Sengupta
Research Reactor Design and Projects Division, BARC, Mumbai, India
e-mail: kotakvk@barc.gov.in

S. Sengupta
e-mail: samsen@barc.gov.in

V. Kotak · S. Sengupta · S. Gopalakrishnan
Homi Bhabha National Institute, Mumbai, India
e-mail: gsugilal@barc.gov.in

S. Bhattacharya
Department of Atomic Energy, Mumbai, India

A. Kumar R. et al. (eds.), *Proceedings of Fluid Mechanics and Fluid Power (FMFP) 2023, Vol. 6*, Lecture Notes in Mechanical Engineering,
https://doi.org/10.1007/978-981-96-9600-0_29

1 Introduction

In jet pumps, kinetic energy of a primary fluid is used to pump a secondary fluid. It consists of four components, viz. nozzle, mixing chamber, diffuser and suction region. Figure 1 shows a typical jet pump along with various flow regions inside the mixing chamber. Primary fluid at high pressure is passed through a converging nozzle, which converts pressure energy into kinetic energy. The fall in pressure at the nozzle exit induces the flow of secondary fluid along with the primary stream. The secondary fluid is sucked at the primary jet periphery and carried forward into a mixing chamber. Due to energy and momentum transfer between primary fluid and secondary fluid through turbulent mixing, the combined jet starts expanding in the mixing chamber. The developing region of fluid flowing through mixing chamber can be divided into various regions: uniform low velocity region, mixing region and uniform high-velocity (potential core) region. In ideal condition, fluid at the exit of mixing chamber almost becomes fully developed. In this process of momentum redistribution, some recovery of static pressure takes place. The fluid from the mixing chamber is discharged through diffuser for further static pressure recovery [1].

For jet pumps used in conventional application, nozzle diameter (d) is generally 5 mm and above and nozzle velocity is up to 25 m/s. However, for some specific applications, miniature size nozzle (1–1.2 mm diameter) along with high nozzle velocity (about 40 m/s) is used. Due to higher velocity and lower nozzle dimension of miniature jet pump, turbulent properties and its distribution inside the mixing chamber can be different than that observed in conventional jet pumps. The present paper describes numerical simulation carried out for a miniature jet pump to establish the mixing behaviour for high velocity application.

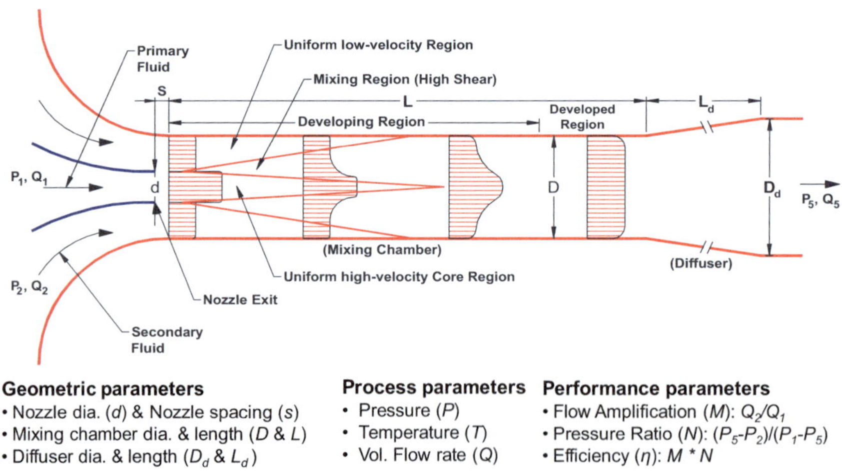

Fig. 1 Sketch of jet pump showing various regions

2 Literature Review and Objective

Several numerical studies on conventional jet pumps using discretized Reynolds Average Navier–Stokes (RANS) equations with closure provided by turbulence model are reported in literature. Hayek and Hammoud [2] carried out numerical simulation of a jet pump using standard k–ε and Reynold Stress turbulent models and predicted velocity and pressure profile inside the jet pump for a maximum nozzle velocity of about 23 m/s. Meakhil and Teaima [3] numerically predicted the velocity profile for different nozzle spacings using standard k–ε turbulent model for a maximum nozzle velocity of about 9 m/s. Yapici and Aldas [4] carried out numerical study for predicting velocity, pressure and turbulent properties using transition SST k–ω turbulent model. The maximum predicted nozzle velocity was about 25 m/s.

Even though generally used RANS equation with turbulent model gives reasonably good and fairly accurate results for most of the industrial problems, advance numerical simulation techniques [e.g. Direct Numerical Simulation (DNS), Large Eddy Simulation (LES)] are very good simulation tools for scientific research by virtue of good computational power. Embedded Large Eddy Simulation (ELES) method is one of the prominent sub-categories of hybrid RANS-LES method, where regions of LES are embedded within a broader RANS domain. This method optimizes the computational power requirement as well as gives more useful and detailed scientific information in the region of interest by modelling the specific region of interest with LES. Holgate et al. [5] carried out review on various studies carried out using Embedded Large Eddy Simulation (ELES) for internal flow. It was summarized that ELES is one of the prospective method to many industrial problems having moderate to high Reynolds number. Zaheer and Masud carried out numerical simulations using ELES for the jet pump with aircraft fuel as a fluid and having nozzle and mixing chamber diameter of 7 mm and 18 mm, respectively [6].

Due to miniature size and different fluid for the present jet pump configuration with reference to the reported literature [6], insight of flow behaviour and turbulent parameter distribution inside the jet pump is the objective of the study here. This paper describes governing equations, grid requirement and numerical setup adopted for ELES simulation for the typical miniature jet pump (shown in Fig. 2). Geometrical dimensions and the process parameters for the miniature jet pump considered for numerical simulation are shown in Table 1. Distribution of velocity, pressure and turbulent properties inside the mixing chamber was predicted using ELES simulation and compared with conventional RANS simulation closed by Standard k–ε turbulent model (RANS-SKE). Flow visualization in terms of typical vortex structure obtained from ELES simulation is also articulated.

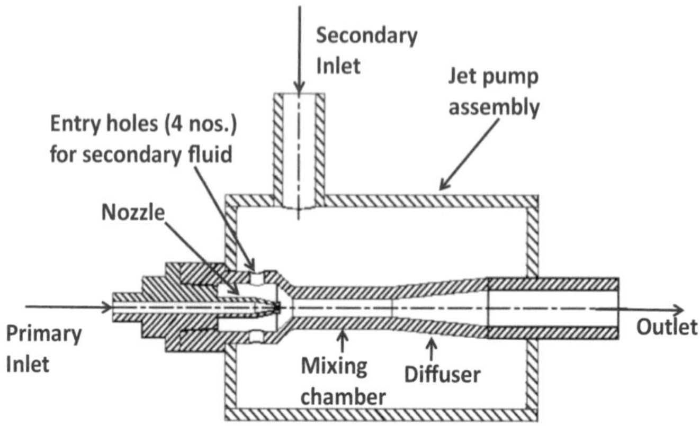

Fig. 2 Schematic diagram of the jet pump assembly

Table 1 Parameters for the miniature jet pump for the present study

Geometrical dimensions	Process parameter
$d = 1.1$ mm	Fluid: Water
$D = 4.5$ mm	M: 1.85
$D_d = 9$ mm	P_5: 2.8 bar
$L = 27$ mm	Q_1: 2.15 lpm
$L_d = 26$ mm	Q_2: 3.98 lpm
$s = 0$ mm	Re_n: 6.7×10^4
Nozzle thickness: 0.7 mm	Re_m: 4.6×10^4
Surface roughness:	T_1: 44 °C
Nozzle: 1.6 μ	T_2: 44 °C
Other parts: 15 μ	

3 Computational Domain, Governing Equations and Numerical Setup

In order to adapt ELES setup, full 3D geometry domain is used for the simulation. The computational domain is shown in Fig. 3. Primary and secondary fluid inlet flow rates (Q_1 and Q_2) are given as inlet boundary condition, and pressure at diffuser outlet (P_5) is given as outlet boundary condition. Since fluid is water, incompressible formulation has been used.

For predicting the gross performance of the jet pump, primary and secondary inlet pressures (P_1 and P_2) are taken as output, which are used to obtain pressure ratio (N) and efficiency (η) for the given flow ratio (M) using the following equations [1]:

$$\text{Flow amplification or flow ratio,} \quad M = \frac{Q_2}{Q_1}, \tag{1}$$

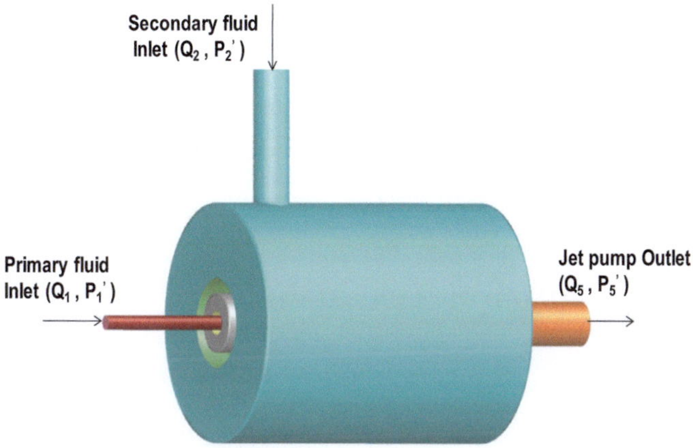

Fig. 3 Computational domain for simulation

$$\text{Pressure ratio,}\quad N = \frac{P_5 - P_2}{P_1 - P_5}, \tag{2}$$

$$\text{Efficiency,}\quad \eta = M * N. \tag{3}$$

3.1 Governing Equations for Numerical Analyses

Navier–Stokes Equations for incompressible flow in absence of body force are shown as Eqs. (4) and (5).

Continuity:

$$\frac{\partial(u_i)}{\partial x_i} = 0. \tag{4}$$

Momentum:

$$\frac{\partial(u_i)}{\partial t} + \frac{\partial\left(u_j u_i\right)}{\partial x_j} = -\frac{1}{\rho}\frac{\partial p}{\partial x} + \frac{1}{\rho}\frac{\partial}{\partial x_j}\left(\mu\frac{\partial u_i}{\partial x_j}\right). \tag{5}$$

Variables (u and p) represent the local instantaneous values, which can be divided in statically time average value (U, P) and fluctuating value as follows:

$$u = U + u', \tag{6}$$

$$p = P + p'. \tag{7}$$

Using the above averaging, the governing equations called as Reynolds Average Navier–Stokes (RANS) equations are shown in Eqs. (8) and (9).

$$\frac{\partial(U_i)}{\partial x_i} = 0, \tag{8}$$

$$\frac{\partial(U_i)}{\partial t} + \frac{\partial(U_j U_i)}{\partial x_j} = -\frac{1}{\rho}\frac{\partial P}{\partial x} + \frac{1}{\rho}\frac{\partial}{\partial x_j}\left(\mu\frac{\partial U_i}{\partial x_j}\right) + \frac{1}{\rho}\frac{\partial}{\partial x_j}(\tau_{ij}), \tag{9}$$

where

$$\tau_{ij} = -\rho\overline{u_j{}'u_i{}'}. \tag{10}$$

Boussinesq hypothesis is employed for estimation of additional Reynolds stress terms (τ_{ij}), and it appears in Eq. (9), which represents it as a function of turbulent viscosity (μ_t) and velocity strain rate (S). For RANS-SKE simulation, the standard k-ε turbulent model is used for evaluating turbulent viscosity which appears in Boussinesq hypothesis and thus used for closure of these equations.

In the Large Eddy Simulation (LES), Eqs. (4) and (5) are filtered as mentioned in Eq. (11) by using grid cell dimension as filtering operator G. The filtering process filters out the eddies whose scales are smaller than grid spacing. The filtered Navier–Stokes (FNS) are shown in Eqs. (12) and (13).

$$\overline{\Phi(x)} = \int \Phi(x')G(x, x')\mathrm{d}x', \tag{11}$$

$$\frac{\partial(\overline{u}_i)}{\partial x_i} = 0, \tag{12}$$

$$\frac{\partial(\overline{u}_i)}{\partial t} + \frac{\partial(\overline{u}_i\overline{u}_j)}{\partial x_j} = -\frac{1}{\rho}\frac{\partial\overline{p}}{\partial x_i} + \frac{1}{\rho}\frac{\partial}{\partial x_j}\left(\mu\frac{\partial\overline{u_i}}{\partial x_j}\right) - \frac{1}{\rho}\frac{\partial}{\partial x_j}(\tau_{\mathrm{sgs}}), \tag{13}$$

where

$$\tau_{sgs} = \rho\overline{u_i u_j} - \rho\overline{u}_i\overline{u}_j. \tag{14}$$

Similar to RANS, Boussinesq hypothesis is used for estimation of subgrid-scale turbulent stress (τ_{sgs}) which appears in Eq. (13). Here, the subgrid-scale turbulent viscosity (μ_{sgs}) was estimated through Wale's model.

By using filtered Navier–Stokes equations, actual flow field and thus, large eddies above grid size are resolved and eddies below grid size are modelled through subgrid viscosity model. Most of the kinetic energy of eddies is carried away by large eddies, which are anisotropic for many real cases. Small eddies are comparatively isotropic and tend to dissipate the kinetic energy. Hence modelling the same through subgrid viscosity model reduces the error of isotropic assumption.

3.2 Grid Requirement and Numerical Setup

For Embedded LES (ELES) implementation for the jet pump, mixing chamber and diffuser, which are zones of interest for detailed simulation, are discretized to very fine mesh required for LES simulation. Other zones are modelled through RANS-SKE turbulent model and have comparatively less refined mesh. Figure 4 shows the half cut of computational domain showing various zones and their interfaces. Inlet boundary condition to LES zone at RANS/LES interface was generated through Vortex method. It generates turbulent structures based on statistically mean turbulent properties evaluated from upstream RANS model [7].

To estimate the mesh size requirement in LES zones, integral length scale l_0 (energy weighted average length of the eddy) of the eddy was evaluated as follows:

$$l_0 \sim \frac{k^{3/2}}{\varepsilon} \tag{15}$$

Based on the initial RANS simulation, the value of turbulent kinetic energy (k) and turbulent dissipation rate (ε) in the mixing chamber region was predicted and l_0 was estimated. For better eddy resolution through LES, at least 80% of the kinetic energy of the eddy shall be actually resolved. As a thumb rule from eddy energy spectrum, this is obtained by using mesh size 5 times finer than the integral length of eddy. Based on this assumption, the LES regions (mixing chamber and diffuser) were

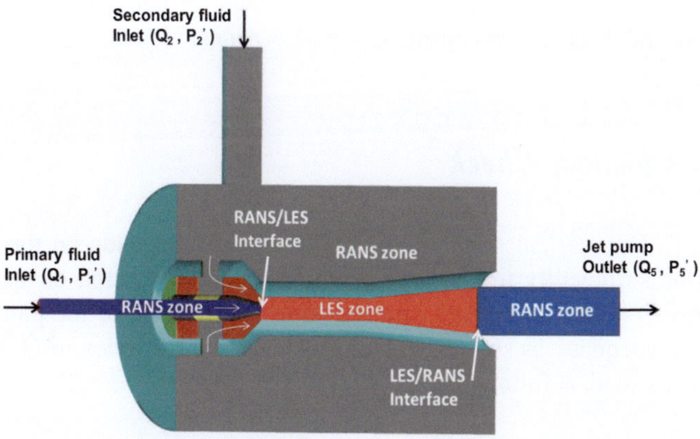

Fig. 4 Half-cut section of computational domain showing various zones for ELES numerical simulation

discretized into 7 million cells. Total number of cells in full computational domain were 11.6 million.

Governing equations mentioned in Sect. 3.1 were discretized through Finite Volume Method (FVM)-based technique available in commercial flow solver.

For spatial discretization of flow variables, first-order upwind differencing scheme was used for RANS-SKE simulation, and central differencing scheme was used for LES zones of ELES simulation. For temporal discretization, implicit transient formulation was used. Initially, analysis was started with steady state, followed by unsteady state simulation using RANS-SKE in all zones. Subsequently, mixing chamber and diffuser zones were changed to analyse through LES. The initial time step was chosen higher (multiple times of corresponding to the required value for Courant Flow limitation) with more iterations per time step. The time step was further reduced to achieve Courant Flow number < 1. The transient analyses were continued till statistically mean value of the output (inlet pressures of primary fluid and secondary fluid) remains unchanged for sufficient time duration. Further to this, the analyses were continued for time greater than the average flow residence time in LES zones to capture the actual flow behaviour.

For comparison with ELES simulation, steady state simulation using RANS-SKE technique in full computational domain was also carried out by using same mesh and boundary conditions.

4 Results and Discussion

The numerical simulations for the parameters mentioned in Table 1 were carried out using both ELES and RANS-SKE technique, and the results are compared and shown here. All the results for ELES were post processed, once convergence of statistically mean values of different parameters are achieved and further running the transient simulation for the fluid residence time in the LES zones.

4.1 Grid Adequacy Check

The accuracy of numerical solution using LES greatly depends on the grid size, which shall be sufficiently fine enough so that significant portion of the turbulent kinetic energy is actually resolved through large eddies. One of the parameters to check the grid adequacy as suggested by Celik et al. is LES Index of Quality (LES IQ), which is defined as follow [8]:

$$\text{LES IQ} = \frac{1}{1 + 0.05\left[\frac{(\mu+\mu_{\text{sgs}})}{\mu}\right]^{0.53}}. \quad (16)$$

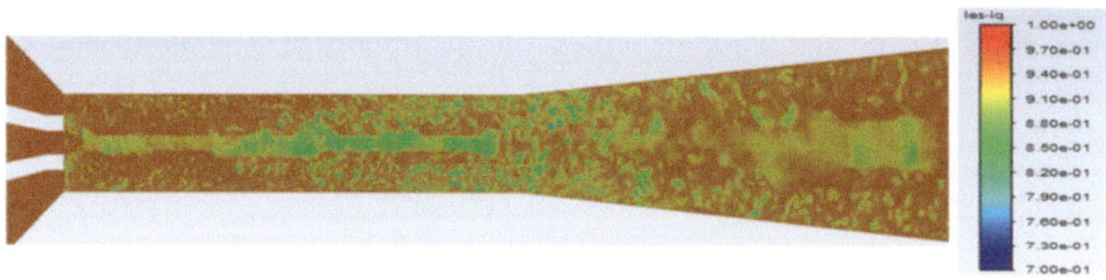

Fig. 5 Contour plot of LES IQ in mixing chamber and diffuser (LES zones) in the symmetry plane

Equation (16) compares modelled subgrid turbulent viscosity (μ_{sgs}) with the laminar viscosity (μ). This index equivalently represents the ratio of resolved to total turbulent kinetic energy. LES IQ varies from 0 to 1 and values greater than 0.8 represents good mesh for LES. The contour of LES IQ in symmetry plane in the LES zones (mixing chamber and diffuser) is shown in Fig. 5. It may be seen that LES IQ is greater than 0.82 in the LES zones, which confirms mesh adequacy for the present study. Another way to ensure grid adequacy is to compare turbulent modelled subgrid viscosity (μ_{sgs}) obtained from LES simulation with the turbulent viscosity (μ_t) obtained from the RANS-SKE simulation. From the contours plot of the viscosity in symmetry plane (Fig. 6), it is clear that turbulent subgrid viscosity obtained from LES simulation is very much less than the turbulent viscosity obtained from the RANS-SKE simulation at any location.

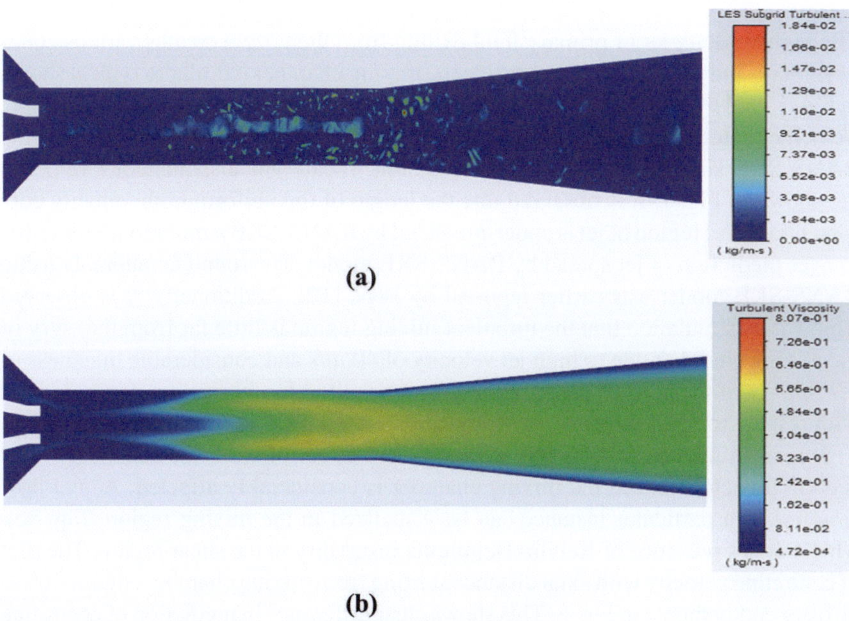

Fig. 6 Contour plot of viscosity in mixing chamber and diffuser in the symmetry plane. **a** Subgrid scale turbulent viscosity from ELES simulation. **b** Turbulent viscosity from RANS-SKE simulation

Table 2 Gross performance validation—comparison of numerical simulation and experimental results

Parameter	Experimental value	RANS simulation	ELES simulation
Pressure ratio (N)	0.0864	0.0843	0.087
Jet pump efficiency (η)	0.16	0.156	0.161

4.2 Experimental Validation

Experiments were carried out for the conditions mentioned in Table 1, and the results were reported by the authors [9]. Using the simulation data, inlet pressure for both the primary and secondary fluids (P_1 and P_2) was estimated. For the given flow ratio ($M =$ 1.85), pressure ratio (N) and efficiency (η) were evaluated from Eqs. (2) and (3). The validation of simulation results through experiments in terms of jet pump efficiency is shown in Table 2. It can be seen from Table 2 that results from both RANS-SKE and ELES simulations are in well agreement with the experimental results. However, prediction from ELES simulation is closer to experimentally observed value.

4.3 Parameter Distribution Inside the Mixing Chamber

The high-velocity jet of primary fluid exiting from the nozzle creates various zones of velocity and pressure gradient inside the mixing chamber (similar to typical shown in Fig. 1) and transfer the momentum to the secondary fluid in the developing region. Velocity contours in the symmetry plane of the mixing chamber obtained from the RANS-SKE simulation and ELES simulation (mean and instantaneous velocity) are shown in Fig. 7. It is observed that the length of the uniform high-velocity core (potential core) region of jet is under-predicted by RANS-SKE simulation. This is due to over prediction of jet spread by RANS-SKE model. The round jet anomaly using RANS-SKE model was earlier reported by Pope [10]. Additionally, it is observed from ELES simulation that the turbulent mixing region is little far from the entry of the mixing chamber due to high jet velocity of 40 m/s and considerable thickness of 0.7 mm for of the nozzle (based on minimum required for fabrication point of view) with respect to other dimensions of the miniature jet pump.

Due to miniature size, nozzle thickness and effect of wall, the secondary fluid suction trajectory inside the mixing chamber is considerably affected. Actual flow structure at a particular instance can be visualized in the mixing region (Fig. 7c), which shows presence of Kelvin–Helmholtz instability in the shear region. The plot of centreline velocity with axial distance starting from mixing chamber entrance up to diffuser end is shown in Fig. 8. This shows clear difference in prediction of centreline velocity variation in mixing chamber between the two techniques. However, in the diffuser region where the mixing is completed, the prediction by both the models are in good agreement.

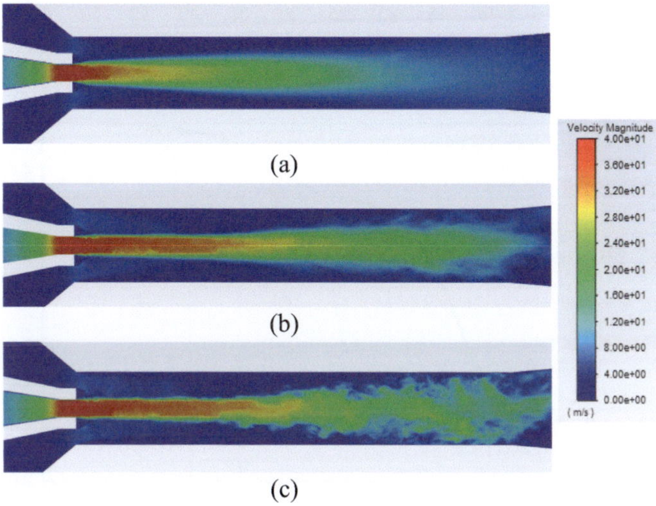

Fig. 7 Velocity contour plot in mixing chamber in symmetry plane. **a** RANS-SKE simulation. **b** Mean velocity, LES simulation, **c** Instantaneous velocity, LES simulation

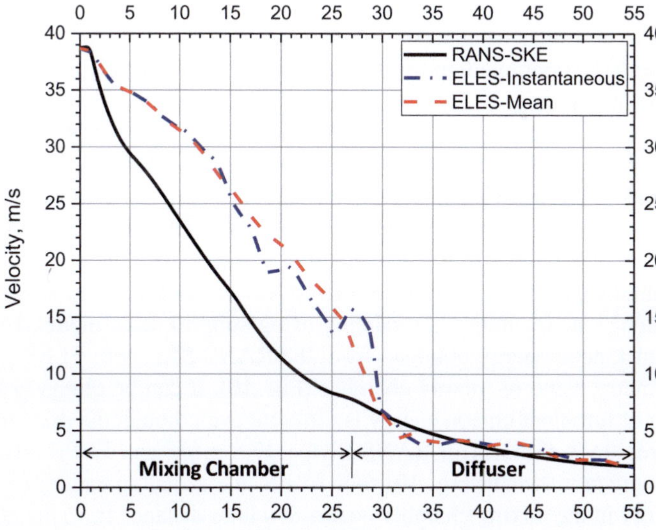

Fig. 8 Axial variation of centreline velocity in the mixing chamber and diffuser

The pressure variation in the mixing chamber and diffuser at the centreline of the jet pump is shown in Fig. 9. Due to momentum redistribution in the developing region, increase in pressure of mixed fluid from RANS-SKE and ELES simulation starts at $x = 5$ mm and $x = 19$ mm, respectively. Pressure at the entrance of the diffuser ($x = 27$), predicted by both techniques, is different. After $x = 35$ mm, both

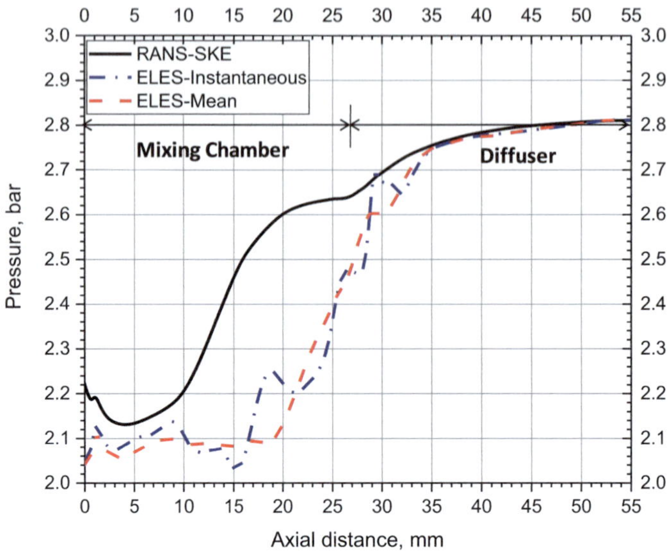

Fig. 9 Axial variation of centreline pressure in the mixing chamber and diffuser

approaches give similar results. This emphasizes that momentum mixing in radial direction might be continued up to certain distance within the diffuser as predicted from ELES simulation.

Another important parameter compared is turbulent kinetic energy (k). The turbulent kinetic energy obtained from RANS-SKE simulation is through solution of transport equations of turbulent kinetic energy (k) and turbulent dissipation rate (ε) along with RANS mass and momentum equations. While in LES, the resolved turbulent kinetic energy is obtained from the fluctuating velocity component of the resolved flow field obtained from the solution of filtered Navier–Stokes equation. The subgrid scale turbulent kinetic energy is modelled using subgrid scale model. Comparing the turbulent kinetic energy obtained from the RANS-SKE with ELES simulation in the symmetry plane of mixing chamber (Fig. 10), it can be observed that axial distribution of turbulent kinetic energy is different even though the maximum value in mixing region is similar (for RANS-SKE—17.8 m^2/s^2 and ELES—16.9 m^2/s^2, both are in green colour in Fig. 10). For ELES, the radial spreading of turbulent kinetic energy in the mixing chamber occurs at a later distance from the nozzle exit than the RANS-SKE simulation. This is evident from results of velocity and pressure distribution shown in Figs. 7, 8 and 9.

From the comparison of the results, it is observed that even though numerical simulation results for both the techniques are in good agreement for predicting the gross parameters (i.e. pressure ratio and efficiency) as shown in Table 2, there is considerable difference in the mixing behaviour inside the mixing chamber. RANS-SKE simulation predicts shorter length required for developing flow and thus, more skin frictional pressure loss occurs at downstream of developing region in the mixing

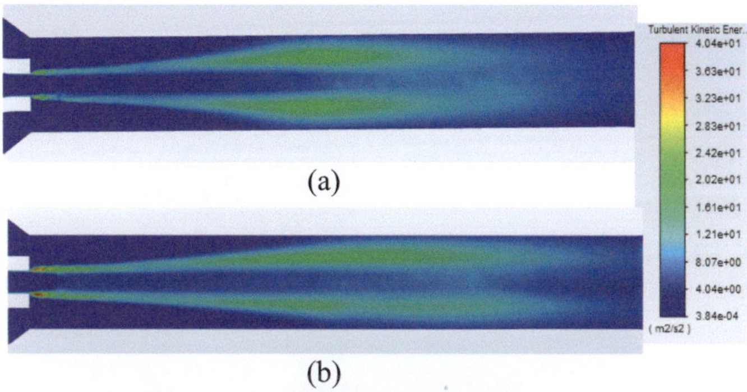

Fig. 10 Turbulent kinetic energy contour plot in mixing chamber in symmetry plane. **a** RANS-SKE simulation. **b** ELES simulation

chamber due to higher wall shear stress. Since frictional loss is more detrimental, the efficiency prediction from RANS-SKE simulation is slightly lesser than the experimental results. For ELES simulation, flow developing length is extended up to certain distance in the diffuser. However, it is well mixed and pressure recovery has occurred before the fluid leaves the diffuser.

4.4 Vortex Visualization

One of the ways to visualize the vortex is by plotting the iso-surfaces of *Q*-criterion. The value *Q*-criterion (QC) is defined in Eq. (17).

$$\mathrm{QC} = \frac{1}{2}\left(\Omega^2 - S^2\right), \tag{17}$$

where S is the Strain rate, and Ω indicates the vorticity rate. The positive value of *Q*-criterion (vorticity rate is higher than the strain rate) indicates the presence of the vortices in the flow domain. Iso-surface for the particular value of *Q*-criterion indicates the particular coherent turbulent structure within the flow domain. The iso-surfaces for a particular value of *Q*-criterion (5% of the maximum found value) in the mixing chamber are shown in Fig. 11. The iso-surface is coloured by velocity magnitude. It is observed that the production of turbulent vortices due to shear layer in the mixing chamber occurs at somewhat far distance from the nozzle exit due to reason explained earlier. The breaking of vortices into smaller sizes just before mixing chamber exits can also be visualized. Absence of vortex just after nozzle exit shows high value of strain rate compared to vorticity rate (very low or negative value of *Q*-criterion).

Additionally, small recirculation zone due to nozzle thickness can be visualized from the velocity vector plot in the symmetry plane of the mixing chamber shown

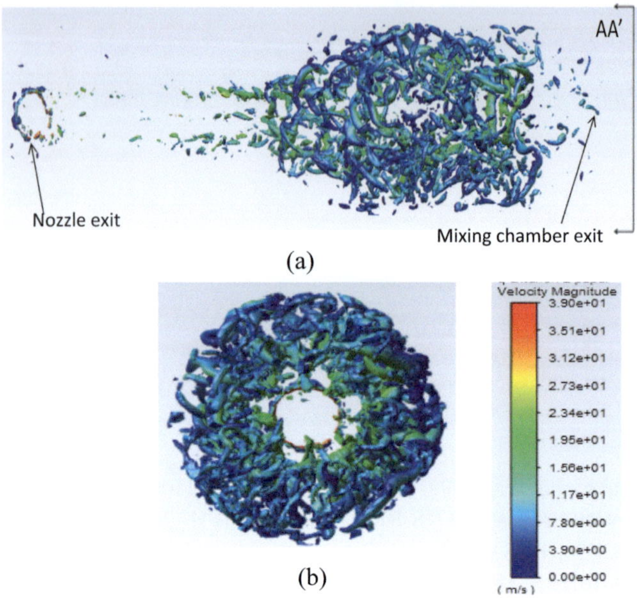

Fig. 11 Iso-surface of Q-criterion (2.27×10^9 1/s^2) in mixing chamber coloured by velocity magnitude obtained from ELES simulation. **a** Front isometric view. **b** View AA′

in Fig. 12. The vector plot also indicates how the nozzle thickness and wall of the miniature jet pump has affected the secondary fluid suction trajectory inside the mixing chamber and delayed the mixing with primary fluid.

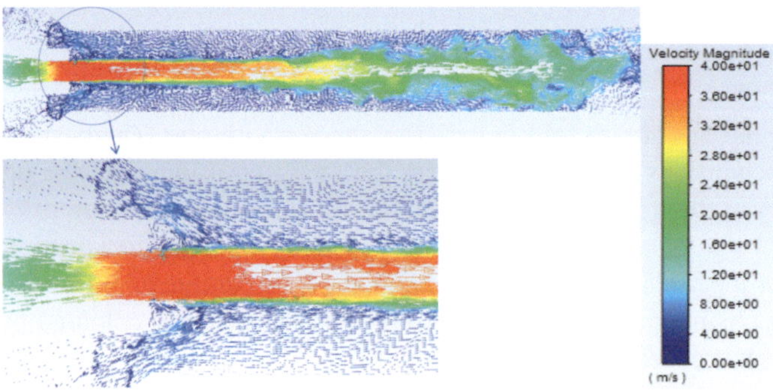

Fig. 12 Velocity vector plot in the symmetry plane of the mixing chamber obtained from ELES simulation

5 Summary and Conclusions

Numerical simulation of a miniature jet pump was studied using an advance numerical technique Embedded Large Eddy Simulation (ELES). It is observed through detailed flow visualization that the developing region of the mixing chamber has different flow regimes: uniform high-velocity (potential core) regime, uniform low velocity regime and high shear (mixing) regime. The ELES simulation was carried out where mixing chamber and diffuser were resolved through LES and other zones of the computational domain were modelled through Reynolds Averaged Navier–Stokes equation with standard k–ε turbulent model (RANS-SKE) technique. The procedure of numerical setup for ELES was described, and the grid adequacy was checked. For comparison, numerical simulation was also carried out using RANS-SKE technique in whole computational domain. The gross results in terms of the pressure ratio and efficiency obtained from both the numerical techniques are in agreement with the experimental results. However, velocity, pressure and turbulent kinetic energy distribution inside the mixing chamber considerably differ between the two simulation techniques. The potential core length of the jet coming out from nozzle is observed to be more in the developing region for ELES than that obtained by RANS-SKE technique. High velocity of primary fluid jet, nozzle thickness and mixing chamber wall shear have considerable effects on the various regimes formed inside the mixing chamber of the miniature jet pump. Further through ELES simulation results, iso-surface of Q-criterion helped in visualization of turbulent coherent structure in the mixing region of mixing chamber.

Nomenclature

d	Primary nozzle tip diameter (mm)
D	Mixing chamber diameter (mm)
G	Filtering operator for Navier–Stokes equations (1/m^3)
k	Turbulent kinetic energy (m^2/s^2)
l	Characteristics dimension (for the current case it is diameter) (m)
L	Mixing chamber length (mm)
l_0	Integral length scale (m)
M	Flow ratio/amplification = Q_2/Q_1 (–)
N	Pressure ratio = $(P_5 - P_2)/(P_1 - P_5)$ (–)
p	Instantaneous pressure (Pa)
P	Average pressure (bar, Pa)
p'	Fluctuating pressure (Pa)
$\overline{p}$	Filtered value pressure (Pa)
Q	Volumetric flow rate (lpm)
QC	Q-criterion (1/s^2)
Re	Reynolds no. = $\rho Vl/\mu$ (–)

S	Nozzle spacing (distance from nozzle to mixing chamber entry) (mm)
S	Strain rate (1/s)
t	Time (s)
T	Temperature (°C)
u	Instantaneous local velocity (m/s)
U	Average velocity (m/s)
u'	Fluctuating velocity (m/s)
$\overline{u}$	Filtered value of velocity (m/s)
V	Area average velocity in jet pump (m/s)
x	Cartesian location (m)

Greek Symbol

η	Jet pump efficiency ($M*N$)
ρ	Density (kg/m^3)
ε	Turbulent dissipation rate (m^2/s^3)
μ	Kinematic viscosity (kg/m s)
μ_{sgs}	Subgrid scale turbulent viscosity (kg/m s)
μ_{t}	Turbulent viscosity (kg/m s)
Ω	Vorticity rate (1/s)
τ_{ij}	Reynold stress (Pa)
τ_{sgs}	Reynold stress (Pa)
$\Phi(x')$	Variable (u,p) in computational domain (m/s, Pa)
$\overline{\Phi(x)}$	Filtered variable ($\overline{u}$, $\overline{p}$) in computational domain (m/s, Pa)

Subscripts

1	Primary inlet
2	Secondary inlet (suction)
3	Mixing chamber inlet
4	Mixing chamber outlet
5	Diffuser outlet
D	Diffuser
m	Mixing chamber
n	Nozzle tip
i,j	Directions

References

1. Eng. Sci. Data Unit (ESDU) (1985) Ejectors and jet pumps—design and performance for incompressible liquid flow. Data item: 85032. ESDU International Ltd., London
2. El Hayek MD, Hammoud AH (2006) Prediction of liquid jet pump performance using computational fluid dynamics. In: Proceeding of the 4th WSEAS International Conference on Fluid Mechanics and Aerodynamics, Greece, Aug. 21–23, pp 148–153
3. Meakhil TA, Teaima IR (2013) A study of the effect of nozzle spacing and driving pressure on the water jet pump performance. Int J Eng Sci Innov Technol 2(5):373–382
4. Yapici R, Aldas K (2013) Optimization of jet pump using numerical simulation. J Power Energy 227(4):438–449
5. Holgate J, Skillen A, Craft T, Revell A (2019) A review of embedded large eddy simulation for internal flows. Arch Comput Methods Eng 26:865–882
6. Zaheer Q, Masud J (2018) Comparison of flow field simulation of liquid ejector pump using standard k–ε and embedded LES turbulence modelling techniques. J Appl Fluid Mech 11(2):385–395
7. Mathley, Cokljat FD, Bertoglio J, Sergent E (2003) Specification of LES inlet boundary condition using vortex method. In: 4th International symposium on turbulence, heat and mass transfer, Antalya
8. Celik I, Cehreli Z, Yavuz I (2005) Index of resolution quality for large eddy simulations. J Fluids Eng 51(27):949–958
9. Kotak VK, Sengupta S, Pathrose A, Sugilal G, Bhattacharya S (2021) Theoretical and experimental study on a miniature jet pump with low area ratio. In: 8th ICTACEM conference, IIT Kharagpur
10. Pope AB (1978) An explanation of the turbulent round-jet/plane jet anomaly. Am Inst Aeronaut Astronaut J 16:279–281

Hydrophobicity Analysis of the Surfaces with Rectangular Micropillars Using Pseudopotential Multiphase Lattice Boltzmann Method

Ganesh S. Meshram, Suman Chakraborty, Partha P. Chakrabarti, Gloria Biswal, and Sasidhar Kondaraju

Abstract Hydrophobicity holds significant importance across a multitude of scientific, technological, and practical domains. We study the surface hydrophobicity with rectangular micropillars using a two-dimensional (2D) pseudopotential multiphase lattice Boltzmann method with a D2Q9 model for various solid–fluid interaction parameter ($\mathbf{G_{ads}}$) varies from − 1.25 to − 2.50. We first analyzed the wettability and hydrophobicity of the smooth surface considering various interaction parameters to scrutinize the accuracy of the present numerical model. We then imposed the micropillars on the bottom wall of the surface by varying the geometrical parameters of micropillars to study the behavior of the surfaces. The fluid domain above the pillar height is filled with water and gaps between the pillars remain with air. The hydrophobicity is analyzed with a droplet of water of radius 30 lattice units in the domain of 200 × 200 lattice units. The present study shows that increasing the solid–fluid interaction parameter of the textured surfaces increases hydrophobicity and reduces wettability because fluid–air interaction is much more dominant than solid–fluid interaction resulting in the reduction in surface area between the fluid droplet and solid wall. We also analyze the effect of solid–fluid interaction on the streamline patterns which shows the generation of vortices inside the water droplet. The impact of geometrical parameters on the wettability and their corresponding states was also studied and when we increase the spacing between micropillars increases the hydrophobicity of the surface.

Keywords Contact angle · Wettability · Hydrophobicity · LBM · Wenzel state · Cassie–baxter state

---

G. S. Meshram (✉) · S. Chakraborty · G. Biswal
Department of Mechanical Engineering, Indian Institute of Technology Kharagpur, West Bengal, India
e-mail: ganeshmeshram.iitkgp@gmail.com

P. P. Chakrabarti
Department of Computer Science & Engineering, Indian Institute of Technology Kharagpur, West Bengal, India

S. Kondaraju
School of Mechanical Sciences, Indian Institute of Technology Bhubaneswar, Odisha, India

A. Kumar R. et al. (eds.), *Proceedings of Fluid Mechanics and Fluid Power (FMFP) 2023, Vol. 6*, Lecture Notes in Mechanical Engineering,
https://doi.org/10.1007/978-981-96-9600-0_30

1 Introduction

It has been a decade since the development of superhydrophobic and hydrophobic surfaces captivated the public's attention because of their wide range of applications in Material Science, Chemical Engineering, Biotechnology, Microfluidics–Nanofluidics, and Nanotechnology for self-cleaning, anti-fouling, drag reduction, anti-icing, oil–water separation, and corrosion [1, 2]. Hydrophobic surfaces can be created by reducing surface energy and changing the hierarchical microstructure. There are various methods for building hierarchical microstructures, including etching, laser ablation, sputtering, spinning, sol-gel, and chemical vapor deposition [3]. In recent decades, the intricate dynamics of wetting on surfaces exhibiting intricate textures have attracted considerable attention from both experimental and theoretical perspectives. Wettability, an intrinsic characteristic, serves as a quantitative measure illuminating the inherent inclination of a liquid to engage in interfacial interactions with diverse liquid and/or solid substrates. This fundamental property assumes paramount importance as it unveils the profound intricacies governing the wetting phenomena during the interplay of solid and liquid phases, thus affording invaluable elucidations into the intricate surface attributes of polymers.

Consequently, the profound understanding of wettability imparts a versatile and indispensable platform for an array of pivotal applications, most notably in the expansive realms of printing and liquid coating. The bedrock for assessing the efficacious nature of wettability lies in the meticulous evaluation of the contact angle, a cardinal metric meticulously employed to discern the tenacity of wetting. A contact angle of diminutive proportions manifests a heightened wetting prowess, signifying the liquid's pronounced predilection to disperse uniformly and conform to the contours of the surface. Conversely, an elevated contact angle signifies a diminishment in wetting potential, underscoring the liquid's inclination to evince a reduced proclivity for interactive engagement with the solid substrate, thereby conferring valuable insights into intermolecular forces and surface energetics.

The fundamental equation for the determination of the contact angle of a liquid droplet on a flat surface was furnished by Young [4]. Later this equation has been modified to calculate the contact angle on patterned or rough surfaces by Wenzel [4] and Cassie–Baxter equations [5]. Typically, liquid substances tend to manifest themselves in either the Cassie–Baxter wetting state or the Wenzel wetting state. In Wenzel's state, a liquid droplet completely wets a rough or textured surface and this phenomenon occurs when the solid–fluid interaction is much stronger than liquid–air interaction, resulting increase in the apparent contact area between liquid and solid due to surface roughness [6]. Within the Cassie–Baxter state, minute air pockets are entrapped within the grooves formed by surface features, thereby giving rise to a composite hydrophobic surface comprising solid and air components and the corresponding contact angles $\mathbf{cos}\theta_{\mathbf{C-B}}$ and $\mathbf{cos}\theta_{\mathbf{W}}$ can be calculated as follows [7].

$$\mathbf{cos}\theta = \left(\frac{\gamma_{\mathbf{SG}} - \gamma_{\mathbf{SL}}}{\gamma_{\mathbf{LG}}}\right) \tag{1}$$

The Wenzel's contact angle is [8]

$$\mathbf{cos\theta_w = (r_f cos\theta)} \tag{2}$$

where $\theta_\mathbf{w}$ is the Wenzel's contact angle, $\mathbf{r_f}$ is the roughness factor of the solid surface, which is the ratio of actual surface area to the projected area, and θ represents Young's contact angle.

$$\mathbf{cos\theta_{CB} = (\phi cos\theta + (1 - \phi))} \tag{3}$$

where θ_{CB} is Cassie–Baxter contact angle and ϕ is the fraction of solid surface that is in contact with the liquid surface.

In recent years, there has been a substantial amount of research devoted to the comprehensive investigation of the wetting transition between these two states. In a recent study conducted by Papadopoulos et al. [9], employing laser scanning confocal microscopy, two distinct transition mechanisms were observed and characterized on surfaces patterned with pillars. Currently, a Lattice Boltzmann Method (LBM) pseudopotential multiphase flow model has been utilized to predict the wetting behavior of droplets on a wall, particularly in cases of large density ratios. Li et al. [10] successfully predicted various contact angles between a wall and droplets by modifying the model by adjusting the coefficient G_w.

In this study, we present the analysis of the wettability and contact angle for different homogeneous and patterned surfaces using the two-dimensional (2D) single-component pseudopotential multiphase lattice Boltzmann method. For two-dimensional simulations, we considered the D2Q9 model for the lattice arrangement which includes nine discrete velocities. For the determination of pressure in the pseudopotential term, the Peng–Robinson equation of state is used in this model. The rest of the paper is organized as follows. In Sect. 2, we introduce the basic governing equations of the lattice Boltzmann method and pseudopotential multiphase model. Then, we present the model validation in Sect. 4, and we then show simulation results, particularly the wetting transitions on the smooth and textured surfaces by changing the geometrical parameters of the pillars. Finally, in the last section, the conclusions from the study are made.

2 Numerical Model-Lattice Boltzmann Method

The lattice Boltzmann equation with single-relaxation-time collision operator can be expressed as [11].

$$\mathbf{f_p(x + e_p\delta t) = f_p(x, t) + \frac{\delta t}{\tau}\left[f_p(x, t) - f_p^{eq}(x, t)\right]} \tag{4}$$

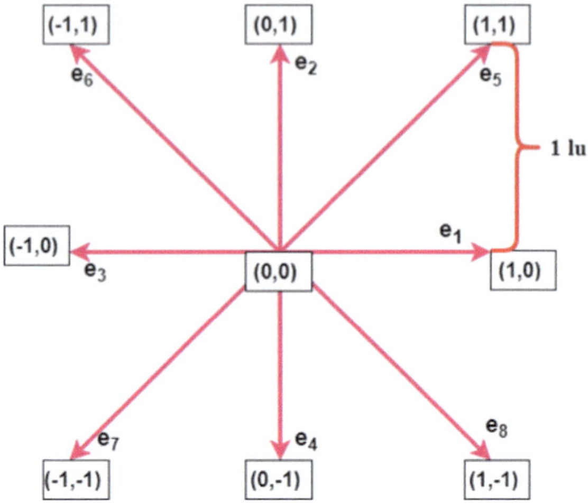

Fig. 1 2D D2Q9 model and their velocity distribution [12]

$$\mathbf{p = 0, 1, 2, 3, \ldots.., n}$$

$$\mathbf{f_p^{eq}(x, t)} = \rho\omega_{\mathbf{p}}\left[\mathbf{1} + \frac{\mathbf{e_p \cdot u}}{\mathbf{c_{ls}^2}} + \frac{(\mathbf{e_p \cdot u})^2}{\mathbf{2c_{ls}^2}} - \frac{\mathbf{u}^2}{\mathbf{2c_{ls}^2}}\right] \tag{5}$$

where $\mathbf{f_p}$ is the density distribution function, $\mathbf{f_p^{eq}}$ is the equilibrium density function, $\mathbf{x}$ is the spatial coordinate, $\mathbf{e_p}$ is the particle velocity along the direction $\mathbf{p}$, $\omega_{\mathbf{p}}$ is the weighting coefficient, $\mathbf{c_{ls}} = \mathbf{c}/\sqrt{\mathbf{3}}$ is the sound speed of a lattice, and $\mathbf{c} = \delta\mathbf{x}/\delta\mathbf{t}$ denotes lattice constant, $\delta\mathbf{x}$ denotes lattice space step, and $\mathbf{u}$ is macroscopic velocity [7].

The D2Q9 model is most suitable for two-dimensional problems and its representation is as shown in Fig. 1. Weighting coefficients are defined as follows:

$$w_k = \begin{cases} 4/9, & k = 0 \\ 1/9, & k = 1, 2, 3, 4 \\ 1/36, & k = 5, 6, 7, 8 \end{cases}$$
$$\mathbf{e}_k = \begin{cases} (0, 0), & k = 0 \\ (\pm 1, 0)c, (0, \pm 1)c, & k = 1, 2, 3, 4 \\ (\pm 1, \pm 1)c, & k = 5, 6, 7, 8 \end{cases} \tag{6}$$

2.1 Shan-Chen Multiphase Model

The Shan and Chen pseudopotential multiphase lattice Boltzmann model was conceptualized to simulate the non-local interaction among particles, as referred in [7, 12, 13]. In the context of single-component multiphase flow, the exerted interaction force upon particles situated at a specific position $\mathbf{x}$ is manifested as follows:

$$\mathbf{F_{int}(x, t) = -G\psi(x, t)\Sigma_p\omega_p\psi(x + e_p\delta t, t)e_p} \tag{7}$$

Here, the $\mathbf{G}$ parameter is used to manipulate the strength of the interaction force. Equation for the Pseudopotential term $\psi(\mathbf{x})$ is defined as

$$\psi = \sqrt{\frac{2(\mathbf{P} - \rho\mathbf{c_{ls}^2})}{\mathbf{Gc_{ls}^2}}} \tag{8}$$

where $\mathbf{P}$ is denoted as fluid pressure.
Peng-Robinson EOS:

$$\mathbf{P} = \frac{\rho\mathbf{RT}}{\mathbf{1 - b}\rho} - \frac{\mathbf{a}\rho^2\varepsilon(\mathbf{T})}{\mathbf{1 + 2b}\rho - \mathbf{b}^2\rho^2} \tag{9}$$

$$\varepsilon(\mathbf{T}) = \left[\mathbf{1} + \varepsilon_0\left(\mathbf{1} - \sqrt{\frac{\mathbf{T}}{\mathbf{T_c}}}\right)\right]^2 \tag{10}$$

$$\varepsilon_0 = \left(\mathbf{0.37464 + 1.54226}\omega - \mathbf{0.26992}\omega^2\right)$$

Here, $\mathbf{R = 1, a = 2/49, b = 2/21}$. The solid–fluid interaction force is [14]

$$\mathbf{F_{ads}} = \left[-\mathbf{G_{ads}}\psi(\mathbf{x, t})\Sigma_p\omega_p\mathbf{s(x + e_p\delta t)e_p}\right] \tag{11}$$

where $\mathbf{G_{ads}}$ is the solid–fluid interaction parameter, $\mathbf{s(x + e_p\delta t)}$ is the indicator function which can be written as

$$\mathbf{s\,(x + e_k\delta t)} = \begin{cases} 0 & \text{if } (\mathbf{x} + \mathbf{e}_k\delta t) \text{ is fluid node} \\ 1 & \text{if } (\mathbf{x} + \mathbf{e}_k\delta t) \text{ is solid node} \end{cases} \tag{12}$$

Total force acting at each particle x

$$\mathbf{F = F_{int} + F_{ads}} \tag{13}$$

$$\Delta\mathbf{f_k(x, t) = f_k^{eq}(}\rho\mathbf{(x, t), u + \Delta u) - f_k^{eq}(}\rho\mathbf{(x, t), u)} \tag{14}$$

The change in velocity due to total force is written as

$$\Delta \mathbf{u} = \frac{\mathbf{F}\delta \mathbf{t}}{\rho} \tag{15}$$

and the actual fluid velocity is as follows:

$$\mathbf{U} = \mathbf{u} + \frac{\mathbf{F}\delta \mathbf{t}}{\mathbf{2}\rho} \tag{16}$$

3 Problem Definition and Boundary Condition

The geometrical description of the problem is shown in Fig. 2. We considered a computational domain (Top) of 200 × 200 [LU] and imposed rectangular micropillars at the bottom wall to analyze the hydrophobicity of the wall, then the bounce-back (no-slip) boundary condition is imposed on the upper wall, micropillars, and lower wall, while periodic boundary condition is taken along the left and right side of the wall. The droplet of radius 25 [LU] is placed in the middle of the bottom wall. We then varied the solid–fluid interaction parameter and geometrical parameters of the micropillars to analyze their impact on the water droplet.

3.1 Grid Independence Test

To achieve accurate results, the mesh independence test is a pivotal consideration in computational fluid dynamics (CFD) modeling. The contact angle was quantified using six different grid sizes (180 × 180, 190 × 190, 200 × 200, and 220 × 220) under identical input conditions, namely a solid–fluid interaction parameter of − 1.50 and a height of triangular micropillars is taken as 50 μm to assess grid independence. To minimize the computational time, we have chosen the grid size of 200 × 200 to expedite further calculations.

3.2 Validation

A code is developed using C Programming for the simulation of water droplets and it is validated with the previously published research article by Pawar et al. [13], where the wettability of the smooth surfaces was analyzed with various solid–fluid interaction parameters. Figure. 3 shows the validation of the present model where the accuracy of the model was found to be 98.5% and hence it is used for further simulations.

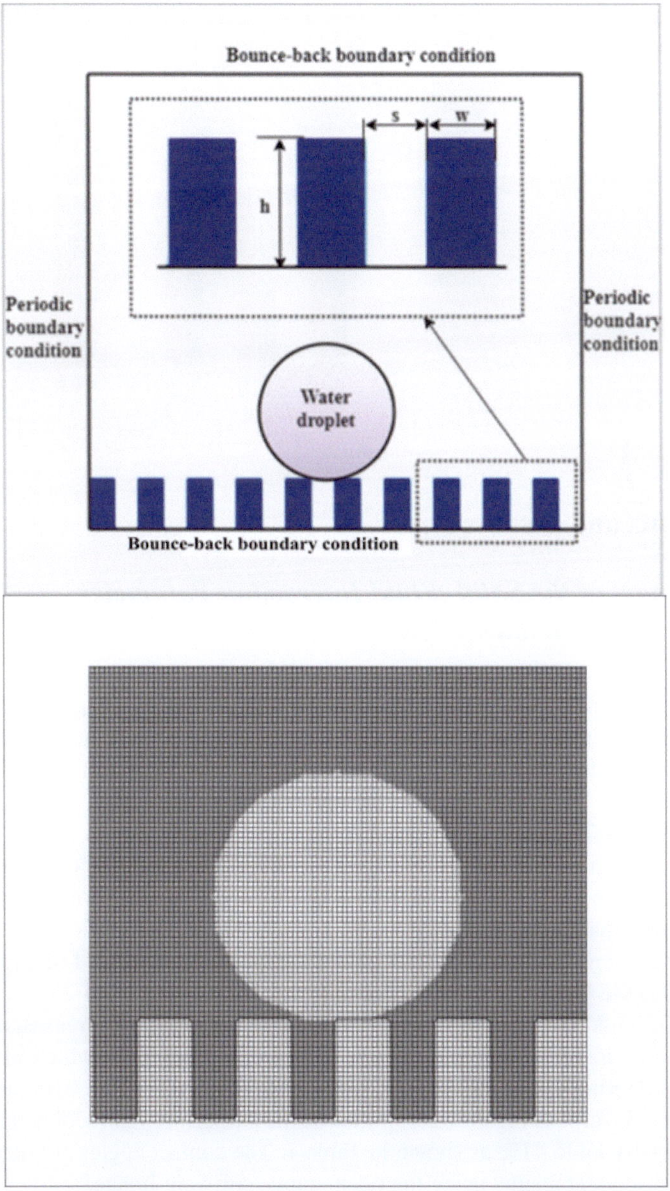

Fig. 2 Problem definition along with BCs (top) and a droplet of radius 25 lu at the center of the bottom wall and mesh (bottom)

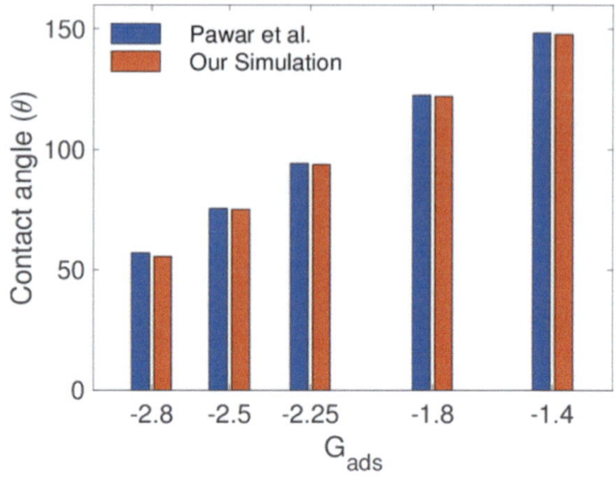

Fig. 3 Validation of the present numerical model

4 Results and Discussion

4.1 Impact of the Solid–Fluid Interaction Parameter on Static Contact Angle

In this section, the impact of solid–fluid interaction parameter ($\mathbf{G_{ads}}$) on the droplets with smooth and textured surfaces is investigated and the contact angle is calculated. Figure 4 shows the effect of $\mathbf{G_{ads}}$ on the smooth surface where the contact angles are measured as 72°, 91°, 103°, 118°, and 138° for the values of $\mathbf{G_{ads}}$ − 2.50, 2.25, 2.00, − 1.75, 1.50, and − 1.25, respectively, as shown in Fig. 7. Then the micropillars with a pillar height of 30 lattice units and a spacing of 5 lattice units (bottom row) are employed on the flat surface to study the effect of $\mathbf{G_{ads}}$ and it has been analyzed that by incorporating the micropillars there is a significant increment in the hydrophobicity. The static contact angles are measured as 104°, 114°, 118°, 134°, and 146° for the $\mathbf{G_{ads}}$ of − 2.50, 2.25, 2.00, − 1.75, 1.50, and − 1.25, respectively. As the solid–fluid interaction parameter decreases, wettability decreases, due to the diminished interaction strength between the solid and the fluid, which hinders the fluid from adhering or wetting the solid surface, making it more cohesive and resistant to spreading. Incorporating the micropillars on the flat surfaces can achieve significant hydrophobicity up to 41% as shown in Table 1. The contact angles are measured in this and subsequent results using the open-source software ImageJ.

Table 1 Physical properties used for the simulation

Property	Value (in LU)
ρ (Density of the liquid)	5.90807
ρ_c (Critical density)	2.54193
ρ_v (Density of vapor)	0.58011
T (Wall temperature)	0.9Tc
T_c (Critical temperature)	0.072922
ν (Lattice viscosity)	1/6

4.2 Effect of the Height and Spacing Between the Pillars on the Surface Wettability

To investigate the effect of spacing between the micropillars on the surface wettability, we have taken three cases where the spacings are taken as $s = 5$, 10, 15, and 20 [lu] with $\mathbf{G_{ads}} = -1.75$ as shown in Fig. 5. Figure 5 shows the temporal evolution of the water droplet on the textured substrate and it has been noticed that the value of contact angle increases with respect to time. The results illustrate the considerable difference between the shape of the droplets on the textured surfaces. For the value $\mathbf{G_{ads}} = -1.75$ with the textures spacing of 5, 10, 15, 20 [lu], the values of mean static contact angles are found approximately to be 115°, 128°, 134°, and 140°, respectively. When the spacing between two pillars is less then the water droplet occurs in Cassie state, i.e., the water droplet completely stays on the top of the micropillars and surfaces behave as hydrophobic in nature. Subsequently, the increasing spacing will change the droplet state from Cassie to Wenzel as shown in Fig. 5. When increasing the spacing from $\mathbf{s} = \mathbf{5}$[lu] to $\mathbf{s} = \mathbf{10}$[lu], the measurement shows that the contact angle decreases by 10.9%. The droplet starts penetrating the micropillars of the textured surface and completely gets penetrated for the spacing of 20 [lu]. The height of the micropillars also seems to be sensitive to the surface wettability and hydrophobicity so the effect of pillar height on the contact angle is studied for the same $\mathbf{G_{ads}}$ value and the temporal evolution of the water droplet is shown in Fig. 6. The height of micropillars is also considered as $\mathbf{h} = \mathbf{5}, \mathbf{10}, \mathbf{15}, \mathbf{20}$[lu], and the corresponding mean static contact angles are found to be 116°, 120°, 132°, and 135°, respectively. It shows that the surface wettability decreases or hydrophobicity increases when the heights of the pillars increase and the water droplet is mostly occurring in Cassie and Cassie–Baxter states. The maximum percentage increase in static contact angle is measured up to 10.242% (Fig. 7).

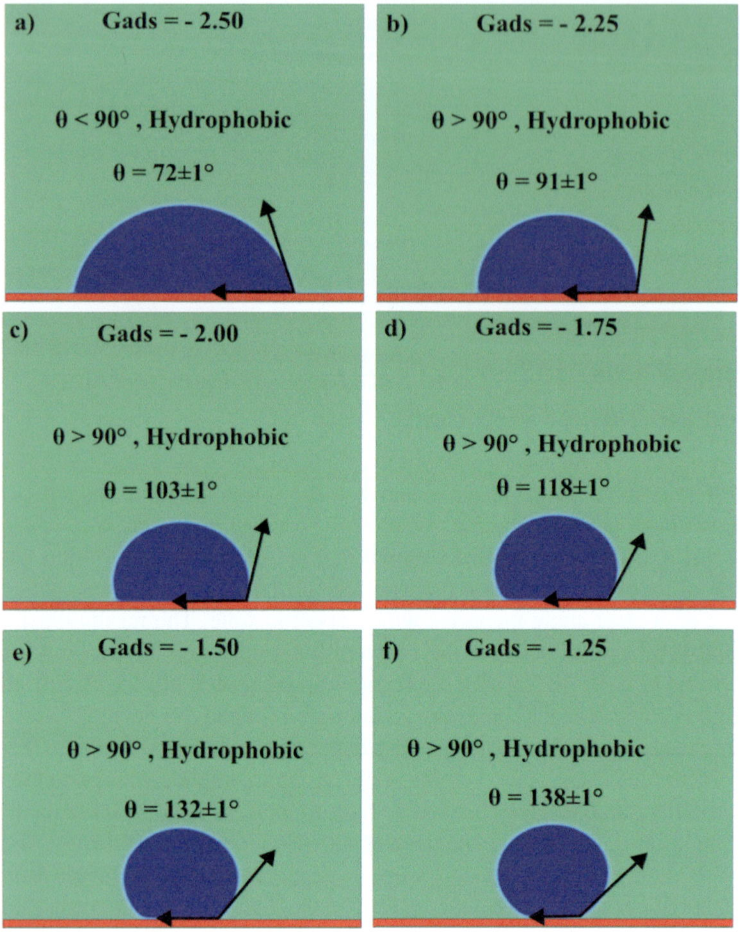

Fig. 4 Comparison of static contact angle on the smooth surfaces with pillar height of 20 lattice units and spacing of 5 lattice units The solid–fluid interaction parameters ($\mathbf{G_{ads}}$) are taken as − 2.00, − 1.75, − 1.50, and − 1.25. The contact angles are measured in this and subsequent results using the open-source software ImageJ

4.3 Comparison of Wettability

The comparison of static contact angle for both smooth and textured surfaces is shown in Fig. 8. Incorporating the micropillars on smooth surfaces can achieve significant hydrophobicity up to 41%. This shows that the surface wettability decreases in the case of textured surfaces as compared to smooth surfaces. Hence, imposing micropillars on smooth surfaces could be more advantageous to fabricate hydrophobic surfaces.

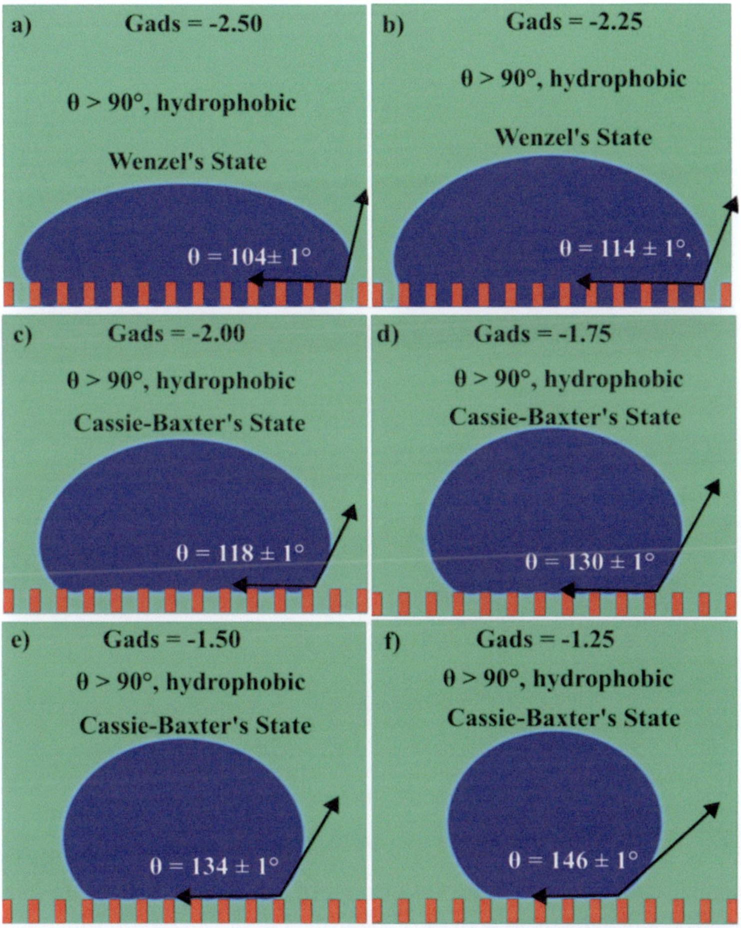

Fig. 5 Comparison of static contact angle on the textured surfaces with pillar height of 20 lattice units and spacing of 5 lattice units. The solid–fluid interaction parameters ($\mathbf{G_{ads}}$) are taken as − 2.00, − 1.75, − 1.50, and − 1.25

5 Conclusions

In this study, we have carried out numerical investigation of hydrophobicity for the pillared surfaces using two-dimensional (2D) pseudo potential single-component multiphase lattice Boltzmann method with D2Q9 model by varying the solid–fluid interaction parameter ($\mathbf{G_{ads}}$) from − 2.50 to − 1.40. The hydrophobicity using the single water droplet placed on the smooth and textured surfaces is investigated for the thickness and depth of the micropillars. The main outcomes and conclusions are summarized as follows:

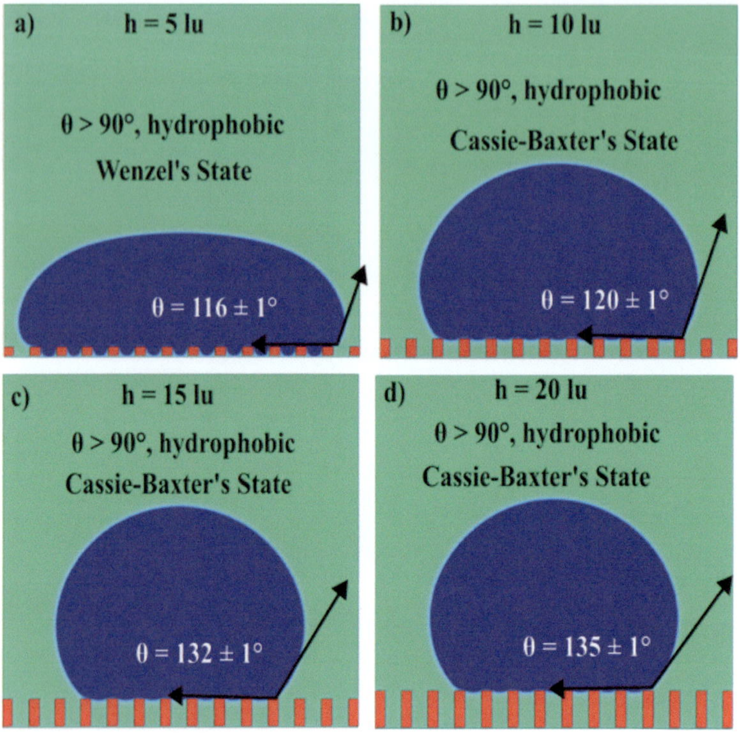

Fig. 6 Effect of height between the micropillars on the static contact angle for solid–fluid interaction parameter ($\mathbf{G_{ads}}$) = $-\mathbf{1.75}$. The heights of the pillar are taken as 5, 10, 15, 20 [lu]

- Simulation results for the static contact angle of both smooth and textured surfaces are in excellent agreement with those published in the literature, which indicates a good degree of accuracy and consistency in the 2D pseudopotential multiphase LBM presented here.
- The contact angle of the textured surfaces compared with smooth surfaces is large and hence the surface wettability decreases which results in an increase in hydrophobicity. The contact angle is increased up to 41.15%.
- Due to the low magnitude of interaction strength, the formation of vortices inside the water droplet is higher.
- At the smaller spacing, the hydrophobicity occurs more as compared to the larger spacing, and the water droplets change from Cassie State to the Wenzel state and the hydrophobicity decreases up to 10.242% which results in an increase in base diameter of the water droplet.
- Due to the larger heights of the micropillars, the surface wettability decreases, and hence this results in increased hydrophobicity.

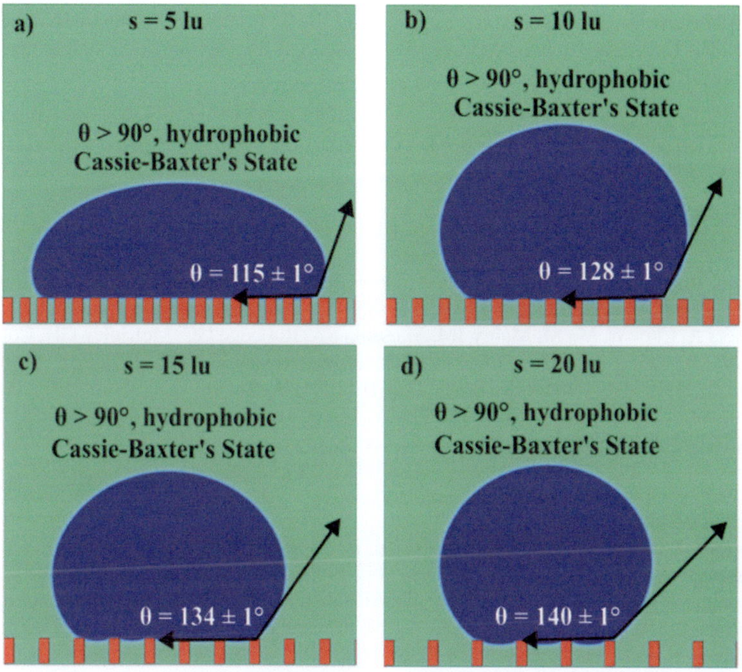

Fig. 7 Effect of spacing between the micropillars on the static contact angle for solid–fluid interaction parameter ($\mathbf{G_{ads}}$) = $-\mathbf{1.75}$. Spacings are taken as 5, 10, 15, 20 [lu]

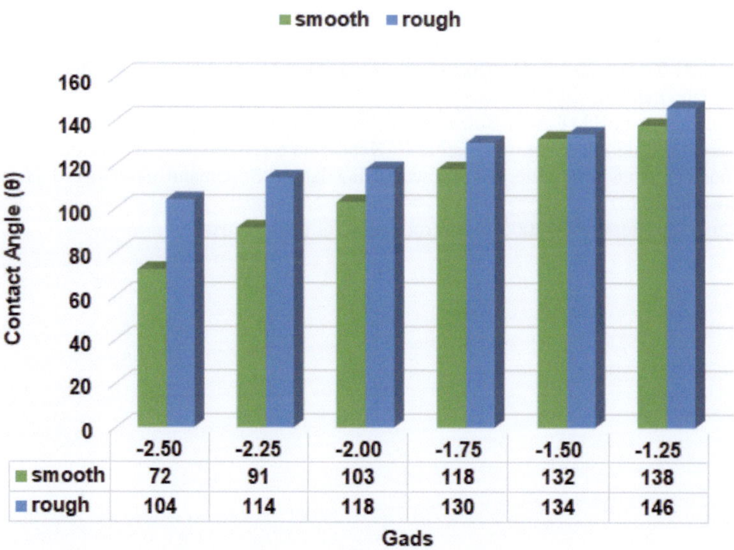

Fig. 8 Comparison of static contact angle for both smooth and textured surfaces

Acknowledgements The authors are very thankful to the Computational Fluid Dynamics and Microfluidics Laboratory in the Department of Mechanical Engineering at the Indian Institute of Technology, Kharagpur.

References

1. Quéré D, Reyssat M (2008) Non-adhesive lotus and other hydrophobic materials. Philos Trans R Soc Math Phys Eng Sci 366(1870):1539–1556
2. Tuteja A, Choi W, Ma M, Mabry JM, Mazzella SA, Rutledge GC, McKinley GH, Cohen RE (2007) Designing superoleophobic surfaces. Science 318(5856):1618–1622
3. Cao L, Jones AK, Sikka VK, Wu J, Gao D (2009) Anti-icing superhydrophobic coatings. Langmuir 25(21):12444–12448
4. Wenzel RN (1936) Resistance of solid surfaces to wetting by water. Ind Eng Chem 28(8):988–994
5. Cassie ABD, Baxter SJToTFS (1944) Wettability of porous surfaces. Trans Faraday Soc 40:546–551
6. Berthier J, Brakke KA (2012) The physics of microdroplets
7. Meshram G, Kondaraju S (2021) Numerical investigation of wettability and its effects on flow through textured micro-channels using lattice Boltzmann method. In: Proceedings of the 26th national and 4th international ISHMT-ASTFE heat and mass transfer conference, IIT Madras, Chennai-600036, Tamil Nadu, India, Begel House Inc
8. Berthier J, Brakke KA, Berthier E (2016) Open microfluidics
9. Papadopoulos P, Lena Mammen X, Deng DV, Butt H-J (2013) How superhydrophobicity breaks down. Proc Natl Acad Sci 110(9):3254–3258
10. Lee C, Nam Y, Lastakowski H, Hur JI, Shin S, Biance A-L, Pirat C, Ybert C et al (2015) Two types of cassie-to-wenzel wetting transitions on superhydrophobic surfaces during drop impact. Soft Matter 11(23):4592–4599
11. Krüger T, Kusumaatmaja H, Kuzmin A, Shardt O, Silva G, Viggen EM (2017) The lattice boltzmann method, vol 10, Springer
12. Mohamad AA (2011) Lattice Boltzmann method, vol 70, Springer
13. Pawar ND, Kale SR, Bahga SS, Farhat H, Kondaraju S (2019) Study of microdroplet growth on homogeneous and patterned surfaces using lattice Boltzmann modeling. J Heat Trans 141(6):062406
14. Sukop MC, Thorne DT (2006) An introduction for geoscientists

Experimental and Numerical Analysis of Plate Heat Exchanger by Optimizing the Cooling Effect for Plates

Satyam Lomate, Pratik Bendre, Gargi Pathare, Akshata Salunke, Yogesh Bhalerao, and Pramod P. Kothmire

Abstract The thermal efficiency and heat transfer rate of a plate heat exchanger (PHE) are primarily influenced by the efficiency of the plates and the heat transfer occurring between them. A higher level of heat transfer between the plates leads to a greater efficiency of the PHE. In this particular study, an analysis of the plates of a plate heat exchanger is conducted using the ANSYS software. The objective is to enhance the heat transfer rate and overall efficiency of the PHE by increasing the length of the side surface of the plate. This elongated surface, located at the boundary of the plate, reduces the plate's temperature and subsequently increases the heat transfer rate and thermal efficiency. Various research studies have been conducted on plate heat exchangers to improve their thermal efficiency and heat transfer rate, including modifications in design, changes in plate geometry, and alterations in plate shape.

1 Introduction

A plate heat exchanger is a type of heat exchanger that utilizes metal plates to facilitate the transfer of heat between two fluids. Unlike conventional heat exchangers, plate heat exchangers expose the fluids to a larger surface area by spreading them out over the plates. This results in a faster temperature exchange and enhances the heat transfer process. These days, plate heat exchangers, particularly compact brazed versions, are becoming increasingly popular and are widely used in millions of combination boilers. A sealed enclosure that consists of pipes and various pathways for fluids is used to circulate liquid using the operating principle of a plate heat exchanger. This allows the liquid to flow through one side while remaining isolated from the other side. Compared to a boiler, heat is transferred more effectively due to

S. Lomate · P. Bendre · G. Pathare · A. Salunke · Y. Bhalerao · P. P. Kothmire (✉)
School of Mechanical Engineering, MIT Academy of Engineering, Alandi, Pune, India
e-mail: ppkothmire@mitaoe.ac.in

Y. Bhalerao
School of Engineering, Faculty of Science, University of East Anglia, Norwich, UK

A. Kumar R. et al. (eds.), *Proceedings of Fluid Mechanics and Fluid Power (FMFP) 2023, Vol. 6*, Lecture Notes in Mechanical Engineering,
https://doi.org/10.1007/978-981-96-9600-0_31

the extremely thin layer of fluid that passes through each plate. The troughs present in each plate further divide the fluids into smaller channels, thereby enhancing the turbulent flow. Experimentally and numerically, a brazed plate's heat exchanger is studied. The intricate plate geometry is closely related to the distinctions between simulations and experiments: modeling with respect to the modified geometry; original geometry modeling; and original geometry modeling with respect to brazing [1]. In this study, we assess the thermal and hydraulic characteristics of three distinct plate geometries in chevron-type Gasketed Plate Heat Exhaust Heat Exchangers. To ensure consistent Reynolds numbers (ranging from 300 to 5000) and Prandtl numbers across all plates with a 30-degree chevron angle, a comprehensive experimental design is employed. The experiments encompass a wide range of plate numbers, flow rates, and input/output temperature values. Based on the experimental data, we derive correlations specific to each plate geometry for the friction factor and Nusselt Number [2]. A three-dimensional numerical model is developed and validated using a finite volume approach. The model focuses on laminar flows and turbulences, specifically with Reynolds numbers ranging from 400 to 10,000. To analyze the turbulences in the heat exchanger, the 2-equation (k-SST) model is employed. The results indicate that pressures per unit length increase as the Reynolds number and corrugation angle of the heat exchanger designs increase. The numerical analysis determines that the heat exchanger operates under single-phase conditions. Cross Cross-corrugated plate heat exchangers (PHEs) are utilized, with different corrugation angles ranging from 30 to 80 degrees. It is observed that the cross-corrugated plate heat exchangers with a corrugation angle of 80 degrees have a friction factor two-times higher than that of a flat PHE [3]. This paper presents a professional approach to enhancing the heat exchange efficiency of plate heat exchangers. The algorithm utilized in this study is based on a screening approach. By applying this approach, the algorithm identifies the optimal configurations for each type of plate, while adhering to specific restrictions. These optimal configurations have the least amount of heat exchange area and therefore the least number of plates required. Furthermore, all the identified combinations exhibit local features that closely resemble the ideal conditions. Additionally, the algorithm identifies the relevant local maxima for each type of plate [4]. For the investigations of the study Computational Fluid Dynamics (CFD) gives accurate results and simulations that are much lesser cost than an experimental setup. This helps to describe temperature distribution and fluid flow through heat plate exchanger. There is a wealth of study on the innovations in plate heat exchanger. For applications of this heat exchanger, many alternatives are chosen and a large amount of research has been done based on their use. This section explains the various research in heat exchangers for solving problems [5]. The following text highlights the scientific analysis of nano-fluid in a plate heat exchanger (PHE), focusing on its experimental characterization, heat transmission, and exergetic performance. In this study, researchers utilized a plate heat exchanger to examine the heat transfer rate between two different types of nanofluids [6]. This paper aims to discuss the research and development endeavors conducted by researchers in the area of corrugated plate heat exchangers. These heat exchangers find extensive usage across multiple engineering sectors and applications. Notably, research reactors heavily depend on corrugated plate heat exchangers due to their advantageous features, including effortless assembly/disassembly and convenient maintenance [7]. This study investigates the existing body of literature regarding the influence of chevron corrugation on the efficiency of plate heat exchangers that possess intricate geometries. Additionally, it explores different methods for enhancing heat transfer in plate heat exchangers, focusing on passive surface techniques and the utilization of nanofluids.

2 Literature Review and Objective

This section explains the various research in heat exchangers for solving problems. Hrnjak et al. [8] this study examines the literature concerning the impact of chevron corrugation on the effectiveness of plate heat exchangers with complex designs. Moreover, it explores various approaches to improving heat transfer in plate heat exchangers, with an emphasis on passive surface techniques and the utilization of nanofluids. Galeazzo et al. [9] in this study, a Computational Fluid Dynamics (CFDs) method was employed to develop a prototype of a plate heat exchanger with four channels and flat plates. Both parallel and series flow configurations were investigated. The results of the experiment were then compared to the heat load predictions obtained from both the 3D CFD model and the 1D plug flow model. Ajeeb et al. [10] nanofluids containing Al_2O_3 were employed in this study, with different low Al_2O_3 concentrations. The base fluids used were distilled water (DW) and a mixture of DW with ethylene glycol. The aim was to investigate the effects of varying the flow rates of low Al_2O_3 concentrations on heat conductivity. The findings demonstrated a noteworthy enhancement in heat conductivity. Zhang et al. [11] experimental research has been carried out on the flow boiling heat transfer and pressure drop in a plate heat exchanger. The primary aim of this research is to establish prediction methods for the heat transfer coefficient and frictional pressure drop. Several fluids were examined during testing, and it was noted that propane and isobutane demonstrate superior heat transfer coefficients compared to other working fluids. Conversely, R236fa exhibited the lowest heat transfer coefficient Wang et al. [12]. The following text describes a professional study on the condensation of R1234ze in a micro-structured plate heat exchanger. The study focuses on investigating the two-phase heat transfer process and presents experimental results of the quasi-local heat transfer coefficient using mixed plates and a hydraulic diameter of 5 mm with R134a. The study examines the effects of mass flux and saturation pressure, and the experimental findings indicate that condensation in the micro-structured plate heat exchanger is primarily controlled by shear forces Saranmanduh et al. [13]. In this study, we designed a heat exchanger with a cross-flow plate and compared it to one with a counter-flow plate in order to improve the cooling performance of an air conditioning cabinet system. The results showed that the system with a cross-flow plate showed better cooling performance and less thermal resistance than the system with a counter-flow plate, especially for large-width air conditioning cabinet cooling systems. We also conducted an experiment using different geometries of heat exchangers to test the performance of the system against IEC standards Stefano De Antonellis et al. [14]. In this study, an experimental investigation was conducted on the construction of Ion Exchange Columns (IECs) using heat exchangers with different geometries. The results indicate that the shape of the plates significantly influences both the surface wettability and the heat transmission rate. It was observed that a reticular plate protrusion can serve as a suitable compromise to enhance the overall effectiveness of the system Young-Jin Baik [15]. The laboratory-scale trials in this study involved evaluating the prototype plate heat exchangers (PHEs). Based on the experimental results, it was found that the double-wave PHE exhibited approximately 50. The above literature survey shows that there is quite fewer research on the extension of the surface of plates in plate heat exchanger which increases its thermal efficiency and rate of heat transfer.

2.1 Literature Gaps

1. Limited Understanding of Plate Extension Effects: There is a lack of comprehensive research on how extending the plates slightly on both sides of a plate heat exchanger affects its thermal efficiency.
2. Comparison between normal plate and extended plate: There is a gap on plate heat exchanger knowledge based on the comparison of normal plate and extended plate in plate heat exchanger. This gap leaves the designers with questions that which plate will be more efficient.
3. Shape and locations of corrugation on surface of extended plates have confined discussions.
4. Optimal Plate Extension Length: There is a gap in knowledge regarding the determination of the optimal length by which plates should be extended on both sides to achieve maximum thermal efficiency. This gap leaves designers without clear guidelines for this parameter.

2.2 Objectives

(1) Investigate the Impact of Plate Extensions: To analyze and quantify the effect of extending plates on both sides of a plate heat exchanger on its thermal performance compared to conventional designs. (2) To compare and find out which type of plate for PHE is better, normal rectangular plate or plate with extension. (3) To compare the extended plates with corrugation on alternate hot plates for different flow rates. (4) Determine Optimal Plate Extension Length: To establish methods for determining the ideal length of plate extensions that maximize thermal efficiency while minimizing potential drawbacks.

2.3 *Scope*

This project intends to contribute to the development of more efficient heat exchanger designs, which can ultimately lead to energy savings and greater sustainability in a variety of industries by investigating plate extension in plate heat exchangers. In addition, practical experiments in the education field get nearby accurate results.

3 Materials and Methods

3.1 *Materials*

The material used in analysis of plate heat exchanger is stainless steel. Stainless steel for a plate heat exchanger offers several advantages due to its unique properties and characteristic such as Corrosion Resistance, Compatibility, Thermal Conductivity, Hygiene and Clean Ability and that makes the stainless steel more useful for many applications of plate heat exchanger.

3.1.1 Geometry

The geometry for heat exchanger comprises of two different models to compare output efficiencies, first the rectangular plate heat exchanger with the effective plate area of 0.018 m2 and thickness of 2 mm outlet pipes of diameter 10 mm. Another model with same dimensions but have extended area along length and the inlet and outlet pipes are now placed in this extended section. Figure 3 shows the design of corrugation on hot plates with analysis model as well as experimental plate with corrugation are shown (Fig. 1).

3.1.2 Mesh Section

Meshing is a technique for breaking up large geometries into smaller pieces, or elements, for both plate heat exchanger geometries the mesh refinement is used. Structured meshes are often applied in regularity and for simpler geometries. For the solid plates, a structured mesh is used with a finer mesh near regions where temperature gradients are expected to be high, such as near the edges where plates are in direct contact. For fluid domains, meshing is applied on both sides of the plates. The boundary layer near the walls is accurately captured by refinement of near-wall which helped capture the velocity and temperature gradients. A finer mesh is employed around regions where flow disturbances, such as bends or obstructions, are present. For accurate heat transfer predictions, the mesh interfaces between solid and fluid domains are aligned and connected properly. Mesh refinement used near

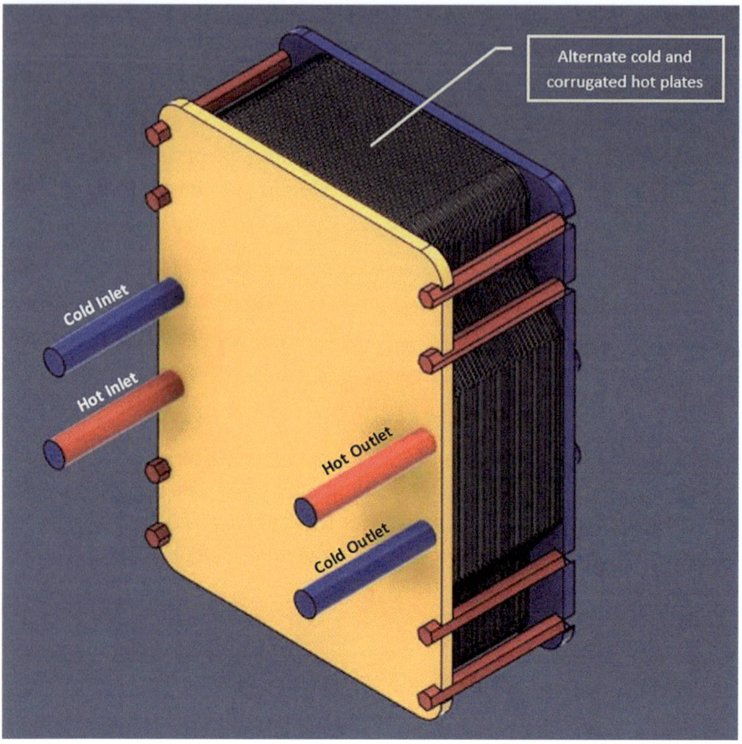

Fig. 1 CAD model

areas of high temperature gradients and where the fluid velocity changes. A dedicated meshing software, ANSYS meshing included with CFD solvers, is used to generate the mesh (Fig. 2).

T1 = Inlet temperature—hot side
T2 = Outlet temperature—hot side
T3 = Inlet temperature—cold side
T4 = Outlet temperature—cold side

3.1.3 Design

Governing Differential Equations

Assuming steady-state heat transfer through the plates and neglecting any heat generation within the plates, Fourier's law is used to trace conduction.

$$q_{\mathrm{cond}} = -kA\mathrm{d}T/\mathrm{d}x$$

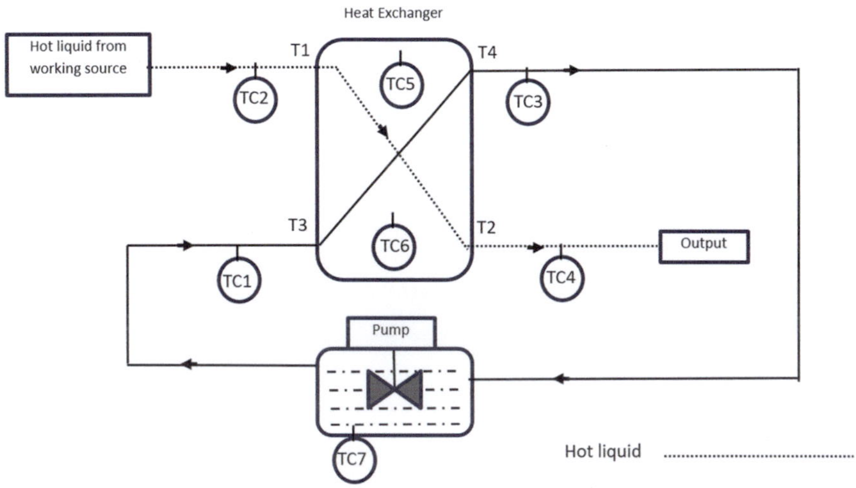

Fig. 2 Layout schematic

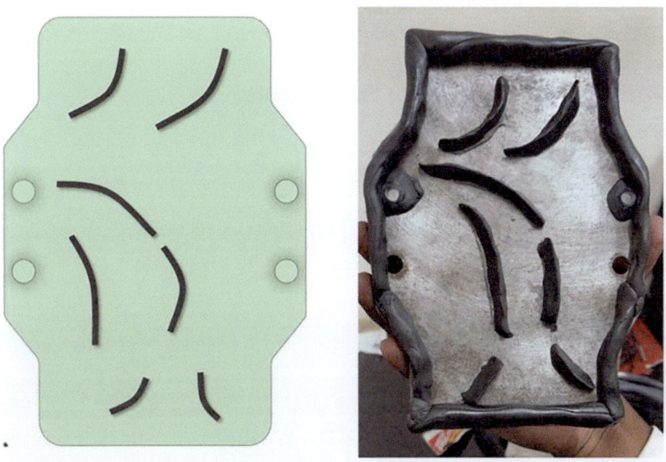

Fig. 3 Corrugation design on hot plates

Convective heat transfer phenomenon is used on both sides of the plates. The heat transfer rate due to convection is described using Newton's law of cooling

$$q_{\text{conv}} = hA\delta T$$

The total heat transfer rate is calculated using the sum of the conduction and convective heat transfer rates,

$$q_{\text{total}} = q_{\text{cond}} + q_{\text{conv}}$$

Conservation of mass and energy equations for liquid flow, the conservation equations for mass and energy describe the flow and heat transfer of the fluids in the heat exchanger and expressed in terms of mass flow rates and temperatures.

For cold fluid at inlet,

$$\text{Mass Conservation: } m_1 = \rho_1 A_1 V_1$$
$$\text{Energy Conservation: } m_1 C_p 1(T_1(\text{inlet}) - T_1) = UA(T_1 - T_2)$$

4 Results and Discussion

This presented analysis shows the comparison between normal plate design plate heat exchanger and plate heat exchanger with extension along length. Ultimately, the heat distribution is more in extended part of plate heat exchanger that increases the heat transfer rate and also increases the efficiency. The surface areas of the extended plate and conventional plate are kept equal. Following are the results shown after the comparison (Figs. 3 and 5).

Figure 4 (a) Rectangular plate (b) Extended plate.

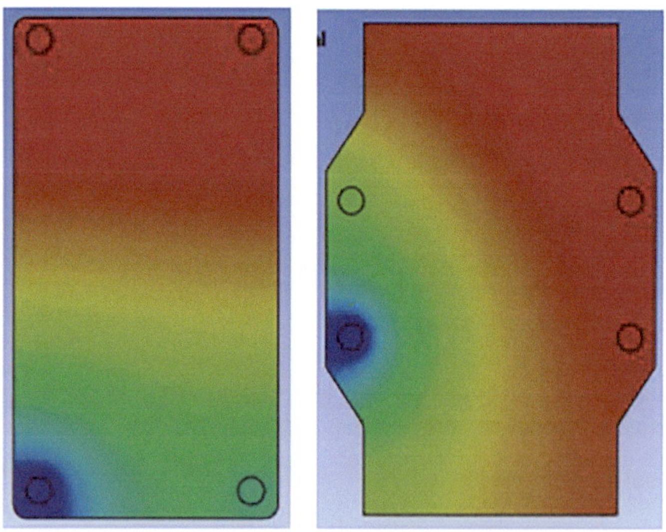

(a) (b)

Fig. 4 Comparison between extended and conventional plate design

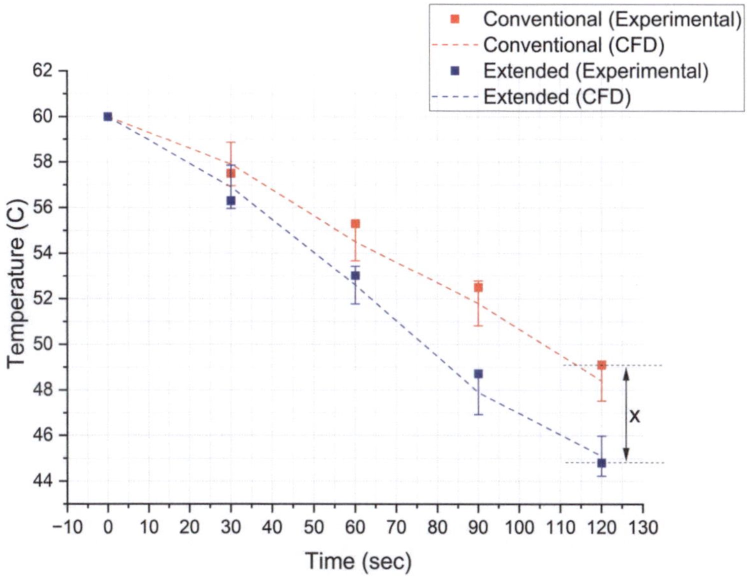

Fig. 5 Comparison between conventional and extended plate design

Figure 4 (a) show the normal rectangular plate with parameters as follows—Cross section area = 0.018 m^2

Cold Inlet Temp = 18 ₒC

Hot Inlet Temp = 90 ₒC.

Max plate surface temp = 203.13 ₒC.

Figure 5 a show the normal rectangular plate with parameters as follows—Cross Sec Area = 0.022 m^2

Hot Inlet Temp = 90 ₒC.

Cold Inlet Temp = 18 ₒC.

Max plate surface temp = 139.12 ₒC.

txt———

Table 1 shows the comparison between rectangular plate heat exchanger and extended plate

Surface heat exchanger

The objective three discusses about the effect of corrugation on alternate hot plates and the various flow rates on the thermal efficiency of PHE. Table 2a, b shows the CFD results. Initially, the extended plate without corrugation and with corrugation for the flow rate of 0.04 kg/s are compared Table 2a. The experimental results along with analysis results obtained are discussed in Table number 2 and the graphs in Fig. 6a, b.

The corrugation on alternate hot plates provides the obstruction and directs the fluid in spiral pathway that increase the pressure inside the hot plate and also decreases the flow rate of hot fluid flowing through plate with respect to cold fluid flowing.

Table 1 Temperature drop comparison between extended and conventional design

S. No.	Time (s)	Temperature oC			
		Conventional (Experimental)	Conventional (CFD)	Extended (Experimental)	Extended (CFD)
1	0	60	60	60	60
2	30	57.5	57.9	56.3	56.9
3	60	55.3	54.5	53.1	52.6
4	90	52.5	51.8	48.72	47.9
5	120	49.1	48.4	44.8	45.1

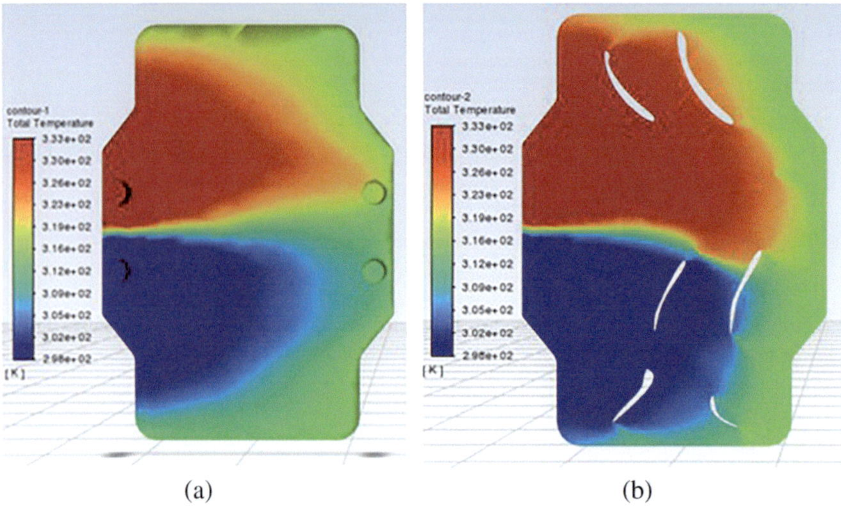

Fig. 6 Extended plates without (**a**) and with (**b**) corrugation

The comparison between flowrates of 0.04, 0.08 and 0.12 kg/s are shown in graph of Fig. 7.

The maximum efficiency, i.e., the maximum cooling rate is found in the condition of maximum flow rate. With flow rate increasing form 0.04 to 0.08 kg/s, the maximum temperature drop form 50 to 47.5 °C with further increase in flowrate to 0.12 kg/s the maximum drop in temperature was observed to be 46 °C, this various flowrate conditions shows that with the increase flowrate the cooling rate increases.

The experimental data generated on extended plates with corrugation shows the similar trend in cooling rate as compared with CFD generated data.

Table 2 Comparison of different flow rates on extended and conventional design

S. No.	Time (s)	Flowrate (kg/s)					
		0.04 (Extended)	0.04 (Conventional)	0.08 (Extended)	0.08 (conventional)	0.12 (Extended)	0.12 (Conventional)
1	0	60	60	60	60	60	60
2	30	57.1	58.21	56.3	57.8	55.4	57.1
3	60	55.4	56.3	53.02	55.3	52.22	54.8
4	90	52.1	53.7	48.72	52.5	47.56	51.83
5	120	48.3	50.4	45.5	49.1	44.7	47.2

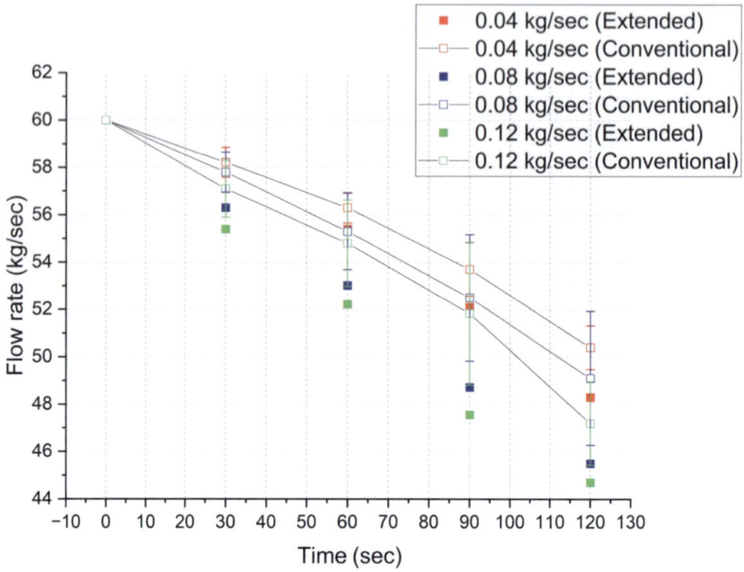

Fig. 7 Different flow rates comparison

5 Conclusions

The current work provides a simple and reasonable scientific method for reaching an appropriate solution of increasing efficiency of the plate heat exchanger. By changing the element, size of the plate can help to reduce the temperature of the plate which significantly increases the efficiency of the overall plate heat exchanger. The comparison between the normal plate heat exchanger that is without extension and with extension can be identified. Plate heat exchanger without extension has more temperature than with extension that shows the temperature of plate with extension reduces and hence ultimately increases the efficiency of the plate heat exchanger. By adding swirl corrugation to alternate hot plates by keeping the surface of cool plates flat the efficiency increases. Plate heat exchanger with swirl corrugation on hot fluid plates and extension has more heat flow than with conventional heat exchangers with same effective plate surface area.

References

1. Jafari A, Sadeghianjahromi A, Wang C (2022) Experimental and numerical investigation of brazed plate heat exchangers—a new approach. Appl Thermal Eng 200
2. Cagin G, Fatih A, Selin A, Nilay SU, Sadik K (2014) Experimental comparison of performances of three different plates for gasketed plate heat exchangers. Int J Therm Sci 75:249–256

3. Saha SK, Khan AH (2020) Numerical study on the effect of corrugation angle on thermal performance of cross corrugated plate heat exchangers. Thermal Sci Eng Prog 20:100711
4. Mota AS, Ravagnani MASS, Carvalho EP (2014) Optimal design of plate heat exchangers. Appl Therm Eng 63(1):33–39
5. Kumar V, Kumar Tiwari A, Ghosh SK (2017) Characterization and performance of nanofluids in plate heat exchanger. In: 5th international conference of materials processing and characterization (ICMPC 2016), 4(2):4070–4078
6. Abou TM, Elmaaty AEK, Mahgoub M (2016) Corrugated plate heat exchanger review. Renew Sustain Energy Rev 70:852–860
7. Zhang J, Zhu X, Mondejar ME, Haglind F (2019) A review of heat transfer enhancement techniques in plate heat exchangers. Renew Sustain Energy Rev 101:305–328
8. Jin S, Hrnjak P (2017) Effect of end plates on heat transfer of plate heat exchanger. Int J Heat Mass Transf 108:740–748
9. Galeazzo F, Miura Y, Gut J, Tadini C (2006) Experimental and numerical heat transfer in a plate heat exchanger. Chem Eng Sci 61(21):7133–7138
10. Ajeeb W, Renato RS, Silva T, Murshed S (2023) Experimental investigation of heat transfer performance of Al_2O_3 nanofluids in a compact plate heat exchanger. Appl Therm Eng 218:119321
11. Zhang J, Haglind F (2021) Experimental analysis of high temperature flow boiling heat transfer and pressure drop in a plate heat exchanger. Appl Therm Eng 196:117269
12. Wang R, Kabelac S (2021) Condensation quasi local heat transfer and frictional pressure drop of R1234ze(E) and R134a in a micro-structured plate heat exchanger. Appl Therm Eng 197:117404
13. Borjigin S, Zhang S, Ma T, Zeng M, Wang Q (2020) Performance enhancement of cabinet cooling system by utilizing cross-flow plate heat exchanger. Energy Convers Manage 213:112854
14. De Antonellis S, Cignatta L, Facchini C, Liberati P (2020) Effect of heat exchanger plates geometry on performance of an indirect evaporative cooling system. Appl Therm Eng 173:115200
15. Kim M, Baik Y-J, Park S-R, Ra H-S, Lim H (2010) Experimental study on corrugated cross-flow air-cooled plate heat exchangers. Exp Thermal Fluid Sci 34(8):1265–1272
16. Camilo Jiménez-García J, Rivera W (2019) Parametric analysis on the experimental performance of an ammonia/water absorption cooling system built with plate heat exchangers. Appl Therm Eng 148:87–95

Numerical Simulation of Electro-coalescence Enhancement via Novel Non-uniform Electric Field Design

Mayur Chaudhari, Tejas Kolte, Ajinkya Jadhav, Mahesh Goudar, and Pramod Kothmire

Abstract This study investigates how different parameters affect the coalescence of water droplets within an oil phase, specifically in a unique concentric semi-elliptic non-uniform electric field configuration. Numerical simulations are used to investigate droplet electro-coalescence systematically. The level set method is applied to investigate binary droplet collisions in the presence of an external electric field. In addition to field geometry, this study explores parameters such as droplet separation distance, boundary design, and initial skew angle. Increased electric field intensity and closer droplet spacing enhance electro-coalescence, according to experimental data that supports electrostatic theories. The study discovered that a larger b/a ratio led to stronger forces and faster coalescence, with a scale factor of 0.5 being ideal. Increasing voltage also decreased the droplet time to attain electric potential. Moreover, alignment accelerates the coalescence process, which is dependent on the skew angle of the electric field. Improving electrode designs for efficient electro-coalescence is one of the research's implications.

M. Chaudhari (✉) · T. Kolte · A. Jadhav · P. Kothmire (✉)
School of Mechanical Engineering, MITAOE, Alandi, Pune, Maharashtra, India
e-mail: mayur.chaudhari@mitaoe.ac.in

P. Kothmire
e-mail: pramodkothmire@iitb.ac.in

T. Kolte
e-mail: tejas.kolte@mitaoe.ac.in

A. Jadhav
e-mail: ajinkya.jadhav@mitaoe.ac.in

M. Goudar
School of Electrical Engineering, MITAOE, Alandi, Pune, Maharashtra, India
e-mail: mdgaudar@mitaoe.ac.in

M. Chaudhari · T. Kolte · A. Jadhav · P. Kothmire
Department of Chemical Engineering, IIT Powai, Mumbai, Maharashtra, India

A. Kumar R. et al. (eds.), *Proceedings of Fluid Mechanics and Fluid Power (FMFP) 2023, Vol. 6*, Lecture Notes in Mechanical Engineering,
https://doi.org/10.1007/978-981-96-9600-0_32

Keywords Electro-coalescence · Skew angle · Non-uniform electric field · Droplet collision

1 Introduction

Despite having a straightforward appearance, droplet coalescence is a significant and useful phenomenon that has received a lot of attention lately. This can be seen in a variety of natural processes, such as the formation of raindrops and clouds, the circulation of atmospheric aerosols, and more [1]. This technique finds applications in various fields, such as microfluidics, fire suppression, and condensation heat transfer. Because of its high coalescence efficiency, the electro-coalescence technique is frequently used to separate water-in-oil emulsions [2].

The efficiency of electro-coalescence is affected by several factors, including electric field strength, oil viscosity, droplet separation distance, and electric field skew angle. Larger electric fields cause droplets to approach and coalesce more quickly. Furthermore, droplets coalesce faster when aligned with the electric field, and the rate of coalescence is influenced by the initial separation distance. For numerical simulations, we use the COMSOL Multiphysics software in this study.

This study investigates the effects of a novel non-uniformly distributed electric field design on droplet coalescence. It considers a number of factors, including the geometry of the field, the initial separation between the droplets, the electric field's strength, viscosity, skew angle, and their effects on the coalescence process.

2 Literature Review

This work examines the behavior of collisions between binary droplets in the presence of an external electric field using numerical simulations. The primary objective is to enhance the design and operation of equipment that employs electric fields to promote droplet coalescence in immiscible liquid phases [2]. In order to describe the coalescence of a water droplet within a dielectric oil phase at the interface with an applied electric field, the Navier–Stokes and charge conservation equations must be solved [3]. The research primarily focuses on understanding the dynamics and kinematics of droplet impacts on superhydrophobic materials. As part of this investigation, the research thoroughly investigates the impact of surface qualities and evaluates the most current scientific discoveries on droplet impact and deformation [4].

The findings of this study indicate the potential for increasing the speed of electro-coalescence. Furthermore, it was discovered that the skew angle of the electric field, oil viscosity, and the initial separation distance of the droplets all had an effect on electro-coalescence. Furthermore, based on participant-level factors, the study shows a link that predicts electro-coalescence kinetics [5]. It is worth noting that, due to the droplet-medium system's properties, the shape of the non-uniform electric field has a substantial influence on coalescence. The study also explores variables such as voltage amplitude, oil viscosity, droplet skew angle, and initial droplet distance. The insights gained from this research contribute to optimizing electrode configurations for more efficient electro-coalescence operations [2].

The study leverages COMSOL Multiphysics, a proprietary software tool, to model the interaction between oil and water using a Level Set methodology. Specific parameters, including grid size, interface thickness, and re-initialization settings, are scrutinized for their influence on the model's sensitivity to input data. The research determines the secondary droplet volume to initial drop volume ratio, correlates it with initial drop size, and compares it to previously published experimental data [3]. According to the results, a non-uniform electric field and the resulting dielectrophoretic effect can hasten the micro-emulsions' phase separation and coalescence [2]. The research considers three different emulsion types: (1) conducting phase in a dielectric medium, (2) leaky dielectric phase in a more conductive phase, and (3) leaky dielectric phase in a less conductive phase. Coalescence in non-uniform electric fields is explored and compared to coalescence in uniform fields [6].

This study investigates the effects of voltage, distance, and frequency in order to evaluate the effectiveness of dissolving water-in-oil emulsions. Numerical results show that water droplets can break between hydrophobic titanium dioxide (TiO_2) coated electrodes spaced 2 mm apart and exposed to an electric field intensity greater than 300 kV/m. Furthermore, the maximum demulsification efficiency of 76% was attained by using an alternating current (AC) voltage of 250 V, a frequency of 200 kHz, and a maintenance of an inter-electrode spacing of 4 mm [6]. This study adds to our understanding of how to improve the efficiency of electrode designs by emphasizing the role of di-electrophoresis with a non-uniform electric field in accelerating coalescence and phase separation in micro-emulsions [2].

3 Objective

This research seeks to address the following key objectives:

1. Investigate the impact of an electric field on droplet electro-coalescence.
2. Assess the effectiveness of external electric field design on binary droplet collision.
3. Examine the effect of the scale factor on droplet electro-coalescence.
4. Investigate how the skew angle affects the electro-coalescence of droplets.

4 Methodology

Start a new 2D model in COMSOL Multiphysics. In the Select Physics tree, select Electrostatics (es) and Laminar Two-Phase Flow, Level Set. Choose the Time Dependent with Phase Initialization study option.

4.1 Geometry

A schematic of the geometry used in the current investigation is shown in Fig. 1a, b. Two concentric semi-ellipses with $a1 = 5$ mm, $b1 = 10$ mm, and $a2 = 10$ mm, $b2 = 20$ mm make up the medium. The radius of the droplet is assumed to be 0.6 mm, with a starting distance of 2 mm.

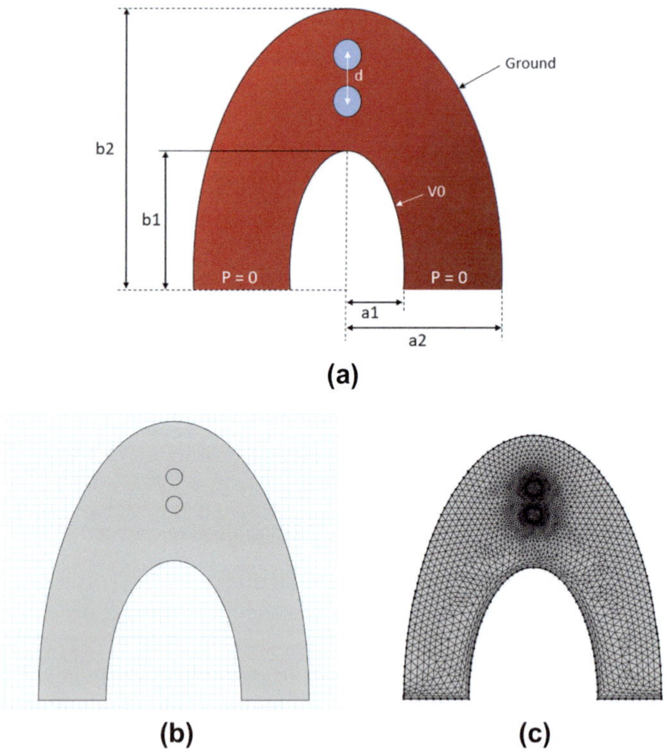

Fig. 1 **a** Rough geometry and electrical boundary conditions in a computational domain. **b** Schematic geometry for $b/a = 2$ and initial droplet distance of 2 mm. **c** Generated physics-controlled fine mesh

4.2 Boundary Condition

Select Initial Values 1 under Electrostatics (es). Enter $y * V0/10$ (mm) in the V text field. On the inner wall, add electric potential $V0$ equal to 1000 V, with the outward curve serving as the ground. Select 1 m as the out-of-plane thickness.

For a two-phase flow level set in surface tension, select the library coefficient for the liquid/liquid interface, and add Olive oil/water. For the section on Material Properties. Click Multiphase Material Add. Add water with a relative permittivity of 80 from the library. In the second phase, fill in the Label text area with Oil and add blank material. Add the following oil properties: permittivity = 2.2, density = 884 kg/m^3, dynamic viscosity = 0.2 Pa-s. Select every domain and apply volume force in Laminar Flow. Assign the droplet a pressure point constraint. Choose the domain material oil as the initial value, fluid 2, under level set (ls). An equal pressure of 0 Pa is applied for the boundary conditions on the designated horizontal lines. Curved lines are additionally regarded as non-slip walls.

4.3 The Governing Equations and Solution Method

The electro-coalescence phenomenon in two-phase flow is mathematically described using momentum, continuity, and electric field equations, with the Navier–Stokes equations solving for velocity and pressure distribution. The following is the expression for the continuity equation of an incompressible fluid:

$$\nabla \cdot u = 0$$

Fluid velocity is represented by u (m/s). Equation of Navier–Stokes is

$$\rho \frac{\partial u}{\partial t} + \rho(u \cdot \nabla)u = -\nabla p + \nabla \cdot \mu\left(\nabla \mathrm{u} + (\nabla u)^T\right) + F$$

The density is represented by ρ (kg/m^3), pressure is indicated by p (Pa), viscosity is indicated by μ (Pa-s), and the total of the external volume forces is indicated by F (N). Surface tension and electric forces combine to create volume forces (F), which are as follows:

$$F = \sigma k \delta n + f_e$$

where f_e (N) is the electric field force, the unit normal vector to the interface points inward toward the droplets; the interface delta function is represented by δ; the curvature of the interface between the two fluids is represented by κ; and the coefficient for surface tension is σ (measured in N/m). This force can be characterized as:

$$f_e = \rho_e E - \frac{1}{2} E^2 \nabla \varepsilon$$

where the electric field is E (V/m), electric permittivity is ε (F/m), and free charge density is ρ_e (C/m^3). Additionally, ρ_e can be obtained using Gauss' law in the following way:

$$\rho_e = \nabla.\varepsilon E$$

and the electric field originates from:

$$E = -\nabla V$$

where electric potential, denoted as V. For perfect-perfect dielectric fluids, the electric field equation under DC electric field conditions is $\rho_e = 0$.

$$\nabla \cdot \varepsilon E = 0$$

As a result, the Navier–Stokes equation's final non-dimensional form is:

$$\rho^* \frac{\partial u^*}{\partial t} + \rho\left(u^* \cdot \nabla\right)u^* = -\nabla p^* + \frac{1}{R_e} \nabla.\mu^*\left(\nabla u^* + (\nabla u)^T\right) - \frac{1}{R_e C_a}(\sigma k \delta_n)$$
$$+ \frac{B_{O_E}}{R_e C_a} \frac{L}{R}\left(-\frac{1}{2} E^{*^2} \nabla \varepsilon\right)$$

In this equation, the dimensionless numbers are:

$$R_e = \frac{\rho_m U R_0}{\mu_m}$$

$$C_a = \frac{\mu_m U}{\gamma}$$

$$B_{O_E} = \frac{\varepsilon_m R_0 E_0^2}{\gamma}$$

In the previously mentioned equations, it is essential to consider the distinct properties of each liquid phase. One of the numerical techniques for distinguishing between different fluid phases in a two-phase fluid flow problem is to track the interface between two immiscible fluids. This technique can be represented by the following equation:

$$\frac{\partial \varphi}{\partial t} + u.\nabla \varphi = \gamma \nabla.\left[\in \nabla \varphi - \varphi(1 - \varphi)\frac{\nabla \varphi}{|\nabla \varphi|}\right]$$

where $\in$ and γ denote the numerical stabilization parameters for determining interface thickness and re-initialization, respectively. The level set function (φ) for one phase is "0" and "1" for the other.

4.4 Meshing

To create a mesh, choose fine for the element size. Figure 1c shows the generated grid surrounding the two droplets.

4.5 Solution

Click Show Default Solver from the study toolbar. Connect the separated groups for electrostatics, velocity (u), and pressure (p). Compute the study. Run the simulation and monitor the outcome. Observe the water droplet's behavior in the fluid volume percentage at various times.

5 Results and Discussion

The schematic representation and boundary conditions employed in this study are illustrated in Fig. 1a, b. The geometry consists of two concentric semi-ellipses, with $b2/a2 = b1/a1 = b/a$. An electrode on the inner curved surface is kept at an electric potential of $V0$, whereas the outer curve is grounded. Figure 2 designated horizontal lines are subjected to a uniform pressure condition ($p = 0$), with the curved lines being handled as non-slip walls. The non-dimensional coalescence time of droplets with a radius of 0.6 mm and an initial spacing of 2 mm was investigated using an average electric field strength of 2.8×10^5 V/m in order to perform a mesh independence study.

Under the influence of the electric field, the droplets undergo polarization, resulting in the generation of charges at both ends. These charges induce an attractive force between the droplets, leading them to move closer until they make contact. At this point, influenced by interfacial tension, they tend to coalesce, culminating in complete coalescence occurring at $t = 0.3$ s, as shown in Fig. 3b. This process neutralizes the different charges on the touching regions of the droplets, resulting in their complete neutralization by the electric field induced by polarization [7]. An electric field threshold of 1000 V is found, below which electro-coalescence cannot be properly initiated. This means that in order to achieve dipole–dipole interaction, the electric field strength must be increased. After coalescence water droplets will move toward the electric potential boundary. Figure 4 demonstrates that an increase in voltage leads to a reduced time for the droplets to reach the electric potential

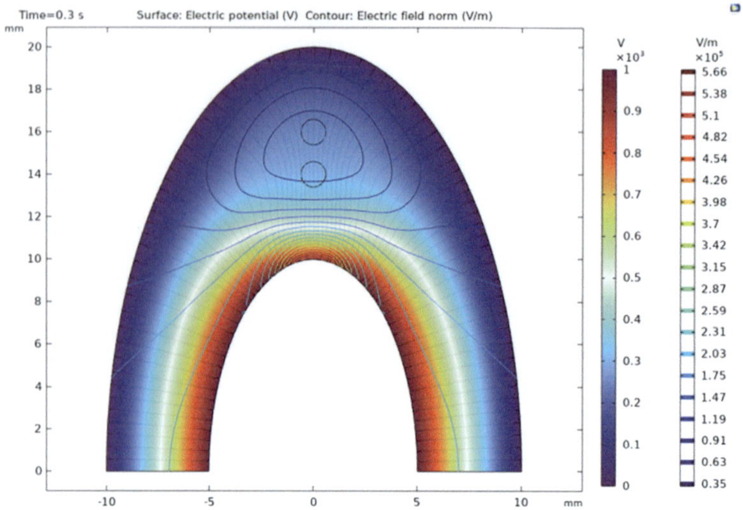

Fig. 2 Electric potential distribution for the geometry where voltage is equal to 1000 V, and *b*/*a* is equal to 2

boundary. Furthermore, the rate of water droplet coalescence increases with higher electric field strength, exceeding 2500 V, causing the water droplets to elongate and dissolve in the direction of the field.

5.1 Effect of b/a *Ratio*

Figure 5a, b show the electric field intensity contours at different time intervals for *b*/*a* ratios of 1 and 5, as well as changes in droplet separation distance. Droplets with a radius of 0.6 mm are initially separated by 2 mm. Table 1 shows the design of the electric field for various *b*/*a* ratios. The viscosity of the oil is 0.2 Pa s, the electric potential is 1000 V, and the scale factor is set to 0.5.

As depicted in Fig. 6, the non-dimensional coalescence period decreases as the *b*/*a* ratio increases. This phenomenon can be attributed to the more potent forces exerted by the non-uniform electric field on the droplets, potentially due to the improved alignment of droplet pairs within the applied non-uniform electric field [8].

5.2 The Influence of the Scale Factor (Sf)

The effects of the outer to inner radii ratio on droplet coalescence in the case of *b*/*a* = 2 are discussed further below. Sf is the "scale factor" ($a1/a2 = b1/b2 = R1/R2$).

Fig. 3 Variations in the distances between the droplets at different points in time when *b/a* ratio equal to 2 and voltage equal to 1000 V

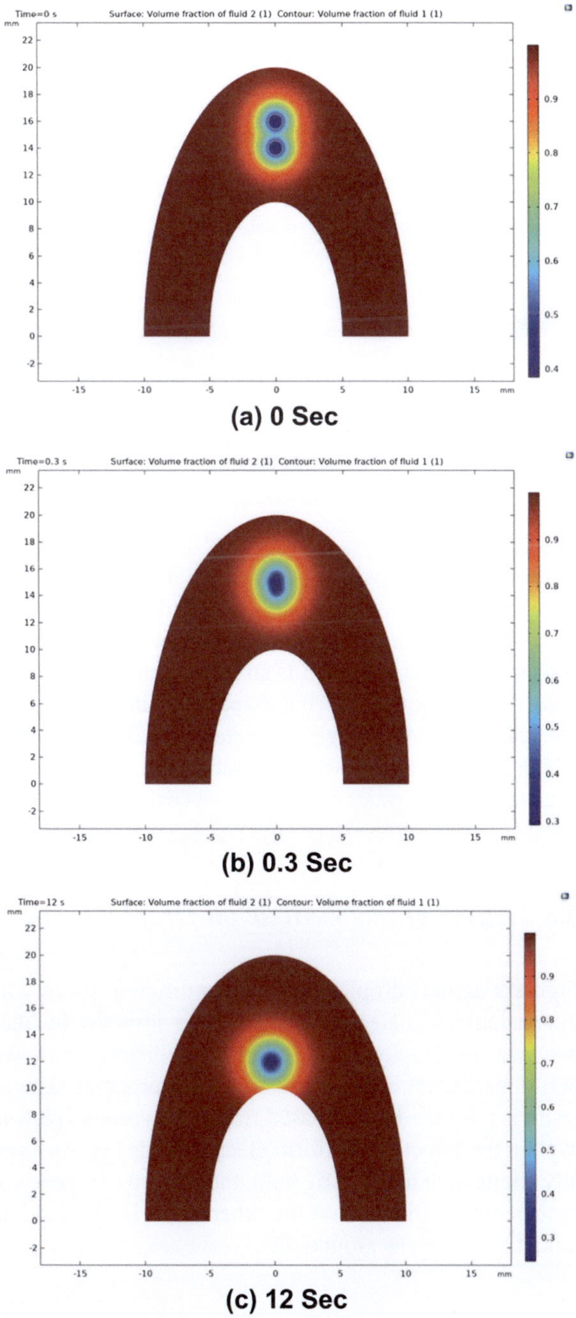

Fig. 4 Time required to move 3 mm for the droplet after coalescence for different voltage

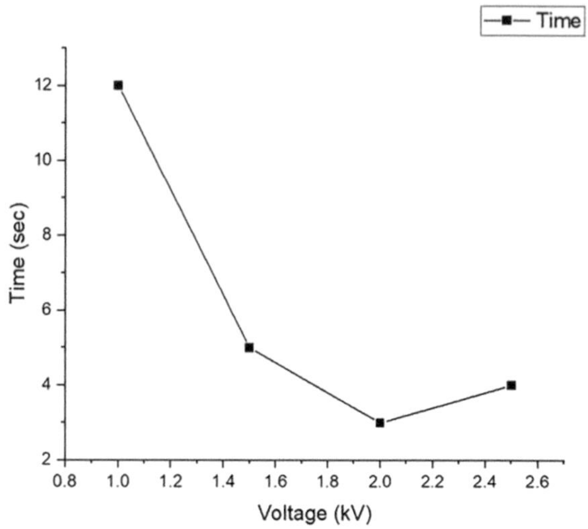

The electric field design is shown in Table 2 for various scale factors. Diagrams for scale factor (Sf) values of 0.2 and 0.6 are shown in Fig. 7a, b. Among these values, Sf = 0.7 results in the longest impingement time, while Sf = 0.2 yields the shortest coalescence time. Sf = 0.5 is an ideal scale factor. Because of non-uniform electric field, the direction in which it is applied to the droplets varies with scale factor. When the scale factor is changed, the electric field's alignment with drop pairs changes, which affects the coalescence duration [8] as shown in Fig. 8. The time required for coalescence of a water droplet with a scale factor of 0.2 is less than that of 0.7.

5.3 Effect of Skew Angle on Droplet Coalescence

Figure 9 depicts droplet positions at different times with a 45° starting angle. The dimensionless coalescence duration is affected by the angle of the tangent line between the droplets at the point of collision, as shown in Fig. 10. The droplets' non-dimensional coalescence time is shortest at skew angles of 12 and 45°. This result is attributed to increased forces between droplets at these angles, which could explain the effects of uniform electric fields [8]. As a result, as the drop pair gradually aligns with the electric field, their approach speeds up significantly. The longest time between droplets, on the other hand, is observed at angles where this force is minimized, such as around 30°.

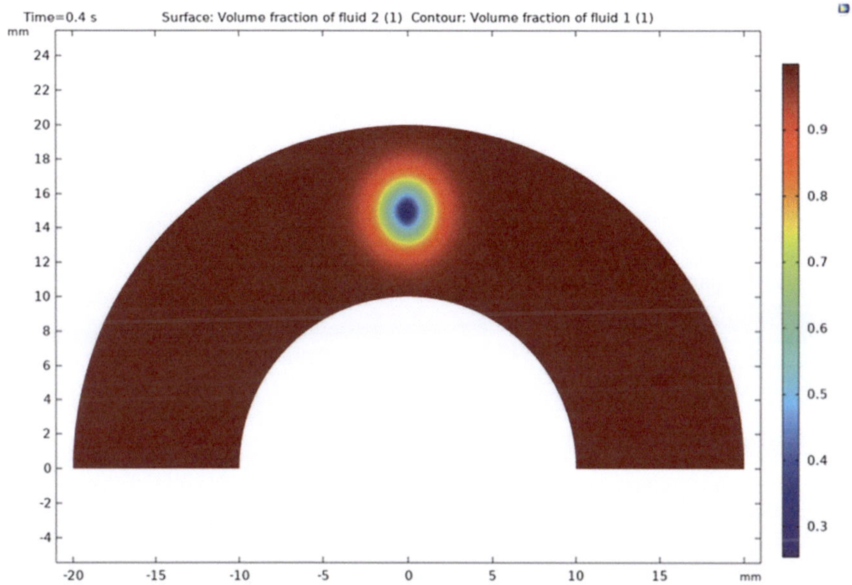

(a)

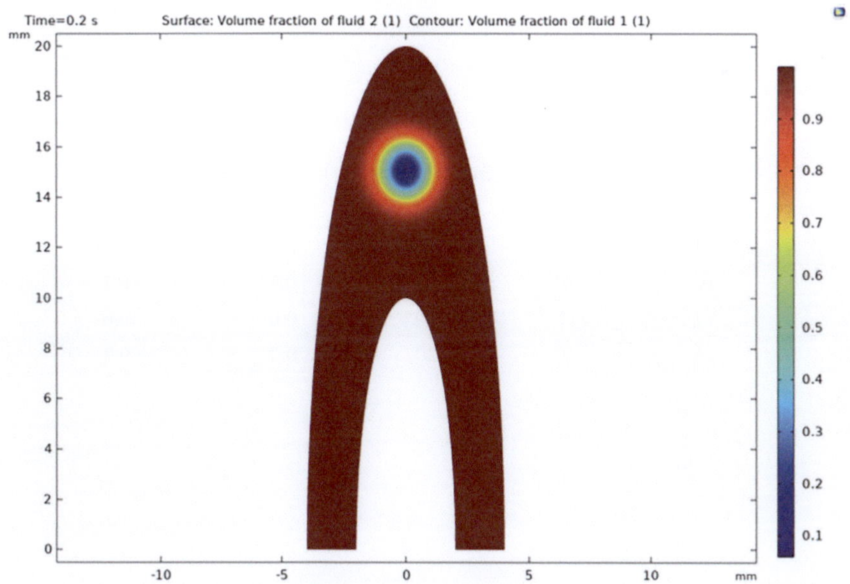

(b)

Fig. 5 **a** Time needed for a droplet to coalesce when the *b/a* ratio is equal to 1. **b** Time needed for a droplet to coalesce when the *b/a* ratio is equal to 5

Table 1 Design of electric field for different b/a ratio

*a*1 (mm)	*b*1 (mm)	*a*2 (mm)	*b*2 (mm)	*b/a*	Time (s)
10	10	20	20	1	0.4
5	10	10	20	2	0.3
3	9	6	18	3	0.25
2	10	4	20	5	0.2

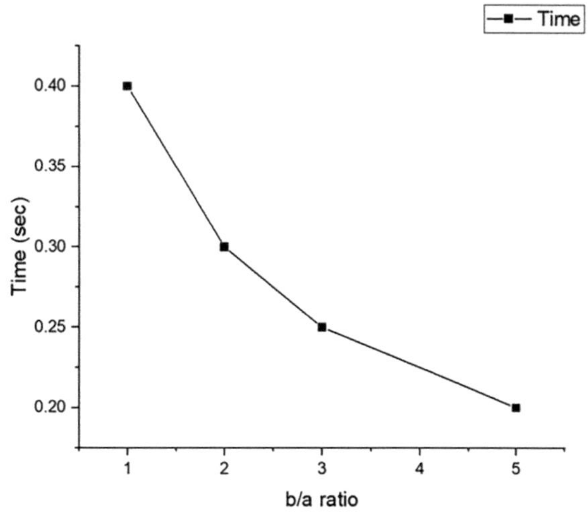

Fig. 6 Time required for coalescence of water droplet for different *b/a* ratio at 1000 V and scale factor 0.5

Table 2 Design of electric field for different scale factor

*a*1 (mm)	*b*1 (mm)	*a*2 (mm)	*b*2 (mm)	Scale factor	Time (s)
2	4	10	20	0.2	0.25
5	10	10	20	0.5	0.3
6	12	10	20	0.6	0.6
7	14	10	20	0.7	0.7

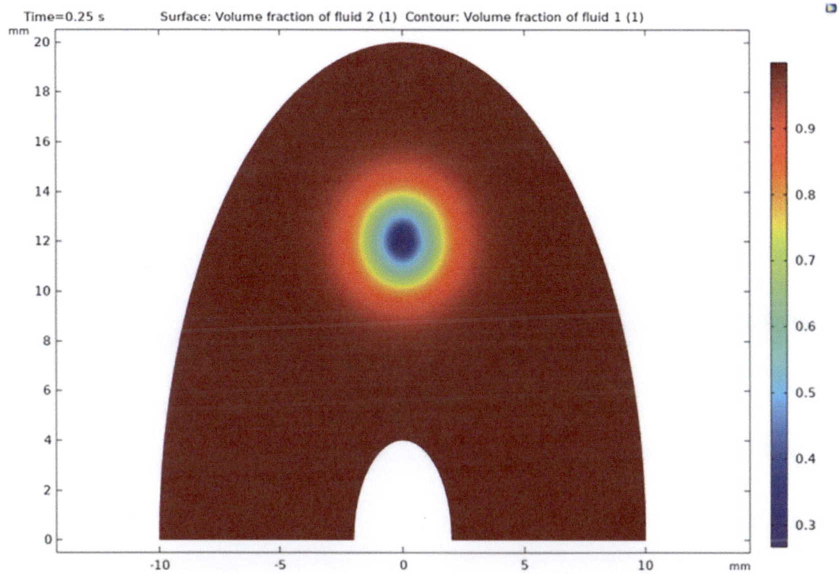

(a)

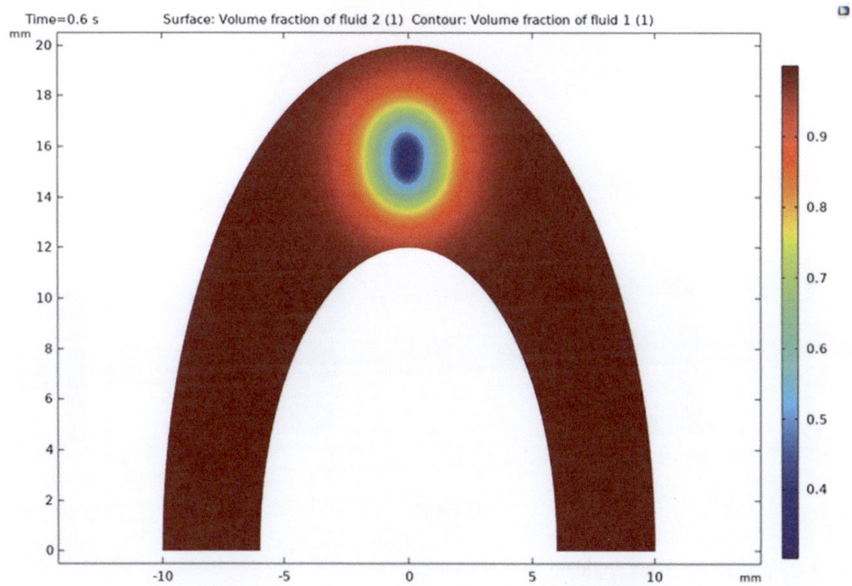

(b)

Fig. 7 **a** Coalescence of droplet with Sf value of 0.2. **b** Coalescence of droplet with Sf value of 0.6

Fig. 8 Time required for coalescence of water droplet for different scale factor at 1000 V and *b/a* ratio 2

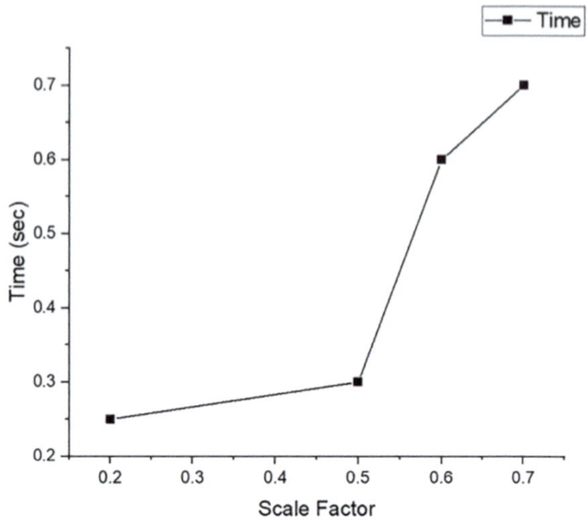

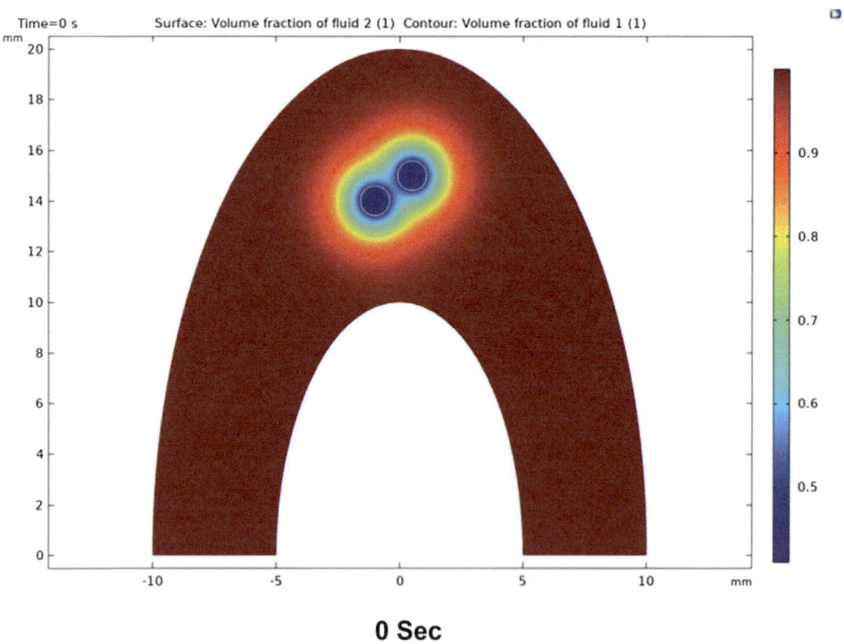

Fig. 9 Impact of the skew angle on droplet coalescence

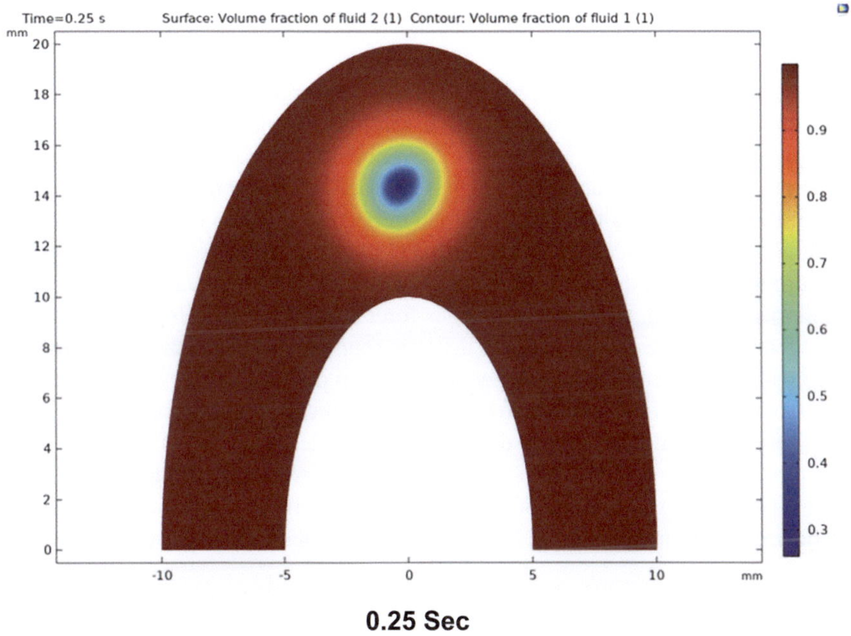

Fig. 9 (continued)

Fig. 10 Time required for coalescence of water droplet for different skew angle at 1000 V and *b/a* ratio 2

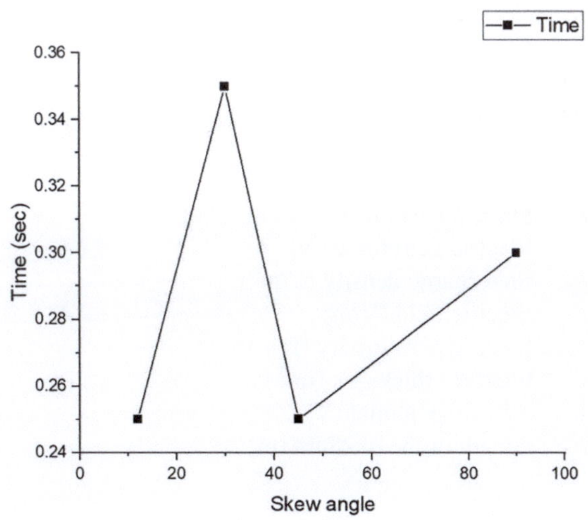

6 Conclusion

This paper examines the coalescence behavior of water droplets within an oil phase under the influence of a novel concentric semi-elliptic non-uniform electric field configuration. It investigates the effect of various parameters on this coalescence phenomenon, providing useful insights. Changes in the geometry of the applied non-uniform fields were found to have a significant impact on electro-coalescence performance. It was discovered that increasing the voltage reduced the time required for a droplet to reach the electric potential boundary. A higher b/a ratio, indicating an elliptical medium, resulted in stronger forces acting on the droplets and faster coalescence. A scale factor of Sf = 0.5 was found to be optimal, indicating that minimizing coalescence time is dependent on drop pair alignment with the electric field. Coalescence time was also found to be significantly influenced by the skew angle between the droplets, with different skew angles leading to faster or slower coalescence. The droplets' coalescence time is shortest for skew angles of 12 and 45°. Further studies into the best ways to use non-uniform electro-coalescence in a variety of medium types and electrode configurations are made possible by this work.

Nomenclature

u	Fluid velocity (m/s)
ρ	Density (kg/ m^3)
p	Pressure (Pa)
μ	Viscosity (Pa-s)
F	External Force (N)
σ	Surface tension coefficient (N/m)
f_e	Electric field force (N)
ρ_e	Free charge density (C/m^3)
E	Electric field (V/m)
ε	Electric permittivity (F/m)
$\in$	Interface thickness (m)
V	Electric potential (V)
L	Length of the medium (m)

References

1. Sun K, Wang T, Zhang P, Law CK (2015) Non-Newtonian flow effects on the coalescence and mixing of initially stationary droplets of shear-thinning fluids. Phys Rev E 91(2):023009

2. Hadidi H, Kamali R, Manshadi MK (2020) Numerical simulation of a novel non-uniform electric field design to enhance the electrocoalescence of droplets. Eur J Mech-B/Fluids 80:206–215
3. Vivacqua V, Ghadiri M, Abdullah AM, Hassanpour A, Al-Marri MJ, Azzopardi B, Hewakandamby B, Kermani B (2016) Analysis of partial electrocoalescence by level-set and finite element methods. Chem Eng Res Des 114:180–189
4. Khojasteh D, Kazerooni M, Salarian S, Kamali R (2016) Droplet impact on superhydrophobic surfaces: a review of recent developments. J Ind Eng Chem 42:1–14
5. Li B, Vivacquab V, Wang J, Wang Z, Sunc Z, Wang Z, Ghadiri M (2019) Electrocoalescence of water droplets in sunflower oil using a novel electrode geometry. Chem Eng Res Des
6. Yasir AT, Hawari AH, Talhami M, Baune M, Thöming J, Du F (2023) The impact of electric field on the demulsification efficiency in an electro-coalescence process. J Electrostat 122:103796
7. Hong W, Ye X, Xue R (2018) Numerical simulation of deformation behavior of droplet in gas under the electric field and flow field coupling. J Dispersion Sci Technol 39(1):26–32
8. Mohammadi M, Shahhosseini S, Bayat M (2014) Electrocoalescence of binary water droplets falling in oil: experimental study. Chem Eng Res Des 92(11):2694–2704
9. Finotello G, Padding JT, Deen NG, Jongsma A, Innings F, Kuipers JAM (2017) Effect of viscosity on droplet-droplet collisional interaction. Phys Fluids 29(6)
10. Luo S, Schiffbauera J, Luo T (2016) Effect of electric field non-uniformity on droplets coalescence. Phys Chem Chem Phys 18:29786

Shape Matter: Investigating Mixing Dynamics in a Rotating Drum

Sunil Kumar and Anshu Anand

Abstract Solid-solid mixing is a fundamental process utilized across numerous industries, yet a comprehensive understanding of granular material behavior remains elusive. The current study employs the Discrete Element Method (DEM) to explore the effect of radial segregation and its relation to variations in particle shape by using the multi-sphere approach. By generating four distinct types of binary mixtures, the investigation emphasizes the significant role of coarse particle shape in governing mixing and segregation dynamics within a rotating drum. The initial homogeneous mixture of coarse and fine particles evolves into radial segregation, marked by the aggregation of fine particles at the drum's core, surrounded by coarser counterparts. Notably, the study unveils that spherical particles, whether act as coarse components in binary mixtures, tend to encourage segregation. Additionally, the study identifies that elongated needle-shaped particles exhibit relatively enhanced mixing behavior compared to spheres. This trend is consistent across both experimental and numerical simulations. The trend involves a sequential decrease in the segregation index value for coarse particles, with the order of shapes being sphere, followed by oblate, then prolate, and finally, elongated needle.

Keywords Mixing · Segregation · DEM · Rotating drum · Multi-sphere

Nomenclature

SI	Segregation Index
C_{AA}	Contacts between particles of type A
C_{BB}	Contacts between particles of type B
C_{AB}	Contacts between particles of type A and B
$p-w$	Particle- Wall
$p-w$	Particle- Particle
$E-Needle$	Elongated Needle

S. Kumar · A. Anand (✉)
Department of Chemical Engineering, Indian Institute of Technology Roorkee, Uttrakhand, India
e-mail: anshu.anand@ch.iitr.ac.in

A. Kumar R. et al. (eds.), *Proceedings of Fluid Mechanics and Fluid Power (FMFP) 2023, Vol. 6*, Lecture Notes in Mechanical Engineering,
https://doi.org/10.1007/978-981-96-9600-0_33

1 Introduction

Granular materials often show segregation during flow, primarily driven by variations in particle characteristics such as cohesion [1], shape [2, 3], size, and density. It's worth highlighting that granular materials comprise a significant portion, around 60%, of blended products in various industries [4]. Moreover, in the pharmaceutical sector, approximately 80% of drugs are formulated in solid dosage forms [5]. The rotating drum is a cost-effective, simple apparatus with extensive utilization in diverse industries. For instance, it is pivotal in tasks such as ball milling for paint and cement production [6], powder manufacturing lines [7], kiln processes [8], and blending operations across sectors like food, cement, refractory materials, fertilizers, and glass ceramics [9].

The Discrete Element Method (DEM) is a powerful numerical technique widely used to analyze the behavior of discrete particles within granular or particulate systems. This method treats individual particles as distinct entities, considering their interactions, motion, and collisions. By simulating the interactions and dynamics of these discrete particles, DEM provides insights into various phenomena such as particle segregation, mixing, flow behavior, and compaction. In this study, all simulations are conducted using the commercial Altair EDEM version 2022.1.0. This software effectively applies the principles of DEM to replicate intricate particle interactions and behaviors within virtual environments.

The phenomenon of segregation, which stands in contrast to mixing, is an intrinsic occurrence in a wide range of flow situations [10]. Accurate measurement of segregation is crucial for enhancing the efficiency of blending procedures. Its detrimental impacts encompass various sectors, giving rise to challenges like uneven compositions in pharmaceutical formulations, inadequate concrete mixing, separation of dry fruits in the food industry, formation of lumps in detergent manufacturing, and inconsistent material discharge from hoppers [10]. Notably, segregation is of particular significance in the context of the pharmaceutical industry, where variations in the composition of active pharmaceutical ingredients (APIs) as a result of segregation can result in irregular dosages during the process of blending. This phenomenon can potentially impact the quality and effectiveness of pharmaceutical products, posing challenges in maintaining uniformity and reliability in dosage administration. Consequently, understanding and mitigating segregation-related issues are crucial in ensuring consistent and accurate medication delivery to patients. Numerous particle and system parameters significantly influence particle segregation dynamics. Particle attributes encompass shape [11–19], size [20–23], density [24], elasticity [25], and cohesiveness [26–28]. Simultaneously, system parameters include drum rotational speed [23], drum length [29], drum side-wall effects [30], fill-level [31], and wall surface roughness [32].

Various techniques have been deployed in previous studies to simulate non-spherical particles. One of the promising techniques is glued sphere [33–36]. Multi-sphere (MS) approximation is a technique in which spheres are glued together to represent other non-spherical shapes [37]. In this approximation, complex non-spherical particles are represented by overlapping spheres. Previous studies have shown that multi-spherical approximation can be successfully and accurately applied to non-spherical shapes such as ellipsoid, [38–41], Prolate [42], cube [43], cylinder [44], rod-like [45], and other complex shapes [46, 47]. Yan et al. [40] employed the MS approximation to represent actual soybean seed shapes. In the study by Tao et al. [48], the multi-sphere (MS) approximation technique is employed to generate particles with a corn-shaped morphology. Their findings suggest that by increasing the number of sub-spheres used in the approximation, there is a noticeable tendency to better mimic the behavior and characteristics of macroscopic particles. This observation underscores the significance of considering the level of detail in particle representation, as a higher number of sub-spheres enhances the accuracy of capturing the macroscopic properties and behavior of the complex-shaped particles. The results become more realistic as the smoothness of the particles increases as the sub-sphere increases. However, this will result in increased computational cost [38]. Similarly, Xu et al. [49] performed a static angle of repose test of soybean seeds with Five, nine, and thirteen sub-spheres and found that a higher number of sub-spheres gave a better result but at a high computational cost.

The aim of this study is to use a combined approach of experimental and DEM techniques to examine the influence of coarse particle shape on radial segregation in binary mixtures. Four distinct particle shapes are generated using the multi-sphere approximation, yielding novel data on various shape combinations. This data holds value for industries aiming to improve segregation control by optimizing coarse particle shapes. Additionally, this work extends our prior experimental study [50] and computational [51] study on rotating drums, confirming and validating the trends observed in that study. The outcome of the work aims to bridge the gap between scientific theory and simulation studies. The significance of this study lies in its adoption of a contact detection approach to quantify the extent of segregation, diverging from the conventional analysis of particle distribution data in spatial dimensions.

2 Methodology

The default model within the framework of the EDEM Software is the Hertz-Mindlin (No Slip) model, as detailed in the EDEM Software manual. This model integrates Hertzian contact theory [52] to ascertain the normal force component in particle

Fig. 1 Particle used in experiments

interactions. Another commonly employed approach is the Mindlin-Deresiewicz model [53, 54], which introduces tangential forces and damping components. In this model, the damping coefficient is closely linked to the coefficient of restitution [55]. The model also accounts for tangential friction forces using the Coulomb law of friction model [56]. Additionally, rolling friction is incorporated by applying a contact-independent directional constant torque model [57]. Collectively, these distinct models provide a comprehensive framework for simulating particle interactions and behavior within the context of discrete element simulations, enabling a more accurate representation of real-world granular systems.

For the purpose of conducting the experimental investigations, a set of specially engineered particles crafted from glass material, characterized by their non-cohesive nature, have been carefully chosen. This selection encompasses both spherical and non-spherical variations of particles, facilitating a comprehensive examination of particle segregation dynamics within the context of a rotating drum. These engineered particles have been meticulously designed and manufactured exclusively for the scope of this study, ensuring their suitability for the intended analysis. The choice of glass particles stems from their advantageous attributes, notably their cost-effectiveness and ready availability. Moreover, the glass particles offer the flexibility of being customized into a range of colors, which significantly enhances the clarity of visualization during the experimental observations.

A schematic representation of the particle utilized in this study is depicted in Fig. 1, providing a visual reference for the particle's shape. Additionally, Table 1 presents a comprehensive compilation of the material properties pertinent to the particles employed in the experimental investigations. This comprehensive data is a valuable resource for understanding the physical characteristics and attributes of the particles used in the study, thereby contributing to the accurate interpretation of experimental results and their implications. Table 2 compiles the parameter values employed in the DEM simulations. The values listed in Table 2 are derived from the properties of glass particles and the acrylic sheet composition of the rotating drum.

Table 1 Material properties and dimension of particles

Particles	Mean radius (mm)			Average mass	Sphericity (Φ)
	a	b	c	(g)	–
Coarse prolate	3	3	6	0.662	0.93
Coarse oblate	6	6	3	1.360	0.91
Coarse elongated	1	1	6	0.128	0.69
needle					
Coarse sphere	6	6	6	2.442	1
Fine sphere	1	1	1	0.011	1

Table 2 Numerical simulation parameters for performing DEM simulation

Properties	Values
Dimensions of drum (m)	L = 0.14 and R = 0.14
Total mass (kg)	3.0
Mass ratio between particle (–)	1:1
Young's modulus (Pa)	10^7
Density of particle (kg/m^3)	2500
Poisson ratio (-)	0.24
Coefficient of restitution [p-p] (–)	0.97
Friction [p-p] (–)	0.3
Coefficient of restitution [p-w] (–)	0.85
Friction [p-w] (–)	0.3
Timestep (s)	10^{-6}
Time period of rotation (s)	6

2.1 Calculation of Segregation Index

This section introduces a robust methodology for quantitatively assessing particle segregation within complex systems. To elucidate this approach, we consider a binary particle system comprising distinct particles categorized as types A and B. These particles interact with one another, forming an intricate network of contacts. The segregation index, a pivotal metric in this analysis, is formulated as the combined fraction of A-type contacts with other A particles and B-type contacts with other B particles. Mathematically, this index is expressed as:

$$SI = \frac{C_{AA}}{C_{AA} + C_{AB}} + \frac{C_{BB}}{C_{BB} + C_{AB}} \tag{1}$$

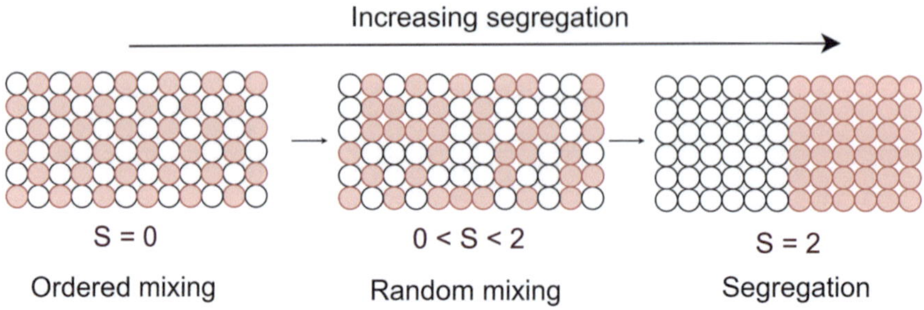

Fig. 2 Schematic depicting the conceptual definition of the segregation index derived from contact numbers

Table 3 Relationship between particle contact and segregation value.

Mixture	C_{AA}	C_{BB}	C_{AB}	Segregation
Ordered mixing	0	0	$\neq 0$	S = 0
Random mixing	$\neq 0$	$\neq 0$	$\neq 0$	0 < S < 2
Complete segregation	$\neq 0$	$\neq 0$	0	S = 2

Here, C_{AA}, C_{BB}, and C_{AB} denote the total number of contacts between particles of the same species (A and B), and different species (A and B), respectively. The schematic representation of the segregation index, visualizing the concept of contact numbers, is illustrated in Fig. 2.

To provide a comprehensive understanding of the relationship between contact numbers and segregation behavior, Table 3 is presented. This table elucidates various mixtures and their corresponding interaction patterns, shedding light on the intricate interplay between contact numbers and the degree of segregation. In scenarios of complete segregation, the absence of C_{AB} or C_{BA} contacts corresponds to a segregation index value of S = 2, signifying a complete separation between particle types. Conversely, when S = 0, it reflects a highly ordered configuration with no contacts between particles of the same type (C_{AA} or C_{BB}). For a deeper exploration of the nuanced relationship between contact numbers and segregation, reference can be made to the comprehensive study conducted by Marigo et al. [58]. This study delves into the intricate dynamics that underscore the segregation phenomena, offering valuable insights into this intricate process.

3 Results and Discussion

To comprehensively investigate the influence of coarse particle shape in binary mixtures, a systematic variation in the shape of coarse particles is undertaken, encompassing four distinct shapes. In this controlled experimentation, the fine particles are

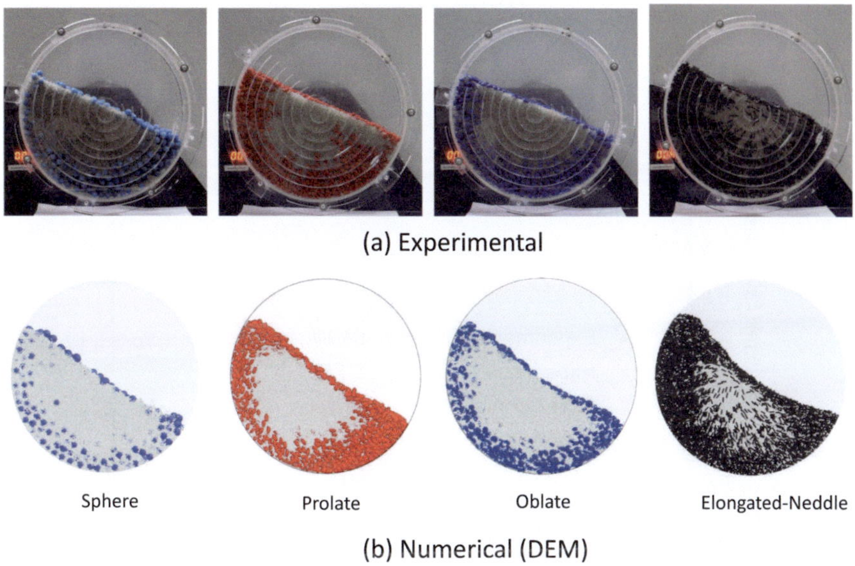

Fig. 3 Post-rotation visual comparison of particle distribution, displaying **a** experimental results and **b** DEM simulation outcomes

fixed to a uniform spherical configuration with a diameter of 2mm. The introduction of particles follows a sequential process, beginning with the finer particles and followed by the coarser counterparts. This initial arrangement ensures a fully segregated state within the system. Adequate settling time is provided for the finer particles to establish their positions before introducing the coarser particles. Subsequently, all particles are granted an additional settling duration before initiating the rotation of the drum.

Illustrated in Fig. 3, a snapshot captures the concluding configuration of the binary system, featuring four diverse shapes of coarse particles while the fine particles remain consistently spherical (highlighted in white). This visual representation depicts the fine particles aggregating at the central core of the drum, encircled by the coarser particles. Notably, this depiction aligns remarkably well with the visual comparison between DEM results and experimental observations.

Both the experimental data and predictions generated through DEM demonstrate a consistent qualitative trend in the normalized fine mass fraction of the final mixed state. This alignment is effectively showcased in Fig. 4, where the normalized fine mass fraction is plotted against the normalized radial position. The concurrence in the trends of the normalized fine mass fraction between the experimental data and DEM predictions strongly indicates the accuracy of DEM simulation in forecasting the normalized fine mass fraction profile. This predictive capability empowers an in-depth exploration of particle mixing behavior within the mixture. The analysis of the

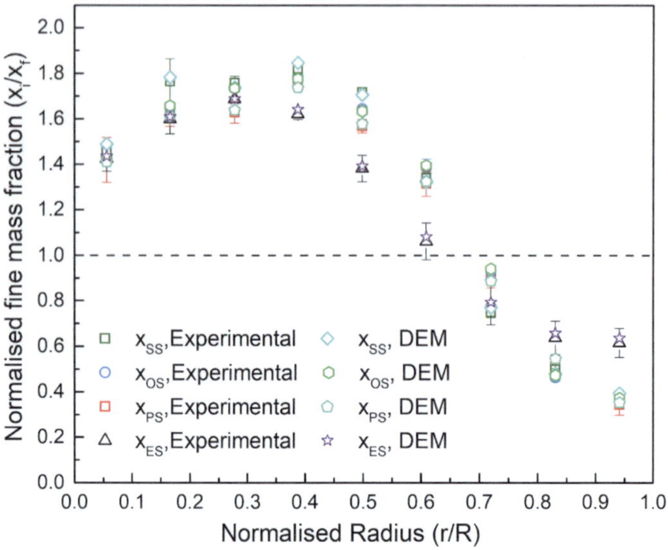

Fig. 4 Comparative analysis of normalized fine mass fraction profiles against normalized radius between DEM results and experimental data

numerical data, acquired through DEM, yields valuable insights into the underlying mechanisms driving particle mixing within the rotating drum.

The evolution of the segregation index over time is vividly depicted in Fig. 5 for a binary system encompassing four distinct coarse particles alongside a fine sphere. Notably, Fig. 5 elegantly illustrates the descending pattern observed in the segregation index (SI) values, specifically: sphere ($SI = 1.11$) > oblate ($SI = 0.97$) > prolate ($SI = 0.94$) > elongated needle ($SI = 0.91$). This finding clearly shows that the relatively most effective mixing is achieved when combining coarse elongated needles with fine prolate particles, whereas the combination of coarse spheres with fine prolate particles yields subpar mixing performance. The study conducted by Kumar et al. [50] not only elucidates this prevailing trend and its underlying mechanism but also establishes a significant correlation between the segregation index and the random packing of monodisperse coarse particles. Our future endeavors will focus on characterizing the flow behavior of these particles within a multiphase flow using water as the medium.

4 Conclusions

This investigation leverages the Discrete Element Method (DEM) to comprehensively explore the evolving dynamics of radial segregation and its interplay with particle shape variations. By generating four distinctive types of binary mixtures,

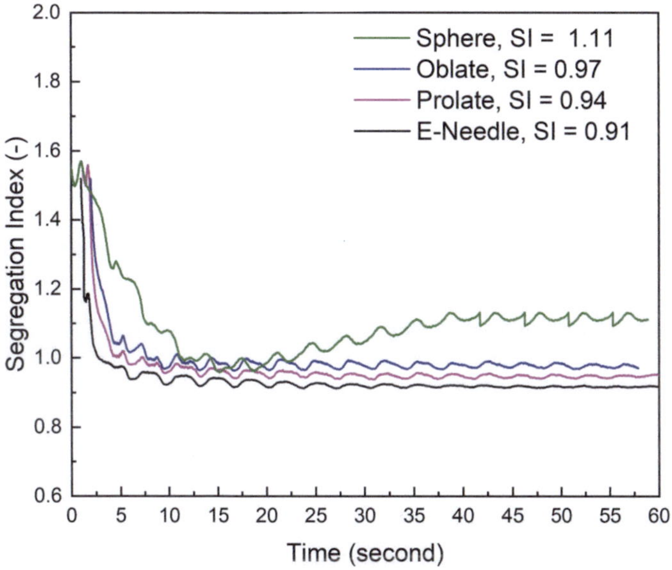

Fig. 5 Temporal evolution of the segregation index for a binary system comprising four coarse particles and a fine sphere

the study delves into the intricate relationship between coarse particle shape and the ensuing mixing and segregation phenomena within a rotating drum. The simulations commence with an initially well-mixed binary amalgamation of coarse and fine particles, gradually transitioning into a configuration characterized by radial segregation. This intriguing transformation culminates in fine particles clustering at the central core of the drum, enveloped by the encircling coarse particles. Notably, the study sheds light on the propensity of spherical particles, particularly when functioning as coarse particles in a binary mixture, to act as catalysts for segregation processes.

Moreover, the outcomes notably unveil the enhanced mixing efficacy associated with elongated needle-shaped particles in contrast to their spherical counterparts. This insight is further validated across both experimental observations and numerical simulations, which consistently reveal a descending trend in segregation behavior. Specifically, the order of segregation propensity among the coarse particles follows the sequence: sphere, oblate, prolate, and elongated needle. The comprehensive exploration of these phenomena provides a profound understanding of how particle shape variation intricately intertwines with mixing and segregation dynamics within the context of a rotating drum.

Acknowledgements The authors would like to express their heartfelt gratitude to the Science & Engineering Board (SERB) for providing financial support. We also want to thank Altair for providing the necessary EDEM Software. Additionally, we would like to thank Ramesha BS and Prasad Avilala from the Altair team for their valuable support.

References

1. Anand A, Curtis JS, Wassgren CR, Hancock BC, Ketterhagen WR (2009) Predicting discharge dynamics of wet cohesive particles from a rectangular hopper using the discrete element method (DEM). Chem Eng Sci 64(24):5268–5275
2. Kumar S, Khatoon S, Dubey P, Yogi J, Anand A (2024) Shape-dependent radial segregation in rotating drum: insights from dem simulations. Powder Technol 432:119134
3. Hernández-Delfin D, Tunuguntla DR, Weinhart T, Hidalgo RC, Thornton AR (2022) Shape matters: competing mechanisms of particle shape segregation. Phys Rev E 106(5):054614
4. Nienow AW, Harnby N, Edwards MF (1992) Mixing in the process industries. In: Mixing in the process industries. Butterworth-Heinemann, Oxford, pp 1–24
5. Huang A-N, Kuo H-P (2014) Developments in the tools for the investigation of mixing in particulate systems-a review. Adv Powder Technol 25(1):163–173
6. Cleary PW (2006) Axial transport in dry ball mills. Appl Math Model 30(11):1343–1355
7. Soliman HMA, Abdel-Rahman HH (2006) The use of rotating cylinder electrode to study the effect of 1, 3-dihydroxypropane on the production of copper powder. J Braz Chem Soc 17(4):705–714
8. Demagh Y, Moussa HB, Lachi M, Noui S, Bordja L (2012) Validation and similarity for an industrial scale kiln. Surface particle motions in rotating cylinders. Powder Technol 224:260–272
9. Ji S, Wang S, Zhou Z (2020) Influence of particle shape on mixing rate in rotating drums based on super-quadric dem simulations. Adv Powder Technol 31(8):3540–3550
10. Anand A, Curtis JS, Wassgren CR, Hancock BC, Ketterhagen WR (2008) Predicting discharge dynamics from a rectangular hopper using the discrete element method (dem). Chem Eng Sci 63(24):5821–5830
11. Khatoon S, Kumar S, Yogi J, Dubey P, Anand A (2024) Influence of aspect ratio of ellipsoidal particles on mixing in vibrated packed bed using discrete element method. AIChE J 70(6):e18405
12. Yogi J, Dubey P, Verma SK, Kumar S, Anand A (2023) Cumulative effect of particle properties on mixing of multi-component mixture in a vibrated packed bed. Powder Technol413:118000
13. Verma SK, Yogi J, Kumar S, Anand A (2022) Experimental investigation of segregation in the dilute phase transport of nonspherical particles through a pneumatic conveying system. Ind Eng Chem Res
14. Yogi J, Dubey P, Khatoon S, Iyer K, Kumar S, Anand A (2023) Effect of non-sphericity of a narrow-sized binary mixture on mixing in convective vibrated packed bed using discrete element method. Powder Technol 118600
15. Yogi J, Verma SK, Kumar S, Anand A (2021) Experimental study of mixing of nonspherical particles in a vibrated packed bed mixer. Ind Eng Chem Res 60(48):17694–17704
16. Shimosaka A, Nousou I, Shirakawa Y, Hidaka J (2013) Effect of particle shape on size segregation of particles. Chem Eng Trans 32:2143–2148
17. He SY, Gan JQ, Pinson D, Zhou ZY (2019) Particle shape-induced radial segregation of binary mixtures in a rotating drum. Powder Technol 341:157–166
18. He SY, Gan JQ, Pinson D, Yu AB, Zhou ZY (2019) Radial segregation of binary-sized ellipsoids in a rotating drum. Powder Technol 357:322–330
19. He SY, Gan JQ, Pinson D, Yu AB, Zhou ZY (2021) Particle shape-induced axial segregation of binary mixtures of spheres and ellipsoids in a rotating drum. Chem Eng Sci 235:116491
20. Dubey P, Yogi J, Kumar S, Khatoon S, Kumari K, Anand A (2024) Shape-dependent size polydispersity: dem investigation of mixing behavior in a vibrating packed bed system. Powder Technol 441:119804
21. Piacenza C, Marconati M, Hare C, Santomaso AC, Ramaioli M (2018) Mixing grains with different elongation in a rotating drums. arXiv preprint arXiv:1809.00153
22. Dury CM, Ristow GH (1999) Competition of mixing and segregation in rotating cylinders. Phys Fluids 11(6):1387–1394

23. Soni RK, Mohanty R, Mohanty S, Mishra BK (2016) Numerical analysis of mixing of particles in drum mixers using dem. Powder Technol 27(2):531–540
24. Wang Q, Feng Y, Lu J, Yin W, Yang H, Witt PJ, Zhang M (2015) Numerical study of particle segregation in a coal beneficiation fluidized bed by a tfm-dem hybrid model: influence of coal particle size and density. Chem Eng J 260:240–257
25. Kuo HP, Hsu RC, Hsiao YC (2005) Investigation of axial segregation in a rotating drum. Powder Technol 153(3):196–203
26. Anand A, Curtis JS, Wassgren CR, Hancock BC, Ketterhagen WR (2010) Segregation of cohesive granular materials during discharge from a rectangular hopper. Granul Matter 12(2):193–200
27. Li H, McCarthy JJ (2006) Cohesive particle mixing and segregation under shear. Powder Technol 164(1):58–64
28. Jarray A, Shi H, Scheper BJ, Habibi M, Luding S (2019) Cohesion-driven mixing and segregation of dry granular media. Sci Rep 9(1):1–12
29. Chand R, Khaskheli MA, Qadir A, Ge B, Shi Q (2012) Discrete particle simulation of radial segregation in horizontally rotating drum: effects of drum-length and non-rotating end-plates. Phys A: Stat Mech Appl 391(20):4590–4596, 2012
30. Jain A, Evrard F, van Wachem B (2023) The effect of side walls on particles mixing in rotating drums. Particuology 72:112–121
31. Arntz MMHD, den Otter WK, Briels WJ, Bussmann PJT, Beeftink HH, Boom RM (2008) Granular mixing and segregation in a horizontal rotating drum: a simulation study on the impact of rotational speed and fill level. AIChE J 54(12):3133–3146
32. Tang P, Puri VM (2007) Segregation quantification of two-component particulate mixtures: effect of particle size, density, shape, and surface texture. Part Sci Technol 25(6):571–588
33. Tao H, Zhong W, Jin B, Ren B (2013) Dem simulation of non-spherical granular segregation in hopper. AIP Conf Proc 1547(1):720–726
34. Maione R, Richter SKD, Mauviel G, Wild G (2017) Axial segregation of a binary mixture in a rotating tumbler with non-spherical particles: experiments and dem model validation. Powder Technol 306:120–129
35. Favier JF, Abbaspour-Fard MH, Kremmer M, Raji AO (1999) Shape representation of axi-symmetrical, non-spherical particles in discrete element simulation using multi-element model particles. Eng Comput 16(4):467–480
36. Soltanbeigi B, Podlozhnyuk A, Kloss C, Pirker S, Ooi JY, Papanicolopulos S-A (2021) Influence of various dem shape representation methods on packing and shearing of granular assemblies. Granular Matter 23(2):1–16
37. Favier JF, Abbaspour-Fard MH, Kremmer M (2001) Modeling nonspherical particles using multisphere discrete elements. J Eng Mech 127(10):971–977
38. Markauskas D, Kačianauskas R, Džiugys A, Navakas R (2010) Investigation of adequacy of multi-sphere approximation of elliptical particles for dem simulations. Granular Matter 12(1):107–123
39. Deng XL, Davé RN (2013) Dynamic simulation of particle packing influenced by size, aspect ratio and surface energy. Granular Matter 15(4):401–415
40. Yan D, Jianqun Yu, Wang Y, Zhou L, Yajun Yu (2020) A general modelling method for soybean seeds based on the discrete element method. Powder Technol 372:212–226
41. Han Y, Zhao D, Jia F, Qiu H, Li A, Bai S (2020) Experimental and numerical investigation on the shape approximation of rice particle by multi-sphere particle models. Adv Powder Technol 31(4):1574–1586
42. Parafiniuk P, Molenda M, Horabik J (2014) Influence of particle shape and sample width on uniaxial compression of assembly of prolate spheroids examined by discrete element method. Physica A 416:279–289
43. Soltanbeigi B, Podlozhnyuk A, Ooi JY, Kloss C, Papanicolopulos S-A (2017) Comparison of multi-sphere and superquadric particle representation for modelling shearing and flow characteristics of granular assemblies. In: EPJ web of conferences, vol 140, p 06015. EDP Sciences

44. Alchikh-Sulaiman B, Alian M, Farhad Ein-Mozaffari, Ali Lohi, and Simant R Upreti (2016) Using the discrete element method to assess the mixing of polydisperse solid particles in a rotary drum. Particuology 25:133–142
45. Lattanzi AM, Stickel JJ (2020) Hopper flows of mixtures of spherical and rod-like particles via the multisphere method. AIChE J 66(4):e16882
46. Maione R, Richter SKD, Mauviel G, Wild G (2015) Dem investigation of granular flow and binary mixture segregation in a rotating tumbler: Influence of particle shape and internal baffles. Powder Technol 286:732–739
47. Ketterhagen WR (2011) Modeling the motion and orientation of various pharmaceutical tablet shapes in a film coating pan using dem. Int J Pharm 409(1-2):137–149
48. Tao H, Jin B, Zhong W, Wang X, Ren B, Zhang Y, Xiao R (2010) Discrete element method modeling of non-spherical granular flow in rectangular hopper. Chem Eng Process 49(2):151–158
49. Jian X, Wang X, Zhang Z, Weibin W (2020) Discrete element modeling and simulation of soybean seed using multi-spheres and super-ellipsoids. IEEE Access 8:222672–222683
50. Kumar S, Khatoon S, Yogi J, Verma SK, Anand A (2022) Experimental investigation of segregation in a rotating drum with non-spherical particles. Powder Technol 117918
51. Kumar S, Khatoon S, Parashar S, Dubey P, Yogi J, Anand A (2023) Effect of aspect ratio of ellipsoidal particles on segregation of a binary mixture in a rotating drum. Powder Technol 118682
52. Hertz H (1881) The contact of elastic solids. J Reine Angew Math 92:156–171
53. Mindlin RD (1949) Compliance of elastic bodies in contact
54. Mindlin RD, Deresiewicz H (1953) Elastic spheres in contact under varying oblique forces
55. Tsuji Y, Tanaka T, Ishida T (1992) Lagrangian numerical simulation of plug flow of cohesionless particles in a horizontal pipe. Powder Technol 71(3):239–250
56. Cundall PA, Strack ODL (1979) A discrete numerical model for granular assemblies. Geotechnique 29(1):47–65
57. Sakaguchi H, Ozaki E, Igarashi T (1993) Plugging of the flow of granular materials during the discharge from a silo. Int J Modern Phy B 7(09n10):1949–1963
58. Marigo M, Cairns DL, Davies M, Ingram A, Stitt EH (2012) A numerical comparison of mixing efficiencies of solids in a cylindrical vessel subject to a range of motions. Powder Technol 217:540–547

The Impact of Air Velocity on the Performance of Pad Evaporative Cooling Systems

Ankitsinh Chauhan, Vimal Patel, and Ashok Parekh

Abstract Evaporative cooling is a prominent solution for indoor climate control, particularly in arid and semi-arid regions. This study delves into the pivotal role of air velocity in the operation and efficacy of pad evaporative cooling systems. The interaction between air velocity and evaporative cooling efficiency is explored through comprehensive experimental analysis and computational simulations. The investigation encompasses diverse air velocities to gauge their influence on temperature reduction and energy consumption. Experimental data reveals a distinct pattern of air temperature drop: 1.7 °C at 5.1 m/s, 1.2 °C at 5.7 m/s, and 1.0 °C at 6.5 m/s. Strikingly, the highest saturation efficiency of 42.5% is attained at an air velocity of 5.1 m/s, surpassing efficiencies of 5.7 m/s by 25.6% and 6.5 m/s by 41.17%—notably, the cooling capacity peaks at approximately 740 W for an air velocity of 1.5 m/s.

Keywords Evaporating cooling · Indoor climate control · Cooling pads · Energy efficiency · Effect of air velocity

1 Introduction

Evaporative cooling, effective in arid regions, harnesses water evaporation for eco-friendly indoor cooling. Air velocity is pivotal among factors affecting pad cooling performance. In a changing climate, demand for efficient cooling grows, as conventional methods prove unsustainable. Evaporative cooling, widely used across residential, commercial, and industrial sectors, offers energy efficiency and environmental

A. Chauhan (✉) · V. Patel · A. Parekh
Department of Mechanical Engineering, SVNIT, Surat, Gujarat, India
e-mail: ankitsinh.svnit@gmail.com

V. Patel
e-mail: vkp@med.svnit.ac.in

A. Parekh
e-mail: adp@med.svnit.ac.in

A. Kumar R. et al. (eds.), *Proceedings of Fluid Mechanics and Fluid Power (FMFP) 2023, Vol. 6*, Lecture Notes in Mechanical Engineering,
https://doi.org/10.1007/978-981-96-9600-0_34

benefits. Cooling efficiency hinges on diverse factors, with air velocity through wet pads playing a key role. Comprehending this influence is vital for optimal evaporative cooling system design and operation.

2 Literature Review and Objective

Extensive research has delved into the transformative effects of evaporative cooling on the efficiency and performance of air conditioning systems. Across a range of studies [1–4], researchers have explored the potential of pre-cooling air as a means to reshape the dynamics of these systems. One prominent study [3] revealed a significant 31% upswing in cooling capacity while concurrently reducing power consumption by 16% through the integration of evaporative cooling methods. This finding underscores the profound impact of leveraging natural processes to augment energy efficiency. Moreover, a detailed investigation into the impact of cooling pad thickness on air conditioning performance was undertaken [4]. This scrutiny unveiled a pivotal correlation between pad thickness and system effectiveness. Impressively, an optimal thickness of five cm emerged, demonstrating its role in optimizing overall system performance and underscoring the importance of meticulous design considerations. Another researcher [5] directed focus toward the condenser element of refrigeration systems. By introducing a five mm cellulose pad for evaporative cooling, a remarkable 50.3% enhancement in performance and a substantial 20.3% reduction in power consumption were achieved. This outcome not only underscores potential energy savings but also highlights the role of innovative approaches in reshaping traditional cooling paradigms. On a regional scale, a study conducted in Sudan by Ahmed et al. [6] compared diverse types of evaporative cooling pads in greenhouse settings. This comparative analysis brought forth intriguing disparities in saturation efficiencies among straw, CELdek, and sliced wood pads, showcasing the influence of pad materials on cooling efficacy. Transitioning to split air conditioners, Thiangchanta et al. [7] embarked on investigating the pre-cooling of condensers through evaporative pads. Their results showcased a notable 74.3% enhancement in heat rejection and an augmented coefficient of performance (COP) compared to conventional methodologies. Likewise, Adarsh et al. [8] harnessed cellulose pads to cool condenser air within air conditioning systems, leading to a substantial COP elevation, attaining an impressive 8.03 in contrast to the conventional 5.98. This advancement holds far-reaching implications for energy-efficient climate control. Chainarog and Doungsong [9] ventured into the realm of evaporative cooling for split air conditioners, uncovering COP improvements spanning from 6 to 48%, alongside a consequential 4–15% reduction in energy consumption. These findings underscore the potential of such methods in the pursuit of sustainable solutions. In a different context, Chang et al. [10] delved into the intricate relationship between material segregation and air velocity within evaporative cooling systems. Their exploration illuminated optimal air velocity effects, paving the way for refined operational strategies. Finally, Ergun's groundbreaking work [11] culminated in a correlation facilitating the understanding of pressure drops in randomly packed beds, forming a cornerstone for comprehending the intricate hydrodynamics of these systems. Collectively, these studies epitomize the potential of evaporative cooling approaches to redefine the efficiency, energy consumption, and overall performance landscape of air conditioning and refrigeration systems.

$$C_T = \frac{T_{\text{turbine}}}{T_{\text{available}}} = \frac{T_{\text{turbine}}}{0.5\rho A V^2 R} \quad (1)$$

$$C_P = C_T \times \text{TSR} \quad (2)$$

3 Experimental Procedure

The experiment study was carried out on water cooler in SVNIT, Surat, and the measurement was taken for month May 2023. For this study, the direct evaporative cooling (DEC), a 65-W axial blower, and an 18-W submersible water pump were employed to move water through the system at a rate of up to 1200 L/h. While air is moving horizontally through the device, water is sprayed on the pad's upper side. A cooling pad made of cellulose with dimensions of 41.5 cm long, 29.5 cm wide, and 5 cm thick was employed as the cooling medium. The effectiveness of the direct evaporative cooling unit at various airflow rates was tested through experiments. An anemometer is used to measure air velocity, dry bulb temperature, and relative humidity of air. The air velocity and pad cross-section area were used to calculate the mass flow rate of air entering the evaporative cooling unit. The evaporative air cooler's axial fan has three speed settings: low, medium, and high. Using a submersible water pump, water from the water tank is transferred to the cooling pad (Figs. 1, 2; Table 1).

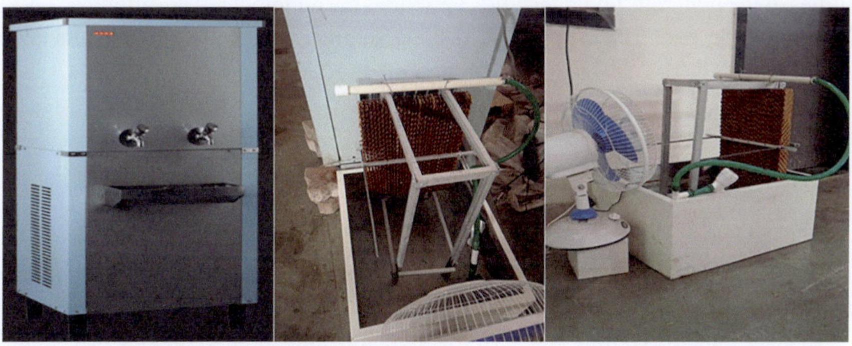

Fig. 1 Pictorial view of experimental setup

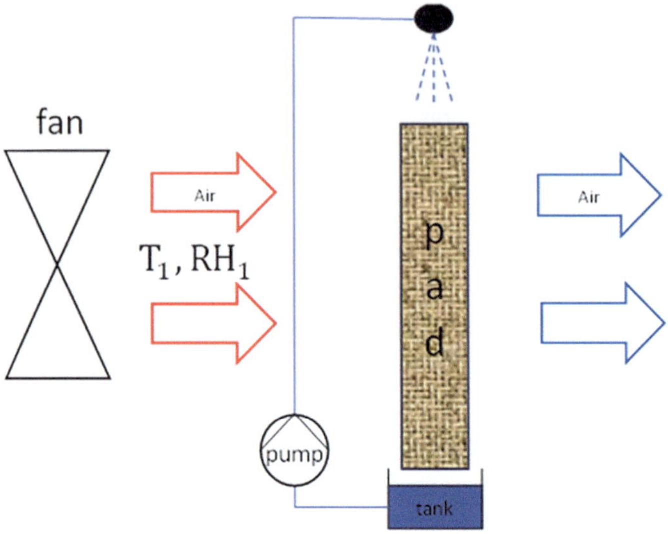

Fig. 2 Schematic diagram of the setup

Table 1 Accuracy of instruments

Independent variables	Accuracy (%)
Air velocity (Range: 0.4–30.0 m/s)	±3
Temperature (T) (K-type) (Range: 0–50 °C)	±1.2
Humidity sensor	±1.2

3.1 Performance Evaluation

The following equations were used to determine the evaporative cooling (EC) capabilities of the materials under test: Eq. 3 provides the air temperature difference (ΔT) throughout the pad. The evaporative cooling saturation efficiency (ε) is calculated via Eq. 4 ASHRAE (2001).

$$\Delta T = T_1 - T_2 \tag{3}$$

$$\varepsilon = (T_1 - T_2)/(T_1 - T_{wb}) \times 100 \tag{4}$$

where T_1 for the inlet temperature, T_2 for outlet temperature and T_{wb} is evaporative inlet wet bulb temperature °C.

To determine the cooling capacity, the mass flow rate of air across the pad should be assessed. It was calculated by dividing the measured air velocity at the outlet V by the average air density and the cross-section area (Ac) of the material sample, as given in Eq. 5. Equation 6 is used to calculate the air mass flow rate. The sensible cooling capacity is calculated using the temperature drop and air mass flow rate as shown in Eq. 6.

$$m_a = \rho V \mathrm{Ac} \tag{5}$$

$$Q_c = m_a C_p (T_1 - T_2) \tag{6}$$

here Q_c is cooling capacity (kJ/s), m_a for air flow rate (kg/s), C_p for specific heat of air (kJ/kg · K).

4 Results and Discussion

The information is displayed as temperature drop, cooling capacity, cooling efficiency, and feasibility index. The highest average temperature drop was recorded at 1.7 °C on 5.1 m/s. The temperature drop was nearly same to 1.2 °C and 1.0 °C, respectively, at greater air speeds of 5.7 m/s and 6.5 m/s. The greatest temperature drop indicates superior cooling capacity, which is 0.74 kJ/s. This finding is fascinating because it demonstrates how important contact parameters and air retention are for the cooling pad. Compared to greater velocity airflow, a slower airflow across the cooling pad allows for more surface contact on the pad surface and improved heat transfer. Less time the air stream is in contact with the wet pad, the greater the air mass flow rate through the cooling surface. Less water evaporates as a result.

5 Conclusions

It was discovered that air velocity does have a significant impact on a direct evaporative system's ability to cool, with the highest cooling capacity being measured at 0.74 kJ/s at the low air velocity from Fig. 5. The setup with the lowest velocity had the highest efficiency, 42.49% from Fig. 4. Slow air velocity encourages longer air retention times on the surface of cooling pads, allowing for more evaporation to occur, and lowers temperature as shown in Fig. 3.

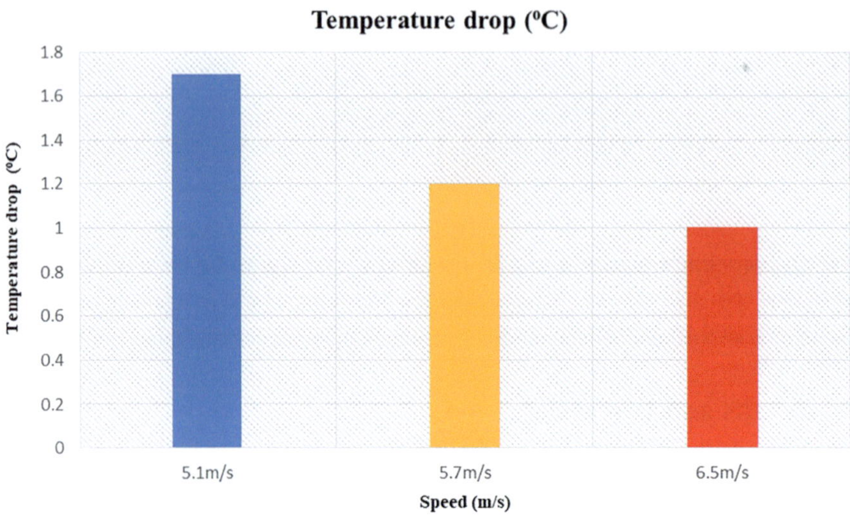

Fig. 3 Temperature drop across the pad at different air velocity

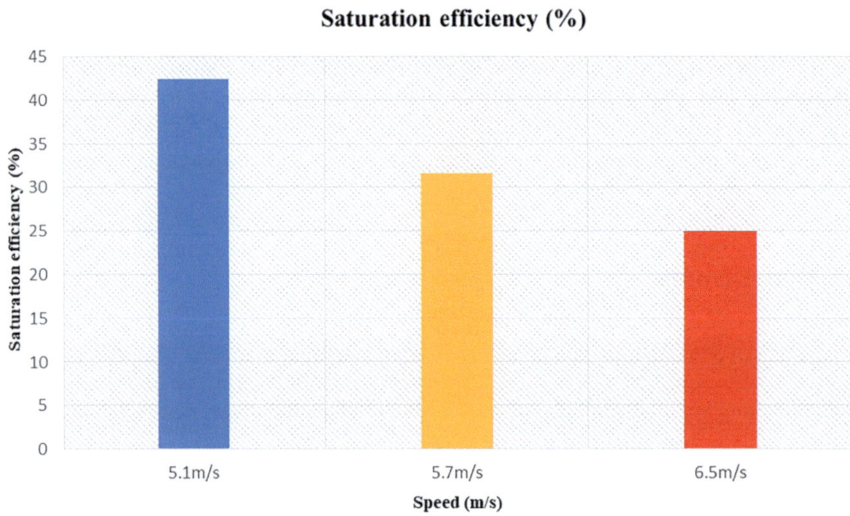

Fig. 4 Saturation efficiency of the pad at different air velocity

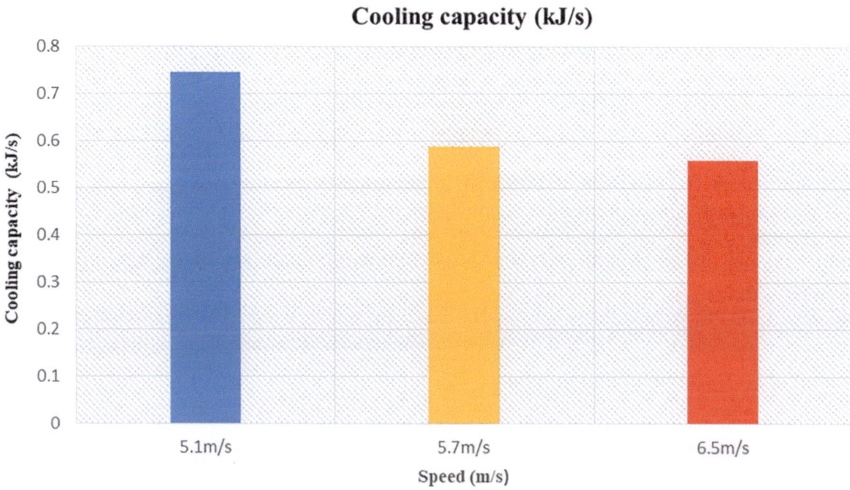

Fig. 5 Cooling capacity of the pad at different air velocity

Nomenclature

Ac	Cross section area (m^2)
C_p	Specific heat at constant pressure ($kJ/kg \cdot K$)
EV	Evaporative cooling
m_a	Mass flow rate of air (kg/s)
Q_c	Cooling Capacity (kJ/s)
T_1	Inlet air temperature (°C)
T_2	Outlet air temperature (°C)
ρ	Density of air (kg/m^3)
ε	Saturation efficiency (%)

References

1. Goswami DY, Mathur GD, Kulkarni SM (1993) Experimental investigation of performance of a residential air conditioning system with an evaporatively cooled condenser
2. Hajidavalloo E (2001) Increasing COP of window air conditioner in very hot weather of Khoozestan. Research Project Report to Management and Programming Organization
3. Hajidavalloo E (2007) Application of evaporative cooling on the condenser of window-air-conditioner. Appl Therm Eng 27(11–12):1937–1943
4. Martínez P, Ruiz J, Cutillas CG, Martínez PJ, Kaiser AS, Lucas M (2016) Experimental study on energy performance of a split air-conditioner by using variable thickness evaporative cooling pads coupled to the condenser. Appl Therm Eng 105:1041–1050
5. Hajidavalloo E, Eghtedari H (2010) Performance improvement of air-cooled refrigeration system by using evaporatively cooled air condenser. Int J Refrig 33(5):982–988

6. Ahmed EM, Abaas O, Ahmed M, Ismail MR (2011) Performance evaluation of three different types of local evaporative cooling pads in greenhouses in Sudan. Saudi J Biol Sci 18(1):45–51
7. Thiangchanta S, Do TA, Suttakul P, Mona Y (2021) Energy reduction of split-type air conditioners using a pre-cooling system for the condenser. Energy Rep
8. Dixit AM, Desai A, Vyas A (2013) Improving efficiency of air conditioner by cellulose pad. Int J Eng Sci Human 3(1):5–9
9. Chaktranond C, Doungsong P (2010) An experimental evaluation of energy saving in a split-type air conditioner with evaporative cooling systems. Int Trans J Eng Manag Appl Sci Technol 1(1):009–018
10. Chang LL, Xu B, Jiang T (2007) Research of air-flow feeding in Sinter. Min Metall Eng 27(2):54–57
11. Zhu H, Hao X, Wen Z, Zhang Y, Chen H (2004) Mathematical model for the thermal process of controlled cooling of wires and its numerical simulation. Int J Miner Metall Mater 11(6):505–510

Experimental Study on the Evaporation of Iron(III) Nitrate Nonahydrate Droplet

A. Ghosh, S. Mikhil, S. Bakshi, and T. N. C. Anand

Abstract Flame spray pyrolysis is a promising method used for the synthesis of nanoparticles which has several applications in the field of bioelectronics, sensors, fuel cells, solar cells and batteries. Droplet heating and evaporation play a vital role in this process. In this work, the heating and evaporation of droplets of the precursor solution of iron (III) nitrate nonahydrate (INN) in ethanol was studied experimentally. The experiments were performed for particular mass fractions of INN in ethanol. The precursor solution droplet was suspended on a glass fibre of 25 μm thickness, and the droplet was exposed to a hot air flow. The temperature of the flow was controlled by the PID controller, and the range of temperature for the experiment was varied from 50 °C to 300 °C. The corresponding air velocities were kept in the range of 1.1 m/s to 3.1 m/s. The effect of temperature on the rate of evaporation was studied, and the corresponding evaporation rates have been reported. The initial phase of droplet evaporation was observed to follow the d^2 law of evaporation, whereas the later phase was characterized by puffing and residue formation.

Keywords Flame spray pyrolysis · Iron(iii) nitrate nonahydrate · Droplet evaporation

A. Ghosh (✉) · S. Mikhil · S. Bakshi · T. N. C. Anand
Department of Mechanical Engineering, Indian Institute of Technology Madras, Chennai, Tamil Nadu, India
e-mail: agnimitra.ghosh10@gmail.com

S. Bakshi
e-mail: shamit@iitm.ac.in

T. N. C. Anand
e-mail: anand@iitm.ac.in

T. N. C. Anand
Department of Mechanical Engineering, Indian Institute of Technology, Palakkad, Kerala, India

A. Kumar R. et al. (eds.), *Proceedings of Fluid Mechanics and Fluid Power (FMFP) 2023, Vol. 6*, Lecture Notes in Mechanical Engineering,
https://doi.org/10.1007/978-981-96-9600-0_35

1 Introduction

Nanoparticle synthesis is the science of creating materials with dimensions at the nanoscale, typically ranging from 1 to 100 nm. These materials exhibit unique properties and behaviours distinct from their bulk counterparts due to their size-dependent quantum effects and increased surface area-to-volume ratio [1]. Nanoparticle synthesis methods aim to precisely control the size, shape and composition of nanoparticles to tailor their properties for specific applications. Flame spray pyrolysis (FSP) is one of the versatile techniques used in nanomaterial synthesis.

In FSP, the process starts with the generation of a flame that acts as a high-temperature reactor. A precursor solution of the metal salt or organometallic compound is sprayed in the form of a fine mist through a capillary into the flame [2, 3]. The process is followed by heating, evaporation and combustion of precursor solution in the gas phase, converting it to nanoparticles. The process involved in FSP determines the final properties of the nanoparticles. The particles are then collected and can be further processed to fine-tune their properties. FSP produces highly uniform and crystalline nanoparticles [4], reducing the need for costly post-synthesis purification steps. The rapid synthesis rate and continuous production capabilities of FSP make it industrially scalable and attractive for large-scale nanomaterial manufacturing. Some of the essential applications of these nanomaterials are in the field of catalysis, optics and photonics, sensors, health care, magnetic material, electroceramics for fuel cells and composite material [3, 5–7]. Therefore, it is of particular interest to control the process, and as a result, the study of droplet evaporation of the precursor solution is important.

Researchers have reported that the evaporation of droplets of precursor solution is governed by two different pathways [8]. For droplet surface temperature T_s below the thermal decomposition temperature T_{th}, the solvent evaporates, leaving a solid particle behind. In the case of T_s equal to or above T_{th}, thermal decomposition occurs, the precursor solution transforms into the gaseous phase, and nanoparticles are formed directly from the gaseous phase. In the context of producing iron oxide nanoparticles, iron(III) nitrate nonahydrate is popularly employed as a precursor [9]. Ethanol is a common solvent used for INN [10]. Narasu et al. [8] conducted a numerical study to understand the behaviour of INN precursor solutions in ethanol in the temperature range of 400 K to 1200 K. Two distinct mass fractions of INN in ethanol were considered, viz. 0.0125 and 0.025. Considering the non-availability of experimental data on the evaporation behaviour of INN, the present work focuses on the evaporation of precursor solution droplets for a mass fraction of 0.0125 at six different temperatures.

2 Experimental Setup

The experimental setup and methodology used in the present study were identical to those used by Surendran et al. [11] to study the evaporation behaviour of UWS droplets under forced convective conditions. Figure 1 shows the experimental setup.

The evaporation rate was studied by suspending the droplet on a micro glass fibre of 25 μm diameter. Micro glass fibre was used to minimize the conduction heat transfer from the suspender to the droplet. The two ends of the micro glass fibre were attached to two posts emerging from a slideable plate. The slideable plate could be slid on the grooves made on a plate mounted at the top of the vertical column (Fig. 2). The column was 1.7 m in height and was used for heating and conditioning the air before it interacted with the droplet. The conditioning of the air was done using a set of coarse and fine metal meshes fixed inside the column. The coarse mesh was kept at the bottom of the column, and the fine mesh was at the top. A 3 kW heating element placed inside the vertical column heated the flowing air and created the desired temperature around the suspended droplet. A K-type thermocouple (thermocouple-B) was fixed 15 cm above the heating element, measured the air temperature, and provided feedback to a PID heater controller. A second K-type thermocouple (thermocouple-A) was set at a point such that the tip of the probe coincided with the point of suspension of the droplet. This thermocouple gave the temperature readings by which the attainment of steady-state conditions was ascertained.

All the experiments were performed with dry air. The airflow rate and the set temperature on the heater, required to obtain the desired temperature around the droplet,

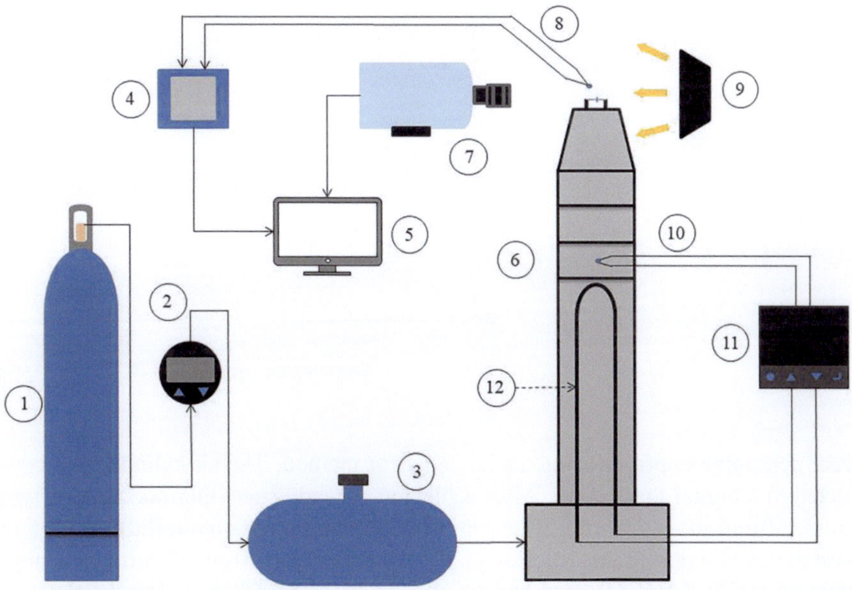

Fig. 1 Schematic of the experimental setup. 1. Compressed air cylinder, 2. Flow metre, 3. Buffer tank, 4. NI data acquisition system, 5. Computer, 6. Verticle column, 7. High-speed camera, 8. K-type thermocouple-A, 9. LED light source, 10. K-type thermocouple-B, 11. PID heater controller, 12. Heating coil

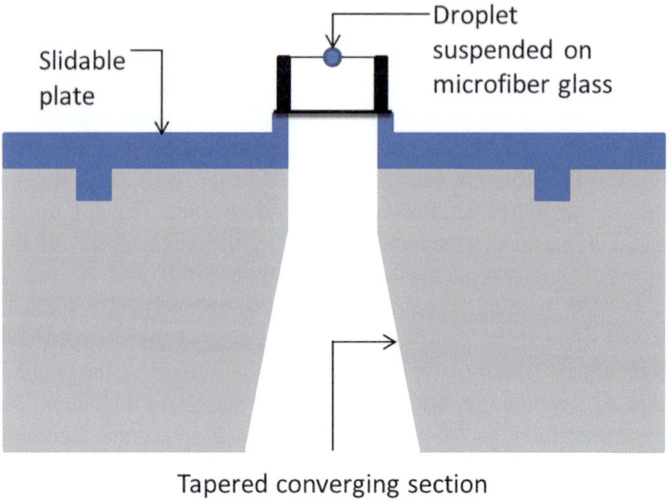

Fig. 2 Droplet suspended on a microfiber glass

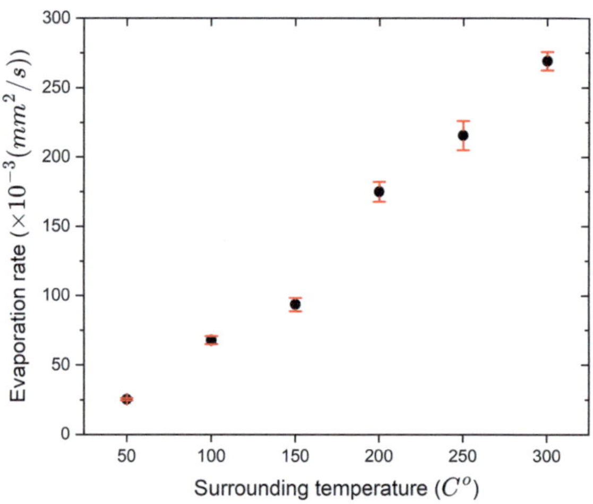

Fig. 3 Evaporation rate versus surrounding temperature of INN/ethanol

were determined upfront using a trial and error method. The air cylinder was connected to a digital flow metre. After achieving a steady state, thermocouple-A was shifted 10 mm above the point of suspension so that the thermocouple did not interfere with the droplet during the experiment. For the experiment, iron(III) nitrate nonahydrate ($Fe(NO_3)_3.9H_2O$) used had a purity of 98% (Srl Chem.). The droplet was suspended by spraying a jet of precursor solution from a 60 mL syringe onto the micro glass fibre. Backlit images of the evaporating droplet were captured using a high-speed CMOS camera (Photron FASTCAM SA4) equipped with a 105 mm macro lens (SIGMA 105 mm 1:2.8 DG MACRO HSM). The required backlighting was obtained using a 50 W LED light source. The resulting images were then processed using an image processing code developed in MATLAB. For each temperature, experiments were performed on at least ten droplets.

3 Results and Discussion

The experimental investigation involved testing the precursor solution of INN in ethanol with a mass fraction of 0.0125 at six different temperatures, namely 50 °C, 100 °C, 150 °C, 200 °C, 250 °C and 300 °C. Each temperature setting required different flow rates, leading to relative velocities between the suspended droplets and the air ranging from 1.1 m/s to 3.1 m/s. The study aimed to understand the behaviour of droplets during evaporation under varying thermal conditions.

$$\left(\frac{d}{d_o}\right)^2 = 1 - \frac{k \cdot t}{d_o^2} \tag{1}$$

The analysis focused on the non-dimensional d^2 versus t plot, which exhibits a characteristic trend known as the d^2 law (1), indicating how droplet size changes with time during evaporation. As anticipated, the droplet's evaporation rate increased with higher surrounding temperatures plotted in Fig. 3. This rate was determined by measuring the slope of the curve, where a linear trend-line was fitted to the data points. The values of the evaporation rate are complied in Table 1. Figure 4 visualizes the evaporation characteristics of the INN/ethanol solution, offering insights into its dynamic behaviour.

Remarkably, at elevated surrounding temperatures, particularly 100 °C and above, the evaporation process displayed prominent peaks at its conclusion, which is notable

Fig. 4 Normalized d^2 versus t plot for evaporating INN/ethanol droplets

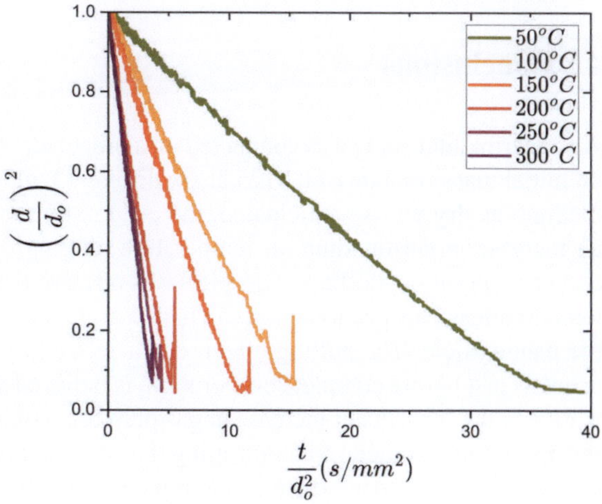

Table 1 Evaporation rates of INN/ethanol for different surrounding temperatures

Temperature °C	Velocity (m/s)	INN-0.0125	
		Evaporation rate x 10^{-3} (mm^2/s)	Std error
50	1.1	25.39	0.056
100	1.5	67.88	0.188
150	1.9	93.54	0.370
200	2.4	174.93	0.452
250	2.8	215.7	0.703
300	3.1	269.14	0.469

in Fig. 4. These peaks signified puffing of the precursor solution, a phenomenon that was not observed at lower temperatures. Puffing refers to the thermally induced rupture of the surface of the droplet, along with the break-up of a secondary single droplet. It occurs when a droplet is heated above its boiling point, and the pressure within the droplet increases to a critical level [12]. The occurrence of puffing in droplets with a mixture of INN as a precursor in a solution of ethanol and 2-ethylhexanoic acid in spray flame synthesis was reported by Jüngst et al. [13]. A similar phenomenon was observed by Haipeng et al. [14] where the gas-phase analysis for the formation of SnO_2 nanoparticles in FSP was conducted experimentally, which revealed instances of surface microexplosions. Figure 5 shows different instances of the droplet during evaporation at 200 °C. The puffing of the droplet could be clearly verified from the image at $t = 5.524$ s (5d). The shell formation and residue formation were prominently visible inside the shell.

4 Conclusions

An experimental study was conducted on a precursor solution of INN in ethanol with an initial mass fraction of 0.0125 at six different temperatures with the surrounding medium as dry air. As anticipated, the evaporation rate of droplets increased with an increase in surrounding air temperature ranging from 50 °C to 300 °C. At the end of evaporation, puffing of droplets was observed, which was not visible at lower temperatures. This phenomenon can have a significant impact on the morphology of the nanoparticle. The puffing effect, characterized by the sudden expansion of the droplets just before complete evaporation, is believed to be the result of rapid vaporization and a subsequent increase in gas pressure within the droplets. Understanding this behaviour is crucial for controlling the size and distribution of the nanoparticles produced in such processes, as it can influence their final properties and applications.

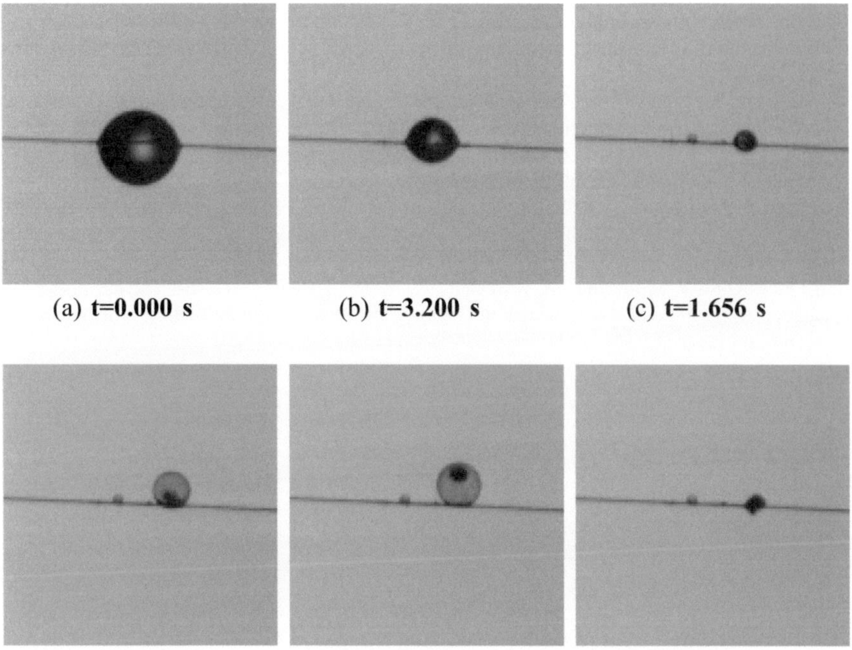

(a) **t=0.000 s** (b) **t=3.200 s** (c) **t=1.656 s**

(d) **t=5.524 s** (e) **t=5.536 s** (f) **5.560**

Fig. 5 Sequence of images of an INN/ethanol droplet evaporating at 200 °C with 0.88 mm initial droplet diameter

References

1. Adlim A (2006) Preparations and application of metal nanoparticles. Indones J Chem 6(1):1–10
2. Schneider F, Suleiman S, Menser J, Borukhovich E, Wlokas I, Kempf A, Wiggers H, Schulz C (2019) Spraysyn—a standardized burner configuration for nanoparticle synthesis in spray flames. Rev Scient Instrum 90(8)
3. Teoh WY, Amal R, Mädler L (2010) Flame spray pyrolysis: an enabling technology for nanoparticles design and fabrication. Nanoscale 2(8):1324–1347
4. Tani T, Mädler L, Pratsinis SE (2002) Homogeneous zno nanoparticles by flame spray pyrolysis. J Nanoparticle Res 4:337–343
5. McNamara K, Tofail SAM (2017) Nanoparticles in biomedical applications. Adv Phys X 2(1):54–88
6. Sharma N, Ojha H, Bharadwaj A, Pathak DP, Sharma RK (2015) Preparation and catalytic applications of nanomaterials: a review. Rsc Adv 5(66):53381–53403
7. Schneider MGM, Martín MJ, Otarola J, Vakarelska E, Simeonov V, Lassalle V, Nedyalkova M (2022) Biomedical applications of iron oxide nanoparticles: current insights progress and perspectives. Pharmaceutics 14(1):204
8. Narasu P, Keller A, Kohns M, Hasse H, Gutheil E (2021) Numerical study of the evaporation and thermal decomposition of a single iron (iii) nitrate nonahydrate/ethanol droplet. Int J Therm Sci 170:107133
9. Morales MP, González-Carreño T, Ocana M, Alonso-Sanudo M, Serna CJ (2000) Magnetic iron oxide/mullite nanocomposite stable up to 1400 °C. J Solid State Chem 155(2):458–462

10. Stodt MFB, Gonchikzhapov M, Kasper T, Fritsching U, Kiefer J (2019) Chemistry of iron nitrate-based precursor solutions for spray-flame synthesis. Phys Chem Chem Phys 21(44):24793–24801
11. Surendran M, Anand TNC, Bakshi S (2020) Experimental investigation of the evaporation behavior of urea-water-solution droplets exposed to a hot air stream. AIChE J 66(2):e16845
12. Shinjo J, Xia J, Ganippa LC, Megaritis A (2014) Physics of puffing and microexplosion of emulsion fuel droplets. Phys Fluids 26(10)
13. Jüngst N, Skenderović I, Südholt BA, Smallwood GJ, Kruis FE, Kaiser SA (2023) High-speed imaging and statistics of puffing and micro-exploding droplets in spray-flame synthesis. Appl Energy Combust Sci 100167
14. Li H, Pokhrel S, Schowalter M, Rosenauer A, Kiefer J, Mädler L (2020) The gas-phase formation of tin dioxide nanoparticles in single droplet combustion and flame spray pyrolysis. Combust Flame 215:389–400

Effect of Viscosity on Shock-Induced Collapse of a Bubble in the Vicinity of a Wall

Dasika Sunder and Ratnesh Shukla

Abstract Cavitation bubbles subjected to incident shock or acoustic waves collapse violently and emit strong blast waves. The emitted blast waves are one of the principal reasons for the damage that collapsing cavitation bubbles cause to underwater hydraulic machinery. The collapse and blast-wave emission processes have been argued to be responsible for the tissue injury in shock and sound wave-based biomedical techniques such as shock wave lithotripsy and burst wave lithotripsy. Since the bubble collapse process is quite violent and short, its detailed experimental characterization is very challenging. Detailed simulations of the bubble collapse process are therefore particularly desirable and can provide the necessary insight. Simulations reveal that viscosity of the medium impacts the collapse and blast-wave emission characteristics significantly. Specifically, the blast-wave-induced peak pressure on the rigid boundary undergoes over a twofold decrease when the viscosity increases by two orders of magnitude. The sensitivity of the peak pressure indicates that the viscosity of the medium can significantly alter the strength of the blast wave and the damage potential of collapsing cavitation bubbles. In addition to peak wall pressure, we also analyze the influence of viscosity of the surrounding liquid on formation of secondary peaks in wall pressure due to impact of re-entrant jet, the shape of the bubble, and movement of the bubble toward the wall during the collapse process.

Keywords Higher-order WENO · Multiphase · Non-Newtonian fluids

Nomenclature

t	Time
z	Axial coordinate
r	Radial coordinate
ρ	Density
u	Axial velocity
v	Radial velocity

D. Sunder (✉) · R. Shukla
Department of Mechanical Engineering, IISc Bangalore, Bengaluru, India
e-mail: dasikasunder@iisc.ac.in

A. Kumar R. et al. (eds.), *Proceedings of Fluid Mechanics and Fluid Power (FMFP) 2023, Vol. 6*, Lecture Notes in Mechanical Engineering,
https://doi.org/10.1007/978-981-96-9600-0_36

p Pressure
E Total energy
α_k Volume fraction of phase k
μ_k Viscosity of phase k
γ_k Specific heat ratio of phase k
π_k Stiffness constant of phase k

1 Introduction

When cavitation bubbles are subjected to incident shock or acoustic waves, they collapse violently and emit strong blast waves. The emitted blast wave could hit a nearby solid surface and cause severe damage to the surface [1]. The pressure peak at the wall due to the blast wave is usually much higher than the peak caused due to incident shock wave hitting the wall. Cavitation is very common in many industrial, medical, and natural phenomena. Many rotating equipments used in marine applications like propellers, rudders, and turbomachinery blades often lead to the formation of cavitation bubbles, which collapse in their vicinity and cause severe damage to their blades. The severity of damage in rotating equipment due to cavitation has been studied extensively in [2–5]. Damage to red blood cells due to artificial heart valves [6] or through acoustic fields [7] is some important physiological examples of damage due to cavitation. Cavitation damage is also used in many productive applications, some of which include cleaning textiles [8] and removing dental plaque by ultrasonic toothbrushes[9]. Bubbles also play an important role in a non-invasive medical procedure known as shock wave lithotripsy, which is used to destroy hardened masses like kidney stones using shock waves. Further, the interaction of the shock wave with the bubbles leads to the formation of very strong blast wave, which could damage the nearby tissues [10, 11]. Light can also be emitted during collapsing bubbles, a phenomenon known as sonoluminescence [12].

Several theoretical, experimental, and numerical studies have been performed to study collapsing bubbles near a wall under various conditions. Equations governing the dynamics of an incompressible bubble collapse in an infinite medium were proposed by Rayleigh and Plesset. [13, 14]. Later Keller and Miksis [15] extended the equations to include the compressibility of the surrounding liquid medium. However, the asymmetry introduced due to the wall and the highly nonlinear nature of shock-bubble interaction makes obtaining theoretical solutions very difficult. Hence, we must rely on numerical or experimental techniques to study the phenomena. Detailed experimental characterization of the collapse process is challenging due to the very short time and spatial scales involved. Johnsen et al. [16] performed detailed numerical studies using higher-order shock and interface capturing finite volume methods to study collapsing bubbles near a wall in an inviscid medium with specific applications to lithotripsy. Nagrath et al. [17] used level-set and finite element methods to study bubble collapse in an incompressible medium. Xiaobin et al. [18] recently performed a three-dimensional bubble collapse near a wall in an inviscid medium using adaptive mesh refinement to study shock formation inside the collapsing bubble in detail. Important experimental studies on collapsing bubbles near solid wall were done by Vogel et al. [19, 20]. Lindau and Laurteborn [21] used modern equipment with up to 100 million frames per second to study the collapse and rebound of a cavitation bubble near a wall.

When the blast wave emitted during the collapse pressure hits the rigid wall in the vicinity, there is an enormous increase in pressure at the wall, which could cause severe damage. Therefore, we will perform detailed studies to understand how the peak in wall pressure changes when the viscosity of the medium changes by two orders of magnitude. In addition to analyzing the peak pressures at the wall, we will also study the effect of viscosity on the shape and movement of the bubble and formation of secondary pressure peaks on the wall caused by the impact of re-entrant jet on the wall.

2 Methodology

Although bubble collapse problems can be solved in three dimensions, the azimuthal symmetry of the problem permits us to use axi-symmetric equations and save a lot of computational cost. The axi-symmetric compressible multiphase Navier–Stokes equations are

$$\frac{\partial \rho}{\partial t} + \frac{\partial(\rho u)}{\partial z} + \frac{\partial(\rho v)}{\partial r} + \frac{\rho v}{r} = 0 \tag{1a}$$

$$\frac{\partial(\rho u)}{\partial t} + \frac{\partial(\rho u^2 + p - \tau_{zz})}{\partial z} + \frac{\partial(\rho uv - \tau_{rz})}{\partial r} + \frac{\rho uv - \tau_{rz}}{r} = 0 \tag{1b}$$

$$\frac{\partial(\rho v)}{\partial t} + \frac{\partial(\rho uv - \tau_{rz})}{\partial z} + \frac{\partial(\rho v^2 + p - \tau_{rr})}{\partial r} + \frac{\rho v^2 - \tau_{rr} + \tau_{\theta\theta}}{r} = 0 \tag{1c}$$

$$\frac{\partial E}{\partial t} + \frac{\partial(u(E+p) - u\tau_{zz} - v\tau_{rz})}{\partial z} + \frac{\partial(v(E+p) - u\tau_{rz} - v\tau_{rr})}{\partial r} + \frac{v(E+p) - u\tau_{rz} + v\tau_{rr}}{r} = 0 \tag{1d}$$

$$\frac{\partial \alpha_1}{\partial t} + u\frac{\partial \alpha_1}{\partial z} + v\frac{\partial \alpha_1}{\partial r} = 0. \tag{1e}$$

The various components of viscous stress tensor and rate tensor are

$$\tau_{zz} = \mu S_{zz}, \quad \tau_{rr} = \mu S_{rr}, \quad \tau_{rz} = \mu S_{rz}, \quad \tau_{\theta\theta} = \mu S_{\theta\theta}$$

$$S_{zz} = 2\frac{\partial u}{\partial z} - \Delta, \quad S_{rr} = 2\frac{\partial v}{\partial r} - \Delta$$

$$S_{rz} = \frac{\partial u}{\partial r} + \frac{\partial v}{\partial z}, \quad S_{\theta\theta} = 2\frac{v}{r} - \Delta$$

$$\Delta = \frac{2}{3}\left(\frac{\partial u}{\partial z} + \frac{v}{r} + \frac{\partial v}{\partial r}\right).$$

In discretizing the above equations, the final terms in Eqs. 1a–1d are treated as source terms. The volume fraction advection Eq. 1e is treated as detailed in [22]. The upwind flux is calculated using rotated HLLC Riemann solver [23]. Spatial discretization of equations is done using fourth-order constrained least squares (CLS)-based adaptive weighted essentially non-oscillatory (WENO) scheme [24]. Temporal discretization is performed using third-order SSP-RK scheme [25]. Stiffened gas equation of state [26] is used to close Eq. 1.

$$e_k = \frac{p_k + \gamma_k \pi_k}{\gamma_k - 1}, \quad k = 1, 2 \tag{2}$$

In the above equation, γ_k is the ratio of specific heats, and π_k is the stiffness constant of the kth component. The mixture equation of state

$$e = \frac{p + \bar{\gamma}\bar{\pi}}{\bar{\gamma} - 1} \tag{3}$$

where $\bar{\gamma}$ and $\bar{\pi}$ are the properties of the mixture given by

$$\frac{1}{\bar{\gamma} - 1} = \frac{\alpha_1}{\gamma_1 - 1} + \frac{\alpha_2}{\gamma_2 - 1}$$

$$\frac{\bar{\gamma}\bar{\pi}}{\bar{\gamma} - 1} = \frac{\alpha_1 \gamma_1 \pi_1}{\gamma_1 - 1} + \frac{\alpha_2 \gamma_2 \pi_2}{\gamma_2 - 1}.$$

The viscosity of the mixture is defined as

$$\mu = \alpha_1 \mu_1 + (1 - \alpha_1)\mu_2$$

where μ_1 is the viscosity of the first fluid, and μ_2 is the viscosity of second fluid.

2.1 Problem Set-Up

The problem set-up is shown in Fig. 1. Initially both the liquid and gas at rest in the domain $(x, y) \in [0, 10] \times [0, 10]$. A bubble of radius $R_0 = 0.5$ is located at a distance H_0 from the wall. The domain is discretized using 2000 cells in each direction. Initially, there are 200 cells across the bubble. A large domain is used to ensure that reflections from the boundaries do not interfere with the bubble collapse process. The distance of the bubble from the wall is varied across the simulations, and the effect of stand-off distance is studied. A left-moving Mach 1.02 shock in liquid moves toward the gas bubble. The side of the bubble closer to the shock is called the proximal or jet side of the bubble. The side of the bubble farther from the shock (closer to the wall) is called the distal side of the bubble. The density of the shocked liquid is $\rho = 1.015$, and pressure is $p = 400$. The velocity of the shocked liquid is $u = -2.407$. The initial stand-off distance is $H_0 = 1.5R_0$, and the shock is initially at a distance $1.5R_0$ from the bubble center. The density and pressure of the liquid in unshocked region are $\rho = 1$ and $p = 1$. The density of the gas bubble is $\rho = 0.001$, and the gas is at the same pressure as the unshocked liquid. The initial speed of sound in the liquid is $c_L = 162.5$.

No-slip adiabatic wall boundary conditions are used on the left wall $z = 0$. The rest of the boundaries are modeled as transmissive boundary conditions. Important parameters used in Eq. 1 are summarized in Table 1. Equation 1 is discretized in space using constrained least squares-based fourth-order accurate adaptive order CLS-WENO-AO(4,3) [24] finite volume method. Three-stage, third-order accurate SSP-RK(3,3) [27] method is used for evolving the solution in time. Rotated HLLC Riemann solver [23] modified for multiphase flows [22] is used for calculating upwind fluxes nec-

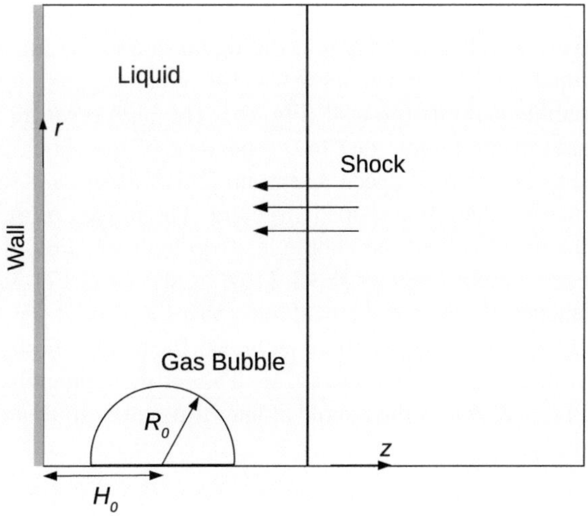

Fig. 1 Computational domain for shock-induced collapse of a gas bubble in liquid. Initially, both the liquid and the gas are at rest

Table 1 Parameters used in Eq. 1

Parameter	High-viscosity fluid	Low-viscosity fluid
γ_1	4.4	4.4
γ_2	1.4	1.4
π_1	6000	6000
π_2	0	0
μ_1	1.0	0.01
μ_2	0.0004	0.0004

essary for convective terms. Viscous fluxes are calculated by central averaging the two extrapolated values of gradients from the cells sharing the given face.

3 Results and Discussion

In this section, we will compare the results of bubble collapse in low-viscosity and high-viscosity liquids with respect to various important parameters.

3.1 Evolution of the Flow

The evolution of flow field with time for collapse in low-viscosity liquid with $\mu_1 = 0.01$ is shown in Fig. 2. After the left-moving shock hits the proximal side of the bubble, a part of the shock wave is transmitted into the bubble, and a rarefaction wave is reflected back into the liquid due to the higher acoustic impedance of the liquid [28], as shown in Fig. 2a. The remaining part of the shock diffracts around the bubble and hits the wall (Fig. 2b). The high pressure induced behind the proximal side of the bubble due to the passage of the shock causes the bubble to involute (Fig. 2c). A high-speed re-entrant jet is formed, and the bubble is further deformed into a kidney-like shaped structure. The impact of the high-speed re-entrant jet on the distal side of the bubble leads to the formation of very strong blast wave (also called water-hammer shock [16]) as shown in Fig. 2d. The blast wave propagates spherically toward the wall and causes a peak in pressure at the wall upon hitting. By this time, the re-entrant jet breaks the bubble through its centerline into a toroidal vortex ring, as seen in Fig. 2e. The jet then catapults the bubble toward the wall. Figure 2f shows the bubble at later times after hitting the wall and rebounding from the wall.

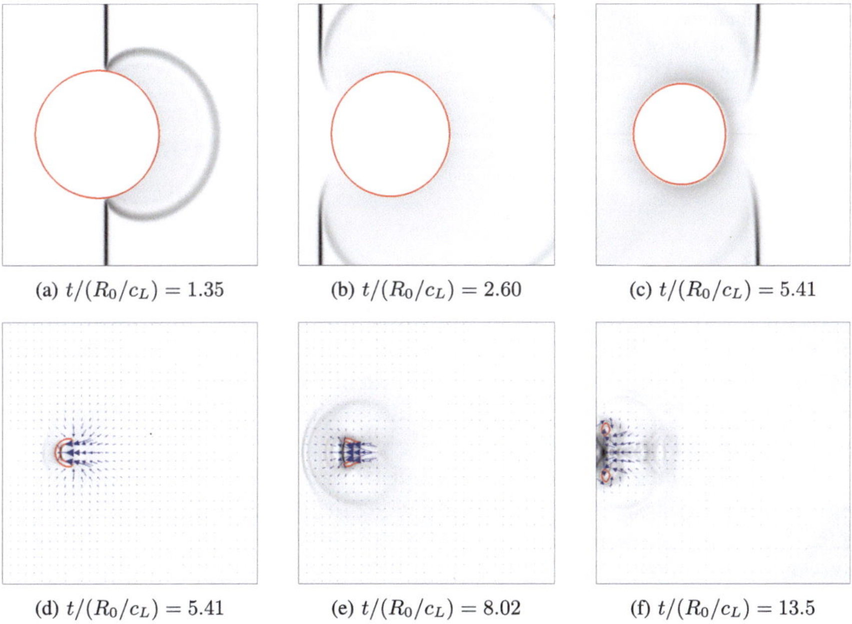

Fig. 2 Numerical pressure Schlieren images showing the evolution of flow. The domain in each figure is $[0, 1] \times \left[0, \frac{1}{2}\right]$. However, to visualize the re-entrant jet and the shape of the bubble more clearly, all images have been reflected along $r = 0$. Contours of $\alpha_1 = 0.01$ (in red color) are also embedded over the Schlieren images

3.2 Peak Wall Pressure

The pressure along the rigid wall is important to monitor, as it indicates the damage caused to the wall due to collapsing bubble. The blast wave hits the wall and causes a peak in wall pressure. The blast wave is formed when the re-entrant jet hits the distal side of the bubble. Figure 3 shows the maximum wall pressure as a function of time with initial stand-off distance $H_0 = 1.5R_0$ for low-viscosity and high-viscosity liquids. In both cases, two prominent peaks in pressure are visible. The first peak at around $t/(R_0/c_L) = 3.54$ is caused due to impact of the incident shock wave on the wall. The second and the larger peak at around $t/(R_0/c_L) = 9.01$ occurs when the blast wave hits the rigid wall. We see that the peak in pressure at the wall in low-viscosity liquid is around 2.2 times higher than the peak in high-viscosity liquid.

To understand the drop in peak wall pressure more thoroughly, we will consider the axial velocity and viscosity profiles at two instants of time. First, at the beginning of re-entrant jet formation, near the proximal side of the bubble, and second, when the re-entrant jet impacts the distal side of the bubble leading to the formation of a strong blast wave.

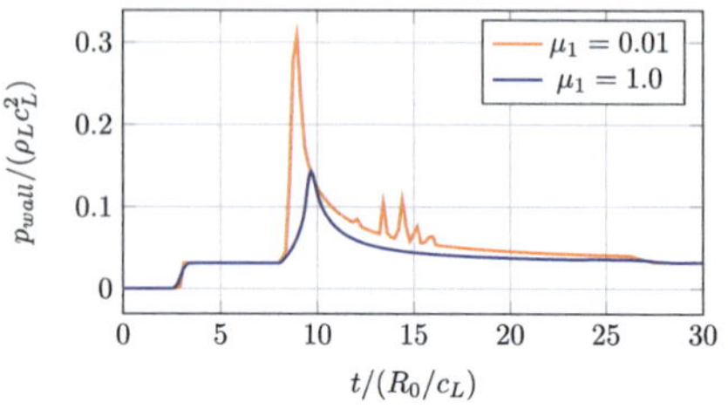

Fig. 3 Maximum wall pressure as a function of time. The initial stand-off distance is $H_0 = 1.5R_0$

Figure 4 shows the velocity along the central axis $r = 0$, just after the re-entrant jet starts forming at the bubble's proximal side. As can be seen from Fig. 4b, the higher viscosity of the fluid leads to a slowdown of the re-entrant jet. Therefore, the peak in re-entrant jet velocity is around 1.4 times higher in low-viscosity liquid than in high-viscosity liquid.

As the re-entrant jet moves toward the distal side of the bubble, it attains very high velocity. Figure 5 shows the velocity along the centerline $r = 0$ when the re-entrant jet is very close to the distal side of the bubble. The peak in re-entrant jet velocity is again higher in low-viscosity liquid than high-viscosity liquid (Fig. 5a and b). However, by this time, the peak in re-entrant jet velocity in low-viscosity liquid is nearly 2.65 times higher than the peak in high-viscosity liquid. The impact of a faster-moving jet on the distal side of the bubble leads to the formation of a stronger blast wave [16] and, therefore, a higher peak in wall pressure. Thus, in addition to dissipation caused due to the higher viscosity of the medium, smaller re-entrant jet velocity is also a major cause for lowering of peak wall pressure with increasing viscosity.

Fig. 4 Axial velocity u along the centerline $r = 0$ at time $t/(R_0/c_L) = 0.52$, just after the formation of re-entrant jet begins

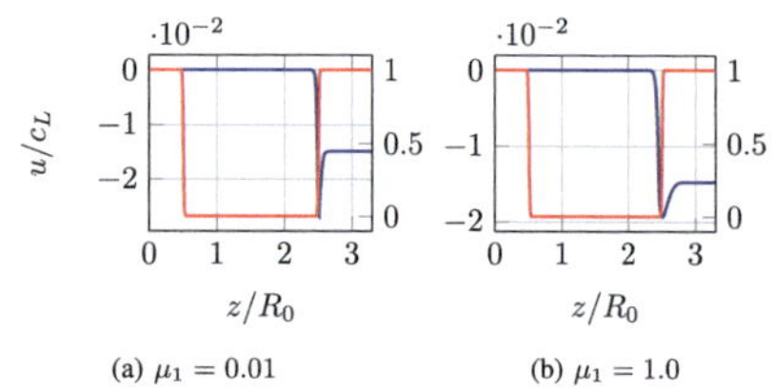

Fig. 5 Axial velocity along the centerline at time $t/(R_0/c_L) = 7.98$, when the jet is close to distal side of the bubble

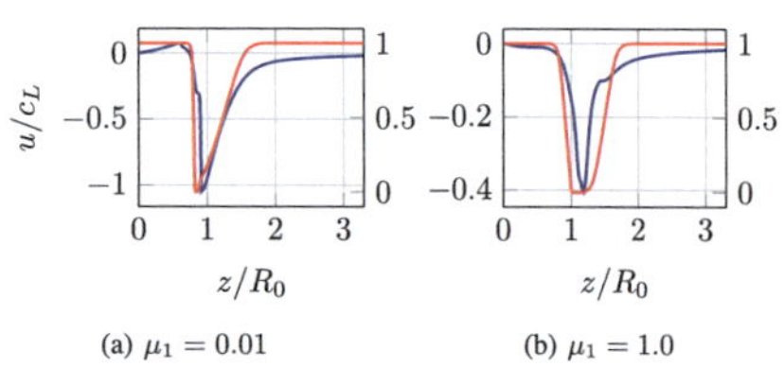

3.3 Secondary Peak Pressures on the Wall

In Fig. 3, apart from the primary peak in wall pressure caused by the impact of the blast wave, secondary peaks in pressure can also be seen in low-viscosity liquid at later times. However, the secondary peaks in pressure are much smaller in magnitude than the primary peak caused by the impact of the blast wave on the rigid wall. As illustrated in Fig. 6a, the secondary peak in pressure is caused due to impact of the re-entrant jet on the wall. However, no such secondary wall pressure peaks are formed in high-viscosity liquid as a result of the reduced velocity of the re-entrant jet due to higher viscosity (Fig. 6b). Figure 7 shows the axial velocity along the centerline $r = 0$ when the peak in re-entrant jet velocity is closest to the wall. In the case of low-viscosity liquid, we can see that the peak in re-entrant jet velocity is more than twelve times higher than that of high-viscosity liquid. Figure 6a also shows that the high re-entrant jet velocity breaks the bubble into smaller toroidal bubbles and propels it toward the wall. Upon hitting, the bubble rebounds back from the wall. In the case of low-viscosity liquid (Fig. 6b), the low-velocity re-entrant jet does not break the bubble into smaller toroidal bubbles. Consequently, the bubble also moves very slowly toward the wall. Until the simulation was run, the bubble did not reach the wall.

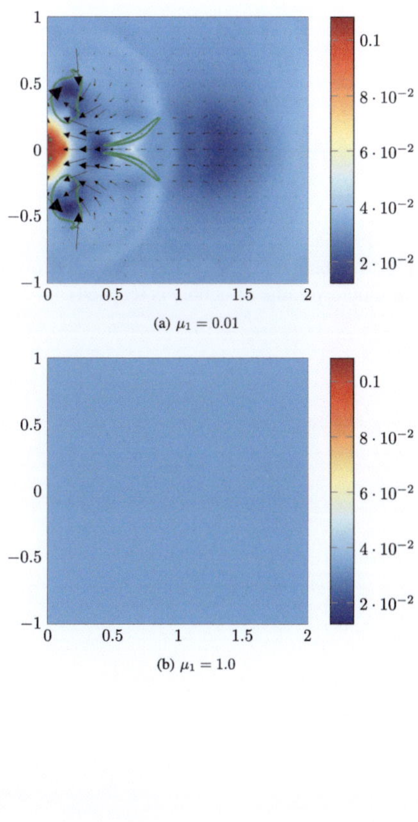

Fig. 6 Pressure field $p/(\rho_L c_L^2)$ at time $t/(R_0/c_L) = 14.3$

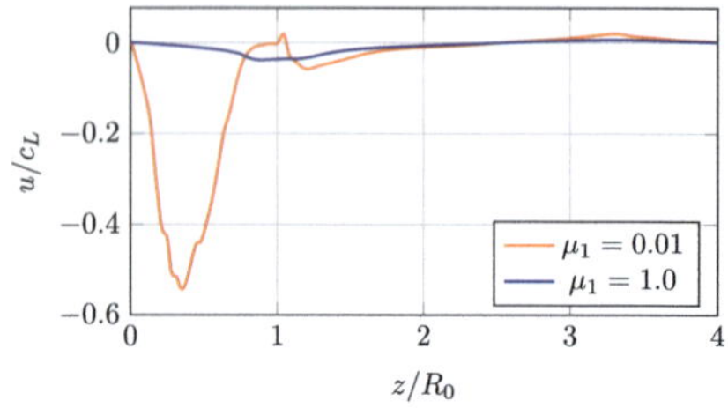

Fig. 7 Axial velocity along the centerline $r = 0$ at $t/(R_0/c_L) = 14.3$, when the peak in re-entrant jet velocity is closest to the wall

3.4 *Bubble Shape and Position*

Figure 8 shows bubble interface ($\alpha_1 = 0.01$) embedded over numerical Schlieren for low-viscosity (Fig. 8a) and high-viscosity (Fig. 8b) liquids at the time of collapse, i.e., when the volume of the bubble is minimum. The shape of the bubble is very distinct in both cases. In the case of low-viscosity liquid, the bubble is on the verge of breaking down into smaller toroidal vortex rings along the centerline. On the other hand, the bubble is much less non-spherical in case of high-viscosity liquid. Figure 9 shows the displacement of the bubble centroid along the z axis. The centroid of the bubble is calculated using its center of volume

$$z_c = \frac{\sum \alpha_2 z}{\sum \alpha_2}$$

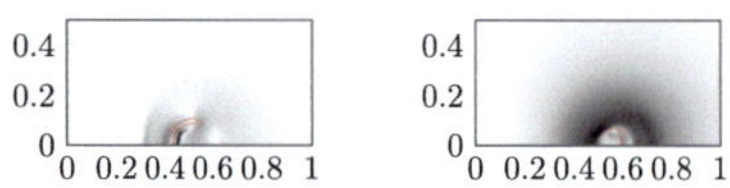

(a) $\mu_1 = 0.01$ (b) $\mu_1 = 1.0$

Fig. 8 Contours of $\alpha_1 = 0.01$, imposed on numerical Schlieren at the time of collapse (when minimum volume of bubble is reached)

Fig. 9 .

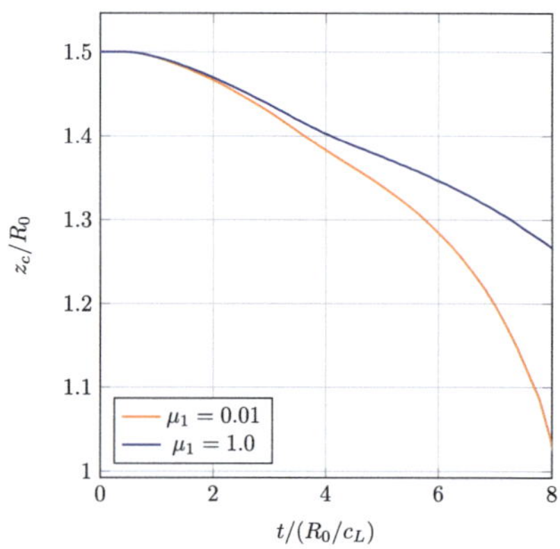

Due to the higher re-entrant jet velocity, the bubble in low-viscosity liquid migrates closer to the wall than the bubble in high-viscosity liquid.

4 Conclusions

In this work, we have analyzed effect of viscosity on shock-induced collapse of a bubble placed near a rigid wall. Axi-symmetric multiphase compressible Navier–Stokes equations were used for solving the problem using fourth-order accurate CLS-WENO-AO(4,3). Simulations were carried out for a Mach 1.02 shock in liquid which corresponds to a pressure ratio of 400 across the shock. Two different liquids having viscosities $\mu_1 = 0.01$ and $\mu_1 = 1.0$ were used for the simulations. Important flow features like the re-entrant jet, water-hammer shock and subsequent peak in pressure at the wall, secondary pressure peaks formed due to jet impact, the bubble's motion and its shape's evolution were studied. Simulations revealed that the viscosity of the medium impacts the collapse and blast-wave emission characteristics significantly. The blast-wave-induced peak pressure on the wall increases by more than two times with decreasing viscosity. Change in re-entrant jet velocity and strength of water-hammer shock due to difference in viscosity were the two significant mechanisms influencing the peak wall pressure. With decreasing viscosity, the higher velocity re-entrant jet can penetrate more through the liquid, forming secondary peaks in pressure. The bubble is also more non-spherical at the time of collapse and moves closer to the wall during the collapse process, leading to a more substantial blast-wave impact.

Acknowledgements We would like to thank SERC, IISc, for providing computational resources (Param-Pravega) necessary for this study.

References

1. Hickling R, Plesset MS (1964) Collapse and rebound of a spherical bubble in water. Phys Fluids 7(1):7–14
2. Knapp RT (1952) Cavitation mechanics and its relation to the design of hydraulic equipment. Proc Inst Mech Eng 166(1):150–163
3. Knapp RT, Daily JW, Hammitt FG (1970) Cavitation. Engineering Societies monographs. McGraw-Hill, New York
4. Li SC (2000) Cavitation of hydraulic machinery. World Scientific
5. Arndt REA (2002) Cavitation in vortical flows. Annu Rev Fluid Mech 34(1):143–175
6. Sotiropoulos F, Trung LB, Gilmanov A (2016) Fluid mechanics of heart valves and their replacements. Annu Rev Fluid Mech 48(1):259–283
7. Daniels S, Kodama T, Price DJ (1995) Damage to red blood cells induced by acoustic cavitation. Ultrasound Med Biol 21(1):105–11
8. Choi J, Kim T-H, Kim H-Y, Kim W (2016) Ultrasonic washing of textiles. Ultrason Sonochem 29:563–567

9. Digel I, Kern I, Geenen E-M, Akimbekov N (2020) Dental plaque removal by ultrasonic toothbrushes. Dentistry J 8:28
10. Lingeman J (2013) Shock wave lithotripsy: state of the art. Springer, New York
11. Freund JB, Colonius T, Evan AP (2007) A cumulative shear mechanism for tissue damage initiation in shock-wave lithotripsy. Ultrasound Med Biol 33(9):1495–1503
12. Brenner MP, Hilgenfeldt S, Lohse D (2002) Single-bubble sonoluminescence. Rev Mod Phys 74:425–484
13. Rayleigh L (1917) On the pressure developed in a liquid during the collapse of a spherical cavity. Lond Edinburgh Dublin Philos Mag J Sci 34(200):94–98
14. Plesset MS (2021) The dynamics of cavitation bubbles. J Appl Mech 16(3):277–282
15. Keller JB, Miksis M (1980) Bubble oscillations of large amplitude. J Acoust Soc Am 68(2):628–633
16. Johnsen E, Colonius T (2009) Numerical simulations of non-spherical bubble collapse. J Fluid Mech 629:231–262
17. Nagrath S, Jansen K, Lahey RT, Akhatov I (2006) Hydrodynamic simulation of air bubble implosion using a level set approach. J Comput Phys 215(1):98–132
18. Yang X, Liu C, Wan D, Hu C (2021) Numerical study of the shock wave and pressure induced by single bubble collapse near planar solid wall. Phys Fluids 33(7):073311
19. Vogel A, Lauterborn W, Timm R (1989) Optical and acoustic investigations of the dynamics of laser-produced cavitation bubbles near a solid boundary. J Fluid Mech 206:299–338
20. Lauterborn W, Bolle H (1975) Experimental investigations of cavitation-bubble collapse in the neighbourhood of a solid boundary. J Fluid Mech 72(2):391–399
21. Lindau O, Lauterborn W (2003) Cinematographic observation of the collapse and rebound of a laser-produced cavitation bubble near a wall. J Fluid Mech 479:327–348
22. Johnsen E, Colonius T (2014) Implementation of WENO schemes in compressible multicomponent flow problems. J Comput Phys 219:715–732
23. Huang K, Wu H, Yu H, Yan D (2011) Cures for numerical shock instability in HLLC solver. Int J Numer Meth Fluids 65:1026–1038
24. Dasika S, Vaghani D, Shukla RK (2022) Constrained least-squares based adaptive-order finite-volume WENO scheme for the simulation of viscous compressible flows on unstructured grids. J Comput Phys 470:111534
25. Gottlieb S, Shu C-W, Tadmor E (2001) Strong stability-preserving high-order time discretization methods. SIAM Rev 43:89–112
26. Harlow FH, Amsden AA (1971) Fluid dynamics: a LASL monograph. National Technical Information Service, U.S. Department of Commerce
27. Gottlieb S, Ketcheson DI, Shu C-W (2009) High order strong stability preserving time discretizations. J Sci Comput 38:251–289
28. Ranjan D, Oakley J, Bonazza R (2011) Shock-bubble interactions. Annu Rev Fluid Mech 43(1):117–140

Thermodynamic Analysis of Various Disc Brake Rotors

Falgun Patil, Poonam Rathod, Girish Patil, Yogesh Gholap, Prachi Chaudhari, and Pramod Kothmire

Abstract The disc brake system in modern passenger cars plays a critical role in converting kinetic energy into thermal energy, necessitating effective heat dissipation for optimal performance. This study employs a combination of experimental and numerical methods, including finite element analysis, to compare different types of brake discs. We assess structural and thermal properties to identify the disc brake rotor type exhibiting the best performance. The research emphasizes the significance of efficient heat dissipation in brake systems to avert potential damage or failure.

Keywords Brake rotor · Experimental analysis · Thermal analysis · Structural analysis

1 Introduction

The disc brake stands as a specialized device, meticulously engineered to decelerate or halt the wheel's rotation. Nonetheless, each act of braking begets heat. The braking system is indispensable for power stability, acting as the conduit through which a vehicle's mechanical energy transforms into a force that brings about its deceleration. Through the medium of dry friction, the brake metamorphoses kinetic energy into thermal energy, which subsequently dissipates into the surrounding environment. To preclude the perils of overheating and ensure unwavering performance, brake cooling emerges as a pivotal concern for automakers striving to fortify their vehicles' safety standards. The brake system constitutes a linchpin in active safety systems, with the disc brake shouldering the role of deftly retarding or arresting wheel motion. When the brake engages, brake pads within the calliper muster frictional force against the disc, transmuting kinetic energy into thermal energy, which demands dispersion through convective cooling. Consequently, the design of brake systems takes on a decisive mantle in guaranteeing the longevity and efficacy of contemporary passenger

F. Patil · P. Rathod (✉) · G. Patil · Y. Gholap · P. Chaudhari · P. Kothmire
School of Mechanical Engineering, MIT Academy of Engineering, Pune, India
e-mail: poonam.rathod@mitaoe.ac.in

A. Kumar R. et al. (eds.), *Proceedings of Fluid Mechanics and Fluid Power (FMFP) 2023, Vol. 6*, Lecture Notes in Mechanical Engineering,
https://doi.org/10.1007/978-981-96-9600-0_37

vehicles. The brake system ranks among the zenith of safety-critical elements in modern vehicles, particularly in its mission to curtail wheel rotation through the pressing of brake pads against rotating wheel discs.

1.1 Basic About the Topic

Sarkar et al. [1] conducted a comprehensive study delving into the temperature distribution of disc brakes. Their methodology hinged on employing Ansys, a finite element simulation tool, to execute static thermal analysis on the disc rotor. Their evaluation of different disc rotor performances hinged on temperature distribution, factoring in cooling mechanisms such as convection and radiation. Parab et al. [2] embarked on an exploration of the structural and thermal facets of brake discs. Their quest involved deploying structural and thermal analysis methodologies to scrutinize disc brakes crafted from three distinct materials: stainless steel, cast iron and carbon-carbon composite. Their mission was to gauge deformation, stress levels, and temperature profiles, with the aim of determining the most suitable material. Adamowicz and Grzes [3] ventured to evaluate the impact of convective cooling on temperature distribution in disc brakes, particularly during repetitive braking scenarios. They harnessed a three-dimensional finite element model for transient thermal analysis, enabling the revelation of temperature profiles on the disc brake's contact surface.

1.2 Parametric Study

Jule et al. [4] embarked on an examination of the structural and thermal characteristics of solid and cross-drilled rotors. Their analysis, rooted in finite element techniques, involved three materials, each endowed with unique properties. This inquiry unveiled total deformation and temperature distribution across the disc brake rotor's surface, offering insights into their repercussions. Belhocine and Bouchetara [5] dived into the realm of the thermal behaviour of full and ventilated brake discs within vehicles. Leveraging the computational prowess of Ansys, their research yielded results that aligned satisfactorily with specialized literature. Their comparative analysis encompassed two disc rotor types employing three distinct cast iron materials. Belhocine and Afzal [6] ventured to evaluate the suitability of various materials for disc rotors. Through numerical simulations, they orchestrated transient thermal and static structural analyses. Their investigation unearthed stress concentration, total deformation, and temperature distribution dynamics over time, ultimately deeming a grey cast iron disc brake rotor as the zenith of performance. Belhocine et al. [7] conducted a thermal and structural analysis of a disc brake assembly during a single-stop braking event. Leveraging Ansys, they unveiled disc temperature dynamics and dissected stress concentration, structural deformation and contact pressure between the brake disc and pads. The study further delved into the multifarious influences encompassing a fixed calliper, variable friction coefficients, and fluctuating disc speeds on stress concentration, structural deformation, and contact pressure. McPhee and Johnson [8] charted a dual course by amalgamating experimental and analytical methods to decode the intricacies of heat transfer and flow within vented brake rotors. Their experimental domain encompassed the assessment of heat transfer and fluid dynamics, with analytical solutions dovetailing where applicable. Notably, their analytical solutions harmonized with the experimental findings. Yevtushenko and Kuciej [9] embarked on an odyssey to unravel the temperature and thermal stresses borne by the pad and disc during the act of braking. An amalgam of analytical and numerical methods was their compass. Their exploration featured a two-element model encapsulating the braking process, comprising a pad (referred to as the strip) and a disc surface (known as the semi-space). This dynamic interplay unfolded under time-dependent velocity and uniform retardation. Dhir [10] delved deep into the thermo-mechanical performance of automotive disc brakes, wielding the potent tool of the Finite Element Method. His mission focused on unravelling the temperature escalation of the brake during operation and its repercussions on disc durability. The findings struck a resonant chord, revealing the substantial influence of temperature elevation on disc durability, with potential ramifications on braking performance and safety. Belhocine [11] conducted a multifaceted exploration, delving into the thermal performance and stresses of an automotive disc brake system. The investigation hinged on a finite element approach, with the prediction of factors such as deformation and von Mises stress serving as heralds of the structural performance of the disc-pad model. To further enhance understanding, a thermos-mechanical analysis entered the fray, encompassing convection, adiabatic elements and heat flux constituents. The veracity of the findings was bolstered through the inclusion of experimental data on contact pressure distributions within brake pads.

1.3 Comparative or Parametric Study

Mathad et al. [12] orchestrated a non-linear analysis scrutinizing the structural and thermal attributes of automotive brake discs. The synergy of SolidWorks for brake disc modelling and Ansys Workbench for analysis kindled their inquiry. The findings reverberated with the supremacy of carbon ceramic disc brakes, surpassing other brake types in structural and thermal domains under similar boundary and loading conditions. Pranta et al. [13] embarked on an expedition into the structural and thermal characteristics of modified disc brake rotors. Their voyage involved the construction of finite element models designed using SolidWorks, subsequently subjected to simulation via Ansys. The results unfurled the ascendancy of modified models, showcasing superior brake force capabilities devoid of cracking or buckling during operation. Cho and Ahn [14] plunged into a transient thermoelastic analysis of a disc brake system. Their voyage navigated the waters of the Fast Fourier Transform and the Finite Element Method. Their findings unveiled the temperature distributions and displacement dynamics within the disc brake system, particularly at low speeds and the emergence of hotspots at speeds transcending the critical threshold.

1.4 Summary of Above Three Paragraphs, Gaps Identified, Objectives and Scope of Future Studies

The depicted landscape of research spans studies that explore the structural and thermal characteristics of disc brake rotors, leveraging both numerical and experimental approaches. A critical focal point encompasses the scrutiny of deformation, temperature distribution, stress concentration, and contact pressure within the brake system. Furthermore, an array of materials and designs have been placed under the analytical spotlight to unravel their performance under a tapestry of operational conditions. This elucidation has uncovered an overarching gap in the absence of studies analysing environmental factors, such as temperature, impacting the thermal behaviour of the brake system. Moreover, the preponderance of studies primarily concentrates on single-stop braking events, failing to address the ramifications of repeated braking episodes.

The core objective of these studies orbits around the meticulous examination of the structural and thermal properties of automotive brake discs. A synergy of experimental and numerical methods, chiefly the instrument of finite element analysis, has been leveraged to explore factors including stress concentration, total deformation and temperature distribution amid the braking process. The comparative evaluations among different types of brake discs strive to anoint the most superior disc brake rotor in terms of structural and thermal properties. The overarching ambition remains the enhancement of performance, reliability and safety within automotive disc brake systems. The envisioned future studies need to bridge these identified gaps and provide a more panoramic understanding of the factors influencing brake disc performance. In particular, a more in-depth exploration of the structural and thermal properties of modified disc brake rotors stands as a beckoning frontier for research and development.

2 Heat Transfer Concept

Disc brake rotors serve as a pivotal element within a vehicle's braking system, and their thermal performance stands as a linchpin in ensuring the vehicle's safe and efficient operation. Heat transfer analysis takes centre stage in the design and optimization of disc brake rotors, assuring their ability to proficiently disperse heat while preserving their functionality. The thermal behaviour of a disc brake rotor is meticulously scrutinized through the prism of conduction and convection heat transfer principles. Conduction heat transfer occurs within the rotor's material itself, channelling heat from the hotter regions near the brake pads to the cooler zones adjacent to the hub. Concurrently, convection heat transfer manifests when the heated rotor surface interfaces with the surrounding air, permitting convective heat transfer to whisk away heat from the rotor. In delving into the thermal behaviour of a disc brake rotor, a medley of parameters steps into the spotlight, encompassing the rotor's material properties, the thermal attributes of the brake pads, the rotor's design, the airflow encircling the rotor, and the specifics of braking conditions such as duration and frequency. To optimize the thermal performance of a disc brake rotor, multifaceted design adjustments can be embraced, including augmenting the rotor's surface area, enhancing airflow surrounding the rotor, and harnessing materials boasting higher thermal conductivity or specific heat capacity.

2.1 Methodology

The methodology employed for disc brake analysis entails a triad of analytical approaches: experimental analysis, static structural analysis and steady-state thermal analysis.

Experimental Analysis: The inaugural step of the methodology involves conducting an experimental analysis of the disc brake. This entails subjecting the brake system to diverse operational conditions to garner data pertaining to its performance. These tests encompass a spectrum of measurements, encompassing temperature and other pertinent parameters.

Static Structural Analysis: The subsequent phase focuses on the execution of a static structural analysis of the disc brake. This analysis is geared towards unveiling areas characterized by high stress or deformation that might precipitate brake system failure.

Steady-State Thermal Analysis: The third phase introduces a steady-state thermal analysis of the disc brake. This analysis uncovers high-temperature regions that might pose a threat of thermal damage or degradation to the brake components.

Integration and Interpretation: In the final segment of the methodology, the results stemming from the experimental analysis, static structural analysis and steady-state thermal analysis coalesce and are scrutinized holistically to gauge the overarching performance of the disc brake.

2.2 Experimental Setup

Within the experimental setup, a disc brake rotor is mounted upon a small-scale go-kart, strategically designed to emulate real-world braking scenarios. The circular

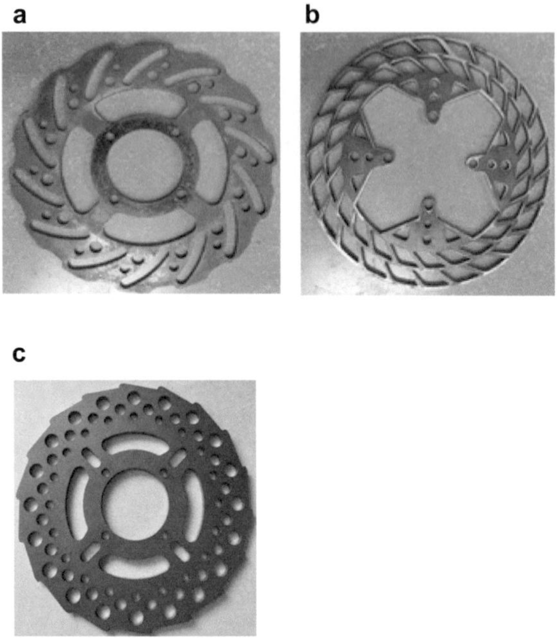

Fig. 1 Disc brake rotor design. **a** Case 1, **b** Case 2 and **c** Case 3

metal disc brake rotor is securely affixed to the wheel hub of the go-kart. The brake system entails brake pads, callipers, and hydraulic lines. When the brake pedal is engaged, it triggers hydraulic pressure generation, compelling the calliper to clamp the brake pads against the disc brake rotor. The ensuing friction between the brake pads and rotor begets deceleration, ultimately halting the go-kart. Assorted parameters, encompassing braking distance, braking efficiency and rotor temperature, are meticulously gauged and scrutinized to evaluate the efficacy and performance of the disc brake system within the go-kart configuration.

The experimental setup is shown in Fig. 1.

Case 1 represents a slotted and drilled rotor configuration, as depicted in Fig. 1a. Case 2 embodies a slotted disc brake rotor, characterized by evenly spaced slots, as illustrated in Fig. 1b. Case 3 portrays the drilled disc-type brake rotor, adorned with small holes, as visualized in Fig. 1c. These slots and holes are ingeniously fashioned to facilitate heat dissipation, enhance braking performance, and remove water, dust and debris. Laser-cut machining processes are employed to craft these rotors, ensuring precise placement of slots and holes.

3 Results and Discussion

The disc brake rotor contends with substantial temperatures, oscillating between a maximum of 116.8 °C and a minimum of 114.63 °C. It contends with a total heat flux of 0.047572 W/mm^2. As evidenced in Fig. 2a, b, the rotor's role is pivotal within the braking system, designed to withstand these intense temperatures and efficiently disperse heat to ensure effective braking performance, as shown in Fig. 2a, b.

The disc brake rotor wrestles with a maximum equivalent stress of 583.21 MPa, signifying its capacity to endure high stress. Simultaneously, it undergoes a total deformation of 0.09571 mm, denoting the extent of deformation or displacement it experiences under load. This is discernible in Fig. 3a, b.

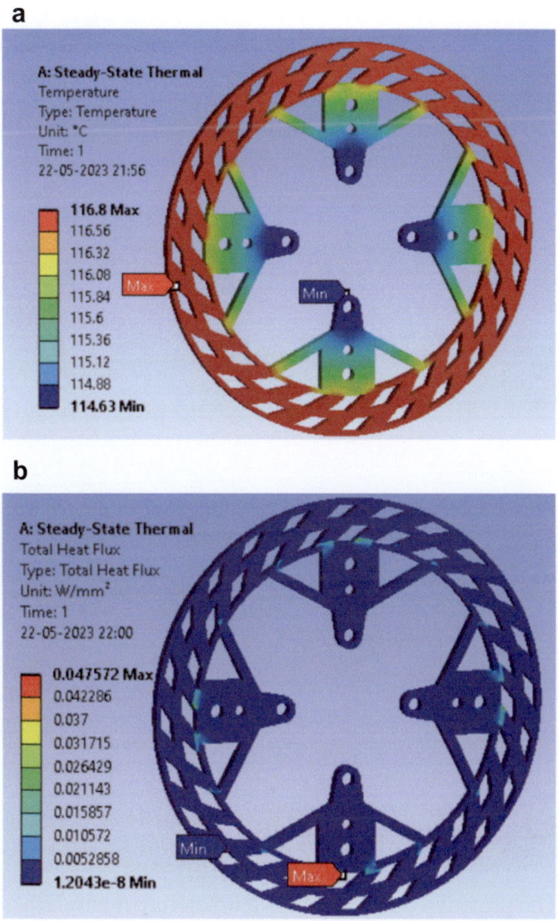

Fig. 2 Steady-state thermal (Case 2). **a** Temperature distribution and **b** heat flux

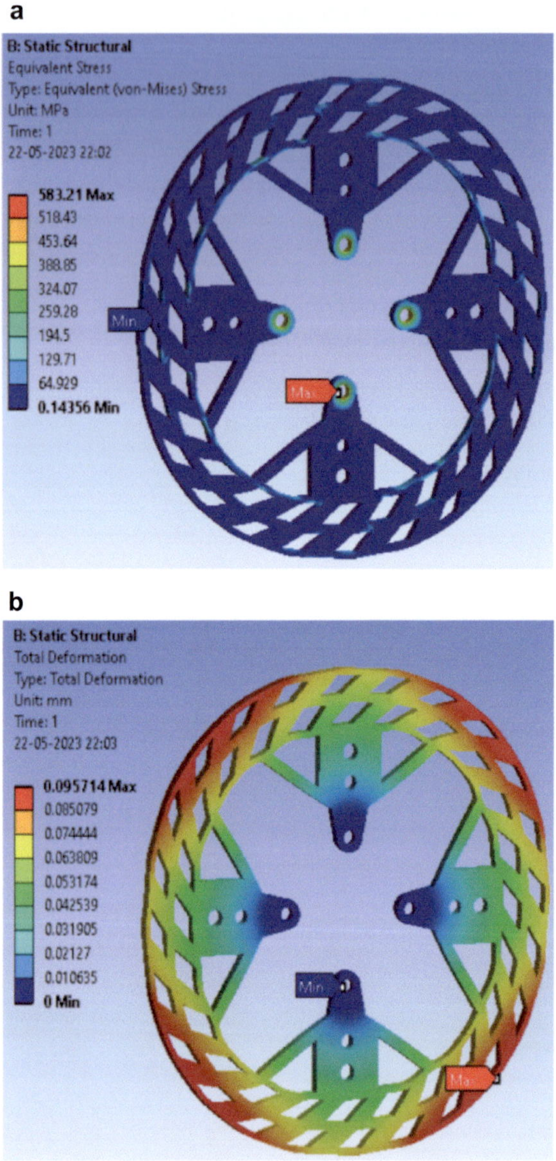

Fig. 3 Static structural (Case 2). **a** Equivalent stress and **b** total deformation

3.1 Comparative Study

The provided dataset unveils temperature measurements for a disc brake rotor under distinct cases, with each case corresponding to particular scenarios or conditions for evaluating the rotor. These temperature records are articulated in °C and are categorized into three cases: Case 1, Case 2 and Case 3. Each case encapsulates

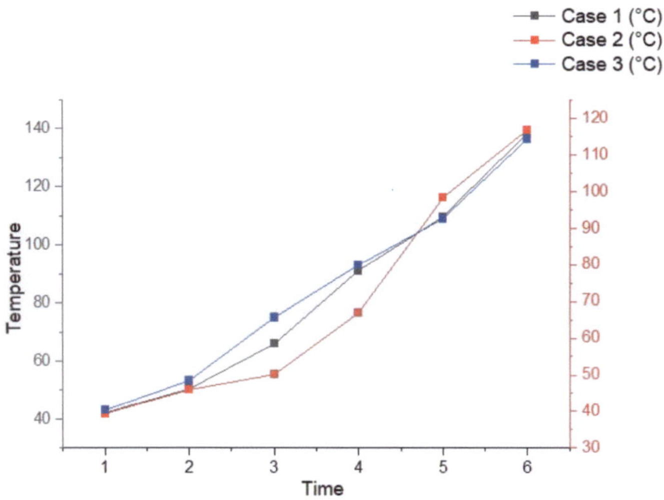

Fig. 4 Actual maximum temperature

multiple temperature measurements, encompassing actual maximum temperature, actual minimum temperature and Ansys minimum temperature.

In Case 2, actual maximum temperatures oscillate between 116.8 and 39.3 °C, while actual minimum temperatures fluctuate between 112.3 and 39.1 °C. The Ansys minimum temperatures in Case 2 vary from 114.4 to 38.98 °C. For Case 3, actual maximum temperatures range from 127.6 to 41.9 °C, while actual minimum temperatures span 121.1 to 40.2 °C. The Ansys minimum temperatures in Case 3 vary from 123.82 to 41.69 °C. In Case 1, actual maximum temperatures extend from 138.4 to 42.1 °C, and actual minimum temperatures range between 130.9 and 40.9 °C. The Ansys minimum temperatures in Case 1 span from 133.27 to 41.851 °C, as illustrated in Figs. 4, 5 and 6.

This dataset serves as a rich trove of information, illuminating the temperature distribution within the disc brake rotor under assorted conditions. It furnishes insights into the rotor's thermal behaviour and aids in the assessment of its performance and potential concerns related to extreme heat or temperature variations. The dataset might have been acquired through experiments or simulations using Ansys software. The values provide a comprehensive snapshot of the temperature distribution and fluctuations within the disc brake rotor across the specified cases, offering a robust foundation for evaluating its performance and thermal behaviour.

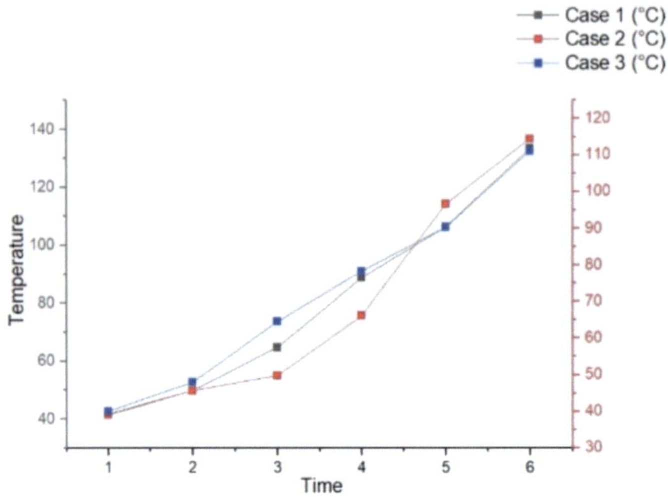

Fig. 5 Actual minimum temperature

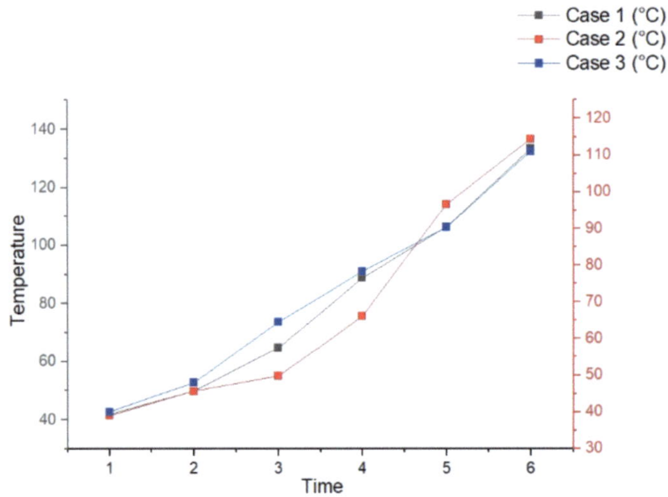

Fig. 6 Ansys minimum temperature

4 Conclusions

A comparative analysis was undertaken, juxtaposing temperature measurements across different cases, namely Case 1, Case 2 and Case 3. The data proffered valuable insights into the thermal behaviour of the disc brake rotor, enabling an appraisal of its performance and identifying potential issues pertaining to excessive heat or temperature fluctuations.

The provided information facilitates an understanding of the thermal nuances of the disc brake rotor, enabling the evaluation of its performance and unveiling issues related to extreme heat or temperature variations. Case 1 experiences temperature differentials ranging from 7.5 to 14.9 °C, with the highest differential noted as 14.9 °C. Case 2 registers temperature variations ranging from 4.5 to 8.5 °C, with the maximum differential recorded as 8.5 °C. Finally, Case 3 showcases temperature differences varying from 6.5 to 10.5 °C, with the highest differential measured at 10.5 °C.

These temperature differentials denote that disc brake rotors face challenges and performance limitations under high-temperature circumstances compared to low-temperature scenarios. These variations in temperature underscore the considerable thermal stress the disc brake rotors endure during operation. The temperature values fluctuate, with Case 1 exhibiting the highest temperatures and Case 2 demonstrating the lowest. These findings underscore the critical importance of effective heat dissipation within disc brake systems to ensure optimal performance and avert potential damage or failure.

References

1. Sarkar S, Rathod PP, Modi AJ (2014) Research paper on modeling and simulation of disc brake to analyse temperature distribution using FEA
2. Parab V, Naik K, Dhale AD (2014) Structural and thermal analysis of brake disc
3. Adamowicz A, Grzes P (2011) Analysis of disc brake temperature distribution during single braking under non-axisymmetric load. Appl Therm Eng 31(6–7):1003–1012. https://doi.org/10.1016/J.Applthermaleng.2010.12.016
4. Jule LT, Krishnaraj R, Nagaprasad N, Stalin B, Vignesh V, Amuthan T (2021) Evaluate the structural and thermal analysis of solid and cross drilled rotor by using finite element analysis
5. Belhocine A, Bouchetara M (2012) Thermal analysis of a solid brake disc. Appl Therm Eng 32:59–67. https://doi.org/10.1016/J.Applthermaleng.2011.08.029
6. Belhocine A, Abu Bakar AR, Bouchetara M (2015) Thermal and structural analysis of disc brake assembly during single stop braking event. https://doi.org/10.1080/14484846.2015.1093213
7. Belhocine A, Afzal A (2020) Computational finite element analysis of brake disc rotors employing different materials. Aust J Mech Eng 1–14. https://doi.org/10.1080/14484846.2020.1733175
8. McPhee AD, Johnson DA (2008) Experimental heat transfer and flow analysis of a vented brake rotor. Int J Therm Sci 47(4):458–467. https://doi.org/10.1016/J.Ijthermalsci.2007.03.006
9. Yevtushenko A, Kuciej M (2010) Temperature and thermal stresses in a pad/disc during braking. Appl Therm Eng 30(4):354–359. https://doi.org/10.1016/J.Applthermaleng.2009.09.015
10. Dhir DK (2018) Thermo-mechanical performance of automotive disc brakes. Mater Today Proc 5(1):1864–1871. https://doi.org/10.1016/J.Matpr.2017.11.287
11. Belhocine A, Ghazaly NM (2015) Effects of material properties on generation of brake squeal noise using finite element method. Lat Am J Solids Struct 12(8):1432–1447. https://doi.org/10.1590/1679-78251520

12. Mathad M, Kohli A, Hombalmath M, Kakol B, Patil AY, Kotturshettar BB, Suresh HK, Patil V (2020) Non-linear structural and thermal analysis of automotive brake disc. https://doi.org/10.1016/J.Matpr.2022.03.426
13. Pranta MH, Rabbi MS, Banik SC, Hafez MG, Chu Y-M (2022) A computational study on structural and thermal behavior of modified disk brake rotors. Alex Eng J. https://doi.org/10.1016/J.Aej.2021.07.013

Thermal Analysis of Exhaust Manifold Using Different Materials

Kaustubh Shinde, Dinesh Puri, Lalit Choudhari, Vaishnavi Kadam, Madhav Murshetwad, and Pramod Kothmire

Abstract When all of the engine's emissions are gathered in one location and fully burned, the exhaust manifold functions as a funnel and directs the emissions into the remaining exhaust system. The creation, development, and manufacturing of a novel exhaust manifold with excellent thermal conductivity is the goal of this project. It also entails not transferring heat to the engine and maintaining a low manifold surface temperature. An analysis of an exhaust manifold system's heat transfer properties is presented in this work. The experimental setup includes a mild steel manifold, SS 304 manifold, and heat wrap. The performance of the system is analysed under different conditions with heat wrap and without heat wrap on MS and SS 304 exhaust manifold. The results show that the system can achieve a maximum thermal conductivity in SS 304 manifold wrapped with glass fibre heat wrap. The study also investigates the effect of different manifolds' thermal conductivity on the system's performance. Overall, this study provides valuable insight into the heat transfer mechanisms of exhaust manifold and their potential to become efficient for single-cylinder engine.

Keywords Exhaust manifold · Heat wrap · Single-cylinder · SS 304 · Glass fibre

1 Introduction

A single-cylinder exhaust manifold is a component used in small internal combustion engines to collect and direct exhaust gases from the engine's single cylinder into the exhaust system. Usually constructed of metal tubing, the exhaust system is mounted to the cylinder head of the engine. The design of single-cylinder exhaust manifold plays a crucial role in the engine's overall performance, particularly in terms of

K. Shinde (✉) · D. Puri · L. Choudhari · V. Kadam · M. Murshetwad · P. Kothmire (✉)
School of Mechanical Engineering, MIT Academy of Engineering, Pune, Maharashtra, India
e-mail: kaustubh.shinde@mitaoe.ac.in

P. Kothmire
e-mail: ppkothmire@mitaoe.ac.in

A. Kumar R. et al. (eds.), *Proceedings of Fluid Mechanics and Fluid Power (FMFP) 2023, Vol. 6*, Lecture Notes in Mechanical Engineering,
https://doi.org/10.1007/978-981-96-9600-0_38

exhaust gas scavenging and back pressure. The manifold's shape, size, and length can affect the flow of exhaust gases and determine the engine's power output.

Heat wrap is often used on exhaust manifolds because it can help to increase exhaust gas velocity, which can improve engine performance by increasing exhaust flow and reducing backpressure. Additionally, by reducing the amount of heat radiated from the exhaust manifold, heat wrap can help to protect other components in the engine compartment from heat damage.

SS 304 material has durability, ease of fabrication, and resistance to corrosion caused by exposure to high-temperature exhaust gases and moisture. Additionally, SS 304 manifold is resistant to oxidation and scaling at high temperatures. Heat wrap provides several benefits to the SS 304 manifold. Firstly, it helps to reduce the amount of heat radiating from the exhaust manifold, which can help to protect surrounding components from excessive heat exposure. Secondly, by retaining more heat within the exhaust system, the heat wrap can improve exhaust gas velocity and scavenging, leading to improved engine performance. Reducing the outside temperature of the exhaust gases leaving an SS 304 manifold using heat wrap can also assist to lessen the impact of heat on the exhaust system components located downstream of the manifold. This can help to improve the durability and lifespan of the exhaust system. The investigation of current exhaust manifold designs in order to gain a deeper comprehension of the importance of the different elements that went into its creation. There is variation in the material and specifications, and the outcomes were compared [1]. Analysis of the engine exhaust manifold's performance to check whether the engine's exhaust manifold has a big impact on how well it runs [2]. The conventional approach to analysis through experimentation is costly and time-consuming. Furthermore, the flow structure is not provided. One effective way to provide flow structure and pressure fluctuation in the flow domain is using CFD simulation with Hyper mesh [3].

A part of a combustion engine that gathers exhaust gases from every cylinder and directs them towards the exhaust system is called an exhaust manifold. It is usually fastened straight to the engine's cylinder head and composed of either cast iron or stainless steel. The main job of the exhaust manifold is to direct exhaust gases entering the exhaust system, so that they may be effectively and safely released from the car, rather than into the cylinders of the engine. Because the exhaust gases are hot and under pressure, the manifold is designed to withstand high temperatures and to resist cracking and warping. The heat wrap provides on the SS 304 manifold reduces the amount of heat radiating from the exhaust manifold, which can help to protect surrounding components from excessive heat exposure. Also retaining more heat within the exhaust system, the heat wrap can improve exhaust gas velocity and scavenging, leading to improved engine performance.

2 Literature Review

Optimizing engine performance requires a thorough understanding of how exhaust manifolds behave in a variety of temperature environments. The thermal properties of exhaust manifolds made of different materials are examined in this examination, along with their loss of heat, structural soundness, and overall efficiency. It seeks to identify the best durable and thermally effective material for these vital engine parts through thorough investigation.

Gopal et al. [4] studied the whole design of the exhaust manifold considering the various factors during finite element analysis. In their study on exhaust manifold design, Manohar and Krishnaraj [5] took into account the thermal strains and deflections that were seen under different operating conditions with different materials and temperatures. The degree of proximity of a few pre-defined physical parameters was taken into consideration when Satish et al. [6] examined the heat transmission in the exhaust manifold. After assessing exhaust energy utilisation and exhaust performance, Liu et al. [7] came to the conclusion that the exhaust pipe's diameter ought to match the manifold's output diameter. According to Reddy and Reddy [8], the development of an exhaust muffler for a diesel engine allows for the management of engine noise level, which explains why the frequency of the Nash Shell Damper muffler has been enhanced. Mogulappa and Ramesh [9] analysed the design of the exhaust manifold by changing its parameters like bend radius and the position of exhaust in the design in the software Pro/Engineer and finally thermal analysis is done in Ansys. Bajpai [10] done the CFD analysis of the exhaust manifold and also compared the back pressure in it considering alternate fuels. The result shows that LPG fuel is the best fuel which gives minimum back pressure. Selvanathan et al. [11] analysed the exhaust manifold by considering the performance of the turbocharger with parameters like velocity vector, static pressure, and turbulent kinetic energy along the manifold and the turbine housing. Puneetha et al. [12] compared two exhaust designs i.e. relative type manifold and resonating type manifold for three different cases for analysing the flow in the exhaust. Li et al. [13] analysed the failure due to cracks in the exhaust manifold for two manifolds parts A and B. Joshi et al. [14] explained the failure analysis of exhaust manifold diffuser plate. Result shows thermal fatigue as the main cause of failure. Masri et al. [15] analysed the design of exhaust manifold for the generator engine for reducing the back pressure.

Although the review of the literature offers a thorough understanding of multiple aspects of exhaust manifold research, there is a significant void in the detailed investigation of thermal analysis for different manifold materials. The cited research examines failure analysis, performance evaluation, back pressure issues, and manifold design. Nevertheless, there is noticeably a lack of a targeted study that carefully contrasts the thermal behaviours of various materials inside exhaust manifolds. Optimizing manifold design for improved durability and performance may be greatly aided by research examining thermal responses—such as distribution of temperatures, stress distribution, along material-specific thermal conductivities—across manifold materials (alloys, composites, etc.) according to different operating conditions. This gap offers a chance to gain a thorough understanding of the thermal responses of various materials inside exhaust manifolds, which can help with selecting materials and design enhancements. but the specific solution must be tailored to the specific application and operating conditions the objectives that are identified are as follows:

1. To evaluate the effectiveness of the exhaust manifold.
2. To identify the appropriate material which can be used to fabricate the exhaust manifold for efficient functioning.
3. To identify the impact of the heat wrap on different materials.

2.1 *Parametric Study*

- Condition 1

Table 1 shows readings of the mild steel exhaust manifold. Reading taken as follows, thermocouples are placed at four locations which is engine casting, first bend of manifold, second bend of manifold, exit of manifold. This shows temperature increases up to 294 °C.

- Condition 2

Table 2 shows a reading of the mild steel exhaust manifold with heat wrap. Reading taken as follows, thermocouples are placed at four locations: engine casting, first bend of manifold, second bend of manifold, and exit of manifold. This shows temperature increases up to 172 °C.

Table 1 MS exhaust manifold reading without heat wrap

Time (s)	Room temperature	T1	T2	T3	T4
0	37	43.5	43	43.1	42.2
0.30	37	72.6	70.8	68.8	54
1.00	37	92	115	124.8	75
1.30	37	95.7	160.7	138.7	95.8
2.00	37	100.1	184.9	155.7	124.5
2.30	37	106.8	218.9	199.6	158
3.00	37	108	250.6	293.9	176.7
3.30	37	104.4	272.1	238.6	203.4
4.00	37	103.4	291.1	230	195.3
4.30	37	99.5	213.3	253.4	191.3
5.00	37	102.3	273.6	263.4	225.4

Table 2 MS exhaust manifold reading with heat wrap

Time (s)	Room temperature	T1	T2	T3	T4
0	37	40.8	33.2	32.1	32.1
0.30	37	52.3	35.4	36.3	33.6
1.00	37	66.6	44.3	45.5	38
1.30	37	92.6	63.2	72	51
2.00	37	99.3	96.9	101.2	74.2
2.30	37	94.5	130.4	126.9	100.6
3.00	37	98.7	144.6	141.8	125.9
3.30	37	104.8	150.8	164.2	138.6
4.00	37	105.7	150.2	161.1	148.7
4.30	37	102.9	146.5	171.9	153.39
5.00	37	97	66	81	159.9

- Condition 3

Table 3 shows readings of the SS 304 austenitic stainless steel exhaust manifold without heat wrap. Reading taken as follows, thermocouples are placed at four locations which is engine casting, first bend of manifold, second bend of manifold, exit of manifold. This shows temperature increases up to 295 °C.

- Condition 4

Table 4 shows readings of the SS 304 austenitic stainless steel exhaust manifold with heat wrap. Reading taken as follows, thermocouples are placed at four locations which is engine casting, first bend of manifold, second bend of manifold, exit of manifold. This shows temperature increases up to 295 °C.

Table 3 SS 304 exhaust manifold reading without heat wrap

Time (s)	Room temperature	T1	T2	T3	T4
0	37	40.8	33.2	32.1	32.1
0.30	37	74.4	93	165.6	56.1
1.00	37	80.9	115.7	177.3	53.3
1.30	37	86.6	113.2	197.5	55
2.00	37	99.3	137.5	223.5	49.7
2.30	37	109.3	176.7	295	70.7
3.00	37	109.3	176.7	244.5	125.9
3.30	37	75.7	181.8	233.9	138.6
4.00	37	96.3	185.7	242.2	148.7
4.30	37	95	184.2	228	153.39
5.00	37	94.4	180.6	220	159.9

Table 4 SS 304 exhaust manifold reading with heat wrap

Time (s)	Room temperature	T1	T2	T3	T4
0	37	40.8	33.2	32.1	32.1
0.30	37	74.4	93	165.6	56.1
1.00	37	66.8	75.3	44.7	64.2
1.30	37	66.9	100.1	54.8	74.6
2.00	37	79.5	120.2	62.7	100.6
2.30	37	81.3	129.7	70.4	109.3
3.00	37	84.1	128.8	78.7	125.9
3.30	37	88.1	126.4	83.8	138.6
4.00	37	88.3	125.2	89.8	148.7
4.30	37	92	155.1	94.9	153.39
5.00	37	95.5	120.1	99	159.9

3 Heat Transfer Concept

The heat transfer to improve the exhaust gas flow, it is also common to use a heat wrap or insulating material around the exhaust manifold. The heat wrap helps to retain heat within the exhaust system, which can enhance exhaust scavenging and potentially improve engine performance. Also retaining more heat within the exhaust system, the heat wrap can improve exhaust gas velocity, leading to improved engine performance. Figure 1 depicts the exhaust manifold's design.

There are a few heat transfer mechanisms that may be used to explain the transfer of heat in an exhaust manifold.

- Convection: This is the transfer of heat through the movement of fluids, such as air. As the exhaust gases flow through the manifold, it transfers heat to exhaust due to convection.
- Conduction: This is the transfer of heat through direct contact. The heat-absorbing surface absorbs heat and transfers it to the air flowing over it through conduction.

A heat-absorbing surface's substance, the exhaust gas flow rate, the manifold's design, and the temperature differential between the surface and the surrounding air all have an impact on the amount of heat transfer in an exhaust manifold. The amount

Fig. 1 Schematic diagram

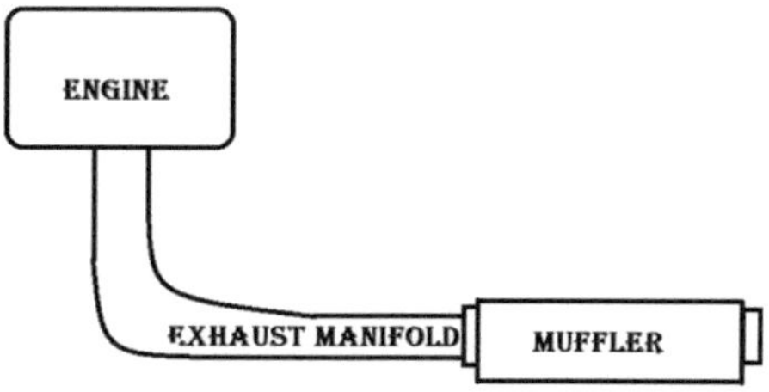

of heat transfer and exhaust manifold efficiency can be increased by increasing the flow of exhaust gases rate and creating a greater temperature differential among the heat-absorbing surfaces and the surrounding air.

4 Thermal Modelling

This section develops and describes an exhaust manifold thermal model. This model depends on the manifold's exhaust gases flowing. By using heat wrap to follow SS, one may calculate the amount of heat transfer from the exhaust gases passing through the manifold.

$$T_1 = 35.8 \quad T_4 = 56.8$$

$$T_3 = T_1 \times r^{(\gamma-1)}$$

$$T_3 = 35.8 \times 8^{(1.4-1)}$$

$$T_3 = 82.246\,^{\circ}\mathrm{C}$$

$$T_2 = T_4 \times r^{(\gamma-1)}$$

$$T_2 = 56.8 \times 8^{(1.4-1)}$$

$$T_2 = 130.4921\,^{\circ}\mathrm{C}$$

$$Q_{\mathrm{add}} = C_{\mathrm{v}} \times (T_2 - T_3)$$

$$Q_{\mathrm{add}} = 0.717 \times (130.4921 - 82.246)$$

$$Q_{\mathrm{add}} = 34.592\,\mathrm{kW}$$

$$Q_{\mathrm{rej}} = C_{\mathrm{v}} \times (T_4 - T_1)$$

$$Q_{\mathrm{rej}} = 0.717 \times (56.8 - 35.8)$$

$$Q_{\mathrm{rej}} = 15.057\,\mathrm{kW}$$

$$\eta = \frac{(Q_{\text{add}} - Q_{\text{rej}})}{Q_{\text{add}}}$$

$$\eta = \frac{(34.592 - 15.057)}{34.592}$$

$$\eta = 56.96\%$$

5 Methodology

5.1 Experimental Setup

An exhaust manifold project includes the following components:

- Mild Steel Manifold: PV solar panels are used to turn solar energy into electricity. The power needs of the cooling system determine the size and output capacity of the solar panel.
- SS 304 Manifold: A battery is used to store extra energy produced by the solar panel so that it may be used when it's dark out. The battery's capacity is determined by how much electricity the cooling system needs and how long it will likely be operating without sunlight.
- Glass Fibre Heat Wrap: The water that will be pumped through the heat exchanger is kept in a water tank. The cooling capability of the cooling system and the anticipated operating time both affect the water tank's size.
- Thermocouple: They are used to monitor the operation of the exhaust system and to ensure that the exhaust gases is at the desired temperature and difference.

The experimental setup is typically housed in a R15 engine, where the system's performance can be tested under different conditions, such as with and without heat wrap, ambient temperature, and maximum temperature. Figure 2 illustrates the exhaust manifold's design. The manifold experimental setup can be seen in Fig. 3.

6 Results and Discussion

Experiments were tested in four different conditions. Using with and without heat wrap as a medium on an exhaust manifold selected to analyse system performance. In the experiment, temperatures were measured at four different locations: at engine casting, first bend of manifold, and second bend of manifold, exit of manifold. At this point, four different thermocouples were introduced, and measurements recorded (Figs. 4, 5, 6 and 7).

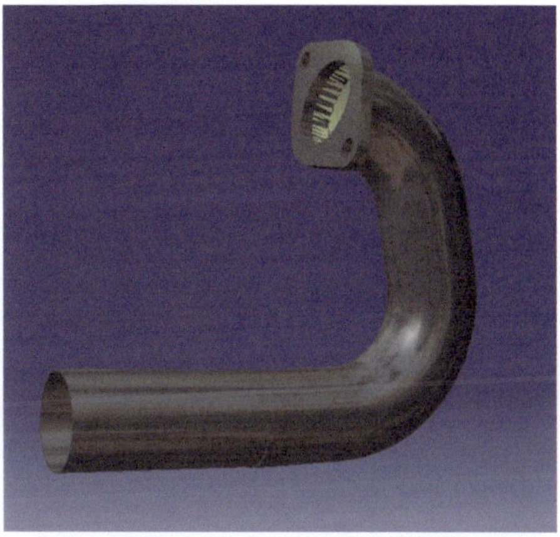

Fig. 2 CAD model of exhaust manifold

Fig. 3 Experimental setup

The experiment was conducted to compare the efficiency of exhaust manifold with and without heat wrap. Four different conditions were used for the experiment, that is MS manifold with and without heat wrap, SS 304 manifold with and without heat wrap. The temperature at surface of manifold was monitored using a thermocouple. The experiment was carried out for a duration of 5 min for each condition with throttle pressing partially.

As a result, the SS 304 manifold with heat wrap was more efficient in reducing the temperature of the engine compartment as compared to the MS with and without heat wrap and SS without heat wrap. If we consider peak temperature during experimentation, The temperature of engine compartment with SS 304 manifold with heat wrap was 13 °C lower than the temperature of MS manifold with heat wrap, The temperature of first bend 2° lower than MS manifold with heat wrap as compared to

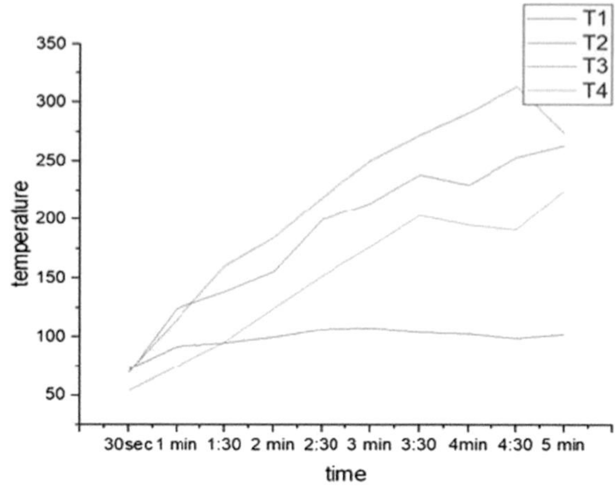

Fig. 4 Condition 1 (MS without heat wrap)

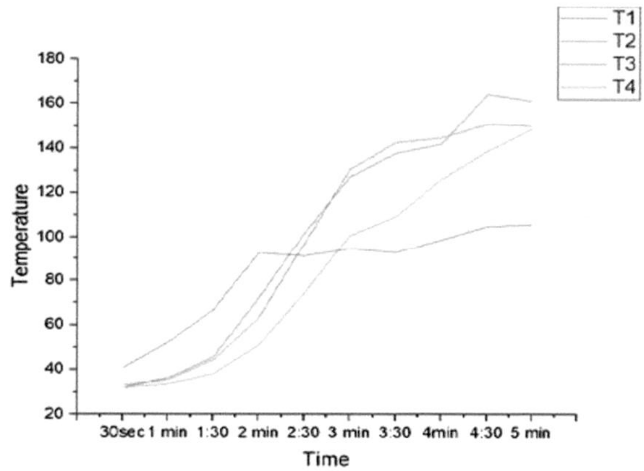

Fig. 5 Condition 2 (MS with heat wrap)

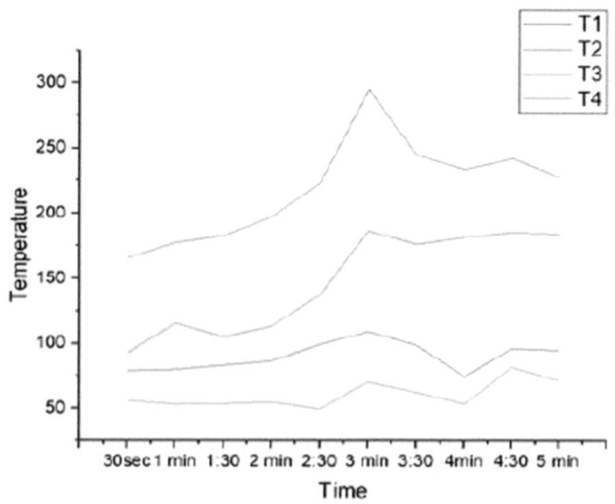

Fig. 6 Condition 3 (SS 304 without heat wrap)

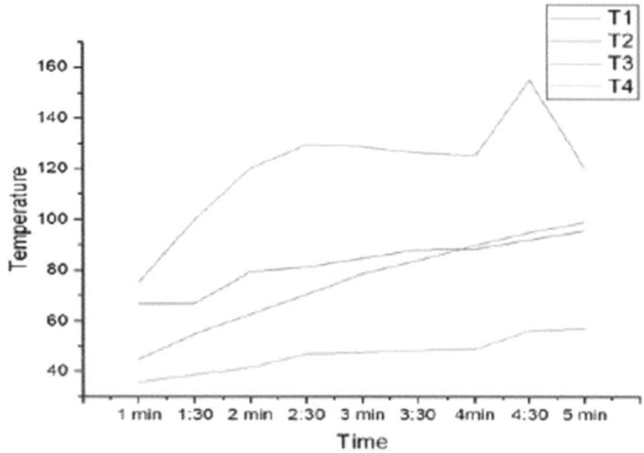

Fig. 7 Condition 4 (SS 304 with heat wrap)

SS 304 manifold with heat wrap and the temperature of second bend of SS manifold with heat wrap was 56.2° lower than MS with heat wrap.

Experiments have shown that using SS 304 manifold with heat wrap is more efficient than other conditions. This is because SS 304 material characteristics and heat wrap. The quantity of heat needed to increase a substance's temperature by 1 °C is known as its heat capacity. SS 304 has a lower heat capacity than MS. So, using SS 304 with will absorb less heat from the exhaust gases, resulting in an increased exhaust gas velocity, which can improve engine performance by increasing exhaust flow and reducing backpressure.

Here is a comparative study of the two systems:

- Design: Exhaust manifold (without heat wrap) typically consists of a MS manifold and SS 304 manifold. An exhaust manifold (with heat wrap) of SS and MS The manifold which transfer exhaust gases form engine to muffler and engine that produces exhaust gases and retain heat within the exhaust system, which can enhance exhaust scavenging and potentially improve engine performance.
- Operation: An exhaust manifold is a component of an internal combustion engine that collects the exhaust gases from each of the engine's cylinders and routes them to the exhaust system.
- Performance: Both the manifold with and without heat wrap can be effective in reducing the temperature of the engine casting and bends of manifold. However, the efficiency and performance of each system may depend on several factors such as the design, size, and operating conditions.

A single-cylinder exhaust manifold is a component used in small internal combustion engines to collect and direct exhaust gases from the engine's single cylinder into the exhaust system. The exhaust manifold is typically made of metal tubing and is located on the engine's cylinder head. The design of a single-cylinder exhaust manifold plays a crucial role in the engine's overall performance, particularly in terms of exhaust gas scavenging and back pressure. The manifold's shape, size, and length can affect the flow of exhaust gases and determine the engine's power output.

Heat wrap is often used on exhaust manifolds because it can help to increase exhaust gas velocity, which can improve engine performance by increasing exhaust flow and reducing backpressure. Additionally, by reducing the amount of heat radiated from the exhaust manifold, heat wrap can help to protect other components in the engine compartment from heat damage.

In summary, both the manifold with and without heat wrap offer an energy-efficient and sustainable way to transfer exhaust gases to muffler. The choice between the two may depend on several factors such as the climate and temperature requirements.

7 Conclusion

The heat transfer through exhaust manifold has been investigated in this research paper. The experimental results have shown that the SS 304 manifold with heat wrap can effectively reduce the temperature of engine compartment and consequently manifold. The manifold consist of SS 304 which is austenitic stainless steel and heat wrap. The rate of heat transfer is dependent on various factors such as the flow rate of exhaust gases, and the design of manifold. The experimental results have demonstrated that the SS 304 manifold with heat wrap can reduce the temperature of engine compartment and first bend of manifold where excessive temperature rise by 13° and 2 °C.

References

1. Raghuwanshi G, Kakirde A, Sharma S (2018) Design and analysis of exhaust manifold comparing different specifications. Int J Eng Trends Technol 62:42–45
2. Zou B, Hu Y, Liu Z, Yan F, Wang C (2013) The impact of temperature effect on exhaust manifold thermal modal analysis. Res J Appl Sci Eng Technol 20:2824–2829
3. Rathnaraj JD (2012) Thermomechanical fatigue analysis of stainless steel exhaust manifolds. IRACST Eng Sci Technol 2(2):265–267
4. Gopal, Varma K, Kumar S (2014) Exhaust manifold design—FEA approach. Int J Eng Trends Technol 17(10):485–489
5. Manohar DS, Krishnaraj J (2018) Modeling and analysis of exhaust manifold using CFD. IOP Conf Ser Mater Sci Eng 012132
6. Satish S, Prithiviraj M, Hari S (2012) Comparison of predictions obtained on an exhaust manifold analysis using conformal and indirect mapped interface. In: International congress on computational mechanics and simulation
7. Liu F, Sun C, Li Y, Shang Y (2021) Performance analysis and optimization design of exhaust system for turbocharging diesel engines. Int J Automot Technol 22:735–745
8. Reddy MR, Reddy KM (2012) Plan and optimization of exhaust muffler in automobiles. Int J Eng Res Appl 2:395–398
9. Mogulappa G, Ramesh G (2017) CFD analysis in exhaust manifold in engines. Int J Eng Sci Adv Technol 7:032–044
10. Bajpai K (2017) CFD analysis of exhaust manifold of SI engine and comparison of back pressure using alternative fuels. IOSR J Mech Civ Eng (IOSR-JMCE) 14:23–29

11. Selvanathan PS, Sudhakaran R, Venkatesh K, Tites CS, Rajamanikandan D (2017) CFD analysis of IC engine exhaust manifold with respect to the performance of a turbocharger. Adv Nat Appl Sci 11:242–249
12. Puneetha CG, Manjunath H, Shashidhar MR (2015) Backpressure study in exhaust muffler of single cylinder diesel engine using CFD analysis. In: Altair technology conference, pp 1–14
13. Li Y, Liu J, Huang W, Wu Y (2023) Failure analysis of a diesel engine exhaust manifold. Int J Metal Cast 17:538–550
14. Joshi B, Shih H, Rose B (2018) Failure analysis and robust optimization of an exhaust manifold diffuser plate. J Fail Anal Prev 18:486–495
15. Masri MM, Aziz AR, Ezrann Z, Zainal A, Mohammad SE (2022) The design and simulation of an exhaust manifold for free piston linear generator engine. In: ICPER 2020, pp 653–659

Hemodynamic Analysis of Local Coil Packing Density in an Anterior Communicating Artery Aneurysm Using Porous Medium Approach

Nisanth Kumar Panneerselvam, B. J. Sudhir, Santhosh K. Kannath, and B. S. V. Patnaik

Abstract The endovascular coiling (EVC) is becoming increasingly popular for treating cerebral aneurysms. Even though EVC has an advantage over conventional surgical clipping, re-treatment is often required within six months. There are several risk factors associated with aneurysm recurrences. Since the aneurysm neck region is highly affected by hemodynamic forces after EVC, studying the influence of local coil packing density at the aneurysm neck volume is valuable for understanding aneurysm recurrence. The present study analyzes the influence of local coil packing density in the anterior communicating artery (AcomA) aneurysm neck volume on the hemodynamics of aneurysm dome volume. A porous medium approach is used to simulate the coil filled inside the aneurysm. From the analysis of flow features (velocity magnitude) and hemodynamic parameters (time-averaged wall shear stress and relative residence time), it was observed that packing density (PD $>$ 20 %) in the aneurysm neck volume reduces the chances of aneurysm recurrence.

Keywords Coil embolization · Intracranial aneurysm · Computational fluid dynamics · Hemodynamics

---

N. K. Panneerselvam (✉) · B. S. V. Patnaik
Department of Applied Mechanics and Biomedical Engineering, Indian Institute of Technology Madras, Chennai, India
e-mail: nh.sp11@gmail.com

B. J. Sudhir
Department of Neurosurgery, Sree Chitra Tirunal Institute for Medical Sciences and Technology, Trivandrum, India

S. K. Kannath
Department of Imaging Sciences and Interventional Radiology, Sree Chitra Tirunal Institute for Medical Sciences and Technology, Trivandrum, India

A. Kumar R. et al. (eds.), *Proceedings of Fluid Mechanics and Fluid Power (FMFP) 2023, Vol. 6*, Lecture Notes in Mechanical Engineering,
https://doi.org/10.1007/978-981-96-9600-0_39

1 Introduction

The endovascular coiling (EVC) is commonly used to treat intracranial aneurysms (IAs) due to the nature of the minimally invasive method [13]. In EVC, Flexible metallic coils are inserted into the aneurysm to block the intra-aneurysmal blood flow; this causes thrombus formation throughout the aneurysm and results in the aneurysm occlusion [16]. Re-treatment of EVC procedure more often results in recanalization of the aneurysm [12]. Compared to surgical clipping, EVC implies a higher possibility of late aneurysm rebleeding and a lesser possibility of complete obliteration [14]. Previous studies suggest that the coil packing density, uniformity of coil distribution inside the aneurysm, local coil density in the neck region, and coil compaction are various critical factors in predicting the chances of aneurysm recurrence [4, 15, 18]. However, comprehensive hemodynamic analysis of the role played by different risk factors is not available in the literature. Since the hemodynamic forces after EVC will affect the aneurysm neck region, studying the effects of different coil packing densities in the aneurysm neck region is of interest in addressing aneurysm recurrence. Hence, the present study analyzes the influence of different packing densities in the aneurysm neck region on the hemodynamics of the aneurysm dome.

2 Methodology

2.1 Construction of Patient-Specific Aneurysm Geometry

The Institute's Ethics Committee approved the acquisition of information on patient-specific aneurysms. The patient-specific anterior communicating artery (AcomA) aneurysm images are obtained using three-dimensional computed tomography angiography (CTA). Digital Imaging and Communications in Medicine (DICOM) images are used to construct a 3D model of an aneurysm, as shown in Fig. 1a

2.2 Porous Medium Approach and CFD

The coil filled inside the AcomA aneurysm is modeled as a homogeneous and isotropic porous medium following Darcy's law [21, 22]; accordingly, the Navier–Stokes equation is modified by adding the source term to account for the coil fill effects as

$$\rho\left(\frac{\partial \boldsymbol{u}}{\partial t}+\nabla\cdot\left(\frac{\boldsymbol{u}\boldsymbol{u}}{\psi}\right)\right)=-\nabla p+\mu\nabla^{2}\boldsymbol{u}-\frac{\mu\psi\boldsymbol{u}}{R} \tag{1}$$

(a)

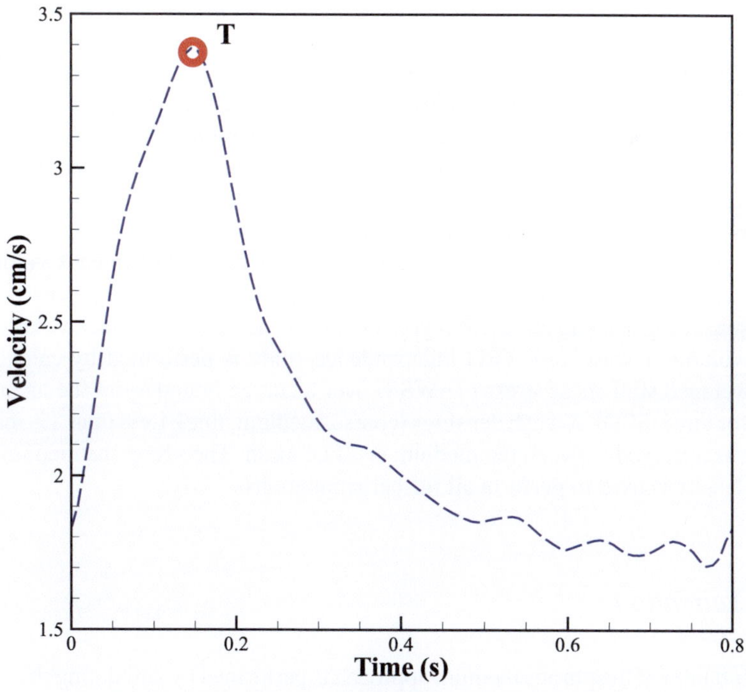

(b)

Fig. 1 **a** Patient-specific AcomA aneurysm model construction from DICOM images. **b** Temporal velocity measurements were obtained from a patient in the AcomA using TCD where ρ is fluid density (kg/m^3), u is superficial velocity (m/s), t is time (s), ψ is porosity, p is pressure (kg/ms^2), g is gravity (m/s^2), μ is dynamic viscosity (kg/ms), and R is permeability (m^2).

$$\psi = 1 - \frac{\text{CPD}}{100}, \quad \text{CPD} = \frac{\text{CV}}{\text{AV}} \tag{2}$$

where CPD is the Coil Packing Density, which is the ratio of Total Coil Volume (TCV in mm^3) to Aneurysm Dome Volume (ADV in mm^3) [20]. The permeability was calculated using the Carman–Kozeny relation [1].

$$R = \frac{\psi^3}{jT^2}, \quad T = \frac{S}{\text{AV}} \tag{3}$$

where j is the Kozeny coefficient, and j is 2 for a cylindrical cross section. S is the porous region's internal surface area. The influence of CPD on intra-aneurysmal hemodynamics is not significant, with the CPD being more than 20% [5, 11].

Navier–Stokes and continuity equations were solved for flow in the arteries. Blood was considered as an incompressible, laminar, and Newtonian fluid [6–8]. The velocity profile at the inlet must be patient-specific in order to precisely match in vivo conditions and CFD models [9]. The Transcranial Doppler (TCD) ultrasonography results obtained from patient-specific AcomA aneurysm is used as an inlet boundary condition, as shown in Fig. 1b [10]. Traction-free exit pressure is imposed at outlets. The aneurysm volume is divided into three equal portions. One-third of the aneurysm volume near the neck is considered as the neck volume, and the region above the neck volume is considered as an aneurysm dome volume. Since the objective of this study is to analyze the influence of local coil density in the aneurysm neck volume on the hemodynamics of the dome, packing density in the aneurysm neck volume is varied from 5 to 20% while the dome volume is kept with 20% packing density in all the cases. The green and red color regions in Fig. 1a are aneurysm neck volume and dome volume, respectively. Grid independence study is performed by calculating time-averaged wall shear stress (TAWSS) over a curved boundary in the aneurysm dome for three different mesh densities (coarse, medium, fine). Less than 2% relative error was observed between the medium and fine mesh. Therefore, the medium size mesh density is used to perform all further simulations.

2.3 Validation

The validation of the porous medium approach is performed by simulating the model aneurysm suggested by, Cha et al. [3]. In this work, simulations are performed on the idealized bifurcation aneurysm geometry with inlet and outlets diameter of 3.6 and 2.55 mm, respectively. The aneurysm diameter is kept as 5.5 mm. The bifurcation angle between the two outlets is 180°, as shown in Fig. 2a. The pulsatile flow rate

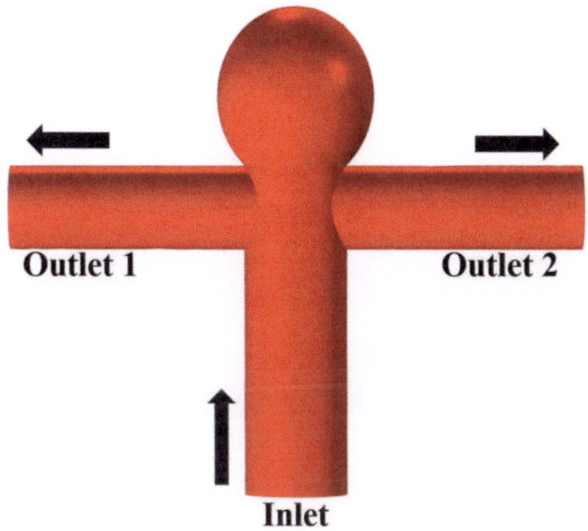

(a)

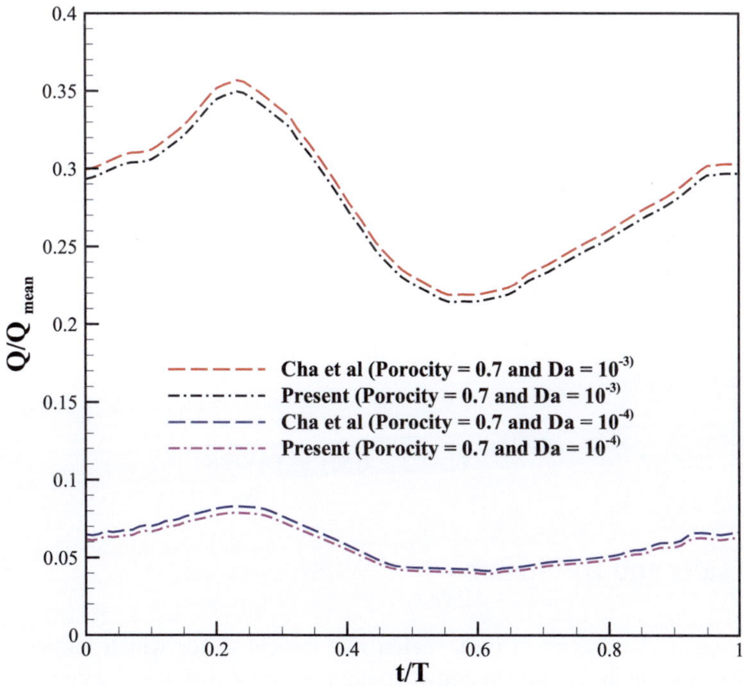

(b)

Fig. 2 **a** Ideal bifurcation aneurysm model. **b** Validation result at the inlet boundary is imposed, and constant pressure is applied on both outlets. Blood is assumed as a Newtonian fluid with a density of 1100 kg/m^3 and a viscosity of 3.4 cP. The CPD and ψ is kept as 30% and 0.7, respectively. Two different cases with different Darcy number (10^{-3} and 10^{-4}) are analyzed with same porosity. The normalized flow rate along the aneurysm neck plane is shown in Fig. 2b. A good comparison maybe noticed.

2.4 Hemodynamic Parameters

IA's mostly occur near the curvatures and bifurcations in the artery, where the hemodynamics are intricate. The hemodynamic parameters that cause the initiation, growth, and rupture of IAs are extremely important which in turn depend on the biomechanical properties [2]. Time-averaged wall shear stress (TAWSS) and relative residence time (RRT) are the most critical hemodynamic parameters which indicate the chances of aneurysm recurrence [17, 19].

The average of wall shear stress (WSS) over a cardiac cycle is called as TAWSS.

$$\text{TAWSS} = \frac{1}{T}\int_0^T \|\tau_{\text{wall}}\|\,\mathrm{d}t \tag{4}$$

RRT reflects the amount of time that a typical erythrocyte spends inside the aneurysm. RRT is calculated using TAWSS and oscillatory shear index (OSI) as,

$$\text{OSI} = \frac{1}{2}\left(1 - \frac{\left\|\frac{1}{4T}\int_0^{4T} \boldsymbol{\tau}_{\text{wall}}\mathrm{d}t\right\|}{\frac{1}{4T}\int_0^{4T} |\boldsymbol{\tau}_{\text{wall}}|\,\mathrm{d}t}\right) \tag{5}$$

$$\text{RRT} = \frac{1}{(1 - 2 \times 0\,\text{SI}) \times \text{TAWSS}} \tag{6}$$

3 Results and Discussion

Coil filling in the aneurysm neck region is a critical factor which influences the hemodynamics in the aneurysm dome. To implement the different packing densities (PD) in the aneurysm neck region, the aneurysm volume is divided into three portions. The region of aneurysm volume next to the neck plane is considered an aneurysm neck volume. The influence of different packing densities in aneurysm neck volume on the hemodynamics of an aneurysm dome is analyzed by varying the PD from 5 to 20% in the aneurysm neck region while keeping the aneurysm dome packing density as 20%. Velocity magnitude (VM) changes in a cross section within the aneurysm

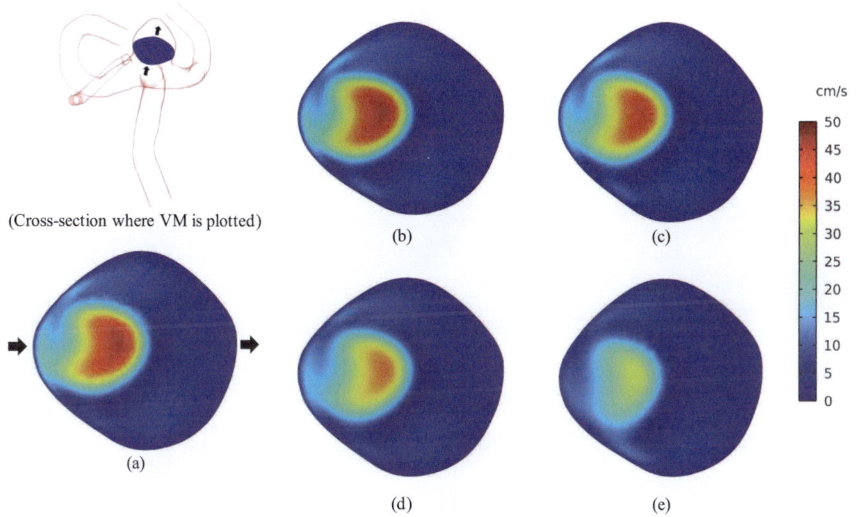

Fig. 3 Velocity magnitude during peak systole. **a** No coil in neck and 20% PD in dome. **b** 5% PD in neck and 20% PD in dome. **c** 10% PD in neck 20% PD in dome. **d** 15% PD in neck 20% PD in dome. **e** 20% PD in neck and dome

dome during peak systole (mentioned as T in Fig. 1b) are shown in Fig. 3. Without coil in neck region, the VM inside the aneurysm dome is higher at the center region. The VM changes are not significant inside the aneurysm dome due to coiling with 5 and 10% PD at the neck region. The resistance to the flow inside the aneurysm dome is much higher above coil PD of 15%. Coil fill with 20% PD allows marginal inflow into the aneurysm dome, which reduces the risk of aneurysm recurrence and initiates flow stasis.

To further investigate the hemodynamic changes in the aneurysm dome due to various packing densities in the aneurysm neck region, TAWSS and RRT plots are shown in Figs. 4 and 5. Without coil in neck region, TAWSS is higher in the aneurysm dome. Varying PD at the aneurysm neck region influences the TAWSS developed in the aneurysm dome. With lower coil fill (PD $\leq$ 10%) at the neck, TAWSS developed in the neck is higher than in the aneurysm dome region. With higher coil fill (PD $\geq$15%) at the neck region almost developed the same range of TAWSS in both the neck and dome region. TAWSS and RRT changes along a curve within the aneurysm dome is shown in Fig. 6a and b. With lower PD in the neck aneurysm region, there are local fluctuations both in TAWSS as well as RRT. However, for higher PD in the neck region, the intensity of fluctuation is smaller. The higher value of TAWSS in the aneurysm wall increases the chances of aneurysm rupture [19]. Since the higher TAWSS region leads the aneurysm to rupture, the higher coil PD in the aneurysm neck region is perhaps helpful.

RRT is an important hemodynamic parameter which provides significant information about the thrombus generation after coiling. Without a coil in the neck region,

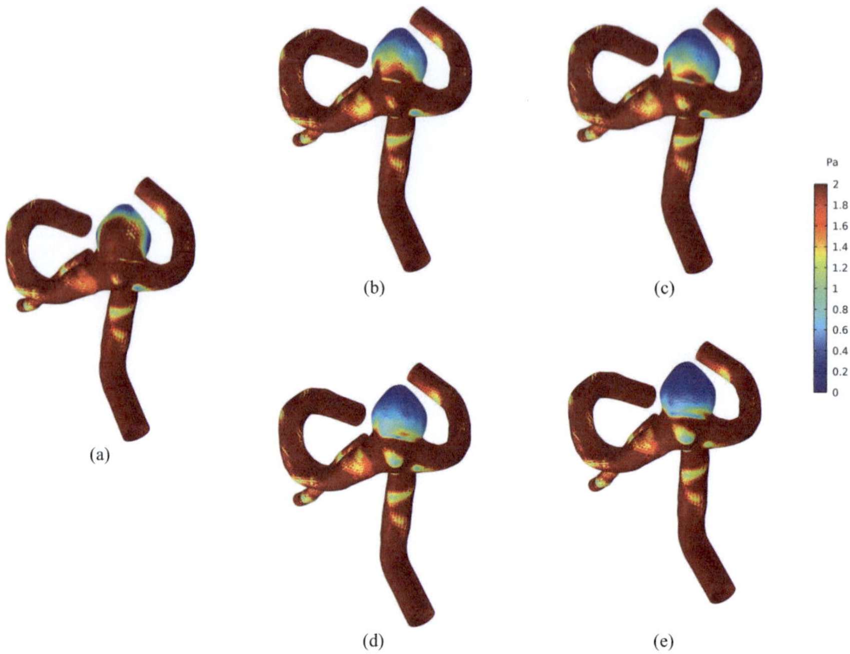

Fig. 4 TAWSS variations for different local coil packing densities. **a** No coil in neck and 20% PD in dome. **b** 5% PD in neck and 20% PD in dome. **c** 10% PD in neck 20% PD in dome. **d** 15% PD in neck 20% PD in dome. **e** 20% PD in neck and dome

a partial portion of the aneurysm dome region develops higher RRT. With lower PD, only the dome region develops higher RRT and the neck region develops lower RRT, which is not sufficient to initiate the thrombus. Due to higher resistance to flow in the higher PD case, higher RRT is developed in both neck and dome regions. The aneurysm wall region with higher RRT perhaps promotes thrombus deposition than the region with lower RRT [17]. Therefore, when RRT is sufficiently high in the entire aneurysm region, thrombus generation within the aneurysm is facilitated.

4 Conclusions

The re-treatment of intracranial aneurysm (IA) is of common concern in EVC, due to the recanalization of the aneurysm. Since the hemodynamic forces after EVC are higher at the aneurysm neck region, it is important to analyze the influence of different local coil packing densities at the aneurysm neck volume. CFD simulations using a porous medium approach are performed for different local coil packing densities (5–20%) at neck volume while keeping the aneurysm dome volume with 20% packing density for all cases.

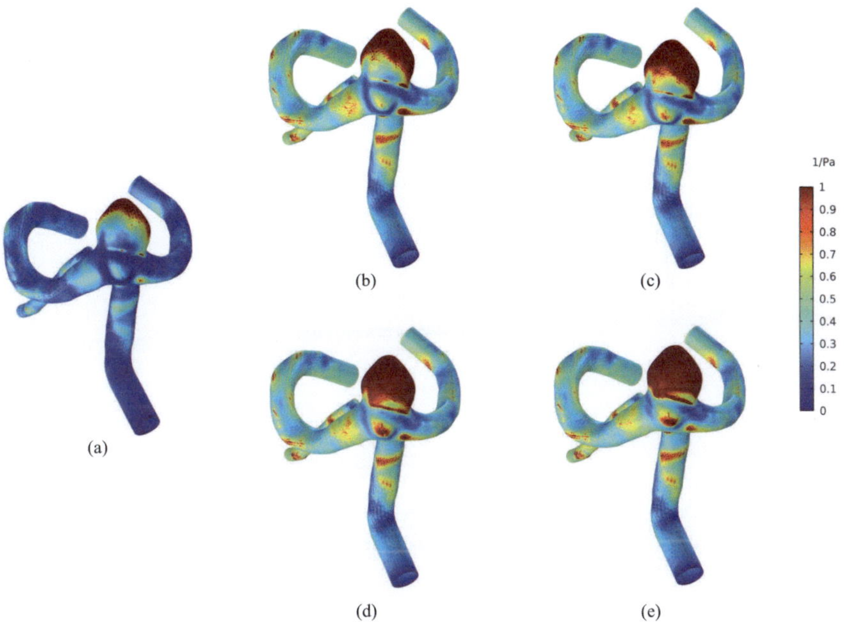

Fig. 5 RRT variations for different local coil packing densities. **a** No coil in neck and 20% PD in dome. **b** 5% PD in neck and 20% PD in dome. **c** 10% PD in neck 20% PD in dome. **d** 15% PD in neck 20% PD in dome. **e** 20% PD in neck and dome

The simulation results of flow feature (VM) and hemodynamic parameters (TAWSS and RRT) show that local packing density (PD $\leq$ 10%) in the aneurysm neck volume has not affected the hemodynamics of aneurysm dome, which leads to the higher chances of aneurysm recanalization. Since the higher local packing density (PD $>$ 20%) in aneurysm neck volume greatly influences the hemodynamics of aneurysm dome volume, as it reduces the chances of aneurysm recurrence.

5 Limitations

The porous medium approach to simulate flow features of coil-filled aneurysms is a simple technique, but it does not account for precise details of coil filling inside the aneurysm. The aneurysm neck volume and dome volume are separated eventually, but in real practice, the coil is unevenly filled in neck volume and dome volume. Several assumptions in the simulation, such as Newtonian blood flow and constant pressure at the exit boundaries are restrictive in nature.

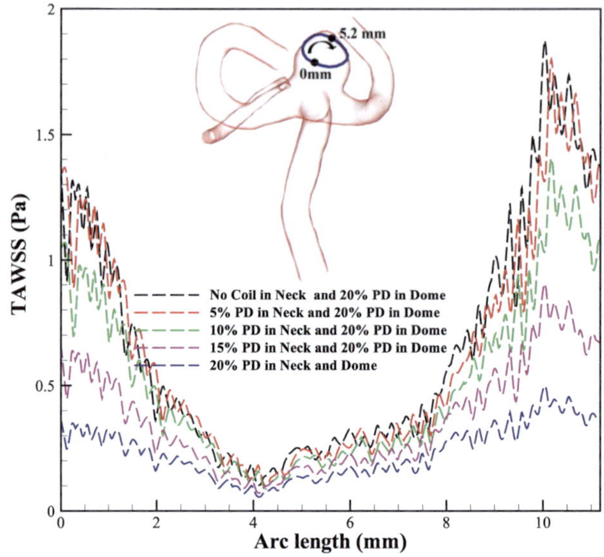

(a)

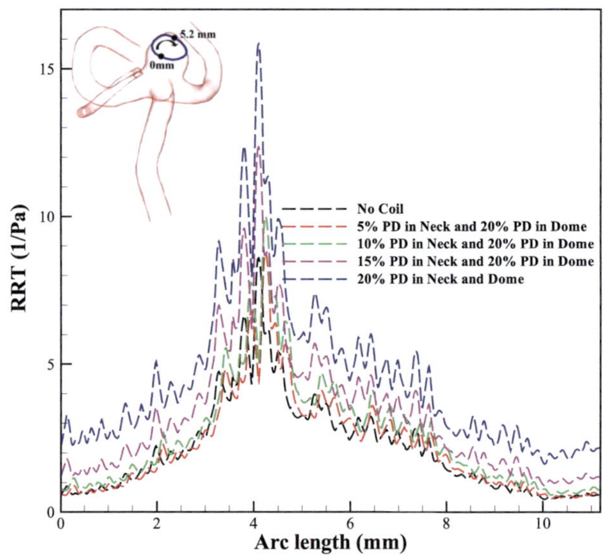

(b)

Fig. 6 **a** TAWSS variation over the curve along aneurysm dome, **b** RRT variation over the curve along aneurysm dome

Acknowledgements This work was supported by grants from Science and Engineering Research Board (SERB), Government of India, under the Scientific and Useful Profound Research Advancement (SUPRA) Scheme (Grant No. SPR/2020/000298) and National Supercomputing Mission (NSM), Government of India (Grant No. DST/NSM/R&D HPC Applications/2021/03.39).

References

1. Carman PC (1956) Flow of gases through porous media. Academic, New York
2. Castro MA (2013) Understanding the role of hemodynamics in the initiation, progression, rupture, and treatment outcome of cerebral aneurysm from medical image-based computational studies. International Scholarly Research Notices 2013
3. Cha KS, Balaras E, Lieber BB, Sadasivan C, Wakhloo AK (2007) Modeling the interaction of coils with the local blood flow after coil embolization of intracranial aneurysms. J Biom Eng 129(6):873–879
4. Chalouhi N, Bovenzi CD, Thakkar V, Dressler J, Jabbour P, Starke RM, Teufack S, Fernando Gonzalez L, Dalyai R, Dumont AS, Rosenwasser R, Tjoumakaris S (2014) Long-term catheter angiography after aneurysm coil therapy: results of 209 patients and predictors of delayed recurrence and retreatment: clinical article. J Neurosurg JNS 121(5):1102–1106
5. Goddard JK, Moran CJ, Cross DWT, Derdeyn CP (2005) Absent relationship between the coil-embolization ratio in small aneurysms treated with a single detachable coil and outcomes. Am J Neuroradiol 26(8):1916–1920
6. Panneerselvam NK, Sudhir BJ, Kannath SK, Patnaik BSV (2023) Hemodynamic analysis of coil filled patient-specific middle cerebral artery aneurysm using porous medium approach. Phys Fluids 35(11):111906–111907. https://doi.org/10.1063/5.0173688
7. Panneerselvam NK, Akade B, Kumar AA, Sudhir BJ, Kannath SK, Patnaik BSV (2024) Hemodynamic investigation of hemorrhagic stroke treatment options for a patient-specific aneurysm using a porous medium model. Phys Fluids 36(7):071915. https://doi.org/10.1063/5.0214784
8. Panneerselvam NK, Sudhir BJ, Kannath SK, Patnaik BSV (2024) Influence of framing coil orientation and its shape on the hemodynamics of a basilar aneurysm model. Med Biol Eng Comput 62(11):3411–3432. https://doi.org/10.1007/s11517-024-03146-4
9. Panneerselvam NK, Mummaneni G, Roncali E (2026) Toward digital twins for optimal radioembolization. PET Clin 21(1):153–167. https://doi.org/10.1016/j.cpet.2025.09.001
10. Maramkandam EB et al (2024) Review of CFD based simulations to study the hemodynamics of cerebral aneurysms. J Indian Inst Sci 104(1):77–110. https://doi.org/10.1007/s41745-024-00431-x
11. Jeong W, Han MH, Rhee K (2013) Effects of framing coil shape, orientation, and thickness on intra-aneurysmal flow. Med Biol Eng Comput 51(9):981–990
12. Lad SP, Babu R, Rhee MS, Franklin RL, Ugiliweneza B, Hodes J, Nimjee SM, Zomorodi AR, Smith TP, Friedman AH et al (2013) Long-term economic impact of coiling vs clipping for unruptured intracranial aneurysms. Neurosurgery 72(6):1000–1013
13. McDonald JS, McDonald RJ, Fan J, Kallmes DF, Lanzino G, Cloft HJ (2013) Comparative effectiveness of unruptured cerebral aneurysm therapies. Stroke 44(4):988–994
14. Molyneux AJ, Kerr RSC, Yu L-M, Clarke M, Sneade M, Yarnold JA, Sandercock P (2005) International subarachnoid aneurysm trial (ISAT) of neurosurgical clipping versus endovascular coiling in 2143 patients with ruptured intracranial aneurysms: a randomised comparison of effects on survival, dependency, seizures, rebleeding, subgroups, and aneurysm occlusion. Lancet 366(9488):809–817
15. Otani T, Shindo T, Ii S, Hirata M, Wada S (2018) Effect of local coil density on blood flow stagnation in densely coiled cerebral aneurysms: a computational study using a Cartesian grid method. J Biomech Eng 140(4):041013

16. Piotin M, Spelle L, Mounayer C, Salles-Rezende MT, Giansante-Abud D, Vanzin-Santos R, Moret J (2007) Intracranial aneurysms: treatment with bare platinum coils-aneurysm packing, complex coils, and angiographic recurrence. Radiology 243(2):500–508 (PMID: 17293572)
17. Rayz VL, Boussel L, Ge L, Leach JR, Martin AJ, Lawton MT, McCulloch C, Saloner D (2010) Flow residence time and regions of intraluminal thrombus deposition in intracranial aneurysms. Ann Biomed Eng 38:3058–3069
18. Sluzewski M, van Rooij WJ, Slob MJ, Bescós JO, Slump CH, Wijnalda D (2004) Relation between aneurysm volume, packing, and compaction in 145 cerebral aneurysms treated with coils. Radiology 231(3):653–658 (PMID: 15118115)
19. Tan FPP, Torii R, Borghi A, Mohiaddin RH, Wood NB, Xu XY (2009) Fluid-structure interaction analysis of wall stress and flow patterns in a thoracic aortic aneurysm. Int J Appl Mech 1(01):179–199
20. Umeda Y, Ishida F, Tsuji M, Furukawa K, Shiba M, Yasuda R, Toma N, Sakaida H, Suzuki H (2017) Computational fluid dynamics (CFD) using porous media modeling predicts recurrence after coiling of cerebral aneurysms. PLoS ONE 12:1–13
21. Wang L, Wang L-P, Guo Z, Mi J (2015) Volume-averaged macroscopic equation for fluid flow in moving porous media. Int J Heat Mass Transf 82:357–368
22. Whitaker S (1986) Flow in porous media I: a theoretical derivation of Darcy's law. Transp Porous Media 1:3–25

A Comparative Analysis of Rhombus-Shaped Ribs in Solar Air Heater

Kashinath Dhamudia and Jnana Ranjan Senapati

Abstract In this article, heat transmission and flow behavior was analyzed in rectangular solar air heater duct. The absorber plate of the solar air heater (SAH) was attached to artificial rough rhombus-shaped ribs with varying rib height ($e = 1$, 1.25, and 1.4 mm). The finding of this problem was done with a computational exercise in Ansys Fluent R22 by varying dimensionless parameters like Reynolds number (4000–18,000), and relative roughness height (e/D_h), 7.14–10 by keeping constant pitch distance ($P = 10$ mm), for this study. A two-dimensional domain was taken to investigate, with uniform heat supplied of 1000 W/m^2 over an absorber plate made up of aluminum. From various turbulence models, the RNG $k - \varepsilon$ model with enhanced wall treatment (EWT) was chosen for simulating the fluid flow equations (mass, momentum, and energy). The result showed that the heat transfer coefficient increases (h) with rise in Reynolds number (Re) and found that maximum heat augment and thermal enhancement factor (TEF) at Re = 10,000.

Keywords SAH · Reynolds number · Rib height · Pitch distance

1 Introduction

In the present scenario, world consumes much energy in every sector, and to fulfill demand, we can't depend upon conventional sources. In our nature, we have ample non-conventional sources, where solar energy can be the primary resource. So, solar energy utilization has become a significant solution in the research community to fulfill the demands of the world. SAH is broadly utilized for applications such as space heating, drying processes, and solar ventilation. A rectangular duct SAH was

K. Dhamudia (✉)
Department of Mechanical Engineering, Parala Maharaja Engineering College, Berhampur, India
e-mail: kisna014@gmail.com

K. Dhamudia · J. R. Senapati
Department of Mechanical Engineering, National Institute of Technology Rourkela, Rourkela, India

A. Kumar R. et al. (eds.), *Proceedings of Fluid Mechanics and Fluid Power (FMFP) 2023, Vol. 6*, Lecture Notes in Mechanical Engineering,
https://doi.org/10.1007/978-981-96-9600-0_40

considered for this analysis, where the absorber plate was kept at the top, and other parts were kept perfectly insulated. As air has a low "h", solar air heaters have low efficiency. In SAH, efficiency can be enhanced by attaching rough surfaces over absorber plate, which enters the viscous sub-layer and creates more swirls in duct, which boosts heat transfer. Still, the pumping power increases with an attachment of ribs. The research community compared rough surfaces with smooth surfaces in SAH and estimated a ratio of Nusselt number with friction factor by varying Reynolds number to optimize enhancement of heat transfer with pumping power.

2 Literature Review and Objective

The researchers used experimental methods to study solar air heaters by attaching roughed struts in absorber plates. Sharma et al. [1] conducted an experimental analysis of roughened SAH with sine wave baffles. The result shows a maximum enhancement of 2.68 in TEF at Reynolds number 9000. Singh et al. [2] used roughed multiple broken transverse ribs and accomplished experiments and CFD analysis in solar air heaters. They found transverse ribs have highest heat removal than non-ribbed surface and square ribs with lesser fiction factors. Sharma and Kalamkar [3] done a study on the SAH with thin ribs, where the blockage effect was maintained at $e/H = 0.1$, it was observed that the optimal TEF were achieved when $8 \leq P/e \leq 10$. Performance outside this range was suboptimal due to reduced heat transfer, which was attributed to low reattachment of the free shear. Gawande et al. [4] executed experiment and CFD investigation with reverse L-shaped ribs at a constant e/D_h of 0.042, and found at Re = 15,000, Nusselt number has maximum augmentation of 2.87 times more w.r.t. smooth duct. Alam et al. [5] used V-shaped perforated blocks in a rectangular duct to experiment by varying (e/H) from 0.4 to 1. They conclude that 0.8 has the maximum TEF. Skullong et al. [6] done a proportional study on rib size arrangement in an SAH for forced convective heat transfer and concluded thin ribs have a higher heat removal rate. By using six different baffles, Sharma et al. [7] carried out a computational with an experimental assessment of SAH. The maximum TEF is 2.05 at Re of 15,000. Mahanand and Senapati [8] have done a comparative study to find the TEF of a SAH with quarter-circular ribs over heated surface at constant $e/D_h = 0.042$. They found at 15,000 Reynolds number, $P/e = 7.14$ has a maximum value of TEF of 1.88. The design of various ribs to enhance the thermo-fluid performance of a solar air heater (SAH) has been developed by Barik et al. [9] and observation were made for T-ribs over a Re range of 5000–18,000, revealing a TEF in range of 1.58–1.7. Ngo and Phu [10] used conic curve profile to examine heat flow and change of pressure in solar air heater with CFD, by varying conic constants in artificial roughness profiles. When the conic constant decreases, Nusselt number increases with a reduction in friction factor, and a topmost TEF of 1.56 is attained at a Reynolds number of 8000. Yadav and Bhagoria [11] carried CFD analysis of heat transfer and fluid flow in a solar air heater was conducted using circular transverse wire rib roughened surfaces on the absorber plate. The study achieved a (TEF) of 1.65 by varying the Reynolds number from 3800 to 18,000. Karwa et al. [12] used chamfered rib roughness on the absorber plate and found about 40% enhancement in thermal efficiency over solar air heater with non-ribbed duct. Ansys is preferred to reduce time and cost compared to experiments. The computational investigation of absorber plate with rough surfaces has been taken by researchers. Mahanand and Senapati [8] and Ngo and Phu [10] analyze heat transfer performance in roughed surfaces with different turbulent models in SAH. They found that RNG $k - \varepsilon$ (EWT) yields the best results in promise with the Dittus-Boelter empirical correlation.

The above survey clear indicates a lot of researchers have used different shapes of artificial tabulators to increase TEF, but none tried a rhombus shape in SAH.

3 Materials and Methods

Figure 1 shows the flow field for CFD simulation of this article. The analysis fluid domain has inlet and exit section with an absorber plate (as test section) where, artificial rhombus ribs are attached as shown in Fig. 1. The sections are chosen of SAH as per ASHRAE 93-2010 standards [13]. The fluid is in the fully developed condition at the absorber plate and better mixing hot and cold air at exit length. For passing air, a height $H = 20$ mm is taken for this model. In the present study, we comprehend results by comparing three different rhombus heights and a smooth duct.

3.1 Boundary Condition and Solution Method

In this study, air at 300 K is used as the analysis fluid, with ambient pressure set as the outlet condition. The Reynolds numbers varied from 4000 to 18,000, considering adiabatic boundary conditions for the walls, except at the inlet, outlet, and absorber plate. The steady-state conformation was achieved by solving basic fluid flow equations, i.e., energy, momentum, and continuity. The (SIMPLE) algorithm was adopted with a second-order upwind scheme. The converging criteria of 10^{-6} were set for continuity, momentum, and turbulence equations, and 10^{-9} chosen energy equation.

Fig. 1 Schematic of two-dimensional computational analysis

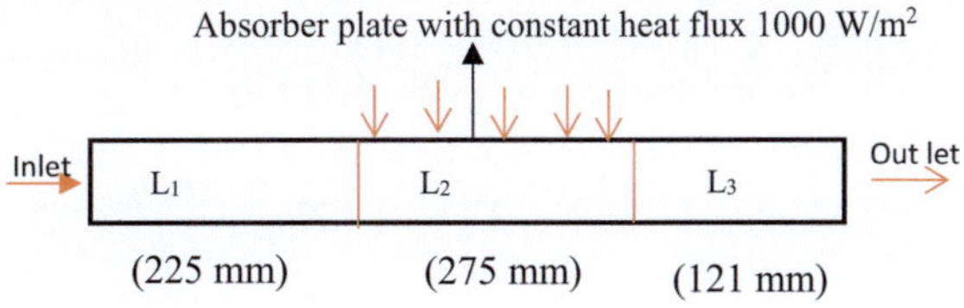

Fig. 2 Meshing rough duct SAH

3.2 Meshing and Grid Independence Test

A non-uniform meshing is adopted for smooth and rough duct, shown in Fig. 2.

The grid independence was achieved by varying nodes from 149,117 to 104,356, and node size 107,278 has the best result with a % difference of 0.083.

3.3 Validation and Model Selection

The computational results for rough duct were estimated by implementing following mathematical relations and the abbreviations are listed in nomenclature.

$$h = \frac{Q}{A_P}/(T_a - T_m) \tag{1}$$

where $\frac{Q}{A_P} = 1000\,\text{W/m}^2$,

$h =$ convective heat transfer coefficient

$T_a =$ average absorber plate temperature

$T_m = \frac{T_i + T_o}{2} =$ mean fluid temperature.

The Nusselt number was calculated by

$$\text{Nu}_r = \frac{hD_h}{k} \tag{2}$$

With the duct's width (W) and depth (H) being 100 mm and 20 mm respectively, the D_h is calculated as follows.

$$D_{\mathrm{h}} = \frac{4W \times H}{2(W + H)}. \tag{3}$$

The 'f_{r}' is determined

$$f_{\mathrm{r}} = \frac{2D_{\mathrm{h}}\mathrm{d}p}{4\rho LV^2} \tag{4}$$

The modified Dittus–Bolter correlation and modified Blasius correlation were used to estimate the Nusselt number and friction factor.

$$\mathrm{Nu} = 0.024\mathrm{Re}^{0.8}\mathrm{Pr}^{0.4} \tag{5}$$

$$f = 0.85\mathrm{Re}^{-0.25} \tag{6}$$

To reach maximum accuracy, five turbulence models were analyzed by numerical simulation in a duct with no rough surface. A rigorous analysis was done on Nusselt number by varying Re from 4000 to 18,000 with turbulence models and Dittus-Bolter correlation. It is observed the average Nusselt number variation is 4.6–14.8%, where RNG $k - \varepsilon$ with EWT has the lowest deviation. For the validation of rough RNG $k - \varepsilon$ with EWT duct turbulence model was compared with the experimental result of square rib from the literature [6] shown in Fig. 3.

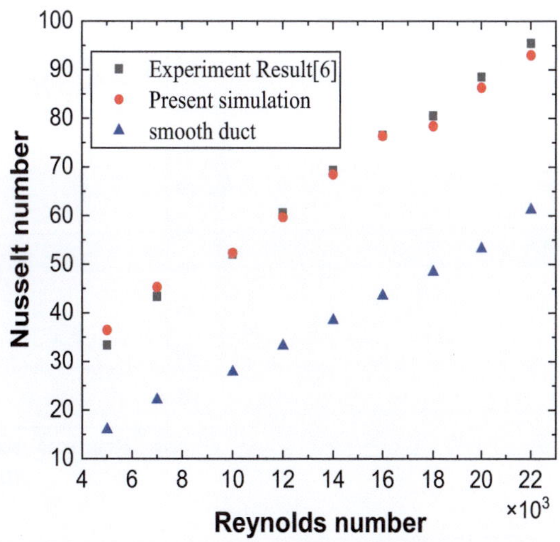

Fig. 3 Validation of numerical model with experimental results of square ribs of $P/e = 6.67$

4 Results and Discussion

A comparison was made by varying "h" versus Re for different cases shown in Fig. 4. This evidently shows that the enhancement in "h" with rise in Reynolds numbers, indicating the Nusselt number also follows the same trend.

Temperature contours at Re = 10,000 are shown in Fig. 5, and illustrates temperature gradually decreased, and at the absorber plate, maximum temperature was achieved. However, for $e = 1$ mm and $e = 1.25$ mm, the maximum temperature was found at 340.22, and 342.54 K. It means the mean temperature absorber plate is less for $e = 1$ mm, hence it has better heat transfer.

When artificial roughness is attached in SAH, the performance parameter is defined as

$$\text{Thermal enhancement factor (TEF)} = \frac{\left(\frac{\text{Nu}_\text{r}}{\text{Nu}_\text{s}}\right)}{\left(\frac{f_\text{r}}{f_\text{s}}\right)^{1/3}}. \tag{7}$$

The comparison of TEF was carried out with Reynolds number (4000–18,000), which is shown in Table 1. The results show that TEF gradually increases with the increase of Reynolds number from 4000 to 10,000 and then decreases from 10,000 to 18,000. The maximum value of 2.17 was achieved at Re = 10,000. It shows higher heat transfer enhancement.

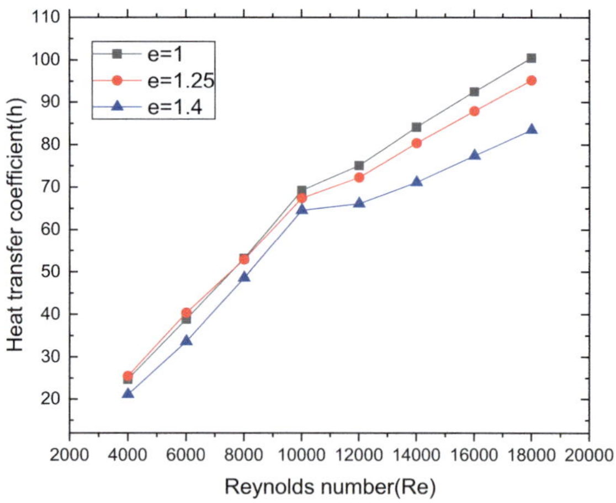

Fig. 4 Comparison of (h) versus Re

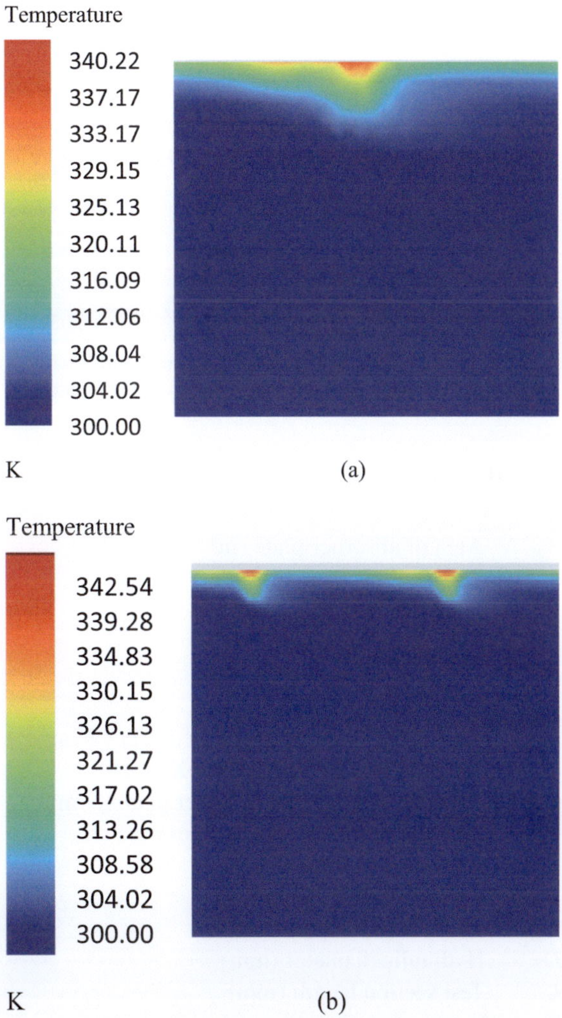

Fig. 5 Temperature contours at Re = 10,000 ($e = 1$, and 1.25 mm)

5 Conclusions

The article presents the variation of rib heights at constant pitch length ($p = 10$ mm) in SAH. The results illustrated the "h" increases with Re. For $e = 1$ mm and maximum TEF of 2.17 at Re = 10,000. It means maximum heat transfer enhancement is achieved in this condition.

Table 1 Variation of (TEF) with Reynolds number at different rib height.

Reynolds number	$e = 1.00$ mm	$e = 1.25$ mm	$e = 1.4$ mm
4000	1.85497	1.76988	1.58424
6000	2.06415	1.97446	1.71626
8000	2.14694	1.98522	1.70125
10,000	**2.17204**	2.08578	1.93631
12,000	2.12123	1.92465	1.61134
14,000	2.09835	1.89828	1.57651
16,000	2.07557	1.8711	1.54543
18,000	2.05428	1.84959	1.51963

Bold numerals correspond to the maximum TEF and the corresponding Reynolds number

Nomenclature

A_p	Area of absorber plate (m^2)
e	Rib height (mm)
P	Pitch distance (mm)
P/e	Relative roughness pitch
H	Depth of duct (mm)
W	Width of duct (mm)
h	Convective heat transfer coefficient (W/m^2 K)
k	Thermal conductivity (W/m K)
dP	Pressure drop across absorber plate (N/mm^2)
T_a	Average absorber plate temperature (K)
T_i	Inlet air temperature (K)
T_o	Outlet air temperature (K)
T_m	Mean fluid temperature (K)
D_h	Hydraulic diameter (mm)
L	Test section length (mm)
Nu_r	Rough duct Nusselt number
Nu	Smooth duct Nusselt number
f_r	Friction factor for rough duct
f	Friction factor for smooth duct
Pr	Prandtl number
Re	Reynolds number
ρ	Air density (kg/m^3)
TEF	Thermal enhancement factor

References

1. Sharma S, Das RK, Kulkarni K (2022) Experimental analysis and thermal management of solar air heater roughened with sine wave baffles. Proc Inst Mech Eng Part A J Power Energy 236(7):1403–1417. https://doi.org/10.1177/09576509221092906
2. Singh I, Vardhan S, Singh S, Singh A (2019) Experimental and CFD analysis of solar air heater duct roughened with multiple broken transverse ribs: a comparative study. Sol Energy 188:519–532. https://doi.org/10.1016/j.solener.2019.06.022
3. Sharma SK, Kalamkar VR (2017) Experimental and numerical investigation of forced convective heat transfer in solar air heater with thin ribs. Sol Energy 147:277–291. https://doi.org/10.1016/j.solener.2017.03.042
4. Gawande VB, Dhoble AS, Zodpe DB, Chamoli S (2016) Experimental and CFD investigation of convection heat transfer in solar air heater with reverse L-shaped ribs. Sol Energy 131:275–295. https://doi.org/10.1016/j.solener.2016.02.040
5. Alam T, Saini RP, Saini JS (2014) Experimental investigation on heat transfer enhancement due to V-shaped perforated blocks in a rectangular duct of solar air heater. Energy Convers Manag 81:374–383. https://doi.org/10.1016/j.enconman.2014.02.044
6. Skullong S, Thianpong C, Promvonge P (2015) Effects of rib size and arrangement on forced convective heat transfer in a solar air heater channel. Heat Mass Transf Stoffuebertragung 51(10):1475–1485. https://doi.org/10.1007/s00231-015-1515-5
7. Sharma S, Das RK, Kulkarni K (2021) Computational and experimental assessment of solar air heater roughened with six different baffles. Case Stud Therm Eng 27. https://doi.org/10.1016/j.csite.2021.101350
8. Mahanand Y, Senapati JR (2021) Thermo-hydraulic performance analysis of a solar air heater (SAH) with quarter-circular ribs on the absorber plate: a comparative study. Int J Therm Sci 161:106747. https://doi.org/10.1016/j.ijthermalsci.2020.106747
9. Barik AK, Mohanty A, Senapati JR, Awad MM (2021) Constructal design of different ribs for thermo-fluid performance enhancement of a solar air heater (SAH). Int J Therm Sci 160:106655. https://doi.org/10.1016/j.ijthermalsci.2020.106655
10. Ngo TT, Phu NM (2020) Computational fluid dynamics analysis of the heat transfer and pressure drop of solar air heater with conic-curve profile ribs. J Therm Anal Calorim 139(5):3235–3246. https://doi.org/10.1007/s10973-019-08709-4
11. Yadav AS, Bhagoria JL (2013) A CFD (computational fluid dynamics) based heat transfer and fluid flow analysis of a solar air heater provided with circular transverse wire rib roughness on the absorber plate. Energy 55:1127–1142. https://doi.org/10.1016/j.energy.2013.03.066
12. Karwa R, Solanki SC, Saini JS (2001) Thermo-hydraulic performance of solar air heaters having integral chamfered rib roughness on absorber plates. Energy 26(2):161–176. https://doi.org/10.1016/S0360-5442(00)00062-1
13. Anon (1977) Methods of testing to determine the thermal performance of solar collectors. ANSI Standard B198. 1, vol 2010, no Ra

Aerodynamic Performance of Symmetric and Cambered Wings Under Ground and Ceiling Proximity

Ravindra A. Shirsath

Abstract The aerodynamic performance of a finite wing is greatly influenced while operating in proximity to the ground and has been extensively researched and reported in the literature, popularly known as Wing-in-Ground (WIG) effect. The present work investigates the same; the influence of ceiling proximity is also studied. The performance of a symmetric wing is compared to that of the cambered wing having the same aspect ratio for the range of angles of attack and particular ground and ceiling clearances. Considerable enhancement in aerodynamic performance is observed in terms of lift coefficient with ceiling effect as compared to ground effect for both the symmetric and cambered wings.

Keywords Finite wing · Cambered wing · Pressure distribution · Symmetric wing · Wing-in-ceiling effect · Wing-in-Ground effect

1 Introduction

The finite wing aerodynamics has been the topic of research for nearly a century. The drag experienced by a finite wing is considerably higher when compared to its constituting airfoil due to the formation of wing tip vortices and the phenomenon of downwash. Hence researchers have tried various methods in the past such as winglets, sharklets, tip fuel tanks, and elliptical planforms in order to control the strength of wing tip vortices that ultimately results in reduced downwash and hence improved aerodynamic performance. Another entirely different approach is to operate the wing in close proximity to the ground such that the ground acts as a blockage to the wing tip vortices and serves the same purpose. In case of ground effect the surface is under the wing, i.e., from pressure side however if the surface happens to be above the wing, i.e., from suction side, the effect needs to be investigated. The common engineering applications of this phenomenon can be found in bi-planes, cascades

R. A. Shirsath (✉)
Academic Development Manager II, Ansys Software Private Limited, Pune, India
e-mail: aero.ravi22@gmail.com

A. Kumar R. et al. (eds.), *Proceedings of Fluid Mechanics and Fluid Power (FMFP) 2023, Vol. 6*, Lecture Notes in Mechanical Engineering,
https://doi.org/10.1007/978-981-96-9600-0_41

of turbine and compressor blades, sport cars and Ground Effect Machines (Gems). Hence in the present work, both the ground and ceiling proximities are considered and compared, in the case of symmetric and cambered wing sections with identical aspect ratio (AR). The investigations include estimating the lift and drag coefficients over a wide range of angles of attack for wing in ceiling effect, out-of-ceiling effect and ground effect. Corresponding pressure distribution is also reported for $\alpha = 0°$ to bring out some of the interesting aspects associated with the phenomenon.

The ground effect can be broadly classified into two categories as chord dominated and span dominated ground effects. The chord dominated ground effect such as aerodynamic characteristics of NACA 0015 airfoil close to ground were studied experimentally by Ahmed and Sharma [1]. Li et al. [2] highlighted the occurrence of enhanced lift and improved L/D ratio only for medium range of angles of attack. Importance of span dominated ground effect was justified as Martinez-Aranda et al. [3] found that low AR and wingtip vortex generation were the primary causes for approximately 40% reduction in lift force regardless of the Re compared to NACA 0012 airfoil, and the edge effect generated induced drag nearly six times greater than the airfoil profile. While approaching the ground, all wings with thick highly cambered airfoils with different AR exhibited enhanced lift-curve slope and lower induced drag resulting in considerable rise in L/D ratio as reported by Fink et al. [4] from wind tunnel investigations determining the ground proximity effect. Dominance of ground effect for $h/c < 1$ was reported by Chawla et al. [5] through wind tunnel study of a finite wing with $AR \approx 2.33$ of a NACA 4415 profile with adjustable flap and removable end and center plates. Aerodynamic characteristics of three dimensional wings in ground effect for Aero-levitation Electric Vehicle were investigated numerically by Moon et al. [6] including different modifications of vertical wing. Accuracy of different boundary conditions, deformation of free surface due to wing tip vortex and inability of inviscid methods were highlighted by Barber [7].

To summarize, most of the research has focused on the aerodynamic analysis of chord-dominated ground effect and lately extended to the span-dominated ground effect. The improvement in the aerodynamic characteristics is reported in both. However, if the wing is operating in the proximity of the ceiling, i.e., if the surface is from the suction side of the wing rather than the pressure side, its influence on the aerodynamic characteristics remains to be investigated. Thus, the present work is aimed at comparing the aerodynamic performance of the wing in ceiling effect to that in the ground effect in terms of force coefficients and pressure distribution.

2 Computational Methodology

This section explains briefly the methodology adapted for steady-state CFD simulations using ANSYS Workbench. Numerical solver is set up with absolute velocity formulation and SIMPLE algorithm is used for pressure–velocity coupling. Least square cell-based method is employed for estimating the gradients while second-order upwind scheme is used for discretizing the equations.

2.1 Geometry and Mesh Generation

The wings investigated in the present work have NACA 0012 (symmetric) and NACA 4412 (cambered) airfoil sections. Both the wings have same aspect ratio, $AR \approx 6.4$, and chord length, $c = 0.063\text{m}$; however, to save computational time, only half of the wing is simulated. The computational domain used for investigating Wing-in-Ground (WIG) effect is shown in Fig. 1a and same for Wing-in-Ceiling (WIC) effect in Fig. 1b. The domain for Wing-out-of-Ceiling (WOC) effect is similar but just doubled in the vertical direction appropriately with the wing at the center.

Fig. 1 Computational domain for WIG and WIC effect

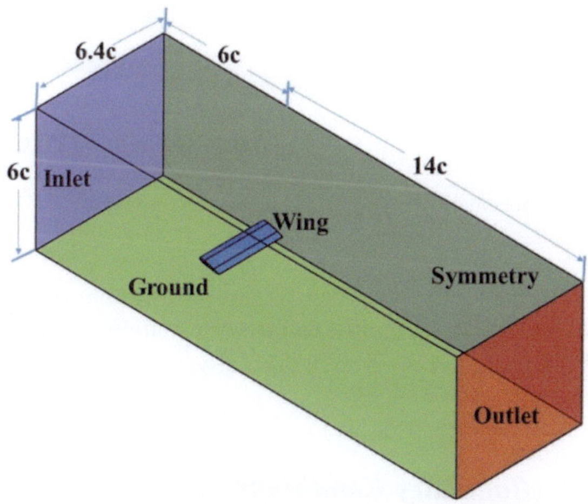

(a) Wing-in-Ground (WIG) effect

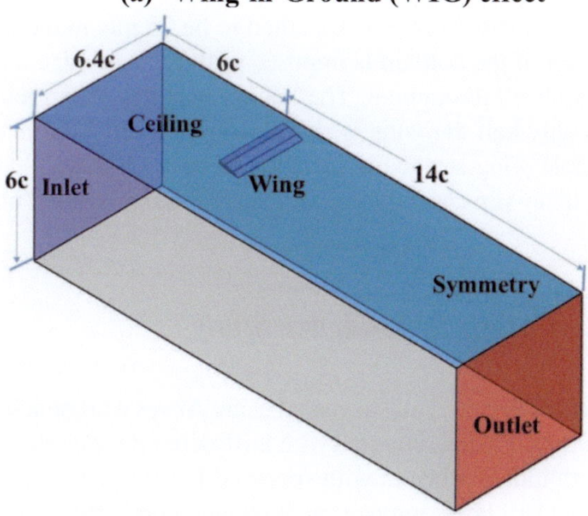

(b) Wing-in-Ceiling (WIC) effect

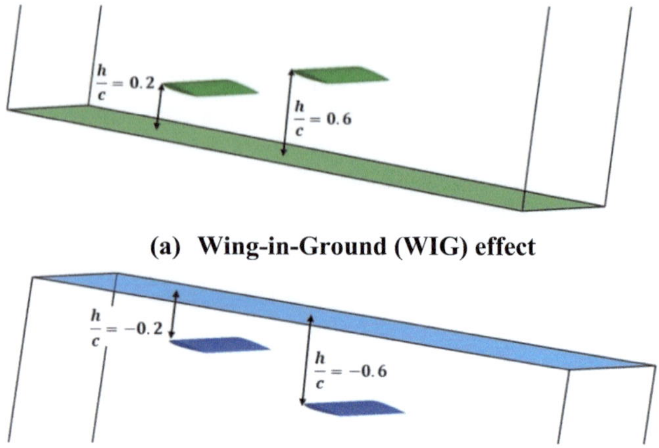

(a) Wing-in-Ground (WIG) effect

(b) Wing-in-Ceiling (WIC) effect

Fig. 2 Schematic of wing-in-ground/ceiling proximity

To distinguish between ground and ceiling effect, the clearance is denoted as positive h/c for ground effect as seen from Fig. 2a and negative h/c for ceiling effect as indicated in Fig. 2b schematically.

2.2 Boundary Conditions

The velocity at inlet is specified to be 25 m/s in the streamwise direction while the outlet of the domain is imposed with the outflow condition. Zero-gauge pressure is initialized throughout. The boundary at the wing root is prescribed to be symmetry as only half the wing is simulated, while the bottom and side are set to walls. The whole wing surface is modeled to be a solid wall. The ground/ceiling is imposed with no-slip condition as schematically represented in Fig. 3.

2.3 Mesh Dependence Study

An unstructured mesh created using Ansys workbench is shown in Fig. 4. While most of the domain is covered with unstructured mesh, the surface of the wing is provided with inflation layers with specified first-layer thickness to capture the formation of boundary layer and near-wall phenomenon efficiently.

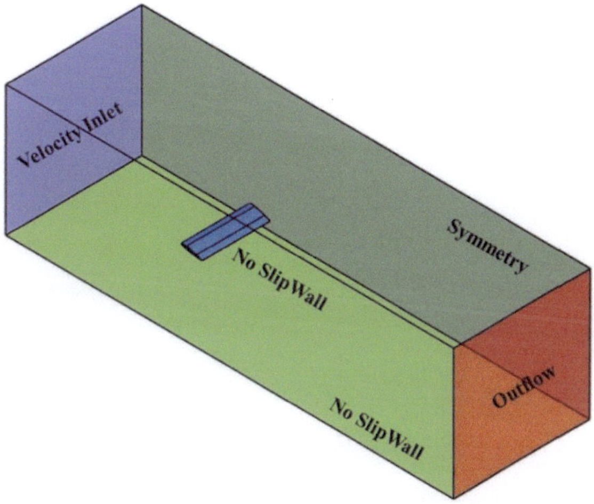

Fig. 3 Boundary conditions (Schematic)

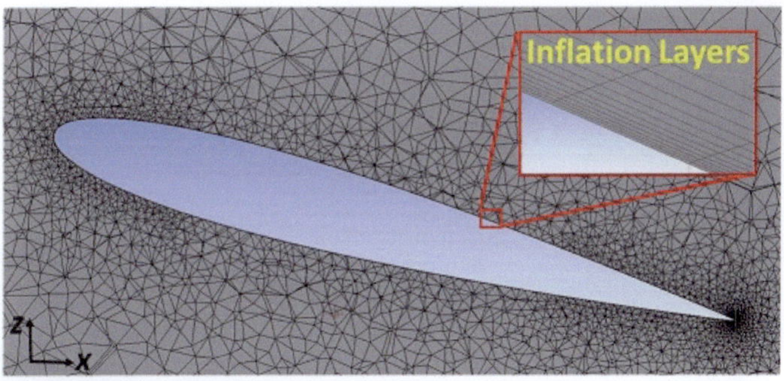

Fig. 4 Unstructured mesh on the cross-section of wing

A mesh dependence study is performed for WOC effect at $\alpha = 6°$. Seven grids are constructed with an increasing number of elements. Figure 5 depicts a grid convergence analysis of C_D of the wing. Eventually the grid with $\approx 3 \times 10^6$ elements is used for the remaining simulations.

2.4 Selection of Turbulence Model

The selection of turbulence model is crucial, especially for the flows at higher angles of attack. As seen from Fig. 6, all the models used are predicting the C_L correctly at smaller angles of attack; however, both the $k - \epsilon$ (Standard) and Spalart–Allmaras models are seen to not only over predict the C_L at larger angles but also much higher stall angle. $k - \omega$ (SST), on the other hand, is seen to predict the stall angle correctly and fairly agreeing with experimental results by Shirsath and Mukherjee [8], in both

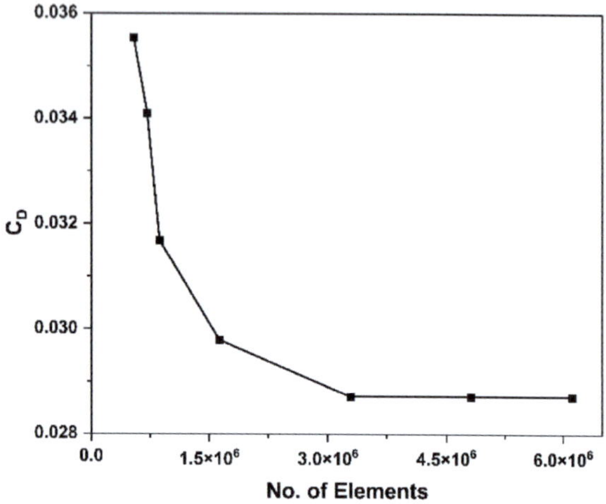

Fig. 5 Grid convergence at $\alpha = 6°$

small and large angles of attack regions though with a marginal over prediction in the latter case. Hence, $k - \omega$ (SST) is used throughout the analysis.

The inflation layers used with specified first layer height over the wing surface ensures that the $y^+ < 1$ over the entire wing as seen from Fig. 7 as required by the $k - \omega$ (SST) model for efficient computations.

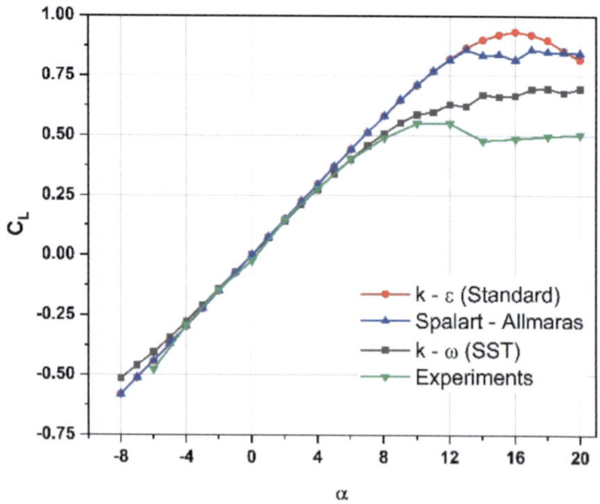

Fig. 6 C_L from different turbulence models

Fig. 7 Comparison of turbulence models and wall y^+

2.5 Convergence

The criteria for convergence is set to be 10^{-5} and additionally the force coefficients on the wing are monitored individually as shown in Fig. 8. It is seen from Fig. 8a that the C_L converges to a constant value fairly quickly for $\alpha < 9°$, though the simulations are continued for a sufficient number of iterations.

However, in the post stall region as seen from Fig. 8b at $\alpha = 17°$, the force coefficient exhibits a dynamic convergence, thus after sufficient number of iterations, the coefficients are averaged once they have reached the dynamic convergence.

3 Results and Discussion

The results are discussed aiming at the comparison of the symmetric wing with that of the cambered wing. The two wings considered for comparison have the same aspect ratio; both are rectangular, operating under the same free-stream conditions except the constituting airfoil sections. The first wing has NACA 0012, i.e., the symmetric airfoil section, while the latter is NACA 4412, i.e., the cambered airfoil section.

The aerodynamic characteristics of the two wings are first investigated without proximity to the ground or ceiling, referred to be the wing-out-of-ceiling (WOC) condition. It is observed that the C_D of a symmetric wing is relatively less than that of the cambered wing for $0° < \alpha < 11°$ while for $11° < \alpha < 18°$ that of the cambered wing is less as seen from Fig. 9a. Interestingly, the C_D of both the wings is identical at $\alpha = 11°$ and $\alpha = 18°$.

The comparison of C_L as seen from Fig. 9b, however, clearly shows that the C_L of cambered wing is always higher than that of the symmetric wing for the entire range of angles of attack investigated. In the range of $0° < \alpha < 10°$, the $\Delta C_L \approx 0.25$ and is nearly constant for the two wings.

Fig. 8 Convergence of lift coefficient with iterations

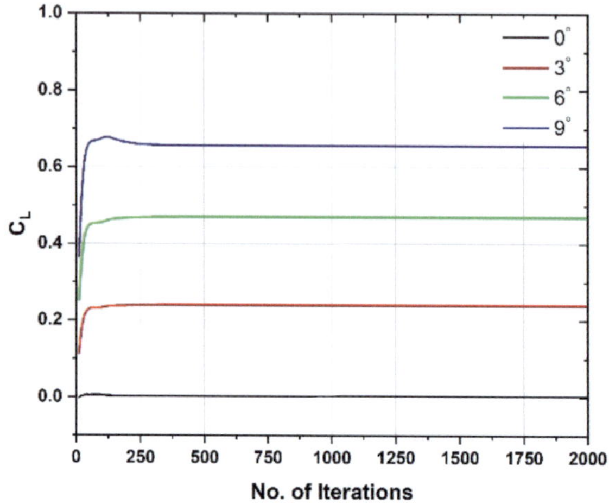

(a) C_L convergence, $\alpha = 0^\circ, 3^\circ, 6^\circ$ *and* 9°

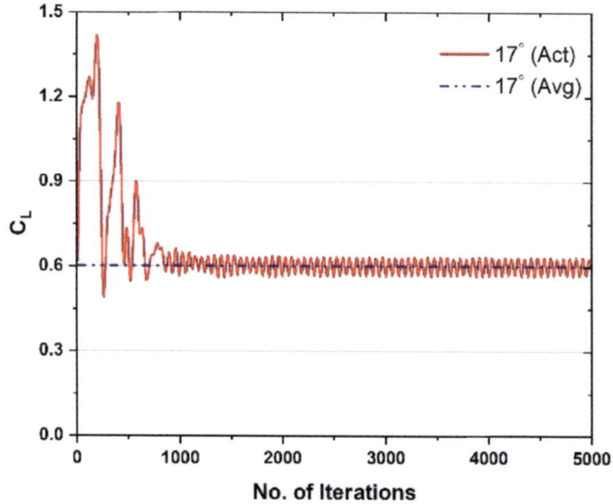

(b) C_L convergence, $\alpha = 17^\circ$

3.1 Aerodynamic Performance Under the Ground and Ceiling Proximity

The effect of surface proximity (both ground and ceiling) along with the airfoil section on the C_D for $h/c = \pm 0.2$ is shown in Fig. 10a. The C_D of a symmetric wing under WIG effect is observed to be marginally lesser than corresponding WIC effect for $\alpha < 9°$, however, beyond this angle the C_D starts increasing drastically and mostly remains higher than the corresponding WIC effect condition. For the cambered wing, however, the C_D in WIG effect condition remains lesser than the WIC effect over a greater range of angles of attack.

Fig. 9 Comparison between NACA 0012 and NACA 4412 wings under WOC effect

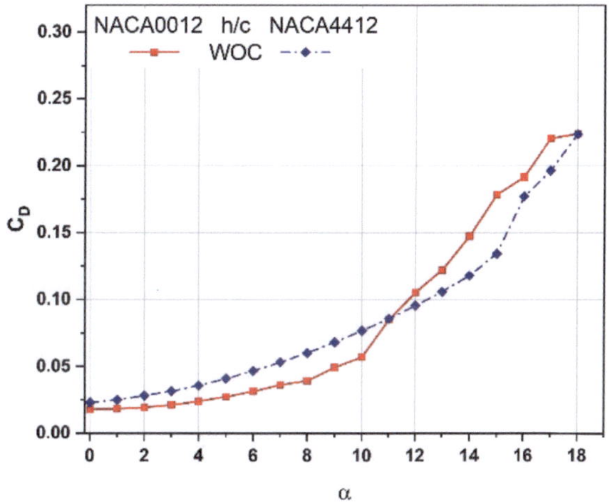

(a) Drag coefficient, C_D vs α

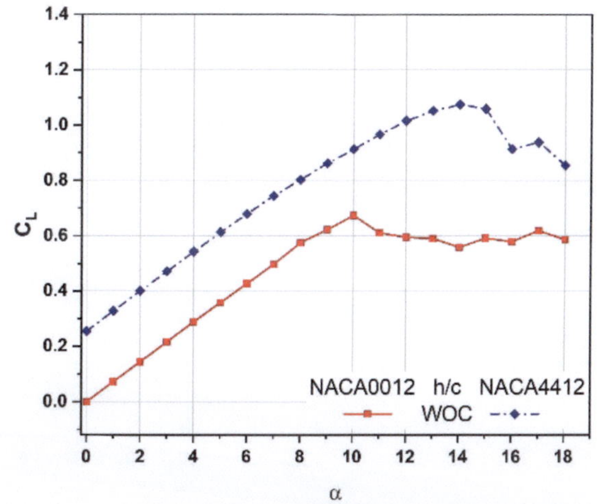

(b) Lift coefficient, C_L vs α

The variation of C_L for corresponding condition is shown in Fig. 10b where the effect of ceiling proximity is seen to be dominating when compared to the effect of ground proximity. Interestingly, for symmetric wing, at $\alpha = 0°$ the $C_L \approx -0.1$ for WIG effect condition as against $C_L = 0$ when in WOC effect. The primary reason for this behavior is the formation of convergent-divergent passage created between the ground and wing as also reported by Ahmed and Sharma [1] for 2D cases. This effect is however confined to very small angles of attack and the wing starts producing positive lift for $\alpha > 1°$. Contrary to this, when in WIC effect condition, similar convergent-divergent passage occurs but now from the upper surface of the wing and thus the reduction in pressure now is rather favoring the production of lift and hence now at $\alpha = 0°$ the $C_L \approx 0.1$. The C_L experienced by the wing in ceiling

Fig. 10 Comparison between NACA 0012 and NACA 4412 wings under WIG and WIC effect ($h/c = \pm 0.2$)

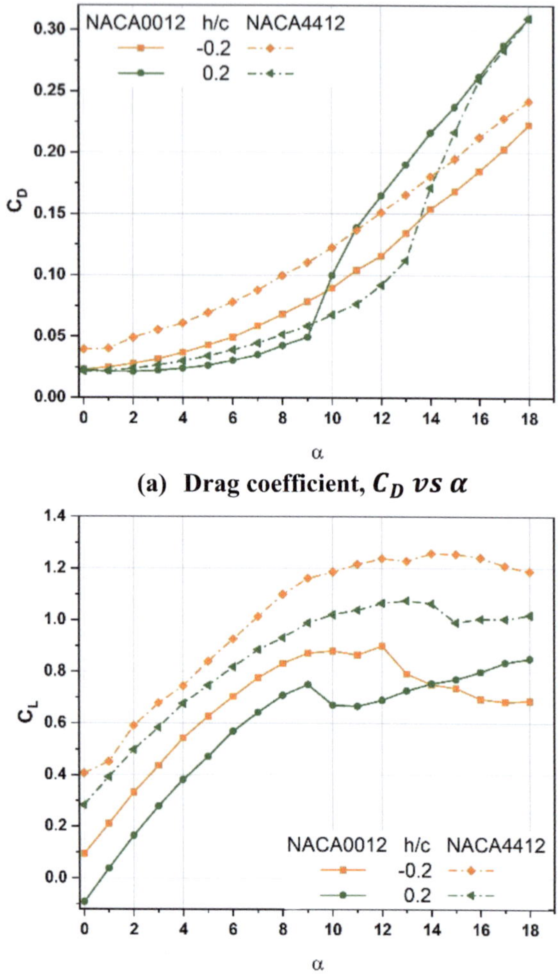

(a) Drag coefficient, C_D vs α

(b) Lift coefficient, C_L vs α

proximity remains greater than that with the ground proximity for most of the range of angles of attack investigated and is notably higher for $9° < \alpha < 14°$, beyond which the trend reversal is observed.

3.2 Pressure Distribution

As observed that at $h/c = \pm 0.2$, the symmetric wing shows a distinctly different behavior, corresponding pressure contours at mid-semi-span location of the wing at $\alpha = 0°$ are shown in Fig. 11. It is observed that at this angle of attack, due to the particular shape of the wing, a convergent-divergent passage is created between the ground and wing surface. This ultimately results in reduced pressure on the lower surface compared to upper surface as seen from Fig. 11a, thereby resulting in negative lift rather than zero lift as also reported by Ahmed and Sharma [1]. While out of ground/ceiling effect, as observed from Fig. 11b, the pressure distribution is exactly identical on the upper and lower surfaces thereby resulting in zero lift as expected. More importantly, when in ceiling effect, the similar phenomenon is observed from Fig. 11c, but now on the upper surface causing considerably higher lift under this condition.

Unlike the symmetric wing, in case of cambered wing no convergent-divergent passage is formed between the ground and wing surface. Thus no region of reduced pressure is observed as seen from pressure contours plot from Fig. 12a at $\alpha = 0°$ even at $h/c = 0.2$, thus corresponding $C_L > 0$ as opposed to symmetric wing where $C_L < 0$ was reported. Further in WOC effect condition, as seen from Fig. 12b, larger pressure difference creating greater lift compared to symmetric wing is seen. More interestingly, at $h/c = -0.2$ the greater reduction in pressure between the upper surface and ceiling as seen from Fig. 12c causes larger than expected lift even at $\alpha = 0°$.

Since the ground effect has been investigated extensively, the present work explores the possibility of potential advantages of wing in ceiling effect. The improvement in aerodynamic characteristics reported lays the foundation for further investigations. The present work opens up an idea of potentially different design of high lift devices that can be more efficient and robust compared to the traditional high lift devices. In this design, the main wing can be considered as a ceiling and a deployable device can be treated as a wing in ceiling effect.

4 Conclusions

The investigations carried out to compare the effect of ground and ceiling proximity on the aerodynamic characteristics of a finite rectangular wings of symmetric and cambered airfoil sections justifies the enhanced aerodynamic performance specifically in terms of increased lift coefficient.

Wing-in-ceiling effect shows even better lift coefficient compared to wing out of ceiling as well as wing-in-ground effect for both the symmetric and cambered wing. Additionally, the cambered wing shows better improvement in lift coefficient at relatively higher angles of attack.

Pressure distribution at mid-semi span clearly indicates adverse effect on symmetric wing-in-ground effect at $\alpha = 0°$ while that in ceiling effect is favorable. In case of cambered wing, however, no such adverse effect is observed.

Thus overall, better aerodynamic performance is observed for both symmetric and cambered wings in ground as well as ceiling proximity and the same is considerably higher in the latter case.

(a) $h/c = 0.2$

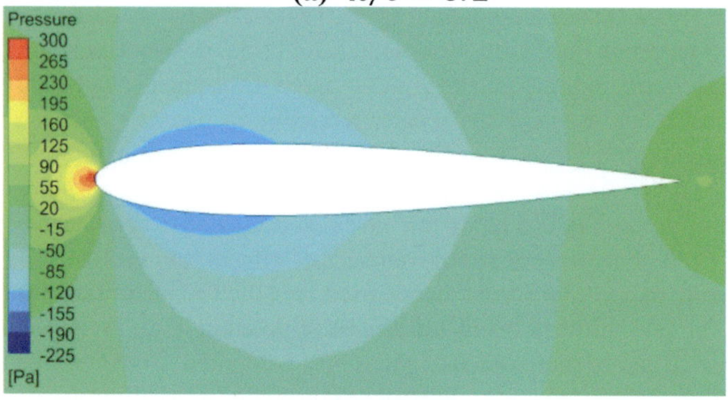

(b) WOC

(c) $h/c = -0.2$

Fig. 11 Pressure contours at mid-section of the symmetric wing under WIG, WOC and WIC effect ($h/c = \pm 0.2$)

(a) $\boldsymbol{h/c = 0.2}$

(b) WOC

(c) $\boldsymbol{h/c = -0.2}$

Fig. 12 Pressure contours at mid-section of the cambered wing under WOC, WIG and WIC effect ($h/c = \pm 0.2$)

Nomenclature

AR	Aspect ratio
CFD	Computational fluid dynamics
C_D	Drag coefficient
C_L	Lift coefficient
NACA	National advisory committee for aeronautics
Re	Reynolds number
SIMPLE	Semi implicit method for pressure linked equations
WIG	Wing-in-ground
WIC	Wing-in-ceiling
WOC	Wing-out-of-ceiling
α	Angle of attack (°)
h/c	Ceiling/ground clearance

References

1. Ahmed M, Sharma S (2005) An investigation on the aerodynamics of a symmetrical airfoil in ground effect. Exp Thermal Fluid Sci 29(6):633–647
2. Li S, Zhou D, Zhang Y, Qu Q (2015) Aerodynamic investigation of RAE 2822 airfoil in ground effect. Proc Eng 126:174–178
3. Martinez-Aranda S, Garcia-Gonzalez A, Parras L, Velazquez-Navarro J, Pino C (2016) Comparison of the aerodynamic characteristics of the NACA 0012 airfoil at low-to-moderate Reynolds numbers for any aspect ratio. Int J Aerospace Sci 4:1–8
4. Fink MJ, Lastinger J (1961) Aerodynamic characteristics of low aspect ratio wings in close proximity to the ground, NASA Technical Note
5. Chawla M, Edwards L, Franke M (1990) Wind tunnel investigation of wing-in-ground effects. J Aircr 27(4):289–293
6. Moon YJ, Oh HJ, Seo JH (2005) Aerodynamic investigation of three dimensional wings in ground effect for aero-levitation electric vehicle. Aerosp Sci Technol 9(6):485–494
7. Barber T (2006) Aerodynamic ground effect: a case study of the integration of CFD and experiments. Int J Veh Des 40(4):299–316
8. Shirsath RA, Mukherjee R (2023) Experimental and computational investigations of aerodynamic characteristics of a finite rectangular wing-in-ground effect. Proc Inst Mech Eng, Part G: J Aerospace Eng 237(5):1007–1024

Effect of Vane Edge Shape on the Performance of Savonius Hydrokinetic Turbines

Vimal Patel and Shaikh Shehzad

Abstract Hydrokinetic turbines, such as Savonius turbine, are used to extract the energy from the flowing water stream, without or with very limited change in the natural path of the water flow. The performance of the Savonius turbine can be improved by geometrical modification of the vanes. In this investigation, three different semi-circular blade shapes have been selected for analysis: (i) Vane with both edges having a semi-circular edge, (ii) Vane with a sharp outer edge and semi-circular inner edge, and (iii) Vane with a sharp inner edge and semi-circular outer edge. The 2D simulations are used in the present investigations. The grid-independent study, domain optimization, and validation of the adopted methodology are carried out before investigations related to the vane shape. The obtained results are indicated as a non-dimensional parameter of the coefficient of power (Cp) and coefficient of torque (Ct). The power coefficient obtained with the optimized design is 0.24 against that of 0.08 with the conventional vane shape.

Keywords Hydrokinetic turbines · Renewable energy · Applied energy · Hydraulic turbines · Savonius turbines

1 Introduction

In most places of India, rivers and canals exhibit regular water currents for most of the year. However, India has not looked closely at growth toward the usage of kinetic energy from river currents [1]. The Savonius turbine is a type of vertical axis turbine that generates electricity from the flow of water in a river or canal without impeding it with a large dam [2]. Savonius turbines have good starting characteristics and straightforward construction features, despite their low power output. Therefore, it is

V. Patel (✉) · S. Shehzad
Department of Mechanical Engineering, SVNIT, Surat, Gujarat, India
e-mail: vimal.iitbombay@gmail.com

S. Shehzad
e-mail: p20tm002@med.svnit.ac.in

A. Kumar R. et al. (eds.), *Proceedings of Fluid Mechanics and Fluid Power (FMFP) 2023, Vol. 6*, Lecture Notes in Mechanical Engineering,
https://doi.org/10.1007/978-981-96-9600-0_42

perfect for producing electricity in isolated areas that are located closer to a water flow resource [3]. The performance of the Savonius hydrokinetic turbine can be improved using flat deflector plates [6], C-shaped deflector plates [7], or flow-diverting blocking plates [8, 9]. Patel and Patel [4] investigated the effect of end arc near the center of the rotor providing a different fillet radius, for a modified vane shape. The effect of the end arc radius was tested using validated numerical simulation. Patel and Patel [5] first time investigated the effect of wavy edge vane on the performance of the Savnonius hydrokinetic turbine. They achieved 37% improvement in coefficient of power compared to the conventional design. However, the relative assessment related to the orientation of the sharp edge and hydrodynamic edge in the vane design is not investigated till now. Hence, in the present investigation, it is decided to investigate the effect of vane edge sharpness on the performance of conventional vanned Savonius hydrokinetic turbine.

In this investigation, three different blade shapes, with semi-circular blade profiles, have been selected for analysis. (1) Design 1: Both side blade edges semi-circular (2) Design 2: Design 2: Sharp outer edge and semi-circular inner edge and (3) Design 3: Sharp inner edge and semi-circular outer edge. All three selected blades are shown in Fig. 1. Keeping sharp edges outside, the flow may separate out at the outer edge of the vane easily, which might lead to creating a larger drag force at the outer edge of the vane. This might lead to increased resultant torque on the turbine vane due to shifting of the center of force toward the outer side of the vane. On the other hand, keeping a semi-circular vane on the outer side may form a hydrodynamic shape of the vane, which leads to minimizing the fluid loss. Hence, it is necessary to investigate the location of sharp and semi-circular vane edge shapes. Hence, the aim of the present investigation is decided to investigate and analyze the vane edge shape on conventional semi-circular vanes of Savonius hydrokinetic turbine with numerical simulations.

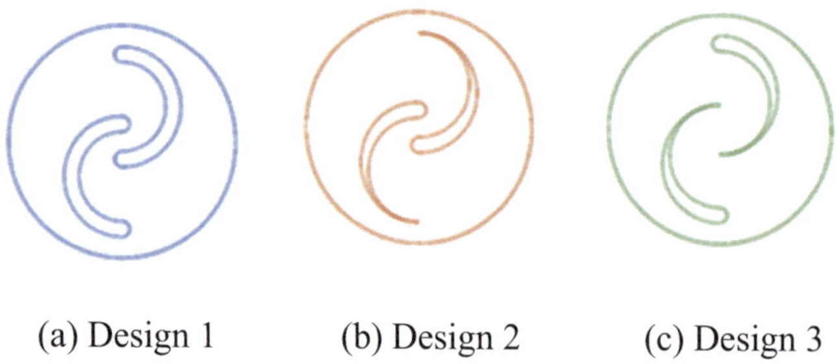

(a) Design 1 (b) Design 2 (c) Design 3

Fig. 1 Investigated vane shapes in the present study

2 Investigated Parameters

In the present investigations, three different impeller designs are considered, as shown in Fig. 1. The investigations are carried out for different angular velocities of the turbine rotor, which is indicated with the non-dimensional numbers as Tip Speed Ratio (TSR). The coefficient of torque and coefficient of power is measured at TSR of 0.1, 0.3, 0.5, 0.7, 0.9. Flow characteristics, velocity, and pressure variation are also analyzed with the help of their contours. All simulations are done considering the constant inlet fluid velocity of 0.3 m/s.

3 Data Reduction

Tip Speed Ratio indicates the non-dimensional number of angular velocity of the turbine impeller. It is the ratio of tangential velocity of the tip of the turbine vane to the free stream velocity of the fluid. It can be calculated using Eq. (1)

$$\mathrm{TSR} = \frac{\omega R}{V} \tag{1}$$

The coefficient of torque indicates the non-dimensional number of available shaft torque. It indicates the ratio of mechanical torque available at the turbine shaft to the hydrodynamic torque supplied to the turbine rotor area. It can be calculated using Eq. (2).

$$C_T = \frac{T_{\mathrm{turbine}}}{T_{\mathrm{avaialble}}} = \frac{T_{\mathrm{turbine}}}{0.5\rho A V^2 R} \tag{2}$$

The coefficient of power indicates the non-dimensional number of power output from the turbine. It indicates the ratio of mechanical power output available at the turbine shaft to the hydrokinetic power supplied to the turbine rotor area. It can be calculated using Eq. (3)

$$C_P = C_T \times \mathrm{TSR} \tag{3}$$

4 Numerical Analysis

The present investigations are done using numerical analysis simulations using Ansys. The 2D simulations are used in the present investigations. The grid-independent study, domain optimization, and validation of the adopted methodology are carried out before investigations related to the vane shape. The detailed discussion is explained in the following topics.

4.1 Geometry and Domain Optimization

The geometry details are shown in Table 1. Two domains are used, namely stationary domain and rotating domain, which are separated by an interface as shown in Fig. 2.

The selected size of the computational domain also affects the results. Too small size of the domain may overestimate the coefficient of torque and, subsequently, coefficient of power, due to the close appearance of the boundary near the turbine rotor. Too large size of the computational domain requires too much time for the computations. Hence, it is required to find the optimum size of the computational domain before carrying out the investigations. The boundary conditions and selected domains used in the present investigations are shown in Fig. 2. To study the turbine, the stationary domain must be such that it has no effect on the rotating domain. To clarify this problem, a domain optimization study has been done and presented by a graph as shown in Fig. 3.

Table 1 Details of geometry

S. No.	Parameter	Value
1	Rotor diameter (D)	0.245 m
2	Gap between two blades (e)	0.045 m
3	Interface diameter	0.3 m
4	Number of blades	2

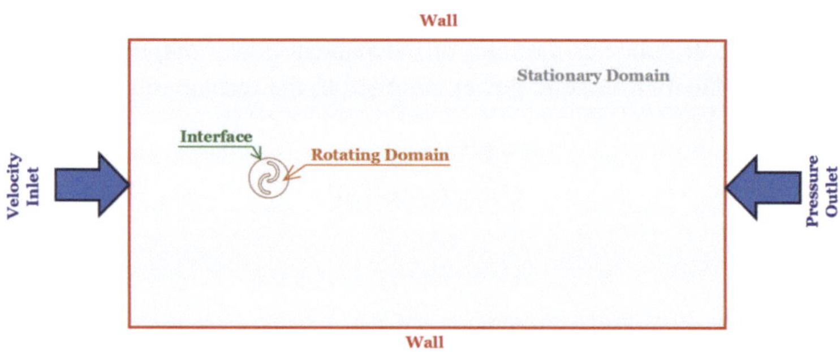

Fig. 2 Boundary conditions used in the present investigations

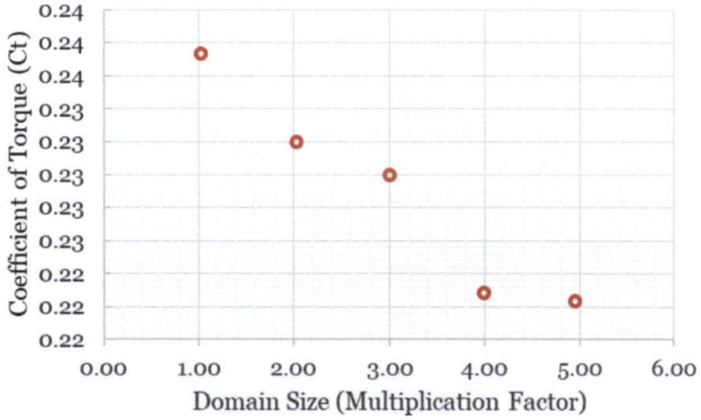

Fig. 3 Domain size optimization

For domain optimization, different sizes of domains are considered. The different size of the domain is formed as different multiplication factors to the basic considered size. Different five sizes of the domains are considered as a multiplication from 1 to 5, respectively, keeping rotor size same. The obtained results are plotted as variations of coefficient of torque with considered multiplication factors as shown in Fig. 3. The results indicate that the variation of obtained coefficient of torque becomes almost same for the size multiplication factor of 4 and 5. Which indicates that the result became independent of the size after multiplication factor of 4. The optimized size is 9D × 16D. All the simulations in the present investigations are done using the same optimized size of the domain.

4.2 Meshing and Grid Independency Study

The meshing of the model has been done using the Ansys meshing module. A triangular elements are used to generate the auto-generate mesh. At the edges, fine mesh is generated, and 15 inflation layers with a 1.12% growth rate are created at blade edges. The number of elements also affects the final results if enough numbers of elements are not considered in numerical simulation. To find out the optimum number of elements, the grid-independent study is carried out before initiating the main study. The variation of the coefficient of torque obtained with different numbers of elements is shown in Fig. 4. From Fig. 4, it is observed that the variation of the coefficient of torque, more than the number of elements 3×10^5, is almost the same. Hence, it can be concluded that, for the selected size of the domain, a minimum of 3×10^5 elements is required to get the results independent from the number of grid elements used. Aspect ratio, skewness, and orthogonality are 1.97, 0.56, and 0.92, respectively. The mesh generated near the turbine vanes and near the edge of the turbine vanes is shown in Fig. 5.

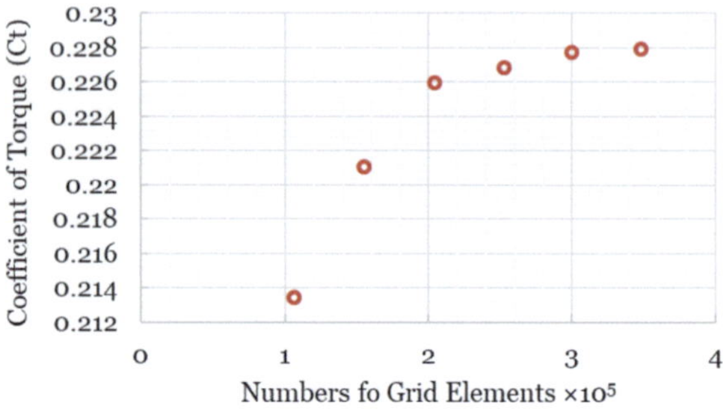

Fig. 4 Grid Independence study

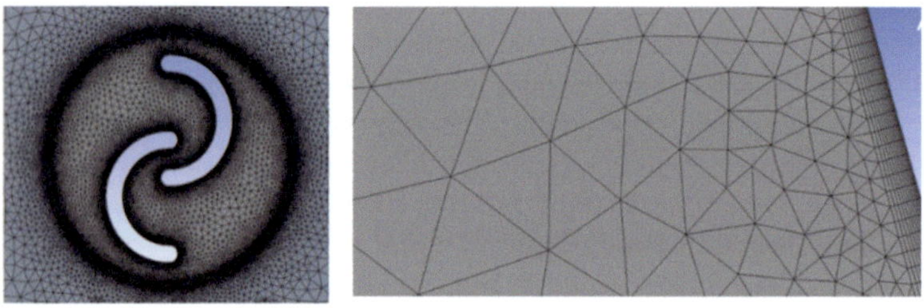

Fig. 5 Meshing

4.3 Fluent Setup

In the present investigations, a 2D, unsteady, and pressure-based solver has been selected, and K-ω turbulent model is considered for the analysis. Water is used as a fluid. Boundary conditions considered in the present investigations are shown in Fig. 2. The time step size is calculated from the considered angular velocity or TSR.

5 Result and Discussion

The aim of the present study is to investigate the best vane design among the three considered shapes of the vanes. The numerical simulations are carried out for all three selected vane designs at different angular velocities of the rotor with validated methodology. The simulations are done at different angular velocities of the turbine runner, and the average coefficient of torque is estimated from the simulations. The obtained values of coefficient of torque (Ct) at different Tip Speed Ratio (TSR) are marked in Fig. 6.

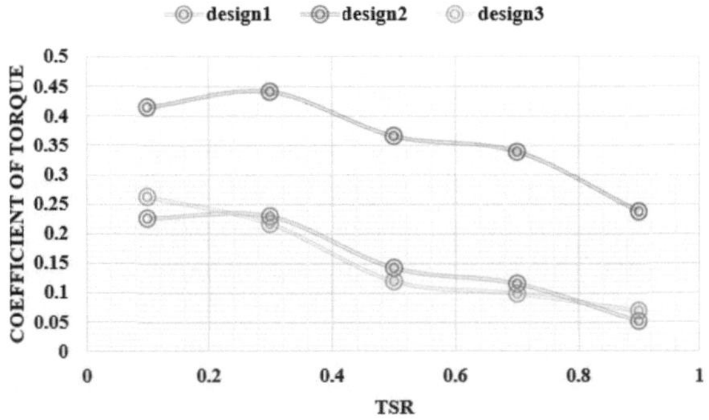

Fig. 6 Variation of coefficient of torque (Ct) at different TSR for 3 selected designs

The general trend in all three designs is very common: the coefficient of torque decreases as TSR increases. Due to the increase in load, torque will increase, and rotational speed will decrease as inlet velocity is constant; this fundamental observation has also been analyzed. Design 1 and design 2 have almost similar values at particular TSR and less than that of design 2. From the graph, it is observed that the coefficient of torque is maximum at 0.3 TSR for design 1 and design 2, and for design 3, it is maximum at 0.1 TSR.

The variation of coefficient of power (Cp) at different Tip Speed Ratio (TSR) is plotted for all three different designs is shown in Fig. 7. The result indicates that design 2, i.e., sharp edge outside the vane, provides the better performance compared to the other two considered designs.

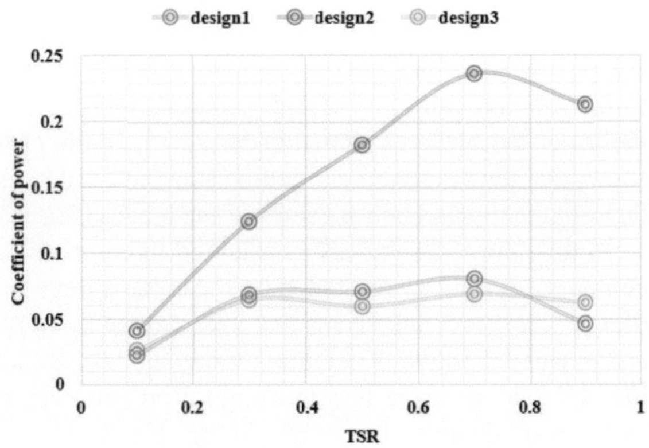

Fig. 7 Comparison of coefficient of power

The power output is dependent on torque and angular velocity. After comparing the coefficient of power at different TSR for all considered designs, it is investigated that the maximum value for all three designs is obtained at 0.7 TSR and the minimum at 0.1 TSR. The maximum value obtained in design 1, design 2, and design 3 are 0.08, 0.24, and 0.069, respectively. It is observed that the maximum coefficient of power obtained using design 2 is approximately 0.24 at TSR of 0.7, against 0.08 at the same TSR for the conventional shape of the vane. Hence, the result indicates that by using the sharp edge outside the vane enhance the performance approximately 3 times the conventional semi-circular vane edge.

When water with velocity 0.3 m/s strikes a Savonius turbine rotor which is placed in such a way that after striking it will start to rotate in a clockwise direction. In this way, fluid kinetic energy is converted into pressure energy and then rotational energy. Figures 8, 9, and 10 indicate the comparison of velocity contours obtained with the different TSR for conventional semi-circular vane edge, sharp edge outside, and sharp edge inside the vanes cases.

In all cases, it is observed that the velocity of the water at the downstream side of the rotor is relatively less. This is due to the fact that the kinetic energy of the water is utilized to convert into mechanical power output from the turbine rotor.

The flow velocity of water doesn't change up to the turbine blades, and after touching the blades, it changes. Due to the rotation of turbine blades, variation of velocity occurs. There are two regions observed in the contours of velocity: one is the high-velocity region at tip of advance blade, and the other is the low-velocity or wake zone downstream of rotor (Figs. 11 and 12).

It can be observed easily that the pressure of the backside of retarding blade increases as TSR increases, and the resultant pressure of the rotor decreases, which will reduce torque. At the tip-advancing blade, a red color is observed in the velocity contour, which means velocity is higher; this can be verified from the pressure contour as it shows pressure loss at that position, and that's why vortices are generated at the tip of the blade as shown in Fig. 4.15. In the pressure contour, the red region shows that fluid particles are getting stagnant, and to maintain flow, fluid has to pass through the top and bottom sides of the blades (Fig. 13).

In this, three designs are considered to compare their performance. In velocity contours, tip velocity presented by red color is more in design 1 and design 3 and in design 2, it is less this means that losses at the tip of blade for design 2 is minimum and power extraction is obtaining more, it's also shown by Cp versus TSR graph in Fig. 4.7. The pressure loss in design 2 pressure contour is found more and hence torque is producing more in this design (Fig. 14).

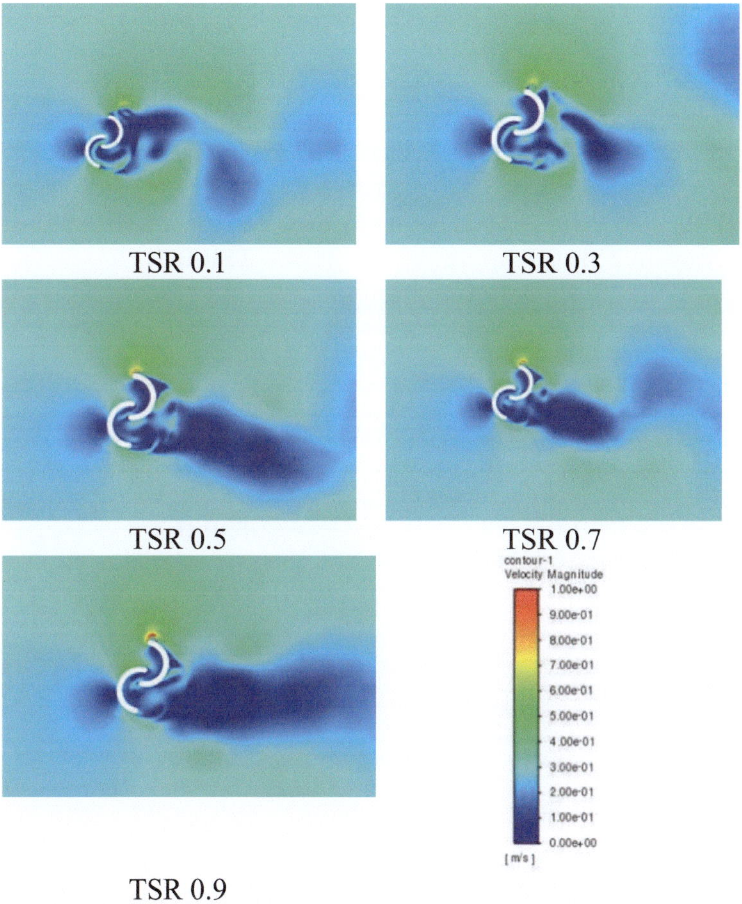

Fig. 8 Velocity contours for design 1

6 Conclusion

In the present investigation, three different vane designs have been considered for analysis. The performance of all selected designs is compared using non-dimensional terms, coefficient of torque (Ct), coefficient of power (Cp), and Tip Speed ratio (TSR). The flow conditions are analyzed using pressure and velocity contours at different Tip Speed Ratio (TSR). From the present investigations, the following conclusions are derived.

- Design 2 is found to be more optimal because the maximum power and maximum torque obtained are quite high as compared to other considered designs.
- The design 2, i.e., sharp edge outside of the vane, enhances the performance by approximately 200% compared to the conventional semi-circular edge shape.

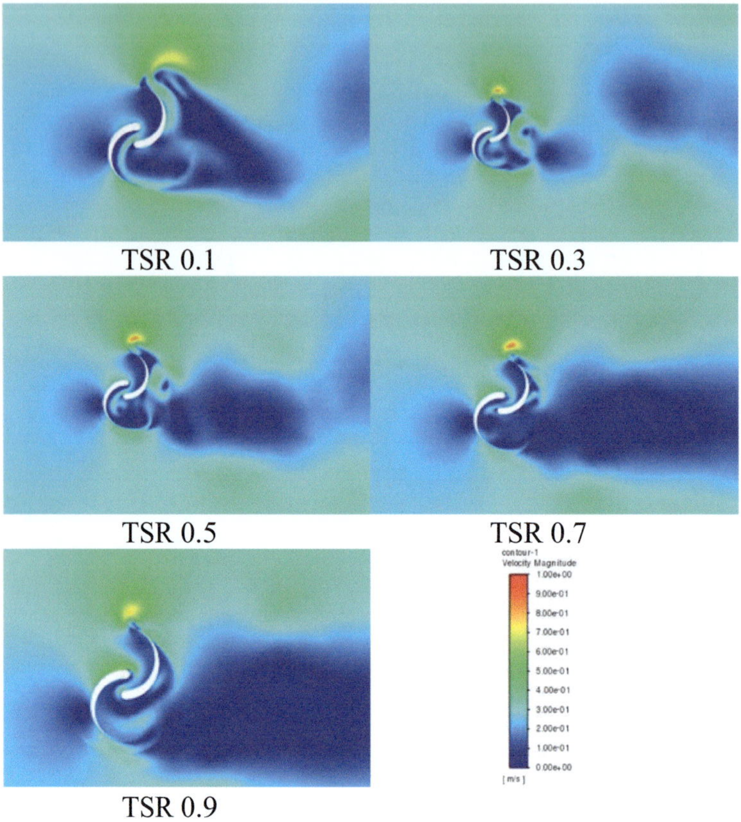

Fig. 9 Velocity contours for design 2

- In all cases, the maximum power coefficient is obtained at 0.7 TSR, and the maximum torque coefficient is at 0.3 TSR for design 1 and design 2, and for design 3, it is at 0.1 TSR. Therefore, the good value of TSR is 0.7

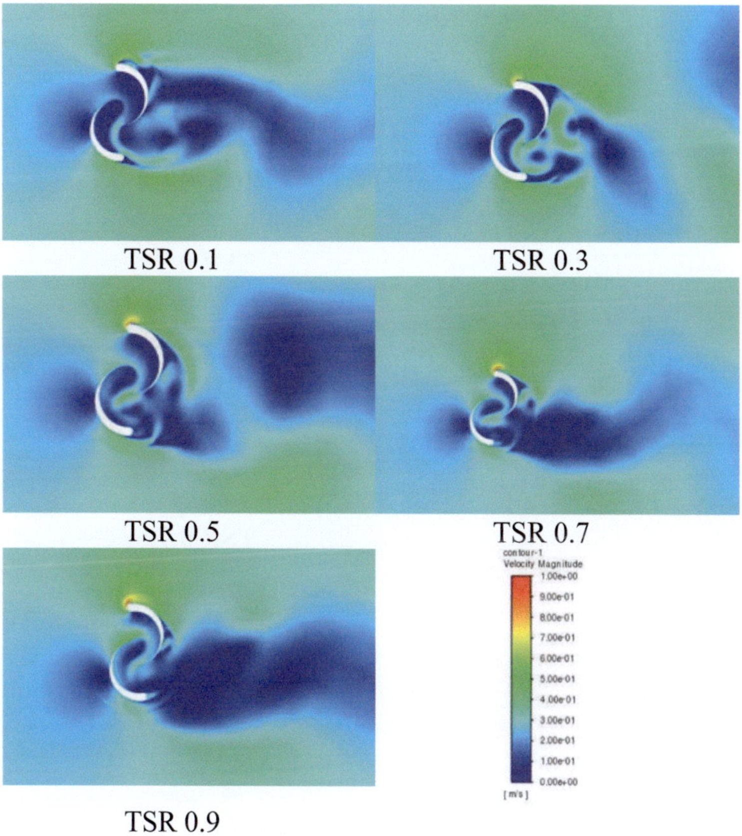

Fig. 10 Velocity contours for design 3

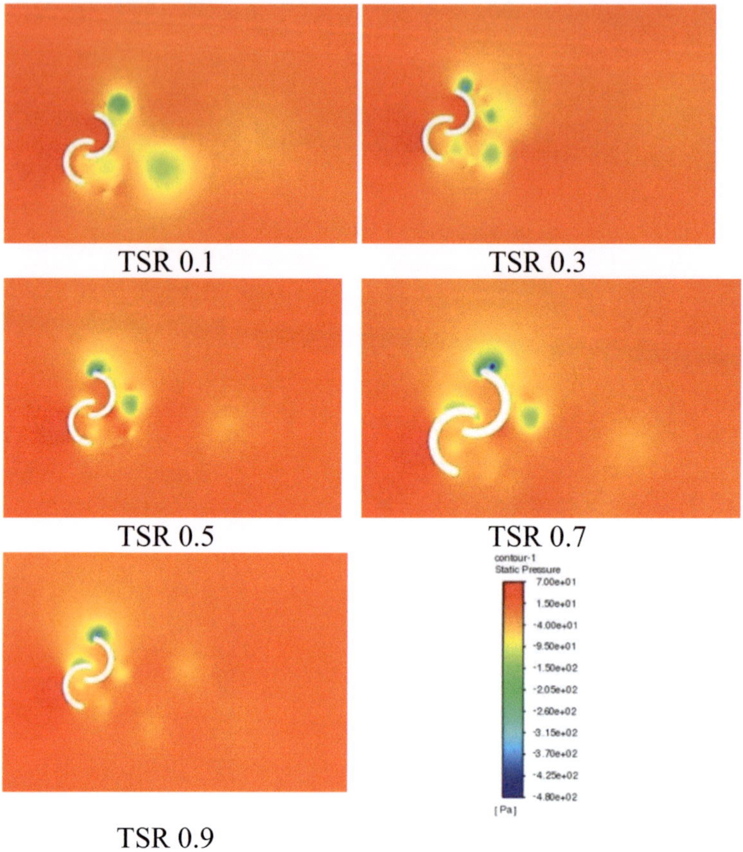

Fig. 11 Pressure contours for design 1

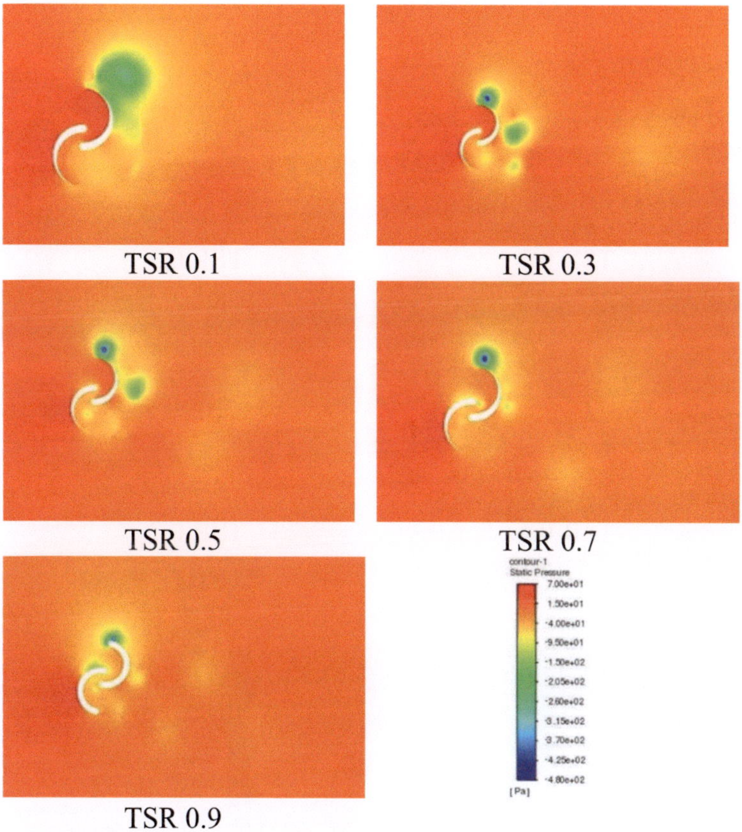

Fig. 12 Pressure contours for design 2

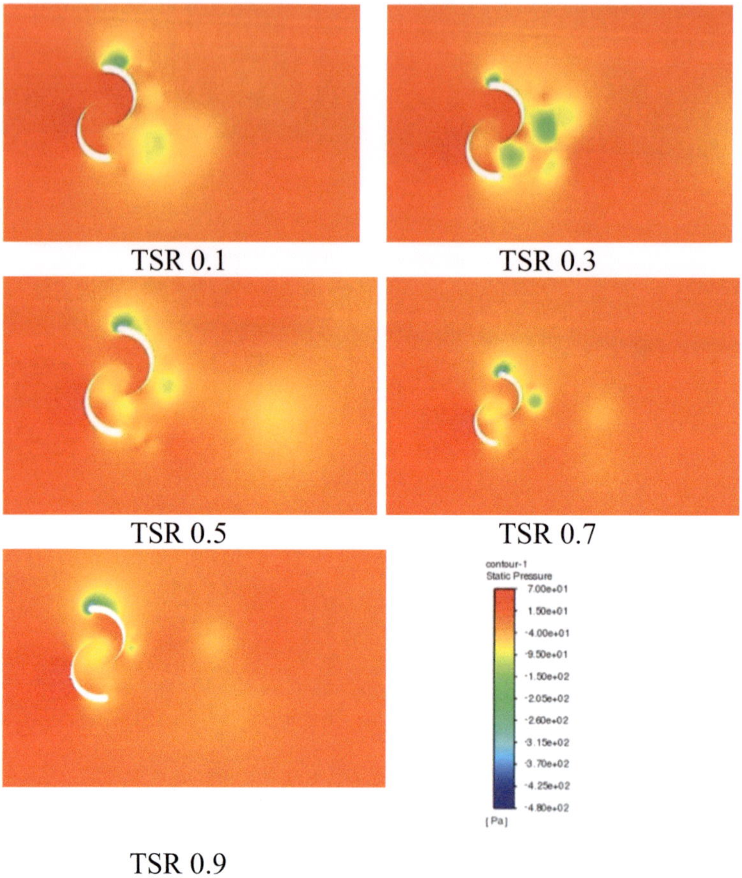

Fig. 13 Pressure contour of design 2

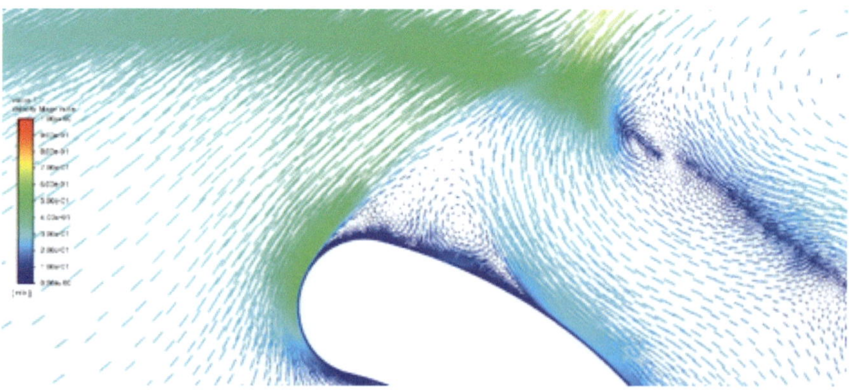

Fig. 14 Vortices at the tip of an advanced blade

Acknowledgements Authors gratefully acknowledge Science and Engineering Research Board (SERB), Department of Science and Technology, Delhi, India, for funding through core research grant for this study. The sanction order number CRG/2020/005420

Nomenclature

D	Rotor diameter (m)
e	Eccentricity (m)
Ct	Coefficient of torque
Cp	Coefficient of power
T	Torque (Nm)
ω	Rotor rotational speed (rad/s)
ρ	Density of fluid (kg/m^3)
A	Area of rotor (m^2)
V	Velocity of fluid (m/s)
R	Radius of rotor (m)

References

1. Patel V, Bhat G, Eldho TI, Prabhu SV (2017) Influence of overlap ratio and aspect ratio on the performance of Savonius hydrokinetic turbine. Int J Energy Res 41(6):829–844
2. Patel V, Eldho TI, Prabhu SV (2018) Theoretical study on the prediction of the hydrodynamic performance of a Savonius turbine based on stagnation pressure and impulse momentum principle. Energy Convers Manage 168:545–563
3. Kerikous E, Thévenin D (2019) Optimal shape of thick blades for a hydraulic Savonius turbine. Renew Energy 134:629–638
4. Patel V, Eldho TI, Prabhu SV (2019) Velocity and performance correction methodology for hydrokinetic turbines experimented with different geometry of the channel. Renew Energy 131:1300–1317
5. Patel V, Patel R (2021) Energy extraction using modified Savonius rotor from Free-flowing water. Mater Today: Proc 45:5190–5196
6. Patel R, Patel V (2023) Effect of waves on leading edge of modified Savonius rotor blades. Ocean Eng 271:113445
7. Patel V, Eldho TI, Prabhu SV (2019) Performance enhancement of a Darrieus hydrokinetic turbine with the blocking of a specific flow region for optimum use of hydropower. Renew Energy 135:1144–1156
8. Patel R, Patel V (2022) Performance analysis of Savonius hydrokinetic turbine using 'C'shaped Deflector. Energy Sources, Part A: Recovery, Utilization, and Environ Effects 44(3):6618–6631
9. Patel R, Patel V (2022) Experimental investigation of Savonius hydrokinetic turbine with different flow diverting blocking plate. Environ Prog Sustainable Energy 41(6):e13908
10. Patel V, Rathod V, Patel C (2023) Optimal utilization of hydrokinetic energy resources through performance improvement of the darrieus turbine using concave and flat blocking plates. Ocean Eng 283:115099

Effect of a Central Shaft on the Performance of Darrieus Hydrokinetic Turbine

Vimal Patel and Rajashekhar Jilla

Abstract Compared to airflow, water flow has a higher energy flow rate per unit flow area. This is because the density of water is higher than that of air. Because of this, hydrokinetic turbines—such as Darrieus turbines—seem to be able to generate more power than wind turbines of the same size. One of the best options for using a hydrokinetic turbine is the Darrieus turbine because of its high coefficient of power. The current investigation examines the impact of the central shaft's diameter, which is crucial in determining how a turbine's performance is affected. The investigations are carried out experimentally using five different diameters of the central shaft. The results indicate that the performance of the turbine continuously deteriorates with increments in the diameter of the central shaft. The maximum coefficient of power of 0.275 is obtained at TSR 1.64 without using a central shaft among all cases of the investigated sizes of the central shafts.

Keywords Hydrokinetic turbines · Renewable energy · Applied energy · Hydraulic turbines · Darrieus turbines

1 Introduction

Wind and hydrokinetic energy are two of the many clean, renewable energy sources that have grown quickly. The kinetic energy of the wind is transformed into mechanical power output by wind turbines. Similar to this, hydrokinetic turbines generate mechanical power by harnessing the kinetic energy of water flowing through man-made water channels, river streams, or tidal currents. Water may extract more power than wind turbines for a given turbine size because of its extremely high-power density per unit flow area [1, 3]. Vertical-axis hydrokinetic turbines provide several advantages, including the elimination of the need for a yaw mechanism, ease of

V. Patel (✉) · R. Jilla
Department of Mechanical Engineering, SVNIT, Surat, Gujarat, India
e-mail: vimal.iitbombay@gmail.com

A. Kumar R. et al. (eds.), *Proceedings of Fluid Mechanics and Fluid Power (FMFP) 2023, Vol. 6*, Lecture Notes in Mechanical Engineering,
https://doi.org/10.1007/978-981-96-9600-0_43

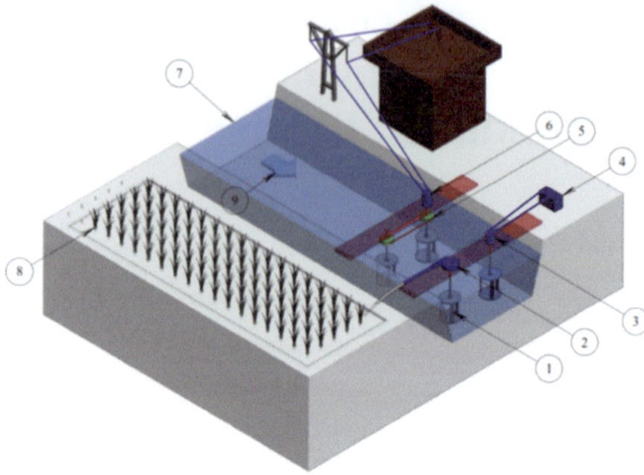

Fig. 1 Applications of the Darrieus turbine [1]

design and fabrication due to the single rotating part. The generator or power output shaft can be brought outside of the water surface [2].

The blocking plate, with its concave and flat design, has been used for the investigation. The two blocking plates have been tested for their various sizes and placement on the turbine's upstream side. It has been determined that the power coefficient of a concave blocking plate improves by 18.45% and that of a flat blocking plate with optimal width and placement, respectively, by 10.92% [2, 4]. The effect of the blocking plate at upstream side is investigated even for the Savonius type turbine also [5, 6]. Even, the effect of upstream side flow diverting plate is also analyzed for vertical axis Savonius type turbine toward its performance enhancement [7]. However, the effect of flow diversion, with the use of central shaft has never been investigated, specifically for the Darrieus type hydrokinetic turbine. Hence, in the present investigation, it is decided to investigate about the effect of different size of the central shaft on the performance of Darrieus hydrokinetic turbine. The main application of hydrokinetic turbine is to develop the power as a standalone unit at remote locations away from power grid. The different applications of the Darrieus turbine are shown in Fig. 1.

2 Conceptual Discussion

The fundamental principle of lift and drag force generation is applicable for the working of the Darrieus turbine. While rotating the turbine in the flow field, the lift and drag force are generated by the turbine vanes, as shown in Fig. 2. The tangential components of these forces decide the magnitude and direction of the resultant torque of the turbine rotor.

Here, it can be thought that, as more water passes from the turbine vanes, then the chances of the higher magnitude of the lift and drag force generation can be enhanced, and subsequently, power output can also be enhanced. The flow of the

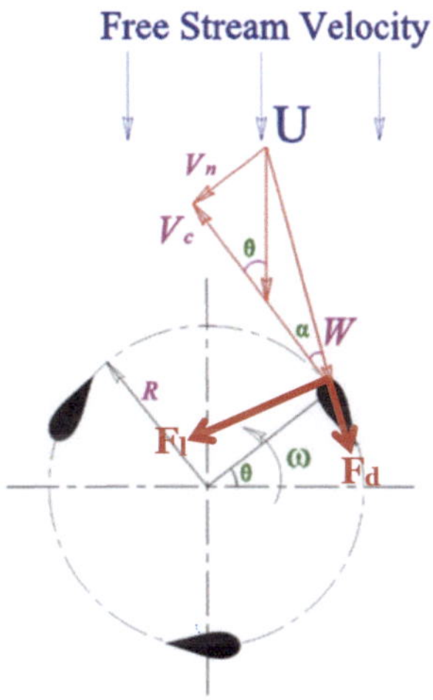

Fig. 2 Working principle of Darrieus turbine

water can be enhanced by using the central shaft, which will deflect the water from the center of the turbine rotor toward the vane side. On the other hand, too large size of the central shaft might create flow resistance for the water to pass from the turbine rotor region. Which subsequently reduces the mass flow rate of water passing from the turbine rotor region. Ultimately, it will lead to a reduction in the power output from the turbine. Hence, it is necessary to investigate the use of a central shaft and optimize the size of the turbine.

3 Data Reduction

3.1 *Solidity*

The ratio of the entire swept area to the total planform area of the rotor is known as its "solidity." To assess a turbine's performance, a specified metric called "solidity" is used.

$$\sigma = \frac{nc}{\pi D} \tag{1}$$

3.2 *Torque*

The torque (T) generated by the turbine rotor may be determined by applying force to the right (T_1) and slack (T_2) sides of the rope that is looped on the shaft.

$$T = (T_1 - T_2) \times R_{\text{Shaft}} \tag{2}$$

3.3 Angular Speed

The angular speed at which the turbine's shaft revolves is a crucial metric to calculate. The following is a definition of angular velocity (ω):

$$\omega = \frac{2\pi N}{60} \tag{3}$$

3.4 Power

The actual power developed by the shaft is given by,

$$\text{Power} = T \times \omega \tag{4}$$

3.5 Tip Speed Ratio (TSR)

Non-dimensional characteristics must also be taken into consideration. The Tip Speed Ratio is the ratio of the speed of the water flow to the speed of the hydrokinetic turbine blade tip. The TSR value of a given turbine is used to measure and verify the turbine's performance.

$$\text{TSR} = \frac{r\omega}{V} \tag{5}$$

3.6 Coefficient of Torque (C_t)

The coefficient of torque (C_T) is the ratio of actual torque generated by a hydrokinetic turbine to the predicted torque computed. These non-dimensional parameters have varied formulas depending on the situation. Open channel flow is used in these equations.

$$C_T = \frac{4T}{\rho H D^2 V^2} \tag{6}$$

3.7 *Coefficient of Power (C_p)*

According to the power coefficient, the hydrokinetic turbine's output may be figured out (C_P). The hydrokinetic turbine's coefficient of power is a measure of how well it turns water energy into electricity. The rotor's surface area and flow velocity are the two most important determining factors.

$$C_p = \frac{2P}{\rho H D V^3} \tag{7}$$

4 Experimental Setup

4.1 *Details of Test Facility*

The experiment is being conducted in the laboratory canal, available in the Advanced Fluid Mechanics and Fluid Power lab, as shown in Fig. 3. The geometry of the canal is elliptical, and there is consistent water circulation throughout the canal. The rotor's construction and frame are both shown in Fig. 3. The details of the canal available in the Advanced Fluid Mechanics and Fluid Power laboratory (AFMFP Lab) of the institute are shown in Table 1.

The rotor assembly was installed in the laboratory canal. On the hump, an experimental setup was erected to gain elevation and obtain high velocity. A 5-horsepower submersible pump is installed in the canal to ensure constant water circulation.

4.2 *Details of Turbine Setup*

For this experiment, the NACA0012 hydrofoil is used for making the three turbine vanes. All three vanes are mounted between the two closed end plates. The arrangement is made such that different sizes of the shaft can be installed and removed easily, without disturbing other parts of the turbines. Figure 4 indicates the turbine used for the present investigations, and Table 2 indicates the design details of the used turbine setups.

In order to attain height and also to obtain high velocity, the complete turbine unit is placed above the top of the hump. Acrylic material is used in the manufacturing of

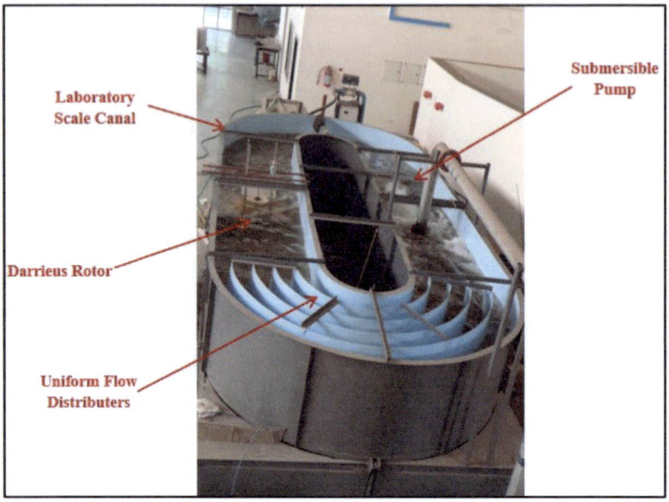

Fig. 3 Test facility available at AFPFP Lab, SVNIT, Surat

Table 1 Details of the laboratory canal

S. No.	Details	Specifications
1	Length	400 cm
2	Width	100 cm
3	Depth	75 cm
4	Water velocity	0.354 m/s
5	Pump capacity	5 HP

both the top and bottom support plates, which also serve as a support structure. The aspect ratio and the solidity of the used turbine rotor are 0.79 and 0.477, respectively. A load cell is attached to the rotor shaft to measure the slack side tension of the rope. The load pan is used, at the tight side of the rope, to apply the different loads on the turbine shaft.

The different central shafts used in the present investigations are shown in Fig. 5.

4.3 Experimental Procedure

The turbine rotor is prepared using a specific diameter of central shaft. The turbine rotor is mounted on the supporting structure of the entire turbine setup. The complete turbine setup is then mounted in the laboratory canal at test section using the supporting frame. The frequency of the pump motor is kept constant throughout the experiment to retain flow velocity constant in the canal. The flow velocity at the test section is measured using a propeller-type velocity current meter available in

Fig. 4 Darrius turbine rotor with center shaft

Table 2 Specifications of the setup

S. No.	Details	Specifications
1	No. of hydrofoil	3
2	Blade profile	NACA 0012
3	Diameter of the rotor	28 cm
4	Height of the rotor	22 cm
5	Hydrofoil chord	14 cm
6	Solidity	0.477
7	Outside diameter of central Shafts	2.2, 3.4, 6.3 and 11.1 cm

the laboratory. The hydrokinetic power available in the turbine region is calculated using this measured flow velocity and area covered by the turbine rotor.

The rope brake dynamometer concept was utilized to measure the torque produced. A weighing pan and a digital load cell are attached to opposite ends of a string that is coiled around the rotor shaft. The tight side of the string is connected with the load pan and the slack side is with the load cell. The net tangential force is calculated by differentiating tight side and slack side tensions. The torque on the turbine shaft is calculated using this net tangential force and shaft radius, where the rope is wounded. The angular velocity of the rotor is measured by tachometer.

Fig. 5 Central shafts used in the present investigations

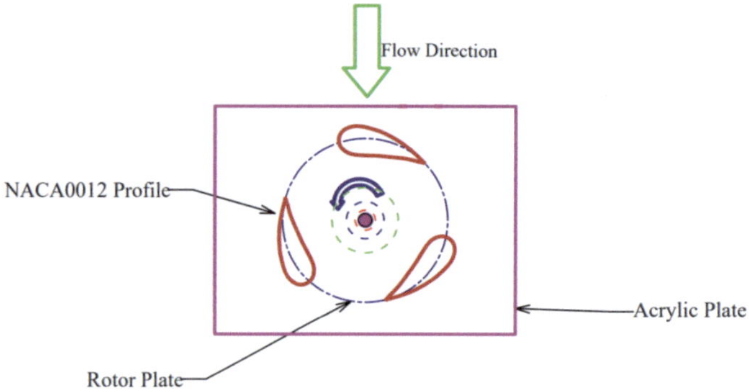

Fig. 6 Conceptual drawing of the turbine setup with different central shaft

The mechanical power output available on the turbine shaft is calculated using the torque on the turbine shaft and measured RPM. The detailed experimental procedure is explained by Patel and Patel [8]. The experiments are repeated by applying different load on the turbine shaft till turbine stops. The same procedure is repeated by installing different central shafts, which are shown in Fig. 5. The conceptual representation is shown in Fig. 6.

5 Validation

The initial experiment was completed without keeping any central shaft. The obtained results without using central shafts have been compared and validated with a result reported by Lee et al. [9]. The comparison of the obtained results is shown in Fig. 7. The close matching of the trend indicates the validity of the experimental measurement and calculation methodology adopted in the present investigations.

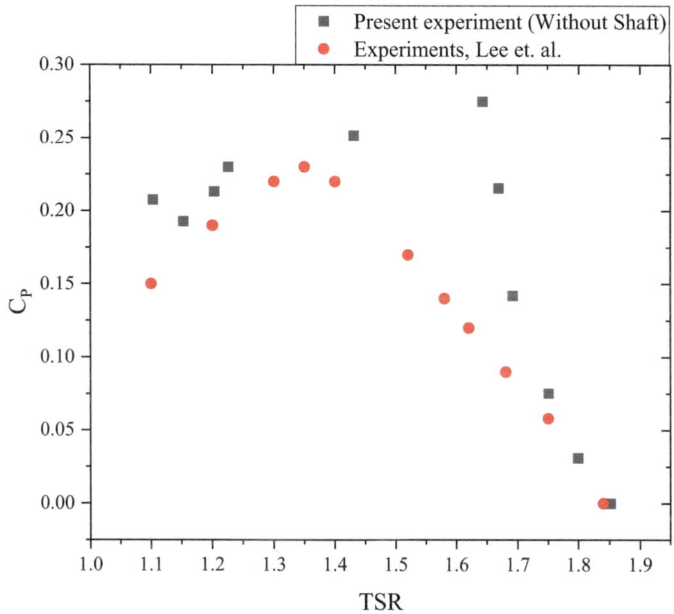

Fig. 7 Validation of used methodology and calculation steps

6 Results and Discussion

The results obtained from the experiments are presented in the form of non-dimensional numbers Coefficient of Power (C_p), Coefficient of Torque (C_t), and Tip Speed Ratio (TSR). For better comparison, the results obtained with different central shaft diameters are plotted on a single chart as shown in Figs. 8 and 9.

The results indicate that initially, by an increase in the load on the turbine shaft, i.e., by decreasing TSR, the coefficient of power continuously increases for all cases of central shaft diameters. However, by further increase in load, the value of coefficient of power subsequently falls continuously. The maximum value of coefficient of power is obtained in the TSR range of 1.3–1.5 for all cases of central shaft diameters. It is further observed that the value of maximum coefficient of power is obtained for the case of no central shaft. By enhancement in the central shaft diameter, the value of maximum coefficient of power obtained is deteriorated subsequently. It is also observed that by increase in the diameter of central shaft, the value of maximum TSR also decreases at no-load condition. The results indicate that the maximum value of the coefficient of power of 0.275 is obtained without the use of the central shaft.

To conclude the effect of the size of the central shaft, the graph of the maximum coefficient of power obtained with different central shaft diameters is presented in Fig. 9. The result indicates that, by using the larger diameter of central shaft, the value of the maximum coefficient of power continuously deteriorates with increment in

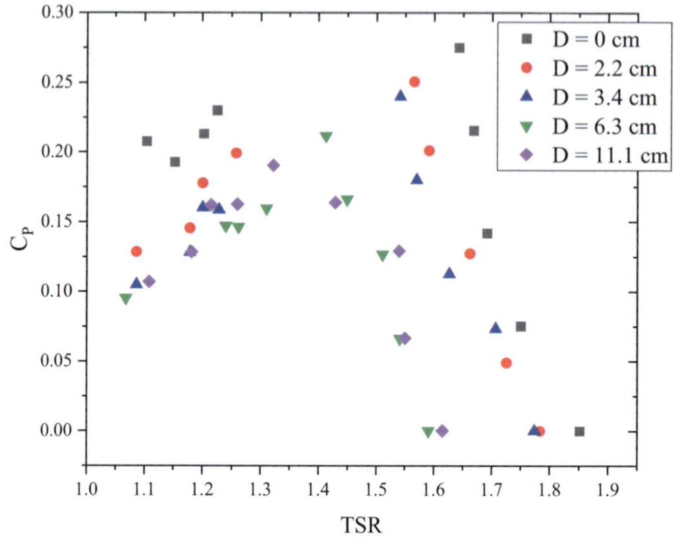

Fig. 8 Variation C_p with TSR with different shaft diameters

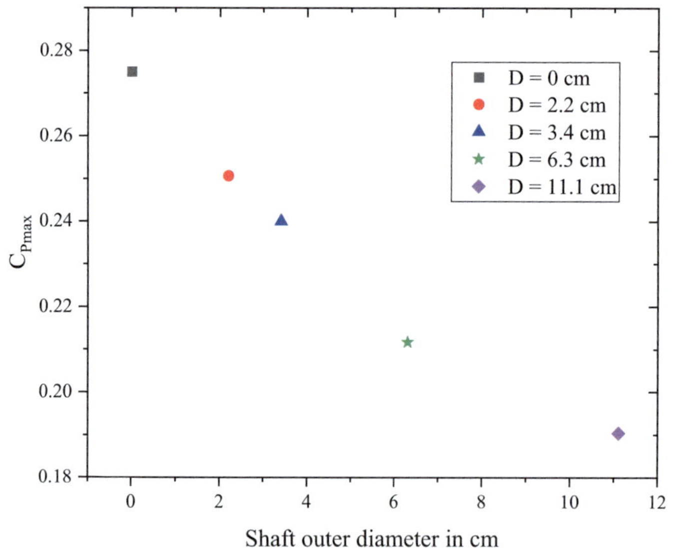

Fig. 9 Variation $C_{P\max}$ according to the central shaft diameters

the size of the shaft diameter. This may be due to the fact that, with the use of the central shaft, the additional resistance to the water flow arises in the turbine rotor. Hence, the water flow may bypass over and out from the turbine rotor region. Hence, the performance of the turbine subsequently decreases. This fact also can be observed in Fig. 8 that with the use of a larger diameter of the central shaft; the no-load TSR (maximum speed case) value continuously decreases with increment in the central shaft diameter. This might be due to the decrement in the flow velocity by enhancement of flow resistance and subsequently approaching flow velocity with enhancement in the size of the central shaft.

7 Conclusion

The experimental investigations are carried out with a three-vane Darrieus turbine with NACA 0012 hydrofoil shape of the vanes. The performance of the turbine is reported experimentally using a total of five different sizes of central shafts, including without the use of a central shaft. From the present investigations, the following conclusions can be obtained.

1. The turbine indicates the best performance, in the form of the highest C_p, without the use of any central shaft.
2. The maximum coefficient of power of 0.275 is obtained at a TSR of 1.64 Tip Speed Ratio without use of a central shaft, which is approximately 2.75 times more than the use of a maximum central shaft diameter of 11.1 mm.
3. By using a larger diameter of the central shaft, the value of the maximum coefficient of power obtained from the turbine rotor continuously deteriorates. Hence, it can be concluded that the turbine without a central shaft provides the best coefficient of power.
4. With the use of a larger diameter of the central shaft, the highest value of TSR also continuously decreases. This fact indicates that, with the use of a central shaft, the approaching velocity of fluid flow reduces due to additional resistance offered by the central shaft. Which leads to bypass the flow over and above from the turbine region. The reduction in the flow from the turbine region leads to deteriorate the coefficient of power and maximum speed of the turbine.

Acknowledgements Authors gratefully acknowledge the Science and Engineering Research Board (SERB), Department of Science and Technology, Delhi, India, for funding through a core research grant for this study. The sanction order number CRG/2020/005420.

Nomenclature

D	Diameter of rotor [mm]
n	Number of blades –
N	Number of revolutions –
D	Diameter of rotor –
C_P	Power coefficient –
C_T	Torque coefficient –
TSR	Tip speed ratio –
D_O	Outer diameter of shaft (mm)
ρ	Density of water [kg/m^3]
ω	Rotor rotational speed [rad/s]

References

1. Patel V, Eldho TI, Prabhu SV (2017) Experimental investigations on Darrieus straight blade turbine for tidal current application and parametric optimization for hydro farm arrangement. Int J Mar Energy 17:110–135
2. Patel V, Eldho TI, Prabhu SV (2019) Performance enhancement of a Darrieus hydrokinetic turbine with the blocking of a specific flow region for optimum use of hydropower. Renew Energy 135:1144–1156
3. Patel MV, Chaudhari MH (2013) Performance prediction of H-Type Darrieus turbine by single stream tube model for hydro dynamic application. Int J Eng Res Technol 2(3)
4. Patel V, Rathod V, Patel C (2023) Optimal utilization of hydrokinetic energy resources through performance improvement of the Darrieus turbine using concave and flat blocking plates. Ocean Eng 283:115099
5. Patel R, Patel V (2022) Performance analysis of Savonius hydrokinetic turbine using 'C' shaped Deflector. Energy Sources Part A Recover Util Environ Eff 44(3):6618–6631
6. Patel R, Patel V (2022) Experimental investigation of Savonius hydrokinetic turbine with different flow diverting blocking plate. Environ Prog Sustain Energy 41(6):e13908
7. Patel VK, Patel RS (2022) Optimization of an angle between the deflector plates and its orientation to enhance the energy efficiency of Savonius hydrokinetic turbine for dual rotor configuration. Int J Green Energy 19(5):476–489
8. Patel R, Patel V (2023) Effect of waves on leading edge of modified Savonius rotor blades. Ocean Eng 271:113445
9. Lee YT, Lim HC (2015) Numerical study of the aerodynamic performance of a 500 W Darrieus-type vertical-axis wind turbine. Renew Energy 83:407–415

Thermodynamic and Transport Property Modeling of High-Pressure LOx-Methane Rocket Engine

Rahul Kumar, Abhishek Sharma, Aravind Vaidyanathan, and T. John Tharakan

Abstract A high-pressure LOx-Methane (LOx-CH_4) rocket engine is characterized by transcritical injection and supercritical combustion phenomena which pose various difficulties during numerical modeling. Near the critical point region, thermophysical properties undergo highly non-linear behavior, which can impact the flow and flame evolution. In this study, a comprehensive thermodynamic modeling of LOx-CH_4 engine is conducted to capture such a non-linear phenomenon. A real gas-based thermodynamic and high-pressure transport framework is invoked to represent transcritical injection and supercritical combustion. Multiple real gas models are tested to evaluate thermophysical properties over the entire range of operations. Our study captured non-linear behavior near critical point operation of LOx-CH_4 engine. We identified a high-pressure thermodynamic and transport modeling framework that can be utilized for computational fluid dynamics (CFD) simulations.

Keywords Transcritical · Supercritical · Thermophysical · Pseudo-boiling · Soft-cryogenic

1 Introduction

Liquid rocket engines have played a pivotal role in modern space exploration. Liquid propellants, like liquid hydrogen (LH_2) and rocket-grade kerosene (RP-1), combined with liquid oxygen (LOx) as oxidizer, offer high specific impulse and controllability. LOx-Methane (LOx-CH_4) has emerged as an optimal propellant combination for interplanetary travel and reusable launch vehicles. The potential of LOx-CH_4

R. Kumar (✉) · A. Vaidyanathan
Department of Aerospace Engineering, Indian Institute of Space Science and Technology, Thiruvananthapuram, Kerala, India
e-mail: rk8682442@gmail.com

A. Sharma · T. J. Tharakan
Liquid Propulsion Systems Centre, ISRO, Thiruvananthapuram, Kerala, India

A. Kumar R. et al. (eds.), *Proceedings of Fluid Mechanics and Fluid Power (FMFP) 2023, Vol. 6*, Lecture Notes in Mechanical Engineering,
https://doi.org/10.1007/978-981-96-9600-0_44

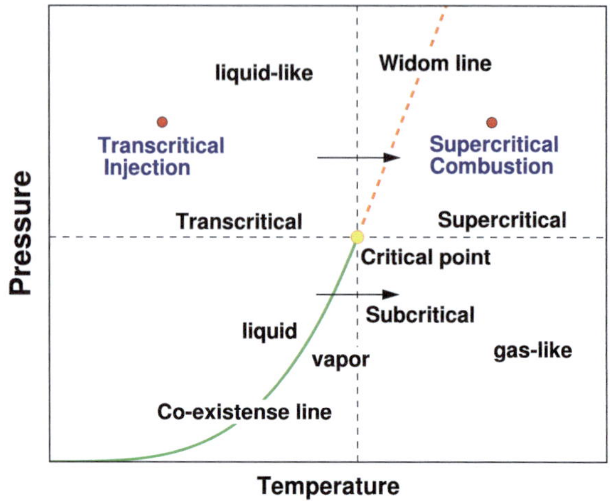

Fig. 1 Phase diagram for LOx-CH_4 engine operation

propulsion is evident from the recent use of the 200 TN thrust class Raptor engine in the SpaceX starship launch vehicle.

The cryogenic fluid state of LOx-methane experiences transcritical injection and supercritical combustion in high-pressure rocket engines as shown in Fig. 1. The propellant undergoes phase transition similar to the subcritical state, but distinct non-linearities are observed in thermodynamic and transport properties near the pseudo-boiling temperature (across Widom line) as described by Banuti [1]. The behavior of cryogenic fluid injection at subcritical and supercritical conditions was experimentally investigated by Mayer et al. and Chehroudi et al. [2, 10]. It was observed that in the vicinity of the critical point, the jet structure changes to gas-like with no detectable interface between vapor and liquid droplets. An apropos study by Singla et al. [13], indicates that droplet formation does not occur in a supercritical environment and mixing is governed by diffusion. The studies indicated the non-linearity associated with transcritical and supercritical regimes needs to be dealt with for accurate numerical modeling. A comprehensive review of the experimental and numerical analysis of LOx-methane combustion is provided by Sharma et al. [12]. It highlights the capability of real gas models in numerical simulations using different reaction mechanisms and combustion models. Kim et al. [5] studied kerosene/LOx combustion analytically, by including real gas thermodynamic models in flamelet equation. They observe discrepancies in combustion characteristics in fuel-rich and oxidizer-rich environments, ensuing real fluid effects.

Accurate modeling of these behaviors, especially during injection and combustion, is essential for efficient rocket system design. Thermodynamic models, like cubic equations of state (EoS) such as Peng-Robinson (PR) and Soave-Redlich-Kwong (SRK), are widely used due to their balance of accuracy and computational efficiency. Understanding fluid behavior in supercritical and transcritical conditions remains critical for advancing rocket technology. Although few studies identified the requirement of real fluid EoS and high-pressure transport models for thermophysical properties, a detailed assessment of all EoS available is required for the LOx-methane propellant combination. In this work, an attempt has been made to reveal the shortcomings and advantages of each model, while deriving a possible new generalized EoS model that can incorporate the advantages of all models and can represent LOx-methane injection and combustion phenomena accurately. This paper presents comprehensive analytical modeling of cubic EoS to investigate thermophysical characteristics of LOx-CH_4 propellants.

2 Methodology

The operating pressure of LOx-CH_4 system is much higher than the critical pressure of O_2 and CH_4. The non-linear variation of thermophysical properties near the critical point, cannot be captured using ideal gas EoS due to the assumptions of weak intermolecular forces and negligible volume occupied by gas. The accurate modeling of thermophysical properties demands invoking real fluid EoS. This section presents important cubic EoSs, briefly highlighting their advantages and shortcomings. A generalized EoS for real fluids is described with the procedure to obtain major thermophysical properties.

2.1 *Real Fluid Equations of State*

The real fluid EoS was first introduced by Van der Waals (VDW), incorporating adjustments for both volume and pressure through correction terms. It helped explain the deviations of real fluids from ideal behavior, especially at high pressures and low temperatures. The equation proposed by Van der Waals is written as follows:

$$P = \frac{RT}{v - b} - \frac{a}{v^2}$$

where, R represents universal gas constant, and v is the molar specific volume. This equation incorporates two parameters: the constraint "a" accommodates the molecular attraction forces, while "b" represents the space occupied by gas. VDW equation fails to accurately predict real fluid effects at low temperature and high-pressure conditions due to its simplified assumptions. Redlich-Kwong modified the VDW equation by including critical temperature (T_c) and pressure (P_c) in the formulation of parameters "a" and "b." Redlich-Kwong (RK) [14] equation for the correct representation of low and high densities in gases is given as:

$$P = \frac{RT}{v - b} - \frac{a}{v(v + b)}$$

where $a = 0.4278\frac{R^2 T_c{}^2}{P_c}$ $b = 0.08664\frac{RT_c}{P_c}$.

The RK equation provides superior prediction in comparison to van der Walls' EoS, but it still performs poorly for the liquid phase and cannot be used to determine vapor-liquid equilibria accurately. Soave modified the RK equation by introducing a temperature correction term "α" for attraction forces. The Soave-Redlich-Kwong (SRK) EoS [14] is given as.

$$P = \frac{RT}{v-b} - \frac{a\alpha}{v(v+b)}$$

The formulation for variables a, b, and α is given in Table 1 . Where k is a constant that depends on the acentric factor ω. After Soave, Peng, and Robinson's [14] provided significant improvements to the existing framework of attraction factor and formulated PR EoS, written as:

$$P = \frac{RT}{v-b} - \frac{a\alpha}{v(v+b)+b(v-b)}$$

The PR EoS has shown good accuracy near the critical point. However, it still lacks in modeling substances with strong hydrogen bonding or associated effects. The details about the formulation of "a," "b," and "α" for SRK and PR EoS are mentioned in Table 1. Combining the advantages of both SRK and PR EoSs, Cismondi, and Mollerup [4] proposed Redlich-Kwong-Peng-Robinson (RK-PR) EoS, introducing a new parameter δ_1, besides "a" and "b." By incorporating δ_1, the prediction of thermophysical properties was improved over the range of critical compressibility factor (Z_c). The parameter δ_1 is expressed as a function of Z_c and modal constants, as given in Table 2. The RK-PR EoS is given by:

$$P = \frac{RT}{v-b} - \frac{a\alpha}{(v+\delta_1 b)(v-\delta_1 b)} \tag{1}$$

formulation for a, b, α are described in Table 1. Equation 1 is also used to represent PR and SRK EoS with appropriate values of δ_1 and δ_2 as described in Table 1.

2.2 Generalized EoS

This study invokes a generalized EoS to determine thermophysical properties of LOx-CH_4 system at transcritical and supercritical conditions. The RK-PR EoS mentioned in Eq. 1, is in a generalized form that can also describe SRK and PR equations. An analytical methodology is developed to solve the generalized equation in terms of compressibility factor (Z), which is further used to determine density. The generalized equation in terms of Z is given as

Table 1 Parameters for the cubic equation of states

Parameter	SRK	PR	RK-PR[a]
δ_1	1	$1+\sqrt{2}$	$d_1+d_2\,(d_3-1.168Z_c)^{d_4} \quad +d_5\,(d_3-1.168Z_c)^{d_6}$
δ_2	0	$1-\sqrt{2}$	$\frac{1-\delta_1}{1+\delta_1}$
a	$0.42747\left(\frac{R^2T_c^2}{p_c}\right)$	$0.42747\left(\frac{R^2T_c^2}{p_c}\right)$	$\frac{3y^2+3yd+d^2+d-1}{(3y+d-1)^2}\left(\frac{R^2T_c^2}{p_c}\right)$
b	$0.08664\left(\frac{RT_c}{p_c}\right)$	$0.07780\left(\frac{RT_c}{p_c}\right)$	$\frac{1}{3y+d-1}\left(\frac{RT_c}{p_c}\right)$
k	0.48508	0.37464	$(C_0+1.168Z_cC_1)$
	$+1.55171\omega$	$+1.54226\omega$	$+(B_0+1.168Z_cB_1)\,\omega$
	$-0.15613\omega^2$	$-0.26992\omega^2$	$+(A_0+1.168Z_cA_1)\,\omega^2$
α	$\left[1+k\left(1-\sqrt{\frac{T}{T_c}}\right)\right]$		$\left(\frac{3}{2+T/T_c}\right)^k$
$\frac{d\alpha}{dT}$	$-k\sqrt{\frac{\alpha}{TT_c}}$		$-\frac{k\alpha}{T_c(2+T/T_c)}$
$\frac{d^2\alpha}{dT^2}$	$\frac{k^2}{2TT_c}+\frac{k}{2}\sqrt{\frac{\alpha}{T^3T_c}}$		$\frac{k(k+1)\alpha}{T_c^2(2+T/T_c)^2}$

[a] $d=(1+{\delta_1}^2)/(1+\delta_1), \quad y=1+[2(1+\delta_1)]^{\frac{1}{3}}+[1/(1+\delta_1)]^{\frac{1}{3}}$

Table 2 Modal constants for RK-PR EoS

δ_1		k	
d_1	0.428363	A_0	0.0017
d_2	18.496215	A_1	– 2.4407
d_3	0.338426	B_0	1.9681
d_4	0.660000	B_1	7.4513
d_5	789.723105	C_0	– 2.7238
d_6	2.512392	C_1	12.5040

$$Z^3 + Z^2(B(\delta_1 + \delta_2 - 1) - 1) + Z(B^2(\delta_1\delta_2 - \delta_1 - \delta_2) - B(\delta_1 + \delta_2) + A) - AB - \delta_1\delta_2(B^3 + B^2) = 0 \quad (2)$$

where $Z = \frac{Pv}{RT}$ $A = \frac{a\alpha}{(RT)^2}$ $B = \frac{bP}{RT}$

The solution of Eq. 2, gives three roots which can be either all real or one real and two complex depending upon the operating conditions. The concept of fugacity is invoked to determine the stable root representing the solution of Eq. 2. Fugacity (f) represents the correction pressure that the fluid exerts if it behaves ideally at a particular value of temperature and volume. The expression for fugacity coefficient ($\phi = f/P$) is given as:

$$\log(\phi) = Z - 1 - \log(Z - B) - \frac{A}{(\delta_1 - \delta_2)B}\log(\frac{Z + \delta_1 B}{Z + \delta_2 B})$$

The fugacity coefficient is determined from the above expression, and depending on which real root has a lower value of ϕ, represents the solution at that condition. In the case of three real roots, that occur in the vapor-liquid-equilibria zone, the first and third real roots are the most significant because they correspond to the liquid and vapor phases, respectively.

2.3 Evaluation of Thermophysical Properties

The major thermodynamic and transport properties such as density, specific heat, enthalpy, entropy, viscosity, and thermal conductivity are important to understand the mixing and combustion processes in rocket engines. As propellants approach their critical points, their thermophysical properties change significantly. Understanding the critical point behavior is important to numerically study transcritical injection and supercritical combustion phenomena in LOx-CH_4 engines. Figure 2 outlines the process of obtaining density. The properties at critical points are given as input, and then Eq. 2 is solved for the compressibility factor Z corresponding to different EoSs. If it gives only one real root, it is further used for calculations of density. If three

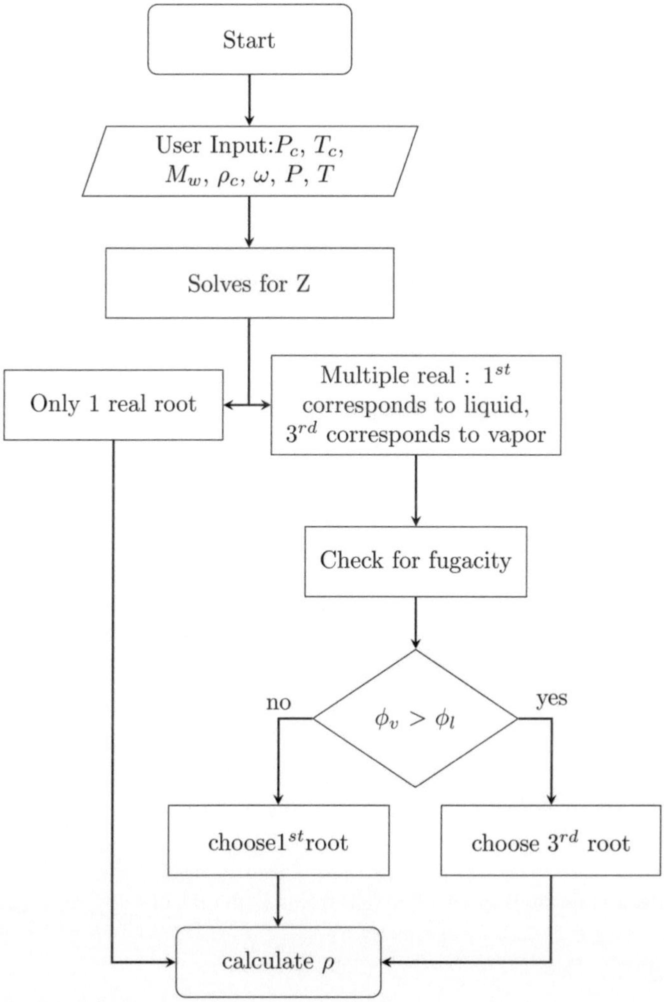

Fig. 2 Flowchart for density calculations of the pure and mixture fluids

real roots are obtained, then the concept of fugacity is invoked, and whichever root has the minimum value of ϕ is selected for density calculations. The thermodynamic properties of O_2 and CH_4 are estimated using ideal gas and departure function values. The departure function accounts for the non-ideal behavior of fluid at low temperature and high pressure, that arises due to deviation of intermolecular forces, molecular size, and polarizability of fluids from the ideal values. The concept of departure function is explained in detail in the book referred [11].

The ideal gas's thermodynamic properties, such as enthalpy, specific heat capacity, and entropy, are evaluated using shomate equations [7]. Shomate relations are considered to be the most accurate, and the coefficients for various compounds are accessible from the NIST Chemistry WebBook [7]. The ideal gas thermodynamic relations are given as:

$$C_v^\circ = C_p^\circ - R$$
$$E^\circ(T) = H^\circ(T) - RT$$
$$S^\circ(T, p) = S^\circ(T, p^\circ) - R\log\frac{p}{p^\circ}$$

where C_p° and C_v° are heat capacity at constant pressure and volume (J/mol K), H° is standard enthalpy (kJ/mol), S° is standard entropy (J/mol K) and E° is internal energy (kJ/mol) The reference state for all these properties is at a temperature of $298K$ and a pressure of 1atm.

The real fluid thermodynamic equation is given as:

$$\text{Prop}_{rf} = \text{Prop}_{ig} + \text{Prop}_{\text{dep}}$$

where, Prop refers to the thermodynamic property of the gas and rf, ig and dep refers to the real fluid, ideal gas, and departure valuesof that property, respectively.

The thermodynamic properties are evaluated using the generalized cubic EoS given in Eq. 1. For deriving the thermodynamic properties, Eq. 1 is differentiated with respect to temperature and volume to obtain the following differential expressions:

$$\left(\frac{\partial p}{\partial T}\right)_v = \frac{R}{v-b} - \frac{(\partial a\alpha/\partial T)}{(v+\delta_1 b)(v+\delta_2 b)}$$
$$\left(\frac{\partial p}{\partial v}\right)_T = -\frac{RT}{(v-b)^2} + \frac{a\alpha\left[2v+(\delta_1+\delta_2)b\right]}{(v+\delta_1 b)^2(v+\delta_2 b)^2}$$

The above expressions are used to determine enthalpy, entropy, and specific heat. The expression for enthalpy is given as

$$h = h^o - RT + K_1\left[a\alpha - T\left(\frac{\partial a\alpha}{\partial T}\right)\right] + pv \quad (3)$$

where, $K_1 = \frac{1}{(\delta_1-\delta_2)b}\ln\left(\frac{v+\delta_2 b}{v+\delta_1 b}\right)$

Similarly the entropy is evaluated using relation:

$$s(T, v) = s^o + R\ln\frac{v-b}{v} - K_1\left(\frac{\partial a\alpha}{\partial T}\right) \quad (4)$$

The Isobaric specific heat capacity is evaluated as

$$c_p = \left(\frac{\partial h}{\partial T}\right)_p = c_p^o - R - K_1 T\left(\frac{\partial^2 a\alpha}{\partial T^2}\right) - T\frac{(\partial p/\partial T)_v^2}{(\partial p/\partial v)_T} \quad (5)$$

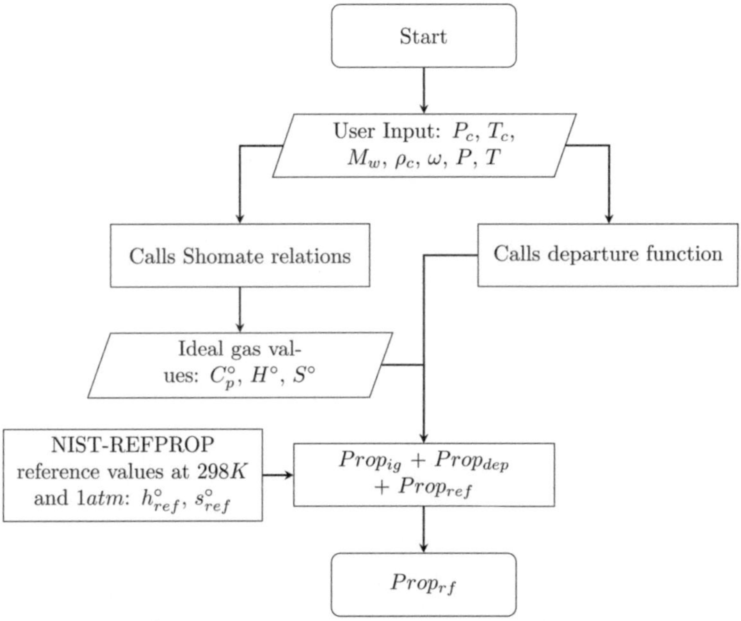

Fig. 3 Flowchart for Thermodynamic property calculations (P refers to Thermodynamic property)

The details of thermodynamic modeling can be referred from Ma thesis [8].

The procedure to obtain thermodynamic properties of O_2 and CH_4 is shown via flowchart in Fig. 3. The critical properties and operating T, P data are given as input to MATLAB code line. The shomate relations are used to obtain ideal gas state values. The departure value of each property is obtained by solving Eqs. 3–5, accounting for the real fluid effect. The ideal gas and departure values are combined with the reference values to obtain real fluid thermodynamic properties. The reference values are obtained from NIST-REFPROP [6].

Transport properties play a crucial role in high-pressure rocket engines. They govern the mixing and transport of physical quantities in LOx-CH_4 which experiences transcritical injection. The method of Chung et al. [3] for the prediction of viscosity and thermal conductivity at high pressure is quite accurate, with an error of less than 5%. This method has been extensively employed in the literature for evaluating dynamic viscosity and conductivity for pure fluids and mixtures [8, 9, 12].

The low pressures dynamic viscosity, η^o (μP, $P = 0.1$Pa s) is given as

$$\eta^0 = 40.785 F_c \frac{\sqrt{M_w T}}{V_c^{2/3} \Omega_v} \tag{6}$$

where M_w indicates the molecular weight of the species (in kg/kmol), V_c is the critical volume (cm^3/mol), Ω_v is the viscosity collision integral of the fluid, and F_c is the correction factor accounting for molecular shape and polarities of dilute gases.

The high-pressure dynamic viscosity η (μP) is expressed as

$$\eta = \eta^o \eta^* \tag{7}$$

where η^o is gas viscosity at low pressure, expressed as:

$$\eta^0 = 36.344 \frac{\sqrt{M_w T_c}}{V_c^{2/3}} \tag{8}$$

and η^* is the correction term accounting for dense fluid effects at high pressures, expressed as

$$\eta^* = \frac{\sqrt{T*}}{\Omega_v} \left[F_c (G_2{}^{-1} + E_6 y) \right] + \eta^{**} \tag{9}$$

The detail of this model and associated variables can be referred from Sharma et al. [12].

The thermal conductivity (λ) of a non-polar fluid like O_2 and CH_4 at high pressure can be estimated using Chung's method [3]. Various other methods for estimating thermal conducting are summarized in a book by Poling [11]. The conductivity is expressed as follows:

$$\lambda = 31.2 \frac{\eta^0 \Psi}{M'_w} \left(\mathcal{G}_2^{-1} + B_6 y \right) + q B_7 y^2 \mathcal{G}_2 \sqrt{T/T_c} \tag{10}$$

where λ ($N\,s/m^2$), is a function of the low-pressure viscosity given by Eq. 8 and M'_w is molecular mass in (kg/mol). The remaining terms are computed from formulations mentioned in Chung's model [3, 12].

3 Results and Discussion

In this section, analytical model results for the LOx-methane rocket engine, operating at a pressure of 6 MPa are presented. The peculiarities associated with transcritical and supercritical operating conditions are discussed in detail. We present the accuracy of different cubic EoS in predicting the thermophysical properties in the entire range of operations.

Figure 4 presents the variation of density with temperature for methane and oxygen. The density variation for different EoSs is plotted and compared with NIST. The reduced pressure (P/P_c) is 1.3043 for CH_4 and 1.19 for LOx. As the operating pressure is above the P_c for methane (4.59 MPa) and oxygen (5.043 MPa), both fluid undergoes a transition from transcritical to supercritical states. It is observed that PR over-predicts the density of both methane and oxygen below their critical temperatures. For SRK and RK-PR EoS, the density prediction is almost of the same accuracy. The prediction capability of SRK and RK-PR is better as compared to PR, with small discrepancies near the critical temperature. The ideal gas shows a large variation from NIST at low temperatures indicating its ineffectiveness in such operating conditions. The relative error in the prediction of density for methane and oxygen with multiple EoSs is shown in Fig. 5. The relative error in PR is less than 5% compared to 15% SRK and RK-PR near the critical temperature. SRK and RK-PR exhibit less relative error than PR, at conditions away from critical point temperature.

The specific heat at constant pressure as predicted by different EOSs versus temperature is shown in Fig. 6 for methane and oxygen. It is observed that the heat capacity increases significantly near the critical temperature, which is not predicted by the ideal gas equation. The rise in heat capacity could be attributed to a drastic density change, as observed in Fig. 4, leading to increased intermolecular interactions. The peak of specific heat occurs at pseudo-boiling temperature, analogous to boiling temperature in subcritical conditions. All three cubic EoS depict good accuracy over a range of temperatures but under-predict near the pseudo-boiling point, where the peak of specific heat occurs at $6MPa$. The relative error of more than

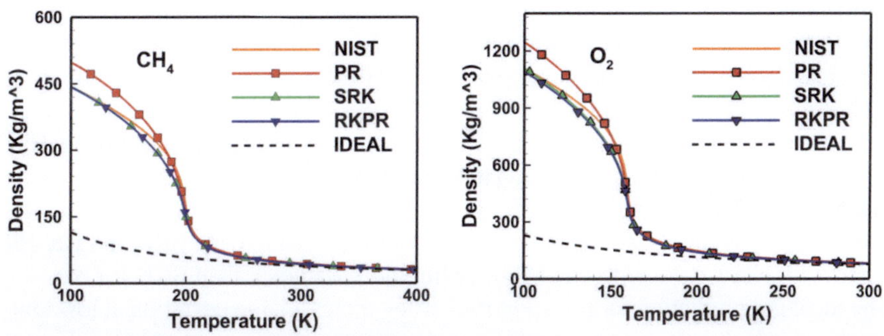

Fig. 4 Density versus temperature variation for CH_4 and O_2, comparison between NIST, RKPR, SRK, PR, and ideal gas

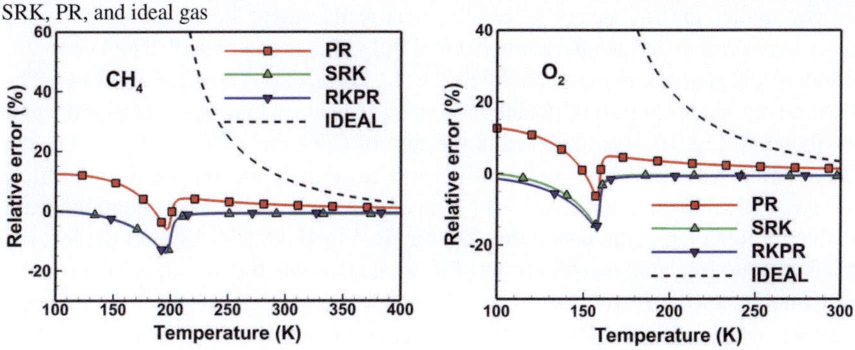

Fig. 5 Relative error of density for CH_4 and O_2 with the temperature predicted by different cubic EoSs and ideal gas

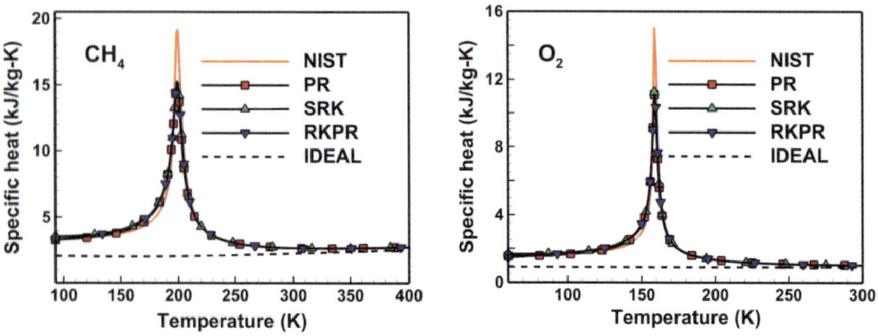

Fig. 6 Specific heat at constant pressure of CH_4 and O_2 versus temperature as predicted by different cubic EoSs, ideal gas and comparison with NIST

15% is observed at pseudo-boiling temperature as shown in Fig. 7. The entropy and enthalpy predictions using different EoSs and comparison with the NIST database are shown in Fig. 8. It is observed that all three cubic equations can predict the enthalpy and entropy with the help of the departure function. The equations are also able to capture a jump in the enthalpy and entropy during the phase transition regime. It is seen that ideal gas EoS does not show any phase change behavior, which brings out the importance of using departure values in the thermodynamic modeling of high-pressure LOx-methane rocket engines. Transport properties like thermal conductivity and viscosity greatly influence the diffusion of fuel during injection and are very important in determining mixing in rocket engines. Figure 9 shows variations in viscosity and thermal conductivity for methane and oxygen as predicted by Chung's Model for different EoSs. For viscosity and thermal conductivity, SRK and RK-PR agree with NIST data, with RK-PR showing more accuracy than SRK for oxygen. The ideal gas prediction for transport properties shows a large deviation at low temperatures, which improves upon temperature increases due to a reduction in the real gas effect at high temperatures. The combustion products of the LOx-methane engine are also studied to simulate conditions like near walls, where the temperatures can be lower than combustion temperature and real fluid effects can prevail. The two major products that are present in stoichiometric CH_4, O_2 reactions are CO_2 and H_2O. The thermodynamic properties of density and specific heat for carbon dioxide and water are shown in Fig. 10. It is observed in the case of CO_2, that unlike CH_4 and O_2, the PR equation was able to predict density more accurately as compared to the SRK and RK-PR cubic equations. At lower temperatures, all three cubic equations were unable to predict specific heat value accurately. For H_2O, RK-PR EoS models the density accurately, where as SRK and PR, underestimate the densities by a significant amount. As the P_c value for water (22.064 MPa) is much greater than operating pressure, water is in subcritical conditions, an abrupt phase change around 550 K is observed for Fig. 10, where liquid-vapor equilibrium exists. The specific heat also shows abrupt changes near-boiling temperatures. The prediction improves as we go near the critical pressure of the species, with some range of error.

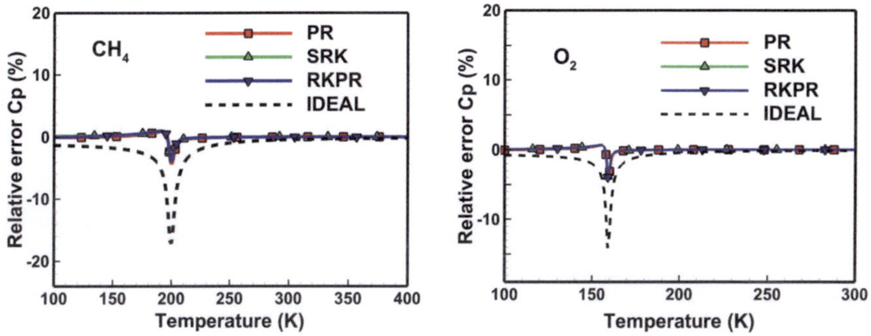

Fig. 7 Relative error of specific heat for CH_4 and O_2 with the temperature predicted by different cubic EoSs and ideal gas

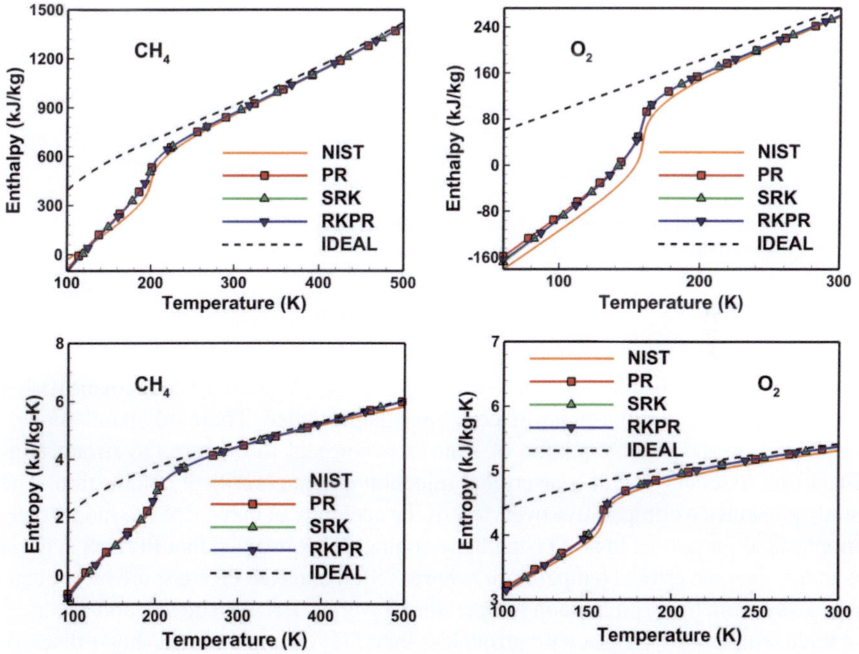

Fig. 8 Enthalpy and enstropy versus temperature variation using Chung's method as predicted by different cubic EoSs, ideal gas and comparison with NIST

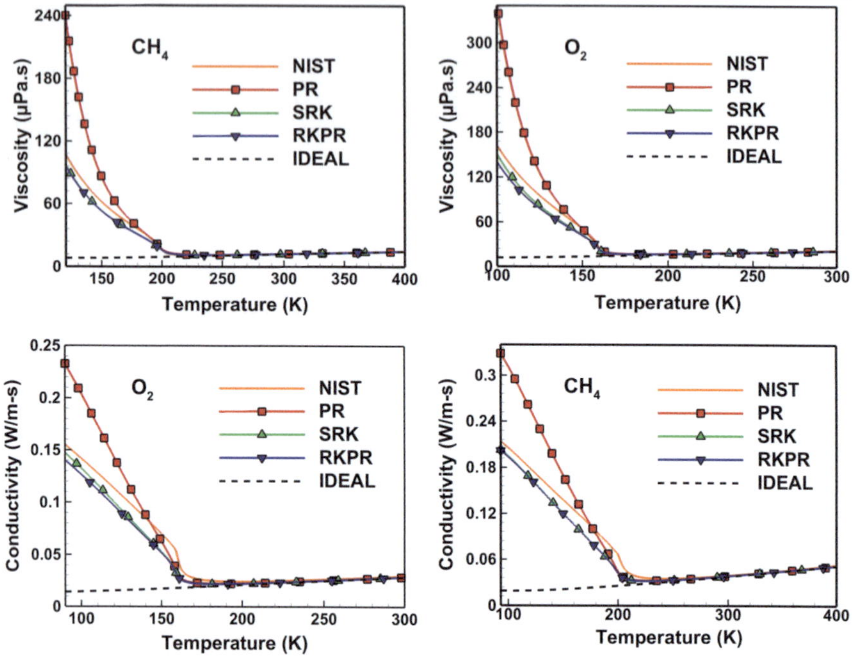

Fig. 9 Viscosity and thermal conductivity versus temperature variation using Chung's method as predicted by different cubic EoSs, ideal gas and comparison with NIST

4 Conclusions

A comprehensive evaluation of the thermophysical properties of LOx-methane rocket engines operating at high-pressure conditions is presented. Thermodynamic modeling based on the cubic equation of state is performed to capture the strong nonlinearities associated with transcritical injection and supercritical combustion. The study presented a comparative overview of the accuracy of cubic EoSs to model thermophysical properties in a LOx-methane engine. It is observed that PR EoS is more accurate near the critical temperature, whereas SRK and RK-PR EoSs are more accurate away from the critical point temperature. Overall, RK-PR shows good accuracy at wide temperature ranges with errors less than 5%. Chung's model shows discrepancies at lower temperatures, which needs further investigation. Our study clearly identifies a high-pressure thermodynamic and transport modeling framework that can accurately represent real fluid effects in injection and combustion of LOx-methane engines.

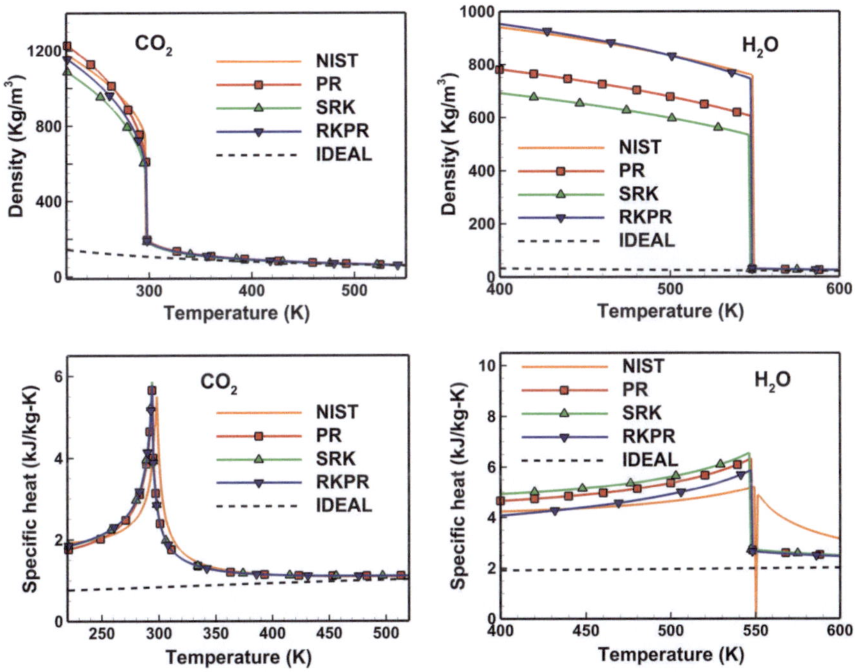

Fig. 10 Thermophysical properties of CO_2 versus temperature at 6MPa and comparison with NIST database

References

1. Banuti DT (2015) Crossing the widom-line-supercritical pseudo-boiling. J Supercritical Fluids 98:12–16
2. Chehroudi B, Talley D, Coy E (2002) Visual characteristics and initial growth rates of round cryogenic jets at subcritical and supercritical pressures. Phys Fluids 14(2):850–861
3. Chung TH, Ajlan M, Lee LL, Starling KE (1988) Generalized multiparameter correlation for nonpolar and polar fluid transport properties. Indus Eng Chem Res 27(4):671–679
4. Cismondi M, Mollerup J (2005) Development and application of a three-parameter RK-PR equation of state. Fluid Phase Equilib 232(1–2):74–89
5. Kim S-K, Choi H-S, Kim Y (2012) Thermodynamic modeling based on a generalized cubic equation of state for kerosene/Lox rocket combustion. Combust Flame 159(3):1351–1365
6. Lemmon EW, Bell IH, Huber ML, McLinden MO (2018) NIST standard reference database 23: reference fluid thermodynamic and transport properties-REFPROP, Version 10.0, National Institute of Standards and Technology
7. Linstrom PJ, Mallard WG (2001) The NIST Chemistry Webbook: a chemical data resource on the internet. J Chem Eng Data 46(5):1059–1063
8. Ma PC (2018) Modeling of turbulent mixing and combustion at transcritical conditions. Ph.D. thesis, Stanford University
9. Madana Gopal JV, Morgan R, De Sercey G, Vogiatzaki K (2023) Overview of common thermophysical property modelling approaches for cryogenic fluid simulations at supercritical conditions. Energies 16(2):885

10. Mayer W, Schik A, Vielle B, Chauveau C, Gokalp I, Talley DG, Woodward RD (1998) Atomization and breakup of cryogenic propellants under high-pressure subcritical and supercritical conditions. J Propul Power 14(5):835–842
11. Poling BE, Prausnitz JM, O'connell JP (2001) Properties of gases and liquids. McGraw-Hill Education
12. Sharma A, De A, Sunil Kumar S (2023) Assessment of statistical and kinetics-based models for LOx-methane green propellant combustion. Int J Energy Clean Environ 24(7)
13. Singla G, Scouflaire P, Rolon C, Candel S (2005) Transcritical oxygen/transcritical or supercritical methane combustion. Proc Combust Inst 30(2):2921–2928
14. Valderrama JO (2003) The state of the cubic equations of state. Indus Eng Chem Res 42(8):1603–1618

Recovering Energy Potential Through ASRU for Agricultural Dependent Hydropower Plant

M. Kurulekar, K. Kumar, and S. Joshi

Abstract Energy supply from electricity has been increased from 19 to 23% from 2010 to 2020, and for this renewable sources has contributed 30% electricity globally. Electricity can be generated from impulse, reaction, axial type of hydropower turbine. Limited turbine capacity, functional restrictions of power units, necessary withdrawal of water from dam and continuously changing weather, demands viable and flexible solution of power generation. Thus, all possible options of power generations should be explored. Canal sites, irrigation dependent hydropower plant, small hydropower plant are the targeted opportunities of power generation. The nexus of such locations and power generation should be emphasized to generate power. Understanding need of hydropower generation, this paper proposes the solution of Archemedes screw recovery unit (ASRU). To explore the use of ASRU, the case study of irrigation dependent power station at Dhom from Wai, Maharashtra, has been considered. This reservoir has two irrigation gates. The two gates are suggested with ASRU. The endeavor of this study is to propose the feasibility of ASRU for power generation. ASRU would able to generate additional power in the range of 455.4–559.3 kW from discharge after two gates for obtained efficiency range of 67–82.25% in addition to 2 MW.

Keywords ASRU · Irrigation dependent · Hydropower plant

M. Kurulekar
Department of Civil Engineering and Department of Engineering Sciences, Vishwakarma Institute of Information Technology and Vishwakarma University, Pune, India
e-mail: maya.kurulekar@vupune.ac.in

K. Kumar (✉)
Hydraulic Division, Central Water and Power Research Station, Pune, India
e-mail: krishnaswamy_kumar@yahoo.co.in

S. Joshi
Department of Civil Engineering, Vishwakarma Institute of Information Technology, Pune, India
e-mail: shardul.joshi@viit.ac.in

A. Kumar R. et al. (eds.), *Proceedings of Fluid Mechanics and Fluid Power (FMFP) 2023, Vol. 6*, Lecture Notes in Mechanical Engineering,
https://doi.org/10.1007/978-981-96-9600-0_45

1 Introduction

Renewable energy and non-renewable energy are the sources of electricity production, and electricity is major contributor toward growth of a nation. The source wise contribution of electricity production by 2012 and 2022 is mentioned in Fig. 1 [1]. Around 30% electricity is supplied by renewable sources showing major shift toward sustainability goal. The power generation by different ways mentioned in Fig. 1 has its own advantages and limitations. Conventional hydropower plants necessitate huge capital and operational expenditure for power generation. Hence, new solution with as minimum as possible investment is expected by the innovators. On the other hand, increasing demand and available site conditions should be accommodated in the solution. Limitation on turbine capacity, functional restrictions of power units, necessary withdrawal of water from dam and continuously changing weather, demands viable and flexible solution of power generation. Also few hydro sites has power generation as secondary objective.

These practical requirements and constraints gave birth to different types of water turbine such as impulse, reaction, and axial. The combination of these power machines with civil works and other hydraulic systems throws ray of hope of power generation at all conditions of site, flow parameters, cost, etc. This study has proposed the solution of the power generation by means of Archemedes screw; for which the case study of irrigation dependent power station at Dhom from Wai, Maharashtra, is considered.

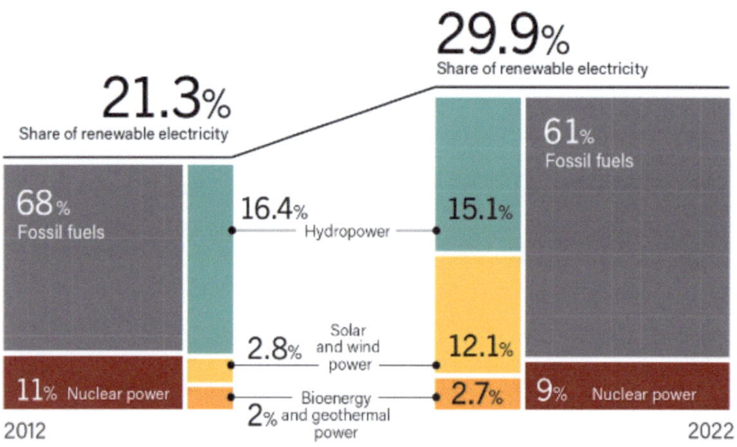

Fig. 1 Source wise contribution of electricity generation [2]

2 Literature Review

Hydropower generation under head as low as 3 m is a challenging problem, and this ultra-low head concept is converting water resources to energy resources to a great extent. Reusing the water from the tail race channel for power generation is possible by incorporating the structures such as head drop, stairways, ramps, weirs, etc., along with ultra-low head turbine. The author has suggested the vertical wall in the flow direction has been damped by introducing the inlet approach profile which is the objective of this literature. Backward facing step profile, stepped bed profile has been reviewed by the Chaulagain et al [3]. The location of turbine was checked by these profiles to know the losses. Inlet approach profile is connecting the turbine with the bed for low approach flow velocity. Vertical, inclined, curve, and combined profile were tested for this objective. Non-rotating model with holes has been used in the experiment. Upstream height and highest velocity has been observed in the curved profile. Further optimized profiles are suggested as future scope. Purece [4] has been provided the performance and implementation of Archemedes screw turbine. It was also mentioned that both low head turbine and Archimedes screw is 10% cheaper than Kaplan cost for 15% more energy production.

Effect of channel geometry on Savonius turbine has been studied by Patel et al. [5] to know the performance of turbine. Effect of cross section, bed, and side wall of the channel has been observed during experiment. Turbine was suggested to place on hump at the channel to improve the performance by 83% power output. Increased velocity has reduced the depth of flow by changing channel bed, and side wall geometry results in increase of power coefficient to 0.45. The review paper by Punys et al. [6] gives overview about historical hydropower resources of EU countries. All the unattended hydropower potential has been summarized. The paper has considered the hydropower plant classification as micro-5 to 100 kW, mini-100 kW to 1 MW and small-1 MW to 10 MW. Thus, paper provides the platform to encourage hydropower generation even at small capacity.

Untapped energy potential of irrigation network is a vast field of opportunities to generate power. The rigorous review shared here strengthens the thought of hydropower generation in water distribution network having energy generation as secondary objective. Action turbines convert hydraulic energy to mechanical energy and generate power due to kinetic energy of flow; whereas the reaction turbine generates power by creating pressure difference. Pérez-Sánchez et al. [7] has suggested the classification of power units used in open channel and close conduit. The type of adapted machines can be used to generate power at the place of dissipated energy like pressure reduction valve or at the place of hydraulic jump. Author has suggested that conventional power units are not viable so pump as turbines (PAT) can be used in open channel having enough civil works.

Unfavorable conditions for the very low head turbine (VLH) has been studied by Wright et al. [8]. The effect of position of turbine model and backward facing step were included along with non-rotating model. VLHT has been developed with 3.5–5.6 m diameter runner, 1.4–4.2 m head, 10–30 m^3/s, 100 to 500 kW. This study will help to reduce the cost of civil work and runner provided in terms of drop structures. Drop structures help to control the grade, reduces erosion, better aqua habitat, etc.

Nuernbergk et al. [9] have investigated the use of AS and attempted the optimal hydraulic design of screws as per availability water inflow. Authors have also mentioned that no pertinent literature is available for hydraulic design of screw as per site conditions. Thus, new analytical model for water inflow has been designed by the authors to obtain the fixed inflow to hydropower plant.

Run-off river flow type hydropower plant is non-polluting and generate revenue if satisfied Green audit. Thus, Adhau et al. [10] have suggested irrigation cum power plant by including large diameter pressurized pipe for desired discharge.

The connection of environment, electrical energy, its cost, and effect are important to zero emission target. Thus, smallest unattended potential of hydropower generation and so solution of low head turbine is area of interest. The turbine with head as low as 2 m reducing the transmission losses in local consuming area were identified. Thus, Darrieus-type turbine has been suggested by Shimokawa et al. [11] with inlet nozzle showing improvement in power generation without inlet nozzle.

Figure 2 from the literature of Yoosef Doost et al. [12] shows the performance of different turbines under different head and flow rate. The requirement of turbine to generate power under low head requires sustainable solution and this graph suggest water wheel, Archimedes screw, Kaplan turbine. The power generated by water wheel is from 100 W to 10 kW, AST 2 kW to 500 kW, and Kaplan 20 kW to 200 MW. Thus, AST can be thought of over water wheel considering large range of power generation even under low head. Kaplan has constraint of large diameter of runner which can be overcome by AST.

The nexus of irrigation and power generation is important at the places where power generation as secondary objective can be achieved. Understanding different approaches and need of hydropower generation, this paper proposes the solution for hydropower generation by suggesting Archemedes screw recovery unit (ASRU) next to irrigation gate. The case study of irrigation dependent power station at Dhom from Wai, Maharashtra, is considered. This reservoir has two irrigation gates. The feasibility of ASRU is obtained with reference to efficiency in power generation as objective of this study [13].

3 Problem Statement and Objective

Due to drought conditions, agricultural dependent power plants required more releases. These irrigation releases do not utilize water energy due to capacity restriction of power plant. Also high rainfall and so less irrigation lead to low head. At this low head and part flow conditions, power turbines become inoperative due to functional restrictions. This catastrophic behavior demands sustainable solution. In line with the objective of the study, Archemedes screw recovery unit (ASRU) is suggested next to two irrigation gates. For the suggested solution, the pilot case study of Dhom

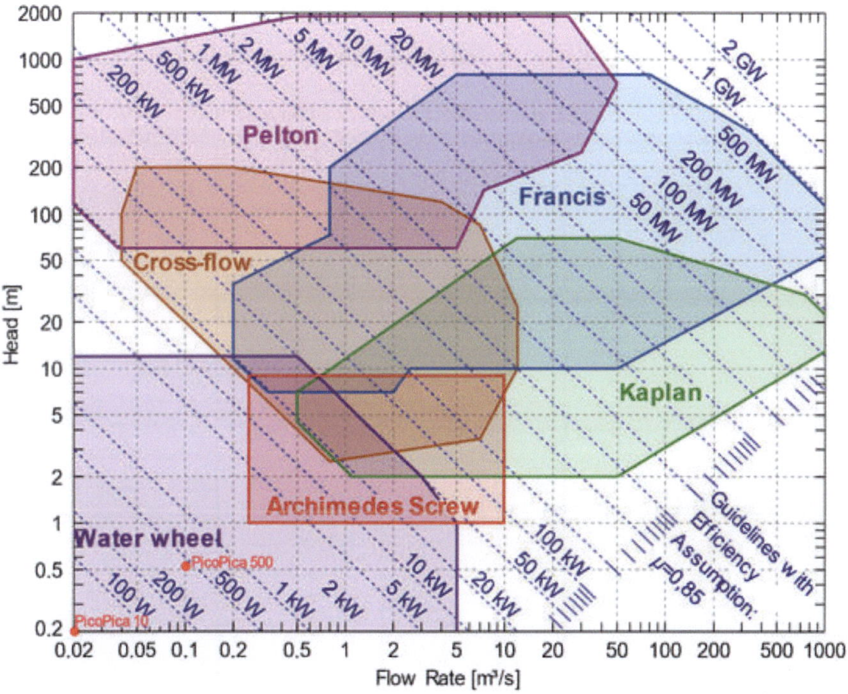

Fig. 2 Performance of various turbine

Hydro Power Station of capacity 2 × 1 MW from Wai of Satara, Maharastra, is taken. The attributes of case study are given in the next section.

3.1 Features of Dhom Hydropower Plant

Dhom Hydro Power Station of capacity 2 × 1 MW from Wai of Satara, Maharastra, is identified as an area of research. Reservoir on right bank of river Krishna has (1) for irrigation discharge (ID) and (2) for energy generation. Left bank canal of Dhom splits from right bank canal on downstream side. The layout of this case study is shown in Fig. 3, and the features of power plants are listed in Table 1.

For irrigation dependent hydropower plant, the power generation is a secondary objective. The Dhom Dam receives an annual average rainfall of about 970 mm. In case of maximum rainfall of about 1725 mm the reservoir had released water of about 108.73 MM^3 toward irrigation without power generation. Also in case of minimum rainfall of about 580 mm, the discharge was about 24.78 MM^3 without generating power. The capacity factor of any plant is a result of energy generation. However, in case of irrigation dependent hydropower plant releases toward irrigation without

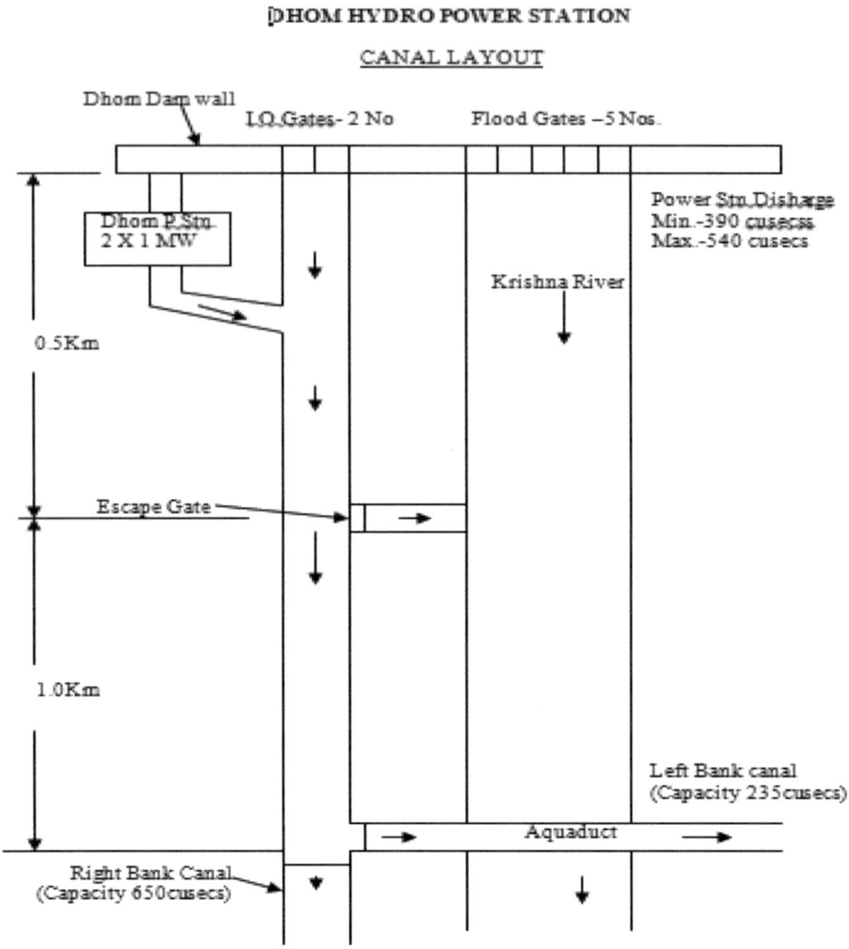

Fig. 3 Layout of Dhom hydropower plant

Table 1 Features of hydropower plant

Installed capacity	2 × 1 MW	Design energy generation/year	10.56 MU
Average rain fall	1500–6250 mm	Average inflow of water	423.91 MM^3
Gross dam storage	382 MM^3	Live storage	331 MM^3
Minimum DD level	733.08 m	Design head	18 m
Design discharge	14 m^3/s	Diameter of penstock	2.5 m

Table 2 Yearly release of water

Year	Irrigation discharge (MCM) (ID)	Water loss (WL) in MU	Year	Irrigation discharge (MCM)	Water loss in MU
1991–92	98.543	3.425	2005–06	30.32	0.968
1992–93	54.346	1.864	2006–07	57.34	2.156
1993–94	58.241	2.432	2007–08	27.85	0.96
1994–95	37.421	1.296	2008–09	154.24	4.671
1995–96	92.357	3.361	2009–10	29.83	0.887
1996–97	61.437	2.683	2010–11	52.21	1.165
1997–98	165.291	5.373	2011–12	90.87	3.246
1998–99	88.541	3.186	2012–13	58.51	1.783
1999–2000	27.651	0.87	2013–14	161.31	6.973
2000–01	59.563	2.453	2014–15	58.76	2.286
2001–02	86.461	3.215	2015–16	58.89	0.96
2002–03	163.435	5.427	2016–17	165.39	4.237
2003–04	56.347	2.543	2017–18	46.23	2.538
2004–05	48.475	1.641	Avg	77.40 MCM/year	

power generation would affect the capacity factor. Table 2 shows the irrigation release in MM^3 and corresponding water loss in MU [14].

The capacity factor of hydropower plant can be increased by suggesting the system of power generation. The probable power generation from irrigation releases is calculated as Eqs. 1 and 2.

$$\text{Average irrigation release} = \frac{77.40 \times 10^6}{12 \times 30 \times 24 \times 3600} = 2.47\ \text{m}^3/\text{s} \quad (1)$$

$$\text{Power generated} = \frac{1000 \times 9.81 \times 14 \times 2.47}{10^6} = 0.339\ \text{MW}. \quad (2)$$

3.2 *Archemedes Screw Recovery Unit (ASRU)*

Yoosef Doost et al. [12] mentioned that run of river has small or no storage of water and so Archimedes screw turbine (AST) of run of river (ROR) type is one of the sustainable solutions for power generation. The threat to climate, wildlife, land, or soil, ecosystem in total and social impacts of ROR can be reduced by adopting AST. At low head also AST shows efficiency from 60 to 80%. Conventional ASTs can operate with heads between 1 and 6 m, and flow rates between 0.1 and 6 m^3/s, producing 1–300 kW of power. The author [11] also mentioned that the improved version of AST can produce around 10 watts of power with 0.01 m^3/s of flow rate and 0.1 m of water head. So the AST can generate power with flowrate of 0.01–10 m^3/s and with head 0.1–10 m. Inaccessible areas, local areas where quick, easy, and cheap supply of electricity is required, areas where construction of large dam is infeasible, existing dam sites for ease of construction or retrofitting, AST is workable solution. The implementation guidelines of AST are discussed in the next section.

3.2.1 Design Study of AST

The design considerations of ASRU have been reviewed and suggested here to know the feasibility of Archemedes screw at the site of case study. There are neither concrete guidelines for the design of ASRU nor the experienced expertise available as per literature [4, 11]. The combination of prediction model and optimization model from [13] is used to calculate efficiency suggested which has been found useful to obtain the feasibility of ASRU. This theoretical study will be helpful for experimental study in future. Figure 4 shows the schematic view of ASRU [4].

1. **Design Principle**

Archimedes screw recovery unit (ASRU) is working mainly working at atmospheric pressure and discharge. Equal area and velocity at inlet and outlet of turbine drive this pressure. Both inlet and outlet are having free surface. The product of force

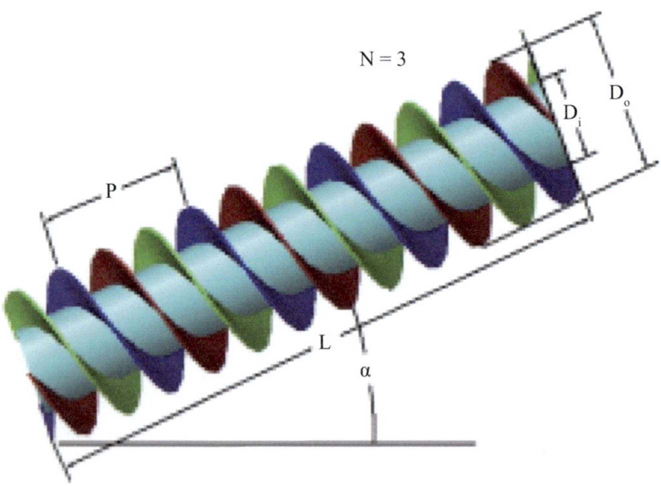

Fig. 4 Schematic view of ASRU. (N = number of bucket, P = pitch, L = length of screw, D_i = inlet diameter, D_o = outlet diameter, α/θ = inclination)

and velocity (speed) of runner give power, and thus, position of turbine creating tangential component of velocity is dominant over weight of water rotating bucket [4].

2. **Installation of Screw**

Since the turbine is working at the atmospheric pressure, it should receive the water energy to run itself. Thus, this unit should be kept at proper inclination. Andrei Dragomirescu [2] has mentioned that inclination from 20° to 35° is sufficient to achieve efficiency. Fundamental principal to select the inclination angle of unit is to reduce the cost by minimizing the length of the screw. High inclination angle will result in spreading of water out of the bucket. Thus to reduce this concurrent loss of water, ASRU can be converted to axial type turbine by shelling in tube and hence efficiency can be increased comparative to propeller as well.

3. **Length of the Screw**

The length of the screw turbine depends on height of turbine and inclination of turbine [12, 13],

$$L = H / \sin\theta \tag{3}$$

Turbine height (H) can be obtained from difference of static head of inlet and outlet of turbine. Inclination angle θ can be set between 20° to 35°.

4. **Number of Blades in the Screw**

The blades of the screw unit make the bucket to hold volume of water. So number of blades criteria should not affect the volume of water flow, weight on the shaft, shock and frictional losses, and the subsequent cost. Functionally, four blades show less efficiency and two blades show great reduction in efficiency; thus, 3 number of blades are suggested by [12, 13].

5. **Diameter Ratio of the Screw**

The torque generated by axial component and tangential components of hydrostatic forces is inversely proportional to inlet to outlet diameter ratio. If this diameter ratio decreases, the torque increases. On the other hand, increasing diameter ratio reduces volume of water buckets which may affect the speed of the runner. Thus diameter ratio (D_i to D_o) is advisable to keep between 0.5 and 0.55.

3.2.2 Design of Archimedes Screw for Case Study

Earlier section covering design considerations for ASRU sets the line for feasibility study. However, the design guidelines available in recent literature of Kazem Shahverdi et al. [15] to design the components of ASRU are used in this paper.

Yoosef Doost et al. [13] has proposed general analytical equation to determine outer (overall) diameter and other geometrical parameters of ASRU. The equation accuracy with correlation 91.80% and mean absolute error 6.61% has been mentioned by literature. The critical design steps were significantly reduced by the author. Present papers have suggested to consider the prediction and optimization model together [13] for the design of Archimedes screw at Dhom dam. The irrigation discharge determined from the available data of Dhom dam is 2.47 m^3/s. The available head at the dam site is 14 m. As per literature review, ASRU is also suitable and efficient at the site of low head and moderate flow. This section covers the calculation of various ASRU components shown in Table 3.

Outer diameter has been calculated and varied by utilizing the range of D_o/H given in prediction model of [13].

$$D_i = 0.01D_o \text{ to } 0.99D_o \tag{4}$$

$$P = 0.01D_o \text{ to } 1.2D_o \tag{5}$$

$$\omega = 50/D_o^{(2/3)} \tag{6}$$

The inner diameter D_i, pitch and angular speed of blades have been calculated as a function of D_o [13]. The feasibility of screw turbine has been obtained for different values of components of ASRU. The efficiency of the screw turbine is observed from the Eq. (7) as given by prediction model in [13],

$$\begin{aligned}\eta = & -169.1 + (722.4 * \sin\theta) + \left(346.8 * \frac{D_o}{H}\right) - (20.7 * N) \\ & - (79.8 * \emptyset) - \left(370.1 * \sin\theta * \frac{D_o}{H}\right) + (18.8 * \sin\theta * N) + (75.1 * \sin\theta * \emptyset) \\ & + \left(9.2 * \frac{D_o}{H} * N\right) + \left(102.4 * \frac{D_o}{H} * \emptyset\right) - (10.4 * N * \emptyset) \\ & - \left(590.9 * (\sin\theta)^2\right) - \left(161 * \left(\frac{D_o}{H}\right)^2\right) + \left(1.6 * N^2\right) - \left(20.4 * \emptyset^2\right)\end{aligned} \tag{7}$$

Table 3 Components of ASRU includes

Components	Symbols	Components	Symbols
Outer diameter of screw in meter	D_o	Diameter ratio	D_R
Inner diameter of screw in meter	D_i	Pitch in meter	P
Length of the screw in meter	L	Number of flights or blades	N
Inclination of screw in degrees	α	Rotation of the screw in rad/s	ω
Discharge through screw in m^3/s	Q	Gap width in meter	G_W
Flow pitch	P_f	Blade thickness	t_b

4 Results and Discussion

This section covers the observations about the feasibility study of ASRU.

1. The required parameters and empirical Eq. (7) to know the efficiency of ASRU are referred from [13].
2. The variations for outer diameter (D_o), length of turbine (L), inclination of turbine (θ) under constant head (H), and discharge (Q) has been done.
3. Tables 4, 5 and 6 show the scenarios varying $\theta = 35°, 28°, 27°, 25°$, length = 24.5, 30, 31, 33, outer diameter = 10.2, 9.8, 9.3 and number of blades (N) 1, 2 3, 4 for constant H and Q.
4. The specific speed φ is calculated as [13]. It has been observed that for the case study of Dhom dam, ASRU can be provided with efficiency from 67 to 82.25%.
5. The screw with 2, 3, 4 blades and angle from 25° to 28° can be reasonably provided at the site next to TWO gates.
6. The scenarios are considered as per relations mentioned in [13] $\sin\theta = \frac{H}{L}$, $\emptyset = \frac{\omega\sqrt{Q}}{(gH)^{3/4}}$, $D_o = s$, $D_i = 0.5D_o$. The range for $\theta = 20°$ to $35°$, $D_o/H = 0.435$ to 0.731, $\varphi = 0.14$ to 0.44 is assumed as limiting parameters to obtain geometry of ASRU.
7. Same range of geometry has been used to know the range of efficiency of ASRU.
8. It has been observed that increasing number of blades will not have significant effect on efficiency. Thus, cost can be saved against efficiency.

Table 4 Scenario 1

sin θ	D_o/H	N	ω	φ	η
0.57	0.73	1.00	10.63	0.42	70.06
0.57	0.73	2.00	10.63	0.42	67.28
0.57	0.73	3.00	10.63	0.42	67.69
0.57	0.73	4.00	10.63	0.42	71.31
0.47	0.73	1.00	10.63	0.42	81.65
0.47	0.73	2.00	10.63	0.42	76.89
0.47	0.73	3.00	10.63	0.42	75.34
0.47	0.73	4.00	10.63	0.42	76.98
0.45	0.73	1.00	10.63	0.42	82.25
0.45	0.73	2.00	10.63	0.42	77.21
0.45	0.73	3.00	10.63	0.42	75.37
0.45	0.73	4.00	10.63	0.42	76.73
0.42	0.73	1.00	10.63	0.42	82.65
0.42	0.73	2.00	10.63	0.42	77.10
0.42	0.73	3.00	10.63	0.42	74.74
0.42	0.73	4.00	10.63	0.42	75.59

Table 5 Scenario 2

sin θ	D_o/H	N	ω	φ	η
0.57	0.70	1.00	10.92	0.43	71.37
0.57	0.70	2.00	10.92	0.43	68.20
0.57	0.70	3.00	10.92	0.43	68.24
0.57	0.70	4.00	10.92	0.43	71.47
0.47	0.70	1.00	10.92	0.43	81.76
0.47	0.70	2.00	10.92	0.43	76.62
0.47	0.70	3.00	10.92	0.43	74.69
0.47	0.70	4.00	10.92	0.43	75.95
0.45	0.70	1.00	10.92	0.43	82.18
0.45	0.70	2.00	10.92	0.43	76.76
0.45	0.70	3.00	10.92	0.43	74.55
0.45	0.70	4.00	10.92	0.43	75.53
0.42	0.70	1.00	10.92	0.43	82.27
0.42	0.70	2.00	10.92	0.43	76.34
0.42	0.70	3.00	10.92	0.43	73.61
0.42	0.70	4.00	10.92	0.43	74.08

Table 6 Scenario 3

sin θ	D_o/H	N	ω	φ	η
0.57	0.66	1.00	11.31	0.44	72.53
0.57	0.66	2.00	11.31	0.44	68.88
0.57	0.66	3.00	11.31	0.44	68.43
0.57	0.66	4.00	11.31	0.44	71.18
0.47	0.66	1.00	11.31	0.44	81.42
0.47	0.66	2.00	11.31	0.44	75.79
0.47	0.66	3.00	11.31	0.44	73.37
0.47	0.66	4.00	11.31	0.44	74.15
0.45	0.66	1.00	11.31	0.44	81.63
0.45	0.66	2.00	11.31	0.44	75.72
0.45	0.66	3.00	11.31	0.44	73.02
0.45	0.66	4.00	11.31	0.44	73.51
0.42	0.66	1.00	11.31	0.44	81.32
0.42	0.66	2.00	11.31	0.44	74.90
0.42	0.66	3.00	11.31	0.44	71.69
0.42	0.66	4.00	11.31	0.44	71.67

5 Conclusions

The Archimedes screw is antiquity not only in the field of pump but also in turbine. It is the need of time to find green, local, feasible, and economical solution for hydropower generation which can satisfy increasing demand of electricity. Use of water turbine has been an antiquity in the field of irrigation dependent hydropower plant. The present study is the sustainable idea to install Archimedes screw recovery unit (ASRU). ASRU is ecosystem friendly due to working at low runner speed. Stochastic nature of weather having high rainfall–low irrigation and low rainfall–high irrigation demands workable solution to address all functional and capacity constraints. These seasonal changes are well capable to handle with ASRU. The present case study of Dhom dam has two irrigation gates which release the discharge of 2.47 m^3/s each. The input power from both the gates would be 680 kW and hydropower plant has been designed for 2 MW. This study for ASRU concludes that ASRU can be viable solution at Dhom dam. The Archemedes screw can be provided next to irrigation gate.

ASRU would able to generate additional power in the range of 455.4–559.3 kW from discharge after two gates for obtained efficiency range of 67–82.25%. The length of screw is another area of concern as increasing in inner diameter increases length results in added cost and lower effective area. More weight of the screw also limits the bearing.

The CFD analysis and experimental analysis have been kept in the scope of this work. Literature surveyed from [4, 12, 13] validates the range of efficiency calculated for the range of geometrical parameters of ASRU.

References

1. https://greenbugenergy.com
2. Dragomirescu A (2021) Design considerations for an Archimedean screw hydro turbine. IOP Conf Ser Earth Environ Sci
3. Chaulagain RK, Poudel L, Maharjan S (2023) Experimental investigation on flow approach profile for comparison of surface velocity at the inlet of the vertical ultra-low head hydro turbine using a non-rotating model. Eng Sci Technol Int J 40:101367
4. Purece C, Corlan L (2021) Archimedean screw as fish-friendly turbines for harnessing hydropower potential. E3S Web Conf 286:02007. https://doi.org/10.1051/e3sconf/202128602007
5. Patel V, Eldho TI, Prabhu SV (2019) Velocity and performance correction methodology for hydrokinetic turbines experimented with different geometry of the channel. Renew Energy 131:1300–1317
6. Punys P, Kvaraciejus A, Dumbrauskas A, Šilinis L, Popa B (2019) An assessment of micro-hydropower potential at historic watermill, weir, and non-powered dam sites in selected EU countries. Renew Energy 133:1108–1123
7. Pérez-Sánchez M, Sánchez-Romero FJ, Ramos HM, López-Jiménez PA (2017) Energy recovery in existing water networks: towards greater sustainability. MDPI, Water
8. Wright MK, Rival DE (2013) An experimental investigation of the approach flow conditions for a non-rotating, very low head water-turbine model. Exp Thermal Fluid Sci 49:105–113

9. Nuernbergk DM, Rorres C (2013) Analytical model for water inflow of an archimedes, screw used in hydropower generation. J Hydraul Eng 139(2): 213–220. https://doi.org/10.1061/(ASCE)HY.1943-7900.0000661
10. Adhau SP, Moharil RM, Adhau PG (2012) Mini-hydro power generation on existing irrigation projects: case study of Indian sites. Renew Sustain Energy Rev 16(7):4785–4795
11. Shimokawa K, Furukawa A, Okuma K, Matsushita D, Watanabe S (2012) Experimental study on simplification of Darrieus-type hydro turbine with inlet nozzle for extra-low head hydropower utilization. Renewable Energy 41:376–382
12. Yoosef Doost A, Lubitz WD (2020) Green and renewable energy generation—a review of potential and design procedures. Sustainability 12(18):7352
13. Yoosef Doost A, Lubitz WD (2021) Archimedes screw design: an analytical model for rapid estimation of archimedes screw geometry. MDPI, Energies
14. https://www.ren21.net
15. Shahverdi K, Najafi G, Mamat R, Ghazali MF, Ei-Shafy AS, Mousa M (2023) Introducing a design procedure for archimedes screw turbine based on optimization algorithm. Energy Sustain Dev 162–172

Development and Validation of the Lattice Boltzmann Method-Based Single-Phase Flow Solver

Tejas Raval and Absar Lakdawala

Abstract Lattice Boltzmann Method (LBM) is widely accepted for solving multi-phase fluid interaction applications.The LBM – a mesoscale method – doesn't require to solve governing equations at molecular level (unlike molecular dynamics-based methods) which makes it computationally lesser costlier. In addition, it is also taking care of the physical properties for both the fluids. LBM with multi-relaxation-time (MRT) model is having higher stability and accuracy, and hence, LBM with MRT-based methodology is implemented in this work. LBM-based multi-phase fluid solver is developed, however in this work only single-phase benchmarking problems are solved in order to validate the proposed solver. 2-D channel flow in backward-facing step and 2-D channel flow over a square cylinder are solved, and results are validated through available numerical and analytical solutions available. The problem of backward-facing step is simulated at $R_e = 800$ and flow over a square cylinder is simulated at two different Reynolds number, i.e., $R_e = 60$ & 100. Efforts are made to showcase the capabilities and benefits of LB-based model in comparison of well-established Navier–Stokes-based solvers. Numerical results of the proposed work are in nice matching with well-established NS-based numerical and analytical solutions available in the literature.

Keywords Lattice Boltzmann Method (LBM) · Multi-relaxation time (MRT) · Single-phase · Flow over square cylinder · Channel flow over backward-facing step

1 Introduction

Now-a-days computational multi-fluid dynamics (CMFD) is very much popular for solving multi-fluid and multi-component fluid interaction applications as its gives cheaper and quite realistic solutions. CMFD totally eliminates the costing of the

T. Raval · A. Lakdawala (✉)
Mechanical Engineering Department, Institute of Technology, Nirma University, Ahmedabad, Gujarat, India
e-mail: absar.lakdawala@nirmauni.ac.in

A. Kumar R. et al. (eds.), *Proceedings of Fluid Mechanics and Fluid Power (FMFP) 2023, Vol. 6*, Lecture Notes in Mechanical Engineering,
https://doi.org/10.1007/978-981-96-9600-0_46

experimental setups and provides approximated results. Navier–Stokes-based solvers fail to encounter effects of physical behavior of particles of the fluids in multi-phase applications [1]. As a result, mesoscale model (LBM based model) is proposed by various researchers for benchmarking problems pertaining to single [2–5] and multi-phase fluids [6]. Different benchmarking problems like flow through 2-D channel [7–9], 2-D flow over cylinder, 2-D channel flow over backward-facing step, etc. [10–12]. In this paper, we proposed solution for 2-D flow over cylinder, 2-D channel flow over backward-facing step [13] for validating our LBM-based methodology. Numerical results of Navier–Stokes (N–S)-based solvers are also available in multiple literatures [3, 4, 6, 14]. Here, the efforts are made to validate the LBM-based solver by comparing the results with the above-mentioned methodologies.

2 Literature Review and Objective

The mesoscale approach of LBM allows researchers to reduce the computational time by considering the effects of physical behavior of each molecule in a group called lattice. Different lattice arrangements are available for computations in the literature. Most of the authors have used $D1Q3$ for 1-D problems, $D2Q9$ for 2-D problems, and $D3Q15$ for 3-D problems [15] as shown in Fig. 1. In LBM collision of particle is first considered between two lattice nodes [16]. Two types of relaxation time defined as single relaxation time (SRT) model [2, 17] and multi-relaxation time (MRT) model [18] are being used by various researchers. In LBM, the collision process deals with the velocity space and can't be accurately executed with SRT model. In MRT model, problems of occurring in SRT model can be handled easily. As a result, MRT model gives higher stability and accuracy than SRT model [8]. Dazhi Yu et al. [17] have suggested the standard LBM to solve the problem of flow over a circular cylinder. In this paper, we have implemented MRT for single-phase problem simulations. As there is no other fluid is involved, we are restricting from interface tracking/capturing methods. In our multi-phase applications related work, we have used Dual grid-based Level Set Method (DGLSM). At first, it was proposed to enhance the performance and to increase the accuracy of the solver. It is to be noted that while going with DGLSM, computational time will slightly increase [19]. Readers may find our other publications for referring the DGLSM and Lattice Boltzmann-based simulations [20].

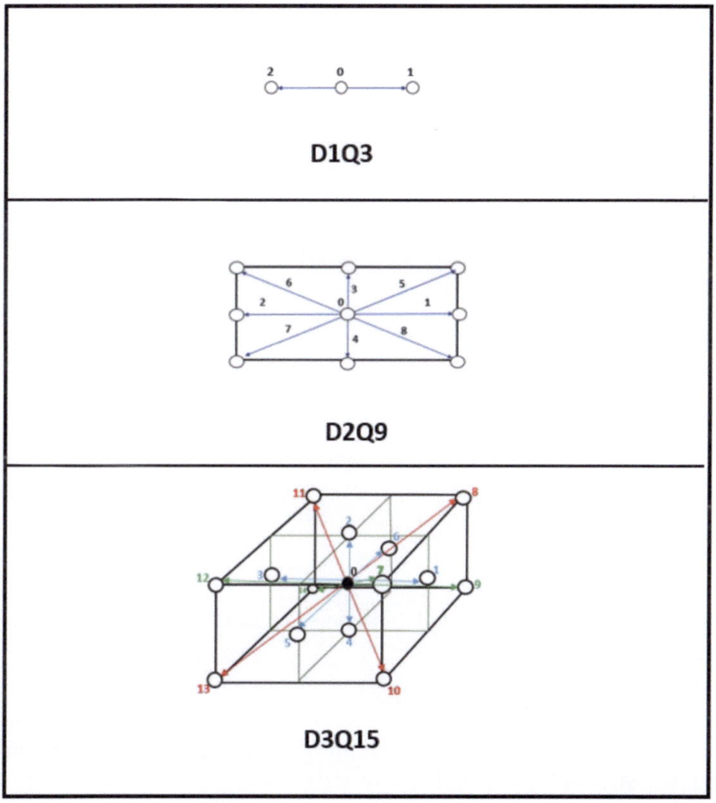

Fig. 1 Different lattice arrangements

3 Mathematical Modeling

$$f(x + c\Delta t, c + F\Delta t, t + t)dxdc - f(x, c, t)dxdc = \Omega(f)dxdcdt, \tag{1}$$

$$f_i(x + c_i\Delta t, t + \Delta t) = f_i(x, t) + \frac{\Delta t}{\tau}\left[f_i^{\text{eq}}(x, t) - f_i(x, t)\right], \tag{2}$$

where

$\Omega = \frac{1}{\tau}(f^{\text{eq}} - f)$.

t = time, i = lattice points, for $D_2Q_9 i = 0, 1, 2 \ldots..8$.

While by considering the effect of the external forces (F), it may be written as,

$$f_i(x + c_i\Delta t, t + \Delta t) = f_i(x, t) + \frac{\Delta t}{\tau}\left[f_i^{\text{eq}}(x, t) - f_i(x, t)\right] + \frac{(c_i - u) \cdot F}{\rho c_s^2} f_i^{\text{eq}} \tag{3}$$

The above equation is known as Lattice Boltzmann Equation (LBE) and is the foundation for solving the multi-phase problems. F is external force which may be modeled by considering various forces like surface tension, pressure, viscous force, buoyant force, etc., depending upon the need of the problem. Macroscopic quantities like density (ρ) and velocity (u) can be calculated as,

$$\rho = \sum_{i}^{n} f_i \tag{4}$$

$$\rho u = \sum_{i}^{n} c_i f_i \tag{5}$$

Further with the help of Taylor series expansion, the equilibrium distribution function for isothermal incompressible flow can be written as,

$$f_i^{\text{eq}} = \rho w_i \left[1 + \frac{c_i \cdot u}{C_s^2} + \frac{1}{2} \frac{\left((c_i.u)^2\right)}{C_s^4} + \frac{1}{2} \frac{u^2}{C_s^2} \right] \tag{6}$$

where w_i = weightage factor and its value are available for D_2Q_9 structure in literature.

Different lattice structures are available in literature for 1-D, 2-D, and 3-D applications. D_2Q_9 is very much popular lattice scheme for 2-D problems in LBM community. At all the lattice nodes the calculation of equilibrium distribution function is done and then after the values will be transferred to next level with the help of advection equation of Level Set function as discussed in next subsection. We may reach to the solution at last by correlating the mesoscale properties to the macroscale ones through Chapman–Enskog (C–E) expansion.

As per the C-E expansion, relaxation time can be correlated with the physical property of the fluid (Kinematic Viscosity) of the fluid as [21]

$$\tau = 3\nu + 0.5 \tag{7}$$

After mapping LB with NS-based equations with C-E expansion, various force terms of F can be calculated.

$$\begin{aligned} F = {} & \text{Surface tension}(F_s) + \text{Viscous force}(F_V) \\ & + \text{Pressure gradient}(F_P) + \text{Buoyancy force}(F_B) \end{aligned} \tag{8}$$

4 Physical Formulation of the Problems

Physical formulation of both the stated problems is shown in this section. They are also nicely represented and explained in details. In addition to this type of boundary conditions applied to achieve results are also discussed. With the current LBM methodology, we got excellent matching of the results in benchmarking problems like Lid driven cavity and 2-D channel flow problems. Readers may refer our previous paper [22].

4.1 Case-I 2-D Channel Flow Over Backward-Facing Step

For validation of the current LBM methodology, at first 2-D channel flow problem is selected first as a benchmarking problem. For the present study, computational domain of length, L and height, H is selected. In the present study, the characteristics length considered is the step height as "$h = H/2$". Entrance height is selected as $L1 = 2h$ and remaining length of the domain is selected as $L2 = 53h$. The mesoscale velocity at inlet is specified as $u_w = 0.1$ m/s and the Reynolds number as is $R_e = 800$. On the eastern wall outlet boundary condition is implemented. In the present simulations, we have kept boundary at sufficiently larger distance in order to avoid formation of the backpressure wave which may deflect the accuracy of results. This kind of pressure waves actually badly affects the solution and leads to poor results. The backward-facing step is chosen in bottom left part as an intricated shape such like that, no fluid flow may happen from it. Actually, a rectangular domain of size $H \times (L2 + L1)$ is selected. Here, equal size grid of $\Delta x = \Delta y = 1$ is generated. Further, the grid in the step of $L1 \times L1$ is blanked (not considered) in the computational domain, and the remaining grid is solved for the governing equation. Finally, a novel approach of blanking is applied in the said problem in which the lattice is formed as it is but certain lattice values are solved in simulations. The wall boundary condition is implemented at the step wall (Fig. 2).

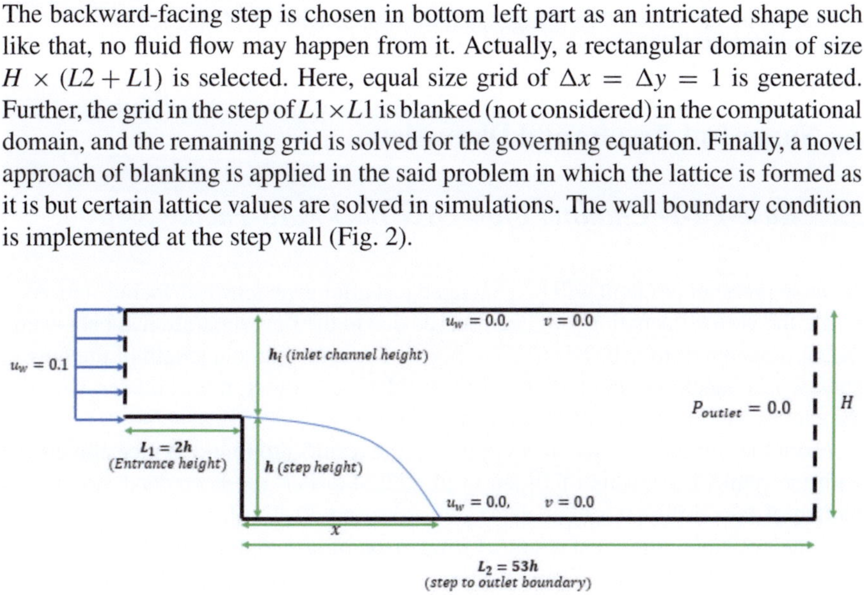

Fig. 2 Geometry of flow over backward-facing step

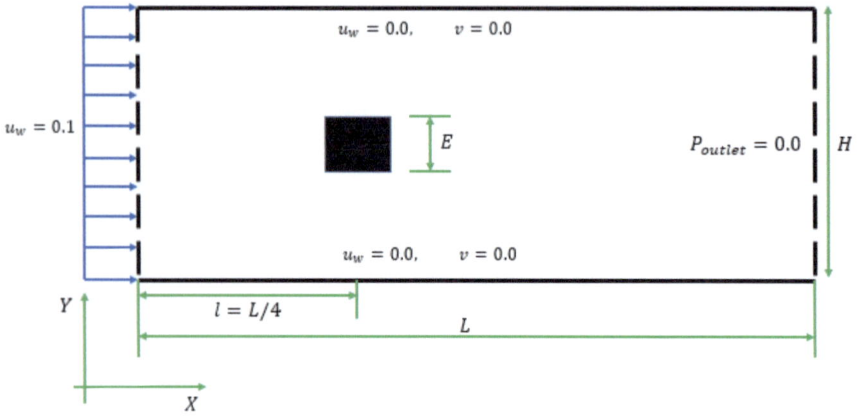

Fig. 3 Geometry of flow over a square cylinder

4.2 Case-II 2-D Flow Over a Square Cylinder

Another problem of external flow over a square cylinder in 2-D channel flow is selected. The total domain length is selected as L and square cylinder is considered at a distance of L/4. The wall boundary condition is applied on the northern and as well as southern walls. On the western wall of the domain, pressure outlet boundary condition is implemented. The ratio of the channel height to diameter of square cylinder (H/E) is selected as 8 suggested by various authors [13] (Fig. 3).

5 Numerical Results and Discussion

5.1 Case-I 2-D Channel Flow Over Backward-Facing Step

Adverse pressure gradient will be generated just after the backward-facing step. As a result, the vortex formation will be observed due to the flow separation at backward-facing step area. It may be observed in Fig. 4. The reattachment length of the flow is chosen as a benchmarking parameter for validation. Further, it is validated with the available literature results of Ercan Ertuk [23].

From the simulations, we may claim that the results are found in nice agreement with the published results of Ercan et al. [23]. Further, the horizontal velocity is plotted at two different values as $x/h = 13.6$ and 30. Further they are validated with the available numerical results. Comparison of horizontal velocity with that of numerical velocity is shown in Figs. 5 and 6, respectively. The RMS error between the present numerical simulation and published results is calculated. The RMS errors for the velocities at $x/h = 13.6$ and 30 are found to be 2% and 5%, respectively.

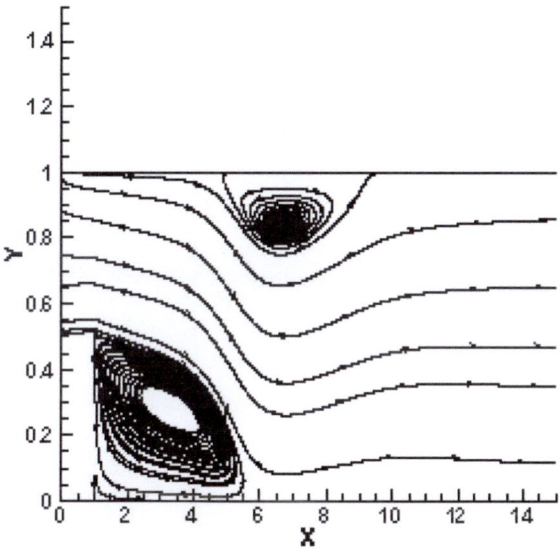

Fig. 4 Streamline patterns in flow over backward-facing step at $R_e = 800$, domain size = 1600×60

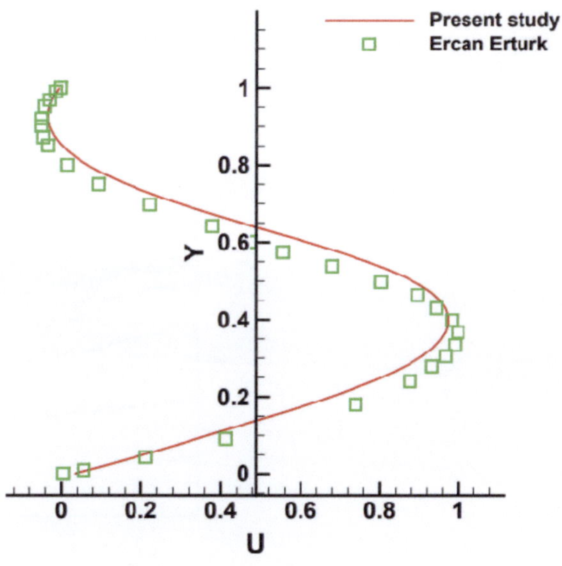

Fig. 5 Horizontal velocity at $^{x}/_{h} = 13.6$

5.2 *Case-II 2-D Flow Over a Square Cylinder*

Velocity streamlines are plotted in 2-D flow over a square cylinder at $R_e = 60\&100$ as shown in Figs. 7 and 8. The vortex shedding can be observed after a square cylinder in both the cases. In Fig. 7, vortex is plotted at $R_e = 60$ and $t = 585$ s while in Fig. 8 vortex is plotted at $R_e = 100$ and $t = 388$ s.

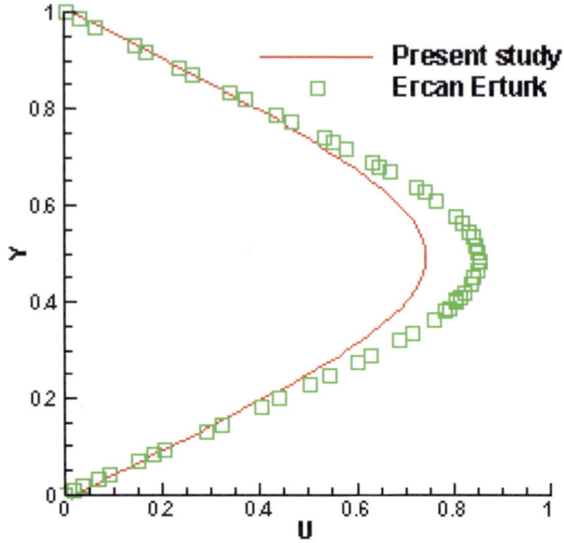

Fig. 6 Horizontal velocity at $x/h = 30$

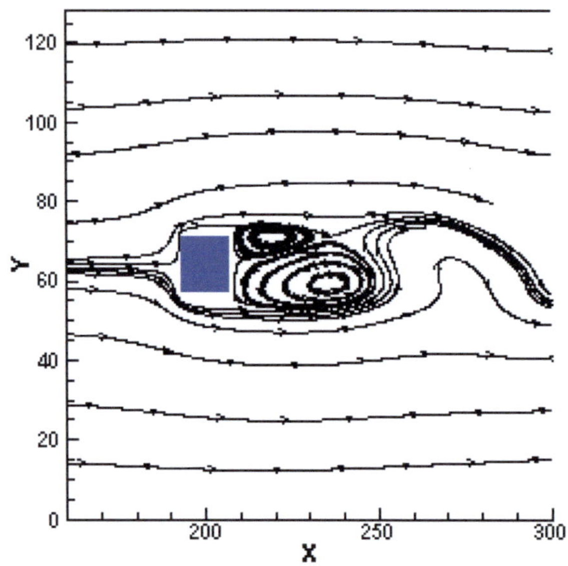

Fig. 7 Velocity streamline patterns for unsteady flow at $R_e = 60$ and time = 585 s

For the problem of flow over a square cylinder, we have selected the critical value of Reynolds number as $R_e = 51$ [10, 11] from the available references. The phenomena become transient just after the critical value of Reynolds number as shown in Fig. 8. The LB-based solver is inherently explicit in nature while NS-based solver is implicite. This is the best advantage of present LB-based solver over Navier-Stokes-based solver.

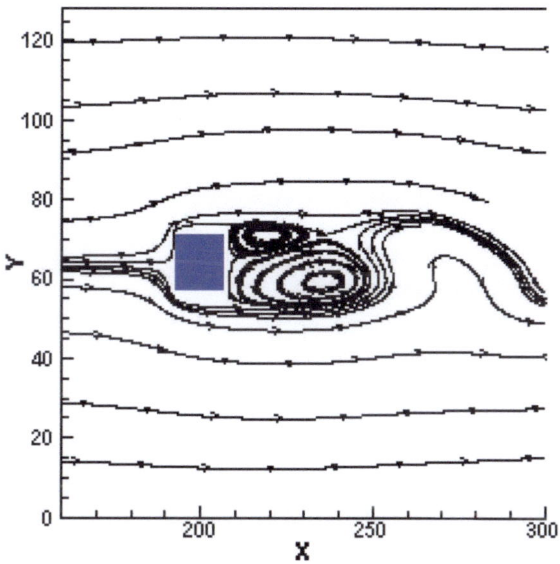

Fig. 8 Velocity streamline patterns for unsteady flow at $R_e = 100$ and time = 388 s

6 Conclusions

To validate our LBM-based solver, we carried out simulations for a single-phase flow benchmarking problems and shown in this present work. Benchmarking problems of 2-D flow in backward-facing step and 2-D flow over a square cylinder are solved. The 2-D flow in backward-facing step problem was solved at $\boldsymbol{R_e} = 800$, domain size = 1600×60 just after the ration of reattachment length to step height. This is validated with available literature results. The results are perfectly matching in the proposed method with available literature results. Adverse pressure gradient near the backward-facing step was observed. Additionally, the horizontal centerline velocity is plotted at $^x/_h$**ratio** = 13.6&30. The proposed LBM results are compared with the results of Ercan Erturk [23]. The simulation results are in good agreement with the published results of Ercan Erturk.

Two different Reynolds number are selected for the problem of flow over a square cylinder. The simulations are carried out at $\boldsymbol{R_e} = 60$ and 100 to measure instabilities of flow at any instant time. Vortices are noted just after the cylinder in both the cases. At higher value of Reynolds Number,$\boldsymbol{R_e} = 100$, the vortices are observed quite earlier (at 388 s) as compared to Reynolds Number, $\boldsymbol{R_e} = 60$ (at 585 s). This is as expected and comparable with analytical as well as published results with NS-based solver. In the present work, LBM-based solver is proposed. Simulation results of proposed LBM-based solver are comparable with published literature results.

Acknowledgements I would like to extend my sincere thanks to Nirma University for helping me in carrying out this work at CFD lab, Mechanical Engineering Department, Institute of Technology, Nirma University, Ahmedabad, Gujarat, India.

Nomenclature

f	Boltzmann distribution function
Ω	Collision operator
x	Position vector
c	Velocity vector
F	External force
C_s	Sound velocity
P	Mean density of the fluid
τ	Relaxation time
ν	Kinematic viscosity

References

1. Mohammad AA (2020) Lattice Boltzmann Method
2. Nemati H, Farhadi M, Sedighi K, Fattahi E, Darzi AAR (2010) Lattice Boltzmann simulation of nanofluid in lid-driven cavity. Int Commun Heat Mass Transfer 37(10):1528–1534. https://doi.org/10.1016/j.icheatmasstransfer.2010.08.004
3. Aidun CK, Clausen JR (2010) Lattice-boltzmann method for complex flows. Annu Rev Fluid Mech 42:439–472. https://doi.org/10.1146/annurev-fluid-121108-145519
4. Ladd AJC (2015) Lattice-Boltzmann methods for suspensions of solid particles. Mol Phys Appear Mol Phys 1–11. https://doi.org/10.1080/0040516YYxxxxxxxx
5. Mishra SC, Poonia H, Vernekar RR, Das AK (2014) Lattice boltzmann method applied to radiative transport analysis in a planar participating medium. In: Heat transfer engineering, pp 1267–1278. https://doi.org/10.1080/01457632.2013.876806
6. Derksen JJ, The Lattice-Boltzmann method for multiphase fluid flow simulations and Euler-Lagrange large-eddy simulations
7. Biswas S, Sharma P, Mondal B, Biswas G (2015) Analysis of mixed convective heat transfer in a ribbed channel using the lattice Boltzmann method. Numeri Heat Transf A Appl 68(1):75–98. https://doi.org/10.1080/10407782.2014.965095
8. Lin LS, Chang HW, Lin CA (2013) Multi relaxation time lattice Boltzmann simulations of transition in deep 2D lid driven cavity using GPU. Comput Fluids 80(1):381–387. https://doi.org/10.1016/j.compfluid.2012.01.018
9. Kefayati GHR, Huilgol RR (2016) Lattice Boltzmann method for simulation of mixed convection of a Bingham fluid in a lid-driven cavity. Int J Heat Mass Transf 103:725–743. https://doi.org/10.1016/j.ijheatmasstransfer.2016.07.102
10. Sewatkar CM, Patel R, Sharma A, Agrawal A (2012) Flow around six in-line square cylinders. J Fluid Mech 710:195–233. https://doi.org/10.1017/jfm.2012.359
11. Sehgal A (2014) Flow past a cylinder using Lattice Boltzmann method
12. Sewatkar CM, Sharma A, Agrawal A (2011) Simulation of flow across a row of transversely oscillating square cylinders. J Fluid Mech 680:361–397. Cambridge University Press. https://doi.org/10.1017/jfm.2011.167
13. Wang H, Zhao H, Guo Z, He Y, Zheng C (2013) Lattice Boltzmann method for simulations of gas-particle flows over a backward-facing step. J Comput Phys 239:57–71. https://doi.org/10.1016/j.jcp.2012.12.032
14. Huang HB, Lu XY, Sukop MC (2011) Numerical study of lattice Boltzmann methods for a convection-diffusion equation coupled with Navier-Stokes equations. J Phys A Math Theor 44(5). https://doi.org/10.1088/1751-8113/44/5/055001

15. Safi MA, Prasianakis N, Turek S (2017) Benchmark computations for 3D two-phase flows: a coupled lattice Boltzmann-level set study. Comput Math Appl 73(3):520–536. https://doi.org/10.1016/j.camwa.2016.12.014
16. Redapangu PR, Chandra Sahu K, Three-dimensional LBM simulations of buoyancy-driven flow using graphics processing units
17. Yu D, Mei R, Luo LS, Shyy W (2003) Viscous flow computations with the method of lattice Boltzmann equation. Prog Aerosp Sci 39(5):329–367. https://doi.org/10.1016/S0376-0421(03)00003-4
18. Safi MA, Turek S (2016) GPGPU-based rising bubble simulations using a MRT lattice Boltzmann method coupled with level set interface capturing. Comput Fluids 124:170–184. https://doi.org/10.1016/j.compfluid.2015.06.001
19. Gada VH, Sharma A (2011) On a novel dual-grid level-set method for two-phase flow simulation. Numer Heat Transf Part B Fundam 59(1):26–57. https://doi.org/10.1080/10407790.2011.540956
20. Lakdawala A, Gada VH, Sharma A (2015) On dual-grid level-set method for computational-electro-multifluid-dynamics simulation. Numer Heat Transf Part B Fundam 67(2):161–185. https://doi.org/10.1080/10407790.2014.949582
21. Li J (2015) Appendix: Chapman-Enskog expansion in the Lattice Boltzmann Method. [Online]. Available: http://arxiv.org/abs/1512.02599
22. Raval T, Lakdawala A (2023) Development and validation of the Lattice Boltzmann method based single-phase flow solver. In: AIP conference proceedings, vol 2872,030007:1–6. https://doi.org/10.1063/5.0164997
23. Erturk E (2008) Numerical solutions of 2-D steady incompressible flow over a backward-facing step, Part I: High Reynolds number solutions. Comput Fluids 37(6):633–655. https://doi.org/10.1016/j.compfluid.2007.09.003

Scaling Law Analysis for Transient Fluid Flow and Heat Transfer Characteristics of Packed Bed with Mono-Sized Pebbles

Pratyush Kumar, Sandip K. Saha, and Atul Sharma

Abstract The present work focuses on the scale law analysis of the thermo-hydraulic behaviour of the packed bed filled with mono-sized pebbles considering the porous media approach. We derived dimensionless time scales and flow and thermal similarity parameters by employing the non-dimensional form of the governing equations for mass, momentum, and local thermal non-equilibrium (LTNE) energy. This work investigates the consistency of the scaling law between the model and prototypes for various mass flow rates and heat generation rates. For the LTNE energy equation, despite the longer time required to reach a steady state due to volumetric heat generation rates, the temperature equation consistently maintains its behaviour over the transient time steps, thus confirming the applicability of the developed scaling law. Furthermore, the dimensionless pressure drops $\left(P^* = \frac{P}{\rho V^2}\right)$ and non-dimensional outlet temperature $\left(T^*_{out} = \frac{T_{out}-T_{\infty}}{T_{in}-T_{\infty}}\right)$ show sustained consistency as the solution approaches steady state. These findings enhance our understanding of transient behaviour in packed bed systems and have broader implications for the design and optimization of porous media systems in engineering applications.

Keywords Transient · Scaling law · LTNE · Porous medium · Convective losses · OpenFOAM

1 Introduction

Packed bed has wide industrial applications, such as in the chemical, metallurgical, and process industries, for various applications in absorption, drying, catalytic reactions and fluidized beds. The advantages of using a packed bed are that it has high-density packing and enhances the heat transfer behaviour. In the present study, we have chosen an application of the packed bed for the tritium breeder blanket concept

P. Kumar (✉) · S. K. Saha · A. Sharma
Department of Mechanical Engineering, IIT Bombay, Mumbai, India
e-mail: pratyushiitb94@gmail.com

A. Kumar R. et al. (eds.), *Proceedings of Fluid Mechanics and Fluid Power (FMFP) 2023, Vol. 6*, Lecture Notes in Mechanical Engineering,
https://doi.org/10.1007/978-981-96-9600-0_47

of the International Thermonuclear Experimental Reactor (ITER) project. Two main objectives of breeder blanket concepts are an adequate supply of tritium and the extraction of high-quality heat [1].

Several numerical and experimental studies have been performed to understand the flow behaviour and heat characteristics of the breeder blanket concepts using a porous medium approach at different scales [2–8]. Bhuyan et al. [2] determined MHD pressure drop in a scaled LLCB TBM. Our previous work involved parametric analysis on a scaled breeder blanket model [3, 4]. Akhalesh et al. [5] studied LLCB TBM flow using CFD-DEM on a 1:20 scaled model. Starke et al. [6] conducted magnetohydrodynamic (MHD) experiments on a scaled HCLL test blanket module mock-up. Satyamurty et al. [7] tested a NaK section of the LLCB TBM and compared results with simulations. Swami et al. [8] validated the TBM shield module's neutronic design with small-scale experiments. Therefore, it is found from literature survey that the behaviour of the breeder blanket has been studied at different scales.

It is important to understand the flow dynamics and heat transfer of the breeder blanket at larger scales before installation. However, conducting large-scale tritium breeder module (TBM) studies demands substantial resources, including computation and infrastructure costs. In such a scenario, scaling law is used to correlate the findings of small-scale with the corresponding large-scale. Bejan [9] proposed a scale analysis approach for porous media transport phenomena. Hariti et al. [10] explored scaling laws for permeability and porosity in porous phase change materials. Fallico et al. [11] investigated scaling laws for hydraulic conductivity and effective porosity in various aquifers, establishing relationships based on slug test injection volumes. However, the analysis of scaling laws for fluid flow and heat transfer behaviour in packed bed porous media models remains limited in the published literature.

The present work on scale law analysis, we conducted a comprehensive scaling analysis of a packed bed filled with mono-sized spherical particles to determine dimensionless flow and thermal similarity parameters linking small-scale models with prototypes, considering both heterogeneous and homogeneous porosity, primarily focusing on assessing the applicability of the scaling law in transient simulations. In this study, considering transient terms, we have developed similarity parameters based on the non-dimensionalized Navier–Stokes equation with Darcy-Forchheimer terms and the local thermal non-equilibrium (LTNE) model. We have also performed parametric studies investigating scaling effects on the flow field and temperature profile for various heat generation and mass flow rates in transient conditions.

2 Problem Definition

In this study, the base model is the cylindrical packed bed having diameter (D) = 161 mm and height (H) = 1490 mm, as shown in Fig. 1a, randomly filled with mono-sized lithium titanate (Li_2TiO_3) pebbles having particle diameter (d = 1 mm). The mid-plane of the packed bed is shown in Fig. 1b. Low-pressure helium flows through

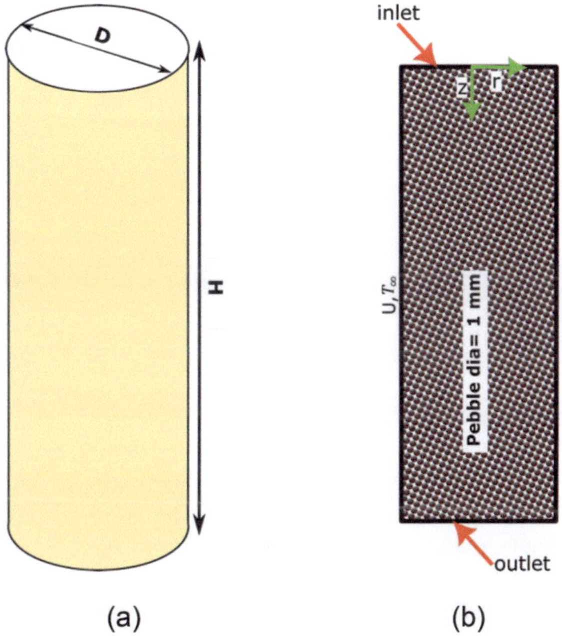

Fig. 1 Schematic of three-dimensioned **a** cylindrical packed bed; **b** the mid-plane filled with pebbles of the packed bed (all dimensions are in mm)

the packed bed to purge tritium from the breeder. Liquid PbLi acts as a coolant around the packed bed, maintaining an ambient temperature of 613 K. To use the Zehner–Bauer–Schlünder (ZBS) model for radiation [3, 4], the thermophysical properties of helium and Li_2TiO_3 are considered from [5].

3 Scaling Law

In present work, we have considered transient, incompressible and viscous flow with Darcy-Forchheimer terms. Assuming the packed bed as porous media, with superficial velocity, V_g at the inlet, and the velocity components u, v and w in the x, y and z directions, respectively, the governing equations can be written as follows:

Continuity equation:

$$\nabla \cdot \rho v = 0, \tag{1}$$

where $\boldsymbol{v}$ is the velocity vector.

Momentum equation:

$$\frac{\partial(\boldsymbol{v})}{\partial t} + (v.\nabla)v = \frac{-\nabla P}{\rho} + \vartheta \nabla^2 v - \left(\frac{\vartheta}{K} + \frac{C_F}{\sqrt{K}}|v|\right).v, \tag{2}$$

where ρ, v, P, ϑ, K and C_F are fluid density, velocity vector, pressure, kinematic viscosity, permeability, and the form-drag coefficient of the porous media, respectively.

LTNE model of the energy equation:

Fluid:

$$\emptyset(\rho c)_f \frac{\partial T}{\partial t} + (\rho c)_f \boldsymbol{v}.\nabla T = \emptyset \nabla.\left(k_f \nabla T\right) + h\left(T_s - T_g\right) + S_f^h. \tag{3}$$

Solid:

$$(1 - \emptyset)(\rho c)_s \frac{\partial T_s}{\partial t} = (1 - \emptyset)\nabla.(k_s \nabla T_s) + h\left(T_g - T_s\right) + S_s^h, \tag{4}$$

where ϕ, ρ, $\boldsymbol{v}$, T_g, T_s, k_g, k_s, h, $c_{p,g}$ and S_s^h are porosity, fluid density, velocity vector, fluid temperature, solid temperature, the thermal conductivity of the fluid, the thermal conductivity of solid, the specific heat of fluid, interfacial heat transfer coefficient and heat generation source term, respectively. The s and f represent solid and fluid, respectively. The Zehner–Bauer–Schlünder (ZBS) model described in our previous works [3] is used for the radiative heat transfer, and the emissivity of the pebbles and the inner walls is based on the study by Sharma et al. [5].

LTE model of the energy equation

Assuming fast heat exchange between solid and gaseous phases and a slow rate of temperature change, both phases' (solid and fluid) temperatures remain nearly constant during the heat transfer, allowing for a model based on local thermal equilibrium (LTE) for heat transfer between them. With these assumptions, we can consider $T_f = T_s = T$. Equations (3) and (4) can be rewritten in combined form as,

$$(\rho c)_m \frac{\partial T}{\partial t} + (\rho c)_f \boldsymbol{v}.\nabla T = \nabla.(k_e \nabla T) + S_e^h, \tag{5}$$

where

$$(\rho c)_m = (1 - \phi)(\rho c)_s + \phi(\rho c)_f, \tag{6}$$

$$k_e = (1 - \phi)k_s + \phi k_f, \tag{7}$$

$$S_e^h = (1 - \phi)S_s^h + \phi S_f^h, \tag{8}$$

where k_f and S^h are the thermal conductivity and volumetric heat generation rate. s *and* f are subscripts for solid and fluid, respectively, and e denotes the equivalent.

3.1 Non-dimensional Parameters

The characteristic length of the packed bed filled with pebbles using the porous medium approach is the equivalent hydraulic diameter (d_h),

$$d_h = d\frac{\emptyset}{1-\emptyset}. \tag{9}$$

Another important parameter is the characteristic velocity defined as,

$$V = \frac{V_g}{\emptyset}, \tag{10}$$

where V_g is the superficial velocity. Using the characteristics parameters, i.e. d_h and V; the characteristics Reynolds number (Re_h) can be defined as,

$$\text{Re}_h = \frac{\rho V d_h}{\mu}. \tag{11}$$

Also, based on the characteristic diameter (d_h) and characteristics velocity (V), the characteristics time scale can be defined as,

$$\tau = \frac{d_h}{V}. \tag{12}$$

Table 1 summarizes the non-dimensional parameters used to non-dimensional the above governing equations (Eqs. 1, 2 and 5).

Table 1 Non-dimensional parameters

Dimensional parameter	Dimensionless parameter	Expression
Time (t)	Characteristics time scale (τ)	$t^* = \frac{t}{\tau}$
Diameter (l)	Characteristic diameter (d_h)	$D^* = \frac{D}{d_h}$
Height (H)	Characteristic diameter (d_h)	$H^* = \frac{H}{d_h}$
Velocity (u, v, w)	Interstitial velocity (V)	$u^* = \frac{u}{V}$, $v^* = \frac{v}{V}$, $w^* = \frac{w}{V}$
Pressure (P)	Density (ρ), interstitial velocity (V)	$P^* = \frac{P}{\rho V^2}$
Permeability (K)	Interstitial velocity (V), kinematic viscosity (ϑ), characteristic diameter (d_h)	$N_K = \frac{KV}{\vartheta d_h}$
Heat generation rate (S_e^h)	Characteristic diameter (d_h), density (ρ), interstitial velocity (V), specific heat capacity (c), temperature difference ($T_\infty - T_{in}$)	$S^{h*} = \frac{S_e^h d_h}{(\rho c)_g V (T_\infty - T_{in})}$

Using the above listed non-dimensional parameters (Table 1), the dimensionless form of the governing equations can be written as,

$$\frac{\partial u^*}{\partial x^*} + \frac{\partial v^*}{\partial y^*} + \frac{\partial v^*}{\partial z^*} = 0, \tag{13}$$

where $x^* = \frac{x}{d_h}$; $y^* = \frac{y}{d_h}$; $z^* = \frac{z}{d_h}$; $u^* = \frac{u}{V}$; $v^* = \frac{v}{V}$; $w^* = \frac{w}{V}$.

X-direction momentum equation:

$$\frac{\partial u^*}{\partial t^*} + u^* \frac{\partial u^*}{\partial x^*} + v^* \frac{\partial u^*}{\partial y^*} + w^* \frac{\partial u^*}{\partial z^*} = -\frac{\partial P^*}{\partial x^*} + \frac{1}{\mathrm{Re}_h}\{\nabla^2 u^*\} - \frac{1}{N_K}\{u^*\} - \frac{1}{N_D}\{u^{*2}\}. \tag{14}$$

For brevity, only x-momentum is derived in non-dimensional form.

Energy equation

The only difference between the LTNE and LTE models is the assumption of local thermal equilibrium. Therefore, the non-dimensional variables obtained by non-dimensionalizing the LTE equations will be consistent with those obtained by non-dimensionalizing the LTNE equations. Therefore, in the present work, we used the LTE form of the equation to derive the scaling parameters for transient heat analysis.

LTE energy equation

$$\frac{\partial T^*}{\partial t^*} + \frac{(\rho c)_f}{(\rho c)_m}\left[u^* \frac{\partial T^*}{\partial x^*} + v^* \frac{\partial T^*}{\partial y^*} + w^* \frac{\partial T^*}{\partial z^*}\right] = \frac{1}{\mathrm{Re}_h \mathrm{Pr}}\{\nabla^2 T^*\} + S^{h*}, \tag{15}$$

where $T^* = \frac{T - T_{\mathrm{in}}}{T_\infty - T_{\mathrm{in}}}$; $\mathrm{Pr} = \frac{\mu c_{pf}}{k}$; $S^{h*} = \frac{S_e^h d_h}{(\rho c)_f V (T_\infty - T_{\mathrm{in}})}$ with T_{in} is the inlet temperature, and T_∞ is the ambient temperature.

In the above equation, $\frac{(\rho c)_f}{(\rho c)_m}$ remains constant during the simulation as we have assumed constant properties and incompressible flow.

3.2 Similarity Parameters

To ensure similarity between the model and the prototype of the cylindrical packed bed filled with mono-sized pebbles, following conditions from Eqs. (13–15) must be satisfied:

$$(t^*)_m = (t^*)_p, \tag{16}$$

$$(\mathrm{Re}_h)_m = (\mathrm{Re}_h)_p, \tag{17}$$

$$(N_K)_m = (N_K)_p, \tag{18}$$

$$(N_D)_m = (N_D)_p, \tag{19}$$

$$(Pr)_m = (Pr)_p, \tag{20}$$

$$\left(S^{h*}\right)_m = \left(S^{h*}\right)_p. \tag{21}$$

Based on d_h, scaling factor (S) can be defined as,

$$\frac{(d_h)_p}{(d_h)_m} = S. \tag{22}$$

Considering Eq. (16), one can obtain the following relation:

$$\frac{(t)_p}{(t)_m} = \frac{(d_h)_p}{(d_h)_m}.\frac{(V)_m}{(V)_p}. \tag{23}$$

From Eq. (17), we will get,

$$\frac{V_m}{V_p} = \frac{\dot{m}_p}{\dot{m}_m}\frac{\O_p}{\O_m} = S. \tag{24}$$

From dimensionless permeability number (N_K), Eq. (18) results in the following expression:

$$\frac{K_p}{K_m} = S^2. \tag{25}$$

For Forchheimer parameter (N_D), we can express Eq. (19) as,

$$(C_F)_m = (C_F)_m. \tag{26}$$

One can get from Eq. (21),

$$\frac{\left(S^{h*}\right)_p}{\left(S^{h*}\right)_m} = \frac{1}{S^2}. \tag{27}$$

Similarly, for particle diameter (d), one can write as,

$$\frac{\left(d\frac{\emptyset}{1-\emptyset}\right)_p}{\left(d\frac{\emptyset}{1-\emptyset}\right)_m} = S. \tag{28}$$

Following Eqs. (22–27), the complete expression for scaling law can be written as,

$$\frac{(t)_p^{1/2}}{(t)_m^{1/2}} = \frac{d_p}{d_m} = \frac{D_p}{D_m} = \frac{H_p}{H_m} = \frac{\dot{m}_p}{\dot{m}_m} = \frac{V_m}{V_p} = \frac{(K)_p^{1/2}}{(K)_m^{1/2}} = \frac{\left(S_s^h\right)_m^{1/2}}{\left(S_s^h\right)_p^{1/2}} = S. \tag{29}$$

4 Numerical Modelling

The C++-based open-source CFD software, i.e. OpenFOAM, is used to solve the governing equations (Eqs. 1–4). The solver is implemented by modifying the pimple Foam, a steady or transient solver based on the PISO-SIMPLE (PIMPLE) algorithm [12]. In the present work, we have adopted the methodology and the algorithm for the LTNE model of energy equation from our previous work for use in the present work [3, 4]. The first-order implicit method is used to discretize the transient term, while central difference approximations are used for advection–diffusion terms. The convergence criteria for velocity, Pressure and Temperature Field is taken as 10^{-6}.

4.1 Simulation Setup

The asymmetric nature of the flow allows the use of a smaller number of cells by applying the wedge type of boundary conditions. Figure 2a shows the asymmetric (wedge) geometric for the simulation setup specified as a wedge with an angle of 5°. Further, Fig. 2b shows the meshing of the specified geometry having 1 cell thick along the symmetry plane.

4.2 Boundary Conditions

No slip condition at the wall, fixed mass flow rates at the inlet and pressure outlet conditions are applied for velocity and pressure field. The inlet temperature is kept at 573 K, while at walls, convection heat transfer is considered with a free stream temperature of $T_\infty = 613$ K. To maintain the consistency between model and prototype, Biot number $\left(\text{Bi} = \frac{h_\infty d_h}{k_o^s}\right)$ should be equal, i.e.

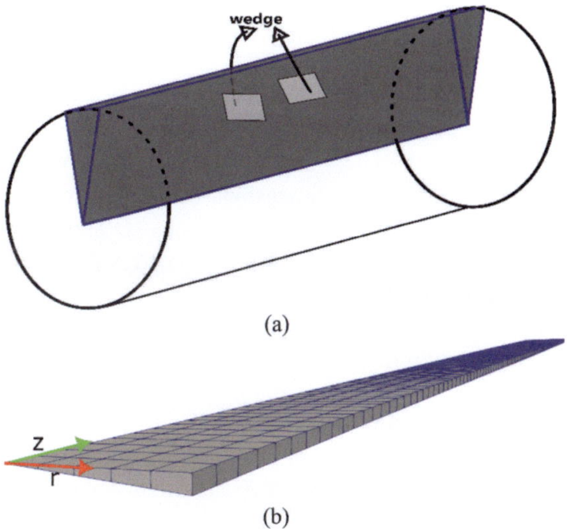

Fig. 2 Schematic of simulation setup for scale S_1; **a** the asymmetric (wedge) domain considered for the present simulation; **b** meshing of the wedge domain

$$(\mathrm{Bi})_m = (\mathrm{Bi})_p. \tag{30}$$

Hence, Eq. (30) is used to determine the scale of the value of the heat transfer coefficient, h, at the wall.

4.3 Grid Independence Test

Grid independence tests are conducted for model S_1 and prototypes, $S_{1/2}$ and $S_{1/3}$, to ensure that the results are independent of mesh sizes. Table 2 shows the grid independence test for the model S_1 for pressure drop and outlet temperature. A mesh with 4768 elements for model S_1 for further simulations is taken. Meshes with 1192 and 495 elements, respectively, are used for prototypes $S_{1/2}$ and $S_{1/3}$ while maintaining the same element size.

Table 2 Grid independence study

Number of elements	T_{out}(K)	Δ (%)	ΔP (kPa)	Δ (%)
1192	592.98	–	3.99	–
4768	592.64	0.057	4.27	7.15
13,419	592.54	0.017	4.47	4.68

5 Results and Discussion

The present study focuses on understanding the transient flow and heat transfer behaviour of cylindrical packed beds when they are scaled. Two scaled-down prototypes at $S_{1/2}$ ($S = 1/2$) and $S_{1/3}$ ($S = 1/3$) are numerically investigated other than the model (S_1) having dimensions 162 (Diameter) × 1490 (Height) mm. Following Eq. 29, the cylindrical bed dimensions (D and H) are calculated for S_1, $S_{1/2}$ and $S_{1/3}$. Using Eq. (29), the particle diameters (d) are considered 1 mm, 0.5 mm and 0.33 mm, respectively, for S_1, $S_{1/2}$ and $S_{1/3}$. The constant porosity is assumed in all the cases.

Considering the model S_1, to study the effect of mass flow rates, different mass flow rates ($\dot{m} \times 10^{-6}$) of 0.56, 1.125, 2.5 kg/s with heat generation rates 320 W are considered. Similarly, different heat generation rates of 320, 560 and 970 W are taken to study the effect of heat generation rates. Note that the different parameters specified above for the model S_1 are calculated using Eq. (29) for the prototypes $S_{1/2}$ and $S_{1/3}$.

For transient analysis, Eq. (16) should be applicable while developing the scale law. As we can also see from Eq. (29) that the correlation for the time scale for model and prototype can be written as,

$$(t)_p^{1/2} = S.(t)_p^{1/2}. \tag{31}$$

Therefore, Eq. (26) is applied for the calculation of time-step (t) of the prototypes $S_{1/2}$ and $S_{1/3}$ based on the time-step (t) of the model, S_1.

As the flow is fully developed, the transient state in pressure and velocity reaches a steady state sharply at $t^* = 0.11$ for S_1, $S_{1/2}$ and $S_{1/3}$, as shown in Fig. 3. This suggests that the scaling law is consistent under transient conditions for velocity and pressure. Furthermore, Fig. 4 shows the dimensionless pressure drop $\left(P^* = \frac{P}{\rho V^2}\right)$ for various Re_h with $S^{h*} = 0.004$. With the increase in Reynolds number, the dimensionless pressure drop decreases. It demonstrates that when the solution reaches a steady state, the dimensionless pressure drop remains consistent for both the model and the prototypes at different characteristic Reynolds numbers (Re_h).

Figure 5 displays the variation of the non-dimensional outlet temperature $\left(T^*_{\text{out}} = \frac{T_{\text{out}} - T_\infty}{T_{\text{in}} - T_\infty}\right)$. It is observed that the temperature equation takes a longer time to reach a steady state due to the volumetric heat generation source. It reaches the steady-state condition at t^* ~ 2000. Nevertheless, the consistency in the plot demonstrates the applicability of the developed scaling law. The temperature consistency is proved from the plot of T^*_{out} versus different heat generation rates ($S^{h*} =$ 0.004, 0.0068, 0.011) as shown in Fig. 6.

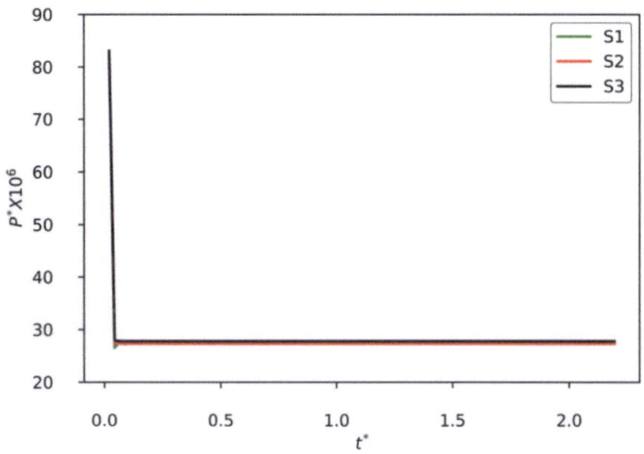

Fig. 3 Variation of non-dimensional pressure drop (P^*) with non-dimensional time-step (t^*) for S_1, $S_{1/2}$ and $S_{1/3}$

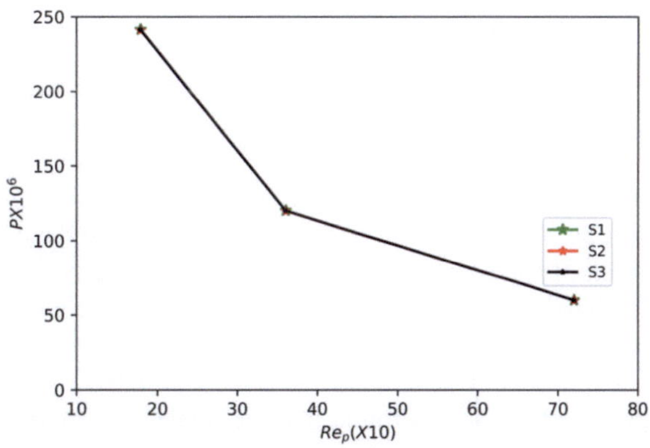

Fig. 4 Variation of non-dimensional pressure drop (P^*) with increasing Re_h for S_1, $S_{1/2}$ and $S_{1/3}$ for cylindrical packed bed filled with mono-sized pebbles

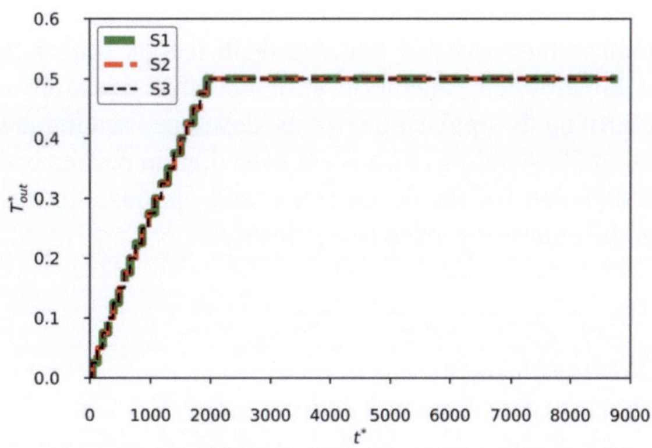

Fig. 5 Variation of non-dimensional outlet temperature (T^*_{out} with non-dimensional time-step (t^*) for S_1, $S_{1/2}$ and $S_{1/3}$

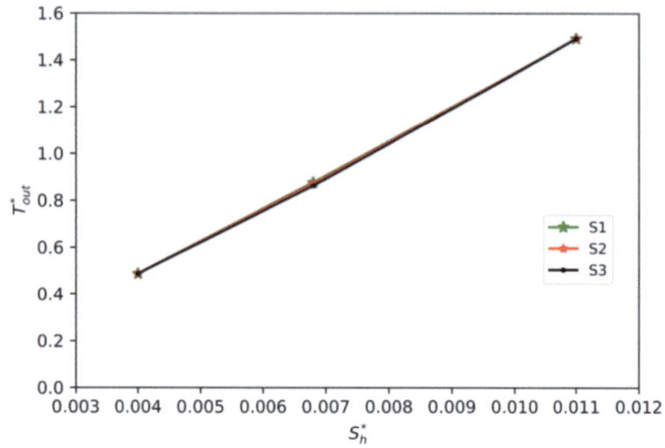

Fig. 6 Variation of non-dimensional outlet temperature (T^*_{out}) with increasing S^{h*} for S_1, $S_{1/2}$ and $S_{1/3}$ for cylindrical packed bed filled with mono-sized pebbles

6 Conclusion

In this study, a comprehensive investigation of transient scaling law using the porous medium approach for fluid flow and heat transfer of packed beds filled with mono-sized pebbles is performed. Therefore, our work investigates the applicability of the scaling law in transient simulations. We derived the dimensionless flow and thermal similarity parameters from the non-dimensional Navier–Stokes equation with Darcy-Forchheimer terms and the local thermal non-equilibrium (LTNE) form of energy considering transient terms. A range of parameters, including different mass flow rates, and heat generation rates are evaluated for the model (S_1) and prototypes ($S_{1/2}$ and $S_{1/3}$). It is found that the transient state in pressure and velocity reached a steady state remarkably quickly, maintaining the consistency of the scaling law over a non-dimensional time scale (t^*). The analysis of the non-dimensional outlet temperature revealed that, although it took longer to reach a steady state due to volumetric heat generation, the temperature equation exhibited sustained consistency, affirming the applicability of the developed scaling law. This research contributes to our understanding of transient behaviour in packed bed systems and provides a solid foundation for the development and optimization of scaled prototypes in various applications of packed bed industries.

Nomenclature

a_{fs}	Specific surface area [m^{-1}]
Bi	Biot number –
C_F	Form-drag coefficient –
c_p	Specific heat [J/kg K]
d	Particle diameter [m]
d_h	Equivalent hydraulic diameter [m]
H	Height [m]

h	Interfacial heat transfer coefficient [W/m^3 K]
h_∞	Overall heat transfer coefficient [W/m^3 K]
$\dot{m}$	Mass flow rate [kg/s]
N_K	Non-dimensional permeability –
N_D	Forchheimer parameter –
P	Pressure [Pa]
$P*$	Non-dimensional pressure $\left(\frac{P}{\rho_g V^2}\right)$ –
Pr	Prandtl number –
Re_h	Equivalent hydraulic diameter –
S	Scale factor –
S_s^h	Heat generation source term –
S^{h*}	Non-dimensional heat generation rate –
T	Temperature [K]
T^*	Non-dimensional temperature [K]
T_{in}	Fluid inlet temperature [K]
T_{out}	Fluid outlet temperature [K]
T_∞	Free stream temperature [K]
u^*	Non-dimensional x component of velocity –
V	Interstitial velocity [m/s]
V_g	Superficial velocity [m/s]

Greek Symbols

ρ	Density [kg/m^3]
μ	Dynamic viscosity [kg/m s]
ϕ	Porosity

Abbreviation

BB	Breeder blanket
CFD	Computational fluid dynamics
Li_2TiO_3	Lithium titanate
LLCB	Lead–lithium cooled ceramic breeder
LTNE	Local thermal non-equilibrium
LTE	Local thermal equilibrium
MHD	Magneto hydrodynamic
TBM	Test blanket modules

References

1. Giancarli LM, Abdou M, Campbell DJ, Chuyanov VA, Ahn MY, Enoeda M, Pan C, Poitevin Y, Rajendra Kumar E, Ricapito I, Strebkov Y, Suzuki S, Wong PC, Zmitko M (2012) Overview of ITER TBM program. Fusion Eng Des 87:1161–1165. https://doi.org/10.1016/j.fusengdes.2011.11.005
2. Bhuyan PJ, Goswami KS (2010) Development of a potential based code for MHD analysis of LLCB TBM. Fusion Eng Des 85:138–145. https://doi.org/10.1016/j.fusengdes.2009.08.008
3. Kumar P, Saha SK, Sharma A (2022) Entropy generation minimisation-based heat transfer and mass transfer study of rectangular packed-bed filled with spherical particles. Fusion Eng Des 184:113295. https://doi.org/10.1016/j.fusengdes.2022.113295
4. Kumar P, Thakur A, Saha SK, Sharma A, Sharma D, Chaudhuri P (2021) CFD investigation of helium gas flow in sphere packed (Pebbled Bed) in a rectangular canister using OpenFOAM. Fusion Eng Des 172:112858
5. Sharma A, Thakur A, Saha SK, Sharma A, Sharma D, Chaudhuri P (2020) Thermal hydraulic characteristics of purge gas in rectangular packed pebble bed of a fusion reactor using DEM-CFD and porous-medium analyses. Fusion Eng Des 160:111848
6. Starke K, Bühler L, Horanyi S (2009) Experimental MHD-flow analyses in a mock-up of a test blanket module for ITER. Fusion Eng Des 84:1794–1798. https://doi.org/10.1016/j.fusengdes.2009.01.047
7. Satyamurthy P, Swain PK, Tiwari V, Kirillov IR, Obukhov DM, Pertsev DA (2015) Experiments and numerical MHD analysis of LLCB TBM Test-section with NaK at 1 T magnetic field. Fusion Eng Des 91:44–51
8. Swami HL, Abhangi M, Sharma S, Tiwari S, Mistry AN, Vasava V, Mehta V, Vala S, Danani C, Chaudhari V (2019) A neutronic experiment to support the design of an Indian TBM shield module for ITER. Plasma Sci Technol 21:065601
9. Bejan A (2004) Convection heat transfer. Wiley, Third
10. Hariti Y, Hader A, Faraji H, Boughaleb Y (2021) Scaling law of permeability and porosity for fluid transport phenomenon in porous media. J Appl Comput Mech 7:84–92
11. Fallico C, Lauria A, Aristodemo F (2020) Porous medium typology influence on the scaling law of confined aquifer characteristic parameters. Water 12(4):1166
12. https://doc.cfd.direct/notes/cfd-general-principles/the-pimple-algorithm. Accessed 30 Aug 2023

Performance Enhancement of NACA 0015 Wing Using Rotating Cylinder Vortex Generator

S. N. Jadhav, Akshay Gupta, Akshaat Singh, Harshvardhan Saxena, S. V. Jain, A. M. Achari, T. N. Raval, and Ath Singhal

Abstract When flow over the upper surface of an airfoil begins to separate, it creates turbulence in which the air flow begins to produce vortices. It results in a decrease in lift and increase in drag. The flow separation can be delayed by imparting momentum to the fluid by providing a vortex generator (VG). Numerical investigations were done on NACA 0015 airfoil without and with VG in the form of a rotating cylinder in the present study. As a VG, ϕ15 mm cylinder with a length of 100 mm was provided at a maximum thickness section of the airfoil. The steady state simulations were carried out in ANSYS Fluent software in the wide range of Reynolds numbers (Re), i.e., 0.29–2.58 lacs at seven different angle of attacks (α), viz. 0°, 3°, 6°, 9°, 12°, 15° and 18°. Numerical analysis was carried out to find an optimum rotational speed of VG. In order to validate the numerical results, NACA 0015 wing was 3D printed and experiments were conducted in a low-speed wind tunnel developed indigenously. The

S. N. Jadhav · A. Gupta · A. Singh · H. Saxena · S. V. Jain (✉) · A. M. Achari · T. N. Raval · A. Singhal
Department of Mechanical Engineering, Institute of Technology, Nirma University, Ahmedabad, India
e-mail: sanjay.jain@nirmauni.ac.in

S. N. Jadhav
e-mail: 19bme507@nirmauni.ac.in

A. Gupta
e-mail: 19bme010@nirmauni.ac.in

A. Singh
e-mail: 19bme007@nirmauni.ac.in

H. Saxena
e-mail: 19bme038@nirmauni.ac.in

A. M. Achari
e-mail: madhusudan.achari@nirmauni.ac.in

T. N. Raval
e-mail: tejas.raval@nirmauni.ac.in

A. Singhal
e-mail: ath.singhal@nirmauni.ac.in

A. Kumar R. et al. (eds.), *Proceedings of Fluid Mechanics and Fluid Power (FMFP) 2023, Vol. 6*, Lecture Notes in Mechanical Engineering,
https://doi.org/10.1007/978-981-96-9600-0_48

flow separation point was delayed by 15–81% and the lift coefficient was increased up to 54% by the provision of VG.

Keywords Airfoil · Flow separation · NACA · Stalling · Vortex generator

1 Introduction

Airflow over an airfoil is slowed down close to a wall due to the viscosity. The speed of the fluid flow in the boundary layer with reference to the airfoil approaches zero resulting in the boundary layer separation. When the boundary layer separates from the wall, it creates vortices that have a big impact on the momentum. The flow reverses in direction near the wall at the point of flow separation and the flow recirculates [1].

In the context of airfoil flow separation, the concept of circulation plays a significant role. Circulation is a fundamental aerodynamic concept that relates to the motion of fluid particles around a closed curve of an airfoil. It determines how much of the flow is deflected around the airfoil. It is important as it influences the boundary layer behaviour and the flow separation. It is directly related to the generation of lift on an airfoil. According to the Kutta-Joukowski theorem [2], the lift produced by an airfoil is proportional to the circulation around it. When circulation is low, the lift generated is insufficient, and the airfoil may experience premature flow separation, leading to reduced lift and increased drag. The circulation distribution along the airfoil surface influences the behaviour of the boundary layer. Higher circulation values can enhance the flow attachment to the surface, delaying or preventing separation. By properly designing the airfoil shape and circulation distribution, engineers can control the separation point and improve the airfoil's performance.

For turbulent flow, the resistance to the separation increases slightly with an increase in Reynolds number (Re) but in laminar flow, the flow separation point isn't affected by Re. Flow separation causes a loss of lift, increases the drag and generates more noise. If the vortex shedding reaches the natural frequency of the system, it may lead to mechanical failure of the systems. The flow controlling methods to delay the flow separation can be classified as active and passive techniques. Active methods require external energy in order to transfer momentum, viz. a rotating cylinder, suction, blowing, plasma actuators oscillation, synthetic jet, etc. Passive methods on the other hand doesn't require external energy to transfer momentum and are simple, viz. vane-type vortex generators, Gurney flap, splitter plate, compliant coating, etc. However, most of these methods are not optimized yet and researchers are working of these methods worldwide.

2 Literature Review and Objectives

The numerical and experimental study of fluid flow over airfoil has been done by several authors. In one of the earlier experimental and analytical study, Modi and Mokhtarian [2] investigated the symmetrical Joukowsky airfoil with circulation control. They studied the boundary layer control under two conditions, viz., a rotating circular cylinder positioned freely with regard to the leading edge and a rotating cylinder swaps out the leading edge of the airfoil. They suggested that although the analytical model predicts proper trends, it is purely qualitative in nature whereas the numerical (surface singularity) method appeared promising. Maximum lift and stall characteristics improved significantly in testing with a leading-edge rotating cylinder. For α as high as 48°, cylinder rotation caused the flow to never completely disengage from the top surface. Higher rotation rates ($U_r/U > 1$) facilitated reattachment of the partially separated flow, resulting in an increase in C_L of up to 150% for $U_r/U = 4$. Thakkar et al. [3], placing the cylinder in the slotted airfoil, numerically investigated and found that increase in diameter and speed of cylinder can delay flow separation.

Martinstetter et al. [4] used a turbulator perpendicular to the flow at 43% of chord length measured from the leading edge resulting in shift of the separation point downstream at low and high Re. It was recommended that the passive technique provides better control over boundary layer separation on low-pressure turbine blade with relatively high loads. In one of the numerical studies to control the boundary layer separation over a NACA 63,218 airfoil, Yahiaoui et al. [5] varied the cylinder speed by 0, 1, 2, 3 and 4 times of inlet velocity (20 m/s). When the cylinder was stationary, it affected the flow negatively due to the gap between airfoil and cylinder body, and the flow became stable with increase in cylinder speed. The fact that the vortices generated by rotor mounted near leading edge re-energize the lower boundary layer and cause separation to be delayed was the conclusion given by Viets et al. [6] in one of their experimental investigations. The separation occurred at $\alpha = 14°$ and it was fully separated at 20°. Mokhtarian et al. [7] concluded that the effectiveness of the leading-edge cylinder improves at lower speed by using scooped configuration than whole cylinder by experimentally testing Joukowsky airfoil at a Re of 2.31×10^5 over different α and different cylinder speed. In a numerical study using Ansys Fluent, Salih and Abdulhussain [8] studied the turbulent airflow over a NACA 0012 airfoil adjusted in tandem with a circular cylinder and concluded that using a circular cylinder at the leading edge caused more pressure drop because of decrease in tandem gap.

Modi et al. [9] experimentally investigated the boundary layer control and achieved enhanced lift and delayed stall by imparting momentum through moving surfaces. The wind tunnel measurements were supplemented by a flow visualization investigation conducted using slit illumination and polyvinylchloride tracer particles. The Joukowsky model showed an increase in C_{lmax} and delayed stall angle by providing a rotating cylinder at the leading edge. Ponnaswami et al. [10] numerically investigated on three variations of NACA 2412, viz. a bare airfoil, with a static cylinder and with a rotating cylinder (cylinder rotation and free-stream velocity ratio range of 1–4). The C_L was maximum for the airfoil with rotating cylinder. The lift to darg ratio (L/D) for an airfoil with a rotating cylinder was maximum justifying it as the most efficient airfoil. Makhija et al. [11] experimentally and numerically investigated a wide range of Re (from 0.55 to 1.12 lakh) and α (from $-6°$ to 20°). According to the results of the research, the value of the C_L increases with α, and stalling occurs at 18°. However, stalling was noticed in the numerical data a bit sooner than $\alpha = 18°$. An average relative error of 18% was found between experimental and CFD results, that might be attributed to the CFD analysis' assumption of a 2-D airfoil.

The objectives of the present work are (i) to carry out numerical and experimental analyses of NACA 0015 airfoil/wing without and with a rotating cylinder VG (ii) to optimize the rotational speed of VG with numerical approach by studying its effects on delay in the flow separation as well as on lift and drag coefficients (C_D) (iii) to validate the numerically optimized results by performing experiments on NACA 0015 wind developed by 3D printing in an indigenously developed wind tunnel over a wide range of Re and α.

3 Numerical Investigations

In the present study, steady state simulations were carried out to reach a stable solution more quickly and with less computational effort compared to transient simulations. Initially, numerical investigations were done on NACA 0015 profile airfoil (without VG) at seven different α, viz. 0°, 3°, 6°, 9°, 12°, 15° and 18° at a velocity of 3, 9, 15, 21 and 27 m/s. The corresponding Re range was 29,000 to 258,000. In the next phase, analysis of NACA 0015 profile was done with a rotating cylinder VG at all the angles and Re mentioned above. As a VG of ϕ15 mm cylinder with a length of 100 mm was provided at a maximum thickness section of the airfoil (i.e. 0.3C, C is the chord length of airfoil i.e. 154 mm). The objective of CFD analysis is to find optimum rotating speed of the VG and to study the effects of VG on the wing performance.

3.1 *Modelling and Grid Generation*

The NACA 0015 wing was generated using around two hundred data points in SolidWorks software. Its wireframe model is shown in Fig. 1a. The 2D drawing of airfoil with a rotating cylinder VG is shown in Fig. 1b. A C-shaped section was constructed for the CFD analysis surrounding the airfoil to resemble the experimental conditions of the wind tunnel. At inlet of C-section, the radius was taken as 1540 mm (10C) and the domain's length was taken as 3080 mm (20C) downstream the airfoil. A multi-block structured grid was generated, with the grid near the airfoil being finer as shown in Fig. 1c. Unstructured mesh was generated adjacent to the cylinder, while structured mesh was used at the airfoil surface with biassing. A 3D model of wing with VG is shown in Fig. 1d.

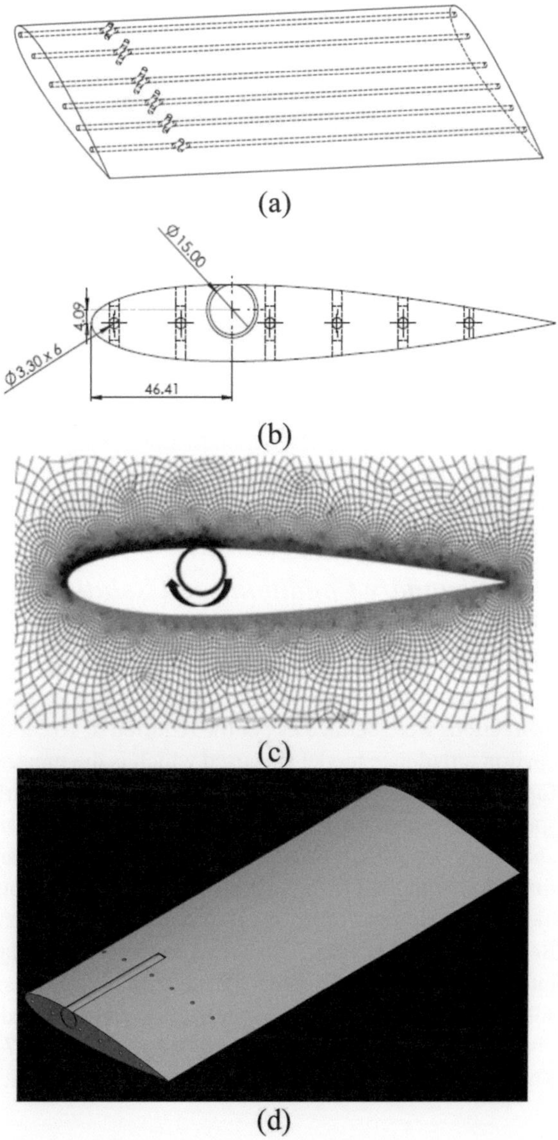

Fig. 1 **a** Wireframe model of NACA 0015 wing without VG, **b** 2D drawing of airfoil with VG, **c** grid with VG, **d** 3D model of wing with VG

3.2 Governing Equations

The 2D mass conservation equation solved for steady incompressible flow is as follows:

$$\frac{\partial u}{\partial x} + \frac{\partial v}{\partial y} = 0. \tag{1}$$

The 2D equations used for momentum conservation (Navier–Stokes equations) in x and y directions are given below:

$$u\frac{\partial u}{\partial x} + v\frac{\partial u}{\partial y} = -\frac{1}{p}\frac{\partial p}{\partial x} + v\left(\frac{\partial^2 u}{\partial x^2} + \frac{\partial^2 u}{\partial y^2}\right) \tag{2}$$

$$u\frac{\partial v}{\partial x} + v\frac{\partial v}{\partial y} = -\frac{1}{p}\frac{\partial p}{\partial y} + v\left(\frac{\partial^2 v}{\partial x^2} + \frac{\partial^2 v}{\partial y^2}\right), \tag{3}$$

where p = pressure (N/m^2), u and v are velocity components in x and y directions (m/s), v = kinematic viscosity (m^2/s).

3.3 Boundary Conditions and Solution Technique

At the inlet of the flow domain, velocity inlet was specified in terms of magnitude and components with an absolute reference frame. The velocity components in x and y directions were defined as per the required α. To capture the turbulence, standard k-epsilon turbulence model was used which is the most commonly used model in CFD for the simulation of mean flow characteristics of turbulent flow. In the turbulence specification method, turbulence intensity and viscosity ratio were specified as 5% and 10, respectively. The wall was considered stationary with no slip shear condition for airfoil without VG. For airfoil with VG, in the momentum condition for the wall motion, moving wall motion with rotational condition (clockwise) was considered. Before CFD solver interface is used to execute calculations, the proper setup for the intended condition was made. The general setting, viscous model, reference values, solution setting, and partial discretization were defined for the same. The density and viscosity of air were taken as 1.1654 kg/m^3 and 1.872×10^{-5} kg/ms, respectively. The residual error for mass, momentum and turbulence was set to 1×10^{-6}.

4 Experimental Investigation

4.1 Wind Tunnel Setup and Calibration

For the experimental work a subsonic wind tunnel was developed indigenously at Mechanical Engineering Department, Nirma University as shown in Fig. 2a. It comprised of inlet section, settling chamber, contraction chamber, test section, diffuser section and induced draft fan at the end. The fan was provided with variable frequency drive to alter its rotational speed and hence flow velocity in the test section. A honeycomb structure was provided in the settling chamber to reduce turbulence. The test section dimensions are 10″ × 15″ ($H \times W$). The major specifications of the wind tunnel are given in Table 1.

Before starting the experiments, the wind tunnel was calibrated at different fan speeds. The wind speed was measured in the test section using pitot-tube along with micro-manometer. In the test section, the wind speed was found in the range of 3–30 m/s at different fan speeds.

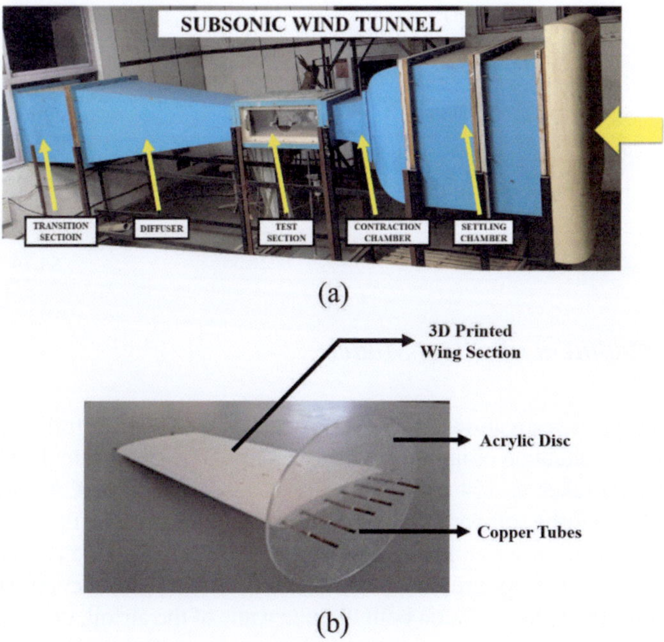

Fig. 2 **a** Subsonic wind tunnel, **b** wing section model

Table 1 Specifications of subsonic wind tunnel

Component	Specifications
Metallic screen	Size: 36″ × 54″, holes: square
Settling chamber	Length: 18″
Contraction chamber	Length: 42″, area ratio: 13
Test section	Size: 10″ × 15″, length: 32"
Diffuser section	Size: 30″ × 30″, length: 85″, diffusing angle: 7°
Transition section	Length: 42″
Fan	Type: induced draft, max. speed: 1500 rpm

Table 2 Delay in point of flow separation

Angle of attack(α)	Reynolds No. (Re)	Delay in flow separation point (%)	% Rise in lift coefficient (C_L)	% Reduction in drag coefficient (C_D)
15°	28,727	39.36	27.63	– 29.19
	86,181	14.62	7.21	– 18.42
	143,635	81.42	13.73	– 24.46
	201,088	31.05	18.85	– 28.54
	258,542	35.95	22.94	– 31.47
18°	28,727	72.89	53.75	– 26.37
	86,181	49.52	22.30	– 27.30
	143,635	52.92	18.90	– 36.69
	201,088	54.85	23.28	– 43.98
	258,542	55.44	27.33	– 44.74

4.2 Development of a Test Model

The experimental investigations were carried out on NACA 0015 airfoil (a test model) without VG for validation of numerical results. The chord length and the span of the test model were taken as 154 mm and 382 mm, respectively. Its CAD model was created in SolidWorks software and then it was 3D printed using PLA (poly lactic acid) white filament material. It was provided with six holes on the top surface and six on the lower surface for pressure measurement. Six numbers of copper tubes (ϕ 3 mm) were inserted in the holes on both the side walls of the airfoil, on which rubber tubes were provided which worked as pressure tapings. A multi-tube manometer with thirteen tubes was used for pressure measurements. Two circular acrylic discs were used to fix the test model in the test section through which the copper tubes were passed. Figure 2b depicts schematic drawing and 3D printed model of NACA 0015 airfoil.

5 Results and Discussion

In this section, some selected results are provided due to the space constraint. The results are analysed to find an optimum rotational speed of the VG. Then, the results obtained for airfoil without and with VG are compared qualitatively in terms of pressure contours, velocity streamlines and velocity vectors as well as quantitatively in terms of pressure coefficient (C_p), lift coefficient (C_L) and drag coefficient (C_D) which were calculated as follows:

$$C_p = \frac{p - p_\infty}{\frac{1}{2}\rho U^2} \tag{4}$$

$$C_L = \frac{\text{Lift force } (L)}{\frac{1}{2}\rho A U^2} \tag{5}$$

$$C_D = \frac{\text{Drag force } (D)}{\frac{1}{2}\rho A U^2}, \tag{6}$$

where p is static pressure (N/m^2), p_∞ is free stream pressure (N/m^2), ρ is density of air (kg/m^3), A is area (m^2), U is freestream air velocity (m/s).

5.1 Grid Independence Test

A grid independence test (GIT) was carried out to monitor the solution mesh independence. This was accomplished by generating a new cell grid and analysing it using a number of obtained alternatives. The number of grid elements were varied from 45,000 to 2 lacs and the value of C_L was found to vary within 1% after around 1.60 lac elements as shown in Fig. 3a [12]. Hence, the further simulations were done with 1.60 lac elements.

5.2 Selection of Rotational Speed of Vortex Generator

In order to determine an optimum speed of VG, here the results are presented for wind velocity $U = 3$ m/s at an α of 15°. The rotational speed of VG (N_r) was varied from 11,500 to 50,000 rpm such that the speed ratio (U_r/U) varied from 3 to 13. The effect of change in U_r/U on C_L is presented in Fig. 3b [12]. It is observed that C_L increases with an increase in U_r/U. However, at $U_r/U = 9$ the C_L suddenly increased, attained the maximum and then decreased. The maximum value of C_L was found at U_r/U = 9. Hence, further simulations were done by keeping $U_r/U = 9$ at different α viz. 0°, 3°, 6°, 9°, 12°, 15° and 18° in the velocity range of 3–27 m/s.

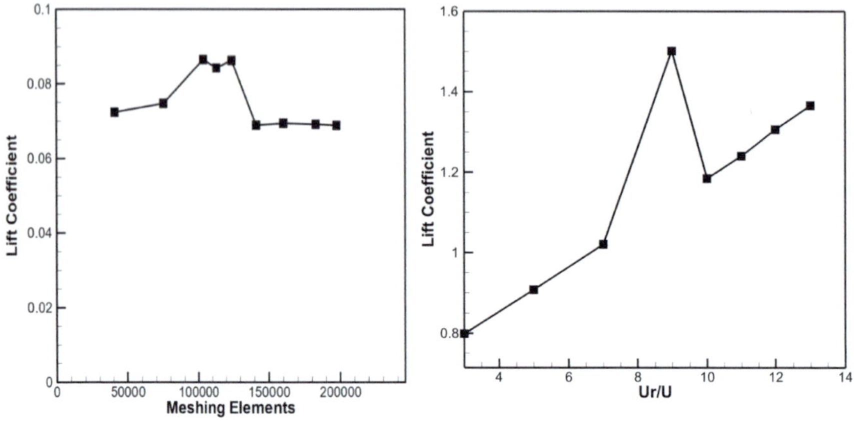

(a) (b)

Fig. 3 **a** Grid independence test, **b** effects of velocity ratio (U_r/U) on lift coefficient (C_L) [12]

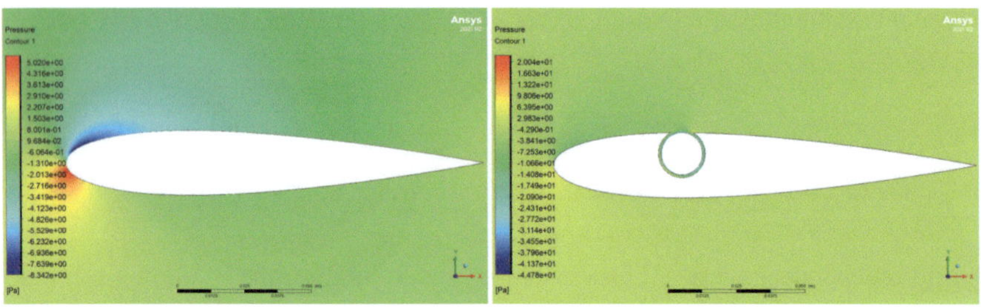

(a) without VG (b) with VG

Fig. 4 Pressure contours for airfoil at Re = 28,727 and $\alpha = 6°$ [13]

5.3 Pressure Contours

Figure 4 shows comparison of pressure contours for airfoil without and with VG at Re = 28,727 and $\alpha = 6°$ [13]. It can be seen that due to VG the pressure difference across the airfoil increased which is due to delay in the flow separation point towards the trailing edge resulting in increase of C_L and decrease of C_D.

5.4 Velocity Streamlines

The comparison of velocity streamlines for airfoil without and with VG at Re = 28,727 and $\alpha = 6°$ is depicted in Fig. 5 [13]. It can be observed that due to VG

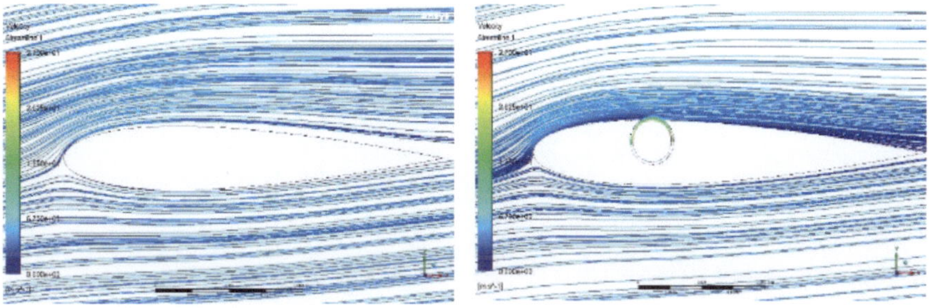

(a) without VG (b) with VG

Fig. 5 Velocity streamlines for airfoil at Re = 28,727 and $\alpha = 6°$ [13]

the flow accelerated in the vicinity of airfoil surface after VG. Thereby the flow separation point is moved towards the trailing edge.

5.5 Identification of Point of Flow Separation

The flow separation or boundary layer separation can be explained as detachment of a boundary layer from a surface into a wake. Separation develops in slowing flow with increasing pressure after passing through the thickest section of the airfoil. Figure 6 shows delay in flow separation point at $U_r/U = 9$ and $\alpha = 15°$ for Re = 143,634 due to provision of VG [13]. It happened due to energising of the flow in the near wall zone downstream of VG. Table 1 shows the percentage change in C_L, C_D and flow separation point due incorporation of rotating VG in airfoil for $\alpha = 15°$ and 18°. It can be seen that, provision of VG led to substantial delay in the flow separation point by 15–81% which resulted in the significant increment in C_L and decrement in C_D. It resulted in 7–54% increment in C_L in the different cases.

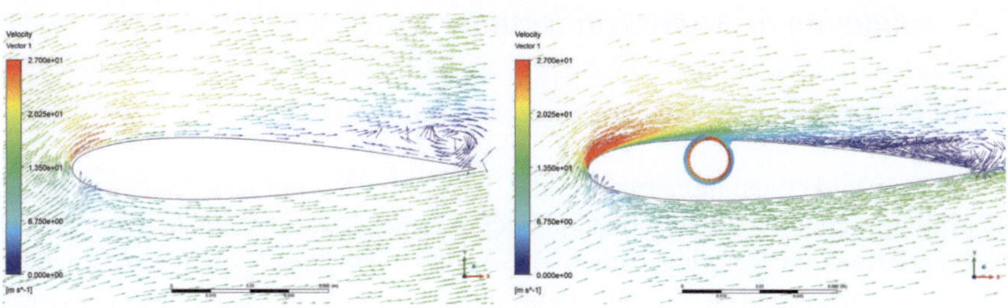

(a) (b)

Fig. 6 Velocity vectors for **a** airfoil without VG, **b** with VG at $U_r/U = 9$, $\alpha = 15°$ for Re = 143,634 [13]

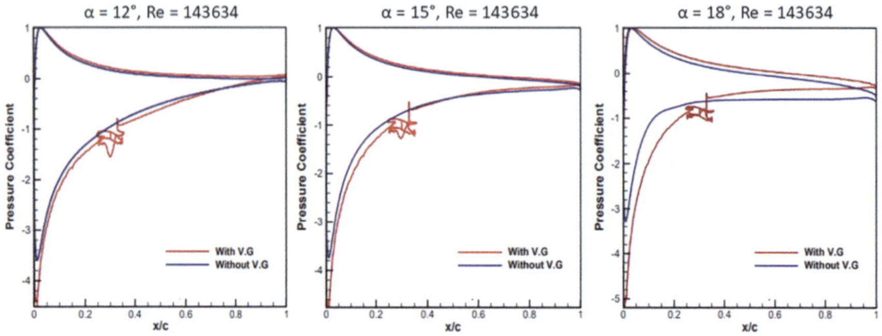

Fig. 7 Comparison of C_p versus x/c plot without and with VG for different α at Re = 143,634

5.6 *Performance Comparison Without and with Vortex Generator*

Figure 7 depicts comparison of C_p versus x/c plot without and with VG for different α at Re $=$ 143,634. It can be seen that the difference in pressures at the lower and the upper surfaces of the airfoil increased in all the cases due to presence of VG. Figure 8 displays a comparison of C_L versus α and C_D versus α plots without and with VG. It can be seen that with increase in α, C_L also increased, reached to maximum and then decreased in all the cases. The sudden decrement in C_L represents stalling characteristics of airfoil. It may be observed that due to VG, C_L and C_D remained nearly same before staling condition. However, after stalling the C_L substantially improved and C_D significantly decreased due to presence of VG at all the Re as shown in Fig. 8a, b respectively [12]. This also resulted in delay in the stalling angle. This suggests that the VG improves the airfoil's ability to generate lift vortex shedding which delays the flow separation and increases the airfoil's effective camber.

5.7 *Validation of Numerical Results*

5.7.1 Pressure Coefficient

In order to validate the numerical results, NACA 0015 wing was developed using 3D printing technology as described in Sect. 4.2 and shown in Fig. 2b. Experiments were performed on the wing without VG in the wind tunnel at different Re. The pressure was measured on top and bottom surfaces of the wing and C_p was evaluated as per Eq. (4). Figure 9 shown comparison of numerical and experimental results for C_p versus X/C plot at different values of α at Re $=$ 143,634 (X is distance from the leading edge) [12]. It can be seen that the numerical results were found in good conformance with the experimental values.

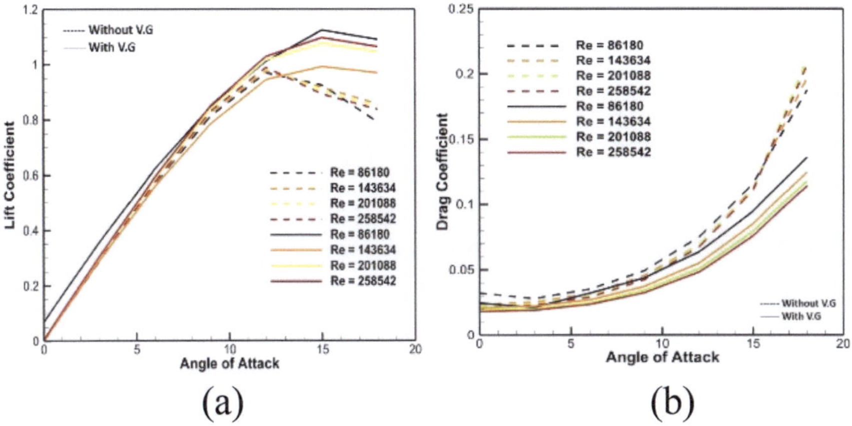

Fig. 8 **a** C_L versus α plot without and with VG, **b** C_D versus α plot without and with VG

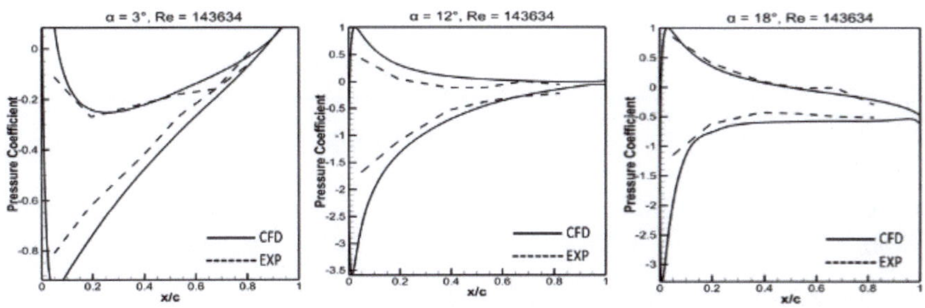

Fig. 9 Comparison of C_p versus x/c plot obtained through experimental and numerical methods [12]

5.7.2 Lift Coefficient

The lift coefficient can be determined by calculating the area bounded by upper and lower surface pressure coefficients ($C_{p,L}$ and $C_{p,U}$). The numerical integration of pressure coefficient will give C_L. The pressure always acts in the normal direction to the surface so only the vertical component of pressure was considered for the calculation of C_L as per Eq. (7) [14]:

$$C_L = \int \left(C_{p,L} * \sin\varphi - C_{p,U} * \sin\varphi\right) d\frac{X}{C}, \tag{7}$$

where φ is angle between normal to the airfoil surface and direction of flow.

Figure 10 shows comparison of C_L versus α plot obtained from numerical and experimental investigations as well as the results published in the NACA report [15]. It can be observed that, the numerical results were found in very good agreement

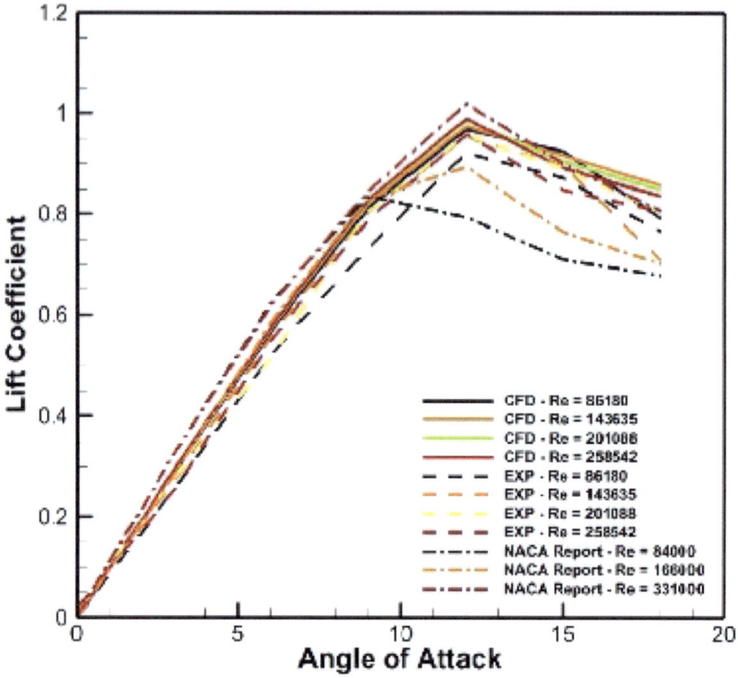

Fig. 10 Numerical and experimental results for C_L versus α at different Reynolds numbers [12]

with the experimental results as well as NACA report except after stalling condition. The reason could be, it may be difficult for the software to predict the actual flow conditions after stalling because of instability created in the flow. Among the previous studies, Zhang et al. [16], Julian et al. [17] also reported the value of lift coefficient in the similar range.

In order to validate the numerical results obtained with VG, it is proposed to perform experiments on NACA 0015 wing by incorporating VG at a maximum thickness point in the wide range of Re and α in the near future.

6 Conclusions

Experimental and numerical study were done on NACA 0015 airfoil without and with VG in the present study. The steady state simulations were carried out in ANSYS Fluent software and the experiments were conducted in the wind tunnel in the wide range of Reynolds numbers i.e. 0.29–2.58 lacs at seven different α viz. 0°, 3°, 6°, 9°, 12°, 15° and 18°. The major conclusions from the study are as under:

- With an increase in the angle of attack, lift coefficient increased, reached to maximum and then decreased at all the Re. The stalling angle was found to be 15° to 18° at different Re.
- The VG comprised of 15 mm dia. was provided at 30% chord length and the analysis was done with a non-dimensional speed (U_r/U) range of 1–13. The optimum value of U_r/U was found to be 9.
- The provision of the VG led to 15–81% delay in flow separation point and resulted in maximum rise in the lift coefficient up to 54%.
- It was concluded that the provision of rotating VG can be efficiently used in the wings of actual aircrafts in order to increase the C_L/C_D ratio by delaying the flow separation point.

Acknowledgements We highly acknowledge the guidance provided by Dr. Kunal Ghosh, Retired Professor and Head, Aerospace Engineering Department, IIT Kanpur and Dr. Chetan Mistry, Assistant Professor, Aerospace Engineering Department, IIT Kharagpur for development of the test facility and conducting the research.

Nomenclature

C_L	Lift coefficient –
C_D	Drag coefficient –
C_p	Pressure coefficient –
U	Free-stream velocity [m/s]
U_r	Rotational speed [m/s]
α	Angle of attack (°)
ρ	Density [kg/m^3]
μ	Viscosity [Ns/m^2]
p	Pressure [Pa]
C	Chord length [mm]

References

1. J. Wang, and L. Feng, Flow Control Techniques and Applications. In: Flow control techniques and applications (Cambridge Aerospace Series, p, I). Cambridge University Press. P: 3, 2018.
2. V. J. Modi, and F. Mokhtarian, Joukowsky Airfoil with Circulation Control. In: 12th Atmospheric flight mechanics conference, 1986.
3. N. Thakkar, A. Lakdawala, and V. Lakhera V. Passive flow separation control on a Cambered Airfoil Using Vortex Generator – a Numerical Approach. In: Proc. Conference on Fluid Mechanics and Fluid Power, Allahabad, UP, 2016.
4. M. Martinstetter, R. Niehuis, and M. Franke, Passive Boundary Layer Control on a Highly Loaded Low Pressure Turbine Cascade. In: Proc. of ASME Turbo Expo: Power for Land, Sea and Air, Glasgow, UK, 2010.

5. T. Yahiaoui, M. Belhenniche, and B. Imine. Effects of Moving Surface on NACA 63218 Aerodynamic Performance. In: EPJ Web of Conferences, 2015.
6. H. Viets, M. Piatt, and M. Ball, Unsteady Wing Boundary Layer Energization. In: 5th Atmospheric flight mechanics conference for future space systems, 1979.
7. F. Mokhtarian, V. J. Modi, and T. Yokomizo, Rotating Air Scoop as Airfoil Boundary Layer Control. Journal of aircraft, 1988.
8. Salih HM, Abdulhussain MA (2019) Drag Investigation Around a Cylinder and Symmetric Airfoil an Tandem with Different Gaps and Angle of Attack. Journal of mechanical engineering research and development 42(1):26
9. Modi VJ, Fernando MSUK, Yokomizo T (1991) Moving surface boundary layer control as applied to two-dimensional and three-dimensional bluff bodies. J Wind Eng Ind Aerodyn 38:83
10. Ponnaswami VMD, Verma A, Reddy CJP, Suri J (2018) Boundary Layer Control of Airfoil using Rotating Cylinder. International Journal of Recent Technology and Engineering 8(6):4742–4750
11. D. Makhija, S. V. Jain, A. M. Achari, and K. Ghosh, Aerodynamic Analysis of Aircraft Model Using Indigenously Developed Wind Tunnel Facility. In: IOP Conference series: Material science and engineering, 2021.
12. S. N. Jadhav, Boundary Layer Control Around Airfoil Using Vortex Generator. BTech Thesis, Institute of Technology, Nirma University – Ahmedabad, 2022.
13. A. Singh, and A. Gupta. Numerical Analysis of a Rotating Cylinder Vortex Generator in an Airfoil. BTech Project, Institute of Technology, Nirma University – Ahmedabad, 2022.
14. J. D. Anderson, and M. L. Bowden, Introduction to Flight, McGraw Hill, 1989.
15. Airfoil section characteristics as affected by variations of the E. N. Jacobs, and A. Sherman, Reynolds number, National Advisor Committee for Aeronautics Report – 586, 1939.
16. Zhang Y, Huang D, Sun X, Wu G (2010) Exploration in Optimal Design of an Airfoil with a Leading Edge Rotating Cylinder. J Therm Sci 19(4):318–325
17. Julian J, Iskandar W, Wahyuni F (2022) Aerodynamics Improvement of NACA 0015 by Using Co-Flow Jet. International Journal of Marine Engineering Innovation and Research 7(4):284–291

Passive Vane-Type Vortex Generators for Aircraft Performance Enhancement: A Review

Devansh Talati, Rushil Palsana, Sourabh Rai, S. N. Jadhav, S. V. Jain, A. M. Achari, T. N. Raval, and Ath Singhal

Abstract A vortex generator (VG) is an aerodynamic device comprised of a small vane attached to a lifting surface like airfoil, aircraft wing, wind turbine blade, etc. The VGs are broadly classified as active and passive device based on the utilization of external energy. This paper provides a comprehensive review of the pivotal role played by vane-type vortex generators in enhancing aerodynamic performance across a range of airfoil designs and practical applications. VGs have proven to be effective tools in mitigating adverse pressure gradients and boundary layer separation, with various studies highlighting their impact on lift coefficients, stall angles, and drag reduction. In conclusion, vane-type VGs are essential tools in boundary layer control, with the potential to significantly improve aerodynamic performance of various airfoils, aircraft wings, wind turbine rotor, etc. In addition, this paper discusses the diverse VG designs and configurations explored in recent researches

D. Talati · R. Palsana · S. Rai · S. N. Jadhav · S. V. Jain (✉) · A. M. Achari · T. N. Raval · A. Singhal
Department of Mechanical Engineering, Institute of Technology, Nirma University, Ahmedabad, India
e-mail: sanjay.jain@nirmauni.ac.in

D. Talati
e-mail: 20bme029@nirmauni.ac.in

R. Palsana
e-mail: 20bme111@nirmauni.ac.in

S. Rai
e-mail: 20bme128@nirmauni.ac.in

S. N. Jadhav
e-mail: 19bme507@nirmauni.ac.in

A. M. Achari
e-mail: madhusudan.achari@nirmauni.ac.in

T. N. Raval
e-mail: tejas.raval@nirmauni.ac.in

A. Singhal
e-mail: ath.singhal@nirmauni.ac.in

A. Kumar R. et al. (eds.), *Proceedings of Fluid Mechanics and Fluid Power (FMFP) 2023, Vol. 6*, Lecture Notes in Mechanical Engineering,
https://doi.org/10.1007/978-981-96-9600-0_49

and shedding light on their varying influences on airfoil performance. The applications of various vane-type VGs such as triangular, rectangular, and gothic, in the modern aircrafts are summarized. It was concluded that the future research will be focused on the combined use of the active and the passive VGs in the upcoming generation of high-performance aircrafts and unmanned aerial vehicles.

Keywords Aerodynamics · Airfoil · Flow control · Vane-type · Vortex generator

1 Introduction

In the early nineteenth century, Sir George Cayley's biomimetic airfoil design inspired aviation. Limited by a lack of theoretical understanding, early aerofoils were highly curved and slender, leading to issues like boundary layer separation. Over the years, aviation field evolved, with a focus on boundary layer control and flow control techniques. These advancements, critical not only in the aviation field but also in areas like turbomachinery, combustion, and transportation, have driven efficiency improvements and innovation in various industries. Ludwig Prandtl, revolutionized fluid dynamics in 1904 by proposing that the effects of friction are felt only extremely close to an object moving through a fluid [1]. During high angles of attack (AOA), the flow tends to get detached from the body and create a wake region within which eddy currents are formed which dissipate energy. This formation of eddy currents increases the pressure drag on the airfoil and in turn decreases the aerodynamic performance. To tackle this problem, researchers have come up with various methods to delay the flow separation [2]. In order to reduce detrimental effects while controlling it for greater aerodynamic efficiency and performance, numerous approaches have been put together since the boundary layer first came into existence. Practically speaking, it is quite valuable to be able to alter a flow field to bring about a desired change.

Delaying boundary layer's conversion from laminar to turbulent has many evident advantages as summarized in Fig. 1. A lower drag translates an object to a longer distance, less fuel consumption, or faster speeds for an aircraft or an aquatic body. Because of this, flow-induced noise is almost non-existent in the laminar region. Flow-induced noise is a consequence of pressure variations in the turbulent boundary layer. The airliner's performance in terms of aerodynamics assurance and ecological impact can potentially be broadened via flow control. A one percent decrease in a large passenger aircraft's drag could result in an increase of approximately 10 additional passengers or a decrease of approximately 1600 kg in the aircraft's empty weight [3]. A study from Boeing company shows that 0.1 increase in lift coefficient (C_L) value leads to a 630 kg excess weight carrying in the empty weight of aircraft at a fixed angle of attack, 1.5% increment in ($C_{L,\max}$) is equivalent to approximately 3000 kg dead weight at operating flying speed and increase in lift to drag ratio by 1% during take-off is same as additional weight advantage of 1270 kg [4].

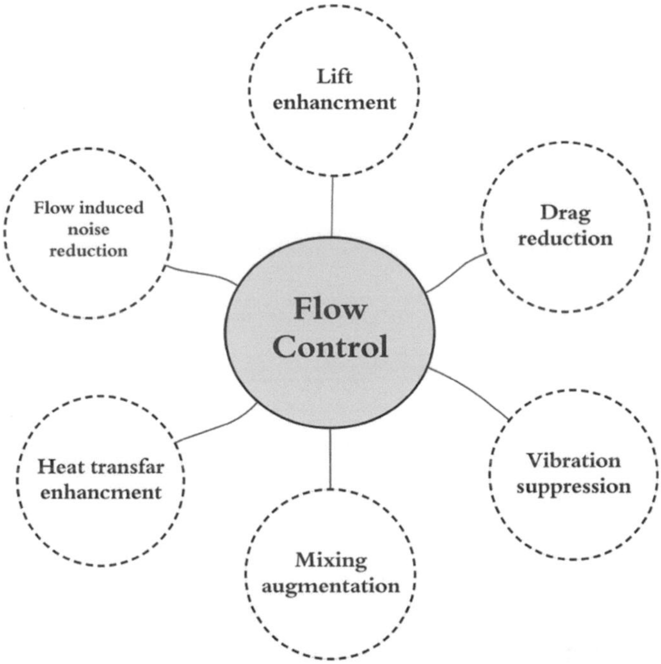

Fig. 1 Advantages of flow control

2 Classification of Flow Control Techniques and Objectives

The most common classification for flow control systems is based on the energy input and is represented in Fig. 2 [3]. Active VGs require external energy in order to transfer momentum, viz., a rotating cylinder VG, which has been widely studied by Mokhtarian et al. [5]. Passive VG on the other hand doesn't require external energy to transfer momentum and are simple, viz., vane-type VG.

The objective of this study is to summarize the passive static vane-type VGs used by different researchers as well as in the field over the years. The current scenario and future scope for research in vane-type VGs are also presented.

3 Vane-Type Passive Vortex Generators

Vane-type VGs are fin like plates that are mounted on the surface which delay boundary layer separation by creating vortices and transferring the momentum from the higher to the lower energized layer near the airfoil surface. VGs can be classified on the basis of their size, shape, direction of rotation of vortices and arrangement

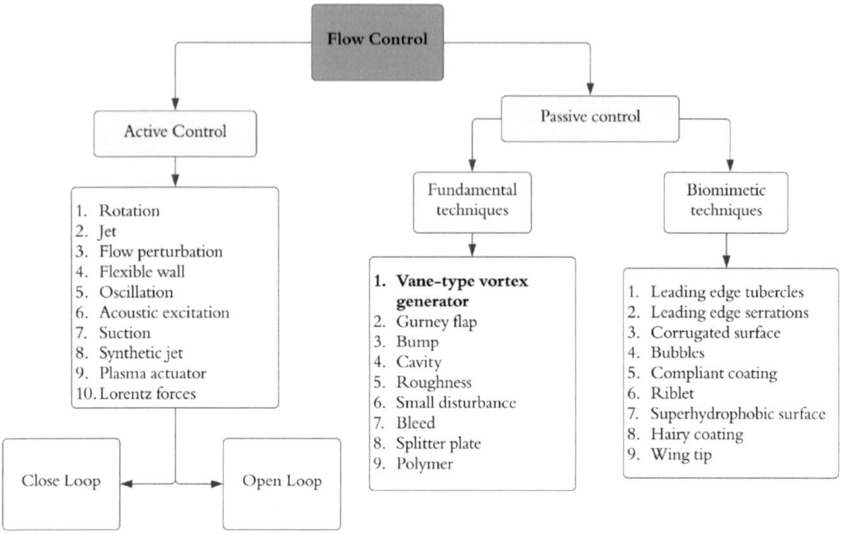

Fig. 2 Different flow control techniques

[6, 7]. Figure 3 shows the classification of various types of passive VGs and Fig. 4 shows different profiles used for vane-type VG.

The vane-type VGs are widely used in different aircrafts as summarized in Table 1.

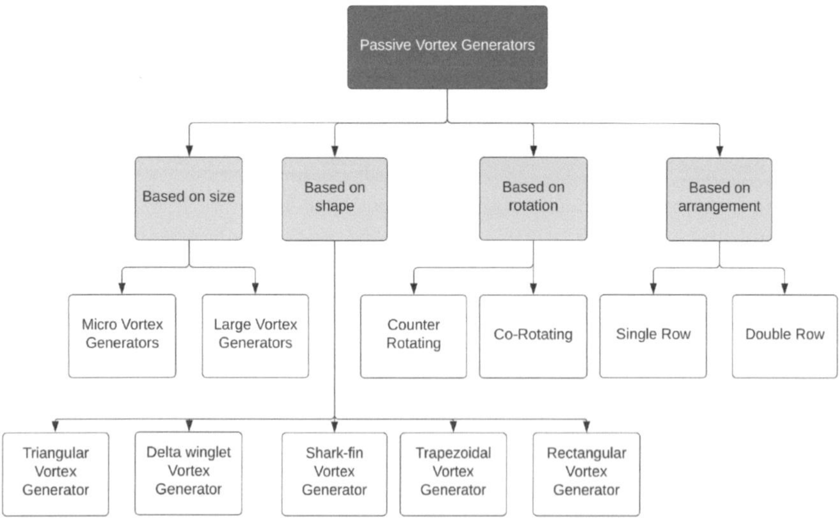

Fig. 3 Classification of passive vortex generators

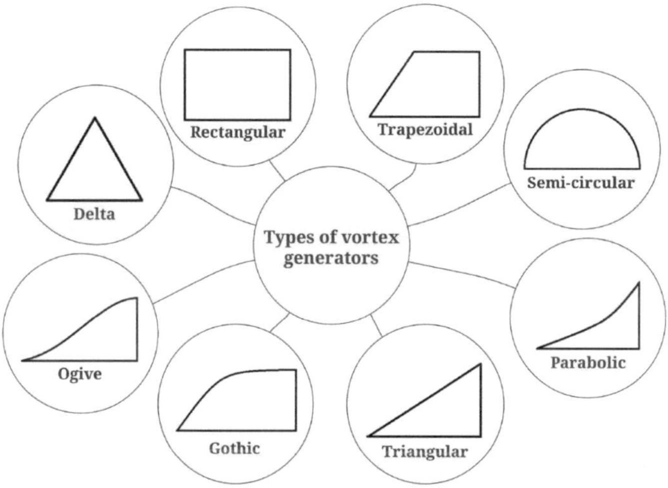

Fig. 4 Different vane-type VG profiles

4 Summary of Most Exemplary Papers

Bragg and Gregorek [16] studied VGs on a canard airfoil model used in the Voyager aircraft. At 3° angle of attack (AOA) lift decreased from 0.9 to 0.55 for the tripped model, worsening with higher AOA. VGs (small, large, delta) reduced $C_{l\max}$ and caused boundary layer reattachment at 0.70c, with the delta VG performing best on the tripped model but slightly increasing drag on the clean model. A rough surface on the airfoil increased drag. Li-shu et al. [17] explored combining VGs and Gurney flaps (GF) on a WA251A airfoil. VGs with a 15° blade incidence angle were optimal. Varying GF heights and locations showed improved lift and reduced drag when closer to the trailing edge. Combining selected VGs and GFs significantly increased lift coefficients, especially in the stall regime.

Sorensen et al. [18] developed a CFD based method to study the impact of VGs on airfoils. They varied VG size, chord-wise distance, aspect ratio, inclination angle, and spacing, using airfoil section FFA-W3-301 (Fig. 5a). Their results indicated a 1° stall angle delay but accompanied by increased drag. They noted an overall increase in C_L for all cases, particularly when VGs were placed near the leading edge, leading to improved lift due to increased pressure differences.

Fouatih et al. [19] engaged in research on the optimization of NACA 4415 airfoil experimentally by equipping counter-rotating configured VG (Fig. 5b) with five congruous parameters, viz., thickness and height of VG, position, orientation angle, spanwise spacing. Triangular-shaped VGs with an incidence angle of 12°, a 3 mm spacing, and positioned at 0.5c (c is chord length) was found to be the most effective in delaying flow separation. They observed a 21% increase in $C_{L,\max}$ and a 17° delay in flow separation with paired VGs.

Table 1 Application of vane-type VGs in aircrafts

Aircraft model	Profile of vane-type VG	Location of VG	Remarks
Boeing's ecoDemonstrater airplane—a large 777-200ER airliner [8]	Gothic	VGs mounted in between 20 and 50% of wing-chord	• "Smart" vortex generators activated with shape-memory alloys • Improved characteristics
ST Aerospace A-4SU Super Skyhawk [9]	Rectangular	Mounted on the leading -edge slat	• Serve to break up the shocks formed at transonic speeds • Improved characteristics
Luncair 235 [10]	Triangular	Mounted just aft of the leading edge	• Improved aerodynamic characteristics
British aerospace Sea Harrier [11]	Trapezoidal	Situated near the leading edge of the swept wings	• Reduced stall speed • Improved stabilator authority, characteristics, controllability
Boeing 737 [12]	Triangular	Aft the leading edge and on nose of fuselage	• Improved stabilator authority, characteristics, controllability
Piper Cherokee Six 300 [13]	Rectangular	Vortex generators attached to the horizontal stabilizer	• Reduced stall speed by 10% • Improved ailerons, characteristics, controllability
Airvan GippsAero GA8 [14]	Gothic	Vortex generators mounted full span on the wings	• Improved elevator authority • Improved characteristics, controllability
Mooney M20M [15]	Rectangular	Mounted just aft of the leading edge	• Reduced stall speed by 8% • Reduced approach speed, better roll rate

Egmond's [20] explored passive vortex generators, specifically internal vortex generators (IVGs), designed to create vortices below the surface. The wedge-type IVG was compared to an external VG on a single-element high lift model. In low-speed wind tunnel tests, the IVG increased maximum lift by 3% compared to baseline, with potential for even greater improvement (up to 33%) when boundary layer tripping was applied, surpassing the 16% lift increase achieved by the external vortex generator.

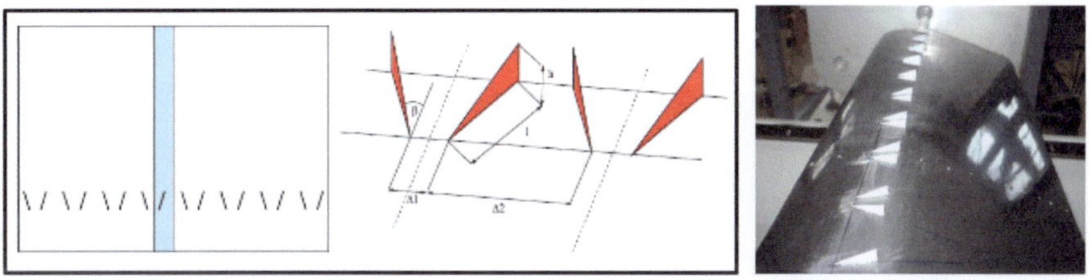

(a) (b)

Fig. 5 **a** Airfoil with VGs [18], **b** experimental setup [19]

Kumar et al. [21] investigated aerodynamic performance of a NACA 0012 airfoil with three types of VGs, viz., triangular, rectangular, and gothic. The VGs were 80 mm long, 90 mm high, and 20 mm thick (with a 70.42 mm fillet radius for gothic VG). All VGs were placed at 0.16c from the leading edge. Tests were conducted at 0° and 10° AOA with a velocity of 250 m/s. At 0°, the triangular VG showed significantly lower drag compared to the others. However, at 10°, the rectangular VG had slightly less drag but not significantly different from the other cases. The study concluded that the triangular VG offered the best aerodynamic efficiency, minimizing drag and delaying boundary layer separation by energizing the flow.

Jumahadi et al. [22] conducted the impact of hybrid micro VGs on a NACA 4415 airfoil. Four models were tested: clean airfoil, active blowing jet airfoil, passive micro VG airfoil, and hybrid micro VG airfoil. The active airfoil showed the most significant increase in lift, exceeding 18% compared to the clean airfoil. Both active and hybrid airfoils effectively delayed stall by over 3° AOA, with a 21.2% increase in $C_{L,\max}$ compared to the clean airfoil. Regarding drag, the active model reduced drag coefficient by 10.4%, while the hybrid model increased it by 11.3%, still outperforming the passive model.

Zulkefli et al. [23] studied the influence of triangular and shark fin VGs on a NACA 4415 airfoil. Both types of VGs had a height of 4 mm and a length of 12 mm, placed at 0.5c. At low AOA, all cases exhibited increased lift coefficients, with the shark fin VG providing the highest lift-to-drag ratio. The differences in drag coefficients were negligible up to 10° AOA.

Agarwal and Kumar [24] explored the effect of VG location on a NACA 4412 airfoil at different AOA. For lower AOA, VGs resulted in minimal lift reduction, with the optimum VG location moving toward the leading edge as AOA increased. At higher AOA, VGs significantly improved lift-to-drag ratio, with a 31% increase observed at 17° AOA when the VG was located at 0.1c. In summary, these studies collectively highlight the potential of VGs and hybrid systems in improving airfoil aerodynamic performance, depending on specific configurations and conditions.

Chillon et al. [25] examined a semi-delta vane-type VG at 18° AOA, using computational fluid dynamics (CFD) at Re = 2,000,000 and a velocity of 46.52 m/s (Fig. 6). They employed RANS for steady-state analysis and DES for AOA from 0° to 20°. The study confirmed wind tunnel results, placing the VG at $x/c = 0.30$ on a DU97W300

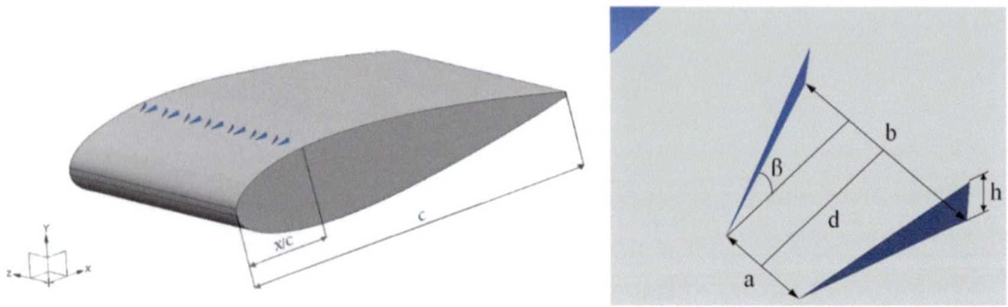

(a) (b)

Fig. 6 Row of vane-type VGs [25]

airfoil ($c = 0.65$ m). Results showed that at higher AOA, the VG's vortex grew, delaying stall and enhancing C_l on the airfoil's suction side.

Few other research studies focused on the impact of VGs on airfoil aerodynamics such as, Ayudia et al. [26] conducted CFD analysis in SolidWorks on a NACA 4412 airfoil with a triangular VG at 24% chord length. Comparing various VG shapes, they found that the triangular VG significantly increased lift with a 13% increase at 5° AOA. Himo et al. [27] examined trapezoidal VGs on an NTUA-t18 airfoil. Numerical results were validated with experiments, showing a lift error of 7% and drag error of 9%. Trapezoidal VGs with a 30° roll angle outperformed triangular VGs, especially in stall conditions. Vortex dynamics analysis revealed differences in vortex behavior between the two VG types.

Hao et al. [28] experimentally investigated the impact of spanwise height distribution of VGs on a NACA4415 wing. Triangular and trapezoidal VGs with varied spanwise heights were compared. Triangular VGs resulted in a 2.5% lift increase, while trapezoidal VGs showed a 1.7% improvement compared to equal-height VGs.

Domel et al. [29] studied three VGs on a NACA 0012 airfoil at 0.26c, Reynolds number (Re) 40,000, and AOAs from 0° to 24°. The shark denticle VG increased lift at all AOA, reduced drag, and improved the C_l/C_d ratio. The continuous shark profile VG had lift similar to the 2D bump VG at low AOA and matched the shark denticle VG at high AOA, with consistently lower drag.

Arunvintham et al. [30] conducted an experiment to study the effects of utilizing a shark scale-based VG on a NACA 0015 airfoil of chord length 100 mm and span of 300 mm. Three different VGs having chord length 0.3c, 0.5c, and 0.8c were tested at a Re of 200,000 with AOA varying from 0° to 24° with increments of 3°. Results showed that the shark scale-based VG deferred stall characteristics and aided in minimizing the adverse effects of flow separation.

Lin et al. [31] investigated VGs on a 22″ chord airfoil with slats and flaps in a landing configuration, testing at Mach 0.2, Re of 5,000,000 and 9,000,000 (Fig. 7). Two VG configurations (delta and trapezoidal) were assessed in counter-rotating and co-rotating setups at 19, 25, and 33% flap chord locations. At 8° AOA, both configurations effectively suppressed boundary layer separation at 25 and 33% positions. The counter-rotating trapezoidal VGs at 25% location showed the best results,

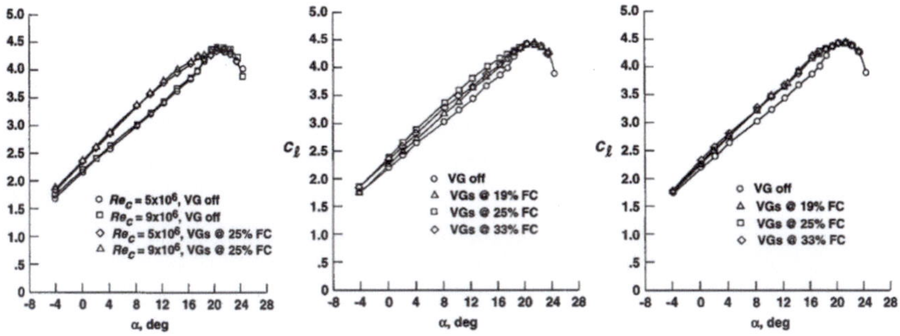

Fig. 7 C_l versus AOA plots for various parameters [31]

increasing lift by 10%, reducing drag by 40%, and improving the lift-to-drag ratio by 80%.

Setyo et al. [32] analyzed NACA 43018 airfoil with VG at $x/c = 10\%$ and 20%. Tests at Re = 7.65×10^5 and 9×10^5 with AOA from 0° to 20° showed minimal differences. It resulted in increase in lift by 5%, drag by 1.5%, and C_L/C_D ratio by 5%, but overall airfoil performance remained similar with or without vortex generators. Delnero et al. [33] tested the effects of VGs on an Eppler 387 airfoil of chord length of 42 cm and span width of 80 cm. The experiments were carried out at a Re of 300,000 at AOA of 0°, 10° and 20°. Two positions for the VGs were studied: 0.1c and 0.2c from the leading edge. The results showed that as AOA increased, $C_{L,\max}$ increased along with the drag. For 0.1c, $C_{L,\max}$ increased by 4.3–15% at different AOA, whereas at 0.2c, $C_{L,\max}$ increased by 8.7–17.4%. There was no change in the stall angle for either case but the stall obtained is smoother. In conclusion, the VGs improved the $C_{L,\max}$ but not the stall angle.

Li et al. [34] studied vortex generators at various installation angles (10° to 30°) and found that higher angles increased negative static pressure on the leeward side and altered axial velocity distribution. VGs delayed stall by 10°, boosted $C_{l,\max}$ to 2.12 and significantly reduced C_d. It resulted in a substantial lift-to-drag ratio increase of 1146.04% at 18° AOA with a 15° installation angle as shown in Fig. 8.

Fig. 8 Plots of C_l/C_d with and without VG [34]

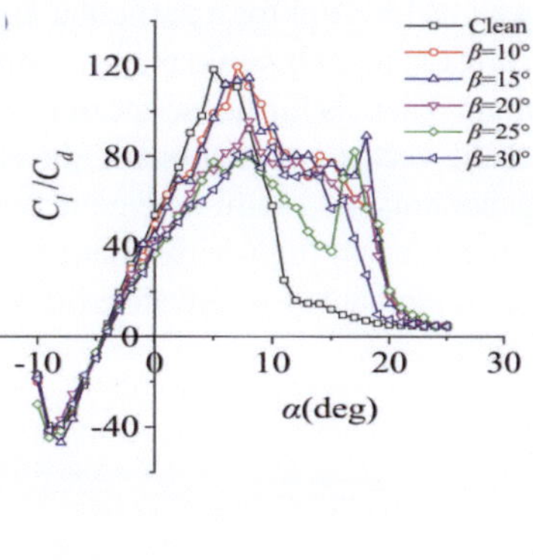

Vinodhini et al. [35] carried out an analytical study on the effects of VGs on the C_D of a NACA0012 airfoil. Three different configurations of VGs were used, viz., gothic, rectangular and delta. The VGs had a length of 0.c to 0.8c and were placed at 0.16c. The tests were carried out at an angle of 0° and a speed of 250 m/s. The results showed that the rectangular VG had the highest C_d value of 0.008797 with the gothic VG having a value of 0.006990 while the delta VG had the lowest value of C_d of 0.006936.

Several other researchers [36, 37] have also implemented passive vortex generators (VGs) on the airfoil of wind turbines to postpone the occurrence of flow separation.

Table 2 shows details of various passive VGs used for different airfoil/aircraft wings.

5 Concluding Remarks

The studies have shown that boundary-layer separation can be efficiently controlled by a passive vane-type VGs. Proper flow control using VG may lead to many benefits, such as flow stabilization, flow reattachment, separation delay, drag reduction, lift increment, aerodynamic performance improvement, increased energy efficiency, and noise reduction.

Different configurations of VGs have been used for different applications such as triangular, rectangular, gothic, and delta. A particular VG configuration optimized for one airfoil profile/wing may not be optimal for another profile of different thickness, camber, or leading-edge radius. In order to design and apply the most suitable flow controlling device, a thorough understanding of flow physics is required, which can be acquired through combined use of numerical and experimental studies. There are numerous examples of flow control methods that have been investigated for decades but have not yet been deployed in real applications due to their insufficient performance or high cost.

VGs are advantageous in many aspects as compared to other high lift augmenting devices such as flaps and slats. Hence, there is a need to optimize VG's geometry as well its location for a particular application. If properly implemented, the vane-type vortex generator could play an important role in many real-life flow-control applications. There is further scope of research on this topic with regards to the arrangement of the vortex generators and the combined use of the active and the passive vortex generators, i.e., semi-active method. It has a lot of potential for use in the upcoming generation of high-performance aircrafts and unmanned aerial vehicles. It can also be implemented in advanced drone technology for better endurance and range of flight.

Table 2 Important performance parameters for the various types of passive VGs from various studies

References	Airfoil profile	Chord length	VG profile	VG dimensions	Remarks
Kumar et al. [21]	NACA 0012	1000 mm	Rectangular, Triangular, Gothic	$l = 80$ mm, $h = 90$ mm, Fillet radius: 70.42 mm (for gothic)	At 0°, triangular had the least drag while at 10°, the rectangular had the least drag
Himo et al. [27]	NTUA-t18	600 mm	Triangular, Trapezoidal	Triangular: $h = \delta$, Roll angle = 20°. Trapezoidal: $h = \delta$ or $3 * \delta$, Roll angle = 30°, 45°, 60°	The L/D ratio for the trapezoidal VG was 10% greater than the triangular
Arunvinthan et al. [30]	NACA 0015	100 mm	Rectangular, Shark skin based	Rectangular: $l = 10$ mm, $h = 6$ mm. Shark skin-based chord: 10, 20, 30 mm	For VG at 0.3c, the lowest drag coefficient is obtained
Lin et al. [31]	McDonnell Douglas Aerospace	22″	Delta wing, Trapezoidal wing	Delta wing: $h = 0.1''$, $h/c = 0.0045$, AOI = 23° Trapezoidal: $h = 0.04''$, $h/c = 0.0018$, AOI = 23°	The optimum location for the VG during landing configuration was found to be 25% of the chord length
Delnero et al. [33]	Eppler 387	420 mm	Triangular	$l = 40$ mm, $h = 10$ mm, $t = 0.5$ mm	The $C_{L,\max}$ increased and the stall curve became smoother
Yang et al. [38]	Modified DU 91-W2-250	600 mm	Triangular	$h = 5$ mm, $l = 17$ mm, AOI = 16.4°, $s = 25$ mm	VG delayed the stall (counter-rotating)
Nickerson Jr [39]	NACA 0024	254 mm	Trapezoidal (co-rotating)	$h = 3.81$ mm, $l = 10.16$ mm, AOI = 45°, $s = 1.58$ mm	L/D ratio increased and the stall angle for the lowest Reynolds number tested
Seshagiri et al. [40]	LS(1)-0417 GA(W)-1	165 mm	Rectangular vane type	$h = 2$ mm, $t = 0.2$ mm (counter and co-rotating)	The VGs were able to increase the $C_{L,\max}$ by 25%

Acknowledgements We highly acknowledge the valuable interactions and guidance provided by Dr. Kunal Ghosh, Retired Professor and Head, Aerospace Engineering Department, IIT Kanpur and Dr. Chetan Mistry, Assistant Professor, Aerospace Engineering Department, IIT Kharagpur.

Nomenclature

C_L	Lift coefficient –
C_D	Drag coefficient –
h	Height of VG [mm]
l	Length of VG [mm]
L	Lift force [N]
D	Drag force [N]
t	Thickness of VG [mm]
s	Space between two VGs [mm]
α	Angle of attack on airfoil ($\circ$)
δ	Boundary layer thickness [mm]

References

1. Anderson JD (2005) Ludwig Prandtl's boundary layer. Phys Today 58(12):42–48
2. Moghaddam T, Neishabouri NB (2017) On the active and passive flow separation control techniques over airfoils. IOP Conf Ser: Mater Sci Eng 248:012009
3. Wang J, Feng L (2018) Flow control techniques and applications, 1st ed. Cambridge University Press
4. Garner P, Meredith P, Stoner R (1991) Areas for future CFD development as illustrated by transport aircraft applications. In: 10th computational fluid dynamics conference, Honolulu, HI, U.S.A, American Institute of Aeronautics and Astronautics
5. Mokhtarian F, Modi VJ, Yokomizo T (1988) Rotating air scoop as airfoil boundary-layer control. J Aircr 25(10):973–975
6. Velte CM, Hansen MOL, Cavar D (2008) Flow analysis of vortex generators on wing sections by stereoscopic particle image velocimetry measurements. Environ Res Lett 3(1):015006
7. Zhu C, Chen J, Wu J, Wang T (2019) Dynamic stall control of the wind turbine airfoil via single-row and double-row passive vortex generators. Energy 189:116272
8. Shape Memory Alloys, National Aeronautics and Space Administration. https://www.nasa.gov/sites/default/files/atoms/files/shape-memory-alloys-activity_0.pdf
9. ST Aerospace A-4SU Super Skyhawk. https://en.wikipedia.org/wiki/ST_Aerospace_A-4SU_Super_Skyhawk
10. Making the Perfect Lancair, Kitplanes. https://www.kitplanes.com/making-the-perfect-lancair/
11. British Aerospace Sea Harrier FA.2, Airplane Pictures. https://www.airplane-pictures.net/type.php?pg=2&p=218&order=views
12. Jirasek A (2006) Vortex generator modeling and its application to optimal control of airflow in inlet. Ph.D. Thesis, Royal Institute of Technology
13. Vortex Generators Boost Performance, Aircraft Owners and Pilots Association. https://www.aopa.org/news-and-media/all-news/2015/june/11/climbing-higher
14. Micro AerodDnamics Inc. https://microaero.com/products/airvan-ga8-airvan-ga8-t320/

15. Mooney M20 Series, Micro Vortex Generator Kit. https://microaero.com/wp-content/uploads/2023/06/mooney-ocs.pdf
16. Bragg MB, Gregorek GM (1987) Experimental study of airfoil performance with vortex generators. J Aircr 24(5):305–309
17. Li-shu H, Chao G, Wen-Ping S, Ke S (2013) Airfoil flow control using vortex generators and a gurney flap. Proc Inst Mech Eng C J Mech Eng Sci 227(12):2701–2706
18. Sorensen NN, Zahle F, Bak C, Vronsky T (2014) Prediction of the effect of vortex generators on airfoil performance. J Phys: Conf Ser 524:012019
19. Fouatih OM, Medale M, Imine O, Imine B (2016) Design optimization of the aerodynamic passive flow control on NACA 4415 airfoil using vortex generators. Eur J Mech B Fluids 56:82–96
20. van Egmond JJ (2015) Passive flow separation control for high lift—an experimental investigation on a novel vortex generator. MS Thesis, Faculty of Aerospace Engineering, Delft University of Technology
21. Kumar GV, Narayanan KS, Aravindhkumar SK, Kishorekumar S (2016) Comparative analysis of various vortex generators for a NACA 0012 Aerofoil. Int J Innov Stud Sci Eng Technol 2(5):3–6
22. Jumahadi MT, Saad MR, Idris AC, Sujipto S, Rahman MRA (2018) The potential of hybrid micro-vortex generators to control flow separation of NACA 4415 airfoil in subsonic flow. In: International conference on engineering and technology (IntCET 2017), Putrajaya, Malaysia, p 020030
23. Zulkefli NF, Ahamat MA, Mohd Safri NF, Mohd Nur N, Mohd Rafie AS (2019) Lift enhancement of NACA 4415 airfoil using biomimetic shark skin vortex generator. Int J Recent Technol Eng 8(4):9231–9234
24. Agarwal A, Kumar P (2016) Numerical investigation of flow field and effect of varying vortex generator location on wing performance. Am J Fluid Dyn 6(1):11–19
25. Chillon S, Uriarte-Uriarte A, Aramendia I, Martínez-Filgueira P, Fernandez-Gamiz U, Ibarra-Udaeta I (2020) JBAY modeling of vane-type vortex generators and study on airfoil aerodynamic performance. Energies 13(10):2423
26. Ayudia SA, Arkundato A, Rohman L (2020) Study of vortex generator effect on airfoil aerodynamics using the computational fluids dynamics method. Comput Exp Res Mater Renew Energy 3(2):23
27. Himo R, Bou-Mosleh C, Habchi C (2021) Aerodynamic performance enhancement of an airfoil using trapezoidal vortex generators. Aircr Eng Aerosp Technol 93(1):76–84
28. Hao L, Hu B, Gao Y, Wei B (2023) Effect of vortex generator spanwise height distribution pattern on aerodynamic characteristics of a straight wing. Adv Aerodyn 5(1):4
29. Domel AG, Saadat M, Weaver JC, Haj-Hariri H, Bertoldi K, Lauder GV (2018) Shark skin-inspired designs that improve aerodynamic performance. J R Soc Interface 15(139):20170828
30. Arunvinthan S, Raatan VS, Nadaraja Pillai S, Pasha AA, Rahman MM, Juhany KA (2021) Aerodynamic characteristics of shark scale-based vortex generators upon symmetrical airfoil. Energies 14(7):1808
31. Lin JC, Robinson SK, McGhee RJ, Valarezo WO (1994) Separation control on high-lift airfoils via micro-vortex generators. J Aircr 31(6):1317–1323
32. Hariyadi SPS, Widodo WA, Junipitoyo B, Suryono W, Supriadi S (2018) Investigasi Perbandingan Posisi rectangular flat plate vortex generator Dengan Posisi Straight Pada wing airfoil NACA 43018. Jurnal Penelitian 3(3):36–43
33. Delnero JS, Leo JMD, Camocardi ME, Martinez MA, Lerner JLC (2012) Experimental study of vortex generators effects on low Reynolds Number airfoils in turbulent flow. Int J Acad Dev 2(1):50
34. Li XK, Liu W, Zhang TJ, Wang PM, Wang XD (2019) Experimental and numerical analysis of the effect of vortex generator installation angle on flow separation control. Energies 12(23):4583
35. Vinodhini PJ, Samuvel TJ, Raj GS (2016) Numerical analysis of drag reduction method using vortex generator on symmetric aerofoil. Int J Eng Trends Technol 35(1):33–36

36. Özden M, Genç MS, Koca K (2023) Passive flow control application using single and double vortex generator on S809 wind turbine airfoil. Energies 16(14):5339. https://doi.org/10.3390/en16145339
37. Chen J, Hu C, Zhang M, Huang B, Zhang H (2022) Experimental investigation into passive control effect of micro vortex generator at various cavitation conditions. Ocean Eng 260:111734. https://doi.org/10.1016/j.oceaneng.2022.111734
38. Yang K, Zhang L, Xu J (2010) Simulation of aerodynamic performance affected by vortex generators on blunt trailing-edge airfoils. Sci China Ser E Technol Sci 53(1):1–7
39. Nickerson J (1986) A study of vortex generators at low Reynolds numbers. In: 24th Aerospace sciences meeting. American Institute of Aeronautics and Astronautics, Reno, NV, U.S.A.
40. Seshagiri A, Cooper E, Traub LW (2009) Effects of vortex generators on an airfoil at low Reynolds Numbers. J Aircr 46(1):116–122

Investigation of Streamline Patterns and Wall Shear Stress Distributions at Different Stages of Carotid Artery Stenosis

Md Yeashir Arafat, Ayesha Siddika, Tasnova Tamanna Silme, and Shashwata Chakraborty

Abstract Carotid artery bifurcations are prone to plaque formation or stenosis due to their complex geometry. In this study, the streamline patterns and wall shear stress at the carotid artery bifurcation have been investigated using a commercial CFD code. The effect of stenosis on the blood flow patterns has been observed at mild (25%), moderate (50%), and severe (75%) stenoses. The steady-state continuity and Navier–Stokes momentum equations with constant blood properties are considered for the numerical analysis. Systole and diastole phases are defined by Reynolds numbers 145 and 40, respectively. The results show increased blood velocity and arterial wall shear stress at the stenosed region.

Keywords Carotid artery bifurcation · Stenosis · Streamline · Wall shear stress · CFD

M. Y. Arafat (✉)
Department of Mechanical Engineering, Bangladesh University of Engineering and Technology, Dhaka, Bangladesh
e-mail: yeashir.arafat45@gmail.com

A. Siddika
Upazila Health Complex, Sakhipur, Tangail, Dhaka, Bangladesh

T. T. Silme
Department of Zoology, University of Dhaka, Dhaka, Bangladesh

S. Chakraborty (✉)
Joint School of Nanoscience and Nanoengineering, North Carolina A&T State University, Greensboro, USA
e-mail: schakraborty@aggies.ncat.edu

A. Kumar R. et al. (eds.), *Proceedings of Fluid Mechanics and Fluid Power (FMFP) 2023, Vol. 6*, Lecture Notes in Mechanical Engineering,
https://doi.org/10.1007/978-981-96-9600-0_50

1 Introduction

Carotid artery stenosis is one of the major causes of stroke and transient ischemic attacks (TIA) in patients [1]. Carotid stenosis is defined as the abnormal narrowing of the arterial blood vessels in the neck that delivers oxygenated blood to the brain and head. Atherosclerosis, the phenomena of plaque formation and progression, is the root cause of carotid stenosis [2]. Plaque is mainly composed of cholesterol, calcium, fatty substances, cellular waste products, and some other substances available in the blood. Plaque formation is accomplished in two different ways. The vulnerable (unstable) plaques are usually surrounded by thin fibrous caps, which are more likely to rupture [3]. Typically, the fissuring of vulnerable plaques is associated with high mechanical stress and blood flow velocities in the artery [4]. On the other hand, the calcified plaques contain thick fibrous caps that make them stable [3]. These plaques continue to harden and narrow the arterial blood, limiting the oxygenated blood flow in the brain [5].

The first and foremost step in the treatment procedures of carotid stenosis is the patient stratification based on the degree of stenosis [6]. The degree of stenosis can be categorized into stages, namely mild (< 30%), moderate (30–69%), or severe (70–99%) [7]. The advent of several vascular imaging techniques, i.e., computed tomography (CT) angiography, magnetic resonance (MR) angiography, and ultrasonography, has paved the way for determining the degree of carotid stenosis in a manifold [6]. However, these advanced techniques are unable to provide complete data on the blood flow domain due to their limited spatial resolution [8]. Therefore, we still lack the lucid perception of the arterial hemodynamic that may add a further dimension to the technology and treatment involved with arterial stenosis. To overcome this limitation, computational fluid dynamics (CFD) models are being widely adopted to predict the flow domain in human arteries [9].

Bifurcations in carotid arteries are more susceptible to the formation of plaque or stenosis and other diseases due to low and altered shear stress distribution [10–12]. This has led to an increasing interest in the blood rheology at the bifurcations in human carotids [13–15]. Computational and experimental studies are being conducted to investigate the use of flow diverters, commonly known as stents, to remove abnormalities as well as for targeted drug delivery to treat diseases at the bifurcating region [16–18]. Hence, blood flow dynamics at stenosed arteries have garnered significant attention in recent years [19–21]. In this study, efforts are aimed at numerically investigating the streamline patterns and wall shear stress at mild, moderate, and severe stages of stenosis and comparing them with those at no stenosis. Blood properties have been considered constant throughout the analysis. Systolic and diastolic phases are simulated by a Reynolds number of 145 and 40, respectively.

2 Mathematical Modeling

The two-dimensional geometry of a carotid artery bifurcation has been adapted from Thomas et al. [22] and presented in Fig. 1. The tuning fork-shaped bifurcating geometry contains a common carotid artery (CCA), which bifurcates into two regions called internal carotid artery (ICA) and external carotid artery (ECA). The inlet diameter at CCA has been considered 5.0 mm, while the outlet diameters at ICA and ECA are 3.6 mm and 2.8 mm, respectively. The bifurcating angle is defined to be 60°. Though blood is a non-Newtonian fluid, its Newtonian viscosity becomes dominant while flowing in larger arteries owing to high shear rates [23]. Since the Newtonian and non-Newtonian models show a slight discrepancy (< 5.0%) while comparing velocity profiles and wall shear stress distributions in a carotid artery, the non-Newtonian viscosity can be neglected to simplify the numerical study [24]. Moreover, the effect of elasticity is negligible (< 2.0%) on the wall shear stress distribution of a carotid artery due to its larger lumen diameter [19]. Hence, the numerical analysis has been conducted considering the walls of the artery to be rigid.

An incompressible, Newtonian, and steady flow of blood has been assumed for the computational fluid dynamics (CFD) study. The numerical study is governed by the continuity and momentum equations stated as follows [25, 26]

$$\frac{\partial u}{\partial x} + \frac{\partial u}{\partial y} = 0 \tag{1}$$

$$\rho\left(u\frac{\partial u}{\partial x} + v\frac{\partial u}{\partial y}\right) = -\frac{\partial p}{\partial x} + \mu\left(\frac{\partial^2 u}{\partial x^2} + \frac{\partial^2 u}{\partial y^2}\right) \tag{2}$$

$$\rho\left(u\frac{\partial v}{\partial x} + v\frac{\partial v}{\partial y}\right) = -\frac{\partial p}{\partial x} + \mu\left(\frac{\partial^2 v}{\partial x^2} + \frac{\partial^2 v}{\partial y^2}\right), \tag{3}$$

where ρ and μ are the density and dynamic viscosity of blood, respectively. The wall shear stress (τ_{wall}) at the artery walls can be defined as follows [25, 26]

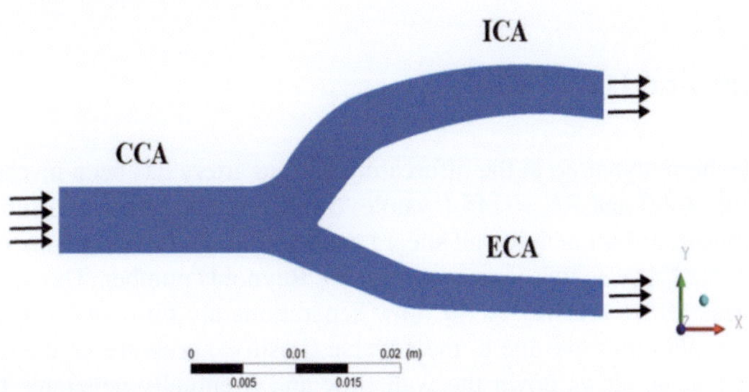

Fig. 1 2-D geometry of carotid artery bifurcation designed in ANSYS design Modeler

Table 1 Grid sensitivity check on the bifurcating geometry with no stenosis at Re = 145

Grid size (mm)	0.4	0.36	0.33	0.3	0.27
Peak wall shear stress (Pa)	0.746	0.78	0.795	0.8	0.802
% |Error|	–	4.558	1.923	0.629	0.25

$$\tau_{\text{wall}} = \mu\left(\frac{\partial u}{\partial y} + \frac{\partial v}{\partial x}\right)\Big|_{y=\pm R}, \tag{4}$$

where R is the radius of the artery lumen. The no-slip boundary condition has been applied at the artery walls. Blood is assumed to enter the inlet (CCA) at Re = 40 during diastole and at Re = 145 during systole. A relative zero pressure is defined at the outlets (ICA and ECA). Blood density (1050 kg/m^3) and viscosity (0.00408 Pa s) are considered constant throughout the study. The length of the geometry has been extended at both the inlet and outlets to ensure that the flow regime of interest is unaffected by the flow conditions at the boundary.

3 Numerical Procedure

The CFD model has been solved in ANSYS Fluent using a pressure-based solver. The pressure and velocity profiles have been coupled by implementing the SIMPLE algorithm. The convective terms in the governing equations have been discretized, adopting the second-order upwind method. Mesh has been generated in ANSYS Meshing. A grid sensitivity test has been performed on the bifurcating geometry with no stenosis for different grid sizes at $Re = 145$, and the results are summarized in Table 1. A slight deviation in the peak wall shear stress is perceived for the grid sizes of 0.3 mm and 0.25 mm. Hence, the simulation has been carried out by selecting the grid size of 0.3 mm. as shown in Fig. 2. Finally, the solutions are converged at a scaled residual $\leq 10^6$.

4 Results and Discussion

Initially, the hemodynamic in the bifurcating carotid artery has been investigated at Re = 40 (diastole) and Re = 145 (systole) to observe the effect of bifurcation on the streamline distribution and wall shear stress. In Fig. 3, the maximum velocity is observed at the CCA, which increases with the Reynolds number. Though the flow regime is less affected at Re = 40, flow separations are observed at Re = 145. These flow separations are due to the adverse (positive) pressure gradient near the bifurcating region, slows down the wall flow and eventually generates backflow. The adverse pressure gradient is resulted from the expansion of the flow area due

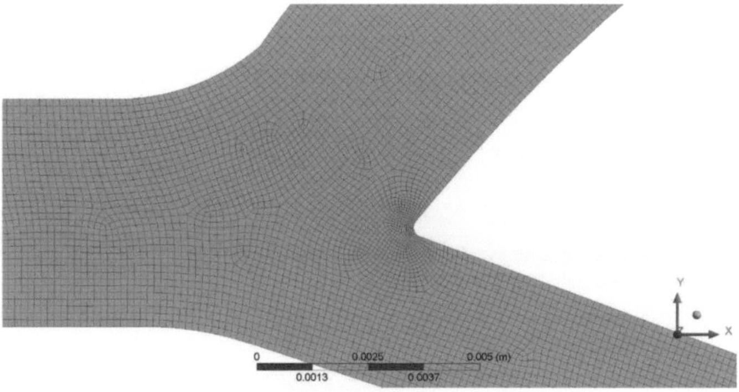

Fig. 2 Meshing surface generated in ANSYS meshing

to bifurcation. Since ICA has a higher diameter than ECA, more flow separation is observed at the junction between CCA and ICA than between CCA and ECA.

The wall shear stress distribution in Fig. 4 also confirms the event of flow separation during diastole. The flow separation causes zero wall shear stress at Re = 145 near the junctions ICA and ECA. The region between the two points with $\tau_{wall} = 0$ Pa at the junction of ICA is larger than that of ECA, which also supports increased flow separation in ICA. At Re = 40, though the values of τ_{wall} reduce at the bifurcating region, they do not reach zero; thus, flow is not separated. The zero-wall shear stress and flow separation at the bifurcating region are responsible for plaque formation, which ultimately leads to stenosis in the long run [8, 27]. The junction at the ICA is more susceptible to plaque formation than that of the ECA, as the flow regime is more affected by the bifurcation.

The effects of stenosis on streamline distribution and wall shear stress at three stages have been presented in Figs. 5, 6, 7 and 8. Mild, moderate, and severe stenoses have been represented by 25%, 50, and 75% blockages, respectively, in ICA near the bifurcation. Streamline distributions in Fig. 5 and shear stress plots in Fig. 6 depict that during diastole (Re = 40).

No separation or zero wall shear stress occurs in the flow regime. Thus, the possibility of plaque accumulation in the bifurcating region is quite low during diastole. However, higher velocities and wall shear stresses are observed with the growth of stenosis in ICA due to the narrowing of the flow channel. Though this increment is negligible in mild conditions (25%), substantial effects are observed for severe stenosis (75%). The ICA stenosis growth has little impact on the flow regime in ECA.

During systole, the flow separation tends to become more prominent with the stenosis growth in ICA, as illustrated by the streamline patterns in Fig. 7 and shear stress plots in Fig. 8. Therefore, ICA becomes more susceptible to plaque accumulation during the systolic phase as the stenosis continues to grow. On the other hand, the flow separation region continues to decrease in ECA and disappears at severe stenosis (75%). This is because, regions with $\tau_{wall} = 0$ Pa has been observed in all the

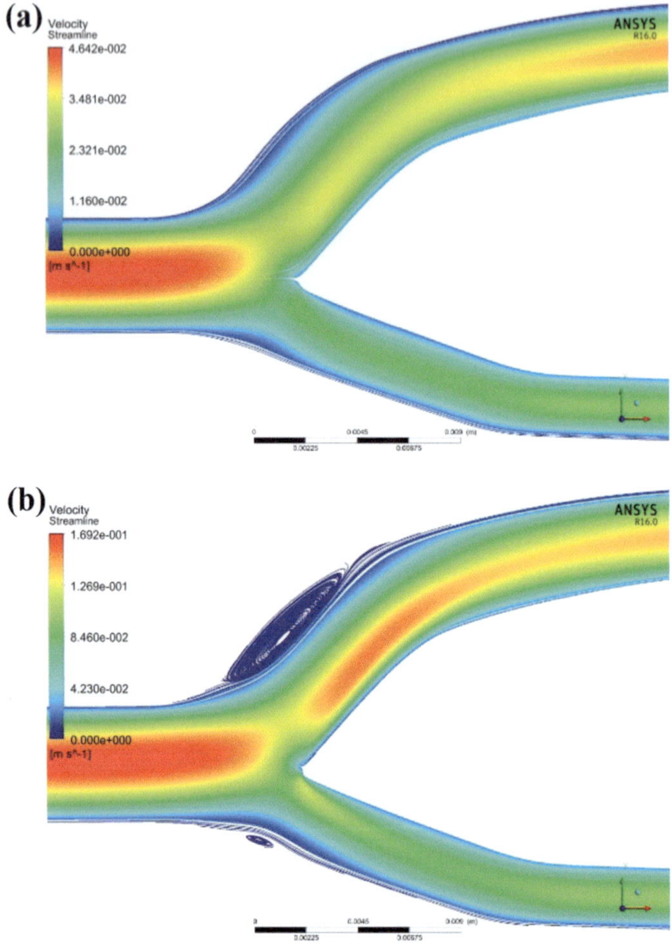

Fig. 3 Streamline distribution in carotid artery bifurcation with no stenosis at **a** Re = 40 (diastole), **b** Re = 145 (systole)

considered cases except for severe stenosis in ECA. With the growth of stenosis, the flow separation tends to occur at the region immediately beyond the lowest diameter, while the separating region just before the lowest diameter continues to disappear. This phenomenon can be explained by the gradual contraction and expansion of the blood vessel due to stenosis. Thus, plaque accumulation will continue in the region immediately after the stenosed region.

Upon narrowing the blood vessel at the ICA, the blood velocity and wall shear stress at the stenosed region continue to increase. At severe stenosis (75%), they reach maximum values considerably higher than those of the artery with no stenosis. The peak velocity and wall shear stress at the carotid artery bifurcation with different

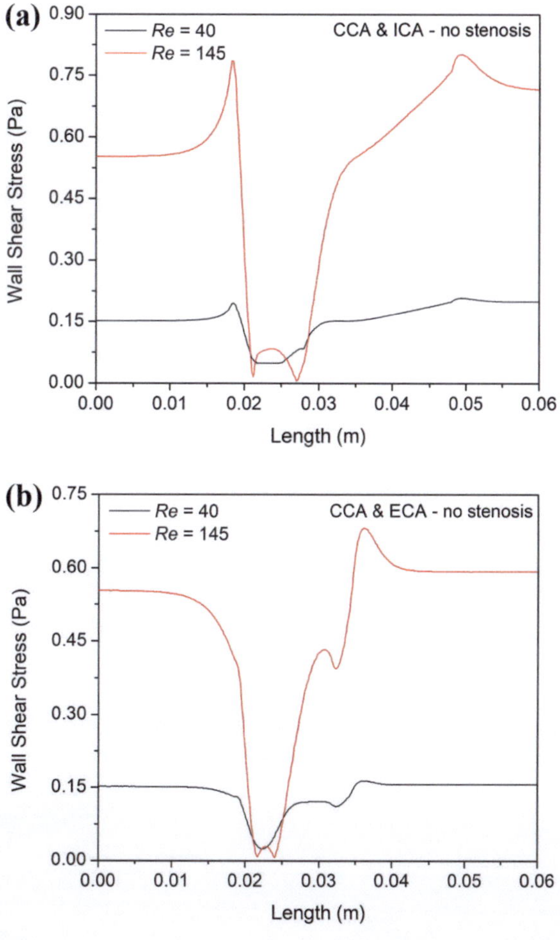

Fig. 4 Wall shear stress distribution in carotid artery bifurcation with no stenosis at **a** CCA and ICA, **b** CCA and ECA

stages of stenosis developed in ICA have been outlined in Table 2 during the systolic and diastolic phases.

When the plaque begins to accumulate near the bifurcating region, the resulting higher velocity and wall shear stress force the vulnerable plaques to break down. These tiny pieces of plaques are carried with the blood toward the artery branches with a lower diameter, thus blocking them [4]. On the other hand, the calcified plaques continue to grow bigger and gradually narrowing down the blood vessel. Though the effect of the narrowed blood vessel is not prominent in mild stenosis, it becomes life-threatening at the moderate to severe stage due to the shortage of the required amount of oxygenated blood supply in the parts distal to the stenosis [5].

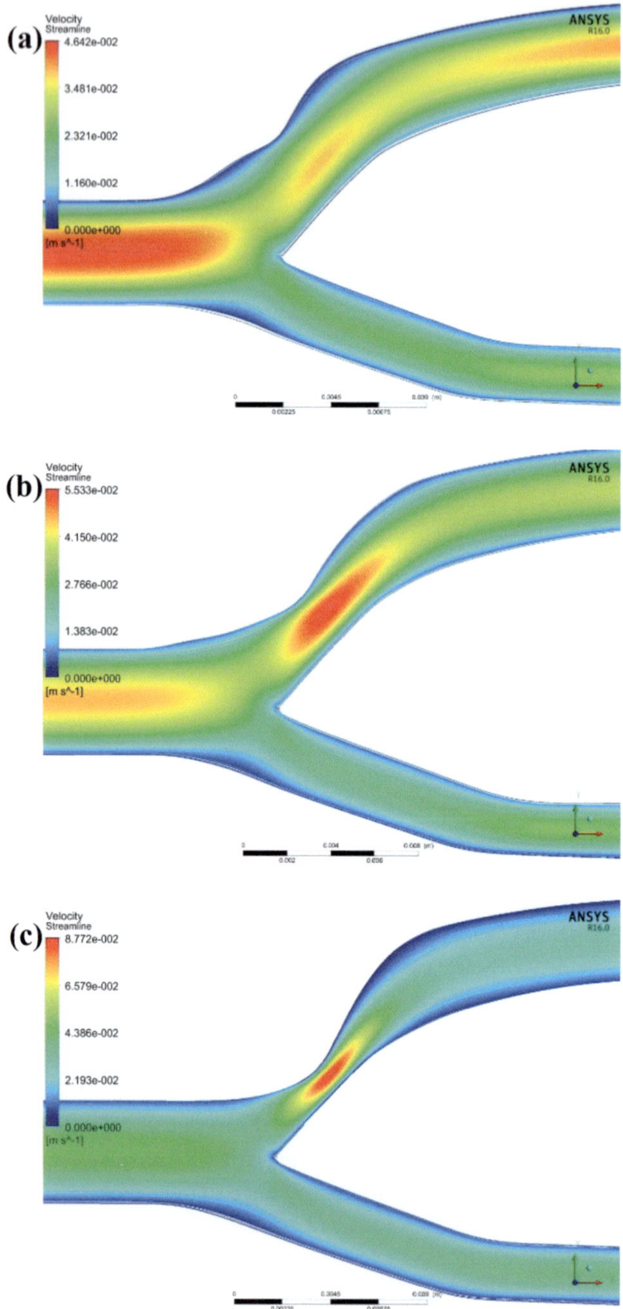

Fig. 5 Streamline distribution in carotid artery bifurcation at Re = 40 with stenosis **a** mild (25%), **b** moderate (50%), **c** severe (75%)

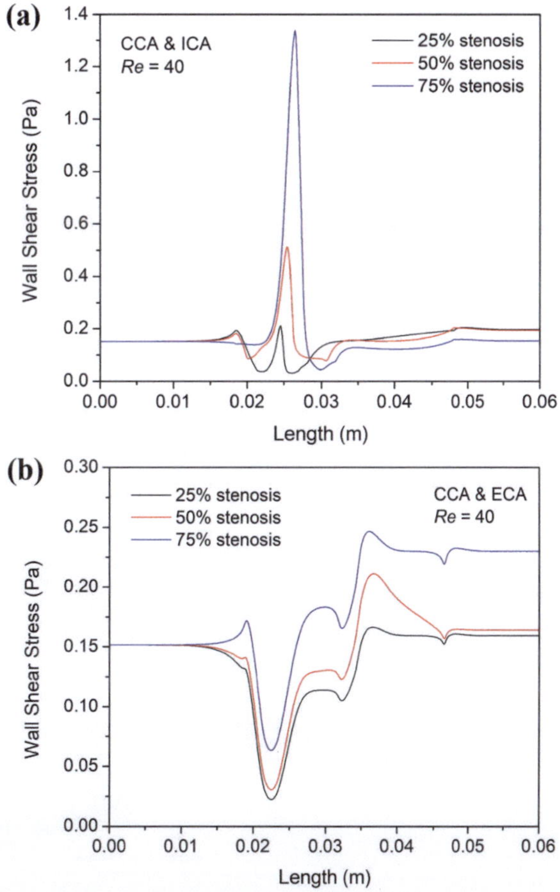

Fig. 6 Wall shear stress distribution in carotid artery bifurcation with stenosis during diastole at **a** CCA and ICA, **b** CCA and ICA

5 Conclusions

A CFD study has been conducted to investigate the streamline patterns and wall shear stress distributions in carotid artery bifurcation. The events of mild, moderate, and severe stenoses have been studied, reducing the ICA diameter by 25%, 50%, and 75%, respectively. The numerical analysis has led to the following outcomes.

i. Reduced wall shear stresses and flow separations are observed near the carotid artery bifurcation due to the expansion of the blood vessel, making it a favorable site for plaque formation and stenosis growth over time. The probability of plaque formation in the ICA junction is greater than that of ECA due to more flow separation and lower wall shear stress.
ii. Though at mild stenosis (25%), the flow regime is less affected by stenosis, higher peak velocity and wall shear stress are observed at moderate stenosis, and noticeable effects are observed at severe stenosis. The increased velocity

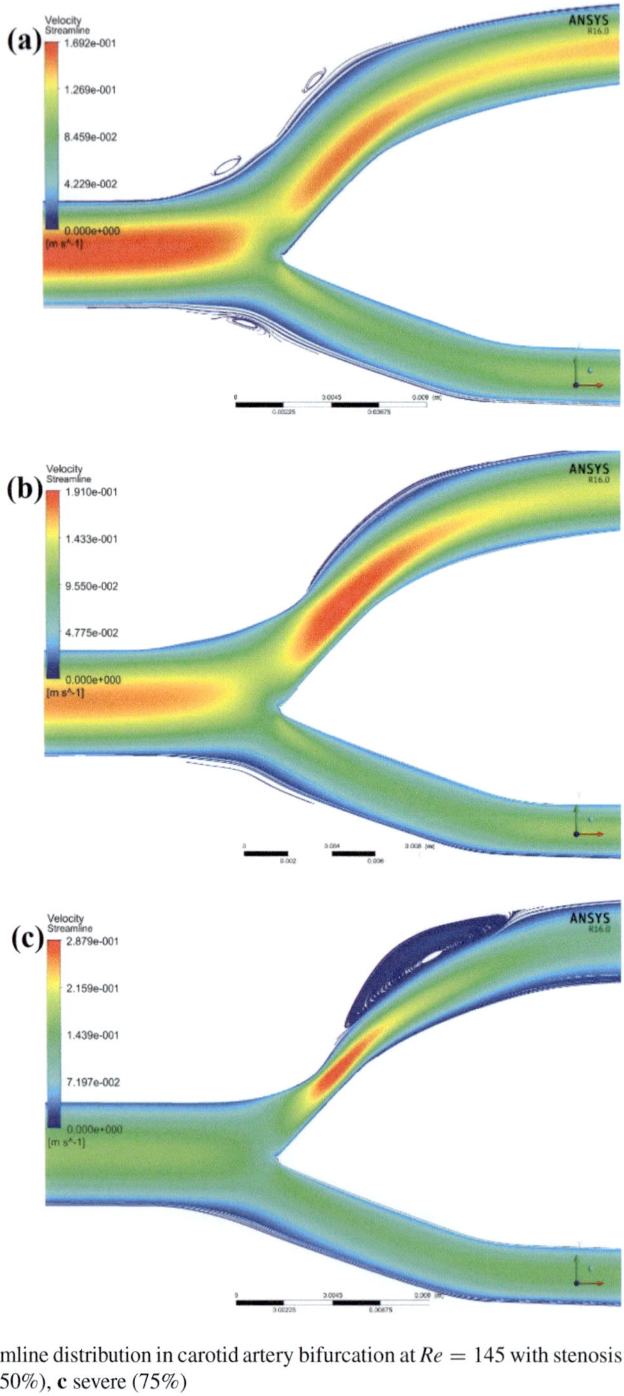

Fig. 7 Streamline distribution in carotid artery bifurcation at $Re = 145$ with stenosis **a** mild (25%), **b** moderate (50%), **c** severe (75%)

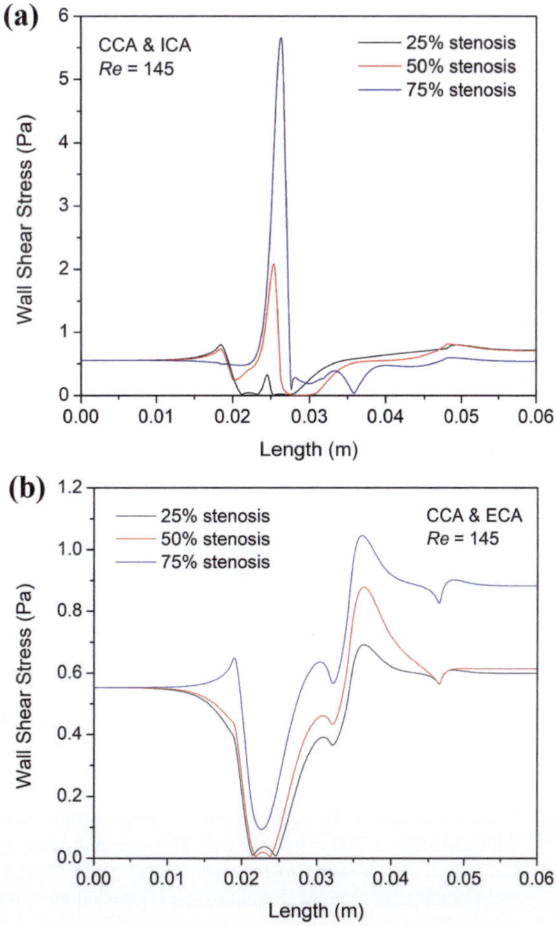

Fig. 8 Wall shear stress distribution in carotid artery bifurcation with stenosis during systole at **a** CCA and ICA, **b** CCA and ICA

Table 2 The peak velocity and wall shear stress at the carotid artery bifurcation

Degree of stenosis	Cardiac cycle	Peak velocity (m/s)	Peak wall shear stress (Pa)
No stenosis	Diastole	0.046	0.207
	Systole	0.169	0.800
Mild (25%)	Diastole	0.046	0.212
	Systole	0.169	0.801
Moderate (50%)	Diastole	0.055	0.512
	Systole	0.191	2.085
Severe (75%)	Diastole	0.088	1.345
	Systole	0.288	5.659

and wall shear stress at the stenosed region cause the vulnerable plaques to rupture. On the other hand, the calcified plaques create notable disturbances in the flow regime and significantly reduce the oxygenated blood supply during severe stenosis.

iii. The results obtained from the CFD study can be further improved by adopting 3D bifurcating geometry and transient flow of blood as a non-Newtonian fluid. Moreover, elasticity can be incorporated at the arterial walls through fluid–solid interfacing (FSI).

References

1. Flaherty ML, Kissela B, Khoury JC, Alwell K, Moomaw CJ, Woo D, Khatri P, Ferioli S, Adeoye O, Broderick JP, Kleindorfer D (2012) Carotid artery stenosis as a cause of stroke. Neuroepidemiology 40(1):36–41
2. Henriques J, Amaro AM, Piedade AP (2023) Understanding atherosclerosis pathophysiology: can additive manufacturing be helpful? Polymers 15(3):480
3. Schmitt C, Soulez G, Maurice RL, Giroux M-F, Cloutier G (2007) Noninvasive vascular elastography: toward a complementary characterization tool of atherosclerosis in carotid arteries. Ultrasound Med Biol 33(12):1841–1858
4. Finn V, Nakano M, Narula J, Kolodgie FD, Virmani R (2010) Concept of vulnerable/unstable plaque. Arterioscler Thromb Vasc Biol 30(7):1282–1292
5. Alfarisi HAH, Mohamed ZBH, Ibrahim MB (2020) Basic pathogenic mechanisms of atherosclerosis. Egypt J Basic Appl Sci 7(1):116–125
6. Brinjikji W, Huston J, Rabinstein AA, Kim G-M, Lerman A, Lanzino G (2016) Contemporary carotid imaging: from degree of stenosis to plaque vulnerability. J Neurosurg 124(1):27–42
7. Warlow C (1991) MRC European carotid surgery trial: interim results for symptomatic patients with severe (70–99%) or with mild (0–29%) carotid stenosis. The Lancet 337(8752):1235–1243
8. Buchmann NA, Yamamoto M, Jermy M, David T (2010) Particle Image Velocimetry (PIV) and Computational Fluid Dynamics (CFD) modelling of carotid artery haemodynamics under steady flow: a validation study. J Biomech Sci Eng 5(4):421–436
9. Lopes D, Puga H, Teixeira J, Lima R (2020) Blood flow simulations in patient-specific geometries of the carotid artery: a systematic review. J Biomech 111:110019
10. Ballambat RP, Zuber M, Khader SMA, Ayachit A, Ahmad KA, Vedula RR, Kamath SG, Shuaib IL (2022) Influence of postural changes on haemodynamics in internal carotid artery bifurcation aneurysm using numerical methods. Vis Comput Ind Biomed Art 5(1):11
11. Zalud NC, Bulusu KV, Plesniak MW (2023) Shear stress metrics associated with pro-atherogenic high-risk anatomical features in a carotid artery bifurcation model. Clin Biomech 105:105956
12. Wild NC, Bulusu KV, Plesniak MW (2023) Vortical structures promote atheroprotective wall shear stress distributions in a carotid artery bifurcation model. Bioengineering 10(9):1036
13. Zohourian T, Hines G (2022) The evolution of current management for carotid artery bifurcation disease. Cardiol Rev. https://doi.org/10.1097/CRD.0000000000000497
14. Zhou X, Yin L, Xu L, Liang F (2020) Non-periodicity of blood flow and its influence on wall shear stress in the carotid artery bifurcation: an in vivo measurement-based computational study. J Biomech 101:109617
15. Hoving M, de Vries EE, Mikhal J, de Borst GJ, Slump CH (2020) A systematic review for the design of in vitro flow studies of the carotid artery bifurcation. Cardiovasc Eng Technol 11(2):111–127

16. Mahmoud M, Farag MA, Farid M, Elserwi A, Abdelsamad A, Guergues W, Hassan F (2020) Application of flow diverters in the treatment of aneurysms in the internal carotid artery bifurcation region. Neuroradiol J 33(4):297–305
17. Hameed MS, Shah AA, Khan MI, Ali A, Hussain I, Bukhari MD (2023) Comparison of blood flow analysis in stenosed and stented carotid artery bifurcation models. Cogent Eng 10(1):2158624
18. Aryan H, Beigzadeh B, Siavashi M (2022) Euler-Lagrange numerical simulation of improved magnetic drug delivery in a three-dimensional ct-based carotid artery bifurcation. Comput Methods Programs Biomed 219:106778
19. Albadawi M, Abuouf Y, Elsagheer S, Sekiguchi H, Ookawara S, Ahmed M (2022) Influence of rigid-elastic artery wall of carotid and coronary stenosis on hemodynamics. Bioengineering 9(11):708
20. Tokugawa J, Kudo K, Mitsuhashi T, Hishii M (2023) Older age, carotid artery stenosis, and female sex as factors correlated with twisted carotid bifurcation based on 457 angiographic studies. Clin Neurol Neurosurg 233:107902
21. Albadawi M, Abuouf Y, Elsagheer S, Ookawara S, Ahmed M (2021) Predicting the onset of consequent stenotic regions in carotid arteries using computational fluid dynamics. Phys Fluids 33(12):123106
22. Thomas JB, Antiga L, Che SL, Milner JS, Steinman DAH, Spence JD, Rutt BK, Steinman D (2005) Variation in the carotid bifurcation geometry of young versus older adults. Stroke 36(11):2450–2456
23. Berger SA, Jou L-D (2000) Flows in stenotic vessels. Annu Rev Fluid Mech 32(1):347–382
24. Boyd J, Buick JM (2007) Comparison of Newtonian and Non-Newtonian flows in a two-dimensional carotid artery model using the lattice boltzmann method. Phys Med Biol 52(20):6215
25. Vimmr J, Jonášová A, Bublík O (2013) Numerical analysis of non-newtonian blood flow and wall shear stress in realistic single, double and triple aorto-coronary bypasses. Int J Numer Methods Biomed Eng 29(10):1057–1081
26. Fan Y, Jiang W, Zou Y, Li J, Chen J, Deng X (2009) Numerical simulation of pulsatile non-newtonian flow in the carotid artery bifurcation. Acta Mech Sin 25(2):249–255
27. Penman D, Ralston SH, Strachan MWJ, Hobson RP (2022) Davidson's principles and practice of medicine international edition, 24th edn. Elsevier, ISBN: 9780702083488

A Numerical Study on Focusing of Micro-particle in Serpentine Microchannels

Bruhad S. Naik, Niraj K. Shah, and Absar M. Lakdawala

Abstract Inertial separation, is one of the promising techniques, separates the particles based on size and without any external force. The 3D simulation along with the particle tracing makes them computationally expensive. To reduce the computation cost, the present study is done using 2D simulations. Parametric investigations are done to get a primary idea of successful focusing of particles. The present study explores the effect of number of turns and Reynolds number on the focusing of 5 and 10 μm microparticles in serpentine microchannel. The study has been carried out for Reynolds number 20, 30, 40, 50, 60 and 80 and number turns of 16, 18, 20, 22 and 24 for the serpentine channel. The number of turns and Reynolds number has significant effect on focusing behaviour of 5 μm particles whereas no significant effect is seen for 10 μm particles. Promising result is seen for Reynolds number 40–60 for 5 μm particles for all turns where as promising results are seen at Reynolds number 20–50 for 10 μm particles irrespective of number of turns. The computational time is significantly reduced for 2D simulation making primary parametric investigation fast and cost-effective compare to 3D simulations.

Keywords Serpentine channel · Effect of number of turns · Effect of Reynolds number · 2D COMSOL simulation

1 Introduction

The lab on chip (LOC) devices are getting more attraction as a possible replacement for conventional diagnostics techniques. Diagnostics of a disease requires the separation of disease cell from the healthy cells. In early stage of cancer in a ml of

B. S. Naik · N. K. Shah (✉) · A. M. Lakdawala
Department of Mechanical Engineering, Institute of Technology, Nirma University, Ahmedabad, India
e-mail: niraj.shah@nirmauni.ac.in

A. M. Lakdawala
e-mail: Absar.lakdwala@nirmauni.ac.in

A. Kumar R. et al. (eds.), *Proceedings of Fluid Mechanics and Fluid Power (FMFP) 2023, Vol. 6*, Lecture Notes in Mechanical Engineering,
https://doi.org/10.1007/978-981-96-9600-0_51

blood, the number of circulating tumour cell (CTC) is 10–15 as compare to millions of red blood cells (RBC) [1]. Inertial separation governed microfluidic LOC device can separate disease particle from whole blood without affecting its cell integrity. A similar device can be used to separate blood constituents such as red blood cell (RBC), white blood cell (WBC) and blood plasma. In inertial separation the separation of particle is achieved by focusing particle in a narrow region and diverting them to the marked outlet. The shape of the channel and cross section of channel along with fluid flow characteristics plays significant role in separation characteristics of particles. The channel can have straight, contraction expansion, serpentine, spiral shapes having square, rectangle, circular or trapezoidal cross sections. The cross section plays a major role in getting number of focusing position, i.e., a square channel may have four focusing position where as a rectangular channel may have two focusing positions [2].

2 Literature Review and Objective

The migration of microparticles at a fix location of 0.6 times radius in a circular tube was first observed by Segre and Silberge [3]. To explain the migration of particle widely accepted theory is that the particle flowing in a Poiseuille flow experiences two lift forces. The parabolic profile of flow creates a force towards the wall pushing the particle away from the centre, called shear induced lift force F_s. The particle near the wall gets pushed with a force away from wall due to presences of wall towards the centre, called wall induced lift force F_w. The direction of both the forces is perpendicular to the direction fluid flow. These opposite forces acting on the particle push the particle in an equilibrium position focusing them in a narrow region. The resultant force is called net lift force F_L. To characterized particles suspended in a Poiseuille flow two non-dimensional numbers channel Reynolds number ($\mathrm{Re}_c = \frac{U_m D_h}{\nu}$) and particle Reynolds number ($\mathrm{Re}_p = \mathrm{Re}_c \left(\frac{d_p}{D_h}\right)^2$) are defined [4]. In curved channel, the presence of curvature pushed maximum velocity away from centre of channel towards the concave surface of the curve. This shift in position of maximum velocity creates two counter rotating vortices called Dean vortices. This phenomenon can be characterized with two non-dimensional numbers Dean number ($\mathrm{De} = \mathrm{Re}_c \left(\frac{D_h}{2*r}\right)^{1/2}$) and curvature ratio ($\delta = \frac{D_h}{2*r}$) [4] (Fig. 1).

The force F_D generated due to dean vortices does not create particle focusing instead it help in reducing the number of focusing position. If $F_D \ll F_L$ no focusing will occur and particle will get mixed due to dominant Dean force. If $F_D \ll F_L$ focusing will occur due to dominant inertial force [5]. The ratio of lift to drag force magnitude scales as $\frac{F_L}{F_D} \sim \frac{1}{\delta}\left(\frac{d_p}{D_h}\right)^3 \mathrm{Re}_c^n (\mathrm{n} < 0)$ suggesting that the smaller particle will remain unfocused where as large particle will focus quickly for the same Reynolds number [6, 7]. The critical condition for particle focusing is that the confidence ratio: $\lambda = \frac{d_p}{D_h} > 0.07$ and particle Reynolds number $Re_p > 1$ [4].

The physics behind the phenomena has been explored by numerous researchers and the lift force formula has been suggested by Brenner (1961), Rubinow and Keller (1961), Saffman (1965), Ho and Leal (1974), Schonberg et al. (1989), McLaughlin et al. (1991), Asomolov (1999), Matas et al. (2009), Hood et al. (2015), Asomolov (2018) [6, 8–17]. All the model has their own limitations due to range of parameters that has chosen and the assumptions that are considered. From all the methods an explicit equation for 2D planar case is only provided by Ho and Leal [12].

$$F_L = K^2 \mathrm{Re}\left[\beta^2 G_1(s) + \beta\gamma G_2(s)\right)] \tag{1}$$

This is simplified for Poiseuille flow as:

$$F_L = \rho U_m^2 K^2 r_p^2 \left[36(1-2s)^2 G_1(s) + 36(1-2s) G_2(s)]\right) \tag{2}$$

$G_1(s)$ is positive for 0 < s < 0.5 and $G_2(s)$ is always positive. Here s = 0 and s = 1 represents bottom and top wall, respectively. In Eq. 1 right hand side, first term represents wall induced lift and second term with γ represents shear induced lift. The Dean drag force is an effect of curvature of geometry. The magnitude of the force can be determined with equation provided by Di Carlo et al. [4, 18].

$$F_D = 3\pi\mu\left(u_f - u_p\right)d_p \tag{3}$$

where u_f, u_p are fluid and particle velocity, respectively, and d_p is particle diameter.

The focusing positions is affected by Re_c, Re_p, d_p, D_h, shape, length and cross section of channel. Combination of these parameter will decide the separation efficiency and accuracy.

A parametric study has been carried out to understand the effect of Reynolds number and number of turns for a serpentine rectangular channel. For different size microparticles this study summarizes the Reynolds number and number of turns that leads to narrow focusing of particles in serpentine microchannel.

3 Materials and Methods

3.1 Details of Computational Domain

Figure 2 represents 2D serpentine computational domain having 18 turns. Five geometries having turns 16, 18, 20, 22 and 24 are created using solidworks.

Figure 3 shows the enlarge view of one turn of geometry of serpentine channel simulated in present work. The amplitude of a turn is equal to 220 μm and pitch is

Fig. 1 Dean flow in a curved channel

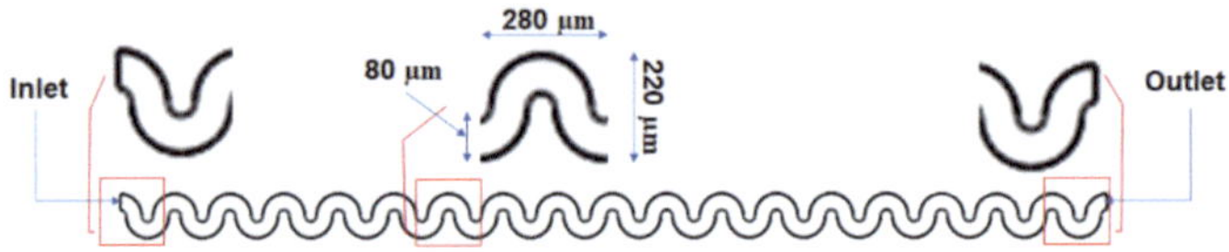

Fig. 2 Computational domain of serpentine channel with 18 turns

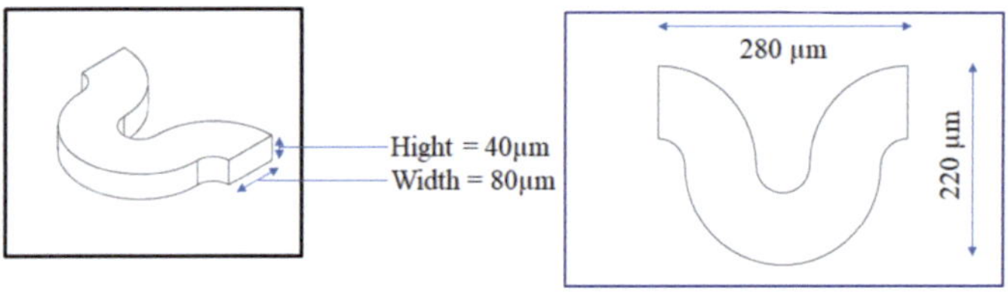

Fig. 3 Schematic diagram of the one turn of channel

280 µm with cross section of 40 µm × 80 µm. For 2D simulations, geometry width is taken as 80 µm.

3.2 Simulation Parameters and Physics

The COMSOL multi-physics is a modular software. In the present work, two modules in COMSOL are used to perform the validation as well as the parametric studies. A Eulerian approach is applied in simulating flow velocity and pressure profile using laminar fluid flow module (LFM). A Lagrangian approach is employed for determining particle trajectories and position. For this particle tracing, for fluid flow module (PTM) is used. The fluid parameter found with LFM is taken as input parameter for determining PTM unsolved parameters.

Flow is assumed to be fully developed at inlet. Flow rate is used as input boundary condition and zero pressure is used as outlet boundary condition during laminar flow simulations. Physics controlled extra fine mesh is used having 6.17 µm maximum and 8.1×10^{-4} µm minimum elements size with growth rate of 1.1 and curvature ratio 0.25. 50 particles are released at random locations of inlet at time $t = 0$. The initial velocities of particles are kept zero. A steady-state solver is used for LFM module and time dependent solver is used for PTM.

The fluid flow parameter solved by LFM steady-state solver is solving Navier–Stokes equations. The steady-state solver configuration is the direct solver: MUMPS direct, fluid flow variables: PARDISO AMG, fluid flow variables (spf): GMRES. The particle trajectories solved by PTM time dependent solver is solving the Newtons second law. The time dependent solver configuration is as direct solver, MUMPS time stepping, generalized alpha iterative and GMRES.

4 Results and Discussion

4.1 Validation

Di Jiang et al. [19] has carried out experimentation and LBM direct numerical simulations for 18 turns rectangular cross section serpentine channel shown in Figs. 2 and 3. The result of this study has taken as reference for validation. The Table 1 shows the flow rate and corresponding Reynolds number used for validation study as well as for parametric investigations.

The Fig. 4 compares the result of experiment, LBM direct numerical simulation and 2D COMSOL simulation. The results of LBM and 2D COMSOL simulation show a similar trend of particle focusing. The position of focusing achieved is almost similar with the LBM simulations. However, the results are having deviation due to the fact that the third wall is not present in the 2D simulations. This result motivates to carry out the further parametric investigation to determined effect of number of turns and Reynolds number on focusing of particles in a serpentine channel.

Table 1 Particle Reynolds number and confidence ratio for 5 μm and 10 μm particles for different Reynolds number

Particle size	5 μm	10 μm
λ	0.0938	0.188
Re_c	$(Re_p)_{5\mu m}$	$(Re_p)_{10\mu m}$
20	0.176	0.703
30	0.264	1.055
40	0.352	1.406
50	0.493	1.758
60	0.527	2.109
80	0.703	2.813

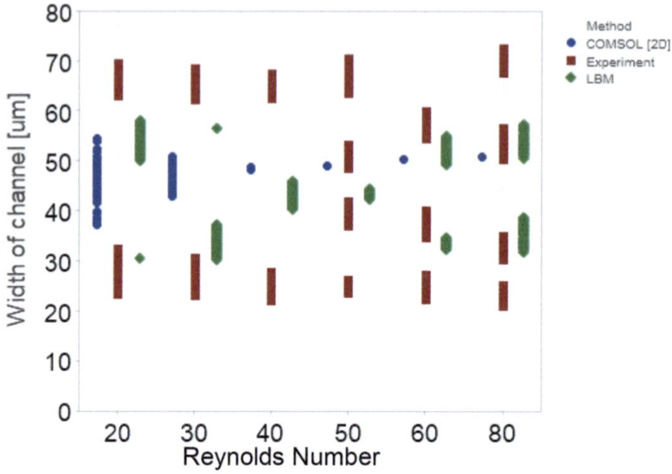

Fig. 4 Comparison of inertial separation results of experimental, LBM simulation and 2D COMSOL simulations of 18 turns serpentine channel at different Reynolds number

4.2 *Effect of Parameters on Focusing Position*

4.2.1 Effect of Reynolds Number

Figures 5 and 7 show the effects of channel Reynolds number on the focusing behaviour of 5 μm and 10 μm particles, respectively.

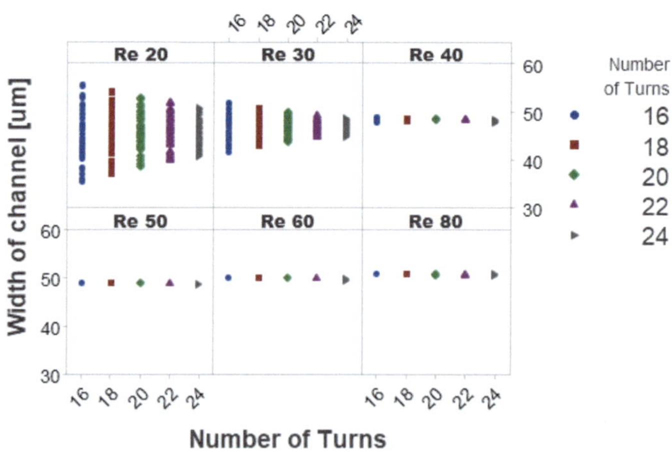

Fig. 5 Effect of number of turns on 5 μm particles for Reynolds number 20, 30, 40, 50, 60 and 80 having number of turns as 16, 18, 20, 22 and 24

It is clearly observed from Fig. 5 that for a particular number of turns, as the Reynolds number increases till 40 focusing region becomes narrower. However, beyond Reynolds number, 40 focusing of particles occurs at same single location irrespective of number of turns of the channel. From Fig. 7, it is evident that number of turns does not affect the focusing position of particles till Reynolds number 50 however after Reynolds number 50 particle starts to unfocused due to influence of Dean forces.

The reasons for such separation are discussed here. The channel Reynolds number is directly associated with particle Reynolds number as $\mathrm{Re}_p = \mathrm{Re}_c\left(\frac{d_p}{D_h}\right)^2$. The limiting properties for 3D channel for focusing of particles depends on $\mathrm{Re}_p > 1$ and $\lambda > 0.07$. Based on particle diameter and geometrical parameter the confidence ratio is found as 0.094 and 0.188 for 5 μm and 10 μm particles, respectively. These values are higher than limiting value of 0.07, making it acceptable for successful particle focusing for given channel design. As shown in Table 2, for 5 μm Re_p values do not satisfy the requirement making it hard to focus at low Reynolds number. However, in case of 2D simulation due to absence of third wall effect, till Reynolds number 40 single point focus is not found. As Reynolds number increases, the inertial lift forces of on particle increases which leads to focusing of particles to a narrow region.

Figure 6 shows the particle trajectories (magenta lines), non-dimensional vorticity contour (coloured one) and velocity vectors (black arrows) for 5 μm particles flowing in 18 turns channel having Reynolds number 20, 30 and 80. Trajectories are found broad for Re 20 and becomes narrow for Re 30 and very narrow for Re 80. It can be observed from Fig. 6a that the vortices strength is low for 20 Reynolds number which makes the particles to scatter in the broad region. Similarly, from Fig. 6c for Reynolds number 80 it can be inferred that the vortices are strong and covering large portion of the channel cross section. This strong vortex makes inertial force stronger and leads to focusing of particles in the narrow region that can be observed based on trajectory line. Overall, as the flow rate increase (Re number increases) vortices are getting stronger and that pushes the particles in the narrow focusing path.

The 10 μm particles focuses at very low Re_c because Re_p values are higher then or near to 1 (refer Table 1) for all the Reynolds numbers mentioned. For 10 μm particles, the focusing start to gets disturbed after $\mathrm{Re}_c > 50$ because of strong Dean forces. The critical Reynolds number is one that supports particle focusing and that is found as 40 and higher than 40 for 5 μm particles and till 50 for 10 μm particles.

4.2.2 Effect of Number of Turns

The effect of number of turns has been studied by changing number of turns as 16, 18, 20, 22 and 24. As number of turns increase the length of channel increases. The Fig. 5 shows the particle focusing positions of 5 μm particle for different number of turns. It can be observed that for Reynolds number 20, 30 and 40 with increase in number of turns the focusing range decreases. For Reynolds number beyond 40 the particles

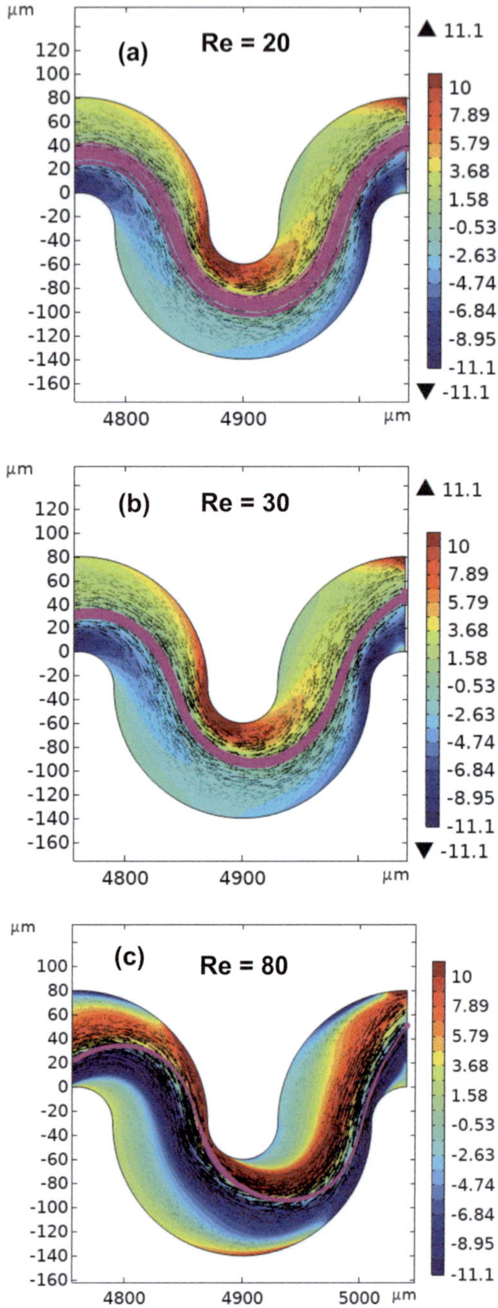

Fig. 6 5 μm particle trajectories and vorticity diagram for 18 turns channel having Reynolds number 20, 30 and 80

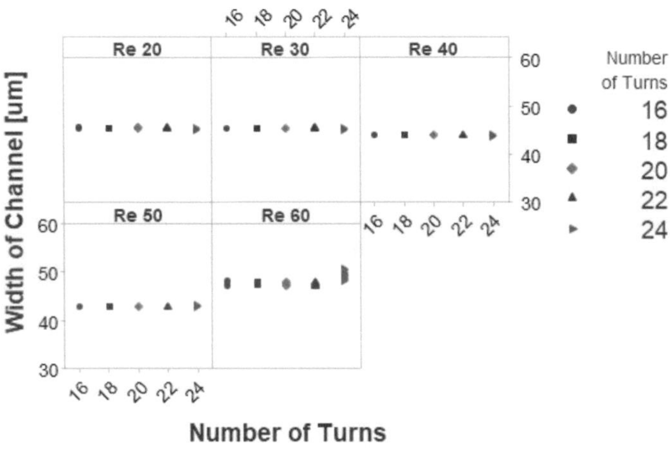

Fig. 7 Effect of number of turns on 10 μm particles for Reynolds number 20, 30, 40, 50 and 60 having number of turns as 16, 18, 20, 22 and 24

focusing is found similar for every number of turns. For a particular Reynolds number, as number of turns increases channel length increases. The increase in length exerts additional lift force on the particles which improves the focusing. However, beyond Reynolds number 40 the Dean force start dominating the particles so the effects of number of turns are not visible. For the 10 μm particle as seen in Fig. 7, there is no effect of number of turns observed. This is evident from Table 1 that 10 μm will focus quickly as the $\lambda = 0.188$ as well as the Re_p is grater then 1 or nearer to 1 for all Reynolds numbers.

In the present study, the 2D simulations are performed to understand the effects of parameters to reduce the time and cost of simulations. For the same configuration and parameters, the 3D simulation is taking 18 h whereas 2D simulation is finished within 2.5 h. Thus, substantial amount of time can be saved for simulations. The idea here is to check the effect of parameter in 2D in order to reduce the effort to find out the superior configuration using 3D simulations. However, it is suggested that comparison study need to be done for 2D and 3D simulation output to practise idea of use of 2D simulations to carry out preliminary study and reduce total simulation time and cost.

5 Conclusions

1. The effect of channel Reynolds number is evident on 5 μm particles as the Reynolds number increases the focusing of particle improves. The channel Reynolds number effect on 10 μm particle is not as evident as on 5 μm particles. This is because of Dean forces affects smaller particle more making them spread more compare to large particles.

2. The channel number of turns affects the 5 μm particles more as compare to 10 μm particles. As number of turns increases the channel length also increases. This leads to increase in the lift force on particles which in turn enhances focusing of the 5 μm particles in the narrow range.
3. The 2D simulation making it easier to perform parametric investigations cost-effective and time saving. The resources consumed also reduces so the lower configuration machine can be used for performing primary analysis.
4. Based on the present 2D simulations, it is found that for a serpentine channel focusing of 5 μm particle are possible after Reynolds number 40 for all turns of channel. The 10 μm particle focusing is found till Reynolds number 50 for any number of turns of channel.

Acknowledgements The work has been carried out with financial support from the research grant (CRG/2021/003349) provided by SERB, New Delhi.

Nomenclature

F_s	Shear induced lift force (N/m^2)
F_w	Wall induced lift force (N/m^2)
F_L	Net lift force on particles (N/m^2)
F_D	Dean force (N/m^2)
Re_c, Re_p	Channel and particle Reynolds number (–)
De	Dean number (–)
d_p, r_p	Particle diameter and radius (m)
D_h	Hydraulic diameter (m)
λ	Confidence ratio (–)
δ	Curvature ratio (–)
r	Radius of curvature (m)
β	Shear rate (N/m^2)
γ	Shear gradient 1/s
U_m	Mean flow velocity (m/s)
$K = \frac{d_p}{2H}$	Characteristics dimension of channel (–)
$s = \frac{y}{H}$	Non dimensional location of particle (–)
y	Vertical co-ordinate of particle (m)
H	Channel height (m)
W	Width of channel (m)
$G_1(s)$, $G_2(s)$	Particle location function. [Provided by Ho and Leal] (m)

References

1. Sajeesh P, Sen AK (2014) Particle separation and sorting in microfluidic devices: a review. Microfluid Nanofluid 17(1):1–52
2. Wyatt Shields Iv C, Reyes CD, López GP (2015) Microfluidic cell sorting: a review of the advances in the separation of cells from debulking to rare cell isolation. Lab Chip 15(5):1230–1249
3. Segré G, Silberberg A (1961) Radial particle displacements in poiseuille flow of suspensions. Nature 189(4760):209–210
4. Di Carlo D, Edd JF, Humphry KJ, Stone HA, Toner M (2009) Particle segregation and dynamics in confined flows. Phys Rev Lett 102(9):094503
5. Nivedita N, Ligrani P, Papautsky I (2017) Dean flow dynamics in low-aspect ratio spiral microchannels. Sci Rep 7(1):1–10
6. Matas JP, Morris JF, Guazzelli É (2004) Inertial migration of rigid spherical particles in Poiseuille flow. J Fluid Mech 515:171–195
7. Martel JM, Toner M (2014) Inertial focusing in microfluidics. Annu Rev Biomed Eng 16:371–396
8. Brenner H (1961) The slow motion of a sphere through a viscous fluid towards a plane surface. Chem Eng Sci 16:242–251
9. Rubinow SI, Keller JB (1961) The transverse force on a spinning sphere moving in a viscous fluid. J Fluid Mech 11(3):447–459
10. Rubinow SI, Keller JB (1972) Flow of a viscous fluid through an elastic tube with applications to blood flow. J Theor Biol 35(2):299–313
11. Saffman PG (1965) The lift on a small sphere in a slow shear flow. J Fluid Mech 22:385–400
12. Ho BP, Leal LG (1974) Inertial migration of rigid spheres in two-dimensional unidirectional flows. J Fluid Mech 65:365–400
13. Hood K, Lee S, Roper M (2015) Inertial migration of a rigid sphere in three-dimensional Poiseuille flow. J Fluid Mech 765:452–479
14. Asmolov ES, Dubov AL, Nizkaya TV, Harting J, Vinogradova OI (2018) Inertial focusing of finite-size particles in microchannels. J Fluid Mech 840:613–630
15. Matas JP, Morris JF, Guazzelli É (2009) Lateral force on a rigid sphere in large-inertia laminar pipe flow. J Fluid Mech 621:59–67
16. Cox R, Brenner H (1968) The lateral migration of solid particles in Poiseuille flow—I Theory. Chem Eng Sci 23:147–173
17. Schonberg JA, Hinch EJ (1989) Inertial migration of a sphere in Poiseuille flow. J Fluid Mech 203:517–524
18. Di Carlo D (2009) Inertial microfluidics. Lab Chip 9(21):3038–3046
19. Jiang D, Tang W, Xiang N, Ni Z (2016) Numerical simulation of particle focusing in a symmetrical serpentine microchannel. RSC Adv 6(62):57647–57657

Graphene Nanoplatelets Powder Versatile Applications in Electronics and Thermal Systems

Rishi Kant Sharma, Rana Pratap Yadav, and Soumen Basu

Abstract Graphene has recently gained considerable attention for its potential role in microwave applications, primarily due to its tunable properties. Through the application of a voltage bias, graphene can shift its resistance values from high to low ranges, enhancing its appeal for various applications. In recent years, significant research interest has been directed toward single-layer carbon films, commonly referred to as graphene, along with modified carbon-based materials such as intercalated graphite, expandable graphite, reduced graphene oxide (RGO), and graphite exfoliation techniques. Producing a few layers of graphene is straight forward, and the cost-effectiveness of graphene does not compromise its tunability. At the midpoint between the microstrip lines, there are few-layer graphene (FLG) flakes, and these lines are connected to a voltage network analyzer at both ends. Utilizing FLG flakes is also applicable in designing tunable patch antennas. Elevating the bias voltage via the bias tees lowers the resistance of the graphene, consequently boosting the transmission across the microstrip line. A prototype has been tested to operate within the frequency range extending from direct current (DC) to 8 GHz. The measured coefficient varies between − 19 dB at 4.5 GHz, achieved by modulating the bias voltage from 2 V (representing minimum attenuation) to 6 V (representing maximum attenuation).

Keywords Graphene · Graphene oxide · FR4 · Microstrip antenna · DC voltage

R. K. Sharma (✉) · R. P. Yadav
Department of Electronics and Communication, Thapar Institute of Engineering and Technology, Patiala, Punjab, India
e-mail: rishikant415@gmail.com

R. P. Yadav
e-mail: ranayadav97@gmail.com

S. Basu
School of Chemistry and Biochemistry, Thapar Institute of Engineering and Technology, Patiala, India
e-mail: soumen.basu@thapar.edu

A. Kumar R. et al. (eds.), *Proceedings of Fluid Mechanics and Fluid Power (FMFP) 2023, Vol. 6*, Lecture Notes in Mechanical Engineering,
https://doi.org/10.1007/978-981-96-9600-0_52

1 Introduction

In recent times, there has been considerable interest surrounding single-layer carbon film, also known as graphene, as well as modified forms of carbon layers such as intercalated graphite, extendable graphite, reduced graphene oxide (RGO), and the process of graphite exfoliation [1]. These techniques encompass various methods including microwave irradiation, thermal reduction, chemical vapor deposition (CVD), electrochemical exfoliation, and bisulfate reduction approaches [2]. These techniques have been explored for their ability to achieve tunability in the microwave frequency range, extending from GHz to THz [2, 3]. Graphene, a single atomic layer of carbon atoms arranged in a hexagonal structure, offers isotropic properties. Notably, its planar sheet features sp^2 carbon bond atoms, and the spacing between the two layers measures 0.0142 nm. In comparison to carbon nanotubes (CNTs), graphene boasts a low density and garners significant attention due to its impressive mechanical and electronic characteristics, including electrical conductivity (~ 20,000 S/cm) and thermal conductivity (~ 5300 W/m K). An intriguing aspect is the alteration in graphene resistance when a bias voltage is applied through the DC voltage supply at the input and output ports. This characteristic enables the manipulation of frequency variation. Additionally, the fabrication of crystal layers on metal strips, known as thin films, stands out due to its simplicity, eco-friendliness (including vacuum and low-contamination conditions), and cost-effectiveness [4].

Basically, graphene is a two-dimensional and unique architecture with excellent properties, which was successfully designed in 2004. At present time widely preparing techniques to design for graphene/graphite nanoplatelets such as chemical oxidation–reduction, CVD (carbon rich compound), and other liquid-phase exfoliation. Graphene is a good conductor because of the no-band gap between covalent and valance bands. In the past few years, graphene has more attraction in scientists and researchers trying to better efforts, and achievements in designing materials. In the energy storage devices flexible in the solar (space program), and other than application in Wi-Fi wireless communication in 5G/6G, radar sensors and nanobots purposely for biomedical and defense applications, etc. The graphite or graphene depends on the > 10 or < 100 layers accordingly like wrinkled 2D and quasi 2D (GNPs) which are created by chemical or physical routes. Graphene nanoplatelets, are easier to determine and have similar characteristics through various applications. GNPs have less acid concentration synthesized for crystalline graphite, which is the ability to be soluble in water and more long-term stable sustain with the support of carbon layer in electronic structure when applied high voltage DC power supply [5].

2 Literature Review and Objective

Above these equations define the size of microstrip width and length patch antennas, also known as microstrip patch antennas, have gained growing significance due to their direct integration onto circuit boards. One of their key advantages is the straightforward fabrication process, contributing to their cost-effectiveness [5]. These antennas, often designed with a low-profile square or rectangular shape, can be conveniently mounted onto flat surfaces

$$\text{Width}(W) = \frac{c}{2f_o\sqrt{(\varepsilon_r + 1)/2}} \tag{1}$$

$$\varepsilon_{\text{eff}} = \frac{\varepsilon_R + 1}{2} + (\varepsilon_{R-1})/2\left[\frac{1}{\sqrt{\left[1 + 12\left(\frac{h}{w}\right)\right]}}\right] \tag{2}$$

$$\text{Length} = \frac{c}{2f_o\sqrt{\varepsilon_{\text{eff}}}} - 0.842h\left(\frac{(\varepsilon_{\text{eff}} + 0.3)\left(\frac{w}{h} + 0.264\right)}{(\varepsilon_{\text{eff}} - 0.358)\left(\frac{w}{h} + 0.8\right)}\right) \tag{3}$$

This paper represents a particular variant of graphene that demonstrates reduced resistance at high voltage vs. frequencies, rendering it suitable for tunable graphene pads operating within the 1–8 GHz frequency range. This property can be connected to minimize gain degradation and induce electronic changes in the microwave domain. The microstrip antenna offers a range of versatile features, including frequency and polarization tuning. When no bias voltage is applied to the graphene, the pads exhibit elevated resistance and function nearly like open circuits. As a result, the transmission of the signal from port 1 to port 2 remains almost complete. As the bias voltage across the graphene increases, its resistance decreases, permitting a higher current flow. Consequently, this current modulation attenuates the signal as it traverses the microstrip line. This letter presents the etching and fabrication of a tunable microstrip attenuator that utilizes FLG and operates within the frequency range spanning from direct current (dc) to 8 GHz. Experimental results are also included.

3 Microstrip Antenna Design by $FECL_3$

The structure shown in Fig. 1 microstrip stub metal patch antenna connected exhibits an input/output impedance depending on variable applied bias voltage to the graphene pad coat on metal 1-mm space.

Based on the stub is analyzed in Fig. 2a consisting of microstrip metal sheet sections: The metal length L_1 and L_2 copper strip which is located between the 2 mm graphene gap and the radiating edge of the patch. The dimension ($g \times w$) of the pad

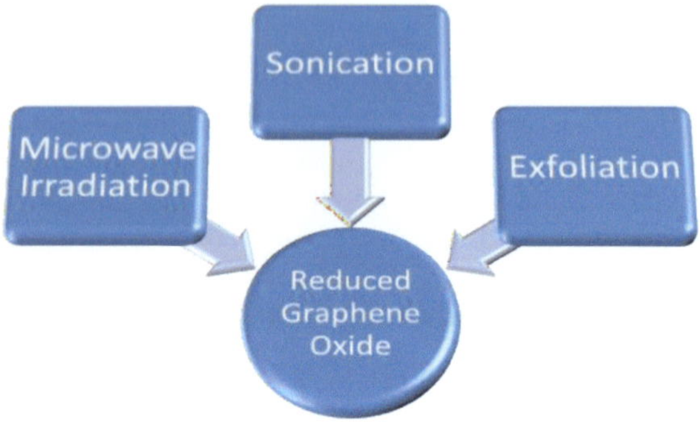

Fig. 1 Schematic diagram of graphene exfoliation process

and metal plate is located between two strip lines. The portions labeled as L_1 and L_2 are connected with thin layers of graphene gel at ports 1 and 2, respectively. Initially designed microstrip line antenna size 3×5 mm ($L = 5$ mm, $W = 3$ mm). A thin conducting filmstrip of $*W$ dimension and located at the center of two wide ground plate separations and the region existing filled with a dielectric material. Although the Microstrip antenna design and fabrication process is straightforward, it does present drawbacks like low gain, limited bandwidth due to dielectric losses, and cross-polar radiation arising from the feeding network. Its popularity stems from its provision of low-profile antenna services for wireless communication applications like GPS, GSM, and Bluetooth. These antennas find utility in reconfigurable systems employing active switches based on various elements such as microstrip line resonators, varactor diodes, and low-cost PIN diodes (typically ranging around 5–5.5 GHz).

The high concentration of $FeCl_3$ mixed with water and dissolved copper transferred electrons from the copper to the ferric ion. The reduction of the concentration rate of the etch process and dissolution of the end. This process is a redox reaction, Fe^{3+} (strong oxidizing agent) and $FeCl_3$ dissolve copper metal fairly. $FeCl_3$ acts as an etchant reaction:

$$Cu + 2Fe^{3+}Cu^{2+} + 2Fe^{2+}$$

This technique various used in industrial PCB designing and analog circuit design boards [6].

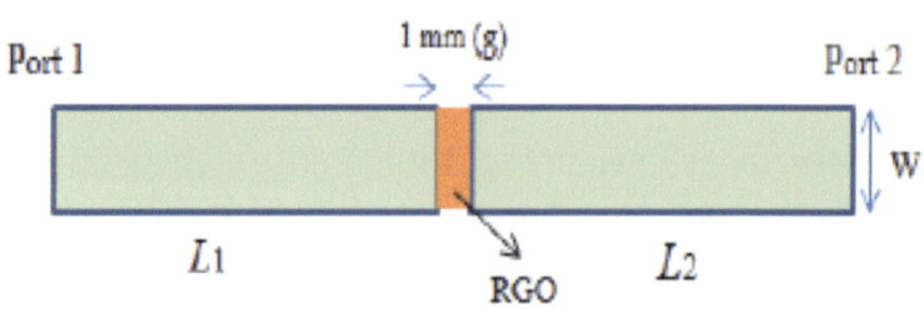

Fig. 2 Schematic diagram of the basic design with graphene fabricated on that place

Fig. 3 Prototype of metal patch film design exfoliates from chemical $FeCl_3$

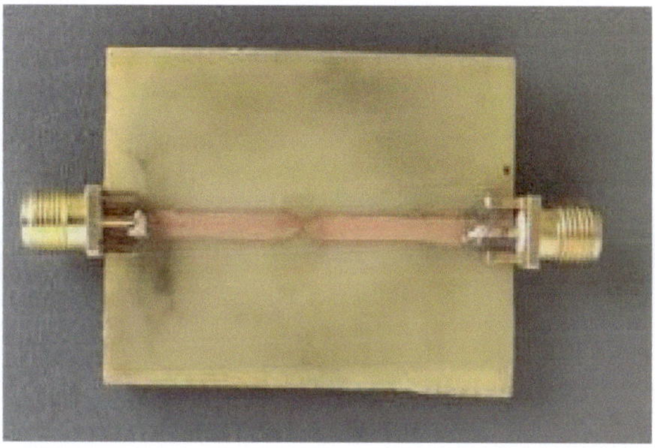

3.1 *Etch of Patch Antenna*

Now we start to discuss the shorted microstrip stub determined to experiment in a basic two phases, which has optimized S*—parameter ***S21***, denoted power leaving at port 2 w. r. t. delivered at port 1 and ***S11*** (dB) reflection coefficient by using a vector network analyzer and simulated physically. Now, we discuss the reflection coefficient of lS11l as a functional frequency based on the preparation of reduced graphene from Anritsu ZVH8 VNA (Vector Network Analyzer) (Fig. 3).

Before experiment, VNA performs calibration by the short, open, load, and through (SOLT) mode. The graphene-patch antenna is DC voltage biasing through the probe tips in substrate layer ground, substrate layer, graphene (G–S–G). The control of all DC voltages is managed through a bias tee (mini-circuits ZX85-12G-S+), which is linked to the VNA via the transmission line. This setup ensures a consistent and adjustable source of voltage or current. The FR4 epoxy property of dielectric constant $\varepsilon_r = 4.4$ and height 1.6 mm of patch antenna with loss tangent loss 0.02. The type of dielectric medium between the internal antenna microstrip feed patch and superstrate has permittivity $\varepsilon_r = 1$ which provides elevated efficiency. The coax probe uses 50 Ω impedance antennas fixed from multilayer metal layer microstrip antenna.

3.2 *Material Processing and Optimization*

Generally, graphite is a good conductor conducting material for metal substrates. In this procedure, if a graphite stick scratches one time then that time that material will not work because without external power that material behaves as a complete insulate. However, if we mix any such soluble the solution performs works as conduction means that conductivity maintains its property otherwise when DC voltage is applied then high voltage is unsuitable and measurement systems can be damaged. Graphene nanoplatelets were prepared by purified natural graphite (98%, which has involved iron 0.4% and particle size 60, mesh) through the mix with isopropanol alcohol (purchased from LOBA Chemie Pvt Ltd). Before taking action, apply heat treatment by microwave and heat at a certain time reason for expand the graphite in the carbon bond and maintain carbon strength. In this experiment, the basic steps to design reduce graphene oxide by thermal treatment to proceed for expansion from graphite powder. After thermal evaporation mix in the IPA solution.

4 Results and Discussion

Graphene biasing was achieved by employing a wide-ranging commercial bias tee, linked between the vector network analyzer (VNA) port and the input microstrip line of the antenna. This setup allowed the application of bias voltage between the ground plane and the patch antenna, ensuring that the voltage was distributed across the graphene pad. The graphene pad increases temperature by 60–80° C when increasing DC via voltage at the port end and potential thermal voltage generated simultaneously. The graphene conduct behaves when increased voltage and partial reflection from the transmit port when voltage changes to 1 V and 6 V and after that the material behaves as insulated.

The transmission and reflection coefficients show that, as the voltage ratio increases up to 5.5 V, the graphene pad exhibits a high red color intensity, indicating enhanced thermal conduction with increasing DC voltage. In contrast, the reflection ratio shows a sudden change at 1 V and 6 V at the same applied voltage, resulting in complete reflection from port 1 to port 1.

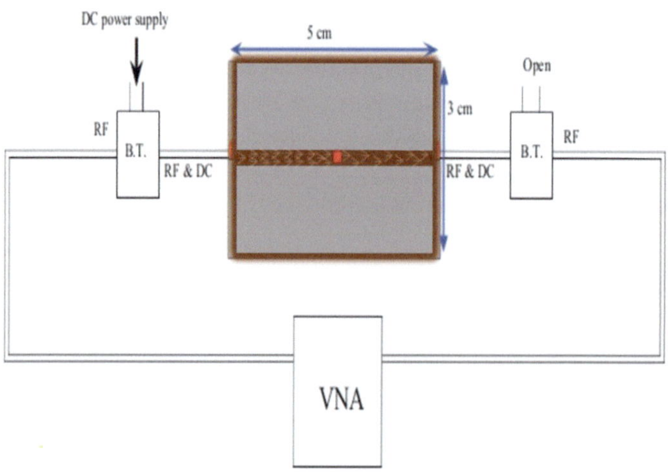

Fig. 4 Complete setup for electronics antenna setup with bias tee and network analyzer

It is important to note that it is theoretically possible to achieve single-layered flakes through additional refinement using chemical exfoliation methods. In contrast to conventional fabrication techniques using exfoliated graphite, our approach avoids significant oxidation, as indicated by the infrared spectra presented in Fig. 4. This preserves the conducting properties of the material (Figs. 5 and 6; Table 1).

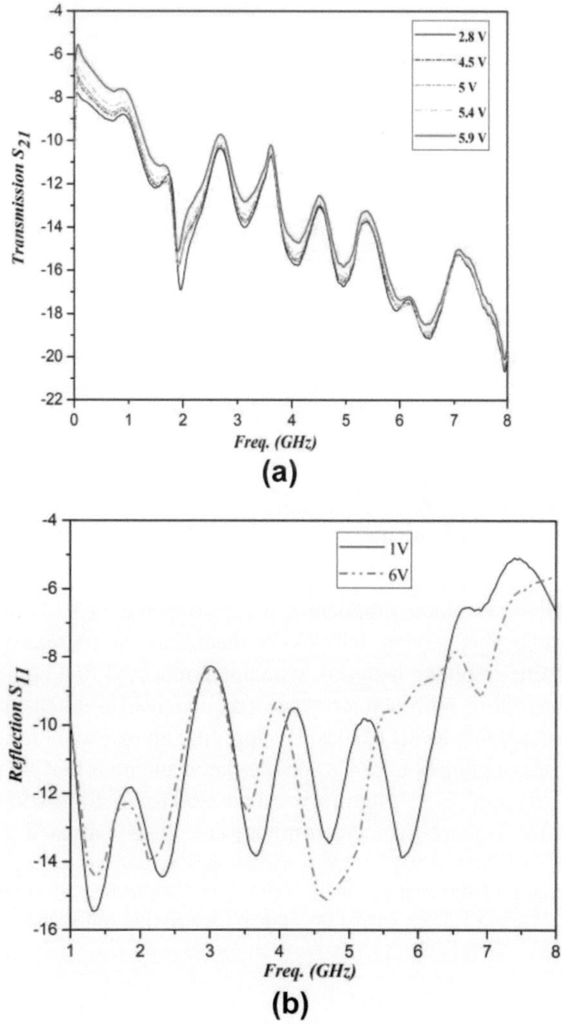

Fig. 5 **a** Transmission and **b** reflection coefficients conductive perform when increases voltage varies at DC bias

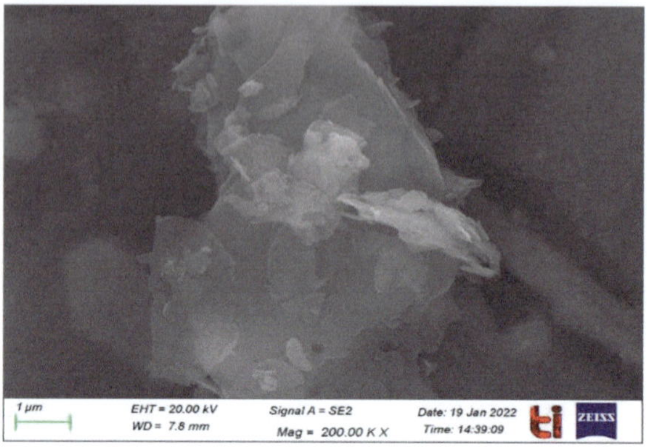

Fig. 6 FESEM analysis of a few layers of carbon nanoflakes

Table 1 Mapping of material quantity

Element	At. No.	Netto	Mass [%]	Mass Norm. [%]	Atom [%]	Abs. error [%] (1 sigma)	Rel. error [%] (1 sigma)
C	6	138,210	90.93	90.93	93.98	4.14	4.56
O	8	2956	7.64	7.64	5.93	0.45	5.83
Au	79	523	1.43	1.43	0.09	0.14	9.85

5 Conclusions

Here examined, graphene is a versatile character of thin film coated whether is a part either mechanical, wireless or electrical, etc. A micro-patch antenna is constructed from graphene material, which is connected to a coplanar transmission line. When operating within microwave frequencies, graphene behaves as a conductor with acceptable levels of loss. By applying an external bias, it becomes possible to adjust the coupling between graphene-patch antennas that are positioned in close proximity. This implies a straightforward method for manipulating the output of the array factor within an arrangement composed of closely situated graphene patches.

Acknowledgements I am thankful to the Department of Electronics and Communication under CEEMs-VT, from Thapar Institute of Engineering and Technology for granting the project fund.

References

1. Halim U, Zheng CR, Chen Y, Lin Z, Jiang S, Cheng R, Huang Y, Duan X (2013) A rational design of cosolvent exfoliation of layered materials by directly probing liquid-solid interaction. Nat Commun 4:2213

2. Xiao W, Zhai X, Ma P, Fan T, Li X (2018) Numerical study on the thermal behavior of graphene nanoplatelets/epoxy composites. Appl Therm Eng 133:330–340
3. Henning K-D (2014) Solvent recycling, removal, and degradation. Carbo Tech Aktivkohlen GmbH, Essen, Germany
4. Moosa AA, Abed MS (2021) Graphene preparation and graphite exfoliation. Mater Today Proc 42:1890–1895
5. Potenza M, Cataldo A, Bovesecchi G, Corasaniti S, Coppa P, Bellucci S (2017) Graphene nanoplatelets: thermal diffusivity and thermal conductivity by the flash method. AIP Advances 7:075214
6. Yu M, Zeng X, Song Q, Liu L, Li J (2016) Examining regeneration technologies for etching solutions: a critical analysis of the characteristics and potentials. J Clean Prod 137:476–488

MIX
Papier aus verantwortungsvollen Quellen
Paper from responsible sources
FSC® C105338

If you have any concerns about our products,
you can contact us on
ProductSafety@springernature.com

In case Publisher is established outside the EU,
the EU authorized representative is:
Springer Nature Customer Service Center GmbH
Europaplatz 3, 69115 Heidelberg, Germany

Printed by Libri Plureos GmbH
in Hamburg, Germany